SCATTERING THEORY OF
MOLECULES, ATOMS AND NUCLEI

Preface

The development of physics during the last century relied heavily on scattering of particles from a variety of targets. There exists in the literature many textbooks and monographs on the subject. Why write a new book? Several recent developments in several areas of research in atomic, molecular and nuclear physics were not covered thoroughly before. This prompted us to write the present monograph. We divided the book into two major parts. The first deals with the fundamentals of collision theory both classical and quantum mechanically for the case of Potential scattering, namely one channel. This contains 8 chapters which are described below. The second part deals with application and generalization of the theory developed in Part I to Many-Body Scattering. This part is composed of 4 Chapters.

The first chapter of Part I, deals with some basic notions such as kinematics, classical scattering theory, and the systems of units used in atomic, molecular and nuclear physics. Chapter 2 develops the partial wave method. In Chapter 3, we treat the scattering from long-range potentials, emphasizing the Coulomb problem. Chapter 4 introduces the important operators of quantum scattering theory, such as the Green's function, the T- and S-matrices. We use these operators to calculate the cross section. We also dwell on the optical theorem. Chapter 5 is a long one that deals with usual approximations employed to treat practical problems. Both perturbation methods and semiclassical methods are fully developed and applied to several cases. In Chapter 6 the spin degree of freedom is introduced and the necessary modification that this ensues is developed. The scattering of identical particles is also discussed in this chapter. The scattering from complex potentials, of paramount importance in nuclear physics, is taken up and developed in Chapter 7. Additional topics in potential scattering, such as the analytical properties of the S-matrix, scattering in one and two dimensions are considered in Chapter 8.

The first chapter of Part II, Chapter 9, introduces the coupled channel formalism for treating collision of composite systems. The Feshbach method used in the treatment of cases involving few channels is developed in Chapter 10. Polarization potentials employed in the one-channel reduction, are calculated both for nuclear and atomic scattering, and the concepts of energy averages and optical potentials are developed. Chapter 11 introduces approximate methods, such as the Distorted Wave Approximation, the Semiclassical method, and the Glauber approximation to treat the many-body scattering. Finally, in Chapter 12, we discuss several selected topics in scattering of atoms and nuclei. Concepts, such as the doorway resonance, the Feshbach resonance in atomic scattering at very low energies, the statistical nuclear reaction theory the three-body scattering problem and others, are introduced and briefly discussed.

We believe that the book is comprehensive enough to benefit a wide range of readers. The material presented in this book requires as prerequisite a first course on quantum mechanics at the graduate level, and some knowledge of atomic, molecular and nuclear physics. The fact that the emphasis of the book is nuclear physics stems from the background of the authors.

Rio de Janeiro and São Paulo, July 2012

Luiz Felipe Canto

Mahir S. Hussein

Acknowledgments

In writing this book we had the privilege of getting help from innumerous colleagues who read one or a few chapters and helped us give a more polished presentation. We would like to thank Akbar Salam, Brett V. Carlson, Daniel Marta, Eduardo Montenegro, Jesus Lubian, Joaquim Lopes Neto, Luiz C. Chamon, Nelson V. C. Faria, Paolo Lotti, Paulo R. S. Gomes, and especially Kouichi Hagino, Pierre Descouvemont, and Raul Donangelo.

Contents

PART 1

Potential Scattering

Chapter 1

Basic Notions

1.1 Introduction

We start this book with the question: *why is it important to study quantum scattering theory?* One could think of a few answers to this question but the most appropriate one is that scattering experiments are the most important tool to investigate the interactions between elementary particles and to probe the internal structure of complex microscopic systems. An early example, before the development of quantum mechanics, was the pioneering work of Rutherford [Rutherford (1911)] on alpha-particle (Helium nuclei) scattering from thin gold foils, which lead to the discovery of the atomic nucleus.

Since direct information on microscopic systems is not available, to investigate the interaction between them one proceeds as follows: a) some hypothesis is made about the nature of the interaction, b) using this hypothesis, predictions for the results of a scattering experiment are made, and c) the experiment is carried out, theoretical predictions are compared with data and the validity of the hypothesis is accordingly checked. In this way, one can investigate the fundamental interactions between elementary particles and also the internal structure of the collision partners (atoms, nuclei, molecules, etc.). Since one is dealing with quantum systems, the predictions for the outcome of the experiment requires the knowledge of the solutions of the Schrödinger equation for a large number of interacting degrees of freedom. However, finding such solutions is a hopeless task. To handle the situation, approximate models are developed. These models must reproduce the main features of the system, while leading to simpler equations of motion. Calling ξ the set of coordinates of the model and $h(\xi)$ the corresponding Hamiltonian, the stationary states, $\varphi_\alpha(\xi)$, are given by the equation

$$h(\xi)\,\varphi_\alpha(\xi) = \varepsilon_\alpha\,\varphi_\alpha(\xi). \tag{1.1}$$

In such situations, it is important to check the validity of the approximations involved. This can be done through a scattering experiment using the system under study - say, the target - and a projectile with simpler internal structure. Neglecting the projectile's intrinsic degrees of freedom, the projectile-target system is described by the set of coordinates $\{\xi, \mathbf{r}\}$, where $\mathbf{r}$ is the projectile-target separation vector, and the total Hamiltonian is

$$H(\mathbf{r}, \xi) = K + h(\xi) + V(\mathbf{r}, \xi). \tag{1.2}$$

Above, K is the kinetic energy associated with the motion of the projectile and the target and $V(\mathbf{r}, \xi)$ is the coupling interaction. It is convenient to describe the collision in the center of mass frame, as will be discussed in section 1.3.1. In this case, the kinetic energy operator is given by

$$K = \frac{\mathbf{p}^2}{2\mu}, \tag{1.3}$$

where $\mathbf{p} = -i\hbar\,\boldsymbol{\nabla}_\mathbf{r}$ is the momentum operator related to $\mathbf{r}$.

Before the collision takes place, the target is in its ground state $\varphi_0(\xi)$. As the projectile approaches the target, the action of $V(\mathbf{r}, \xi)$ leads to excitations, such that there are finite probabilities for finding the target in different final states $\varphi_\alpha(\xi)$. These probabilities are measured and compared with the theoretical predictions based on the approximation used for the target dynamics. These predictions depend strongly on matrix elements of the interaction between pairs of intrinsic states $\{\varphi_\alpha, \varphi_\beta\}$. In this way, the approximate model can be tested. Thus, the scattering experiment plays a fundamental role in the development of approximate models for the quantum dynamics of nuclei, atoms and molecules. Of course, this testing procedure relies on quantitative relations between the model wave functions and observable quantities. Such relations are provided by quantum scattering theory.

Frequently, the collision partners are composite systems and their intrinsic structures have strong influence on the scattering processes. The projectile and the target, which are usually in their ground states before the collision, may end up in several different intrinsic states. It is then necessary to develop a quantum theory of many-body scattering, which is the subject of part II of this book. On the other hand, processes in which the intrinsic states of the projectile and the target are kept frozen throughout the collision are much easier to describe. The ground state densities are used to generate an average projectile-target potential and the intrinsic degrees of freedom do not have to be included explicitly in the scattering theory. The problem is then equivalent to the scattering of an effective particle with mass $\mu = m_{\rm P}\, m_{\rm T}/(m_{\rm P} + m_{\rm T})$ from the average potential. This theory, which is known as *potential scattering*, is studied in chapters 1 to 8 (part I) of this book. In fact, actual nuclear, atomic or molecular collisions can hardly be reduced to a problem of potential scattering. However, potential scattering is very important for two reasons. The first is that several techniques used in many-body scattering are well illustrated in potential scattering. The second is that potential scattering is very useful in the study of elastic scattering. As we will show in chapter 10, it is always possible to reduce many-body formulations of the elastic cross section to a problem of potential scattering. Although this reduction requires the calculation of a very complicated potential, it is suitable for the development of useful approximations.

1.2 Several definitions

In this section, we introduce several concepts used in collision theory[1]. Figure 1.1 represents schematically a typical scattering experiment. Projectiles are accelerated, go through a collimator and collide with a target. Some projectiles are scattered by the target and reach the detector. The particles in the beam could be protons, electrons, positrons, pions, ions, ionized molecules, etc.. According to the nature of the projectile, the required collision energy, energy resolution, beam intensity, etc., a variety of accelerators may be used. Ideally, the beam should be sharp in momentum space, i.e., should have good energy resolution and be well collimated. Usually, good experiments require high count rates. This is attained with an intense beam and/or a thick target. However, if the beam it so intense that the projectiles are close enough to interact, considerable complications appear in the theoretical description. Theoretical difficulties also arise when the target is so thick that multiple scattering becomes relevant.

The detectors can simply acknowledge the arrival of a particle or it can give additional information like the particle charge or energy.

[1]For further details see refs. [Segrè (1977); McDowell and Coleman (1970); Friedrich (1998)]

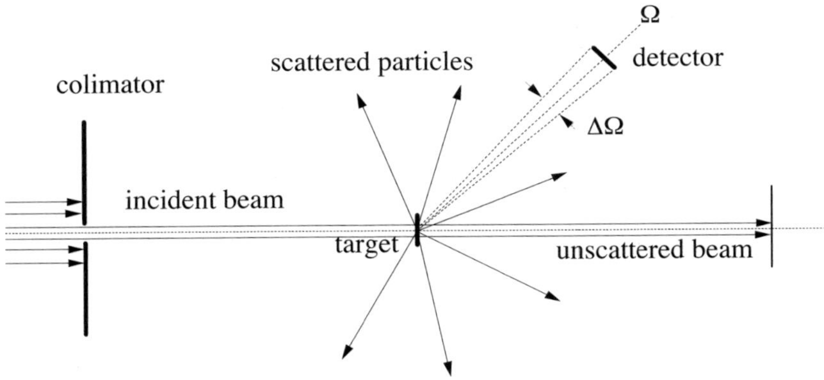

Fig. 1.1 Schematic representation of a scattering experiment.

1.2.1 *Types of collisions — reaction channels*

When one of the collision partners (or both) have relevant internal degrees of freedom, the intrinsic states of the projectile-target system may change from $\varphi_0(\xi)$ to a different final state $\varphi_\alpha(\xi)$. Therefore, the characterization of the final state of the system requires the knowledge of the relative wave function and also the intrinsic quantum numbers expressed by the label α. Such states are called reaction channels ("channel-α"). The most important one is the *elastic* or *entrance* channel, corresponding to the ground state, $\alpha = 0$. Let us list below some frequently studied channels.

a) *elastic channel* - in this channel, the collision does not produce intrinsic excitation. Since the collision process conserves the total energy, the kinetic energy of the relative motion is also unchanged. A simple example of elastic scattering is[2]

$$n + {}^{16}\mathrm{O} \rightarrow n + {}^{16}\mathrm{O}$$

with ^{16}O remaining in the ground state.

b) *inelastic channels* - inelastic channels are those in which projectile and target maintain their identities but one (or both) is (are) excited to a higher intrinsic energy ε_α. For example,

$$n + {}^{16}\mathrm{O} \rightarrow n + \left({}^{16}\mathrm{O}\right)^*,$$

[2]We use the standard Nuclear Physics notation where ^{16}O stands for an oxygen nucleus with mass number 16.

where the notation $(\ )^*$ is used to indicate excitation. In inelastic collisions, part of the kinetic energy of the projectile-target relative motion goes into intrinsic excitation. *Inelastic scattering* can be characterized by a reduction of the relative kinetic energy or detection of the de-excitation product (usually photons).

c) *reaction channels* - the name reaction channel is used for states of the system where the projectile or/and the target change identity. The population of such channels is called a *reaction*. A few examples are given below

$$\begin{aligned} n + {}^{16}\mathrm{O} &\rightarrow p + {}^{16}\mathrm{N} \\ e^+ + \mathrm{H} &\rightarrow e^+ + p + e \\ p + p &\rightarrow p + n + \pi^+ + \pi^0. \end{aligned}$$

The coupling interaction V has symmetries that cannot be violated in the collision process. Only channels can be excited for which there is conservation of total energy, electric charge, baryonic and leptonic numbers, etc.. For this reason, a reaction like

$$e^+ + \mathrm{H} \rightarrow e^+ + p + e^+$$

cannot occur owing to charge conservation.

1.2.2 *Q-values, threshold, open and closed channels*

An important quantity in inelastic and two-body reaction channels is the intrinsic energy difference,

$$Q = \varepsilon_0 - \varepsilon_\alpha, \tag{1.4}$$

which is called the Q-value of the channel. Energy conservation relates the Q-value to the initial, K_0, and final, K_α values of the kinetic energy of the relative motion. That is,

$$Q = K_\alpha - K_0. \tag{1.5}$$

Eq. (1.5) means that the Q-value expresses the variation of the kinetic energy in the collision process.

A collision process in which the intrinsic energy decreases ($Q > 0$), converting the difference into kinetic energy of the relative motion, is called *exothermic*. In the opposite situation, the process is called *endothermic*. Energy conservation does not lead to any restriction on exothermic processes. On the other hand, it leads to restrictions on endothermic processes.

In a scattering experiment with collision energy E (measured in the center of mass frame), a channel with $-Q > E$ can never be excited. The reason is that the maximal kinetic energy available to be transformed into intrinsic excitation is E, which is not high enough. As an example, the inelastic process

$$e + \mathrm{H} \;\rightarrow\; e + (\mathrm{H})^*,$$

with an electron beam of $E = 10$ eV is forbidden by energy conservation, because the lowest excited state in H has the energy $\varepsilon_1 = 10.2$ eV. The lowest collision energy that allows the excitation of an endothermic channel is referred to as its *threshold* energy, and it is usually denoted by E_{th}. In the above example, $E_{\mathrm{th}} = 10.2$ eV. When the collision energy is below the threshold for a given channel, one says that this channel is *closed.* Otherwise, it is said to be *open.*

1.3 Cross sections

Let us consider an experiment such as that represented in figure 1.1, in which we measure the count rate of events leading to the population of channel-α, $N_\alpha(\Omega, \Delta\Omega)$. A few remarks should be made:

- If the detector aperture is small, $N_\alpha(\Omega, \Delta\Omega)$ should be proportional to $\Delta\Omega$.
- Assuming that multiple scattering can be neglected, $N_\alpha(\Omega, \Delta\Omega)$ should be proportional to the number n of target particles per unit area in the region attained by the beam.
- Assuming that interactions between beam particles can be neglected, the count rate $N_\alpha(\Omega, \Delta\Omega)$ should be proportional to the incident flux J.

The count rate, $N_\alpha(\Omega, \Delta\Omega)$, can then be written

$$N_\alpha(\Omega, \Delta\Omega) = (\Delta\Omega \cdot n \cdot J)\; \frac{d\sigma_\alpha(\Omega)}{d\Omega} \;.$$

The constant of proportionality,

$$\frac{d\sigma_\alpha(\Omega)}{d\Omega} = \frac{N_\alpha(\Omega, \Delta\Omega)}{\Delta\Omega \cdot n \cdot J} \,, \tag{1.6}$$

is called the differential cross section for channel-α. This quantity is very useful, since it does not depend on experimental details (detector size, incident flux, target thickness and density). It depends exclusively on the physics of the projectile and target particles. The elastic scattering cross section plays a unique role, since it appears in any collision.

Frequently, one is interested in the integrated (over orientation) cross section. As an example, the (total) elastic cross section is given as

$$\sigma_{\rm el} = \int d\Omega \left[\frac{d\sigma_{el}(\Omega)}{d\Omega} \right] . \tag{1.7}$$

1.3.1 *Center of mass and laboratory frames*

For comparison between theory and experiment, it is necessary to have cross sections in the same reference frame. Of course, the measured cross section is obtained in the laboratory frame, where the target is at rest. From the theoretical point of view, however, it is important to take advantage of the translational invariance of the projectile-target Hamiltonian, introducing the center of mass frame (CM). Calling $\mathbf{r}_{\rm P}$ and $\mathbf{r}_{\rm T}$ the position vectors of the projectile and target in the laboratory frame respectively, and $\mathbf{p}_{\rm P}$ and $\mathbf{p}_{\rm T}$ their associated momenta, one introduces the transformations

$$\mathbf{r}_{\rm P};\, \mathbf{r}_{\rm T} \;\rightarrow\; \mathbf{R}_{\rm c.m.} = \frac{m_{\rm P}\,\mathbf{r}_{\rm P} + m_{\rm T}\,\mathbf{r}_{\rm T}}{M}; \quad \mathbf{r} = \mathbf{r}_{\rm P} - \mathbf{r}_{\rm T}, \tag{1.8}$$

and

$$\mathbf{p}_{\rm P};\, \mathbf{p}_{\rm T} \;\rightarrow\; \mathbf{P}_{\rm c.m.} = \mathbf{p}_{\rm P} + \mathbf{p}_{\rm T}; \quad \mathbf{p} = \frac{m_{\rm T}\,\mathbf{p}_{\rm P} - m_{\rm P}\,\mathbf{p}_{\rm T}}{M}, \tag{1.9}$$

where $m_{\rm P}$ and $m_{\rm T}$ are respectively the projectile and target masses and $M = m_{\rm P} + m_{\rm T}$.

Let us discuss the consequences of this transformation on the Hamiltonian, in a collision of two structureless particles interacting through a conservative potential. In this case, there are no intrinsic degrees of freedom and the modulus of the relative velocity is conserved. Calling $\mathbf{v}$ and $\mathbf{v}'$ the relative velocities before and after the collision, we can write $|\mathbf{v}| = |\mathbf{v}'|$. With this transformation, the Hamiltonian,

$$H = \frac{\mathbf{p}_{\rm P}^2}{2m_{\rm P}} + \frac{\mathbf{p}_{\rm T}^2}{2m_{\rm T}} + V(\mathbf{r}_{\rm P} - \mathbf{r}_{\rm T}),$$

separates into a center of mass and a relative term

$$H = H(\mathbf{R}_{\rm c.m.}, \mathbf{P}_{\rm c.m.}) \;+\; H(\mathbf{r}, \mathbf{p}).$$

These terms are,

$$H(\mathbf{R}_{\rm c.m.}, \mathbf{P}_{\rm c.m.}) = \frac{\mathbf{P}_{\rm c.m.}^2}{2M} \quad \text{and} \quad H(\mathbf{r}, \mathbf{p}) = \frac{\mathbf{p}^2}{2\mu} + V(\mathbf{r}), \tag{1.10}$$

where $\mu = m_{\rm P}\, m_{\rm T}/M$ is the reduced mass.

The new set of coordinates turns out to be very convenient both in classical mechanics (see next section) and quantum mechanical descriptions of the scattering problem. In the latter case, the system wave function may be written as the product

$$\psi(\mathbf{R}_{\text{c.m.}}, \mathbf{r}, t) = \psi(\mathbf{R}_{\text{c.m.}}, t) \times \psi(\mathbf{r}, t),$$

where the center of mass wave function is a trivial plane wave which can be ignored. The relative wave function, which contains all the relevant information, satisfies the Schrödinger equation

$$\left[-\frac{\hbar^2}{2\mu}\nabla_{\mathbf{r}}^2 + V(\mathbf{r})\right] \psi(\mathbf{r},t) = i\hbar \frac{\partial\psi(\mathbf{r},t)}{\partial t}. \tag{1.11}$$

Inspecting the above equation, we conclude that the transformation into center of mass and relative coordinates has a very important practical consequence: the number of variables in the Schrödinger equation is reduced from six (components of $\mathbf{r}_{\text{P}}$ and $\mathbf{r}_{\text{T}}$) to 3 (components of $\mathbf{r}$).

For the comparison of data with theoretical cross sections, it is necessary to relate the collision energies, the angles and the cross sections in the two reference frames (CM and laboratory). The collision energies[3] in these frames are given by[4]

$$E_{\text{lab}} = \frac{\mathbf{p}_{\text{P}}^2}{2m_{\text{P}}}, \quad E_{\text{c.m.}} = \frac{\mathbf{p}^2}{2\mu}.$$

Using Eq. (1.9), with $\mathbf{p}_{\text{T}} = 0$, and the definition of the reduced mass, we find the relation

$$E_{\text{c.m.}} = \frac{m_{\text{T}}}{M} E_{\text{lab}}. \tag{1.12}$$

When the target is much heavier than the projectile, $E_{\text{c.m.}}$ is very close to E_{lab}. Otherwise, it can be much smaller.

To relate cross sections in the CM and in the laboratory frames, one should have in mind that the count rate in the detector, $N_\alpha(\Omega, \Delta\Omega)$, is independent of the reference frame. The same happens with the incident flux, which is given by the product of the beam density with the velocity of the projectiles with respect to the target. The desired transformation then

[3]Note that $E_{\text{c.m.}}$ is the energy of the relative motion, which corresponds to the total collision energy in the CM frame. It is **not** the energy of the center of mass motion in the laboratory frame, as this notation could suggest.

[4]We are assuming that the target is at rest in the laboratory frame. This assumption is not valid in experiments involving colliding beams.

depends exclusively on the orientation Ω ($\Omega \equiv \theta, \varphi$) and the elementary solid angle $\Delta\Omega$. We can write

$$\frac{d\sigma_{\text{el}}(\Omega_{\text{lab}})}{d\Omega_{\text{lab}}} = \frac{d\sigma_{\text{el}}(\Omega_{\text{c.m.}})}{d\Omega_{\text{c.m.}}} \times \frac{d\Omega_{\text{c.m.}}}{d\Omega_{\text{lab}}}, \tag{1.13}$$

where $d\Omega_{\text{c.m.}}/d\Omega_{\text{lab}}$ is a trivial geometrical factor, which can be calculated by elementary classical mechanics. It is evaluated below, for the case of elastic scattering.

The initial velocities in the laboratory frame are $\mathbf{v}_{\text{P}}$ and $\mathbf{v}_{\text{T}} = 0$ and after the collision they become respectively $\mathbf{v}'_{\text{P}}$ and $\mathbf{v}'_{\text{T}}$. The initial and final velocities in the $\{\mathbf{R}_{\text{c.m.}}, \mathbf{r}\}$ coordinate system are given by the derivatives of Eq. (1.8)). We find,

$$\mathbf{V}_{\text{c.m.}} = \frac{m_{\text{P}}}{M}\,\mathbf{v}_{\text{P}} \quad \text{and} \quad \mathbf{v} = \mathbf{v}_{\text{P}}, \tag{1.14}$$

and

$$\mathbf{V}'_{\text{c.m.}} = \frac{m_{\text{P}}}{M}\,\mathbf{v}'_{\text{P}} + \frac{m_{\text{T}}}{M}\,\mathbf{v}'_{\text{T}} \quad \text{and} \quad \mathbf{v}' = \mathbf{v}'_{\text{P}} - \mathbf{v}'_{\text{T}}. \tag{1.15}$$

Owing to momentum and energy conservations, these velocities have the following properties:

$$\mathbf{V}'_{\text{c.m.}} = \mathbf{V}_{\text{c.m.}} \quad \text{and} \quad |\mathbf{v}'| = |\mathbf{v}|\,. \tag{1.16}$$

From the above equations, we get the relation

$$\mathbf{v}'_{\text{P}} = \mathbf{V}'_{\text{c.m.}} + \frac{m_{\text{T}}}{M}\,\mathbf{v}', \tag{1.17}$$

which can be written also as

$$\mathbf{v}'_{\text{P}} = \mathbf{V}'_{\text{c.m.}} + \mathbf{u}'_{\text{P}}, \tag{1.18}$$

where

$$\mathbf{u}'_{\text{P}} = \frac{m_{\text{T}}}{M}\,\mathbf{v}' \tag{1.19}$$

is the final velocity of the projectile measured in the CM frame.

Since the center of mass velocity is assumed to be along the z-axis, the transformation will keep the polar coordinate, φ, unchanged so that

$$\frac{d\Omega_{\text{c.m.}}}{d\Omega_{\text{lab}}} = \frac{\sin\theta_{\text{c.m.}}\ d\theta_{\text{c.m.}}}{\sin\theta_{\text{lab}}\ d\theta_{\text{lab}}}. \tag{1.20}$$

The angles θ_{lab} and $\theta_{\text{c.m.}}$ are respectively the angles between the z-axis and the final velocities in the laboratory and CM frames, $\mathbf{v}'_{\text{P}}$ and $\mathbf{u}'_{\text{P}}$. They are given by

$$\cos\theta_{\text{lab}} = \frac{\mathbf{v}'_{\text{P}} \cdot \hat{\mathbf{z}}}{v'_{\text{P}}}. \tag{1.21}$$

and

$$\cos\theta_{\rm c.m.} = \frac{\mathbf{u}'_{\rm P}\cdot\hat{\mathbf{z}}}{u'_{\rm P}}. \tag{1.22}$$

Having in mind that the projectile's velocity in the CM frame is parallel to the relative velocity, the above equation can be put in the alternative form,

$$\cos\theta_{\rm c.m.} = \frac{\mathbf{v}'\cdot\hat{\mathbf{z}}}{v}. \tag{1.23}$$

Replacing

$$\mathbf{v}'_{\rm P}\cdot\hat{\mathbf{z}} = V_{\rm c.m.} + \frac{m_{\rm T}}{M}\,\mathbf{v}'\cdot\hat{\mathbf{z}} = V_{\rm c.m.} + \frac{m_{\rm T}}{M}\,v\cos\theta_{\rm c.m.}$$

and

$$v'_{\rm P} = \sqrt{V_{\rm c.m.}^2 + \left(\frac{m_{\rm T}}{M}\right)^2 v^2 + 2\,\frac{m_{\rm T}}{M}\,V_{\rm c.m.}\,v\,\cos\theta_{\rm c.m.}}$$

in Eq. (1.21), we get

$$\cos\theta_{\rm lab} = \frac{V_{\rm c.m.} + \frac{m_{\rm T}}{M}\,v\cos\theta_{\rm c.m.}}{\sqrt{V_{\rm c.m.}^2 + \left(\frac{m_{\rm T}}{M}\right)^2 v^2 + 2\frac{m_{\rm T}}{M}\,V_{\rm c.m.}\,v\,\cos\theta_{\rm c.m.}}}. \tag{1.24}$$

A simpler formula can be derived if we divide both the numerator and the denominator by $V_{\rm c.m.}$. Calling

$$\lambda = \left(\frac{m_{\rm T}}{M}\,\frac{v}{V_{\rm c.m.}}\right)^{-1} = m_{\rm P}/m_{\rm T}\,,$$

we obtain

$$\cos\theta_{\rm lab} = \frac{\lambda + \cos\theta_{\rm c.m.}}{\sqrt{1+\lambda^2+2\,\lambda\cos\theta_{\rm c.m.}}}. \tag{1.25}$$

The above equation indicates that, if the projectile is heavier than the target ($\lambda > 1$) it is always deflected into the forward hemisphere. Setting $d\cos\theta_{\rm lab}/d\cos\theta_{\rm c.m.}=0$, we find that the maximum value of $\theta_{\rm lab}$ is

$$[\theta_{\rm lab}]_{\rm max} \equiv \theta_0 = \sin^{-1}\left(\frac{1}{\lambda}\right) = \sin^{-1}\left(\frac{m_{\rm T}}{m_{\rm P}}\right), \tag{1.26}$$

which corresponds to

$$\cos\theta_{\rm c.m.} = -1/\lambda \quad\rightarrow\quad \theta_{\rm c.m.} = [\theta_{\rm lab}]_{\rm max} + \frac{\pi}{2}. \tag{1.27}$$

Differentiating both sides of Eq. (1.24) and using Eqs. (1.20) and (1.13), we obtain the desired transformation

$$\frac{d\sigma_{\rm el}(\Omega_{\rm lab})}{d\Omega_{\rm lab}} = \frac{d\sigma_{\rm el}(\Omega_{\rm c.m.})}{d\Omega_{\rm c.m.}}\left[\frac{\left(1+2\lambda\cos\theta_{\rm c.m.}+\lambda^2\right)^{3/2}}{|1+\lambda\cos\theta_{\rm c.m.}|}\right]. \tag{1.28}$$

In summary, to change from the center of mass frame to the laboratory frame, the energy is given by the inverse of Eq. (1.12),

$$E_{\rm lab} = \frac{M}{m_{\rm T}}\ E_{\rm c.m.}, \tag{1.29}$$

$\theta_{\rm lab}$ is related to $\theta_{\rm c.m.}$ by Eq. (1.25), and the cross section in the laboratory frame is expressed in terms of its value in the CM frame through Eq. (1.28).

Now we consider the inverse problem: converting variables given in the laboratory frame to the center of mass frame. Since the transformation of the collision energy has already been given in Eq. (1.12), we concentrate in the transformation of θ and $d\sigma/d\Omega$. To convert $\theta_{\rm lab} \to \theta_{\rm c.m.}$, we have to solve Eq. (1.25) for $\cos\theta_{\rm c.m.}$ and it is necessary to consider separately the $\lambda < 1$ and $\lambda > 1$ cases. We get,

$$\cos\theta_{\rm c.m.} = -\lambda\sin^2\theta_{\rm lab} + \cos\theta_{\rm lab}\sqrt{1-\lambda^2\sin^2\theta_{\rm lab}}, \quad \text{for} \quad \lambda < 1 \tag{1.30}$$

and

$$\cos\theta_{\rm c.m.} = -\lambda\sin^2\theta_{\rm lab} \pm \cos\theta_{\rm lab}\sqrt{1-\lambda^2\sin^2\theta_{\rm lab}}, \quad \text{for} \quad \lambda > 1 \tag{1.31}$$

These results are very interesting. First, we notice that there is always a solution for $\lambda < 1$, and this solution is unique. On the other hand, for $\lambda > 1$ the situation is more complicated. If $\sin\theta < 1/\lambda$, the quantity within the square root is positive. Thus, one finds two real values of $\theta_{\rm c.m.}$ as one picks the positive or the negative sign. On the other hand, for $\sin\theta > 1/\lambda$, the quantity within the square root becomes negative. In this case, Eq. (1.31) has no physically acceptable solution.

The situation can be better understood in the graphical representation of figure 1.2. It shows the final velocity of the projectile in the laboratory frame as the vector sum $\mathbf{v}'_{\rm P} = \mathbf{V}_{\rm c.m.} + \mathbf{u}'_{\rm P}$, for $\lambda < 1$ (panel (a)) and $\lambda > 1$ (panel (b)). Since $|\mathbf{u}'_{\rm P}| = v\, m_t/M$ (owing to Eq. (1.19) and energy conservation), the extremity of this vector should be on a circle of radius vm_t/M, centered at the tip of $\mathbf{V}_{\rm c.m.}$ (dashed lines in figures 1.2 (a) and (b)). We know that the vector $\mathbf{v}'_{\rm P}$ is oriented as in the figure (forming an angle $\theta_{\rm lab}$ with the z-axis) but, in principle, we do not know its modulus. The vector $\mathbf{u}'_{\rm P}$ is then determined by the intersection of a line passing by the basis of $\mathbf{V}_{\rm c.m.}$, and the direction of $\mathbf{v}'_{\rm P}$, with the dashed circle (see figure1.2). If $\lambda < 1$, $|\mathbf{V}_{\rm c.m.}| < |\mathbf{u}'_{\rm P}|$. Thus, $\mathbf{V}_{\rm c.m.}$ is contained within the circle, as in panel (a). In this case, the interception leads to a single angle $\theta_{\rm c.m.}$ within the interval $\{0, 180^{\rm o}\}$.

For $\lambda > 1$ (panel (b)), the situation is quite different. In this case, $|\mathbf{V}_{\rm c.m.}| > |\mathbf{u}'_{\rm P}|$, so that the basis of $\mathbf{V}_{\rm c.m.}$ is outside the circle. Now the

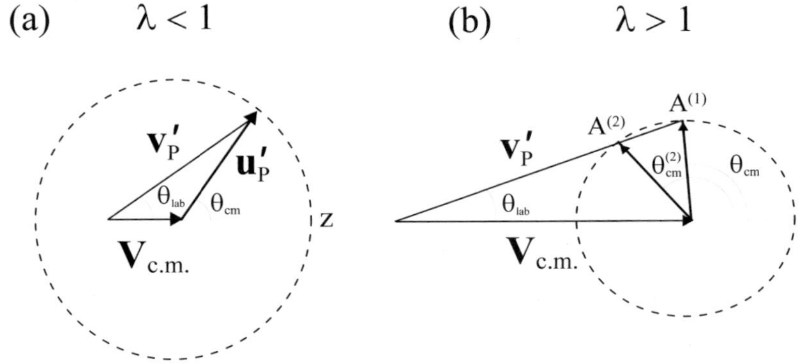

Fig. 1.2 (a) Final velocities in the laboratory frame for $\lambda \equiv m_{\rm P}/m_{\rm T} < 1$ and (b) $\lambda > 1$. For details see the text.

line parallel to $\mathbf{v}'_{\rm P}$ may intercept the circle or not. In the example of panel (b), there are two interceptions, which give rise to two acceptable values of $\theta_{\rm c.m.}$: $\theta^{(1)}_{\rm c.m.}$ and $\theta^{(2)}_{\rm c.m.}$. However, for $\theta_{\rm lab} = 1/\lambda$ the line would be tangent to the circle, so that the two solutions would coincide, and for larger $\theta_{\rm lab}$ there would be no interception. If the experiment measures both $\theta_{\rm lab}$ and $|\mathbf{v}'_{\rm P}|$ it is possible to eliminate the ambiguity in the determination of $\theta_{\rm c.m.}$. However, if one only knows $\theta_{\rm lab}$, the measured cross section will be a mixture of contributions from different $\theta_{\rm c.m.}$. This kind of data would not be very useful to help understanding the physics involved in the collision.

Finally, the transformation for the cross section can be obtained inverting Eq. (1.28):

$$\frac{d\sigma_{\rm el}(\Omega_{\rm c.m.})}{d\Omega_{\rm c.m.}} = \frac{d\sigma_{\rm el}(\Omega_{\rm lab})}{d\Omega_{\rm lab}} \left[\frac{|1 + \lambda \cos\theta_{\rm c.m.}|}{(1 + 2\lambda\cos\theta_{\rm c.m.} + \lambda^2)^{3/2}} \right]. \tag{1.32}$$

This equation cannot be used directly, since it is expressed in terms of the angle in the center of mass frame. However, it can be written in terms of $\theta_{\rm lab}$ with the help of Eq. (1.30).

1.4 Classical scattering

We first study the cross section from the classical mechanics point of view (see, for example [Goldstein (1980)]). We consider the relative motion of the projectile-target system with the Hamiltonian $H(\mathbf{r}, \mathbf{p})$ of Eq. (1.10). If the potential $V(r)$ is conservative and spherically symmetric, the energy E

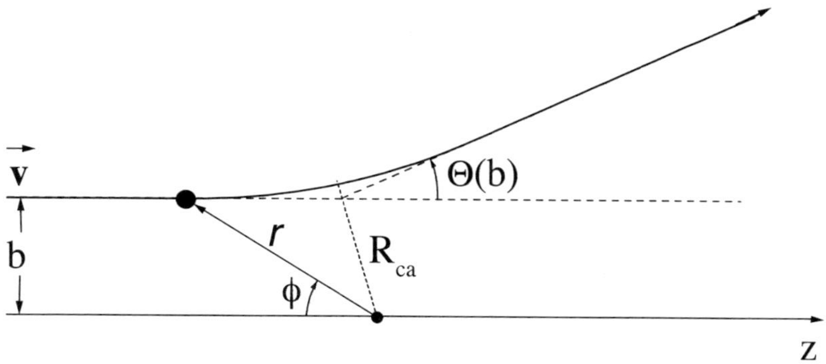

Fig. 1.3 Schematic representation of the scattering of a classical particle with reduced mass μ from a potential.

and the angular momentum vector **L** are conserved. Conservation of the orientation of **L** implies that the trajectory remains on the plane determined by the initial velocity, **v**, and the target. A typical situation is represented in figure 1.3. Choosing the z-axis parallel to the incident velocity, the trajectory is fully determined by the absolute value of the initial velocity

$$v = \sqrt{2E/\mu} \tag{1.33}$$

and the impact parameter

$$b = \frac{L}{\mu\, v}. \tag{1.34}$$

We describe the motion in terms of the coordinates r and ϕ in the collision plane, indicated in figure 1.3. Using conservation of energy and angular momentum, we get

$$E = \frac{1}{2}\mu\left(\dot{r}^2 + r^2\,\dot{\phi}^2\right) + V(r) \tag{1.35}$$

and

$$L = \mu\, vb = \mu\; r^2\,\dot{\phi} \rightarrow \dot{\phi} = \frac{vb}{r^2}. \tag{1.36}$$

Taking $\dot{\phi}$ from Eq. (1.36) and replacing in Eq. (1.35), we get

$$\dot{r} = \pm\, v\sqrt{1 - \frac{b^2}{r^2} - \frac{V(r)}{E}}, \tag{1.37}$$

where the minus and plus signs apply for the incoming and outgoing branches of the trajectory, respectively. We now replace $\dot{\phi} = \dot{r}\ [d\phi/dr]$ in Eq. (1.36) and get

$$\dot{r} = \frac{vb}{r^2}\ \left[\frac{d\phi}{dr}\right]^{-1}. \tag{1.38}$$

Using Eqs. (1.37) and (1.38), we obtain the differential equation

$$\frac{d\phi}{dr} = \pm \frac{b}{r^2 \sqrt{1 - b^2/r^2 - V(r)\,/\,E}}\,.$$

Integrating over the incoming and outgoing branches of the trajectory and using the appropriate sign in each case, we get the asymptotic value of ϕ for a collision with impact parameter b

$$\phi(b) = 2\int_{R_{\rm ca}}^{\infty} \frac{b\;dr}{r^2 \sqrt{1 - b^2/r^2 - V(r)\,/\,E}}\,. \tag{1.39}$$

The lower limit of the integral is the distance of closest approach, $R_{\rm ca}$, determined by the condition

$$1 - b^2/R_{\rm ca}^2 - \frac{V(R_{\rm ca})}{E} = 0\,. \tag{1.40}$$

The deflection of the projectile is determined from $\phi(b)$ (see figure 1.3), through the expression $\Theta(b) = \pi - \phi(b)$, or explicitly

$$\Theta(b) = \pi - 2\int_{R_{\rm ca}}^{\infty} \frac{b\;dr}{r^2 \sqrt{1 - b^2/r^2 - V(r)\,/\,E}}\,. \tag{1.41}$$

$\Theta(b)$ is called the *deflection function.* It plays a very important role in the classical and in semiclassical (see chapter 5) descriptions of the differential elastic cross section. Frequently, the deflection function is given in terms of the angular momentum L, instead of b. In this case, we have

$$\Theta(L) = \pi - 2\int_{R_{\rm ca}}^{\infty} \frac{L\,dr}{r^2 \sqrt{2\mu\,[E - L^2/2\mu r^2 - V(r)]}}\,, \tag{1.42}$$

or

$$\Theta(L) = \pi - 2\int_{R_{\rm ca}}^{\infty} \frac{L\,dr}{r^2 \sqrt{2\mu\;[E - V_L(r)]}}\,, \tag{1.43}$$

where $V_L(r)$ is the effective potential (containing the centrifugal term)

$$V_L(r) = V(r) + \frac{L^2}{2\mu r^2}\,. \tag{1.44}$$

1.4.1 *The classical cross section*

To evaluate $d\sigma(\Omega)/d\Omega$, we consider a beam of projectiles approaching a single target particle with random impact parameter, with intensity of J projectiles per unit area per unit time. If the potential is spherically symmetric, the cross section has axial symmetry and, therefore, $N(\Omega, \Delta\Omega) \to N(\theta, 2\pi \sin\theta\; \Delta\theta)$. In this way, $N(\theta, \Delta\Omega)$ corresponds to the

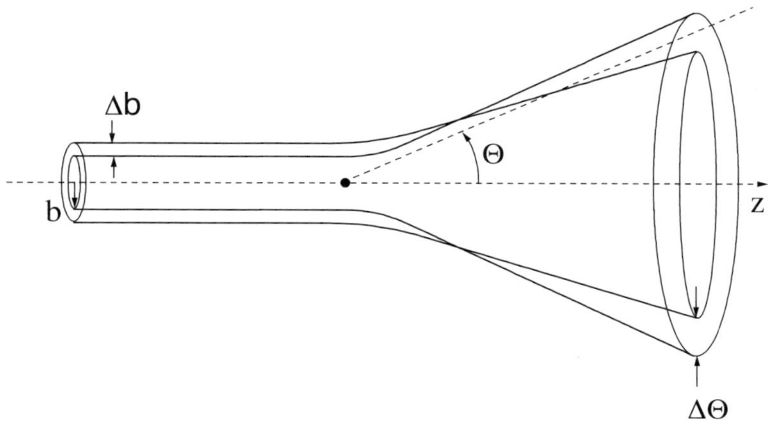

Fig. 1.4 The elementary solid angle $\Delta\Omega$ and the associated area $2\pi\, b\, db$.

number of projectiles scattered per unit time interval between two cones with apertures θ and $\theta + \Delta\theta$, indicated in figure 1.4. Using the inverse of the deflection function[5], we can determine the impact parameters b and $b + \Delta b$ which lead to the deflections θ and $\theta + \Delta\theta$, respectively. The number of particles scattered within the solid angle $\Delta\Omega$ is then given by the number of incident particles in the area between the circles with radii b and $b + \Delta b$, $\Delta S = 2\pi\, b\, \Delta b$. That is,

$$N(\theta, \Delta\Omega) = J\, \Delta S = 2\pi\, b\, \Delta b\, J\,.$$

Inserting this result in Eq. (1.6), with $n = 1$, we get

$$\frac{d\sigma(\theta)}{d\Omega} = \frac{b}{\sin\theta} \left| \frac{\Delta b}{\Delta\theta} \right|, \tag{1.45}$$

where the absolute value is used to guarantee positive cross sections. Using the deflection function, we can express $\Delta\theta$ in terms of Δb. Replacing,

$$\Delta\theta = \frac{d\Theta(b)}{db}\, \Delta b$$

in Eq. (1.45), we obtain the classical differential cross section

$$\frac{d\sigma(\theta)}{d\Omega} = \frac{b}{\sin\theta} \left| \frac{d\Theta(b)}{db} \right|^{-1}. \tag{1.46}$$

[5]Note that the inverse of the deflection function may be multivalued. In some situations, different impact parameters result in the same deflection angle. In this case the cross section is given by the sum of contributions from these impact parameters. On the other hand, some deflection angles can be reached by no impact parameter. The classical cross section then vanishes for this particular angle.

In most situations this expression is inappropriate for comparison with experiment. However, this classical treatment gives good insight into the problem and is helpful as a starting point for approximations. If different values of the impact parameter, $b_1, b_2, .., b_n$ lead to the same angle θ, the classical cross section is generalized as

$$\frac{d\sigma(\theta)}{d\Omega} = \sum_{i=1}^{n} \frac{b_i}{\sin\theta} \left| \frac{d\Theta(b)}{db} \right|_{b=b_i}^{-1} . \tag{1.47}$$

In fact, the quantum mechanical result is rather different. The above formula has the important shortcoming of neglecting interference between the different trajectories leading to the same scattering angle. In chapter 5, we show that this problem can be eliminated in semiclassical approximations for potential scattering.

1.4.2 *Orbiting, rainbow and glory scattering*

Inspecting the deflection function (Eqs. (1.41) or (1.43)), one realizes that the integrand is singular at the lower integration limit. One should therefore investigate the influence of this behavior on the deflection function. For this purpose, we write the integral appearing in Eq. (1.43),

$$I(L) = \int_{R_{\rm ca}}^{\infty} \frac{L\,dr}{r^2\sqrt{2\mu\left[E - V_L(r)\right]}},$$

as $I(L) = I_1(L) + I_2(L)$, where

$$I_1(L) = \int_{R_{\rm ca}}^{R_{\rm ca}+\Delta} \frac{L\,dr}{r^2\sqrt{2\mu\left[E - V_L(r)\right]}} \tag{1.48}$$

and

$$I_2(L) = \int_{R_{\rm ca}+\Delta}^{\infty} \frac{L\,dr}{r^2\sqrt{2\mu\left[E - V_L(r)\right]}} .$$

Since Δ is finite, $I_2(L)$ is bound[6]. In this way, only $I_1(L)$ could give rise to singularities. To investigate this integral, we change $r \to x = r - R_{\rm ca}$, introduce the notation

$$F(x, L) = 2\mu\left[E - V_L(x + R_{\rm ca})\right] \tag{1.49}$$

and express $I_1(L)$ as the limit

$$I_1(L) = L \lim_{\epsilon\to 0} \left[\int_{\epsilon}^{\Delta} \frac{dx}{(x + R_{\rm ca})^2\sqrt{F(x)}} \right] . \tag{1.50}$$

[6]In any relevant situation, the potential asymptotically goes to zero, at least as fast as $1/r$ (Coulomb case). As shown in chapter 3, this integral converges, in spite of the infinite upper limit.

Choosing a small Δ, we can expand

$$F(x,L) = F'(0,L)\,x + F''(0,L)\,\frac{x^2}{2} + O\left[x^3\right], \tag{1.51}$$

in the full interval $[\epsilon, \Delta]$. Above, primes and double-primes stand for first and second derivatives with respect to x. If $F'(0,L) \neq 0$, we obtain

$$I_1(L) = \lim_{\epsilon \to 0} \left[\frac{2L}{R_{\rm ca}^2 \sqrt{F'(0,L)}} \sqrt{x}\,\left(1 + O\left[x\right]\right) \right]_\epsilon^\Delta = \frac{2L}{R_{\rm ca}^2 \sqrt{F'(0,L)}} \sqrt{\Delta}\,. \tag{1.52}$$

We conclude that $I_1(L)$ is finite and therefore the deflection function is well defined.

The situation is rather different when $F'(0,L)$ vanishes, which can be expressed by the condition

$$V_L'(R_{\rm ca}) = 0\,. \tag{1.53}$$

Since $R_{\rm ca}$ depends on both E and L, this equation establishes the *orbiting condition*[7]. For a given value of the collision energy, E, the orbiting condition corresponds to an equation for the orbiting angular momentum, $L_{\rm O}(E)$. Note that this equation does not necessarily have a real solution. In a similar way, given the value of the angular momentum, L, Eq. (1.53) corresponds to an equation for the orbiting energy, $E_{\rm O}(L)$.

For repulsive potentials, $V'(r) < 0 \Longrightarrow V_L'(r) < 0$ for any r and L, so that the above condition is never met. However, for some attractive potentials this condition may be satisfied. Let us consider the orbiting condition in the case of a potential $V(r)$ that is attractive everywhere. This is illustrated in figure 1.5, which shows the effective potentials for the angular momenta[8] L_1, L_2 and L_3. The orbiting energy in each case is determined by the condition that the classical turning point is located at the potential maximum. This situation is indicated in the figure for the angular momenta L_1 and L_2. There is a critical value of the angular momentum, $L_{\rm crit}$, above which the effective potential has no maximum. Above this limit no orbiting occurs. An example of this situation is the effective potential for L_3 in figure 1.5.

---

[7]The reason for this name will be explained later in this section.

[8]Although the potential is attractive everywhere, the centrifugal term is repulsive. If $V(r)$ is limited at the origin and has a finite range, the centrifugal terms dominate at small distances and beyond the range of the potential. Thus, in these regions the effective potentials are positive.

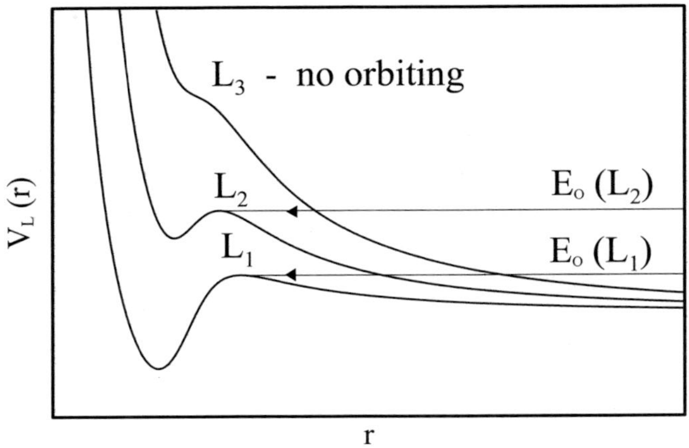

Fig. 1.5 The effective potentials and the orbiting energies for different angular momenta.

When the pair of variables $\{E, L\}$ satisfies the orbiting condition, the dominant term in the expansion of Eq. (1.51) is of second order and it leads to the integral

$$I_1(L) = L\sqrt{\frac{2}{F''(0,L)}}\lim_{\epsilon\to 0}\left[\int_\epsilon^\Delta \frac{dx}{x}\right] \to \infty\,.$$

In this case, $I(L)$ has a logarithmic divergence and we can write

$$\lim_{L\to L_O} \Theta(L) \to -\infty. \tag{1.54}$$

The meaning of this equation is that in a collision with energy E and angular momentum L_O the projectile is caught in an orbit with constant radius $r = R_{ca}$ and it does not emerge. In collisions with $L \simeq L_O$ the projectile goes around the target several times before coming out.

In figure 1.6 we show deflection functions presenting orbiting. In (a), the potential is attractive everywhere, so that the deflection angle is negative for any angular momentum. In (b), the potential is the superposition of a short-range attractive part with a long-range repulsion. This situation is discussed in chapter 3, from the quantum mechanical point of view. The orbiting angular momentum, L_O, is indicated in each case.

In the scattering from a potential which is attractive at small separations and repulsive at large distances, as the one used in figure 1.6 b, other

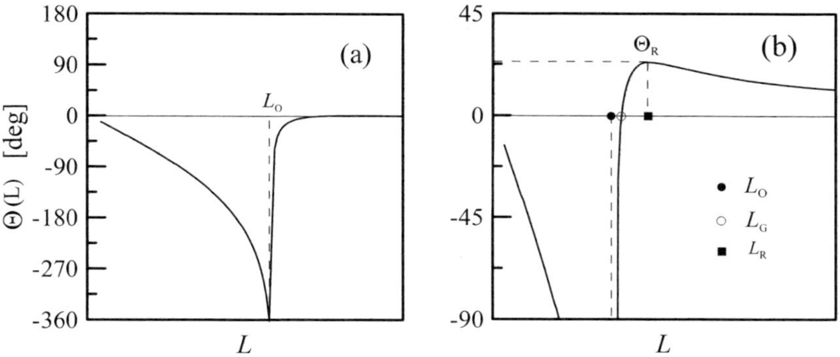

Fig. 1.6 Deflection functions for attractive potentials. In (a) the effective potential V_L (r) is attractive everywhere. The deflection angle is then negative. In (b) V_L (r) is attractive at small separations, which are reached in collisions with small angular momenta, and is repulsive at large distances. The orbiting angular momentum, L_O, is indicated in each case. In (b), the rainbow and glory angular momenta are also indicated.

interesting phenomena can occur[9]. The first one is *rainbow scattering*, arising for the angular momentum L_R (see figure 1.6 b). At this angular momentum, the deflection function goes through a maximum. This has two important consequences. The first one is that, according to classical mechanics, no particle of the incident beam is deflected to angles $\theta > \Theta_R$. The second consequence is that the classical scattering cross section at this angle diverges. This could be checked inserting $[d\Theta/db]_{b=b_R} = [d\Theta/dL]_{L=L_R} = 0$ in Eq. (1.46). In fact, the predictions of quantum mechanics, which are consistent with observation, are not the same. The cross section remains finite at $\theta = \Theta_R$ and it does not vanish at larger angles. However, the classical results are qualitatively correct, in the sense that the observed cross section presents a maximum near the rainbow angle and it shows a shadow region beyond this angle, where $d\sigma/d\Omega$ decreases rapidly. This situation is analogous to the rainbow phenomenon, that occurs in the scattering of light from liquid drops contained in clouds (for an accessible review article we refer to [Nussenzveig (1977)]). In section 5.2.3, we develop a semiclassical approximation for rainbow scattering with no singularities, and illustrate it with applications in atomic physics, nuclear physics and in the meteorological rainbow.

[9]Similar phenomena can occur for interactions that are attractive at long distances and repulsive at short distances. This is the case of the Lennard-Jones potential, discussed in chapter 5 (see Eq. (5.368)).

The classical cross section also diverges when the projectile is scattered at $\theta = 0$ or $\theta = \pi$, in a collision with finite impact parameter, since in such cases $\sin\theta = 0$. This can happen in the scattering from potentials with attractive and repulsive regions, such as the one used in figure 1.6 b. For some particular angular momentum, a combination of attraction and repulsion along the trajectory may lead to scattering to the forward or the backward direction. In 1.6 b, this behavior occurs at the angular momentum $L_{\rm G}$, for which the scattering angle vanishes. An analogous phenomenon, called the *glory effect*, occurs in Optics. There, a *halo* is formed around the shadow of an object, which is produced by light scattering to forward angles. As in the case of the rainbow scattering, wave mechanical effects eliminate the divergence of the cross section.

1.5 Stationary scattering of a plane wave

We now consider the cross section from the quantum mechanical point of view. We first study the stationary states in the scattering of a plane wave from a short range potential, $V(\mathbf{r})$. Although this problem does not seem to correspond to the scattering of a beam of particles considered in section 1.3, we will show in the next section that it can be a good approximation to it.

In this case, the wave function $\psi(\mathbf{r},t)$ must satisfy the Schrödinger equation (1.11). A stationary state with energy E can be written

$$\psi(\mathbf{r},t) = \psi(\mathbf{r})\; e^{-\frac{i}{\hbar}E\,t} \equiv \psi(\mathbf{r})\; e^{-i\,\omega\,t}. \tag{1.55}$$

Using Eq. (1.55), Eq. (1.11) can be put in the form

$$\left(\nabla^2 + k^2\right)\; \psi(\mathbf{r}) = U(r)\; \psi(\mathbf{r}). \tag{1.56}$$

Above,

$$\omega = \frac{E}{\hbar}, \quad k = \frac{\sqrt{2\mu E}}{\hbar} \quad \text{and} \quad U(r) = \frac{2\mu}{\hbar^2}\; V(r). \tag{1.57}$$

We look for a solution of Eq. (1.56) corresponding to the distortion of the plane wave

$$\phi(\mathbf{k};\mathbf{r}) = A\, e^{i\mathbf{k}\cdot\mathbf{r}}, \tag{1.58}$$

where A is an arbitrary normalization constant. Treating the RHS of Eq. (1.56) as a source, we write the desired wave function in the form

$$\psi^{(+)}(\mathbf{k};\mathbf{r}) = \phi(\mathbf{k};\mathbf{r}) + \psi^{\rm sc}(\mathbf{k};\mathbf{r}), \tag{1.59}$$

where $\psi^{\rm sc}(\mathbf{k};\mathbf{r})$ is a particular solution which sets the boundary conditions. In our case, it should be an outgoing spherical wave, produced by the scatterer. For this solution, we use the notation $\psi^{(+)}(\mathbf{k};\mathbf{r})$, where the upper script $^{(+)}$ indicates the outgoing character of the spherical wave. Near the detector, $r \to \infty$ and $\psi^{\rm sc}(\mathbf{k};\mathbf{r})$ should have the asymptotic behavior[10]

$$\psi^{\rm sc}(\mathbf{k};\mathbf{r}) \to A\, f(\Omega)\, \frac{e^{ikr}}{r}. \tag{1.60}$$

Above, $f(\Omega) \equiv f(\theta, \varphi)$ is the scattering amplitude, which plays a major role in calculations of the cross section. It is responsible for the orientation dependence of the strength of the outgoing spherical wave. The factor $1/r$ keeps the flux constant through portions of spherical surfaces intercepting a cone with vertex on the scatterer. This will become clear below, when we evaluate the elastic cross section.

Using Eqs. (1.60) and (1.58) in Eq. (1.59) we get the asymptotic form of the wave function

$$\psi^{(+)}(\mathbf{k};\mathbf{r}) \to A\, \left(e^{i\mathbf{k}\cdot\mathbf{r}} \;+\; f(\Omega)\, \frac{e^{ikr}}{r} \right). \tag{1.61}$$

Eq. (1.61) corresponds to the superposition of the incident plane wave with an outgoing spherical wave, both propagating with speed (phase velocity)

$$v_{\rm P} = \frac{\omega}{k} \; = \frac{1}{2} \left(\frac{\hbar k}{\mu} \right). \tag{1.62}$$

We remark that this speed, which is obtained from the condition of constant phase, $d(kr - \omega t)/dt = 0$, is half of that corresponding to a classical particle with momentum $\hbar k$. This is not a real problem as the plane wave cannot represent a localized particle, which should be given by a wave packet. The velocity of the classical particle then corresponds to the group velocity of the wave packet and there is no inconsistency.

We now turn to the calculation of the differential scattering cross section, defined in Eq. (1.6). First, we must obtain the count rate $N(\Omega, \Delta\Omega)$. This quantity is given by the flux of the scattered current $\mathbf{j}^{\rm sc}$,

$$\mathbf{j}^{\rm sc} = \frac{\hbar}{2\mu i} \; \left[\left(\psi^{\rm sc}(\mathbf{k};\mathbf{r})\right)^* \; \nabla\psi^{\rm sc}(\mathbf{k};\mathbf{r}) \;-\; \psi^{\rm sc}(\mathbf{k};\mathbf{r})\; \nabla \left(\psi^{\rm sc}(\mathbf{k};\mathbf{r})\right)^* \right], \tag{1.63}$$

[10]Note that the spherical wave below, e^{ikr}/r, does have the desired outgoing nature. This is clear if one looks at its global phase (including the time dependence), $\alpha(r,t) = kr - \omega t$. Since the propagation velocity is given by the motion of the wave front, setting $\dot\alpha = 0$ we get the radial propagation velocity $\dot r = \omega/k$. Since it has a positive sign, the wave propagates away from the scatterer.

through the detector surface

$$\mathbf{\Delta S} = \Delta S\ \hat{\mathbf{r}} = \Delta\Omega\ r^2\ \hat{\mathbf{r}}, \tag{1.64}$$

where $\hat{\mathbf{r}}$ is the unit vector along the radial direction, pointing from the scatterer outwards. We get,

$$N(\Omega, \Delta\Omega) = \mathbf{j}^{\rm sc} \cdot \mathbf{\Delta S} = \left(\Delta\Omega\ r^2\right)\ \hat{\mathbf{r}} \cdot \mathbf{j}^{\rm sc}. \tag{1.65}$$

Writing the gradient in polar coordinates,

$$\nabla = \hat{\mathbf{r}}\ \frac{\partial}{\partial_r} + \hat{\theta}\frac{1}{r}\frac{\partial}{\partial_\theta} + \hat{\varphi}\ \frac{1}{r\sin\theta}\ \frac{\partial}{\partial\varphi}, \tag{1.66}$$

using the asymptotic form of $\psi^{\rm sc}$ (Eq. (1.60)) in Eq. (1.63), and inserting the result in Eq. (1.65), we obtain

$$\begin{aligned} N(\Omega, \Delta\Omega) &\simeq |\ f(\Omega)|^2\ \ |A|^2\ \ \Delta\Omega\ r^2 \\ &\times \left\{ \frac{\hbar}{2\mu i}\ \left[\frac{e^{-ikr}}{r}\ \frac{d}{dr}\left(\frac{e^{ikr}}{r}\right) - \frac{e^{ikr}}{r}\ \frac{d}{dr}\left(\frac{e^{-ikr}}{r}\right)\right]\right\}. \end{aligned}$$

Carrying out the derivation with respect to r, we get

$$N(\Omega, \Delta\Omega) = v\ \ |\ f(\Omega)|^2\ \ |A|^2\ \ \Delta\Omega, \tag{1.67}$$

where $v = \hbar k/\mu$ is the velocity of a classical particle with momentum $\hbar k$. To calculate the cross section, we need the beam intensity, J, given by the current associated with the incident plane wave, $\mathbf{j}^{\rm in}$:

$$J = \hat{\mathbf{z}} \cdot \mathbf{j}^{\rm in} = \frac{\hbar}{2\mu i}\ \left[e^{-ikz}\frac{d}{dz}\left(e^{ikz}\right) - e^{ikz}\frac{d}{dz}\left(e^{-ikz}\right)\right] = v\ \ |A|^2. \tag{1.68}$$

Substituting Eqs. (1.67) and (1.68) in Eq. (1.6), we obtain the quantum mechanical differential cross section

$$\frac{d\sigma_{el}(\Omega)}{d\Omega} = |\ f(\Omega)|^2\ . \tag{1.69}$$

Note that the total current,

$$\begin{aligned} \mathbf{j} &= \frac{\hbar}{2\mu i}\ \left[(\psi^{(+)}(\mathbf{k};\mathbf{r}))^*\ \nabla\psi^{(+)}(\mathbf{k};\mathbf{r}) - \psi^{(+)}(\mathbf{k};\mathbf{r})\ \nabla\left(\psi^{(+)}(\mathbf{k};\mathbf{r})\right)^*\right] \\ &= \mathbf{j}^{\rm in} + \mathbf{j}^{\rm sc} + \mathbf{j}^{\rm int}, \end{aligned} \tag{1.70}$$

contains the cross term

$$\mathbf{j}^{\rm int} = \frac{\hbar}{\mu i}\ \mathrm{Re}\left\{\phi^*(\mathbf{k};\mathbf{r})\ \nabla\psi^{\rm sc}(\mathbf{k};\mathbf{r}) + (\psi^{\rm sc}(\mathbf{k};\mathbf{r}))^*\ \nabla\phi(\mathbf{k};\mathbf{r})\right\},$$

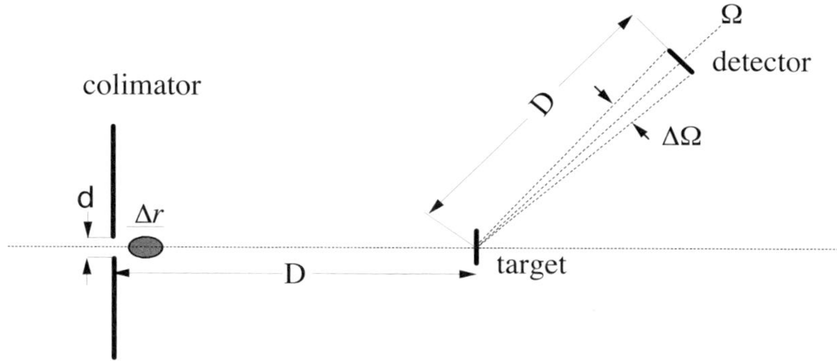

Fig. 1.7 Characteristic lengths in the collision of a wave packet.

which has been disregarded. This procedure is justified by the oscillating behavior of $\mathbf{j}^{\text{int}}$ at any $\theta \neq 0$. Since any experiment has a finite angular resolution, the average over the detector aperture leads to destructive interference. Therefore, its contribution can be neglected, except at very forward angles. For a detailed discussion of this point, we refer to [Joachain (1983)].

1.6 Scattering of a wave packet

We now consider the scattering of a beam of wave packets, which is the appropriate description of a scattering experiment in quantum mechanics. First we discuss the typical dimensions involved. The collision of one wave packet with a single target is represented in figure 1.7. The wave packet goes through a collimator with aperture d, of the order of a few millimeters. The size of the wave packet, Δr, should have the same order of magnitude; $\Delta r \sim d \sim 1$ mm. To reach the detector, it traverses a distance D_1 - from the collimator to the target - and a distance D_2 - from the target to the detector. Typically, these distances are of the order $D \sim 10\,\text{cm} - 1\,\text{m}$. Before the collision, the wave packets propagate freely. They are deformed as the collision takes place, and then move on, propagating freely again.

1.6.1 *Propagation of a free wave packet*

Let us first consider the propagation of a free wave packet. Since the wave packet is a solution of the Schrödinger equation with the free particle

Hamiltonian, it can be written as the superposition of plane waves,

$$\Phi(\mathbf{k}_0;\mathbf{r},t) = \int d^3\mathbf{k}\ A(\mathbf{k}-\mathbf{k}_0)\ \phi(\mathbf{k};\mathbf{r},t) \tag{1.71}$$

$$= \frac{1}{(2\pi)^{3/2}} \int d^3\mathbf{k}\ A(\mathbf{k}-\mathbf{k}_0)\, e^{i(\mathbf{k}\cdot\mathbf{r}-\omega_k t)}\ , \tag{1.72}$$

where $\omega_k = \hbar k^2/2\mu$. Above, $A(\mathbf{k}-\mathbf{k}_0)$ is the wave packet in momentum space. It is centered at $\mathbf{k}_0$ and has a width Δk, given by the uncertainty relation $\Delta k \sim 1/\Delta r$. The wave packet $A(\mathbf{k}-\mathbf{k}_0)$ can be found by evaluating the inverse Fourier transform of $\Phi(\mathbf{k}_0;\mathbf{r},t)$ at $t=0$. $\Phi(\mathbf{k}_0;\mathbf{r},t)$ and $A(\mathbf{k}-\mathbf{k}_0)$ are normalized as

$$\int\ |\Phi(\mathbf{k}_0;\mathbf{r},t)|^2\ d^3\mathbf{r} = \int\ |A(\mathbf{k}-\mathbf{k}_0)|^2\ d^3\mathbf{k} = 1\,. \tag{1.73}$$

It is important to establish some relations involving the characteristic sizes: *i*) the macroscopic lengths D and d, *ii*) the size of the packet Δr, *iii*) the range $\bar{R}$ of the interaction (a microscopic quantity), and *iv*) the wavelength associated with the dominant wave number $\lambdabar_0 = \lambda_0/2\pi = 1/k_0$. In order that only scattered particles reach the detector, it is necessary that the uncertainty in the transverse momentum be much smaller than the dominant momentum. This condition implies that $\Delta k \ll k_0$ or, equivalently, $\lambdabar_0 \ll \Delta r$. On the other hand, the range of the interaction is $\bar{R} \sim 10^{-15}$ m for nuclei and $\bar{R} \sim 10^{-10}$ m for atoms. Thus, we can write

$$D \gg d \sim \Delta r \gg \lambdabar_0. \tag{1.74}$$

To study the evolution of the free packet, we separate the modulus of $A(\mathbf{k}-\mathbf{k}_0)$ from its phase, writing

$$A(\mathbf{k}-\mathbf{k}_0) = |A(\mathbf{k}-\mathbf{k}_0)|\ e^{i\alpha(\mathbf{k})}\,, \tag{1.75}$$

and insert it into Eq. (1.71). We get

$$\Phi(\mathbf{k}_0;\mathbf{r},t) = \frac{1}{(2\pi)^{3/2}} \int d^3\mathbf{k}\ \ |A(\mathbf{k}-\mathbf{k}_0)|\ e^{i\beta}\ , \tag{1.76}$$

where

$$\beta \equiv \beta(\mathbf{k},\mathbf{r},t) = \mathbf{k}\cdot\mathbf{r} - \omega_k\, t + \alpha(\mathbf{k})\,.$$

Since β tends to change very rapidly with $\mathbf{k}$ around $\mathbf{k} = \mathbf{k}_0$, which is the region where $|A(\mathbf{k}-\mathbf{k}_0)|$ is relevant, the integrand of Eq. (1.76) oscillates strongly. Therefore, the integral is negligible, except in the vicinity of some point $\mathbf{R}_t$ where $\beta(\mathbf{k},\mathbf{R}_t,t)$ becomes stationary at $\mathbf{k} = \mathbf{k}_0$. This condition is expressed as

$$[\nabla_\mathbf{k}\ \beta(\mathbf{k},\mathbf{R}_t,t)]_{\mathbf{k}=\mathbf{k}_0} = 0\,. \tag{1.77}$$

Using the results

$$\nabla_{\mathbf{k}} (\mathbf{k} \cdot \mathbf{r}) = \mathbf{r} \tag{1.78}$$

$$[\nabla_{\mathbf{k}} (\omega_k \, t)]_{\mathbf{k}=\mathbf{k}_0} = \frac{t \, \hbar}{2\mu} \; [\nabla_{\mathbf{k}}(k^2)]_{\mathbf{k}=\mathbf{k}_0} = \mathbf{v}_0 \, t \, ,$$

where $\mathbf{v}_0 = \hbar \mathbf{k}_0/\mu$, and defining (this notation will be soon justified)

$$[\nabla_{\mathbf{k}} \; \alpha(\mathbf{k})]_{\mathbf{k}=\mathbf{k}_0} = - \, \mathbf{r}_0 \, ,$$

Eq. (1.77) reduces to the condition

$$\mathbf{R}_t = \mathbf{r}_0 + \mathbf{v}_0 t \, . \tag{1.79}$$

Therefore, the wave function is concentrated around the center of the packet, $\mathbf{R}_t$, which moves according to Eq. (1.79). Note that at $t = 0$, $\mathbf{R}_t = \mathbf{r}_0$. This justifies the notation $[\nabla_{\mathbf{k}} \; \alpha(\mathbf{k})]_{\mathbf{k}=\mathbf{k}_0} = - \, \mathbf{r}_0$. It is convenient to choose the z-axis along the beam direction and set the time scale such that at $t = 0$ the center of the wave packet is on the x-y plane ($z_0 = 0$). With these choices, $\mathbf{r}_0$ corresponds to the impact parameter vector $\mathbf{b}$ and we have

$$[\nabla_{\mathbf{k}} \, \alpha(\mathbf{k})]_{\mathbf{k}=\mathbf{k}_0} = -\mathbf{r}_0 = -\mathbf{b}; \tag{1.80}$$

$$\alpha(\mathbf{k}) \simeq \alpha(\mathbf{k}_0) - \mathbf{b} \cdot (\mathbf{k} - \mathbf{k}_0) \, . \tag{1.81}$$

Eq. (1.80) indicates that the impact parameter information is fully contained in $\mathbf{k}$-dependence of the phase of $A(\mathbf{k} - \mathbf{k}_0)$ in the vicinity of $\mathbf{k} = \mathbf{k}_0$. Eq. (1.79) then becomes

$$\mathbf{R}_t = \mathbf{b} + \mathbf{v}_0 \, t \, . \tag{1.82}$$

Near the center of the wave-packet, i.e. $\mathbf{r} \simeq \mathbf{R}_t$, the stationarity condition is not exactly fulfilled. In this case, $[\nabla_{\mathbf{k}} \; \beta(\mathbf{k}, \mathbf{r}, t)]_{\mathbf{k}=\mathbf{k}_0}$ is small but does not vanish. The wave function $\Phi(\mathbf{k}_0; \mathbf{r}, t)$ can then be calculated expanding $\beta(\mathbf{k}, \mathbf{r}, t)$ around $\mathbf{k} = \mathbf{k}_0$. Assuming that second and higher order terms are negligible, we can write

$$\beta(\mathbf{k}, \mathbf{r}, t) \simeq [\alpha(\mathbf{k}_0) + \mathbf{k}_0 \cdot \mathbf{r} - \omega_0 \, t] + \; [\nabla_{\mathbf{k}} \; \beta(\mathbf{k})]_{\mathbf{k}=\mathbf{k}_0} \cdot (\mathbf{k} - \mathbf{k}_0) \, , \tag{1.83}$$

where $\omega_0 = \omega_{k_0}$. It is convenient to choose the global phase of the wave-packet $A(\mathbf{k} - \mathbf{k}_0)$ so that

$$\alpha(\mathbf{k}_0) = 0.$$

Using Eq. (1.83), Eq. (1.76) takes the form

$$\Phi_{\mathbf{b}}\left(\mathbf{k}_0;\mathbf{r},t\right) \simeq e^{i(\mathbf{k}_0\cdot\mathbf{r}-\omega_0\, t)}\; \chi\left(\mathbf{r}-\mathbf{R}_{\mathrm{T}}\right)\,, \tag{1.84}$$

where

$$\chi(\mathbf{r}-\mathbf{R}_t) = \frac{1}{(2\pi)^{3/2}}\int d^3\mathbf{q}\;|A(\mathbf{q})|\; e^{i(\mathbf{r}-\mathbf{R}_t)\cdot\,\mathbf{q}}\,, \tag{1.85}$$

with

$$\mathbf{q} = \mathbf{k} - \mathbf{k}_0\ .$$

In Eq. (1.84), we use the notation $\Phi_{\mathbf{b}}$ to emphasize the impact parameter dependence. Note that the wave packet has been written as a product of two factors. The first corresponds to a plane wave for the dominant wave vector $\mathbf{k}_0$. The second gives the shape of the packet, which is independent of $\mathbf{k}_0$. Eq. (1.73) yields, at any t,

$$\int d^3\mathbf{r}\;|\chi(\mathbf{r}-\mathbf{R}_t)|^2 = 1\ . \tag{1.86}$$

The validity of the expansion of Eq. (1.83) requires that second order terms be negligible. A necessary condition is that for $q_i \sim \Delta k_i;\; q_j \sim \Delta k_j$

$$\begin{aligned}\frac{1}{2}&\left|\sum_{i,j=1}^{3}\left[\frac{\partial^2\beta(\mathbf{k},\mathbf{r},t)}{\partial k_i\,\partial k_j}\right]_{\mathbf{k}=\mathbf{k}_0} q_i\, q_j\right| \\ &= \frac{1}{2}\left|\sum_{i,j=1}^{3}\left[\frac{\partial^2\alpha(\mathbf{k})}{\partial k_i\,\partial k_j} - \frac{\partial^2\left(t\omega_k\right)}{\partial k_i\,\partial k_j}\right]_{\mathbf{k}=\mathbf{k}_0} q_i\, q_j\right| \ll 1.\end{aligned} \tag{1.87}$$

Since the incident wave packet is sharp in momentum space, the contribution from the second derivatives of $\alpha(\mathbf{k})$ should be small. However, the contribution from $t\omega_k$ grows with t, so that the above condition cannot be satisfied as $t\to\infty$. Eq. (1.87) then leads to the requirement

$$\frac{1}{2}\sum_{i,j=1}^{3}\left[\frac{\partial^2\left(t\omega_k\right)}{\partial k_i\,\partial k_j}\right]_{\mathbf{k}=\mathbf{k}_0} q_i\, q_j = \frac{\hbar(\Delta k)^2}{\mu}\; t \ll 1.$$

Since the relevant values of t correspond to the time needed for the packet to go from the collimator to the detector, we should take $t\simeq 2D/v_0$. Using that $\Delta k \approx 1/\Delta r \approx 1/d$, we get the condition

$$\sqrt{\lambda D} \ll d,$$

which is met in most relevant situations.

1.6.2 *Collision of the wave packet with a target*

We now consider the wave packet given by the same $A(\mathbf{k}-\mathbf{k}_0)$ of the previous section, colliding with a target particle represented by the potential V. The wave function $\Psi^{(+)}(\mathbf{k}_0;\mathbf{r},t)$ should then be given by Eq. (1.71), with the replacement

$$\phi(\mathbf{k};\mathbf{r}) = \frac{\mathbf{1}}{(2\pi)^{3/2}}\, e^{i\mathbf{k}\cdot\mathbf{r}} \quad \to \quad \psi^{(+)}(\mathbf{k};\mathbf{r}).$$

That is,

$$\Psi_{\mathbf{b}}^{(+)}(\mathbf{k}_0;\mathbf{r},t) = \int d^3\mathbf{k}\ \left|A\left(\mathbf{k}-\mathbf{k}_0\right)\right|\ e^{i\alpha(\mathbf{k})}\, e^{-i\,\omega_k\,t}\ \psi^{(+)}(\mathbf{k};\mathbf{r}),$$

where $\psi^{(+)}(\mathbf{k};\mathbf{r})$ is the stationary state with outgoing wave boundary condition, discussed in section 1.5. Expanding $\alpha(\mathbf{k})$ to lowest order around $\mathbf{k}=\mathbf{k}_0$ and dropping the constant phase $\alpha\left(\mathbf{k}_0\right)$, the above equation becomes

$$\Psi_{\mathbf{b}}^{(+)}(\mathbf{k}_0;\mathbf{r},t) = \int d^3\mathbf{k}\ \left|A\left(\mathbf{k}-\mathbf{k}_0\right)\right|\ e^{-i(\mathbf{k}-\mathbf{k}_0)\cdot\mathbf{b}}\ e^{-i\,\omega_k\,t}\,\psi^{(+)}(\mathbf{k};\mathbf{r}). \tag{1.88}$$

Writing $\psi^{(+)}(\mathbf{k};\mathbf{r})$ as in Eq. (1.59),

$$\psi^{(+)}(\mathbf{k};\mathbf{r}) = \phi(\mathbf{k},\mathbf{r}) + \psi^{\rm sc}(\mathbf{k};\mathbf{r}),$$

$\Psi_{\mathbf{b}}^{(+)}(\mathbf{k}_0;\mathbf{r},t)$ splits into two wave-packets, in the form

$$\Psi_{\mathbf{b}}^{(+)}(\mathbf{k}_0;\mathbf{r},t) = \Phi_{\mathbf{b}}(\mathbf{k}_0;\mathbf{r},t) + \Psi_{\mathbf{b}}^{\rm sc}(\mathbf{k}_0;\mathbf{r},t)\,. \tag{1.89}$$

Above, $\Phi_{\mathbf{b}}(\mathbf{k}_0;\mathbf{r},t)$ corresponds to the incident wave-packet propagating freely and $\Psi_{\mathbf{b}}^{\rm sc}(\mathbf{k}_0;\mathbf{r},t)$ is a scattered packet, produced by the interaction with the target.

Let us consider the behavior of $\Psi_{\mathbf{b}}^{(+)}(\mathbf{k}_0;\mathbf{r},t)$ at the $t\to\pm\infty$ limits. In these limits, the packet is far away from the potential center and we can use the asymptotic form of the stationary state (Eq. (1.61) with $A=(2\pi)^{-3/2}$)

$$\psi^{(+)}(\mathbf{k};\mathbf{r}) \to \frac{\mathbf{1}}{(2\pi)^{3/2}}\left(e^{i\mathbf{k}\cdot\mathbf{r}} + f_{\mathbf{k}}(\Omega)\ \frac{e^{ikr}}{r}\right), \tag{1.90}$$

or,

$$\psi^{\rm sc}(\mathbf{k};\mathbf{r}) \to \frac{\mathbf{1}}{(2\pi)^{3/2}}\, f_{\mathbf{k}}(\Omega)\ \frac{e^{ikr}}{r}\,.$$

With this procedure, the scattered packet becomes

$$\Psi_{\mathbf{b}}^{\rm sc}(\mathbf{k}_0;\mathbf{r},t) = \frac{1}{(2\pi)^{3/2}}\ \frac{1}{r}\int d^3\mathbf{k}\ \left|A\left(\mathbf{k}-\mathbf{k}_0\right)\right|\ \ f_{\mathbf{k}}(\Omega)\ e^{i\gamma(\mathbf{k})}. \tag{1.91}$$

Above, we have introduced the notation

$$\gamma(\mathbf{k}) = kr - \omega_k t + \alpha(\mathbf{k}) \ .$$

Using Eq. (1.80) with the constant phase $\alpha(\mathbf{k}_0)$ neglected, $\gamma(\mathbf{k})$ becomes

$$\gamma(\mathbf{k}) \simeq kr - \omega_k t - (\mathbf{k} - \mathbf{k}_0) \cdot \mathbf{b}. \tag{1.92}$$

Let us now look for stationary points in the integrand. In typical situations, $f_{\mathbf{k}}(\Omega)$ is a slowly varying function[11] of $\mathbf{k}$ within the range of relevant values of $|A(\mathbf{k} - \mathbf{k}_0)|$, so that only $\gamma(\mathbf{k})$ must be considered. We should then expand this phase to first order around $\mathbf{k} = \mathbf{k}_0$. Since

$$[\nabla_{\mathbf{k}}(kr)]_{\mathbf{k}=\mathbf{k}_0} = r\,\hat{\mathbf{k}}_0 = r\;\hat{\mathbf{z}}; \quad [\nabla_{\mathbf{k}}(t\omega_k)]_{\mathbf{k}=\mathbf{k}_0} = v_0\,t\,\hat{\mathbf{z}}; \quad [\nabla_{\mathbf{k}}(\mathbf{k}\cdot\mathbf{b})]_{\mathbf{k}=\mathbf{k}_0} = \mathbf{b}\,,$$

a stationary point at $\mathbf{k} = \mathbf{k}_0$ requires that

$$\Big[\nabla_{\mathbf{k}}\,\gamma(\mathbf{k})\Big]_{\mathbf{k}=\mathbf{k}_0} = (r - v_0\,t)\;\hat{\mathbf{z}} - \mathbf{b} = 0. \tag{1.93}$$

Since we are considering the scattered wave at asymptotic distances, we should analyze the condition for stationary phase long before and after $t = 0$, which is the time of closest approach. That is, we should look at $t \to \pm\infty$. As $t \to -\infty$, $r \to \infty$, so that $r - v_0\,t = r + v_0\,|t|$ is a very large positive definite quantity and the condition of Eq. (1.93) cannot be satisfied. This implies that $\Psi^{\text{sc}}(\mathbf{r}, t)$ is vanishingly small and $\Psi^{(+)}_{\mathbf{b}}(\mathbf{k}_0; \mathbf{r}, t)$ reduces to the free wave packet $\Phi_{\mathbf{b}}(\mathbf{k}_0; \mathbf{r}, t)$. Therefore, we have

$$\Psi^{(+)}_{\mathbf{b}}(\mathbf{k}_0; \mathbf{r}, t \to -\infty) = \Phi_{\mathbf{b}}(\mathbf{k}_0; \mathbf{r}, t \to -\infty). \tag{1.94}$$

This is not surprising since at $t \to -\infty$ the wave packet had not reached the range of the interaction and the collision had not started. Therefore, there can be no scattered wave.

We now consider the $t \to +\infty$ limit, which is the one relevant for the scattering cross section. Eq. (1.93) cannot be exactly satisfied except for a head-on collision ($\mathbf{b} = 0$). However, an approximate stationary point occurs under two conditions:

(1) First, the $\hat{\mathbf{z}}$-component of Eq. (1.93) vanishes. That is,

$$r = v_0 t. \tag{1.95}$$

Eq. (1.95) is satisfied for an outgoing spherical wave propagating with the same speed as the incident packet. Thus, $\Psi^{(+)}_{\mathbf{b}}(\mathbf{k}_0; \mathbf{r}, t \to -\infty)$ is the superposition of the incident wave packet with a radial scattered packet centered at $r = v_0 t$.

[11] The situation is different if there is a resonance near the energy $E_0 = \hbar^2 k_0^2 / 2\mu$. This case will be considered in section 2.9.1

(2) The second condition is that the size of the wave packet in the coordinate space, $\Delta r = 1/\Delta k$, be much larger than the range of the interaction, $\bar{R}$. In this case, the scattering of the wave packet is equivalent to the diffusion of a plane wave[12]. To derive this condition, we point out that $[\nabla_{\mathbf{k}}\,\gamma(\mathbf{k})]_{\mathbf{k}=\mathbf{k}_0}$ does not have to vanish identically. It is sufficient that $\gamma(\mathbf{k})$ changes slowly within the range of the wave packet in momentum space, Δk. Expanding $\gamma(\mathbf{k})$ to first order around $\mathbf{k}_0$ and using the notation $\mathbf{q} = \mathbf{k} - \mathbf{k}_0$, one should have

$$\left| \left(\nabla_{\mathbf{k}}\,\gamma(\mathbf{k})\right)_{\mathbf{k}=\mathbf{k}_0} \cdot\ \mathbf{q} \right| \ll 1. \tag{1.96}$$

This condition should be satisfied for any $|\mathbf{q}| \lesssim \Delta k$, where the integrand of Eq. (1.91) has appreciable values. Owing to Eqs. (1.93) and (1.95), the above equation reduces to

$$\left| \ \mathbf{b} \cdot \mathbf{q} \ \right| \ll 1. \tag{1.97}$$

Since the projectiles with impact parameter larger than the range of the interaction, $\bar{R}$, are not scattered[13], and the integrand is negligible for $|\mathbf{q}| > \Delta k$, the most unfavorable situation to use the first order expansion of $\gamma(\mathbf{k})$ is when $|\mathbf{b}|$ and $|\mathbf{q}|$ have their maximal values: $\bar{R}$ and Δk, respectively. The above equation then becomes

$$\bar{R}\ \Delta k \ll 1. \tag{1.98}$$

Or, using the uncertainty relation for the wave packet,

$$\Delta r \gg \bar{R}. \tag{1.99}$$

Let us consider the scattered wave packet in further detail. For this purpose, we expand the phase $\gamma(\mathbf{k})$ to first order around $\mathbf{k} = \mathbf{k}_0$, and obtain

$$\begin{aligned} \gamma(\mathbf{k}) &\simeq [k_0 r - \omega_0\, t] + [(r - v_0\, t)\,\hat{\mathbf{z}} - \mathbf{b}] \cdot \mathbf{q} \\ &= [k_0 r - \omega_0\, t] + (r\hat{\mathbf{z}} - \mathbf{R}_t)\ \cdot \mathbf{q}\,. \end{aligned} \tag{1.100}$$

Eq. (1.91) then becomes

$$\Psi^{\text{sc}}_{\mathbf{b}}(\mathbf{k}_0; \mathbf{r}, t \to \infty) = f(\Omega)\ \frac{e^{i(\,k_0 r - \omega_0 t)}}{r}\ \chi\left(r\hat{\mathbf{z}} - \mathbf{R}_t\right), \tag{1.101}$$

[12] This will be shown at the end of this section.

[13] A quantum mechanical argument to neglect contributions from impact parameter larger than $\bar{R}$, or equivalently angular momenta larger than $l = k\bar{R}$, will be presented in section 2.6.

Fig. 1.8 Schematic representations of $\Psi_{\mathbf{b}}^{(+)}(\mathbf{k}_0;\mathbf{r},t)$ in the $t \to -\infty$ (left panel) and in the $t \to +\infty$ (right panel) limits. In this representations, we take a negligible impact parameter.

where we have introduced the short-hand notation, $f(\Omega) \equiv f_{\mathbf{k}_0}(\Omega)$. Above, $\chi(\mathbf{s})$ is a wave packet with the general form of Eq. (1.85). However, here its argument is $\mathbf{s} = r\hat{\mathbf{z}} - \mathbf{R}_t$, instead of $\mathbf{s} = \mathbf{r} - \mathbf{R}_t$. These arguments are very different. In the present case χ is spherically symmetric, since it depends only on r. Thus, the orientation dependence of $\Psi_{\mathbf{b}}^{\rm sc}(\mathbf{k}_0;\mathbf{r},t\to\infty)$ comes exclusively from $f(\Omega)$.

In figure 1.8 we give a schematic representation of the time-dependent wave function in the distant past (left panel) and in the distant future (right panel)[14]. This illustration corresponds to an incident wave packet with $\mathbf{b} = 0$. Whereas as $t \to -\infty$ the wave function is equal to the incident wave packet, as $t \to +\infty$ it is the superposition of the incident packet with the radial scattered packet.

We now turn to the evaluation of the differential cross section. The first step is to evaluate the probability, $P(\Omega, \Delta\Omega)$, that the projectile reaches a detector placed at the orientation Ω ($\Omega \equiv \{\theta,\phi\}$) with aperture $\Delta\Omega$. This probability is obtained by integrating the density $|\Psi_{\mathbf{b}}^{\rm sc}(\mathbf{k}_0;\mathbf{r},t)|^2$ over the volume of a cone with vertex at the potential center, axis with the orientation Ω and solid angle $\Delta\Omega$. If $\Delta\Omega$ is sufficiently small, we can write

$$P(\Omega,\Delta\Omega) = \Delta\Omega \cdot \int dr\, r^2\, |\Psi_{\mathbf{b}}^{\rm sc}(\mathbf{k}_0;\mathbf{r},t)|^2$$
$$= \Delta\Omega \cdot |f(\Omega)|^2 \cdot \int_0^\infty dr \Big|\chi\left[(r - v_0 t)\,\hat{\mathbf{z}} - \mathbf{b}\right]\Big|^2 . \qquad (1.102)$$

[14]Although it is not represented explicitly, the strength of the spherical scattered packet has an orientation dependence arising from $f(\Omega)$.

At this stage, it is convenient to change the integration variable,

$$r \to z = r - v_0\, t \Rightarrow \int_0^\infty dr \to \int_{-v_0 t}^\infty dz.$$

In the $t \to \infty$ limit we can replace

$$\int_0^\infty dr \to \int_{-\infty}^\infty dz$$

and the expression for $P(\Omega, \Delta\Omega)$ can be written

$$P(\Omega, \Delta\Omega) = \Delta\Omega \; \cdot |f(\Omega)|^2 \; \cdot \int_{-\infty}^\infty dz \; |\chi\,(z\hat{\mathbf{z}} - \mathbf{b})|^2 . \tag{1.103}$$

In a typical scattering experiment, the incident beam is described by a bunch of non-overlapping wave packets. Therefore, the number of particles reaching the detector per unit time, $N(\Omega, \Delta\Omega)$, must be given by the incoherent sum of contributions in the form of Eq. (1.103), one for each projectile. The number of projectiles incident upon an area $d^2\mathbf{b}$ is $n_p = J\, d^2\mathbf{b}$, where J is the incident flux. For unit flux, $N(\Omega, \Delta\Omega)$ is

$$N(\Omega, \Delta\Omega) = \Delta\Omega \; \cdot |f(\Omega)|^2 \; \cdot \int d^2\mathbf{b} \int_{-\infty}^\infty dz \; |\chi\,(z\hat{\mathbf{z}} - \mathbf{b})|^2 .$$

Changing variables: $b_x \to x = -b_x$; $b_y \to y = -b_y$, we get

$$N(\Omega, \Delta\Omega) = \Delta\Omega \; \cdot |f(\Omega)|^2 \; \cdot \int d^3\mathbf{r} \; |\chi\,(\mathbf{r})|^2 .$$

Using the normalization of the wave packet (Eq. (1.86)) and the definition of the differential elastic cross section, we obtain

$$\frac{d\sigma}{d\Omega} = |f(\Omega)|^2 . \tag{1.104}$$

This result justifies the treatment of the scattering problem as the diffusion of a stationary plane wave by the potential V.

1.6.3 *The Optical Theorem*

In this section we derive an important relation involving the integrated elastic scattering cross section and the scattering amplitude at forward angles: the *Optical Theorem*. Although its derivation is more straightforward with partial-wave expansion, the derivation given below has the advantage of stressing that it arises from probability conservation.

As a starting point, we consider the norm of the state[15] $\Psi(\mathbf{r}, t)$,

$$\mathcal{N} = \int d^3\mathbf{r} \; |\Psi(\mathbf{r}, t)|^2 . \tag{1.105}$$

[15] For simplicity of notation, we drop those indices which are not relevant to the discussion in this section.

Since $H = H^\dagger$, scalar products are conserved. Thus, the norm of $\Psi_{\mathbf{b}}^{(+)}$ does not depend on t. Using Eq. (1.89), the norm can be written

$$\mathcal{N} = \int d^3\mathbf{r}\ |\Phi(\mathbf{r},t)|^2 + X_1(t) + X_2(t), \tag{1.106}$$

where

$$X_1(t) = \int d^3\mathbf{r}\ |\Psi^{\rm sc}(\mathbf{r},t)|^2 \tag{1.107}$$

and $X_2(t)$ is the interference term,

$$\begin{aligned} X_2(t) &= \int \left[(\Psi^{\rm sc}(\mathbf{r},t))^*\ \Phi(\mathbf{r},t) + \Psi^{\rm sc}(\mathbf{r},t)\,\Phi^*(\mathbf{r},t) \right] d^3\mathbf{r} \\ &= 2\,\mathrm{Re}\left\{ \int \Psi^{\rm sc}(\mathbf{r},t)\ \Phi^*(\mathbf{r},t)\ d^3\mathbf{r} \right\}. \end{aligned} \tag{1.108}$$

Since the free wave-packet is normalized to one (Eq. (1.73)), we can write

$$X_1(t) + X_2(t) = \mathcal{N} - 1. \tag{1.109}$$

To find the value of $\mathcal{N}$, we consider Eq. (1.105) as $t \to -\infty$. According to Eq. (1.94), in this limit $\Psi \to \Phi$ so that its normalization must be the same as that of Φ. Therefore $\mathcal{N} = 1$ and we have

$$X_1(t) = -X_2(t), \tag{1.110}$$

or, explicitly,

$$\int |\Psi^{\rm sc}(\mathbf{r},t)|^2\, d^3\mathbf{r} = -2\,\mathrm{Re}\left\{ \int \Psi^{\rm sc}(\mathbf{r},t)\ \Phi^*(\mathbf{r},t)\ d^3\mathbf{r} \right\}. \tag{1.111}$$

In the $t \to \infty$ limit (i.e., after the collision), this equation has a straightforward meaning: probability conservation requires that the scattered flux be compensated by a reduction of the forward flux. This reduction results from destructive interference between the scattered and the incident wave packets, producing a "shadow" behind the target.

Let us consider in further detail the $t \to \infty$ limit. As $t \to \infty$, the center of the incident wave packet moves away from the scatterer according to Eq. (1.82), so that its z-component becomes much larger than the impact parameter. Therefore, $\Psi^{\rm sc}(\mathbf{r},t)$ and $\Phi(\mathbf{r},t)$ only overlap at very forward angles and we can approximate

$$\Psi^{\rm sc}(\mathbf{r},t) \simeq f(0)\ \frac{e^{i(k_0 r - \omega_0 t)}}{r}\ \chi(r\hat{\mathbf{z}} - \mathbf{R}_t)$$

in the integrand of Eq. (1.108). Since $\chi\,(r\hat{\mathbf{z}} - \mathbf{R}_t)$ depends only on the modulus of $\mathbf{r}$, the angular dependence of the product $\Psi^{\rm sc}\,\Phi^*$ comes exclusively from Φ^* and we can write[16]

$$X_2 = 2\,\mathrm{Re}\left\{f(0)\int r\;dr\;e^{i(k_0 r-\omega_0\,t)}\chi\,(r\hat{\mathbf{z}} - \mathbf{R}_t)\,I^*(r)\right\}, \tag{1.112}$$

where

$$I(r) = \int d\Omega\;\Phi(\mathbf{r},t)\;.$$

With the help of Eq. (1.71), we obtain

$$I(r) = \frac{1}{(2\pi)^{3/2}}\int d^3\mathbf{k}\;A(\mathbf{k}-\mathbf{k}_0)\,e^{-i\omega_k\,t}\int d\Omega\,e^{i\mathbf{k}\cdot\mathbf{r}}\;. \tag{1.113}$$

Calling $x = \cos(\alpha)$, where α is the angle subtended by the vectors $\mathbf{k}$ and $\mathbf{r}$, the angular integration can easily be performed,

$$\int d\Omega\,e^{i\mathbf{k}\cdot\mathbf{r}} = 2\pi\int_{-1}^{1} dx\,e^{ikrx} = \frac{2\pi}{ikr}\left[e^{ikr} - e^{-ikr}\right].$$

Using this result in Eq. (1.113), we get

$$\begin{aligned} I(r) &= \frac{2\pi}{ik_0}\left[\frac{1}{(2\pi)^{3/2}}\int d^3\mathbf{k}\;A(\mathbf{k}-\mathbf{k}_0)\,\frac{e^{i(kr-\omega_k\,t)}}{r}\right] \\ &\quad - \frac{2\pi}{ik_0}\left[\frac{1}{(2\pi)^{3/2}}\int d^3\mathbf{k}\;A(\mathbf{k}-\mathbf{k}_0)\,\frac{e^{-i(kr+\omega_k\,t)}}{r}\right]. \end{aligned} \tag{1.114}$$

Above, we used the fact that $A(\mathbf{k}-\mathbf{k}_0)$ is a sharp packet in momentum space, so that $1/k$ is slowly varying and can be moved out of the integral with the value $1/k_0$. As pointed out in the previous section, the phase $kr + \omega_k t$ oscillates rapidly because it cannot have a stationary point at $\mathbf{k} = \mathbf{k}_0$. Therefore, the second line of Eq. (1.114) vanishes. The integral within square brackets in the first line of Eq. (1.114) is the same as the scattered packet of Eq. (1.91), except for the absence of the scattering amplitude. Therefore, it should be given by Eq. (1.101) divided by the scattering amplitude. Using this result in Eq. (1.114) and taking the conjugate, we obtain

$$I^*(r) = \left(\frac{2\pi i}{k_0}\right)\;\frac{e^{-i\,(k_o r-\omega_0 t)}}{r}\;\chi^*\,(r\,\hat{\mathbf{z}} - \mathbf{R}_t)\,. \tag{1.115}$$

[16] Above and in the remaining part of this section we use the shorthand notation: $X_1(t\to\infty) = X_1,\;\; X_2(t\to\infty) = X_2$.

Using Eq. (1.115) in Eq. (1.112), we get

$$X_2 = -\frac{4\pi}{k_0}\mathrm{Im}\left\{\mathrm{f}(0)\int \mathrm{dr}\ \left|\chi\left(\mathrm{r}\,\widehat{\mathbf{z}} - \mathbf{R}_{\mathrm{t}}\right)\right|^2\right\}$$
$$= -\left(\frac{4\pi}{k_0}\mathrm{Im}\left\{\mathrm{f}(0)\right\}\right)\cdot\int dr\ \left|\chi\left(r\,\widehat{\mathbf{z}} - \mathbf{R}_{\mathrm{T}}\right)\right|^2. \qquad (1.116)$$

Now we turn to the evaluation of X_1. In the $t \to \infty$ limit, we can use the asymptotic form of Ψ^{sc} (Eq. (1.101)) and Eq. (1.107) becomes

$$X_1 = \left(\int d\Omega\ \left|f(\Omega)\right|^2\right)\cdot\int dr\ \left|\chi\left(r\hat{\mathbf{z}} - \mathbf{R}_{\mathrm{T}}\right)\right|^2, \qquad (1.117)$$

where the angular integral within the round brackets is promptly identified as the integrated elastic scattering cross section σ_{el}.

Using Eqs. (1.117) and (1.116) in (1.110) we get the Optical Theorem

$$\sigma_{\mathrm{el}} = \frac{4\pi}{k_0}\ \mathrm{Im}\left\{f(0)\right\}. \qquad (1.118)$$

Further aspects of the Optical Theorem will be considered in forthcoming chapters. At the end of section 2.3, it will be derived using the partial-wave expansion for stationary scattering and in section 4.2.1 it will be presented as an equivalent relation for the transition operator

1.7 Systems of units

In the scattering of microscopic systems, typical values of lengths, masses, charges, energies, etc. are extremelly small when expressed in scales of the International System of units (SI) – used in the present text – or any other macroscopic system of units. Therefore, it is convenient to introduce appropriate microscopic units. We discuss this problem below, in two situations where different scales are used: a) the scattering of nuclei at non-relativistic energies and, b) the scattering of atoms, ions and molecules.

1.7.1 *Nuclear physics*

In the scattering of nuclear systems, the typical length is the nuclear radius: $R \sim 10^{-15} - 10^{-14}$ m. One then uses as unit of length the Fermi[17] (fm), defined by the relation: 1 fm $= 10^{-15}$ m. This leads naturally to the

[17] This unit is also known as fentometer. For this reason, one frequently uses the notation 'fm' - instead of 'F'.

cross section unit of fm^2; 1 $\text{fm}^2 = 10^{-30}$ m^2. However, the cross sections are usually expressed in barns; $1\text{b} = 10^{-28}\text{m}^2$, milibarns; 1 mb $= 10^{-3}$ b, microbarns; 1 μb $= 10^{-6}$ b, nanobarns; 1 nb $= 10^{-9}$ b or picobarns; 1 pb $= 10^{-12}$ b.

In nuclear physics, electromagnetic forces play a major role. In such systems, the natural unit of charge is the modulus of the electronic charge e, which has the value $e = 1.6 \times 10^{-19}\,\text{C}$ (or $e = 4.803\,\text{esu}$ in Gaussian units). Usually, a scattering experiment uses a beam of charged projectiles accelerated by electromagnetic forces. In an electrostatic acelerator the kinetic energy of a projectile of charge q accelerated by the potential ΔU is $E_{\text{lab}} = q\,\Delta U$. This leads to the use of the *electron Volt*, defined as the kinetic energy acquired by an electron through the acceleration by a potential of 1 Volt. It can be easily checked that 1 eV$= 1.602 \times 10^{-19}$ J. For investigations of the internal structure of nuclei, the projectile's energy is usually high enough to surmount the potential barrier (resulting from the interplay between short range attractive forces and Coulomb repulsion) and excite the intrinsic states of interest. This amounts to energies of the order of MeV (10^6 eV) or GeV (10^9 eV), depending on the system's charge. Nowadays, the beam energy is frequently expressed by the ratio, ϵ, of the collision energy in the laboratory frame, E_{lab}, to the number of nucleons in the projectile, A_{P}. That is $E_{\text{lab}} = \epsilon \cdot A_{\text{P}}$. It can be readly checked that $E_{\text{c.m.}} = \epsilon \cdot A_{\text{P}} A_{\text{T}} / (A_{\text{P}} + A_{\text{T}})$.

Projectile's and target's masses are frequently multiplied by c^2 and given in MeV. Some examples are:

$$\begin{aligned}
m_{\text{P}}\, c^2 &= 938.256\,\text{MeV} \\
m_n\, c^2 &= 939.550\,\text{MeV} \\
m_e\, c^2 &= 511.006\,\text{keV} \\
m_N\, c^2 &= 931.480\,\text{MeV} \\
M(^{12}\text{C})\, c^2 &= 12 \times m_N\, c^2,
\end{aligned}$$

where the labels p, n and e stand respectively for proton, neutron and electron, m_N is the average mass of a bound nucleon in ^{12}C and $M(^{12}\text{C})$ is the mass of the $A = 12$ carbon isotope.

The Coulomb potential energy in a collision between nuclei with atomic numbers Z_{P} and Z_{T} can be writen

$$V(r) = \frac{e^2}{4\pi\varepsilon_0} \frac{Z_{\text{P}}\, Z_{\text{T}}}{r}\,.$$

In order to express it in MeV and r in fm, we must evaluate $e^2/4\pi\varepsilon_0$ accordingly. Writting

$$\frac{e^2}{4\pi\varepsilon_0} \equiv \frac{e^2}{4\pi\varepsilon_0\hbar c}\ \hbar c\,,$$

using the explict values of the fine structure constant

$$\alpha = \frac{e^2}{4\pi\varepsilon_0\hbar c} = \frac{1}{137.04}\,,$$

and the constant

$$\hbar c = 197.33\ \text{MeV}\ \times\ \text{fm}\,,$$

we get

$$\frac{e^2}{4\pi\varepsilon_0} = 1.44\ \text{MeV}\ \times\ \text{fm}\,.$$

1.7.2 *Atomic and molecular physics*

Atoms and molecules are about 10^5 times larger than nuclei. Their radii are frequently measured in Angstroms (Å), where $1\text{Å} = 10^{-10}\text{m} = 10^5$ fm. On the other hand their appropriate energy scale is much smaller. Typical excitation energies are of the order of the keV, eV or a fraction of eV. The constants $\hbar c$ and e^2 can be expressed in terms of these units as

$$\hbar c = 1973.29\ \text{eV}\ \times\ \text{Å}\,,$$

$$\frac{e^2}{4\pi\varepsilon_0} = 14.4\ \text{eV}\ \times\ \text{Å}\,.$$

However, in practical applications Hartree's atomic units (a.u.) are used. In Hartree's atomic unit system, the unit of length is the radius of the classical orbit of the electron in a hydrogen atom. According to Bohr's atomic model, this radius is

$$a_0 = \frac{4\pi\varepsilon_0\hbar^2}{m_e e^2} = 0.529 \times 10^{-10}\,\text{m} = 0.529\,\text{Å}\,.$$

An important remark is that, despite the use of atomic units for lengths, cross sections in atomic and molecular physics are usually given in cm^2. The units of mass and charge are the mass and the charge of the electron. That is, m_e and e. To complement the unit system, one adopts $\hbar$ as the unit of action and, as in the Gaussian system, sets $\varepsilon_0 = 1/4\pi$, $\mu_0 = 4\pi/c^2$. In this way, in a.u. we have: $m_e = e = a_0 = \hbar = 4\pi\varepsilon_0 = \mu_0 c^2/4\pi = 1$. The atomic units for other physical quantities are defined in terms of these basic units. We give below a few examples.

a) The a.u. of velocity is that of an electron in Bohr's orbit, v_0. It can be readily checked that $v_0 = \alpha\, c = 2.18769 \times 10^6$ m/s, so that $c = 137.036$ a.u.
b) The a.u. of energy, the Hartree (H), is twice the ionization energy of the hydrogen atom. That is: $1\text{H} = e^2/4\pi\varepsilon_0 a_0 = 27.2116$ eV,
c) The a.u. of time is the time needed to a particle with velocity v_0 traverse the distance a_0. That is, $a_0/v_0 = 2.41889 \times 10^{-17}$ s.

In the case of electromagnetic radiation, the photon energy can also be specified by its frequency or by its wavelength. The spectroscopic tradition uses MHz (mega Hertz) for photon frequencies in the infrared and microwave regions and wave numbers in the visible and ultraviolet regions. With this in mind, the energy of 1 a.u. corresponds to a frequency of $\nu = 6.57968 \times 10^9$ MHz, wave length $\lambda = 455.633$ Å and wave number $k = 219475$ cm^{-1}. This energy can also be expressed as a temperature (in units of the Boltzmann constant, $\kappa = 8.617 \times 10^{-5}$ eV/K), 1 a.u. $= 3.15777 \times 10^5$ K.

Exercises

(1) In a scattering experiment a beam of ^{9}Be ($Z_\text{P} = 4$) nuclei with energy $E_\text{lab} = 19$ MeV impinges on a thin solid target of ^{64}Zn ($Z_\text{T} = 30$). The scattered particles are measured by a set of five detectors distributed on a circumference at the angles (referred to the beam direction) $\theta_1, ..., \theta_5$. The yields in the detectors, $N(\theta)$, are listed in the table below. The experimental setup is such that the product $J \cdot n \cdot \Delta\Omega$ is 2.9×10^{27} cm^{-2}. Transform these angles to the CM-frame and obtain the corresponding experimental cross sections. Plot the experimental points in comparison with the Rutherford cross section.

$$\left(\frac{d\sigma}{d\Omega}\right)_\text{Ruth} = \frac{a^2}{4}\left[\frac{1}{\sin^4(\theta/2)}\right]$$

with

$$a = \frac{1}{2}\left(\frac{e^2 Z_\text{P}\, Z_\text{T}}{4\pi\varepsilon_0 E}\right),$$

where e is the absolute value of the electronic charge and $v = \sqrt{2E_\text{lab}/M_\text{P}}$.

θ(deg)	30	60	90	120	150
$N(\theta)$	32983±181	2296±50	545±10	181±7	82±3

(2) Check Eq. (1.41) applying it to the trivial case of a free particle ($V(r) \equiv 0$) and showing that it leads to the expected result: $\Theta(b) \equiv 0$.

(3) Consider the scattering of a particle of mass μ and energy E from the spherically symmetric potential $V(r) = \lambda/r^2$. Show that the deflection function is

$$\Theta(b) = \pi \left[1 - \frac{b}{\sqrt{b^2 + \lambda/E}} \right],$$

where b is the impact parameter. Then prove that the classical cross section can be written (see [Goldstein (1980)])

$$\frac{d\sigma}{d\Omega} = \frac{\lambda}{E\pi \sin(\beta\pi)} \left[\frac{1-\beta}{\beta^2 \left(2-\beta\right)^2} \right]$$

with $\beta = \theta/\pi$.

(4) A particle of mass m and energy E is scattered from a potential with the form $V(r) = -A/r^4$, where A is a positive constant. Find the orbiting angular momentum in this case.

(5) A beam of neutrons with energy $E_{\rm lab} = 10.0 \pm 0.1$ keV goes through a a collimator with radius $d = 2$ mm and impinges on a target of He atoms, located 1.5 m away from the collimator. The scattered particles are detected by a set of detectors placed between 10 and 170 degrees. Show that the elastic cross section measured in this experiment can be treated as the diffusion of a plane wave by a short-short range potential.

Chapter 2

The Partial Wave Expansion Method

As we have shown in the previous section, the differential cross section for elastic scattering is given in terms of the scattering amplitude $f(\Omega)$, which appears in the asymptotic form of the stationary wave function. Finding this cross section then amounts to solving the Schrödinger equation,

$$H\psi = E\psi. \tag{2.1}$$

Above, E is the collision energy and H is the Hamiltonian,

$$H = K + V, \tag{2.2}$$

with K representing the kinetic energy operator and V a short-range potential[1]. We are interested in the solution $\psi^{(+)}(\mathbf{k};\mathbf{r})$, which has the asymptotic form of Eq. (1.61),

$$\psi^{(+)}(\mathbf{k};\mathbf{r}) \to A\left(e^{i\mathbf{k}\cdot\mathbf{r}} + f(\Omega)\,\frac{e^{ikr}}{r}\right), \tag{2.3}$$

where A is an arbitrary normalization constant.

There are different methods to solve this problem. In this chapter, we discuss in detail the method of partial-waves. In this method, the scattering amplitude is written as a sum of contributions from different angular momenta, given in terms of phase shifts. The phase shifts are then obtained through the solution of the radial equation. The method is very convenient for spherically symmetric potentials ($V(\mathbf{r}) = V(r)$) at low collision energies (this condition will be quantitatively presented in section 2.6).

The first step is to write the kinetic energy operator in spherical coordinates,

$$K = -\frac{\hbar^2}{2\mu}\left[\frac{1}{r^2}\frac{\partial}{\partial r}\left(r^2\frac{\partial}{\partial r}\right) + \frac{1}{r^2\sin^2\theta}\frac{\partial}{\partial\theta}\left(\sin\theta\frac{\partial}{\partial\theta}\right) + \frac{1}{r^2\sin^2\theta}\frac{\partial^2}{\partial\varphi^2}\right],$$

[1]This excludes the case of Coulomb potentials, which will be considered in chapter 3.

where $\{r, \theta, \varphi\}$ are the spherical coordinates corresponding to the vector $\mathbf{r}$. Having in mind the explicit form of the orbital angular momentum operator [Edmonds (1974); Merzbacher (1970); Messiah (1999)] L^2,

$$L^2 = -\frac{\hbar^2}{\sin^2\theta}\left[\frac{\partial}{\partial\theta}\left(\sin\theta\frac{\partial}{\partial\theta}\right) + \frac{\partial^2}{\partial\varphi^2}\right],$$

we can write

$$K = -\frac{\hbar^2}{2\mu}\left[\frac{1}{r^2}\frac{\partial}{\partial r}\left(r^2\frac{\partial}{\partial r}\right) - \frac{L^2}{\hbar^2 r^2}\right]. \tag{2.4}$$

For spherically symmetric potentials, which is the most frequent case, Eq. (2.4) guarantees that the Hamiltonian commutes with the operators L^2 and L_z. Therefore, one can express the solutions of Eq. (2.1) as linear combinations of the simultaneous eigen-functions of the operator set $\{H, L^2, L_z\}$,

$$\psi_{klm}(\mathbf{r}) = R_l(k,r)\, Y_{lm}(\hat{\mathbf{r}}). \tag{2.5}$$

Above, Y_{lm} is the usual spherical harmonic, $\hat{\mathbf{r}}$ is a shorthand notation for the angles $\{\theta, \varphi\}$ and $R_l(k,r)$ is the radial part of the wave function. Expanding the scattering wave function in this basis and taking scalar product with[2] $(Y_{lm}|$, we get the radial equation

$$-\frac{\hbar^2}{2\mu}\left[\frac{1}{r^2}\frac{d}{dr}\left(r^2\frac{d}{dr}\right) - \frac{l(l+1)}{r^2}\right] R_l(k,r) + V(r)\, R_l(k,r) = E\, R_l(k,r)\,. \tag{2.6}$$

In the argument of the radial wave function, we use the wave number $k = \sqrt{2\mu E/\hbar^2}$ to characterize the collision energy, instead of the quantum number E. Note that this equation was derived under the assumption that the potential is spherically symmetric. If that is not the case, instead of a single radial equation one would obtain a set of coupled equation, as in multi-channel scattering (see chapter 9).

Frequently, one uses the modified radial wave function $u_l(k,\rho)$, defined by the relation

$$R_l(k,r) = \frac{u_l(k,r)}{kr}\,. \tag{2.7}$$

[2]We use round brackets to represent scalar products in the angular space.

In this case, the wave function is expanded as[3]

$$\psi^{(+)}(\mathbf{k};\mathbf{r}) = A \sum_{lm} C_{lm}(\hat{\mathbf{k}})\ Y_{lm}(\hat{\mathbf{r}})\ \frac{u_l(k,r)}{kr} \tag{2.8}$$

and the radial equation takes the simpler form

$$-\frac{\hbar^2}{2\mu}\left[\frac{d^2}{dr^2} - \frac{l(l+1)}{r^2}\right] u_l(k,r) + V(r)\ u_l(k,r) = E\ u_l(k,r)\,. \tag{2.9}$$

Eq. (2.9) corresponds to a one-dimensional Schrödinger equation with a repulsive potential with infinite strengh at negative r-values plus the effective l-dependent potential

$$V_l(r)\ = V(r) + V_l^{\rm cent}(r). \tag{2.10}$$

Above, $V_l^{\rm cent}(r)$ is the centrifugal potential

$$V_l^{\rm cent}(r) = \frac{\hbar^2}{2\mu}\ \frac{l(l+1)}{r^2}\,. \tag{2.11}$$

It is convenient to introduce the dimensionless variable $\rho = kr$ and rewrite Eqs. (2.6) and (2.9) as

$$\left[\frac{d^2}{d\rho^2} + \frac{2}{\rho}\frac{d}{d\rho} + \left(1 - \frac{l(l+1)}{\rho^2}\right)\right] R_l(k,\rho) = U(\rho)\ R_l(k,\rho)\,, \tag{2.12}$$

where

$$U(\rho) = \frac{V(\rho/k)}{E}, \tag{2.13}$$

and

$$\left[\frac{d^2}{d\rho^2} + 1 - \frac{l(l+1)}{\rho^2}\right] u_l(k,\rho) = U(\rho)\ u_l(k,\rho)\,. \tag{2.14}$$

---

[3]Since we are not making any assumption about the direction of the incident wave, the expansion coefficients should carry a dependence on $\hat{\mathbf{k}}$. This dependence is dropped if one assumes that the incident wave is parallel to the z-axis. In this case, the wave function becomes axially symmetric. Accordingly, only $m = 0$ terms should be included in the partial-wave expansion.

2.1 Free particle in spherical coordinates

In the case of a free particle, $V = 0$ and the RHS of Eqs. (2.12) and (2.14) vanish. Eq. (2.12) then reduces to the spherical Bessel equation. Its solutions can be expressed as linear combinations of the independent pair of real solutions $j_l(\rho)$ e $n_l(\rho)$. The former is the spherical Bessel function, which is regular at the origin. The latter is the spherical Neumann function, which diverges at $\rho = 0$. At the limiting values of the argument, $\rho \to 0$ and $\rho \to \infty$, they have the behaviors

$$j_l(\rho \to 0) \sim \frac{\rho^l}{(2l+1)!!}, \quad j_l(\rho \to \infty) \sim \frac{\sin(\rho - l\pi/2)}{\rho}, \tag{2.15a}$$

$$n_l(\rho \to 0) \sim \frac{(2l-1)!!}{\rho^{l+1}}, \quad n_l(\rho \to \infty) \sim \frac{\cos(\rho - l\pi/2)}{\rho}. \tag{2.15b}$$

These equations show that for large values of the argument both functions oscillate with decreasing amplitudes, within the envelope $1/\rho$.

The regular solutions satisfy the orthogonality relation

$$\int_0^\infty r^2\, j_l(kr)\, j_l(k'r)\, dr = \frac{\pi}{2k^2}\, \delta(k - k')\,. \tag{2.16}$$

The properties of $j_l(\rho)$ e $n_l(\rho)$ can be found in many books (see, for example, [Abramowitz and Stegun (1972); Merzbacher (1970)], or Appendix C of [Joachain (1983)]) and numerical values are given in subroutines available in standard computer libraries (see [Press *et al.* (1992)], for example). For $l = 0$ and $l = 1$, they have the simple forms[4]

$$j_0(\rho) = \frac{\sin\rho}{\rho}, \quad j_1(\rho) = \frac{\sin\rho}{\rho^2} - \frac{\cos\rho}{\rho}, \tag{2.17a}$$

$$n_0(\rho) = \frac{\cos\rho}{\rho}, \quad n_1(\rho) = \frac{\cos\rho}{\rho^2} + \frac{\sin\rho}{\rho}\,. \tag{2.17b}$$

For higher l, j_l e n_l can be obtained with the help of the recursion relations [Abramowitz and Stegun (1972)]

$$w_l(\rho) = (2l+1)\, \frac{w_{l+1}(\rho)}{\rho} - w_{l-1}(\rho)\,, \tag{2.18}$$

where w stands for j or n. They can also be obtained through successive derivations with respect to the argument,

$$j_l(\rho) = (-\rho)^l \left(\frac{1}{\rho}\frac{d}{d\rho}\right)^l j_0(\rho); \quad n_l(\rho) = (-\rho)^l \left(\frac{1}{\rho}\frac{d}{d\rho}\right)^l n_0(\rho)\,,$$

[4] Different sign conventions are used for $n_l(\rho)$. We follow the convention used by [Messiah (1999)]. Other authors, like Joachain [Joachain (1983)] and Roman [Roman (1965)], define the Neumann function with a minus sign with respect to our definition.

The solutions of the spherical Bessel equation can also be expressed in terms of the complex functions $h_l^{(+)}(\rho)$ and $h_l^{(-)}(\rho)$, called Haenkel functions. They are defined as[5]

$$h_l^{(\pm)}(\rho) = n_l(\rho) \pm i\, j_l(\rho)\,. \tag{2.19}$$

Asymptotically, $h_l^{(+)}$ e $h_l^{(-)}$ behave respectively as an outgoing and an incoming spherical wave[6]. That is,

$$h_l^{(\pm)}(\rho \to \infty) \sim \frac{e^{\pm\, i(\rho - l\pi/2)}}{\rho}. \tag{2.20}$$

Similarly, free particle solutions can be found for Eq. (2.14). In this case, we have the Ricatti-Bessel equation. The solutions corresponding to j_l e n_l are, respectively, the Ricatti-Bessel function, $\hat{\jmath}_l$, and the Ricatti-Neumann function, $\hat{n}_l$, given as [Taylor (2006)]

$$\hat{\jmath}_l(\rho) = \rho\, j_l(\rho); \quad \hat{n}_l(\rho) = \rho\, n_l(\rho). \tag{2.21}$$

As $\rho \to 0$ and $\rho \to \infty$, $\hat{\jmath}_l(\rho)$ and $\hat{n}_l(\rho)$ behave as[7]

$$\hat{\jmath}_l(\rho \to 0) \sim \frac{\rho^{l+1}}{(2l+1)!!}\ ; \quad \hat{\jmath}_l(\rho \to \infty) \sim \sin(\rho - l\pi/2), \tag{2.22a}$$

$$\hat{n}_l(\rho \to 0) \sim \frac{(2l-1)!!}{\rho^l}\ ; \quad \hat{n}_l(\rho \to \infty) \sim \cos(\rho - l\pi/2). \tag{2.22b}$$

These functions are illustrated in figure 2.1, for two partial-waves. At small ρ, the regular $\hat{\jmath}_l(\rho)$ converges to zero while $\hat{n}_l(\rho)$ diverges. For large values of ρ, both functions oscillate with constant amplitudes. The transition between these trends occurs at the points

$$\rho_t = \sqrt{l(l+1)}\ (\simeq l, \text{ for large } l), \tag{2.23}$$

indicated by circles in figure 2.1. The meaning of these points can be better understood if we get back to the variable r and consider the Riccatti-Bessel equation in the form of Eq. (2.14), with $U = 0$. In this case, it

[5]Usually, these functions are denoted $h_l^{(1)}$, which corresponds to $h_l^{(+)}$, and $h_l^{(2)}$, which corresponds to $h_l^{(-)}$. We choose the notation $h_l^{(\pm)}$ in order to stress the incoming ($-$) or outgoing ($+$) nature of these functions. One should also point out that some authors, like Joachain [Joachain (1983)], use a different phase convention for the Haenkel functions ($h_l^{(1)} = -i\, h_l^{(+)}$ and $h_l^{(2)} = i\, h_l^{(-)}$)

[6]These behaviors are clear in the assymptotic phases, $\varphi^{(\pm)}$, of the wave functions $h_l^{(\pm)}$: $\varphi^{(\pm)} = \pm\rho - \omega t$. Since for $h_l^{(+)}$ the condition $d\varphi/dt = 0$ leads to a positive phase velocity ($v_p = d\rho/dt = \omega$), this spherical wave moves away from the origin. The opposite occurs for $h_l^{(-)}$.

[7]To extend Eq. (2.22b) to $l = 0$, we adopt the convention $(-1)!! = 1$.

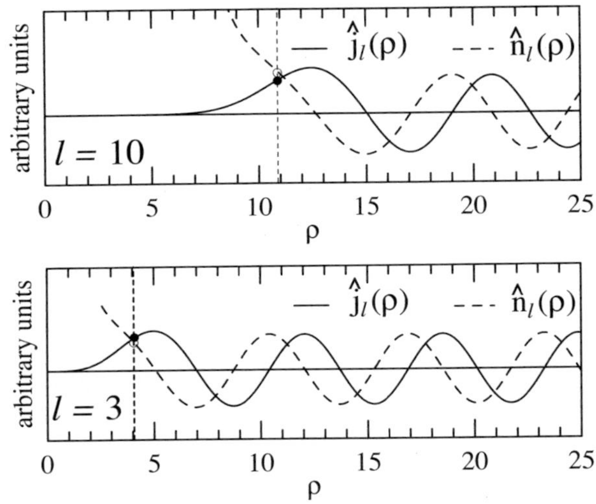

Fig. 2.1 The functions $\hat{\jmath}_l(\rho)$ and $\hat{n}_l(\rho)$ for $l = 3$ and 10. The vertical dotted line represents the boundary between the classically allowed and forbidden regions.

corresponds to a one-dimensional Schrödinger equation with an effective potential $V_l(r) = V_l^{\rm cent}(r)$. The radial distance

$$r_t = \frac{\rho_t}{k} = \frac{\sqrt{l(l+1)}}{k}$$

is then the classical turning point, given by the condition

$$V_l(r_t) = \frac{\hbar^2}{2\mu}\,\frac{l(l+1)}{r_t^2} = E\,.$$

Therefore, the transition points are the boundaries between the classically allowed and the classically forbidden regions.

The Ricatti-Haenkel functions, $\hat{h}_l^{(\pm)}$, are defined analogously to the Haenkel functions,

$$\hat{h}_l^{(\pm)}(\rho) \equiv \rho\, h_l^{(\pm)}(\rho) = \hat{n}_l(\rho) \pm i\,\hat{\jmath}_l(\rho), \tag{2.24}$$

and they have the asymptotic behavior

$$\hat{h}_l^{(\pm)}(\rho \to \infty) \sim e^{\pm\, i(\rho - l\pi/2)}\ . \tag{2.25}$$

Since the Ricatti-Bessel functions are regular solutions with the same kind of boundary condition, they form an orthogonal set. From Eqs. (2.16) and (2.21) we get the relation

$$\int_0^\infty \hat{\jmath}_l(kr)\,\hat{\jmath}_l(k'r)\,dr = \frac{\pi}{2}\,\delta(k-k'). \tag{2.26}$$

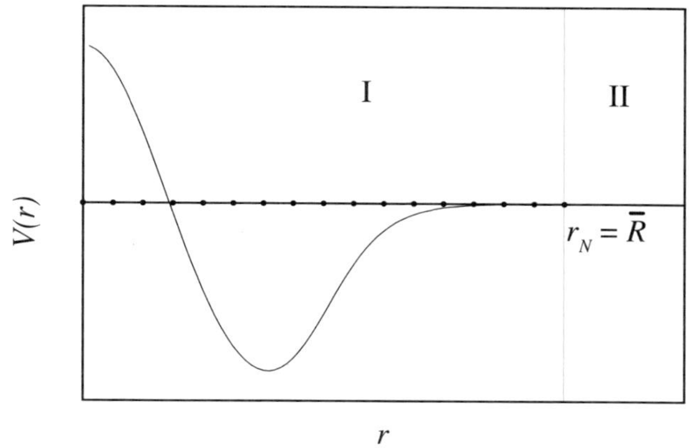

Fig. 2.2 Schematic representation of the potential and the mesh points for the numerical calculation.

2.2 Numerical solutions of the radial equation

The radial wave functions $u_l(k,r)$ can be obtained through numerical integration of Eq. (2.9). Let us consider a potential $V(r)$ with range[8] $\bar{R}$, as represented in figure 2.2. The point $r = \bar{R}$ divides the r-axis into region I, where the potential is different from zero, and region II, where the particle moves freely. In region II, $u_l(k,r)$ satisfies the Ricatti-Bessel equation and, therefore, can be put in the general form

$$u_l^{\rm II}(k,r) = \alpha_l \left[\hat{\jmath}_l(kr) + \beta_l\,\hat{n}_l(kr)\right] , \tag{2.27}$$

where α_l is an arbitrary normalization constant. It will soon become clear that α_l is irrelevant to the scattering amplitude. The effects of the potential are contained in the constant β_l, which vanishes when $V = 0$. The dimensionless logarithmic derivative at the boundary of the external region is then expressed in terms of β_l as

$$\mathcal{L}^{\rm II} = \bar{R}\,\frac{\left[du_l^{\rm II}(k,r)/dr\right]_{r=\bar{R}}}{u_l^{\rm II}(k,\bar{R})} = k\bar{R}\left[\frac{\hat{\jmath}_l'(k\bar{R}) + \beta_l\;\hat{n}_l'(k\bar{R})}{\hat{\jmath}_l(k\bar{R}) + \beta_l\;\hat{n}_l(k\bar{R})}\right] , \tag{2.28}$$

where primes stand for derivatives with respect to the argument.

To determine β_l, one should match Eq. (2.28) with the logarithmic derivative at the boundary, calculated with the radial wave function in

[8] In this chapter we deal with short-range potentials, which vanish or can be neglected for $r > \bar{R}$. Long-range potentials will be considered in chapter 3.

the internal region. For this purpose, we replace the r-axis by a mesh consisting of the points $\{r_0, r_1, ..., r_N\}$, with $r_0 = 0$ and $r_N = \bar{R}$, and approximate the derivatives of the radial equation by finite differences[9]. We then start integrating the differential equation from the origin. In order to keep $\psi^{(+)}$ finite at the origin (see Eq. (2.8)), one must set $u_l^{\rm I}(k, 0) \equiv y_0 = 0$. The initial value of the derivative, $\bar{R}\ [du_l^{\rm I}(k, r)/dr]_{r=0} \equiv y_0'$, can be chosen arbitrarily (1 for example). This choice will only affect α_l, which cancels out when the logarithmic derivative is calculated. With the help of a numerical procedure, e.g. the Numerov algorithm [Press *et al.* (1992)], the radial function and its first derivative can be obtained at all mesh points. The logarithmic derivative is then calculated from the results at the last mesh point, $r_N = \bar{R}$,

$$\mathcal{L}^{\rm I} = \bar{R}\left(\frac{y_N'}{y_N}\right), \tag{2.29}$$

with the notation $y_i \equiv u_l^{\rm I}(k, r_i)$ and $y_i' \equiv [du_l^{\rm I}(k, r)/dr]_{r=r_i}$.

Since $\mathcal{L}^{\rm I}$, $\hat{\jmath}_l'(k\bar{R})$, $\hat{n}_l'(k\bar{R})$, $\hat{\jmath}_l(k\bar{R})$ and $\hat{n}_l(k\bar{R})$ are known quantities, the matching condition,

$$\mathcal{L}^{\rm II} = \mathcal{L}^{\rm I}, \tag{2.30}$$

provides an equation for β_l. For the calculation of the scattering amplitude, it is convenient to express β_l in terms of a new parameter δ_l, defined by the relation

$$\beta_l = \tan\delta_l\,. \tag{2.31}$$

Note that Eq. (2.31) only determines δ_l modulo π. That is, it does not distinguish among δ_l, $\delta_l \pm \pi$, $\delta_l \pm 2\pi$, $\delta_l \pm 3\pi$, etc. However, this indetermination leads to no serious difficulty, since any value satisfying Eq. (2.31) is suitable for our purposes. Eq. (2.27) then takes the form

$$u_l^{\rm II}(k, r) = \alpha_l\ \left[\hat{\jmath}_l(kr) + \tan\delta_l\ \hat{n}_l(kr)\right]\,. \tag{2.32}$$

Using Eq. (2.31) in Eq. (2.28) and the continuity of the logarithmic derivatives (Eq. (2.30)), we obtain the explicit result

$$\tan\delta_l = -\left[\frac{k\bar{R}\ \hat{\jmath}_l'(k\bar{R}) - \hat{\jmath}_l(k\bar{R})\ \mathcal{L}^{\rm I}}{k\bar{R}\ \hat{n}_l'(k\bar{R}) - \hat{n}_l(k\bar{R})\ \mathcal{L}^{\rm I}}\right]. \tag{2.33}$$

[9] In fact, the matching radius does not have to be exactly $\bar{R}$. Any $r > \bar{R}$ would be suitable.

The meaning of δ_l becomes clear if we look at the asymptotic behavior of $u_l(k,r)$. Taking the limits $\hat{j}_l(kr \to \infty)$ and $\hat{n}_l(kr \to \infty)$ (Eqs. (2.22a) and (2.22b)) in Eq. (2.27), we get

$$\begin{aligned} u_l(k, r \to \infty) &= \alpha_l' \, [\sin(kr - l\pi/2)\cos\delta_l + \sin\delta_l \, \cos(kr - l\pi/2)] \\ &= \alpha_l' \, \sin(kr - l\pi/2 + \delta_l), \end{aligned} \tag{2.34}$$

where $\alpha_l' = \alpha_l / \cos\delta_l$ is a new normalization constant. In the absence of potential, the argument of the sinus in the second line of the above equation reduces to the asymptotic phase of the Ricatti-Bessel function: $kr - l\pi/2$. Therefore, the potential V leads to the shift δ_l in this phase (*phase shift*).

Frequently, the solutions of (2.14) in region II are expressed in terms of the Ricatti-Haenkel functions, in the form

$$u_l(k, r > \bar{R}) = \gamma_l \left[\hat{h}_l^{(-)}(kr) - S_l \, \hat{h}_l^{(+)}(kr)\right], \tag{2.35}$$

where,

$$S_l \; = e^{2i\delta_l}, \tag{2.36}$$

and $\gamma_l = i\, e^{-i\delta_l}\alpha_l' \,/\, 2$ is a new choice of the normalization of the radial wave function. It will be shown in chapter 4 that S_l corresponds to the matrix-element of the scattering operator in the angular momentum representation. That is, the partial-wave projected S-matrix.

We have shown that the normalization of the radial wave function is irrelevant for calculations of the scattering amplitude through the numerical solution of the radial equation. However, some equations in Scattering Theory (see, e.g., Eq. (2.70) or the l-projected Lippmann-Schwinger equations of chapter 4) are derived assuming a particular normalization. In such cases, one should make sure that one keeps the same normalization all the way. To avoid ambiguities, we will adopt the normalization

$$\gamma_l = \frac{i}{2}, \tag{2.37}$$

unless otherwise specified. This normalization corresponds to

$$\alpha_l' = e^{i\delta_l} \tag{2.38}$$

and leads to the asymptotic form

$$u_l(k, r \to \infty) = e^{i\delta_l} \, \sin(kr - l\pi/2 + \delta_l) \tag{2.39a}$$

$$\equiv \frac{i}{2} \left[e^{-i(kr - l\pi/2)} - S_l \, e^{i(kr - l\pi/2)}\right]. \tag{2.39b}$$

We should mention the alternative choice

$$\bar{\gamma}_l = \frac{i}{2}\, e^{-i\delta_l} \text{ (or equivalenty, } \alpha_l' = 1), \tag{2.40}$$

which is adopted by many authors. It leads to a radial wave function with the asymptotic form

$$\bar{u}_l(k, r \to \infty) = \sin\,(kr - l\pi/2 + \delta_l) \tag{2.41a}$$

$$\equiv \frac{i}{2}\, e^{-i\delta_l} \left[e^{-i(kr-l\pi/2)} - S_l\, e^{i(kr-l\pi/2)} \right]. \tag{2.41b}$$

To distinguish the radial wave function with the asymptotic form of Eq. (2.39b) from that with the asymptotic form of Eq. (2.41a), we use the notation $\bar{u}_l$ for the latter. Inspecting Eqs. (2.39b) and (2.41b), we get to the relation

$$\bar{u}_l(k, r) = e^{-i\,\delta_l}\, u_l(k, r). \tag{2.42}$$

We point out that both $\bar{u}_l(k, r)$ and $u_l(k, r)$ converge to $\hat{\jmath}_l(kr)$ in the $V \to 0$ limit.

The normalization of $\bar{u}_l$ has the nice feature of keeping the radial wave function real for real potentials[10]. However, we prefer the normalization of u_l, since it can easily be extended to multi-channel scattering. In such problems there is no incident wave in the inelastic channels (see chapter 9), so that the asymptotic forms of the corresponding radial wave functions cannot be given by a sinus function.

2.3 Scattering amplitude and cross section

The scattering amplitude is obtained comparing the asymptotic form of the partial-wave expanded wave function (Eq. (2.8)) with Eq. (2.3). Both are written in terms of incoming and outgoing spherical waves, and the incoming and outgoing contents of each equation are required to be equal. This procedure is described in detail below.

First, we carry out the partial-wave expansion of the free-particle state $\phi\,(\mathbf{k}, \mathbf{r})$. For this purpose, we use the well known Bauer's expansion [Merzbacher (1970)]

$$e^{i\mathbf{k}\cdot\mathbf{r}} = \sum_{lm} 4\pi\, Y^*_{lm}(\hat{\mathbf{r}})\, Y_{lm}(\hat{\mathbf{k}})\, i^l\, j_l(kr) \tag{2.43a}$$

$$= \sum_{l} (2l+1)\, P_l(\cos\theta)\, i^l\, j_l(kr), \tag{2.43b}$$

[10]This advantage disappears in the scattering from a complex potentials (see chapter 7).

where P_l are the Legendre Polynomials [Abramowitz and Stegun (1972)] and $\theta = \cos^{-1}(\mathbf{k}\cdot\mathbf{r}\,/\,kr)$. To go from Eq. (2.43a) to Eq. (2.43b), we have used the addition theorem [Messiah (1999)],

$$(2l+1)\,P_l(\cos\theta) = 4\pi \sum_m Y^*_{lm}(\hat{\mathbf{k}})\,Y_{lm}(\hat{\mathbf{r}})\,.$$

Adopting the usual normalization, $A = (2\pi)^{-3/2}$ (see Eq. (2.3)), we obtain

$$\phi(\mathbf{k};\mathbf{r}) = \frac{1}{(2\pi)^{3/2}} \sum_{lm} 4\pi\, Y^*_{lm}(\hat{\mathbf{k}})\; Y_{lm}(\hat{\mathbf{r}})\, i^l\, \frac{\hat{\jmath}_l(kr)}{kr} \tag{2.44a}$$

$$= \frac{1}{(2\pi)^{3/2}} \sum_{l} (2l+1)\, P_l(\cos\theta)\, i^l\, \frac{\hat{\jmath}_l(kr)}{kr}. \tag{2.44b}$$

Eq. (2.44a) corresponds to Eq. (2.8) with $A = (2\pi)^{-3/2}$, $u_l(k,r) \to \hat{\jmath}_l(kr)$, and the coefficients

$$C_{lm}(\hat{\mathbf{k}}) = 4\pi\, Y^*_{lm}(\hat{\mathbf{k}})\, i^l\,. \tag{2.45}$$

Taking the $r \to \infty$ limit in Eq. (2.44a) and replacing the Ricatti-Bessel function by its asymptotic form (Eq. (2.22a)) we get

$$\phi(\mathbf{k};\mathbf{r}) \to \frac{1}{(2\pi)^{3/2}} \left[\frac{e^{-ikr}}{r}\, Z_- + \frac{e^{ikr}}{r}\, Z_+\right], \tag{2.46}$$

with[11]

$$\begin{aligned} Z_- &= \sum_{lm} C_{lm}\, Y_{lm}(\hat{\mathbf{r}})\, i^l \left(-\frac{i^l}{2ik}\right) \\ &= -\frac{1}{2ik} \sum_{lm} 4\pi\, Y^*_{lm}(\hat{\mathbf{k}})\, Y_{lm}(\hat{\mathbf{r}})\; i^{2l} \end{aligned} \tag{2.47}$$

and

$$\begin{aligned} Z_+ &= \sum_{lm} C_{lm}\, Y_{lm}(\hat{\mathbf{r}})\, i^l \left(\frac{i^{-l}}{2ik}\right) \\ &= \frac{1}{2ik} \sum_{lm} 4\pi\, Y^*_{lm}(\hat{\mathbf{k}})\, Y_{lm}(\hat{\mathbf{r}})\,. \end{aligned} \tag{2.48}$$

Eq. (2.46) means that the plane wave corresponds to the superposition of an incoming spherical wave $\left(e^{-ikr}/r\right)$ with amplitude Z_- with an outgoing spherical wave $\left(e^{ikr}/r\right)$ with amplitude Z_+.

[11] To simplify the notation, we omit the explicit dependence of the coefficients C_{lm} on $\hat{\mathbf{k}}$.

Now, we switch to the scattering wave function. First, we write it in the form

$$\psi^{(+)}(\mathbf{k};\mathbf{r}) = \phi(\mathbf{k};\mathbf{r}) + \psi^{sc}(\mathbf{k};\mathbf{r})\,. \tag{2.49}$$

The scattered wave $\psi^{sc}(\mathbf{k};\mathbf{r})$, which is defined by the above equation, must have the asymptotic form

$$\psi^{sc}(\mathbf{k};\mathbf{r}) \to \frac{1}{(2\pi)^{3/2}} \left[f(\theta)\, \frac{e^{ikr}}{r} \right],$$

so that Eq. (2.3) is satisfied. Taking the $r \to \infty$ limit of $\psi^{(+)}(\mathbf{k};\mathbf{r})$ we obtain

$$\psi^{(+)}(\mathbf{k};\mathbf{r}) \to \frac{e^{-ikr}}{r}\, Z_- + \frac{e^{ikr}}{r}\, \bar{Z}_+\,. \tag{2.50}$$

Since the ingoing wave arises exclusively from $\phi(\mathbf{k};\mathbf{r})$, the amplitude Z_- is still given by Eq. (2.47). However, the outgoing amplitude should be modified, to add the contribution from ψ^{sc}. That is

$$\begin{aligned} Z_+ \to \bar{Z}_+ &= Z_+ + f(\theta) \\ &= \frac{1}{2ik} \sum_{lm} 4\pi\, Y^*_{lm}(\hat{\mathbf{k}})\, Y_{lm}(\hat{\mathbf{r}}) \;+\; f(\theta). \end{aligned} \tag{2.51}$$

To express the scattering amplitude in terms of phase shifts, or equivalently the l-components of the S-matrix, we take the $r \to \infty$ limit in Eq. (2.8), write the radial wave function in this limit (Eq. (2.39b)), and insert the coefficients[12] of Eq. (2.45). The scattering wave function then takes the general form of Eq. (2.50). The amplitude of the incoming wave is the same as in Eq. (2.47) while the one of the outgoing wave is given by

$$\tilde{Z}_+ = \frac{1}{2ik} \sum_{lm} 4\pi\, Y^*_{lm}(\hat{\mathbf{k}})\; Y_{lm}(\hat{\mathbf{r}})\, S_l. \tag{2.52}$$

Setting $\bar{Z}_+ = \tilde{Z}_+$ we obtain the partial-wave expansion of the scattering amplitude

$$f(\theta) \equiv f(\hat{\mathbf{k}}', \hat{\mathbf{k}}) = \frac{1}{2ik} \sum_{lm} 4\pi\, Y^*_{lm}(\hat{\mathbf{k}})\, Y_{lm}(\hat{\mathbf{k}}')\ [S_l \,-\, 1] \tag{2.53a}$$

$$\equiv \frac{1}{k} \sum_{lm} 4\pi\, Y^*_{lm}(\hat{\mathbf{k}})\, Y_{lm}(\hat{\mathbf{k}}')\ e^{i\delta_l} \sin\delta_l\,, \tag{2.53b}$$

[12]Note that the partial-wave expansions of $\psi^{(+)}(\mathbf{k};\mathbf{r})$ and $\phi(\mathbf{k};\mathbf{r})$ have the same coefficients C_{lm}. This results from the facts that these expansions have the same incoming wave.

where $\hat{\mathbf{k}}'$ is a unitary vector parallel to $\mathbf{r}$ and $\theta = \cos^{-1}\left(\hat{\mathbf{k}}'\cdot\hat{\mathbf{k}}\right)$. Using the addition theorem, the scattering amplitude can be written

$$f(\theta) = \frac{1}{2ik}\sum_l (2l+1)\, P_l(\cos\theta)\ [S_l\ - 1] \tag{2.54a}$$

$$\equiv \frac{1}{k}\sum_l (2l+1)\, P_l(\cos\theta)\ e^{i\delta_l}\sin\delta_l\,. \tag{2.54b}$$

A useful expression for $f(\theta)$ can be derived multiplying both sides of Eq. (2.54a) by $\cos\theta$ and using the recursion relations for the Legendre polynomials [Abramowitz and Stegun (1972)]

$$x\ (2l+1)\ P_l(x) = (l+1)\ P_{l+1}(x) + l\, P_{l-1}(x). \tag{2.55}$$

After some algebra, one obtains (see, e.g. [Brink (1985)])

$$(1-\cos\theta)\ f(\theta) = \frac{1}{2ik}\sum_l P_l(\cos\theta)\left[(2l+1)\, S_l - l\, S_{l-1} - (l+1)\, S_{l+1}\right]. \tag{2.56}$$

This equation is particularly useful when dealing with cases with strong absorption, to be treated more fully in chapter 7. In the strong absorption, black disk, limit, the S-matrix in the angular momentum representation is given by,

$$S_l = \Theta(l-\bar{L}),$$

where Θ is the step function and $\bar{L}$ is the orbital angular momentum in a grazing grazing collision. Then, Eq. (2.56) gives immediately,

$$f(\theta) = \frac{\bar{L}}{2ik\ (1-\cos\theta)}\left[P_{\bar{L}}(\cos\theta) - P_{\bar{L}-1}(\cos\theta)\right].$$

The above exact expression for the scattering amplitude in the black disk limit can be further reduced to a form which is used to describe Fraunhofer diffraction, treated extensively in chapter 7.

We recall that the above expressions were obtained under the assumption that the potential is spherically symmetric. Some model calculations are based on axially-symmetric or a triaxial potentials, which violates the rotational invariance of the Hamiltonian. As a result, the scattering amplitude is given by a more complicated expression, depending also on the angle φ.

Inserting in Eq. (2.45) the explicit form of the coefficients C_{lm} (Eq. (2.45)), we get the partial-wave expansion of the wave function

$$\psi^{(+)}(\mathbf{k};\mathbf{r}) = \frac{1}{(2\pi)^{3/2}} \sum_{lm} 4\pi\, Y^*_{lm}(\hat{\mathbf{k}})\, Y_{lm}(\hat{\mathbf{r}})\, i^l\, \frac{u_l(k,r)}{kr} \tag{2.57a}$$

$$= \frac{1}{(2\pi)^{3/2}} \sum_{l} (2l+1)\, P_l(\cos\theta)\, i^l\, \frac{u_l(k,r)}{kr}\,. \tag{2.57b}$$

In terms of radial wave functions normalized as in Eq. (2.41b), the above equation becomes

$$\psi^{(+)}(\mathbf{k};\mathbf{r}) = \frac{1}{(2\pi)^{3/2}} \sum_{lm} 4\pi\, Y^*_{lm}(\hat{\mathbf{k}})\, Y_{lm}(\hat{\mathbf{r}})\, i^l\, e^{i\delta_l}\, \frac{\bar{u}_l(k,r)}{kr} \tag{2.58a}$$

$$= \frac{1}{(2\pi)^{3/2}} \sum_{l} (2l+1)\, P_l(\cos\theta)\, i^l\, e^{i\delta_l}\, \frac{\bar{u}_l(k,r)}{kr}\,. \tag{2.58b}$$

The elastic cross section is obtained from the scattering amplitude through Eq. (1.69),

$$\frac{d\sigma}{d\Omega} = |f(\theta)|^2\,. \tag{2.59}$$

An expression for the total elastic cross section, σ_{el}, is obtained through angular integration of Eq. (2.59). Using Eq. (2.54b) and the orthogonality relation for the Legendre polynomials,

$$\int P_l(\cos\theta)\; P_{l'}(\cos\theta)\; d(\cos\theta) = \frac{2}{2l+1}\,\delta_{ll'}, \tag{2.60}$$

we get

$$\sigma_{\text{el}} = \frac{\pi}{k^2} \sum_l (2l+1)\, \left[1 + |S_l|^2 - 2\,\text{Re}\,\{S_l\}\right] \tag{2.61a}$$

$$= \frac{4\pi}{k^2} \sum_l (2l+1)\; \sin^2\delta_l. \tag{2.61b}$$

From the partial-wave expansions for σ_{el} and for $f(\theta=0)$, one can trivially derive the Optical Theorem, introduced in section 1.6.3. Since at $\theta = 0$, $P_l(\cos\theta) \equiv P_l(1) = 1$, the imaginary part of Eq. (2.54b) can be written

$$\text{Im}\,\{f(\theta=0)\} = \sum_l \frac{(2l+1)}{k}\sin^2\delta_l\,. \tag{2.62}$$

Comparing Eqs. (2.61b) and (2.62) we obtain the Optical Theorem,

$$\sigma_{\text{el}} = \frac{4\pi}{k}\; \text{Im}\,\{f(0)\}\;.$$

2.4 Wronskian relations

We derive below a very useful relation involving linearly independent solutions of the Ricatti-Bessel equation. First, we write separately the Ricatti-Bessel equations for the solutions $\hat{\jmath}_l(\rho)$ and $\hat{n}_l(\rho)$,

$$\hat{\jmath}_l''(\rho) + \left(1 - \frac{l(l+1)}{\rho^2}\right) \hat{\jmath}_l(\rho) = 0 \tag{2.63a}$$

$$\hat{n}_l''(\rho) + \left(1 - \frac{l(l+1)}{\rho^2}\right) \hat{n}_l(\rho) = 0. \tag{2.63b}$$

where double primes stand for second order derivatives with respect to ρ. Multiplying Eq. (2.63b) by $\hat{\jmath}_l(\rho)$, Eq. (2.63a) by $\hat{n}_l(\rho)$ and subtracting the resulting equations we get,

$$\hat{n}_l''(\rho)\, \hat{\jmath}_l(\rho) - \hat{\jmath}_l''(\rho)\, \hat{n}_l(\rho) = 0.$$

We now integrate this equation from zero to some arbitrary value $\tilde{\rho}$. Integrating by parts, we obtain

$$\Big[\hat{n}_l'(\rho)\, \hat{\jmath}_l(\rho) - \hat{\jmath}_l'(\rho)\, \hat{n}_l(\rho)\Big]_0^{\tilde{\rho}} = 0,$$

or

$$\hat{n}_l'(0)\, \hat{\jmath}_l(0) - \hat{\jmath}_l'(0)\, \hat{n}_l(0) = \hat{n}_l'(\tilde{\rho})\, \hat{\jmath}_l(\tilde{\rho}) - \hat{\jmath}_l'(\tilde{\rho})\, \hat{n}_l(\tilde{\rho}).$$

This result indicates that the Wronskian $W\{\hat{n}_l, \hat{\jmath}_l\}$, defined as

$$W\{\hat{n}_l, \hat{\jmath}_l\} = \hat{n}_l(\rho)\, \hat{\jmath}_l'(\rho) - \hat{n}_l'(\rho)\, \hat{\jmath}_l(\rho)\,, \tag{2.64}$$

is independent of ρ. In particular, it can be evaluated at a large $\tilde{\rho}$, where the asymptotic forms of Eqs. (2.22a) and (2.22b) can be used. We get[13]

$$W\{\hat{n}_l, \hat{\jmath}_l\} = \cos^2(\rho - l\pi/2) + \sin^2(\rho - l\pi/2) = 1\,. \tag{2.65}$$

The generalization of the above result to the regular and irregular solutions of Eq. (2.14) is straightforward. In this case, we take the regular, $\bar{u}_l(k,r)$, and irregular, $\bar{w}_l(k,r)$, solutions normalized such as to have the asymptotic forms

$$\bar{u}_l(k, r \to \infty) = \sin(kr - l\pi/2 + \delta_l)\,,$$
$$\bar{w}_l(k, r \to \infty) = \cos(kr - l\pi/2 + \delta_l)\,,$$

[13] Note that if $\hat{n}_l(\rho)$ is defined with the phase-convention of [Joachain (1983)], i.e. with a minus sign with respect to the convention adopted here, the Wronskian will have opposite sign.

and follow the same procedures as above. We obtain

$$W\{\bar{w}_l, \bar{u}_l\} \equiv \frac{d\bar{u}_l(k,r)}{dr} \ \bar{w}_l(k,r) - \frac{d\bar{w}_l(k,r)}{dr} \ \bar{u}_l(k,r) = k\,. \tag{2.66}$$

Note that the factor k appears because we chose to consider derivatives with respect to r, instead of derivatives with respect to ρ, as we had done in the case of Ricatti-Bessel functions.

If we adopt for the radial wave functions our standard normalization,

$$u_l(k, r \to \infty) = e^{i\delta_l} \ \sin(kr - l\pi/2 + \delta_l)\,,$$
$$w_l(k, r \to \infty) = e^{i\delta_l} \ \cos(kr - l\pi/2 + \delta_l)\,,$$

Eq. (2.67) becomes

$$W\{w_l, u_l\} \equiv \frac{du_l(k,r)}{dr} \ w_l(k,r) - \frac{dw_l(k,r)}{dr} \ u_l(k,r) = e^{2i\delta_l}\, k\,. \tag{2.67}$$

2.5 Integral formulae for the phase shifts

The phase shifts can also be evaluated by an integral involving radial wave functions. Here, it is convenient to use the normalization of Eq. (2.41a).

We start by multiplying Eq. (2.14) by $\hat{\jmath}_l(\rho)$, Eq. (2.63a) by $\bar{u}_l(\rho)$, and subtracting the resulting equations. We obtain

$$\bar{u}_l''(k,\rho)\ \hat{\jmath}_l(\rho) - \bar{u}_l(k,\rho)\ \hat{\jmath}_l''(\rho) = U(\rho)\ \hat{\jmath}_l(\rho)\ \bar{u}_l(k,\rho), \tag{2.68}$$

where we have used the notations: $d^2\bar{u}_l(k,\rho)/d\rho^2 = \bar{u}_l''(k,\rho)$ and $U(\rho) = V(\rho/k)/E$. We now integrate both sides over ρ from zero to ∞ and carry out the integration by parts on the LHS. We get

$$\left[\bar{u}_l'(k,\rho)\ \hat{\jmath}_l(\rho) - \bar{u}_l(k,\rho)\ \hat{\jmath}_l'(\rho)\right]_0^\infty = \int_0^\infty d\rho\ \hat{\jmath}_l(\rho)\ U(\rho)\ \bar{u}_l(k,\rho). \tag{2.69}$$

Since both $\hat{\jmath}_l(\rho)$ and $\bar{u}_l(k,\rho)$ are regular at the origin, the LHS of Eq. (2.69) has only contribution from the upper limit, where $\bar{u}_l(k,\rho)$ can be replaced by its asymptotic form (Eq. (2.41a))

$$\bar{u}_l(k, \rho \to \infty) = \sin\left(\rho - l\pi/2 + \delta_l\right).$$

Using this result together with the asymptotic form of $\hat{\jmath}_l(\rho)$ (Eq. (2.22a)) in Eq. (2.69), we get[14],

$$\sin\delta_l = -\int d\rho\; \hat{\jmath}_l(\rho)\, U(\rho)\; \bar{u}_l(k,\rho). \tag{2.70}$$

or

$$\sin\delta_l = -\frac{1}{k}\left(\frac{2\mu}{\hbar^2}\right)\int dr\; \hat{\jmath}_l(kr)\, V(r)\; \bar{u}_l(k,r). \tag{2.71}$$

For the exact evaluation of phase shifts, the above equation is not an improvement over Eq. (2.33). This is because the radial wave function $\bar{u}_l(k,r)$ is not known. However, it is useful for the development of approximations. If the potential is weak ($U(\rho) \ll 1$ for any ρ), the radial wave function cannot be very different from the Ricatti-Bessel function and one can approximate,

$$\sin\delta_l \simeq -\int d\rho\; U(\rho)\, \hat{\jmath}_l^2(\rho) = -\frac{1}{k}\left(\frac{2\mu}{\hbar^2}\right)\int dr\; V(r)\, \hat{\jmath}_l^2(kr)\,. \tag{2.72}$$

This equation is similar to the Born approximation, which will be discussed in chapter 5. Since $\hat{\jmath}_l^2(\rho) \geq 0$, one concludes that the sign of the phase shift is determined by the sign of the potential $V(r)$. Attractive potentials ($V(r) < 0$) give rise to positive phase shift while repulsive potentials lead to negative phase shifts. This property played an important role in the discussion on the existence of a repulsive core in the strong interaction between two protons, which took place some decades ago [Brink (1965)]. The fact that the experimental s-wave phase shift[15] in proton-proton collisions changed from positive to negative at the collision energy of 250 MeV indicated that the picture of a fully attractive nuclear interaction was not consistent with the data.

2.6 Convergence of the partial-wave expansion

The partial-wave expansion of Eq. (2.54a) is very useful when the series can be truncated after a few terms. To determine in which circumstances this occurs, we use a semiclassical argument. Let us consider a collision with

[14]If a different normalization for the radial wave function is adopted, Eq. (2.70) takes a different form. If, for example, one uses $\alpha_l = 1$ in Eq. (2.32), as in [Joachain (1983)], the LHS of Eq. (2.70) changes to $\tan\delta_l$.

[15]We mean the phase shift associated with the short-range interaction. As we will show in chapter 3, in the scattering from a Coulomb plus short-range potential the Coulomb phase shift is considered separately.

energy E (in the CM frame) and impact parameter b. The corresponding wave number is $k = \sqrt{2\mu E/\hbar^2}$ and the angular momentum (in $\hbar$ units) is $l = kb$. The partial-waves which contribute to the elastic cross section should be those for which the projectile comes into the reach of the potential. Calling $\bar{R}$ the potential range, the important partial-waves satisfy the condition

$$l \lessapprox k\bar{R} = \bar{R}\sqrt{2\mu E/\hbar^2}\,.$$

The convergence of the series is, therefore, particularly fast when $k\bar{R} \sim 1$. In terms of the collision energy, this condition can be written

$$E \sim \frac{\hbar^2}{2\mu\bar{R}^2}\,.$$

The same conclusion can be reached in quantum mechanics, through an inspection of Eq. (2.72). For the integrand to have appreciable values, it is necessary that the potential $U(\rho)$ overlaps with $\hat{\jmath}_l^2(\rho)$. Since $U(\rho)$ vanishes for $\rho > \bar{\rho} = k\bar{R}$ and $\hat{\jmath}_l^2(\rho)$ decreases rapidly inside the classically forbidden region (see figure 2.1), given by $\rho < \rho_t \simeq l$, the condition for strong overlap is that

$$l \lessapprox \bar{\rho} = k\bar{R}\,.$$

2.7 Hard sphere scattering

An interesting problem which can easily be handled is the scattering from an infinitely repulsive square well potential. This potential gives an approximation to hard-cores in the nuclear interaction. The hard sphere potential is,

$$\begin{aligned} V(r) &= 0, && \text{for } r > \bar{R}\,, \\ V(r) &\to +\infty && \text{for } r < \bar{R}\,. \end{aligned} \tag{2.73}$$

In this case, the wave function vanishes in region I and its continuity at the boundary requires that[16] $u_l^{\text{II}}(k,\bar{R}) = 0$. Using its explicit form, as given in Eq. (2.32), this condition leads to

$$\tan\delta_l = -\frac{\hat{\jmath}_l(k\bar{R})}{\hat{n}_l(k\bar{R})}. \tag{2.74}$$

This result becomes particularly simple for s-waves. Using Eqs. (2.17a), (2.17b) and (2.21), we get

$$\tan\delta_0 = -\tan(k\bar{R}) \quad \to \quad \delta_0 = -k\bar{R}. \tag{2.75}$$

[16]Note that for infinite potentials the continuity of the derivative $u_l'(k,r)$ is not required. In this particular case, it is discontiuous at $\rho = \bar{\rho}$.

2.8 Absolute phase shifts — Levinson Theorem

As we mentioned in section 2.2, the phase shifts are defined modulo π. This means that one cannot distinguish among δ_l, $\delta_l \pm \pi$, $\delta_l \pm 2\pi$,..., etc. This indefinition can be eliminated if we replace Eq. (2.33) by

$$\delta_l = \tan^{-1}\left[-\frac{k\bar{R}\,\hat{\jmath}_l'(k\bar{R}) - \hat{\jmath}_l(k\bar{R})\,\mathcal{L}^{\mathrm{I}}}{k\bar{R}\,\hat{n}_l'(k\bar{R}) - \hat{n}_l(k\bar{R})\,\mathcal{L}^{\mathrm{I}}}\right]. \tag{2.76}$$

Now δ_l is uniquely defined, since the function $\tan^{-1}$ is defined in the interval $\{-\pi/2, +\pi/2\}$. However, this procedure has the disadvantage of leading to phase shifts which are not continuous functions of the collision energy. As the energy varies, a phase shift would have a jump of π whenever it reaches $\pm\pi/2$.

A search for an absolute definition of δ_l should begin with a study of its behavior in the limiting situations $E \to 0$ and $E \to \infty$. In the high energy limit, the phase shift should satisfy the condition $\tan\delta_l \to 0$. If the potential is limited everywhere and has a finite range[17], it becomes negligible as compared to E in the Schrödinger equation. The dynamics then reduces to that of a free particle. On the other hand, it is straightforward to show that $\tan[\delta_l(k \to 0)] = 0$ (see section 2.11). However, the condition $\tan\delta_l = 0$ is satisfied when the phase shift is equal to any multiple of π, not necessarily 0. To eliminate ambiguities, we set

$$\delta_l(k \to \infty) = 0, \quad \text{for any } l, \tag{2.77}$$

and require that the phase shifts be continuous as the energy is reduced. In this way, it is necessary to add (subtract) π to the phase shift at the energies where $\tan^{-1}$ jumps from $\pi/2$ to -$\pi/2$ (from -$\pi/2$ to $\pi/2$). Following this procedure one can find that $\delta_l(k \to 0) \neq 0$, although[18] $\tan[\delta_l(k \to 0)] = 0$. According to Levinson theorem [Levinson (1949)], the value of $\delta_l(k \to 0)$ is related to the number of bound states of the potential with the same angular momentum, n_l, through the expression,

$$\delta_0(k \to 0) = n_l\,\pi. \tag{2.78}$$

Below, we discuss Levinson theorem in the simple case of s-wave scattering from the square well potential[19]

$$V(r) = 0, \qquad \text{for } r > \bar{R}, \tag{2.79}$$

$$V(r) = -V_0, \quad \text{for } r < \bar{R}. \tag{2.80}$$

[17]Note that this discussion is not applicable to the Coulomb potential.

[18]As we will see in this section, there are anomalous situations where $\tan[\delta_0(k \to 0)] \neq 0$.

[19]A detailed proof of the Levinson Theorem can be found in [Levinson (1949)] or in [Taylor (2006)].

We consider the low energy limit of the s-wave phase shift as the depth V_0 is varied. The radial wave function inside the well is known analytically:

$$u_0^{\mathrm{I}}(k,r) = A\sin(Kr)\,, \tag{2.81}$$

where

$$K = \frac{\sqrt{2\mu(V_0+E)}}{\hbar}\,. \tag{2.82}$$

The internal logarithmic derivative at the boundary $\bar{R}$ then is

$$\mathcal{L}^{\mathrm{I}} \equiv \bar{R}\,\frac{[du_0^{\mathrm{I}}(k,r)/dr]_{r=\bar{R}}}{u_0^{\mathrm{I}}(k,\bar{R})} = K\bar{R}\,\cot(K\bar{R})\,. \tag{2.83}$$

Using Eq. (2.83) and the explicit forms of $\hat{\jmath}_0(kr)$, $\hat{n}_0(kr)$, $\hat{\jmath}_0'(kr)$, $\hat{n}_0'(kr)$,

$$\begin{aligned} \hat{\jmath}_0(kr) &= \sin(kr)\,; \quad & \hat{\jmath}_0'(kr) &= \cos(kr)\,,\\ \hat{n}_0(kr) &= \cos(kr) \quad & \hat{n}_0'(kr) &= -\,\sin(kr)\,, \end{aligned}$$

in Eq. (2.33), we get

$$\begin{aligned} \tan\delta_0 &= \frac{k\cos(k\bar{R}) - K\,\sin(k\bar{R})\,\cot(K\bar{R})}{k\sin(k\bar{R}) + K\,\cos(k\bar{R})\,\cot(K\bar{R})}\\ &= \frac{\frac{k}{K}\tan(K\bar{R}) - \tan(k\bar{R})}{1+\frac{k}{K}\tan(K\bar{R})\,\tan(k\bar{R})}\,. \end{aligned} \tag{2.84}$$

Calling

$$\alpha = \tan^{-1}\left[\frac{k}{K}\tan(K\bar{R})\right]; \quad \beta = k\bar{R},$$

we use in Eq. (2.84) the trigonometric identity

$$\frac{\tan\alpha - \tan\beta}{1+\tan\alpha\,\tan\beta} = \tan(\alpha-\beta).$$

We get

$$\delta_0 = \tan^{-1}\left[\frac{k}{K}\tan(K\bar{R})\right] - k\bar{R}\,. \tag{2.85}$$

At this stage, we leave open the possibility of summing or subtracting integer multiples of π to δ_0. The need for such a procedure becomes clear as we investigate the low energy behavior of δ_0 as a function of the potential depth, V_0.

Initially, we consider a very weak potential so that,

$$K_0\bar{R} = \left(\frac{\sqrt{2\mu V_0}}{\hbar}\right)\bar{R} \ll 1\,.$$

In the low energy limit, $k \to 0$ and we can approximate

$$\delta_0 \simeq k\bar{R}\left(\frac{\tan(K_0\bar{R})}{K_0\bar{R}} - 1\right). \tag{2.86}$$

Therefore, the phase shift of Eq. (2.86) goes to zero proportionally to k. This is consistent with the ideia vanishing phase shifts for a free particle.

We now let the potential strength grow until it satisfies the condition

$$K_0\bar{R} = \left(\frac{\sqrt{2\mu V_0}}{\hbar}\right)\bar{R} = \frac{\pi}{2}\,. \tag{2.87}$$

In this case, the approximation of Eq. (2.86) breaks down and Eq. (2.85) must be used. Replacing $K_0\bar{R} = \pi/2$ and then taking the limit $k \to 0$, we find the anomalous (in the sense that $\tan\left[\delta_0(k \to 0)\right] \neq 0$) phase shift

$$\lim_{k\to 0} \delta_0 = \frac{\pi}{2}\,.$$

Let us study this potential in further details. Inserting the value $K\bar{R} = \pi/2$ in Eq. (2.83) we find that $\mathcal{L}^{\mathrm{I}} = 0$. In this way, the internal wave function has an horizontal tangent at the matching radius. It is the limiting case where the internal wave function can be matched to a decaying exponential in the external region. It corresponds to a bound state with zero binding energy. Therefore, Eq. (2.87) indicates that the minimal depth for a square well to bind an s-state is

$$V_0 = \frac{\pi^2\hbar^2}{8\mu\bar{R}^2}\,. \tag{2.88}$$

Let us now consider a deeper potential such that

$$\frac{3\pi}{2} > K_0\bar{R} > \frac{\pi}{2}\,. \tag{2.89}$$

In this case, Eq. (2.86) can again be used and δ_0 goes to zero in the $k \to 0$ limit. However, now it approaches zero from negative values. For the phase shift to grow monotonically with the potential depth, it is necessary to sum the constant phase π to Eq. (2.86). That is,

$$\delta_0 \simeq \pi + k\bar{R}\left(\frac{\tan(K\bar{R})}{K\bar{R}} - 1\right). \tag{2.90}$$

If we let the potential get deeper, a new anomaly will occur when $K_0\bar{R} = 3\pi/2$. In this case,

$$\lim_{k\to 0} \delta_0 = \frac{3\pi}{2}.$$

It corresponds to the situation where the potential has two bound states. The one it had in the previous case plus a second bound state with one node (besides the one at the origin) and zero energy. As the depth exceeds this value, again a constant phase π has to be added to Eq. (2.90).

Following the same procedures successively, we conclude that the anomaly occurs whenever the depth of the potential reaches the value

$$V_0 = \frac{\pi^2\hbar^2}{2\mu\bar{R}^2}\left(n_0 - \frac{1}{2}\right)^2, \tag{2.91}$$

where $n_0 = 1, 2, \ldots$, and it corresponds to the condition for the potential to bind n_0 S-states. Using the same arguments as above, we get

$$\delta_0(k \to 0) = n_0\,\pi\ , \tag{2.92}$$

which is the result predicted by Levinson theorem (Eq. (2.78)).

As an illustration, we derive absolute s-wave phase shifts for a square well potential with $K_0\bar{R} = 10$. In this case, the potential can bind 3 s-states ($K\bar{R} = \pi/2,\ 3\pi/2,$ and $5\pi/2$). We have performed numerical calculations starting at $E \gg V_0$, where $\delta_0 \simeq 0$, and gradually reduced the energy until $E \simeq 0$. In order to have a continuous function of E, we have eliminated the jumps of $\tan^{-1}$ adding or subtracting π, as required. The results are presented in figure 2.3. Figure 2.3 (a) shows the discontinuous phase shifts, defined in Eq. (2.76). Figure 2.3 (b) shows the absolute phase shifts obtained with the above described procedure. The phase shifts are normalized with respect to π and are plotted as functions of E/V_0. We note that the absolute phase shift of figure 2.3 (b) is a smooth function of the energy. In accordance with the Levinson Theorem, $\delta_0(k \to 0) = 3\pi$.

To conclude this section, we point out an interesting feature of low energy scattering from attractive potentials. Looking at figure 2.3 (b), we see that the phase shift can take the values π and 2π for finite k-values. When this occurs, $S_0 = 1$ and the contribution from $l = 0$ to the scattering amplitude vanishes. If this happens at a very low energy, the partial-wave series can be truncated at $l = 0$ and the elastic cross section becomes very small. At this energy, the target will be transparent to the beam. This phenomenon, known as the Ramsauer-Townsend effect, has been observed in collisions of 0.7 eV electrons with A, Kr, and Xe targets.

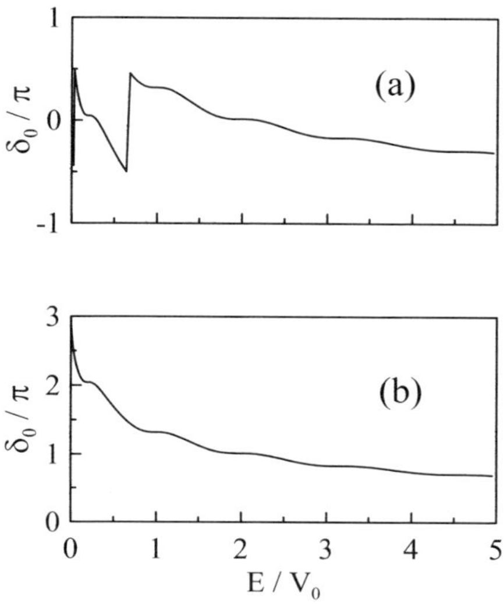

Fig. 2.3 The phase shifts defined in the $\{-\pi/2, \pi/2\}$ interval (a) and the absolute phase shifts (b). For details, see the text.

2.9 Resonances

Although the phase shifts are usually slowly varying functions of the collision energy, there may be situations where a particular phase shift δ_l changes rapidly with E. In this case, the corresponding partial wave tends to give important contribution to the cross section. This behavior is called a resonance. In many important examples, the description of the resonance requires the consideration of intrinsic degrees of freedom. This is, for example, the case of the compound nucleus resonances in Nuclear Physics (see, e.g. [Fe92]). However, a deep attractive potential can also give rise to resonances. In this section we study resonances in potential scattering in the simple case of an s-wave.

As a starting point, we write the asymptotic wave function as in Eq. (2.35),

$$u_0^{\rm II}(k,r) = e^{-ikr} - S_0 \, e^{ikr}, \tag{2.93}$$

where $S_0 = e^{2i\delta_0}$ and we set $\gamma_l = 1$. We now use this equation to calculate the logarithmic derivative

$$\mathcal{L}^{\rm II} = \bar{R} \; \frac{[du_0^{\rm II}(k,r)/dr]_{r=\bar{R}}}{u_0^{\rm II}(k,\bar{R})} = -ik\bar{R} \; \left[\frac{1 + S_0 \, e^{2ik\bar{R}}}{1 - S_0 \, e^{2ik\bar{R}}} \right].$$

It is convenient to express S_0 in terms of the logarithmic derivative and use the matching condition: $\mathcal{L}^{\mathrm{II}} = \mathcal{L}^{\mathrm{I}}$. We get,

$$S_0 = -e^{-2ik\bar{R}} \left[\frac{k\bar{R} - i\mathcal{L}^{\mathrm{I}}}{k\bar{R} + i\mathcal{L}^{\mathrm{I}}} \right]. \tag{2.94}$$

Now we write separatately the real and the imaginary parts[20] of $\mathcal{L}^{\mathrm{I}}$,

$$\mathcal{L}^{\mathrm{I}} = a - b\,i, \tag{2.95}$$

and rewrite Eq. (2.94) as

$$S_0 = -e^{-2ik\bar{R}} \left[\frac{(k\bar{R} - b) - i\,a}{(k\bar{R} + b) + i\,a} \right]. \tag{2.96}$$

Using Eq. (2.96) in the s-wave contribution for the scattering amplitude $f_0(\theta) \equiv f_0$,

$$f_0 = \frac{1}{2ik} \left(S_0 - 1 \right), \tag{2.97}$$

we get

$$f_0 = \frac{1}{2ik} \left\{ -e^{-2ik\bar{R}} \left[\frac{(k\bar{R} - b) - i\,a}{(k\bar{R} + b) + i\,a} \right] - 1 \right\}.$$

Adding and subtracting $e^{-2ik\bar{R}}$ within the curly brackets, f_0 takes the form

$$f_0 = f_{\mathrm{pot}} + f_{\mathrm{res}}, \tag{2.98}$$

with

$$f_{\mathrm{pot}} = \frac{1}{2ik} \left(e^{-2ik\bar{R}} - 1 \right), \tag{2.99}$$

and

$$f_{\mathrm{res}} = \frac{e^{-2ik\bar{R}}}{k} \left[\frac{k\bar{R}}{a - i(k\bar{R} + b)} \right]. \tag{2.100}$$

Eq. (2.99) gives the s-wave term of Eq. (2.54a) with the phase shift

$$\delta_{\mathrm{pot}} = -k\bar{R}.$$

Therefore, it corresponds to the scattering from a hard sphere with radius $\bar{R}$, as given in Eq. (2.75). The resonant behavior has then to be associated with f_{res}. A resonance occurs when the real part of $\mathcal{L}^{\mathrm{I}}$ vanishes. In this way, the condition

$$a(E_r) = 0 \tag{2.101}$$

[20]Real potentials lead to real logarithmic derivatives. However, we take it as a complex number so that the theory can be generalized to complex potentials (see chapter 7).

is equivalent to an equation for the resonance energy, E_r. To study the cross section in the neighborhood of the resonance, we expand $a(E)$ around $E = E_r$,

$$a(E) \simeq \left[\frac{\partial a(E)}{\partial E}\right]_{E=E_r} (E - E_r) \equiv a'(E_r)\ (E - E_r)\ , \tag{2.102}$$

and introduce the notation

$$\begin{aligned} \Gamma_{\rm e} &= -\frac{2k\bar{R}}{a'(E_r)}, \\ \Gamma_a &= -\frac{2b}{a'(E_r)} \\ \Gamma &= \Gamma_{\rm e} + \Gamma_a. \end{aligned} \tag{2.103}$$

The quantities $\Gamma_{\rm e}$, Γ_a and Γ are known as the scattering width, the absorption (or reaction) width and the total width, respectively.

Using the expansion for $a(E)$, the resonant amplitude can be written

$$f_{\rm res} \simeq -\frac{e^{-2ik\bar{R}}}{k} \left[\frac{\Gamma_e/2}{(E - E_r) + i\,\Gamma/2}\right]. \tag{2.104}$$

For a strong and isolated resonance, the contribution from $f_{\rm res}$ to the cross section is so large that $f_{\rm pot}$ can be neglected (see exercise 4). In this case, the integrated elastic cross section becomes[21]

$$\sigma_{\rm el} = \frac{\pi}{k^2} \left[\frac{\Gamma_{\rm e}^2}{(E - E_r)^2 + \Gamma^2/4}\right]. \tag{2.105}$$

This result is known as the Breit-Wigner formula. The cross section is maximal at $E = E_r$ and has the half-width $|E - E_r| = \Gamma/2$. For real potentials, as assumed in this chapter, the logarithmic derivative is real and, therefore, $b = 0$. In this case the reaction width vanishes and $\Gamma_{\rm e} = \Gamma$.

2.9.1 *Time delay*

Near a resonance, the phase of $f_{\mathbf{k}}(\theta)$ changes rapidly with k and some modifications in the discussion of wave packet scattering of section 1.6.2 are needed. This can be clearly seen in the simple case of low-energy ($k\bar{R} \ll 1$)

[21] In the case of complex potentials (see chapter 7), the resonant behavior also appears in the absorption cross section. It is given by $\sigma_a \equiv (\pi/k^2)\left[1 - |S_0|^2\right] = (\pi/k^2)\ \Gamma_e\Gamma_a \,/\, \left[(E - E_{\rm res})^2 + \Gamma^2/4\right]$.

s-wave resonances in a real potential. In this case, the logarithmic derivative is real, $\mathcal{L}^{\text{I}} = a$, so that $b = 0$, $\Gamma_a = 0$ and $\Gamma = \Gamma_e$. Using the low-energy approximation, $e^{-2ik\bar{R}} \simeq 1$, f_{pot} vanishes, so that $f_0(\mathbf{k}) = f_{\text{res}}$. Eq. (2.104) then becomes

$$f_{\text{res}}(E) \simeq -\frac{1}{k}\left[\frac{\Gamma/2}{(E-E_r)+i\,\Gamma/2}\right]. \tag{2.106}$$

We now define the resonant phase shift δ_{res} by the condition

$$\frac{1}{2ik}(e^{2i\delta_{\text{res}}}-1) = f_{\text{res}}(E). \tag{2.107}$$

Using Eq. (2.106) and the trigonometrical relation

$$\tan(2\delta) = \frac{2\tan\delta}{1-\tan^2\delta}, \tag{2.108}$$

we get

$$\delta_{\text{res}} = \tan^{-1}\left[\frac{-\Gamma/2}{(E-E_r)}\right]. \tag{2.109}$$

Inspecting Eq. (2.109), we see that, as $E \to E_r$, the absolute phase shift goes through $n\pi + \pi/2$, where the integer n depends on the potential strength (Levinson's theorem). This behavior has important implications for wave packet scattering (see section 1.6.2). We have shown that the scattered wave packet can be written as (Eq. (1.91))

$$\psi^{sc}_{\mathbf{b},\mathbf{k}_0}(\mathbf{r},t) = \left(\frac{1}{2\pi}\right)^{3/2}\frac{1}{r}\int d\mathbf{k}^3\ |A(\mathbf{k}-\mathbf{k}_0)|\ f_{\mathbf{k}}(\theta)\ e^{i\gamma(\mathbf{k})}\ ,$$

where $\gamma(\mathbf{k})$ is the phase

$$\gamma(\mathbf{k}) = kr - t\,\omega_k - (\mathbf{k}-\mathbf{k}_0)\cdot\mathbf{b}\,.$$

In typical situations, $f_{\mathbf{k}}(\theta)$ is a slowly varying function of $\mathbf{k}$ in the range of relevant values of $|A(\mathbf{k}-\mathbf{k}_0)|$. In section 1.6.2, it was moved out of the integral with the value $f_{\mathbf{k}_0}(\theta)$ and this approximation lead to the conclusion that the stationary phase condition is satisfied by an outgoing spherical wave, moving according to the equation

$$R_t = v_0\,t.$$

In the case of a resonance, this approximation is no longer valid. In a scattering experiment to study the resonance shape, the wave packets should have good energy resolution. The energy uncertainty of the wave packets must satisfy the condition

$$\Delta E = \frac{\hbar^2 k}{2\mu}\,\Delta k \ll \Gamma.$$

This condition guarantees that $|f(E)|$ can be approximated by $|f(E_0)|$, where $E_0 = \hbar^2 k_0^2/2\mu$. In this way, it can be moved out of the integral, as was done in section 1.6.2. However, the same cannot be done with the phase of $f_{\rm res}(E)$. According to Eq. (2.107), the resonant amplitude contains the term $e^{2i\delta_{\rm res}}/2ik$, which has the rapidly varying phase $2\delta_{\rm res}$. It should then be added to $\gamma(\mathbf{k})$ in the search of stationary points. That is,

$$\gamma(\mathbf{k}) \to \tilde{\gamma}(\mathbf{k}) = kr - t\,\omega_k - (\mathbf{k} - \mathbf{k}_0)\cdot\mathbf{b} + 2\delta_{\rm res}(E)\,.$$

This leads to an extra term in the series expansion around $\mathbf{k} = \mathbf{k}_0$, namely

$$\begin{aligned} 2\delta_{\rm res}(E) &\simeq 2\,\delta_{\rm res}(E_0) + 2(\mathbf{k} - \mathbf{k}_0)\cdot\Big[\nabla_{\mathbf{k}}\delta_{\rm res}(E)\Big]_{\mathbf{k}=\mathbf{k}_0} \\ &= 2\,\delta_{\rm res}(E_0) + 2\hbar\,v_0\;\hat{\mathbf{z}}\cdot(\mathbf{k} - \mathbf{k}_0)\;\frac{d\delta_{\rm res}(E)}{dE_0}, \end{aligned} \qquad (2.110)$$

where $v_0 = \hbar k_0/\mu$, and we used the short hand notation

$$\frac{d\delta_{\rm res}}{dE_0} = \left[\frac{d\delta_{\rm res}(E)}{dE}\right]_{E=E_0}.$$

Including the above term in the expansion of the global phase $\tilde{\gamma}(\mathbf{k})$, the stationary phase condition takes the form

$$R_t = v_0\,(t - \Delta_t), \qquad (2.111)$$

where

$$\Delta_t = 2\hbar\left(\frac{d\delta_{\rm res}}{dE_0}\right). \qquad (2.112)$$

Thus, owing to the resonance, the scattered wave emerges with the time delay Δt.

When a sharp resonance occurs, the cross section increases abruptly. It exhibits a maximum at the resonance energy, where the absolute phase shift reaches $(n + 1/2)\,\pi$. In such cases, the time delay may be quite large. Examples of sharp resonances for an attractive square well potential are given in the next section (see table 2.2 and figure 2.4).

We remark that the phase shift can also reach a value $(n + 1/2)\pi$ as it decreases with the collision energy. However, this does not correspond to a resonance. In such cases, the phase shifts are slowly varying and there is no time-delay. Since $d\delta_l/dE$ is negative, the potential accelerates the incident wave, instead of slowing it down.

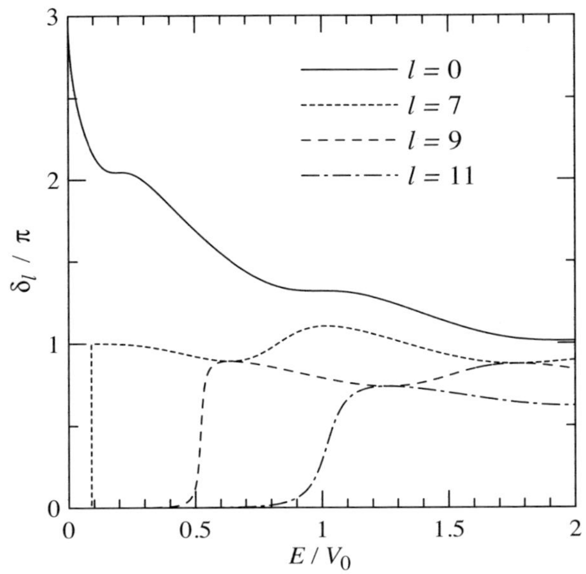

Fig. 2.4 Absolute phase shifts (according to Levinson's theorem) for a few partial-waves in the scattering from a square well potential with characteristic parameter $K\bar{R} = 10$.

2.10 Scattering from a square-well

Let us now calculate phase shifts in the scattering from a square-well of depth V_0 and range $\bar{R}$, for an arbitrary partial-wave. In this case, the radial wave function in the internal region becomes

$$u_l^{\mathrm{I}}(k,r) = A\,\hat{\jmath}_l(Kr)$$

and the logarithmic derivative is

$$\mathcal{L}^{\mathrm{I}} = K\bar{R}\;\frac{\hat{\jmath}_l'(K\bar{R})}{\hat{\jmath}_l(K\bar{R})}. \tag{2.113}$$

Inserting Eq. (2.113) in Eq. (2.33), we get

$$\delta_l = \tan^{-1}\left[-\left(\frac{k\;\hat{\jmath}_l'(k\bar{R})\,\hat{\jmath}_l(K\bar{R}) - K\;\hat{\jmath}_l(k\bar{R})\,\hat{\jmath}_l'(K\bar{R})}{k\;\hat{n}_l'(k\bar{R})\,\hat{\jmath}_l(K\bar{R}) - K\;\hat{n}_l(k\bar{R})\,\hat{\jmath}_l'(K\bar{R})}\right)\right]. \tag{2.114}$$

Absolute phase shifts can be obtained through the procedure described in section 2.8. As an illustration, we consider a strongly attractive square well with $K_0\bar{R} = 10$. We start from a very high energy and progressively reduce it, adding or subtracting π as required to eliminate the discontinuities. Note that the phase shifts depend exclusively on the characteristic

Table 2.1 Energies of the bound states with angular momentum l, normalized with respect to the depth of the well, $\varepsilon = E_{n,l}/V_0$. The quantum number n corresponds to the number of nodes of the radial wave function (excluding the zero at the origin).

l	n	ε
0	0	-0.92
	1	-0.68
	2	-0.29
1	0	-0.83
	1	-0.52
	2	-0.08
2	0	-0.73
	1	-0.34
3	0	-0.60
	1	-0.14
4	0	-0.46
5	0	-0.29
6	0	-0.11

Table 2.2 Locations and widths of the resonances at the partial-waves $l = 7, 8, 9, 10$ and 11.

l	ε	γ
7	0.09	0.000007
8	0.30	0.002
9	0.52	0.012
10	0.77	0.026
11	1.04	0.57

parameter of the well, $K_0\bar{R}$, and on the wave number k. The results are shown in figure 2.4, for a few partial-waves. Again, the x-axis gives the normalized scattering energy $\varepsilon = E/V_0 = k^2/K_0^2$. The behavior of $\delta_l(\varepsilon)$ can be better understood in connection with the bound states of the potential. In table 2.1 they are given as functions of the angular momentum in $\hbar$-units, l, and the number of nodes of the wave function, n. We see that the potential binds three states for each of the angular momentum values $l = 0, 1$, two states for $l = 2, 3$ and a single state for $l = 4, 5$ and 6. There are no bound state for higher angular momenta. According to Levinson's theorem, the phase shifts should have the limits $\delta_l(k \to 0) = 3\pi$ for $l = 0$ and 1, $\delta_l(k \to 0) = 2\pi$ for $l = 2$ and 3, and $\delta_l(k \to 0) = \pi$ for $l = 4$, 5 and 6. At higher partial-waves the phase shifts should go to zero in the $k \to 0$ limit. This behavior is confirmed in figure 2.4. For $l = 7$ the bound state ocurrying at $l = 6$ (see Table 2.1) becomes a sharp resonance. For higher angular momenta, the resonance becomes progressively broader and gradually disappears. Fitting Breit-Wigner shapes, we get the resonance parameters indicated in table 2.2. Similarly to the energies in figure 2.4, the widths Γ are normalized with respect to the potential depth, as $\gamma = \Gamma/V_0$. In section 8.1, we will discuss this point in further details, connecting bound states and resonances to the poles of the S-matrix.

2.11 Low energy scattering

In this section we develop approximate formulae for low-energy scattering. First, we point out that in the $k \to 0$ limit the radial wave function becomes an energy independent function, $z_l(r)$, which satisfies the equation

$$-\frac{\hbar^2}{2\mu}\left[\frac{d^2}{dr^2} - \frac{l(l+1)}{r^2}\right] z_l(r) \; + \; V(r)\; z_l(r) = 0. \tag{2.115}$$

Eq. (2.115) is obtained from Eq. (2.9), setting $E = 0$. Accordingly, the logarithmic derivative $\mathcal{L}^{\text{I}}$ takes the energy independent value L_l, given by

$$L_l = \lim_{k \to 0} \mathcal{L}^{\text{I}} = \bar{R}\; \frac{[dz_l(r)/dr]_{r=\bar{R}}}{z_l(\bar{R})}. \tag{2.116}$$

The phase shifts of Eq. (2.33) are also given by a simpler expression in the low energy limit. Using the small argument limits of the Bessel and Neumann functions (Eqs. (2.22a) and (2.22b)) and their derivatives in Eq. (2.33) and replacing $\mathcal{L} \to L_l$, one gets

$$\tan\delta_l = \frac{\left(k\bar{R}\right)^{2l+1}}{(2l+1)!!\;(2l-1)!!}\left[\frac{(l+1-L_l)}{l+L_l}\right]. \tag{2.117}$$

This expression becomes particularly simple for $l = 0$. This case is very important and will be discussed in detail below.

2.11.1 *The scattering length*

At low energies, the scattering amplitude is dominated by the contribution from the s-wave (see section 2.6). Thus, the phase shift δ_0 becomes particularly important. Inserting the explicit forms of $\hat{\jmath}_0$ and $\hat{n}_0$ (Eqs. (2.17a) and (2.17b)) and their derivatives in Eq. (2.33), one gets

$$\tan\delta_0 = \frac{x\cos x - \mathcal{L}^{\text{I}}\sin x}{x\sin x + \mathcal{L}^{\text{I}}\cos x}, \tag{2.118}$$

where the short-hand notation $x = k\bar{R}$ has been used. To lowest order in x, one can approximate $\cos x \simeq 1$, $\sin x \simeq x$ and replace $\mathcal{L}^{\mathrm{I}} \to L_{l=0} \equiv L$. Eq. (2.118) becomes

$$\tan\delta_0 = -\,k\bar{R}\,\left[\frac{L-1}{L}\right]. \tag{2.119}$$

For the study of low-energy scattering, it is useful to introduce the scattering length, a, defined as the limit

$$a \equiv -\lim_{k\to 0}\left(\frac{\tan\delta_0}{k}\right) = \bar{R}\left[\frac{L-1}{L}\right]. \tag{2.120}$$

Thus, in the $k \to 0$ limit, one has the relation[22]

$$\delta_0 \simeq \tan\delta_0 \simeq -\,ka. \tag{2.121}$$

Comparing Eq. (2.121) with Eq. (2.75), we conclude that the scattering length can be interpreted as the radius of an *effective hard-sphere* that produces the same low-energy s-wave phase shift.

The scattering length is qualitatively related to the potential in a simple way. To show this relation, we consider the radial wave function outside the potential well, as given in Eq. (2.34). We set $l = 0$ and adopt the short-hand notation: $\alpha'_l = A$. Taking the $k \to 0$ limit and using Eq. (2.121), one obtains

$$u_0^{\mathrm{II}}(k,r) = A\,\sin(kr+\delta_0) \simeq A\,(kr+\delta_0) \simeq A\,k\,(r-a)\,. \tag{2.122}$$

Eq. (2.122) indicates that the external wave function is linear near $r \simeq \bar{R}$. Extending the function $u_0^{\mathrm{II}}(k,r)$ into region I (where it is not the solution of the radial equation), the scattering length is given by the coordinate of the point where this straight line intercepts the horizontal axis. This establishes a qualitative relation between the potential and the scattering length. First we consider a repulsive potential, as represented in the upper panel of figure 2.5a. Since region I ($0 < r < \bar{R}$) is classically forbidden, the wave function $u_0^{\mathrm{I}}(k,r)$ grows very rapidly (exponentially for a square well) with r. The continuity of the logarithmic derivative requires that u_0^{I} and u_0^{II} match as indicated in the bottom panel of figure 2.5a. The intersection of the dashed line with the horizontal axis occurs at a point between the origin and the matching radius. Therefore, the scattering length is positive and smaller than $\bar{R}$.

[22]We point out that the approximation $\tan\delta_0 \simeq \delta_0$ is not valid for absolute phase shifts of an attractive potential with bound states. According to the Levinson theorem, $\lim_{k\to 0}[\delta_0] = n\pi$, where n is the number of bound states with $l = 0$. However, this does not lead to any relevant change in this section.

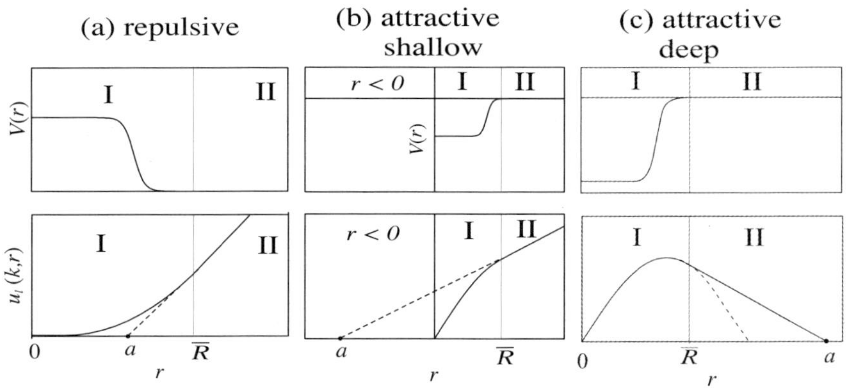

Fig. 2.5 The scattering length for repulsive and attractive potentials. The upper panels represent different potentials. The lower panels show the corresponding radial wave functions. The dashed lines are the extrapolations of the external radial functions into the internal region.

We now turn to attractive potentials. We first consider a shallow potential well, not deep enough to have a bound state (upper panel in figure 2.5b). In this case, the phase of the $u_0^{\mathrm{I}}(k,r)$ is small at $r = \bar{R}$. It does not reach its first maximum in the internal region, as indicated in the bottom panel in figure 2.5b. The intersection of the dashed line with the horizontal axis is then at the left of $r = 0$, so that the scattering length is negative.

Finally, we consider a deeper potential well (upper panel of figure 2.5c), such that the internal wave function is strongly bent in the internal region. The phase at $r = \bar{R}$ is then between $\pi/2$ and π. This corresponds to the condition for the well to have a single bound state. In this case, $u_l^{\mathrm{II}}(k,r)$ matches a decreasing wave function at $r = \bar{R}$, as represented in figure 2.5c. The scattering length is then positive and larger than the matching radius. This analysis is summarized below.

$$
\begin{array}{ll}
V(r) > 0 \text{ (repulsive)} & \bar{R} > a > 0 \\
V(r) < 0 \text{ (attractive) - 0 BS} & a < 0 \\
V(r) < 0 \text{ (attractive) - 1 BS} & a > \bar{R} > 0.
\end{array}
$$

The case of deep attractive potentials with more than one bound state is more complicated. The scattering length is either negative (when $\mathcal{L}^{\mathrm{I}} > 0$) or positive and larger than $\bar{R}$ (when $\mathcal{L}^{\mathrm{I}} < 0$).

We now look for low energy approximations for the scattering amplitude. For this purpose, we use the identity

$$f = \frac{e^{i\delta_0}}{k} \sin\delta_0 = \frac{1}{k\cot\delta_0 - ik}, \tag{2.123}$$

which can be easily checked. The next step is to write the inverse of the scattering amplitude,

$$W(k) = \frac{1}{f} = k\cot\delta_0 - ik, \tag{2.124}$$

and consider approximations of zero, first and second order in k. To lowest order, one obtains

$$W^{(0)} = \lim_{k\to 0}\left[k\cot\delta_0 - ik\right] = -\frac{1}{a} \tag{2.125}$$

and thus

$$f^{(0)} = -a\ . \tag{2.126}$$

Since the above scattering amplitude is isotropic, the elastic cross section can be trivially evaluated. One gets

$$\sigma_{\rm el} = \int d\Omega\ |f|^2 = 4\pi a^2. \tag{2.127}$$

The expansion of $W(k)$ to first order in k can easily be derived. Since the power series for $\sin x$ ($\cos x$) has only odd (even) terms, it can be easily checked that $k\cot\delta_0$ does not have a first order term. In this way, the only first order term in $W(k)$ is $-ik$. Therefore, we can write

$$W^{(1)}(k) = -\frac{1}{a} - ik \quad\Longrightarrow\quad f^{(1)} = \frac{1}{-\frac{1}{a} - ik}. \tag{2.128}$$

2.11.2 *The effective range formula*

Now we derive an improved approximation for the scattering amplitude, taking into account second order terms in $W(k)$. The result is known as the effective range formula. Since $(-ik)$ is of first order, it is clear that the second order correction to $W(k)$ arises from the first term at the RHS of Eq. (2.124). That is, from

$$g(x) = k\cot\delta_0 = \frac{x}{\bar{R}}\cot\delta_0, \tag{2.129}$$

where $x = k\bar{R}$. Using the inverse of Eq. (2.118), the above equation becomes

$$g(x) = \frac{x}{\bar{R}}\left[\frac{x\sin x + \mathcal{L}^{\rm I}\cos x}{x\cos x - \mathcal{L}^{\rm I}\sin x}\right]. \tag{2.130}$$

One should then calculate $\sin x$, $\cos x$ and $\mathcal{L}^{\rm I}$ more accurately than in the previous section. We use the truncated expansions

$$\sin x \simeq x - \frac{x^3}{6}; \quad \cos x \simeq 1 - \frac{x^2}{2}, \tag{2.131}$$

and take into account second order corrections to $\mathcal{L}^{\rm I}$. For this purpose, we write the differential equations for $u_0(k,r)$ and for $z_0(r)$ (Eqs. (2.9) and (2.115) with $l = 0$),

$$u_0''(k,r) - U(r)\, u_0(k,r) = -\, k^2\, u_0(k,r); \tag{2.132}$$

$$z_0''(r) - U(r)\, z_0(r) = 0, \tag{2.133}$$

where, $U(r) = 2\mu V(r)/\hbar^2$. Carrying out the operation

$$z_0(r) \times \text{ Eq. (2.132)} - u_0(k,r) \times \text{ Eq. (2.133)},$$

we obtain

$$\frac{d}{dr}\Big(z_0(r)\, u_0'(k,r) - u_0(k,r)\, z'(r) \Big) = -k^2\, u_0(k,r)\, z_0(r).$$

Integrating the above equation between $r = 0$ and $r = \bar{R}$, and using the regularity of u_0 and z_0, we get

$$z_0(\bar{R})\, u_0'(k,\bar{R}) - u_0(k,\bar{R})\, z_0'(\bar{R}) = -\, k^2 \int_0^{\bar{R}} dr\; u_0(k,r)\, z_0(r).$$

Now, we multiply both sides by $\bar{R}/\left[z_0(\bar{R})\, u_0(k,\bar{R}) \right]$. Using the definitions of $\mathcal{L}^{\rm I}$ and L (see Eq. (2.116)), we obtain

$$\mathcal{L}^{\rm I} - L = -\, k^2\, \frac{\bar{R}}{u_0(k,\bar{R})\, z_0(\bar{R})} \int_0^{\bar{R}} dr\; u_0(k,r)\, z_0(r).$$

Since the difference $u_0(k,r) - z_0(r)$ leads to corrections of order k^3 or higher, we can replace $u_0 \to z_0$. The logarithmic derivative then becomes

$$\mathcal{L}^{\rm I} \simeq L - x^2\, \Delta, \tag{2.134}$$

with[23]

$$\Delta = \frac{1}{\bar{R}\; z_0^2(\bar{R})} \int_0^{\bar{R}} dr\; z_0^2(r). \tag{2.135}$$

[23] Note that this result is independent of the normalization of $z_0(r)$.

Inserting Eqs. (2.131) and (2.134) in Eq. (2.130), and keeping terms up to x^2, we get the second order approximation for $g(x)$,

$$g^{(2)}(x) = \frac{1}{\bar{R}} \left[\frac{L + x^2 (1 - \Delta - L/2)}{1 - L + x^2 (L/6 + \Delta - 1/2)} \right]$$
$$= -\frac{1}{a} + \frac{k^2}{2} \left[2\bar{R} \left(\frac{1 - L - \Delta + L^2/3}{(L-1)^2} \right) \right],$$

or

$$k \cot \delta_0 \simeq g^{(2)}(x) = -\frac{1}{a} + \frac{1}{2} r_{\text{eff}} \, k^2 \tag{2.136}$$

where r_{eff} is the effective range, given by[24]

$$r_{\text{eff}} = 2\bar{R} \left[\frac{1 - L - \Delta + L^2/3}{(L-1)^2} \right]. \tag{2.137}$$

The effective range can be expressed in terms of the range of the potential and the scattering length. Using Eq. (2.120) we get

$$r_{\text{eff}} = 2\bar{R} \left[\left(1 - \frac{\bar{R}}{a} \right) (1 - \Delta) + \frac{1}{3} \left(\frac{\bar{R}}{a} \right)^2 \right]. \tag{2.138}$$

We point out here some interesting properties of a and r_{eff} for very weak potentials. Let us consider the $U \to 0_\pm$ limit, where 0_+ and 0_- are used for repulsive and attractive potentials, respectively. The discussion of figure 2.5 leads to the conclusion

$$\lim_{U \to 0_\pm} [a] = 0_\pm. \tag{2.139}$$

Therefore, in this limit the scattering length approaches zero from positive values, for repulsive potentials, or negative values, for attractive potentials. Inspection of Eq. (2.138) shows that the effective range diverges in this limit. However, it can be shown that the product $(a \, r_{\text{eff}})$ remains finite and is given by

$$a \, r_{\text{eff}} = \pm 2 \, \bar{R}^2/3,$$

with + and - for repulsive and attractive potentials, respectively.

[24]Usually, text books (e.g. [Joachain (1983)]) give an alternative expression for r_{eff}, in terms of integrals involving $z_0(r)$ and an auxiliary function which does not vanish at the origin. It can be easily shown that their expression is identical to Eq. (2.137).

Eq. (2.136) leads to an improved expression to the scattering amplitude. Replacing this equation in Eq. (2.123), we obtain

$$f \simeq f^{(2)} = \frac{1}{g^{(2)} - ik} \tag{2.140}$$

where

$$g^{(2)} = -\frac{1}{a} + \frac{1}{2} r_{\text{eff}}\, k^2 . \tag{2.141}$$

Eq. (2.136) yields also an improved expression for the s-wave phase shift. It can be written

$$\tan\delta_0 = -ka \left[\frac{1}{1 - \frac{1}{2}k^2\,(a\, r_{\text{eff}})} \right] . \tag{2.142}$$

As an application of the above discussed low-energy approximations, we consider the s-wave phase shifts in the scattering from a square well with depth V_0 and range $\bar{R}$. In this case,

$$z_0(r) = A\,\sin\left(K_0 r\right) ,$$

where $K_0 = \sqrt{2\mu V_0}/\hbar$, and we get

$$\begin{aligned} L &= K_0\bar{R}\,\cot\left(K_0\bar{R}\right) \\ \Delta &= \frac{1}{2}\left[\frac{1}{\sin^2\left(K_0\bar{R}\right)} - \frac{\cot\left(K_0\bar{R}\right)}{K_0\bar{R}}\right] . \end{aligned}$$

Figure 2.6, shows s-wave phase shifts in the scattering from a square well, in comparison with the first order approximation (Eq. (2.121)) and the effective range formula for $\tan\delta_0$ (Eq. (2.142)). The results are plotted against $k\bar{R}$, in the range $2 > k\bar{R} > 0$. Above $k\bar{R} = 1$, these approximations are not expected to be reliable. In figure 2.6a, we used the deep potential of the section 2.8, with $K_0\bar{R} = 10$. In this case, the scattering length is positive and $\lim_{k\to 0}\delta_0 = 3\pi$. The exact phase shifts are the same as in figure 2.3, except that only a very small part of the figure is shown. In the present figure, the limit of the x-axis $\left(k\bar{R} = 1\right)$ corresponds to the $E/V_0 \equiv k^2/K_0^2 = 0.01$. In figure 2.6b, we used a much shallower potential, with $K_0\bar{R} = 1.5$. It has been chosen such as to be just below the limit to have a bound state $\left(K_0\bar{R} = \pi/2\right)$. In this case, the scattering length is negative and very large. As a result, the phase shift changes more rapidly with k. The figure shows that the two approximations are accurate at $k\bar{R} \lesssim 0.2$ and get poorer above this value. The second order approximation (effective range formula) is clearly better.

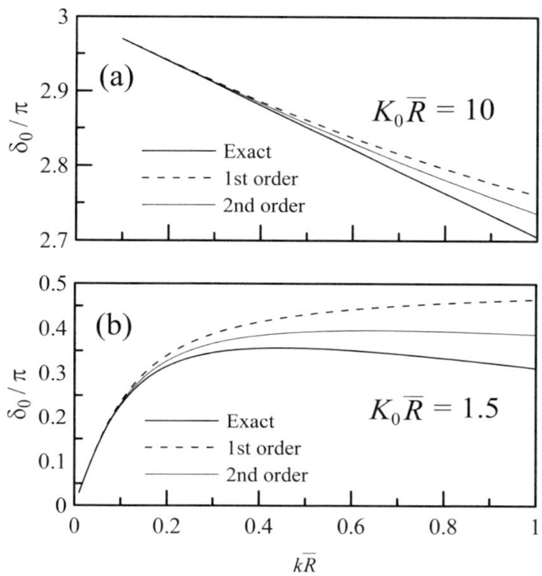

Fig. 2.6 Exact vs. approximate low-energy s-wave phase shifts in the scattering from a square well. For details, see the text.

The effective range expansion can be improved further taking into account the fourth order term. This is discussed in section 8.3, and also in section 8.4 (in the context of Calogero's equation).

Exercises

(1) Derive the expression (Eq. (2.56)),

$$(1 - x)\ f(\theta) = \frac{1}{2ik}\ \sum_l\ P_l(x)\ [(2l + 1)\, S_l - l\, S_{l-1} - (l + 1)\, S_{l+1}]\,,$$

where $x = \cos\theta$.

(2) Consider the square well potential of depth V_0 and range $\bar{R}$,

$$V(r) = -V_0,\ \text{for}\ r \leq \bar{R} \quad \text{and} \quad V_0 = 0,\ \text{for}\ r > \bar{R}.$$

Using qualitative arguments involving the radial wave functions inside and outside the well, find the minimum value of V_0 for a fixed $\bar{R}$ for this potential to have a bound state.

(3) Calculate the Wronskian $W\left\{\hat{h}_l^{(-)}(\rho), \hat{h}_l^{(+)}(\rho)\right\}$, where $\hat{h}_l^{(\pm)}(\rho)$ are the Ricatti-Haenkel functions discussed in the text.

(4) Consider a low energy resonance in the scattering from a real potential with range $\bar{R}$. Assume that

$$k\bar{R} \ll 1 \quad \text{where} \quad k = \frac{\sqrt{2\mu E_r}}{\hbar},$$

with E_r standing for the resonance energy. Show that the amplitude f_{pot} of Eq. (2.99) can be neglected in comparison with the resonant amplitude of Eq. (2.104), and thus one can approximate,

$$\sigma_{\text{el}} \simeq 4\pi \left|f_{\text{res}}\right|^2 = \frac{4\pi}{k^2}.$$

(5) Derive an integral expression for the phase shifts in the scattering from a potential with range $\bar{R}$ in terms of the radial wave function that outside the range of the potential behaves as

$$\tilde{u}_l(k, r > \bar{R}) = \hat{\jmath}_l(kr) + \tan\delta_l \; \hat{n}_l(kr).$$

(6) A particle of mass μ and energy E is scattered off the delta-like potential barrier,

$$V(r) = C\,\delta(r - a).$$

Show that the s-wave phase shift is given by the expression[25]

$$\delta_0 = \tan^{-1}\left[\frac{ka\,\tan(ka)}{ka + \alpha\tan(ka)}\right] - ka,$$

where $k = \sqrt{2\mu\, E}/\hbar$ and α is the constant

$$\alpha = \frac{2\mu\, C}{\hbar^2}\, a.$$

[25] *Hint*: Since the potential barrier is infinite at $r = a$, $u_0'(r)$ is not continuous. The sudden change of the derivative in the neighborhood of a is

$$\lim_{\epsilon\to 0}\left[u_0'(a+\epsilon) - u_0'(a-\epsilon)\right] = \int_{a-\epsilon}^{a+\epsilon} u_0''(r)\, dr$$

and $u_0''(r)$ is given by the Schrödinger equation.

Chapter 3

Coulomb Scattering

A necessary condition for the scattering wave function to have the asymptotic behavior

$$\psi^{(+)}(\mathbf{k};\mathbf{r}) \to A\left(e^{i\mathbf{k}\cdot\mathbf{r}} + f(\Omega)\,\frac{e^{ikr}}{r}\right) \tag{3.1}$$

is that $V(r \to \infty)$ goes to zero faster than $1/r$. Otherwise, the incident wave function is distorted even at asymptotic distances. This is the case of the Coulomb potential, in collisions of charged particles[1]. If the dimensions of the collision partners can be neglected, the Coulomb potential is

$$V_{\mathrm{C}}(r) = \frac{1}{4\pi\varepsilon_0}\,\frac{q_{\mathrm{P}}q_{\mathrm{T}}}{r}, \tag{3.2}$$

where q_{P} and q_{T} are respectively the projectile and target charges. Therefore, Coulomb scattering requires a special treatment.

3.1 Classical mechanics description of Coulomb scattering

Let us first study Coulomb scattering using Classical Mechanics. It is convenient to introduce the Sommerfeld parameter

$$\eta = \frac{1}{4\pi\varepsilon_0}\,\frac{q_{\mathrm{P}}\,q_{\mathrm{T}}}{\hbar v} \tag{3.3}$$

and the half distance of closest approach in a head-on collision

$$a = \frac{1}{4\pi\varepsilon_0}\,\frac{q_{\mathrm{P}}\,q_{\mathrm{T}}}{2E}. \tag{3.4}$$

These quantities are related as

$$\eta = ka\,. \tag{3.5}$$

[1] We use here the SI unit system.

In chapter 1, we derived an expression for the classical deflection function, $\Theta(b)$, in the scattering from spherical potentials. It was written as

$$\Theta(b) = \pi - \phi(b), \tag{3.6}$$

with $\phi(b)$ given by Eq. (1.41). In the case of Coulomb scattering, the potential is given by Eq. (3.2) and $\phi(b) \equiv \phi_{\rm C}(b)$ becomes

$$\phi_{\rm C}(b) = 2b \int_{R_{\rm ca}}^{\infty} \frac{dr}{r\sqrt{r^2 - 2ar - b^2}}. \tag{3.7}$$

Before evaluating the integral, we must find the distance of closest approach, $R_{\rm ca}$, as a function of b. Setting in Eq. (1.37) the closest approach condition, $\dot{r} = 0$, and using Eq. (3.4), we get

$$R_{\rm ca} = a + \sqrt{a^2 + b^2}. \tag{3.8}$$

Sometimes it is convenient to express $R_{\rm ca}$ in terms of the Sommerfeld parameter and the angular momentum measured in $\hbar$ units[2], $l = L/\hbar = kb$. Using Eq. (3.5) and replacing $b = l/k$, we obtain

$$R_{\rm ca} = \frac{1}{k}\left[\eta + \sqrt{\eta^2 + l^2}\right]. \tag{3.9}$$

The integral of Eq. (3.7), which can be found in standard integral tables (see, for example [Gradshteyn and Ryzhik (1965)]), is

$$\phi(b) = \left[2\sin^{-1}\left(-\frac{ar + b^2}{r\sqrt{a^2+b^2}}\right)\right]_{R_{\rm ca}}^{\infty} = \pi - 2\sin^{-1}\left[\frac{a}{\sqrt{a^2+b^2}}\right]. \tag{3.10}$$

Comparing Eqs. (3.10) and (3.6), we get

$$\frac{\Theta_{\rm C}}{2} = \sin^{-1}\left[\frac{a}{\sqrt{a^2+b^2}}\right].$$

The above equation implies that

$$\sin(\Theta_{\rm C}/2) = \frac{a}{\sqrt{a^2+b^2}} = \frac{\eta}{\sqrt{\eta^2+l^2}}, \tag{3.11a}$$

$$\cos(\Theta_{\rm C}/2) = \frac{b}{\sqrt{a^2+b^2}} = \frac{l}{\sqrt{\eta^2+l^2}}, \tag{3.11b}$$

$$\tan(\Theta_{\rm C}/2) = \frac{a}{b} = \frac{\eta}{l}. \tag{3.11c}$$

[2]Although using $\hbar$ as the angular momentum unit is a strange procedure in a classical theory, it is convenient to derive the semiclassical limits of chapter 5.

The deflection function can be specified by any of these equations. The most frequently used is

$$\Theta_C(b) = 2\,\tan^{-1}\left(\frac{a}{b}\right) = 2\,\tan^{-1}\left(\frac{\eta}{l}\right). \tag{3.12}$$

Eq. (3.12) leads to an expression for the distance of closest approach as a function of the deflection angle. From Eq. (3.12) we can write $b = a\,\cot(\Theta_C/2)$. Inserting this result in Eq. (3.8), we get

$$R_{ca} = a\left[1 + \frac{1}{\sin(\Theta_C/2)}\right]. \tag{3.13}$$

The classical cross section is given by Eq. (1.46),

$$\frac{d\sigma_C(\Omega)}{d\Omega} = \frac{b}{\sin\theta}\left|\frac{d\Theta_C}{db}\right|^{-1}.$$

Evaluating the derivative of Eq. (3.12),

$$\frac{d\Theta_C}{db} = -\frac{2a}{a^2+b^2},$$

we get

$$\frac{d\sigma_C(\Omega)}{d\Omega} = \frac{b}{\sin\theta}\left[\frac{a^2+b^2}{2a}\right] = \frac{a^2}{2\sin\theta}\left[\frac{a}{b}\left(\frac{a^2}{a^2+b^2}\right)\right]^{-1}.$$

Writing $\sin\theta = 2\sin(\theta/2)\,\cos(\theta/2)$ and using Eqs. (3.11a), (3.11b) and (3.11c), we get the cross section for Coulomb scattering, usually called the *Rutherford cross section*,

$$\sigma_{\mathrm{Ruth}}(\theta) \equiv \frac{d\sigma_C(\Omega)}{d\Omega} = \frac{a^2}{4}\left[\frac{1}{\sin^4(\theta/2)}\right]. \tag{3.14}$$

We introduced above the usual notation for the Rutherford cross section, $\sigma_{\mathrm{Ruth}}(\theta)$. This is a very interesting result, since it remains the same when evaluated by quantum mechanics (see next section). Note that the above expression diverges as $\theta \to 0$, which is a consequence of the infinite range of the Coulomb interaction. Owing to this feature, even collisions with infinite impact parameters are influenced by the potential and they contribute to scattering to $\theta \simeq 0$.

3.2 Quantum mechanical description

The wave function in Coulomb scattering satisfies the Schrödinger equation

$$-\frac{\hbar^2}{2\mu}\nabla^2\,\phi_C(\mathbf{k};\mathbf{r}) + \frac{1}{4\pi\varepsilon_0}\,\frac{q_P q_T}{r}\,\phi_C(\mathbf{k};\mathbf{r}) = E\,\phi_C(\mathbf{k};\mathbf{r}), \tag{3.15}$$

which can be put in the form,

$$\left[\nabla^2 + k^2 - \frac{2\eta k}{r}\right] \phi_{\rm C}(\mathbf{k};\mathbf{r}) = 0\,. \tag{3.16}$$

Owing to the long range of the Coulomb field, the wave function does not have the asymptotic form of Eq. (3.1). To solve the above equation, we take the incident beam along the $z-$axis and try the ansatz

$$\phi_{\rm C}(\mathbf{k};\mathbf{r}) = C\, e^{ikz}\, g(r-z)\,, \tag{3.17}$$

where $g \to 1$ if the strength of the Coulomb field goes to zero ($\eta \to 0$). Inserting Eq. (3.17) in Eq. (3.16) and calling $\xi = r - z$, we get[3]

$$\xi\, g''(\xi) + (1 - ik\xi)\, g'(\xi) - \eta k\, g(\xi) = 0\,, \tag{3.18}$$

where primes stand for derivatives with respect to the argument. We then change variables:

$$\begin{aligned} \xi &\to s = ik\xi \\ g(\xi) &\to w(s) = g(\xi), \end{aligned} \tag{3.19}$$

and use the notation

$$d = -i\eta, \quad c = 1. \tag{3.20}$$

Eq. (3.18) takes the form

$$s\, w''(s) + (c - s)\, w'(s) - d\, w(s) = 0. \tag{3.21}$$

We identify above the Kumer-Laplace equation, which has the confluent hypergeometric function, ${}_1F_1(d;c;s)$, as regular solution. Replacing d, c, and s according to Eqs. (3.19) and (3.20) in the argument of ${}_1F_1$ and using Eq. (3.17), we get

$$\phi_{\rm C}(\mathbf{k};r,z) = C\, e^{ikz}\ {}_1F_1(-i\eta;1;ik(r-z))\,. \tag{3.22}$$

The function ${}_1F_1(-i\eta;1;ik(r-z))$ can be written as (see, for example, appendix B of [Messiah (1999)] or [Abramowitz and Stegun (1972)])

$${}_1F_1(-i\eta;1;ik(r-z)) = W_1(-i\eta;1;ik(r-z)) + W_2(-i\eta;1;ik(r-z))\,,$$

---

[3] To derive Eq. (3.18), one writes the laplacian in terms of parabolic coordinates $\{\xi,\,\zeta,\,\varphi\}$, where φ is the azimuthal angle, $\xi = r - z$ and $\zeta = r + z$, and looks for a solution in product form, $\psi_{\rm C}(\mathbf{r}) = g(\xi)\cdot f(\zeta)$. It is easy to show that $f(\zeta) = \exp(-ik\zeta/2)$ satisfies the Schödinger equation, if $g(\xi)$ satisfies Eq. (3.18). The details of this procedure are given in [Joachain (1983)].

where W_1 and W_2 are integrals in the complex plane. The wave function $\phi_C(r,z)$ is then given by the sum

$$\phi_C(\mathbf{k};\, r,z) = \psi_i(\mathbf{k};r,z) + \psi_s(\mathbf{k};r,z)\,, \tag{3.23}$$

where

$$\psi_i(\mathbf{k};r,z) = C\,e^{ikz}\; W_1(-i\eta;1;ik(r-z))\,, \tag{3.24}$$

$$\psi_s(\mathbf{k};r,z) = C\,e^{ikz}\; W_2(-i\eta;1;ik(r-z))\,. \tag{3.25}$$

Writing the normalization constant as

$$C = A\;\Gamma(1+i\eta)\,e^{-\pi\eta/2}\,,$$

where A is a constant and Γ is the gamma-function (see, e.g., [AS72]), and using the asymptotic forms of W_1 and W_2, we obtain the limits

$$\psi_i(\mathbf{k};r,z)_{|r-z|\to\infty} \to A\,e^{ikz}\cdot\left[e^{-i\eta\,\ln(k(r-z))}\right] \tag{3.26}$$

$$\psi_s(\mathbf{k};r,z)_{|r-z|\to\infty} \to A\,f_C(\theta)\,\frac{e^{ikr}}{r}\cdot\left[e^{-i\eta\,\ln(2kr)}\right]\,. \tag{3.27}$$

Inserting Eqs. (3.24) and (3.25) in Eq. (3.23), we get

$$\phi_C(\mathbf{k};r,z)_{|r-z|\to\infty} \to A\left\{e^{ikz}\;\left[e^{-i\eta\,\ln(k(r-z))}\right] + f_C(\theta)\,\frac{e^{ikr}}{r}\;\left[e^{-i\eta\,\ln(2kr)}\right]\right\}, \tag{3.28}$$

where $f_C(\theta)$ is the Coulomb scattering amplitude,

$$f_C(\theta) = -\frac{a}{2}\,e^{2i\sigma_0}\;\frac{e^{-i\eta\ln(\sin^2\theta/2)}}{\sin^2(\theta/2)}. \tag{3.29}$$

Above, σ_0 is the s-wave Coulomb phase shift

$$\sigma_0 = \arg\left\{\Gamma(1+i\eta)\right\}. \tag{3.30}$$

The behavior of σ_0 as a function of η is illustrated in figure 3.1. It is negative below $\eta \simeq 2$ and grows monotonically for $\eta \gtrsim 2$. Note that Eqs. (3.26)-(3.28) are not valid at $\theta = 0$. In this case, $|\,r-z| = 0$, so that the condition $|r-z| \to \infty$ is not satisfied. It is worthwhile to mention that the point charge Coulomb scattering amplitude, Eq. (3.29), can be rewritten in the following form,

$$f_C(\theta) = f_C(\pi)\;\frac{e^{-i\eta\ln(\sin^2\theta/2)}}{\sin^2\theta/2}, \tag{3.31}$$

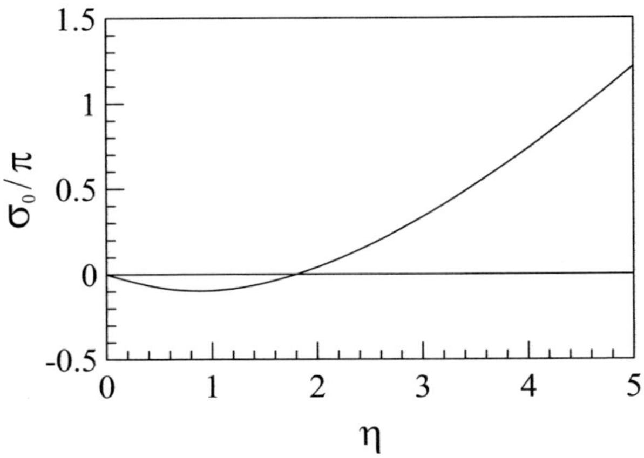

Fig. 3.1 The s-wave phase shift (normalized with respect to π) in Coulomb scattering, as a function of the Sommerfeld parameter.

which, after calling $\cos\theta = x$, can be further reduced to

$$f_{\rm C}(\theta) = f_{\rm C}(\pi)\left(\frac{1-x}{2}\right)^{-1-i\eta}. \tag{3.32}$$

Inspecting Eqs. (3.26) and (3.27), we conclude that ψ_i and ψ_s correspond respectively to the incident plane wave and to the scattered wave. However, due to the long range of the interaction, these waves have logarithmic distortions given by the factors within square brackets. In this way, Eq. (3.28) is the Coulomb analog of the scattering boundary condition for short-range potentials, given in Eq. (3.1). To be consistent with chapter 2, we then switch to the notation

$$\phi_{\rm C}(\mathbf{k};\mathbf{r}) \to \phi_{\rm C}^{(+)}(\mathbf{k};\mathbf{r})\,.$$

3.2.1 *The quantum mechanical cross section in Coulomb scattering*

Now we evaluate the elastic cross section in the case of Coulomb scattering. Since the derivation given in section 1.5 was based on the asymptotic form of Eq. (3.1), which is no longer valid, it should be re-examined here.

A natural attempt to get rid of the asymptotic distortions of Eq. (3.28) would be to modify the Coulomb potential, multiplying it by a screening function. This function would only affects the potential at asymptotic distances, making it vanish faster them $1/r$. Such an attempt leads, however, to disappointing results. The wave function turns out to depend on the details of the screening procedure[4]. Since the Coulomb potential is an idealization which is never found in nature, it is not clear wether the solution of Coulomb scattering can be used in actual scattering problems. In any realistic situations, the Coulomb potential is screened by the physical environment in the macroscopic scale of the experiment. In the case of nuclear scattering, for example, the Coulomb potential is screened by the atomic electrons as the radial distance reaches the order of the atomic radius. However, the situation is not as bad as it looks. Although pure and asymptotically screened Coulomb potentials yield different wave functions, they lead to the same incident and scattered currents, and hence to the same cross section. To prove this point, we first calculate the incident current, J, which is associated with ψ_i. Using Eq. (3.26), we get

$$J = v \left| A \right|^2 \left[1 - \frac{a}{z} \right], \tag{3.33}$$

where a is given by Eq. (3.4). We now evaluate the flux of the scattered current through an elementary surface located on a sphere of large radius r, at the orientation Ω and subtending the solid angle $\Delta\Omega$. Using Eq. (3.27) and proceeding as in section 1.5, we obtain

$$N(\Omega, \Delta\Omega) = v \left| A \right|^2 \left| f_{\rm C}(\theta) \right|^2 \Delta\Omega \left[1 - \frac{a}{r} \right]. \tag{3.34}$$

The cross section is then given by Eq. (1.6), with J and $N(\Omega, \Delta\Omega)$ evaluated asymptotically ($z \to -\infty$ and $r \to \infty$). In this way, a/z and a/r vanish and we obtain the same result of chapter 1. Therefore, the scattering cross is given by the expression

$$\frac{d\sigma_{\rm C}(\Omega)}{d\Omega} = \left| f_{\rm C}(\theta) \right|^2 .$$

Using Eq. (3.29), we get

$$\frac{d\sigma_{\rm C}(\Omega)}{d\Omega} = \frac{a^2}{4} \left[\frac{1}{\sin^4(\theta/2)} \right]. \tag{3.35}$$

A very interesting feature of Coulomb scattering is that the above quantum mechanical cross section is identical to the classical cross section of Eq. (3.14).

[4]A simple illustration of this fact can be found in [Taylor (2006)].

3.3 Partial wave expansion

In the case of pure Coulomb scattering, partial-wave expansions lead to difficulties. Since the range of the Coulomb field is larger than that of the centrifugal potential, Coulomb phase shifts do not tend to zero as $l \to \infty$ (see figure 3.2 of section 3.3.2). In this way, a converging partial-wave expansion for the scattering amplitude, as that given in chapter 2, does not exist. However, this is not a serious problem since the scattering amplitude is known analytically. On the other hand, the partial-wave expansion of the wave function is very useful when the potential contains a short-range term, in addition to the Coulomb interaction. This situation is frequently found in nuclear physics, where the nucleus-nucleus potential is the sum of the Coulomb repulsion with a short ranged term, arising from nuclear forces. It is also the case of collisions between ions or elementary charged particles with ions.

We first consider the case of pure Coulomb scattering. The wave function $\phi_{\mathrm{C}}^{(+)}(\mathbf{k};\mathbf{r})$ can be expanded in partial waves similarly to Eq. (2.8),

$$\phi_{\mathrm{C}}^{(+)}(\mathbf{k};\mathbf{r}) = \frac{1}{(2\pi)^{3/2}} \sum_{lm}^{\infty} C_{lm}\, Y_{lm}(\hat{\mathbf{r}})\, \frac{u_{\mathrm{C},l}(k,r)}{kr}\,, \tag{3.36}$$

where $u_{\mathrm{C},l}(k,r)$ satisfies the radial equation

$$-\frac{\hbar^2}{2\mu}\left[\frac{d^2}{dr^2} - \frac{2\eta k}{r} - \frac{l(l+1)}{r^2}\right] u_{\mathrm{C},l}(k,r) = E\, u_{\mathrm{C},l}(k,r)\,.$$

Making the change of variable: $r \to \rho = kr$ the radial equation takes the form

$$u''_{\mathrm{C},l}(\rho) + \left[1 - \frac{l(l+1)}{\rho^2} - \frac{2\eta}{\rho}\right] u_{\mathrm{C},l}(\rho) = 0\,. \tag{3.37}$$

One identifies above the Coulomb wave equation [Abramowitz and Stegun (1972)]. Its solutions can be expressed as linear combinations of the independent pair of real solutions $F_l(\eta,\rho)$ and $G_l(\eta,\rho)$. $F_l(\eta,\rho)$ is the regular solution and $G_l(\eta,\rho)$ diverges at the origin (except for $l = 0$). The main properties of these functions are discussed in standard text books (e.g. [Messiah (1999)], appendix B). They are summarized below.

- Weak interaction limit: $V_C \to 0 \Rightarrow \eta \to 0$

$$F_l(\eta \to 0, \rho) \to \hat{j}_l(\rho)\,, \tag{3.38}$$

$$G_l(\eta \to 0, \rho) \to \hat{n}_l(\rho)\,. \tag{3.39}$$

Eqs. (3.38) and (3.39) are not surprising since Eq. (3.37) reduces to the Ricatti-Bessel equation when $\eta = 0$.

- $\rho \to 0$ limit $\left(\rho \ll \eta + \sqrt{\eta^2 + l\,(l+1)}\right)$

$$F_l(\eta, \rho \to 0) \to A_l \; \rho^{l+1} \tag{3.40}$$

$$G_l(\eta, \rho \to 0) \to \frac{1}{(2l+1)\,A_l}\; \rho^{-l} \tag{3.41}$$

where

$$A_0 = \left(\frac{2\pi\eta}{e^{2\pi\eta}-1}\right)^{1/2}, \tag{3.42}$$

$$A_{l>0} = \frac{A_0}{(2l+1)!!} \cdot \Pi^l_{m=1}\left(1+\frac{\eta^2}{m^2}\right)^{1/2}. \tag{3.43}$$

- $\rho \to \infty$ limit

$$F_l(\eta, \rho \to \infty) \to \sin\left[\rho - \frac{l\pi}{2} - \eta\ln(2\rho) + \sigma_l\right] \tag{3.44}$$

$$G_l(\eta, \rho \to \infty) \to \cos\left[\rho - \frac{l\pi}{2} - \eta\ln(2\rho) + \sigma_l\right]. \tag{3.45}$$

In Eqs. (3.44) and (3.45) we have introduced the Coulomb phase shifts for the lth partial-wave, which is given by

$$\sigma_l = \arg\Big\{\Gamma(1+l+i\eta)\Big\}. \tag{3.46}$$

Usually, one evaluates

$$\sigma_0 = \arg\Big\{\Gamma(1+i\eta)\Big\}$$

and obtain Coulomb phase shifts for higher partial-waves from the series

$$\sigma_l = \sigma_0 + \sum_{m=1}^{l} \tan^{-1}\left(\frac{\eta}{m}\right). \tag{3.47}$$

As we will show in the next subsection, the Coulomb phase shifts for large η and large l can be evaluated in closed form.

The solutions of the Coulomb wave equation can also be expressed as linear combinations of incoming and outgoing Coulomb waves, respectively denoted by $H_l^{(-)}(\rho)$ and $H_l^{(+)}(\rho)$. They are defined as

$$H_l^{(\pm)}(\eta, \rho) = G_l(\eta, \rho) \pm iF_l(\eta, \rho) . \tag{3.48}$$

Using Eqs. (3.44) and (3.45) we find their asymptotic forms

$$H_l^{(\pm)}(\eta, \rho \to \infty) \to e^{\pm\, i\left[\rho - \frac{l\pi}{2} - \eta \ln(2\rho) + \sigma_l\right]}. \tag{3.49}$$

As the strength of the Coulomb field goes to zero ($\eta \to 0$), $H_l^{(\pm)}$ converge to the Ricatti-Haenkel functions,

$$H_l^{(\pm)}(\eta \to 0, \rho) \to \hat{h}_l^{(\pm)}(\rho). \tag{3.50}$$

3.3.1 *Approximate Coulomb phase shifts — asymptotic series*

Here we derive useful expressions based on asymptotic expansions and uniform asymptotic series. The first such application is the derivation of the asymptotic form of the Coulomb phase shift. First, we rewrite Eq. (3.46) as

$$e^{2i\sigma_l} = \frac{\Gamma(1 + l + i\eta)}{\Gamma(1 + l - i\eta)} \tag{3.51}$$

and use the integral form of the gamma function,

$$\Gamma(1 + z) = \int_0^\infty dt\ t^z\, e^{-t}, \tag{3.52}$$

where $z = l + i\eta$. Writing $t^z = e^{z \ln t}$, we can use the saddle point method to evaluate the integral

$$\Gamma(1 + z) = \int_0^\infty dt\ \ e^{-t + z\ln(t)}. \tag{3.53}$$

Expanding the exponent of the integrand around its extremum, at $t_0 = z$ and keeping terms up to second order,

$$-t + z\ln(t) \simeq -z + z\ln(z) - \frac{1}{2z}(t - z)^2, \tag{3.54}$$

Eq. (3.53) becomes

$$\Gamma(1 + z) \simeq e^{-z} z^z \int_0^\infty dt\ e^{-(t-z)^2/2z}. \tag{3.55}$$

For Re $z = l \gg 1$ and Im $z = \eta > 0$, the gaussian integral can be easily evaluated as

$$\int_0^\infty dt\ e^{-(t-z)^2/2z} = \int_{-\infty}^\infty dt\ e^{-(t-z)^2/2z} = \sqrt{2\pi z} \tag{3.56}$$

and using this result in the previous equation, we obtain

$$\Gamma(1+z) = \sqrt{2\pi}\, e^{-z}\, z^{z+1/2}. \tag{3.57}$$

The approximation of Eq. (3.57) is just the first term in what is known as Stirling's formula,

$$\Gamma(z) = \sqrt{2\pi}\, e^{-z}\, z^{z-1/2} \left[1 + \frac{1}{12\,z} + \frac{1}{288\,z^2} + \cdots\right]. \tag{3.58}$$

This becomes clear if we use the identity of Gamma functions (Eq. (6.1.15) of [Abramowitz and Stegun (1972)])

$$\Gamma(z+1) = z\,\Gamma(z). \tag{3.59}$$

However, the Stirling's formula leads to more accurate estimates of the phase shifts when the value of l is not sufficiently large. We present below a detailed discussion of Eq. (3.58).

We begin by introducing the following parabolic functional mapping,

$$\omega(t) = [t - z\ln t] = A + \frac{s^2}{2}, \tag{3.60}$$

where $A = z - z\ln z$. Then the formula for the Γ function can be written as,

$$\Gamma(z+1) = e^{z - z\ln z} \int_{-\infty}^{\infty} e^{-s^2}\, \frac{dt}{ds}\, ds. \tag{3.61}$$

The derivative dt/ds can be written as a power series in s^2 using the mapping equation,

$$\frac{dt}{ds} = \sum_{k=0}^{\infty} a_{2k}\, s^{2k}. \tag{3.62}$$

It is a simple matter to evaluate the coefficients a_k, by repeated differentiation of the mapping equation. We calculate below the first two terms, a_0, and a_2 (higher order terms can be derived similarly). We write

$$\frac{d\omega(t)}{ds} = \left(1 - \frac{z}{t}\right) \frac{dt}{ds} = s \tag{3.63}$$

$$\frac{d^2\omega(t)}{ds^2} = \left(1 - \frac{z}{t}\right) \frac{d^2t}{ds^2} + \frac{z}{t^2} \left(\frac{dt}{ds}\right)^2 = 1 \tag{3.64}$$

$$\frac{d^3\omega(t)}{ds^3} = \left(1 - \frac{z}{t}\right) \frac{d^3t}{ds^3} + \frac{2z}{t^2}\, \frac{d^2t}{ds^2}\, \frac{dt}{ds} - \frac{2z}{t^3} \left(\frac{dt}{ds}\right)^3 = 0 \tag{3.65}$$

$$\frac{d^4\omega(t)}{ds^4} = \left(1 - \frac{z}{t}\right)\frac{d^4t}{ds^4} + \frac{4z}{t^2}\frac{d^3t}{ds^3}\frac{dt}{ds} + \frac{3z}{t^2}\left(\frac{d^2t}{ds^2}\right)^2$$
$$- \frac{12z}{t^3}\frac{d^2t}{ds^2}\left(\frac{dt}{ds}\right)^2 + \frac{6z}{t^4}\left(\frac{dt}{ds}\right)^4. \qquad (3.66)$$

We now evaluate these equations at the extremum point, defined by the condition

$$\left[\frac{d\omega(t)}{dt}\right]_{t_0} = 0, \qquad (3.67)$$

which is $t_0 = z$ and $s = 0$. The above four equations lead to the results

$$\left[\frac{dt}{ds}\right]_{t=t_0} = \sqrt{z}, \quad \left[\frac{d^2t}{ds^2}\right]_{t=t_0} = \frac{2}{3} \quad \text{and} \quad \left[\frac{d^3t}{ds^3}\right]_{t=t_0} = \frac{1}{6\sqrt{z}}. \qquad (3.68)$$

Thus the coefficients a_{2k} above are,

$$a_0 = \left[\frac{dt}{ds}\right]_{t=t_0} = \sqrt{z} \qquad (3.69)$$

$$a_2 = \frac{1}{2!}\left[\frac{d^3t}{ds^2}\right]_{t=t_0} = \frac{1}{12\sqrt{z}}. \qquad (3.70)$$

Using the Gaussian integral formula,

$$\int_{-\infty}^{\infty} ds\; e^{-s^2/2}\; s^n\, ds = \sqrt{2\pi}\;(n-1)!!, \qquad (3.71)$$

the full integral in the Γ function can be written down in the form of the Sterling formula above, namely,

$$\Gamma(z+1) = \sqrt{2\pi}\; e^{-z}\, z^z \sum_{k=0}^{\infty}(2k-1)!!\; a_{2k}. \qquad (3.72)$$

Thus, evaluating the coefficients a_{2k} and inserting above, we get

$$\Gamma(z+1) = \sqrt{2\pi z}\; e^{-z}\, z^z \left[1 + \frac{1}{12\,z} + \frac{1}{288\,z^2} + \cdots\right]$$
$$= \sqrt{2\pi}\;\; e^{-z}\; z^{z+1/2}\left[1 + \frac{1}{12\,z} + \frac{1}{288\,z^2} + \cdots\right]. \qquad (3.73)$$

A more precise series representation of the Γ function can be found, which incorporates the above Stirling series. We develop this below.

Let us briefly describe the ideas behind Gudermann's corrections and present Stirling's series. Since Euler's gamma function satisfies $\Gamma(z+1) = z\Gamma(z)$, we get from (3.57) the approximation

$$\Gamma(z) \simeq \sqrt{2\pi}\, e^{-z}\, z^{z-1/2}. \qquad (3.74)$$

However, (3.74) contrasts with the expression obtained from (3.57) itself by replacing z by $z-1$:

$$\Gamma(z) \simeq \sqrt{2\pi}\, e^{-(z-1)}\, (z-1)^{z-1/2}. \tag{3.75}$$

Since both expressions (3.74) and (3.75) are only valid for $|z|$ very large, there is no practical difference between them. Nevertheless, one can better deal with this situation by seeking an exact representation for Γ for all values of z. This is done by first writing,

$$\Gamma(z) = \sqrt{2\pi}\, e^{-z}\, z^{z-1/2} e^{\mu(z)}, \tag{3.76}$$

and fixing the correction factor $e^{\mu(z)}$ by imposing the relation $\Gamma(z+1) = z\Gamma(z)$. A simple computation reveals that this condition implies that μ has to satisfy the functional equation

$$\mu(z) - \mu(z+1) = \left(z + \frac{1}{2}\right) \ln\left(1 + \frac{1}{z}\right) - 1. \tag{3.77}$$

Moreover, the validity of (3.74) for "large" $|z|$ leads to the condition $\lim_{|z|\to\infty} \mu(z) = 0$. This allows for a unique solution of (3.77). Indeed, it follows immediately from (3.77) that for any positive integer n one has

$$\mu(z) - \mu(z+n) = \sum_{m=0}^{n-1} \left[\left(z + m + \frac{1}{2}\right) \ln\left(1 + \frac{1}{z+m}\right) - 1\right]. \tag{3.78}$$

Hence, the condition $\lim_{|z|\to\infty} \mu(z) = 0$ implies, in particular, that $\lim_{n\to\infty} \mu(z+n) = 0$ and we get

$$\mu(z) = \sum_{m=0}^{\infty} \left[\left(z + m + \frac{1}{2}\right) \ln\left(1 + \frac{1}{z+m}\right) - 1\right]. \tag{3.79}$$

This series is known as Gudermann series. It can be expressed as,

$$\mu(z) = \sum_{n=1}^{\infty} \left[\frac{B_{2n}}{(2n-1)2n}\right] \frac{1}{z^{2n-1}}, \tag{3.80}$$

where B_k is the kth Bernoulli number. The first terms of (3.80) are

$$\mu(z) = \left[\frac{1}{12}\right] \frac{1}{z} - \left[\frac{1}{360}\right] \frac{1}{z^3} + \left[\frac{1}{1260}\right] \frac{1}{z^5} + \cdots. \tag{3.81}$$

For the correcting factor $e^{\mu(z)}$, this gives

$$e^{\mu(z)} = 1 + \left[\frac{1}{12}\right] \frac{1}{z} + \left[\frac{1}{288}\right] \frac{1}{z^2} - \left[\frac{139}{51840}\right] \frac{1}{z^3} + \cdots. \tag{3.82}$$

Therefore, we may write

$$\Gamma(z) = \sqrt{2\pi}\, e^{-z}\, z^{z-1/2} \left\{1 + \left[\frac{1}{12}\right] \frac{1}{z} + \left[\frac{1}{288}\right] \frac{1}{z^2} - \left[\frac{139}{51840}\right] \frac{1}{z^3} + \cdots \right\}, \tag{3.83}$$

which is the aforementioned Stirling series. In the following derivation of the Coulomb phase shift, however, we use the more exact Gudermann representation of the gamma function, Eqs. (3.76) and (3.81).

According to Eq. (3.51), the phase shifts can be written

$$e^{2i\sigma_l} = \frac{\Gamma(z)}{\Gamma(z^*)}, \tag{3.84}$$

with $z = l + 1 + i\eta$. To lowest order in $1/z$, one keeps only the first term within the curly brackets in Eq. (3.83). The resulting approximation for the phase shifts, $\sigma_l^{(0)}$, is given by the relation

$$e^{2i\sigma_l^{(0)}} \simeq \frac{e^{-z}\; z^{z-1/2}}{e^{-z^*}\;(z^*)^{z^*-1/2}} = \frac{e^{-z+(z-1/2)\ \ln\ z}}{e^{-z^*+(z^*-1/2)\ \ln\ z^*}} \tag{3.85}$$

or

$$2i\sigma_l = -z + (z - 1/2)\ \ln\ z + z^* - (z^* - 1/2)\ \ln\ z^*. \tag{3.86}$$

Replacing $z = l + 1 + i\eta$ in the above equation, we get

$$\sigma_l^{(0)} = -\eta + \left(l + \frac{1}{2}\right)\ \tan^{-1} \frac{\eta}{l+1} + \eta\ \ln\left(\sqrt{(l+1)^2 + \eta^2}\right). \tag{3.87}$$

The above formula is valid for $|z| \gg 1$, which means that we can use it both when $l \gg 1$ and $\eta \ll 1$, or, $\eta \gg 1$ and $l \ll 1$. Accordingly, the leading term in the $l = 0$ phase shift, has the following form,

$$\sigma_0^{(0)} = \frac{1}{2} \tan^{-1} \eta + \eta\ [\ln(\eta) - 1], \tag{3.88}$$

which for $\eta \gg 1$ reduces to,

$$\sigma_0^{(0)} = \frac{\pi}{4} + \eta\ [\ln(\eta) - 1]. \tag{3.89}$$

The case of $\eta \ll 1$ but $l \gg 1$ is important as it is encountered in the scattering of charged particles at high energies, which will be dealt with in Chapter 5, under the title of the eikonal approximation. The formula for the Coulomb phase shift can be read off from Eq. (3.87), by considering $l \gg 1$ and $\eta/l \ll 1$, and using the series representation

$$\tan^{-1}(x) = \left[x - \frac{x^3}{3} + \frac{x^5}{5} + \cdots\right]. \tag{3.90}$$

One finally obtains [Barata *et al.* (2011)],

$$\sigma_l = \eta \left[\ln (l) - \gamma \right], \tag{3.91}$$

where γ is the Euler Constant, which has the value $\gamma = 0.57721$. The above asymptotic formulae will be discussed further in Chapter 5, within the low-energy, large-η, large-l case of the WKB approximation, and the high-energy, small-η, large-l case of the eikonal approximation.

A first order correction to Eq. (3.87) can easily be evaluated. For this purpose, we rewrite Eq. (3.84) in the form

$$e^{2i\sigma_l} = e^{2i\sigma_l^{(0)}} \, F(z), \tag{3.92}$$

with

$$F(z) = \frac{\exp\left[\mu(z)\right]}{\exp\left[\mu(z^*)\right]} \tag{3.93}$$

or

$$F(z) = \frac{\exp\left[\frac{1}{12\,z} + \cdots\right]}{\exp\left[\frac{1}{12\,z^*} + \cdots\right]}. \tag{3.94}$$

The lowest order approximation of Eq. (3.87) corresponds to setting $F(z) \simeq F^{(0)}(z) = 1$. The first order approximation for $F(z)$ is

$$F^{(1)}(z) = \exp\left[\frac{1}{12\,z} - \frac{1}{12z^*}\right]. \tag{3.95}$$

Evaluating the above expression, we find

$$F^{(1)}(z) = e^{2i\ \Delta\sigma_l^{(0)}}, \tag{3.96}$$

with

$$\Delta\sigma_l^{(0)} = -\frac{1}{2}\left(\frac{\mathrm{Im}\,\{z\}}{6\ |z|^2}\right). \tag{3.97}$$

Replacing $z = l + 1 + i\eta$, we get

$$\Delta\sigma_l^{(0)} = -\left[\frac{\eta}{12\ \left[(l+1)^2 + \eta^2\right]}\right]. \tag{3.98}$$

The first order approximation to the Coulomb phase shifts then is

$$\sigma_l^{(1)} = \sigma_l^{(0)} + \Delta\sigma_l^{(0)}, \tag{3.99}$$

with $\sigma_l^{(0)}$ and $\Delta\sigma_l^{(0)}$ given respectively by Eqs. (3.87) and Eqs. (3.98).

3.3.2 *Some numerical results*

3.3.2.1 *Coulomb phase shifts*

Figure 3.2 shows Coulomb phase shifts versus l, for two values of the Sommerfeld parameter. In contrast to the case of short range potentials, the Coulomb phase shifts do not converge to zero as $l \to \infty$. Therefore, the partial-wave expansion method is not appropriate for the evaluation of the Coulomb scattering amplitude[5].

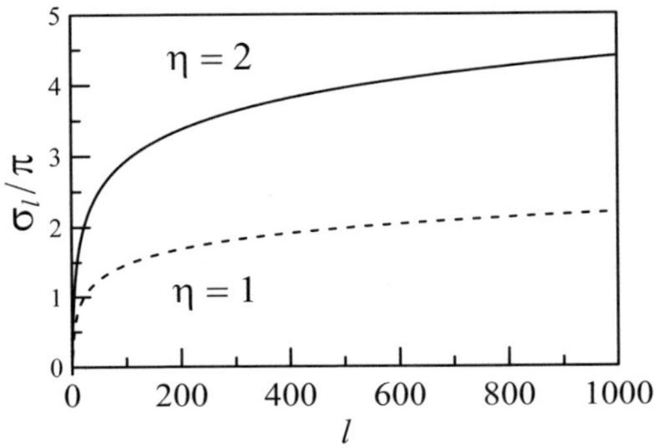

Fig. 3.2 Coulomb phase shifts as functions of l. Results normalized with respect to π are given for two values of the Sommerfeld parameter.

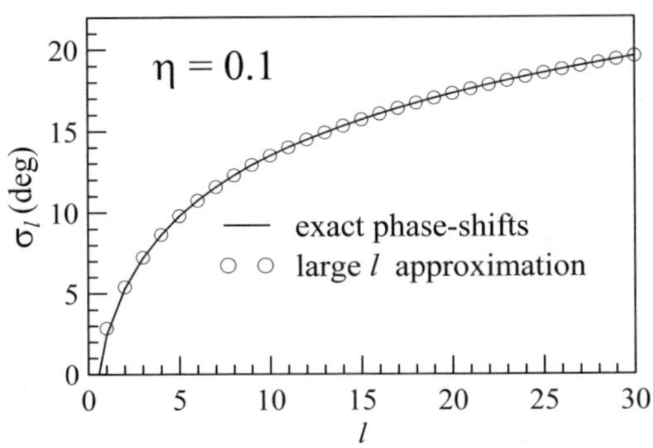

Fig. 3.3 Exact and the approximate Coulomb phase shifts for $\eta = 0.1$, as functions of l. For details see the text.

[5]Although the partial-wave series for the scattering amplitude diverges, it has been shown to exist as a distribution [Taylor (2006)].

Table 3.1 Exact and approximate Coulomb phase shifts as functions of l, for two values of the Sommerfeld parameter.

η	l	$\sigma_l^{(0)}$ (deg.)	$\sigma_l^{(1)}$	σ_l^{exact}
0.1	0	26.08	-3.44	-3.28
	1	2.87	2.39	2.43
	2	5.41	5.29	5.29
	3	7.25	7.20	7.20
	4	8.66	8.63	8.63
1.0	0	-12.30	-17.03	-17.28
	1	30.06	27.68	27.72
	2	55.22	54.27	54.28
	3	73.19	72.71	72.72
	4	87.03	86.75	86.75

The lowest order approximations (in $1/z$) of the previous section are supposed to work for $|\, z \,| \gg 1$, which means $l \gg 1$ and/or $\eta \gg 1$. They are, in fact, very accurate, even when these conditions are not well satisfied. A first illustration of this fact is presented in figure 3.3, where we compare the exact phase shifts (solid line) with the lowest order approximation $\sigma_l^{(0)}$ (open circles) for a small η, as functions of l. They are very close. The only exception is the case of $l = 0$, where they are quite different (see table 3.1). For larger values of η, the agreement is much better. This can be seen in table 3.1, where we compare the two approximations of the previous section with the exact Coulomb phase shifts. Note that the $\sigma_l^{(1)}$ is very accurate even for $l = 0$ and $\eta = 0.1$.

This point can be seen more clearly in figure 3.4, where the exact s-wave phase shifts are compared with the approximations $\sigma_l^{(0)}$ and $\sigma_l^{(1)}$. The comparison indicates that the usual large-l approximation, $\sigma_l^{(0)}$, is not very good. The situation is completely different with the improved approximation, $\sigma_l^{(1)}$, which includes the first order correction to Stirling's formula. In this case, one gets accurate results for any value of the Sommerfeld parameter.

3.3.2.2 *Coulomb wave functions*

Now we illustrate the behavior of the Coulomb wave functions. In figure 3.5 we present $F_l(\eta, \rho)$ and $G_l(\eta, \rho)$ as functions of ρ, in the case $\eta = 5$, for a few partial waves. Quantitatively, they are similar to the corresponding

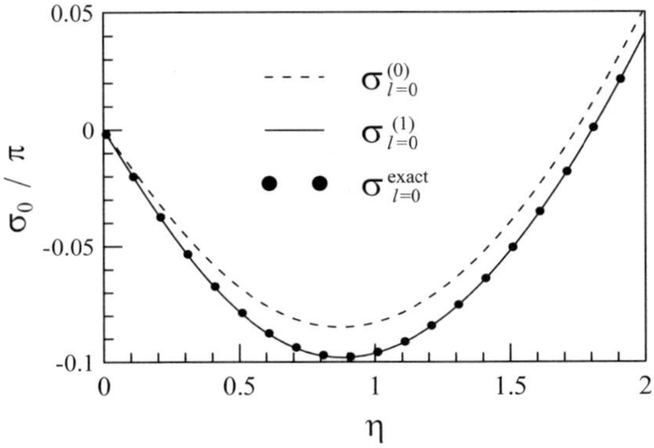

Fig. 3.4 Coulomb s-wave phase shifts as a function of the Sommerfeld parameter. Exact values are compared with the approximations discussed in the text.

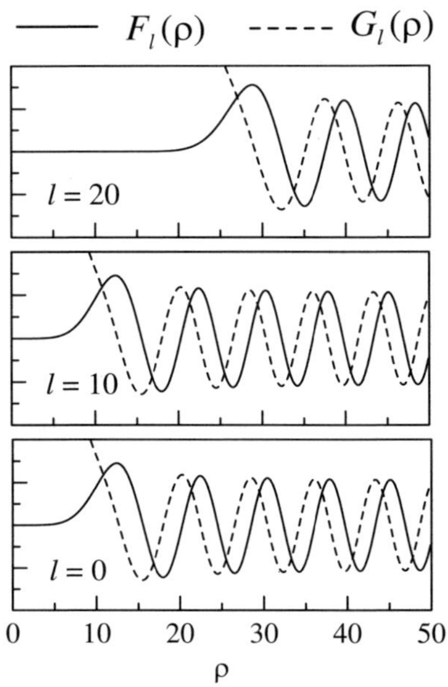

Fig. 3.5 Coulomb wave functions for $\eta = 5$ and $l = 0$, 10 and 20. We remark that $G_{l=0}$ remains finite at $\rho = 0$, although it is very large (see Eq. (3.42)). This cannot be seen above, owing to the scale used in the figure.

Ricatti-Bessel functions, shown in figure 2.1. For large values of ρ, $F_l(\eta,\rho)$ and $G_l(\eta,\rho)$ oscillate with constant amplitude, out of phase with respect to one another. For small ρ, $F_l(\eta,\rho)$ converges to zero while $G_l(\eta,\rho)$ diverges. The transition between these behaviors occurs at the classical turning points, $\rho_t = kR_{\rm ca}$. Using Eq. (3.8), we can write

$$\rho_t = k\left(a+\sqrt{a^2+b^2}\right) = \eta + \sqrt{\eta^2+l^2}. \tag{3.100}$$

Note that the s-wave Coulomb turning point does not vanish, in contrast with that for the Ricatti-Bessel function (Eq. (2.23)). This is because in the present case the s-wave effective potential contains the Coulomb repulsion.

3.3.3 *Partial-wave expansion of* $\phi_{\rm C}^{(+)}(\mathbf{k};\mathbf{r})$

The partial-wave expansion of the wave function $\phi_{\rm C}^{(+)}(\mathbf{k};\mathbf{r})$ is given by Eq. (3.36), with $u_{{\rm C},l}(k,r) = F_l(\eta,\rho)$ and with the coefficients[6]

$$C_{lm} = 4\pi\, Y_{lm}^*(\hat{\mathbf{k}})\, i^l\, e^{i\sigma_l}. \tag{3.101}$$

That is

$$\phi_{\rm C}^{(+)}(\mathbf{k};\mathbf{r}) = \frac{1}{(2\pi)^{3/2}} \sum_{lm} 4\pi\, Y_{lm}^*(\hat{\mathbf{k}})\, Y_{lm}(\hat{\mathbf{r}})\, i^l\, e^{i\sigma_l}\, \frac{F_l(\eta,\rho)}{kr} \tag{3.102a}$$

$$= \frac{1}{(2\pi)^{3/2}} \sum_{l} (2l+1)\, P_l(\cos\theta)\, i^l\, e^{i\sigma_l}\, \frac{F_l(\eta,\rho)}{kr}. \tag{3.102b}$$

In the $\eta \to 0$ limit, $\sigma_l \to 0$, $F_l(\eta,\rho) \to \hat{\jmath}_l(\rho)$ and the above equations reduce to the partial-wave expansion of the plane wave (Eqs. (2.44a) and (2.44b), with $A = (2\pi)^{-3/2}$).

It is also useful to write the partial-wave expansion in terms of incoming and outgoing Coulomb waves. Using Eq. (3.48), Eq. (3.102a) becomes

$$\phi_{\rm C}^{(+)}(\mathbf{k};\mathbf{r}) = \frac{1}{(2\pi)^{3/2}} \sum_{lm} 4\pi\, Y_{lm}(\hat{\mathbf{k}})\, Y_{lm}^*(\hat{\mathbf{r}})\, i^l\, e^{i\sigma_l} \times \frac{i}{2}\left[\frac{H_l^{(-)}(\eta,\rho) - H_l^{(+)}(\eta,\rho)}{kr}\right] \tag{3.103}$$

[6]The proof that the coefficients C_{lm} are given by Eq. (3.101) is not trivial. It is based on the integral representation of the confluent hypergeometric function in Eq. (3.22). For a detailed proof, we refer to [Messiah (1999)], section XI-10 .

or

$$\phi_{\rm C}^{(+)}(\mathbf{k};\mathbf{r}) = \frac{1}{(2\pi)^{3/2}} \sum_l (2l+1)\, P_l(\cos\theta)\; i^l\, e^{i\sigma_l} \times \frac{i}{2}\left[\frac{H_l^{(-)}(\eta,\rho) - H_l^{(+)}(\eta,\rho)}{kr}\right]. \tag{3.104}$$

In the $r \to \infty$ limit, we can use Eq. (3.49) in Eq. (3.103) and $\phi_{\rm C}^{(+)}(\mathbf{k};\mathbf{r})$ takes the form

$$\phi_{\rm C}^{(+)}(\mathbf{k};\mathbf{r}) \to \frac{1}{(2\pi)^{3/2}}\left[\frac{e^{-i\Theta(r)}}{r}\, Z_- + \frac{e^{i\Theta(r)}}{r}\, Z_+\right], \tag{3.105}$$

with

$$Z_- = -\frac{1}{2ik}\sum_{lm} 4\pi\, Y_{lm}^*(\hat{\mathbf{k}})\; Y_{lm}(\hat{\mathbf{r}})\; i^{2l} \tag{3.106}$$

and

$$Z_+ = \frac{1}{2ik}\sum_{lm} 4\pi\, Y_{lm}^*(\hat{\mathbf{k}})\; Y_{lm}(\hat{\mathbf{r}})\, e^{2i\sigma_l}\;. \tag{3.107}$$

Above, $\Theta(r)$ is the Coulomb distorted aymptotic phase

$$\Theta(r) = kr - \eta \ln 2kr\,. \tag{3.108}$$

3.4 Coulomb plus short-range potentials

Let us consider the scattering from the potential

$$V(r) = V_{\rm C}(r) + \overline{V}(r), \tag{3.109}$$

where $V_{\rm C}(r)$ is the Coulomb potential of Eq. (3.2) and $\overline{V}(r)$ is a short-range potential, which vanishes or can be neglected for $r > \bar{R}$. In this case the Schrödinger equation can be written

$$\left[\nabla^2 + k^2 - \frac{2\eta k}{r}\right]\psi(\mathbf{k};\mathbf{r}) = \frac{2\mu}{\hbar^2}\,\overline{V}(r)\;\psi(\mathbf{k};\mathbf{r}). \tag{3.110}$$

Following the procedures of section 1.5, we look for a solution in the form

$$\psi^{(+)}(\mathbf{k};\mathbf{r}) = \phi_{\rm C}^{(+)}(\mathbf{k};\mathbf{r}) + \bar{\psi}^{sc}(\mathbf{k};\mathbf{r}), \tag{3.111}$$

where the Coulomb wave function, $\phi_{\rm C}^{(+)}(\mathbf{k};\mathbf{r})$, plays the role of the "homogeneous" solution and $\psi^{sc}(\mathbf{k};\mathbf{r})$ is a spherical outgoing wave, resulting from the action of $\overline{V}(r)$. In the present case, however, the asymptotic form of $\bar{\psi}^{sc}(\mathbf{k};\mathbf{r})$ is distorted by the Coulomb potential, in the same way as the Coulomb scattered wave (Eq. (3.27)). That is,

$$\bar{\psi}^{sc}(\mathbf{r})_{|r-z|\to\infty} \to \frac{1}{(2\pi)^{3/2}}\left[\bar{f}(\theta)\,\frac{e^{i\Theta(r)}}{r}\right], \tag{3.112}$$

where $\Theta(r)$ is the asymptotic phase of Eq. (3.108) and is $\bar{f}(\theta)$ is the correction to the Coulomb scattering amplitude arising from the short-range potential.

To find $\bar{f}(\theta)$, we write the partial-wave expansion,

$$\psi^{(+)}(\mathbf{k};\mathbf{r}) = \frac{1}{(2\pi)^{3/2}}\sum_{lm}^{\infty} C_{lm}\,Y_{lm}(\hat{\mathbf{r}})\,\frac{u_l(k,r)}{kr}, \tag{3.113}$$

where the radial wave functions $u_l(k,r)$ are the solutions of the equation

$$-\frac{\hbar^2}{2\mu}\left[\frac{d^2}{dr^2}-\frac{l(l+1)}{r^2}-\frac{2\eta k}{r}\right]u_l(k,r)+\overline{V}(r)\,u_l(k,r) = E\,u_l(k,r). \tag{3.114}$$

For $r > \bar{R}$, $\overline{V}(r)$ vanishes and Eq. (3.114) reduces to the Coulomb wave equation. Therefore, $u_l(k, r > \bar{R})$ can be expressed as a linear combination of $H_l^{(-)}$ and $H_l^{(+)}$. Using a convenient normalization, we can write

$$u_l(k,r>\bar{R}) = \frac{i}{2}\left[H_l^{(-)}(\eta,\rho) - \bar{S}_l\,H_l^{(+)}(\eta,\rho)\right]. \tag{3.115}$$

Above, $\bar{S}_l = e^{2i\bar{\delta}_l}$ is the lth partial-wave component of the S-matrix associated with $\overline{V}$, and $\bar{\delta}_l$ is the corresponding phase shift. The normalization of u_l is such that it reduces to $F_l(\eta,\rho)$ as $\overline{V}\to 0$. On the other hand, in the $\eta\to 0$ limit, $u_l(k, r>\bar{R})$ is given by Eq. (2.35) (with $\gamma_l = i/2$).

To derive the partial-wave expansion of $\bar{f}(\theta)$, we proceed similarly to section 2.3 and the previous section. First, we remark that the partial-wave expansion of Eq. (3.113) is the same as that for pure Coulomb scattering (Eq. (3.36)), except for a different radial wave function. That is, replacing $u_{{\rm C},l}(k,r)\equiv F_l\,(\eta,\rho)\to u_l(k,r)$. For $r>\bar{R}$, $u_l(k,r)$ is given by Eq. (3.115) while in the previous section we had

$$F_l\,(\eta,\rho) = \frac{i}{2}\left[H_l^{(-)}(\eta,\rho) - H_l^{(+)}(\eta,\rho)\right]. \tag{3.116}$$

Comparing Eqs. (3.115) and (3.116) we conclude that they have the same incoming waves ($H_l^{(-)}$ terms) while their outgoing waves ($H_l^{(+)}$ terms) differ only by the factor $\bar{S}_l$. Therefore, the asymptotic ($r \to \infty$) form of $\psi^{(+)}$ is similar to Eq. (3.105), and we can write

$$\psi^{(+)}(\mathbf{k};\mathbf{r}) \to \frac{1}{(2\pi)^{3/2}} \left[\frac{e^{-i\Theta(r)}}{r} Z_- + \frac{e^{i\Theta(r)}}{r} \bar{Z}_+ \right]. \tag{3.117}$$

Above, Z_- given by Eq. (3.106) and the outgoing amplitude is

$$\bar{Z}_+ = \frac{1}{2ik} \sum_{lm} 4\pi \, Y_{lm}^*(\hat{\mathbf{k}}) \, Y_{lm}(\hat{\mathbf{r}}) \, e^{2i\sigma_l} \, \bar{S}_l \, . \tag{3.118}$$

Eq. (3.118) is identical to Eq. (3.107), except for the factor $\bar{S}_l$.

The next step is to take the $r \to \infty$ limit in Eq. (3.111). Since the asymptotic form of $\bar{\psi}^{sc}$ is purely outgoing (see Eq. (3.112)), the incoming part of $\psi^{(+)}$ is the same as that of $\phi_C^{(+)}$. On the other hand, the outgoing part of $\psi^{(+)}$ is the sum of contributions from $\phi_C^{(+)}$ and from $\bar{\psi}^{sc}$. Taking into account Eqs. (3.107) and (3.112), we can write

$$\psi^{(+)}(\mathbf{k};\mathbf{r}) \to \frac{1}{(2\pi)^{3/2}} \left[\frac{e^{-i\Theta(r)}}{r} Z_- + \frac{e^{i\Theta(r)}}{r} \tilde{Z}_+ \right], \tag{3.119}$$

with Z_- given by Eq. (3.106) and

$$\tilde{Z}_+ = \bar{f}(\theta) + \frac{1}{2ik} \sum_{lm} 4\pi \; Y_{lm}^*(\hat{\mathbf{k}}) \, Y_{lm}(\hat{\mathbf{r}}) \, e^{2i\sigma_l}. \tag{3.120}$$

Setting $\tilde{Z}_+ = \bar{Z}_+$, we get

$$\bar{f}(\theta) = \frac{1}{2ik} \sum_{lm} 4\pi \, Y_{lm}^*(\hat{\mathbf{k}}) \, Y_{lm}(\hat{\mathbf{r}}) \, e^{2i\sigma_l} \left(\bar{S}_l - 1 \right) \tag{3.121a}$$

$$= \frac{1}{2ik} \sum_{l} (2l+1) \; P_l(\cos\theta) \; e^{2i\sigma_l} \left(\bar{S}_l - 1 \right). \tag{3.121b}$$

To get the cross section, we point out that the wave function can be written, as in Eq. (3.28) with the replacement

$$f_C(\theta) \to f(\theta) = f_C(\theta) + \bar{f}(\theta). \tag{3.122}$$

Therefore, the cross section is

$$\frac{d\sigma_C(\Omega)}{d\Omega} = \left| f_C(\theta) + \bar{f}(\theta) \right|^2, \tag{3.123}$$

with $f_C(\theta)$ given by Eq. (3.29) and $\bar{f}(\theta)$ by Eq. (3.121b). Usually one is interested in investigating the action of the short-ranged potential. For this purpose, the cross sections are normalized with respect to their pure Coulomb (or Rutherford) counterparts.

The evaluation of the phase shifts $\bar{\delta}_l$ is carried out in the same way as in the absence of Coulomb interaction. The boundary $\bar{R}$ divides the r-axis into an internal and an external region. The radial equation is solved numerically in the internal region and the logarithmic derivative, $\mathcal{L}^{\mathrm{I}}$, is calculated. Matching it to the one calculated with the radial wave function at the external region, one finds an expression for the phase shift $\bar{\delta}_l$. This expression is identical to Eq. (2.33) with the replacement of Ricatti-Bessel and Ricatti-Neumann functions by F_l and G_l, respectively. It becomes

$$\tan\bar{\delta}_l = -\left[\frac{k\bar{R}\,F_l'(\eta,k\bar{R}) - F_l(\eta,k\bar{R})\,\mathcal{L}^{\mathrm{I}}}{k\bar{R}\,G_l'(\eta,k\bar{R}) - G_l(\eta,k\bar{R})\,\mathcal{L}^{\mathrm{I}}}\right]. \tag{3.124}$$

As usual, primes stand for derivatives with respect to argument $\rho\ (= k\bar{R})$.

We now consider the scattering wave function. Since the wave functions $\phi_C^{(+)}$ and $\psi^{(+)}$ have identical incoming contents, the expansions of Eqs. (3.36) and (3.113) must have the same coefficients, C_{lm}. Using Eq. (3.101) in Eq. (3.113), we get

$$\psi^{(+)}(\mathbf{k};\mathbf{r}) = \frac{1}{(2\pi)^{3/2}}\sum_{lm} 4\pi\,Y_{lm}^*(\mathbf{k})\,Y_{lm}(\hat{\mathbf{r}})\,e^{i\sigma_l}\,i^l\,\frac{u_l(k,r)}{kr} \tag{3.125a}$$

$$= \frac{1}{(2\pi)^{3/2}}\sum_{l}(2l+1)\,P_l(\cos\theta)\,i^l\,e^{i\sigma_l}\,\frac{u_l(k,r)}{kr}. \tag{3.125b}$$

As pointed out in chapter 2, different normalizations for the radial wave function are found in the literature. Some authors normalize the radial wave function so that for $r > \bar{R}$ it takes the form

$$\bar{u}_l(k,r) = \frac{i}{2}\left[H_l^{(-)}(\eta,\rho) - \bar{S}_l\,H_l^{(+)}(\eta,\rho)\right]e^{-i\bar{\delta}_l} = u_l(k,r)\,e^{-i\bar{\delta}_l}. \tag{3.126}$$

In the $r \to \infty$ limit the above expression gets the simpler form

$$\bar{u}_l(k,r) = \sin\left(kr - \eta\ln 2kr - l\,\pi/2 + \sigma_l + \bar{\delta}_l\right). \tag{3.127}$$

The scattering state is then written as

$$\psi^{(+)}(\mathbf{k};\mathbf{r}) = \frac{1}{(2\pi)^{3/2}}\sum_{lm} 4\pi\,Y_{lm}^*(\mathbf{k})\,Y_{lm}(\hat{\mathbf{r}})\,i^l\,e^{i(\sigma_l+\bar{\delta}_l)}\,\frac{\bar{u}_l(k,r)}{kr} \tag{3.128a}$$

$$= \frac{1}{(2\pi)^{3/2}}\sum_{l}(2l+1)\,i^l\,P_l(\cos\theta)\,e^{i(\sigma_l+\bar{\delta}_l)}\,\frac{\bar{u}_l(k,r)}{kr}. \tag{3.128b}$$

3.4.1 *An illustration: $\alpha - \alpha$ scattering*

To finish this section we study the elastic scattering of two α-particles. The $\alpha - \alpha$ potential is approximated as in Eq. (3.109),

$$V(r) = V_{\rm C}(r) + \overline{V}(r)\,,$$

with the attractive gaussian potential

$$\overline{V}(r) = -\,V_0\, e^{-r^2/\bar{R}^2} \tag{3.129}$$

and

$$\begin{aligned} V_{\rm C}(r) &= \frac{1}{4\pi\varepsilon_0}\,\frac{4e^2}{r}, \quad \text{for } r \geq R_{\rm C} \simeq R_{\rm P} + R_{\rm T} = 2R_\alpha\,, \\ &= \frac{1}{4\pi\varepsilon_0}\,\frac{4e^2}{2R_{\rm C}}\left(3 - \frac{r^2}{R_{\rm C}^2}\right), \quad \text{for } r < R_{\rm C}. \end{aligned} \tag{3.130}$$

The above Coulomb potential, $V_{\rm C}$, takes into account the finite size of the α-particles. If they do not overlap ($r > R_{\rm C}$) it corresponds to the point charge interaction of Eq. (3.2). If they overlap, it is approximated by the potential of a point charge $q_{\rm P} = 2e$ inside a uniformly charged sphere of radius $R_{\rm C}$ and total charge $q_{\rm T} = 2e$. For the numerical calculations, we used typical nuclear scattering parameters: $V_0 = 60$ MeV, $R_{\rm C} = \bar{R} = 4.5$ fm ($= 4.5 \times 10^{-15}$ m).

The effective l-dependent potentials (see Eqs. (2.10) and (2.11)),

$$V_l(r) = V(r) + \frac{\hbar^2}{2\mu}\; l(l+1)/r^2,$$

are shown in figure 3.6, for a few partial waves. Details of the Coulomb barrier are displayed in the inset. The barrier height is $V_B = 0.42$ MeV.

In figure 3.7, we show the phase shifts of the short-range potential, for the partial waves[7] $l = 0, 2, ..., 10$. We plot absolute phase shifts using Levinson's theorem[8], as discussed in section 2.8. The s-wave phase shift decreases monotonically from the value 4π, as the collision energy increases. A similar behavior is found for $l = 2$, except for its smaller low-energy limit, $\bar{\delta}_2(E = 0) = 3\pi$. According to Levinson's theorem, this means that the

[7] As we discuss in chapter 6, the $\alpha - \alpha$ scattering amplitude has only contribution from even partial waves. This is a consequence of the fact that the projectile and the target are identical particles with an even number of fermions.

[8] In this case one uses a generalization of Levinson's theorem for clusters of identical particles, according to which the phase shift at $E \to 0$ is given by $N\pi$, where N is the sum of the numbers of bound and redundant states. The latter are solutions of the radial equation that generate vanishing many-body states (see, e.g. [Friedrich (1981)]).

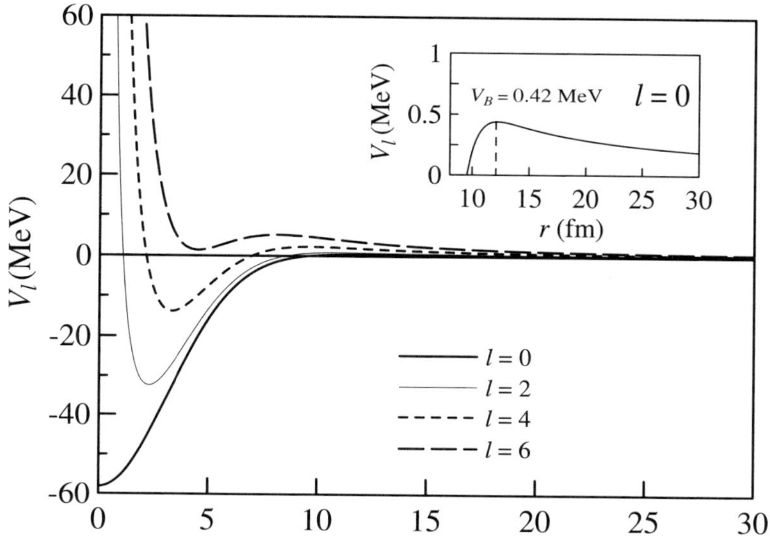

Fig. 3.6 Effective potentials in $\alpha - \alpha$ scattering for a few partial waves. The inset shows details of the Coulomb barrier ($V_{l=0}(r)$).

potential has four bound states with $l = 0$ and three with $l = 2$. For $l = 4$, the phase shift has the value π at $E = 0$, indicating the existence of a single bound state. As the energy increases, $\bar{\delta}_4$ grows slowly until the energy $E = 1.60$ MeV, where it exhibits a sharp resonance. It then falls to 0 as $E \to \infty$. The phase shifts for higher partial waves vanish at $E = 0$. Therefore, the potential has no bound states for angular momentum above $l = 4$. $\bar{\delta}_6$ shows a broader resonance at 4.90 MeV and $\bar{\delta}_8$ and $\bar{\delta}_{10}$ have broad maxima before going to zero asymptotically.

In Table 3.2, we give the barrier heights of the effective potential, $V_{\mathrm{B}}(l)$, and the resonance energies, $E_r(l)$, for $l = 4$ and 6. These quantities play a central role in the resonance widths. The width Γ_l is related to the life-time τ_l through the uncertainty relation, $\Gamma_l = \hbar/\tau_l$, and the life-times are inversely proportional to the corresponding tunelling probability, T_l. Therefore, Γ_l is proportional to T_l. Since the tunelling probability decreases rapidly with $\Delta_l \equiv V_{\mathrm{B},l} - E_r(l)$, the same should occur with Γ_l. This trend is confirmed in figure 3.7. From Table 3.2 we find that $\Delta_6 = 0.34$ MeV $< \Delta_4 = 0.70$ MeV. Accordingly, in figure 3.7 Γ_6 is clearly larger than Γ_4.

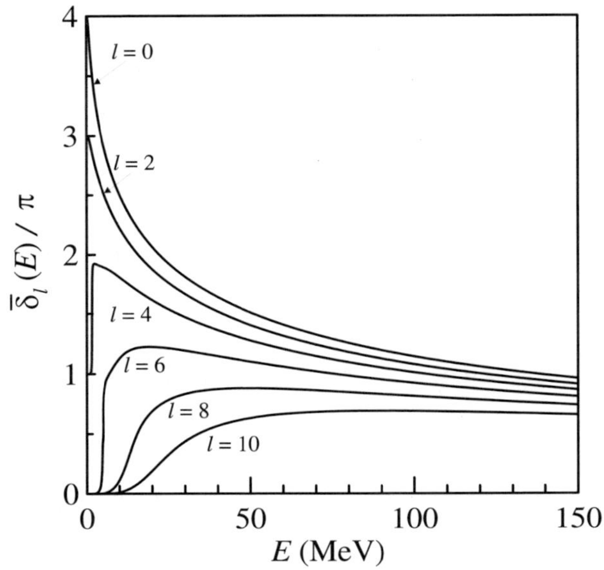

Fig. 3.7 Phase shifts of the short-range potential in $\alpha - \alpha$ collision.

Table 3.2 Resonance energies and the barrier heights for $l = 4$ and 6.

l	$V_B(l)$ [MeV]	$E_r(l)$ [MeV]	$\Delta_l = V_{B,l} - E_r(l)$ [MeV]
4	2.30	1.60	0.70
6	5.24	4.90	0.34

One important experimental fact which is relevant to our discussion of $\alpha + \alpha$ scattering is that the only bound state of this system is the ground state of ^{8}Be and it is unstable. It decays into two α particles, with a life time of $7 \cdot 10^{-17}$ s.

Exercises

(1) Heavy ion collisions at above-barrier energies are characterized by intense Coulomb field and strong absorption. At small impact parameters the large number of open channels yields strong absorption of the incident flux while at large impact parameters the collision is similar to

a collision of two point charges. This situation gave rise to Fresnel models for heavy ion scattering [Frahn (1966, 1971); Frahn and Venter (1963)], discussed in chapter 7. In its simplest form, the S-matrix is approximated as

$$\begin{aligned} S_l &= 0; \qquad \text{for } l < l_g \\ S_l &= e^{2i\sigma_l}; \quad \text{for } l > l_g\,, \end{aligned}$$

where l_g is the angular momentum in a grazing collision. According to this model, the grazing angle, θ_g (associated with l_g through the Rutherford trajectory), is identified by the condition $\sigma(\theta_g)/\sigma_{\mathrm{Ruth}}(\theta_g) = 1/4$ (see section 7.5.2). Using the experimental data of the figure below, make an estimate of the nuclear radius parameter, $r_0 = (R_{\mathrm{T}} + R_{\mathrm{P}})/(A_{\mathrm{T}}^{1/3} + A_{\mathrm{P}}^{1/3})$.

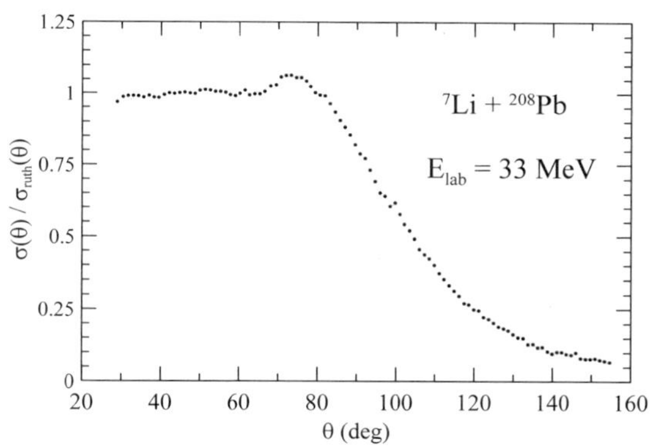

(2) Show that the approximate Coulomb potential of Eq. (3.130) corresponds to the interaction potential between a point charge of value $q_1 = 2e$ and a spherically symmetric uniform distribution of the total charge $q_2 = 2e$ within a sphere of radius $R_{\mathrm{P}} + R_{\mathrm{T}}$.

(3) Consider the scattering of a particle of mass μ with energy E from a Coulomb plus short-rangel potential. Derive the integral formula for the phase shift $\bar{\delta}_l$, in terms of the Sommerfeld parameter, η, the range of the potential, $\bar{R}$, and the radial wave function $\bar{u}_l(k,r)$, of Eq. (3.127). Show that this formula reduces to Eq. (2.71) in the $\eta \to 0$ limit.

(4) Prove that the regular and irregular Coulomb wave functions, $F_l(\eta,\rho)$ and $G_l(\eta,\rho)$, where $\rho = kr$, satisfy the Wronskian relation (primes mean derivatives with respect to ρ)

$$W\left\{G_l(\eta,\rho), F_l(\eta,\rho)\right\} \equiv G_l'(\eta,\rho)\,F_l(\eta,\rho) - F_l'(\eta,\rho)\,G_l(\eta,\rho) = 1.$$

Chapter 4

Green's Functions, T- and S-Matrices

The previous two chapters were devoted to the partial-wave expansion method in scattering theory. Here, we discuss a new approach: the *Green's function method.* This method leads to the Lippmann-Schwinger equations, which have integro-differencial form in the coordinate representation. Although exact solutions of these equations are very hard to find, they are suitable for the derivation of perturbative approximations. In the discussion of the Lippmann-Schwinger equations, we introduce the concepts of Green's functions, Møller operators and the S- and the T-matrices, which play very important roles in scattering theory.

Since we have shown (see chapter 1) that the realistic time-dependent wave-packet treatment of the scattering problem is equivalent to the stationary diffusion of a plane wave, we base most of the present chapter on the simpler time-independent approach. In most situations one deals with real potentials, which leads to a series of important properties. However, sometimes one uses complex potentials to simulate coupled-channel effects on the elastic channel. Throughout this chapter, we assume that the interaction potential is real and spherically symmetric. The scattering from complex potentials will be discussed in chapters 7 and 10.

4.1 Lippmann-Schwinger equations

We start with the Schrödinger equations for a free particle with wave vector $\mathbf{k}$ and energy $E_{\mathbf{k}} \equiv E_k = \hbar^2 k^2/2\mu$, and the corresponding equation when this particle is submitted to the action of a short-range potential, which vanishes for $r > \bar{R}$. Using Dirac's notation, we can write

$$(E_k - H_0) \left|\phi\left(\mathbf{k}\right)\right\rangle = 0 \tag{4.1}$$

and

$$(E_k - H) \, |\psi(\mathbf{k})\rangle = 0. \tag{4.2}$$

Above, $H_0 = K$ is the kinetic energy operator and $H = H_0 + V$. Since $H_0^\dagger = H_0$ and $H^\dagger = H$, the states $|\phi(\mathbf{k})\rangle$ and $|\psi(\mathbf{k})\rangle$ can be normalized so as to satisfy the relations

$$\langle \phi(\mathbf{k}') | \, \phi(\mathbf{k}) \rangle = \delta(\mathbf{k} - \mathbf{k}'), \tag{4.3}$$

and

$$\langle \psi(\mathbf{k}') | \, \psi(\mathbf{k}) \rangle = \delta(\mathbf{k} - \mathbf{k}') \tag{4.4a}$$

$$\langle \psi_m | \, \psi(\mathbf{k}) \rangle = 0 \tag{4.4b}$$

$$\langle \psi_m | \, \psi_n \rangle = \delta_{mn}. \tag{4.4c}$$

Above, m and n stand for negative energy states which the Hamiltonian H may possibly have, and δ_{mn} and $\delta(\mathbf{k} - \mathbf{k}')$ are respectively the usual Kronecker's and Dirac's deltas. The identity operator in the Hilbert spaces spanned by $|\phi(\mathbf{k})\rangle$ and $|\psi(\mathbf{k})\rangle$ can be written

$$\int |\phi(\mathbf{k})\rangle \, d^3\mathbf{k} \, \langle \phi(\mathbf{k})| = \mathbf{1} \tag{4.5}$$

and

$$\int |\psi(\mathbf{k})\rangle \, d^3\mathbf{k} \, \langle \psi(\mathbf{k})| + \sum_n |\psi_n\rangle \, \langle \psi_n| = \mathbf{1}\,. \tag{4.6}$$

The free particle's and the full Green's operators are respectively defined as,

$$G_0(E) = \frac{1}{E - H_0} \tag{4.7}$$

$$G(E) = \frac{1}{E - H}. \tag{4.8}$$

To find a relation between G and G_0, we use the operator identity

$$A^{-1} = B^{-1} + B^{-1}(B - A)A^{-1}.$$

Setting $A = E - H \equiv E - (H_0 + V)$ and $B = E - H_0$, we get

$$G(E) = G_0(E) + G_0(E) \, V \, G(E), \tag{4.9}$$

or, setting $A = E - H_0$ and $B = E - (H_0 + V)$, we get

$$G(E) = G_0(E) + G(E) \, V \, G_0(E). \tag{4.10}$$

The state $|\psi(\mathbf{k})\rangle$ satisfies equations involving the Green's operators. To derive these equations, we put Eq. (4.2) in the form

$$(E - H_0)\,|\psi(\mathbf{k})\rangle = V\,|\psi(\mathbf{k})\rangle \tag{4.11}$$

and write $|\psi(\mathbf{k})\rangle$ as the sum of the incident and the scattered waves,

$$|\psi(\mathbf{k})\rangle = |\phi(\mathbf{k})\rangle + |\psi^{\rm sc}(\mathbf{k})\rangle\,. \tag{4.12}$$

Inserting Eq. (4.12) into Eq. (4.11) and using Eq. (4.1), we obtain

$$(E_k - H_0)\,|\psi^{\rm sc}(\mathbf{k})\rangle = V\,|\psi(\mathbf{k})\rangle\,. \tag{4.13}$$

Multiplying from the left with $G_0(E)$ and replacing $|\psi^{\rm sc}(\mathbf{k})\rangle = |\psi(\mathbf{k})\rangle - |\phi(\mathbf{k})\rangle$, we get the desired equation,

$$|\psi(\mathbf{k})\rangle = |\phi(\mathbf{k})\rangle + G_0(E_k)\,V\,|\psi(\mathbf{k})\rangle\,. \tag{4.14}$$

A similar equation can be obtained in terms of the full Green's function. For this purpose, we use Eq. (4.12) in the RHS of Eq. (4.13) and re-arrange the equation in the form

$$(E_k - H_0 - V)\,|\psi^{\rm sc}\rangle = V\,|\psi(\mathbf{k})\rangle\,.$$

Multiplying from the left with $G(E_k)$ we get

$$|\psi^{\rm sc}\rangle = G(E_k)\,V\,|\phi(\mathbf{k})\rangle\,.$$

Adding $|\phi(\mathbf{k})\rangle$ to both sides of the above equation and using Eq. (4.12), we obtain

$$|\psi(\mathbf{k})\rangle = |\phi(\mathbf{k})\rangle + G(E_k)\,V\,|\phi(\mathbf{k})\rangle = (1 + G(E_k)\,V)\,|\phi(\mathbf{k})\rangle\,. \tag{4.15}$$

Eqs. (4.14) and (4.15) are usually called Lippmann-Schwinger (LS) equations.

4.1.1 *The free particle Green's function*

We now evaluate the matrix elements of the free Green's operator in the coordinate representation, called the *free-particle Green's Function.* We start from the spectral decomposition of G_0,

$$\begin{aligned} G_0(E_k) &\equiv \int d^3\mathbf{q}\;\frac{|\phi(\mathbf{q})\rangle\,\langle\phi(\mathbf{q})|}{E_k - E_q} \qquad (4.16)\\ &= -\left(\frac{2\mu}{\hbar^2}\right)\int d^3\mathbf{q}\,\frac{|\phi(\mathbf{q})\rangle\,\langle\phi(\mathbf{q})|}{q^2 - k^2}\,, \end{aligned}$$

and take matrix elements in the coordinate representation. We get

$$G_0(E_k;\mathbf{r},\mathbf{r}') = -\left(\frac{2\mu}{\hbar^2}\right)\int d^3\mathbf{q}\,\frac{\phi(\mathbf{q};\mathbf{r})\,\phi^*(\mathbf{q};\mathbf{r}')}{q^2-k^2}. \tag{4.17}$$

Next, we use the normalized free particle's wave functions

$$\phi(\mathbf{q};\mathbf{r}) = \frac{1}{(2\pi)^{3/2}}\,e^{i\mathbf{q}\cdot\mathbf{r}};\quad \phi^*(\mathbf{q};\mathbf{r}') = \frac{1}{(2\pi)^{3/2}}\,e^{-i\mathbf{q}\cdot\mathbf{r}'}, \tag{4.18}$$

introduce the notation

$$\mathbf{R} = \mathbf{r}-\mathbf{r}',$$

and evaluate the integral over $\mathbf{q}$ in spherical coordinates, with the z-axis along the $\mathbf{R}$-direction. Owing to the axial symmetry, the integration over the azimuthal angle produces the trivial factor 2π and we get[1]

$$G_0(E_k;\mathbf{r},\mathbf{r}') = -\left(\frac{2\mu}{\hbar^2}\right)\frac{1}{(2\pi)^2}\int\frac{q^2\,dq}{q^2-k^2}\int_{-1}^{1}d(\cos\theta_{\mathbf{q}})\,e^{iqR\cos\theta_{\mathbf{q}}}\,, \tag{4.19}$$

where $\theta_{\mathbf{q}}$ is the angle between the vectors $\mathbf{q}$ and $\mathbf{R}$. Performing the integration over $\theta_{\mathbf{q}}$, the above integral takes the form

$$G_0(E_k;\mathbf{r},\mathbf{r}') = \left(\frac{2\mu}{\hbar^2}\right)\frac{i}{(2\pi)^2 R}\,(I_1+I_2)\,, \tag{4.20}$$

with

$$I_1 = \int_0^\infty dq\,\frac{q\,e^{iqR}}{q^2-k^2}\quad\text{and}\quad I_2 = -\int_0^\infty dq\,\frac{q\,e^{-iqR}}{q^2-k^2}. \tag{4.21}$$

Changing $q\to -q$, the second integral in Eq. (4.21) becomes

$$I_2 = \int_{-\infty}^{0} dq\,\frac{q\,e^{iqR}}{q^2-k^2}\,,$$

so that

$$I_1+I_2 = \int_{-\infty}^{\infty} dq\,\frac{q\,e^{iqR}}{q^2-k^2}. \tag{4.22}$$

Using Eq. (4.22) and the identity

$$\frac{q}{q^2-k^2} = \frac{1}{2}\left[\frac{1}{q-k}+\frac{1}{q+k}\right]$$

in Eq. (4.20), we get

$$G_0(E_k;\mathbf{r},\mathbf{r}') = \left(\frac{2\mu}{\hbar^2}\right)\frac{i}{2\,(2\pi)^2 R}\int_{-\infty}^{\infty}dq\,\left[\frac{1}{q-k}+\frac{1}{q+k}\right]e^{iqR}. \tag{4.23}$$

[1]The fact that the Green's function depends on $\mathbf{r}$ and $\mathbf{r}'$ only through the distance $R = |\mathbf{r}-\mathbf{r}'|$ is a consequence of the translational and rotational invariances of H_0.

The above integral is not defined owing to the poles of the integrand at $q = \pm k$. However, meaningful Green's functions $G_0^{(+)}(E_k;\mathbf{r},\mathbf{r}')$ and $G_0^{(-)}(E_k;\mathbf{r},\mathbf{r}')$ can be obtained moving the poles into the complex-plane, up or down the real axis, according to the prescriptions,

$$\pm k \to \pm\,(k + i\,\epsilon') \quad (G_0(E_k;\mathbf{r},\mathbf{r}') \to G_0^{(+)}(E_k;\mathbf{r},\mathbf{r}')) \tag{4.24}$$

and

$$\pm k \to \pm\,(k - i\,\epsilon') \quad (G_0(E_k;\mathbf{r},\mathbf{r}') \to G_0^{(-)}(E_k;\mathbf{r},\mathbf{r}')). \tag{4.25}$$

These procedures correspond to redefining the Green's operator as

$$G_0(E_k) \longrightarrow G_0^{(\pm)}(E_k) = \lim_{\epsilon\to 0}\left[\frac{1}{E_k - H_0 \pm i\,\epsilon}\right], \tag{4.26}$$

where $\epsilon = \left(\hbar^2 k/\mu\right)\epsilon'$ is an infinitesimal quantity. It is equivalent to the analytical continuation of the variable E_k onto the complex plane, where it takes the values $E_k \pm i\,\epsilon$. In a similar way, for the full interacting Green's function, we define

$$G^{(\pm)}(E_k) = \lim_{\epsilon\to 0}\left[\frac{1}{E_k - H \pm i\,\epsilon}\right]. \tag{4.27}$$

It can be easily checked that the matrix-elements of the free and the full Green's operators in the coordinate representation, $G_0^{(\pm)}(E_k;\mathbf{r},\mathbf{r}')$ and $G^{(\pm)}(E_k;\mathbf{r},\mathbf{r}')$, respectively satisfy the equations

$$\left[E_k + \frac{\hbar^2}{2\mu}\nabla_{\mathbf{r}}^2\right] G_0^{(\pm)}(E_k;\mathbf{r},\mathbf{r}') = \delta(\mathbf{r}-\mathbf{r}') \tag{4.28}$$

and

$$\left[E_k + \frac{\hbar^2}{2\mu}\nabla_{\mathbf{r}}^2 - V(r)\right] G^{(\pm)}(E_k;\mathbf{r},\mathbf{r}') = \delta(\mathbf{r}-\mathbf{r}'). \tag{4.29}$$

Eqs. (4.26) and (4.27), together with the hermiticity of the Hamiltonians, lead to the properties

$$\left[G_0^{(\pm)}(E_k)\right]^\dagger = G_0^{(\mp)}(E_k) \tag{4.30}$$

$$\left[G^{(\pm)}(E_k)\right]^\dagger = G^{(\mp)}(E_k). \tag{4.31}$$

Using the standard result (see, e.g., [Rodberg and Thaler (1967)])

$$\lim_{\epsilon\to 0}\left[\frac{1}{x \pm i\epsilon}\right] = \mathcal{P}\left(\frac{1}{x}\right) \mp i\pi\,\delta(x), \tag{4.32}$$

where $\mathcal{P}$ stands for the Cauchy's principal value, $G_0^{(\pm)}(E_k)$ is split as

$$G_0^{(\pm)}(E_k) = \mathcal{P}\left(\frac{1}{E_k - H_0}\right) \mp i\pi\,\delta(E_k - H_0). \tag{4.33}$$

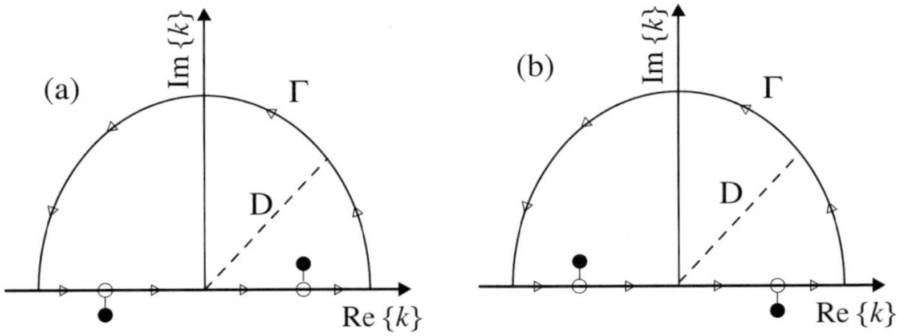

Fig. 4.1 The closed contour on the complex plane and the shifts of the poles away from the real axis (solid circles). The options represented in (a) and (b) lead respectively to $G^{(+)}_{E_i}$ and $G^{(-)}_{E_i}$.

The first term at the RHS of Eq. (4.33) corresponds to a standing wave Green's function [Rodberg and Thaler (1967)] and the second term is the on-the-energy-shell part. Usually, these terms are called respectively the *off-shell* and the *on-shell* parts of the Green's operator. Similar results hold for the full Green's operator.

Shifting the poles as in Eq. (4.24) or as in Eq. (4.25), Eq. (4.23) leads to two different Green's functions, $G_0^{(+)}$ and $G_0^{(-)}$. They can be written as

$$G_0^{(\pm)}(E_k;\mathbf{r},\mathbf{r}') = \left(\frac{2\mu}{\hbar^2}\right) \frac{i}{2\,(2\pi)^2\,R}\, I^{(\pm)}\,, \tag{4.34}$$

where

$$I^{(\pm)} = \int_{-\infty}^{\infty} dq\, \left\{\frac{1}{q-(k\pm i\,\epsilon')} + \frac{1}{q+(k\pm i\,\epsilon')}\right\}\, e^{iqR}\,. \tag{4.35}$$

These integrals can be evaluated on the complex plane, over the closed contour represented in figure 4.1. In the $D \to \infty$ limit, it corresponds to the real axis plus a semi-circle, Γ, of infinite radius. The exponential factor in the integrand of Eq. (4.35) can be written

$$e^{iR(\mathrm{Re}\{q\}+i\,\mathrm{Im}\{q\})} = e^{iR\,\mathrm{Re}\{q\}} \times e^{-R\,\mathrm{Im}\{q\}}\,.$$

Considering that $R = |\mathbf{r}-\mathbf{r}'|$ is a positive definite quantity and that $\mathrm{Im}\{k'\} > 0$ over the path Γ, the factor $e^{-R\,\mathrm{Im}\{q\}}$ goes to zero as $D \to \infty$. Therefore, the integral over Γ vanishes and the integral over the closed contour reduces to that over the real axis, as in Eq. (4.35). In this way, it is given by the residue of the pole enclosed by the contour. That is, $k + i\,\epsilon'$

for $I^{(+)}$ and $-(k - i\,\epsilon')$ for $I^{(-)}$. Taking the contour in the counterclockwise sense, we get

$$I^{(\pm)} = 2\pi i \times \text{Res}\left\{q = \pm\left(k \pm i\,\epsilon'\right)\right\} = 2\pi i\; e^{\pm ikR}\,. \tag{4.36}$$

Inserting Eq. (4.36) into Eq. (4.34) and replacing $R = |\mathbf{r} - \mathbf{r}'|$, we obtain the free particle Green's function:

$$G_0^{(\pm)}(E_k; \mathbf{r}, \mathbf{r}') = -\left(\frac{2\mu}{\hbar^2}\right)\frac{1}{4\pi}\;\frac{e^{\pm\, ik\,|\mathbf{r}-\mathbf{r}'|}}{|\mathbf{r} - \mathbf{r}'|}\,. \tag{4.37}$$

We remark that some authors write the free particle's Green's function without the factor within round brackets at the RHS of Eq. (4.37). This happens when the starting equation is written in the form

$$\left[\nabla^2 + k^2\right]|\psi(\mathbf{k})\,\rangle = U\,|\psi(\mathbf{k})\,\rangle,$$

with $U = 2\mu V/\hbar^2$. In this case, the Green's function is defined as $G_0^{(\pm)}(E_k) = \left(\nabla^2 + k^2\right)^{-1}$. Following the same procedures as above, one gets Eq. (4.37) but without the factor $2\mu/\hbar^2$. However, the resulting Lippmann-Schwinger equation is unaffected since the potential V must be replaced by U and in this way the factor $2\mu/\hbar^2$ is recovered.

A practical consequence of the above discussion is that different wave functions are generated by the Green's functions $G_0^{(+)}$ and $G_0^{(-)}$. One should distinguish the states $|\psi^{(+)}(\mathbf{k})\rangle$ and $|\psi^{(-)}(\mathbf{k})\rangle$, given by the Lippmann-Schwinger equations

$$|\psi^{(\pm)}(\mathbf{k})\rangle = |\phi(\mathbf{k})\rangle + G_0^{(\pm)}(E_k)\,V\,|\psi^{(\pm)}(\mathbf{k})\rangle\,. \tag{4.38}$$

This point will be further discussed in the next section.

Equivalent Lippmann-Schwinger equations are generated by the full Green's functions $G^{(\pm)}(E_k)$, namely

$$|\psi^{(\pm)}(\mathbf{k})\rangle = |\phi(\mathbf{k})\rangle + G^{(\pm)}(E_k)\,V\,|\phi(\mathbf{k})\rangle \equiv \left(1 + G^{(\pm)}(E_k)\,V\right)|\phi(\mathbf{k})\rangle\,. \tag{4.39}$$

4.1.2 *The scattering amplitude*

Although $\psi^{(+)}(\mathbf{k})$ and $\psi^{(-)}(\mathbf{k})$ are solutions of the same equation with the same energy, they have different boundary conditions. To clarify this point, we evaluate the asymptotic forms of these wave functions. In the coordinate representation, the Lippmann-Schwinger equations are written[2]

$$\psi^{(\pm)}(\mathbf{k};\mathbf{r}) = \phi(\mathbf{k};\mathbf{r}) + \int d^3\mathbf{r}'\,G_0^{(\pm)}(E_k;\mathbf{r},\mathbf{r}')\,V(\mathbf{r}')\,\psi^{(\pm)}(\mathbf{k};\mathbf{r}') \tag{4.40}$$

[2]Here, we are assuming the locality of the potential V. Non-local potentials will be discussed in section 4.4.

or, using the Eq. (4.37),

$$\psi^{(\pm)}(\mathbf{k};\mathbf{r}) = \phi(\mathbf{k};\mathbf{r}) - \frac{2\mu}{4\pi\hbar^2}\int d^3\mathbf{r}' \, \frac{e^{\pm ik|\mathbf{r}-\mathbf{r}'|}}{|\mathbf{r}-\mathbf{r}'|} V(\mathbf{r}')\,\psi^{(\pm)}(\mathbf{k};\mathbf{r}')\,. \quad (4.41)$$

We now take the $|\mathbf{r}| \to \infty$ limit. Since V vanishes when $\mathbf{r}'$ is outside the potential range ($|\mathbf{r}'| > \bar{R}$), $|\mathbf{r}'| / |\mathbf{r}| \to 0$ and one can replace

$$\frac{1}{|\mathbf{r}-\mathbf{r}'|} \to \frac{1}{r}. \quad (4.42)$$

The factor $\exp(\pm ik\,|\mathbf{r}-\mathbf{r}'|)$ must be handled more carefully. It is necessary to expand to first order in r'/r,

$$k\,|\mathbf{r}-\mathbf{r}'| \equiv k\left(r^2 + r'^2 - 2\,\mathbf{r}\cdot\mathbf{r}'\right)^{1/2} \simeq kr - k\,\hat{\mathbf{r}}\cdot\mathbf{r}'.$$

Accordingly, the exponential factor is approximated as

$$e^{\pm ik\,|\mathbf{r}-\mathbf{r}'|} = e^{\pm\, ikr} \times e^{\mp\, ik\,\hat{\mathbf{r}}\cdot\mathbf{r}'}, \quad (4.43)$$

where $\hat{\mathbf{r}}$ is the unit vector along $\mathbf{r}$. Inserting Eqs. (4.42) and (4.43) into Eq. (4.41) and defining the final wave vector $\mathbf{q} = k\,\hat{\mathbf{r}}$, we obtain

$$\psi^{(\pm)}(\mathbf{k};\mathbf{r}) \to \phi(\mathbf{k};\mathbf{r}) + \frac{e^{\pm ikr}}{r}\left(-\frac{2\mu}{4\pi\hbar^2}\int d^3\mathbf{r}'\, e^{\mp\, i\,\mathbf{q}\cdot\mathbf{r}'}\, V(\mathbf{r}')\,\psi^{(\pm)}(\mathbf{k};\mathbf{r}')\right).$$

Using Eq. (4.18), and switching to Dirac's notation, the above equation becomes

$$\psi^{(\pm)}(\mathbf{k};\mathbf{r}) \to \frac{1}{(2\pi)^{3/2}}\left[e^{i\mathbf{k}\cdot\mathbf{r}} + \frac{e^{\pm ikr}}{r}\left(-2\pi^2\left(\frac{2\mu}{\hbar^2}\right)\langle\phi(\pm\mathbf{q})\,|\,V\,|\psi^{(\pm)}(\mathbf{k})\,\rangle\right)\right]. \quad (4.44)$$

Inspecting the above equation for $\psi^{(+)}$, we conclude that the Green's function $G_0^{(+)}(E_k;\mathbf{r},\mathbf{r}')$ generates a state which asymptotically behaves as the sum of the incident plane wave with an outgoing spherical wave. This is the desired scattering solution. A comparison with Eq. (2.3) (with the normalization constant $A = (2\pi)^{-3/2}$) leads to the relation

$$f(\theta) = -2\pi^2\left(\frac{2\mu}{\hbar^2}\right)\left\langle\phi(\mathbf{q})\,\middle|\,V\,\middle|\psi^{(+)}(\mathbf{k})\right\rangle. \quad (4.45)$$

Above, θ is the observation angle which corresponds to the angle between the initial ($\mathbf{k}$) and the final ($\mathbf{q}$) wave vectors. The meaning of the solution generated by $G_0^{(-)}(E_k;\mathbf{r},\mathbf{r}')$ is not straightforward. However, it is very useful for calculations, as will be discussed later in this chapter.

4.1.3 Orthonormality relation for scattering states

Since the solutions $|\psi^{(+)}\rangle$ have the same boundary condition, they form an orthonormal set. The set of states $|\psi^{(-)}\rangle$ have the same property. The orthonormalization relations of Eqs. (4.4a) and (4.4c) should then be rewritten as

$$\langle\psi^{(\pm)}(\mathbf{k}')|\psi^{(\pm)}(\mathbf{k})\rangle = \delta(\mathbf{k}-\mathbf{k}'), \tag{4.46}$$

$$\langle\psi_m|\psi^{(\pm)}(\mathbf{k})\rangle = 0, \tag{4.47}$$

$$\langle\psi_m|\psi_n\rangle = \delta_{m,n}. \tag{4.48}$$

These orthonormality relations can be formally derived from the Lippmann-Schwinger equations. Let us, for example, evaluate the scalar product of Eq. (4.46) in the case of outgoing states. We first use the Hermitian conjugate version of the Lippmann-Schwinger equation (Eq. (4.39)) for $|\psi^{(+)}(\mathbf{k}')\rangle$, writing the explicit form of the full Green's function (Eq. (4.27)). We get

$$\begin{aligned}\langle\psi^{(+)}(\mathbf{k}')|\psi^{(+)}(\mathbf{k})\rangle &= \langle\phi(\mathbf{k}')|\psi^{(+)}(\mathbf{k})\rangle \\ &\quad + \left\langle\phi(\mathbf{k}')\middle| V\frac{1}{E_{\mathbf{k}'}-H-i\epsilon}\middle|\psi^{(+)}(\mathbf{k})\right\rangle.\end{aligned} \tag{4.49}$$

Applying the operator $(E_{\mathbf{k}'}-H+i\epsilon)^{-1}$ on $|\psi^{(+)}(\mathbf{k})\rangle$ in the second term on the RHS of Eq. (4.49) we obtain

$$\begin{aligned}\langle\psi^{(+)}(\mathbf{k}')|\psi^{(+)}(\mathbf{k})\rangle &= \langle\phi(\mathbf{k}')|\psi^{(+)}(\mathbf{k})\rangle \\ &\quad + \frac{1}{E_{\mathbf{k}'}-E_{\mathbf{k}}-i\epsilon}\left\langle\phi(\mathbf{k}')\middle| V\middle|\psi^{(+)}(\mathbf{k})\right\rangle.\end{aligned}$$

Now, we use Eq. (4.38) for $|\psi^{(+)}(\mathbf{k})\rangle$ in the first term on the RHS of the above equation and it becomes

$$\begin{aligned}\langle\psi^{(+)}(\mathbf{k}')|\psi^{(+)}(\mathbf{k})\rangle &= \langle\phi(\mathbf{k}')|\phi(\mathbf{k})\rangle + \left\langle\phi(\mathbf{k}')\middle|\frac{1}{E_{\mathbf{k}}-H_0+i\epsilon}V\middle|\psi^{(+)}(\mathbf{k})\right\rangle \\ &\quad + \frac{1}{E_{\mathbf{k}'}-E_{\mathbf{k}}-i\epsilon}\left\langle\phi(\mathbf{k}')\middle| V\middle|\psi^{(+)}(\mathbf{k})\right\rangle.\end{aligned} \tag{4.50}$$

Letting the operator $(E_{\mathbf{k}}-H_0+i\epsilon)^{-1}$ act on the bra state $\langle\phi(\mathbf{k}')|$, the second term on the RHS of Eq. (4.50) exactly cancels the third. The above equation is then reduced to

$$\langle\psi^{(+)}(\mathbf{k}')|\psi^{(+)}(\mathbf{k})\rangle = \langle\phi(\mathbf{k}')|\phi(\mathbf{k})\rangle. \tag{4.51}$$

Since our free states form an orthonormal set, Eq. (4.46) is proved. An analogous proof can be carried out for the estates $|\psi^{(-)}(\mathbf{k})\rangle$.

4.1.4 The Møller wave operators

It is convenient to introduce the Møller wave operators, $\Omega^{(\pm)}$, which transform free waves into scattering states. That is,

$$\Omega^{(\pm)} \left|\phi(\mathbf{k})\right\rangle = \left|\psi^{(\pm)}(\mathbf{k})\right\rangle . \tag{4.52}$$

These operators are a very useful tool in the derivation of formal expressions in scattering theory. Inserting Eq. (4.52) into Eq. (4.38) we obtain

$$\Omega^{(\pm)} \left|\phi(\mathbf{k})\right\rangle = \left[1 + G_0^{(\pm)}(E_k)\, V\, \Omega^{(\pm)}\right] \left|\phi(\mathbf{k})\right\rangle .$$

This equation means that the operators $\Omega^{(\pm)}$, or equivalently $\left[1 + G_0^{(\pm)}(E_k)\, V\, \Omega^{(\pm)}\right]$, transform a free state with energy E_k into the scattering state with the same energy. This can be expressed by the Lippmann-Schwinger equation

$$\Omega^{(\pm)}(E_k) = 1 + G_0^{(\pm)}(E_k)\, V\, \Omega^{(\pm)}(E_k) . \tag{4.53}$$

We emphasize that the Møller wave operators, as defined in Eq. (4.52) and also in the forthcoming Eq. (4.101), are energy-independent. However, their Lippmann-Schwinger equations only hold in the space of states with the energy appearing in the argument of the Green's function, E_k. To emphasize this point, we use the notation $\Omega^{(\pm)}(E_k)$. A similar equation in terms of $G^{(\pm)}(E_k)$ can be derived. Comparing Eqs. (4.52) and (4.39) we get

$$\Omega^{(\pm)}(E_k) = 1 + G^{(\pm)}(E_k)\, V. \tag{4.54}$$

It can easily be checked that the operators $\Omega^{(\pm)}$ have the properties:

$$\Omega^{(\pm)} = \int d^3\mathbf{k}\, \left|\psi^{(\pm)}(\mathbf{k})\right\rangle \left\langle\phi(\mathbf{k})\right| . \tag{4.55}$$

$$\Omega^{(\pm)\dagger}\, \Omega^{(\pm)} = \mathbf{1} , \tag{4.56}$$

$$\Omega^{(\pm)}\, \Omega^{(\pm)\dagger} = \mathbf{1} - \sum_n \left|\psi_n\right\rangle \left\langle\psi_n\right| . \tag{4.57}$$

Several useful identities involving the Møller operators can be readily derived. We have

$$\Omega^{(\pm)\dagger} = 1 + \Omega^{(\pm)\dagger} V G_0^{(\pm)\dagger} = 1 + \Omega^{(\pm)\dagger} V G_0^{(\mp)}, \tag{4.58}$$

which can be solved for $\Omega^{(\pm)\dagger}$ and $\Omega^{(\pm)}$, with the results

$$\Omega^{(\pm)\dagger} = \frac{1}{1 - V G_0^{(\mp)}} \tag{4.59}$$

$$\Omega^{(\pm)} = \frac{1}{1 - G_0^{(\pm)} V} . \tag{4.60}$$

4.2 The transition and the scattering operators

Eq. (4.45) indicates that the scattering amplitude is proportional to the matrix-element of the potential between a final free state with wave vector along the observation direction, and the full scattering state. However, these states do not belong to the same orthogonal set. Thus, it is convenient to use the operator $\Omega^{(\pm)}$ to express the scattering amplitude in terms of matrix-elements between free states with the initial and the final wave vectors. Using Eq. (4.52), we obtain

$$f(\theta) = -2\pi^2 \left(\frac{2\mu}{\hbar^2}\right) \left\langle \phi\left(\mathbf{k}'\right) \middle| V\, \Omega^{(+)} \middle| \phi\left(\mathbf{k}\right) \right\rangle. \tag{4.61}$$

The transition operator is defined as

$$T = V\,\Omega^{(+)}. \tag{4.62}$$

The set of matrix-elements of T between free states, represented as

$$\begin{aligned} T_{\mathbf{k}',\mathbf{k}} &\equiv \langle \phi\left(\mathbf{k}'\right)|\, T\, |\phi\left(\mathbf{k}\right)\rangle = \langle \phi\left(\mathbf{k}'\right)|\, V\,\Omega^{(+)}\, |\phi\left(\mathbf{k}\right)\rangle \\ &= \left\langle \phi\left(\mathbf{k}'\right) \middle| V \middle| \psi^{(+)}\left(\mathbf{k}\right) \right\rangle, \end{aligned} \tag{4.63}$$

is called the T-matrix. Its matrix-elements on-the-energy-shell, i.e. with $|\mathbf{k}'| = |\mathbf{k}|$, give the probability amplitude for the transition from the incident momentum $\hbar\mathbf{k}$ to the final momentum $\hbar\mathbf{k}'$, through the action of the potential V. Thus, in the absence of potential the transition matrix vanishes identically. Inserting Eq. (4.63) into Eq. (4.45), the scattering amplitude becomes[3]

$$f(\theta) = -2\pi^2 \left(\frac{2\mu}{\hbar^2}\right) T_{\mathbf{k}',\mathbf{k}}. \tag{4.64}$$

Therefore, the T-matrix elements lead directly to the scattering amplitude and give the cross section.

Eq. (4.63) is frequently written as a power series. Using recursively the Lippmann-Schwinger equation for $|\psi^{(+)}\left(\mathbf{k}\right)\rangle$ (Eq. (4.38)), we get

$$T_{\mathbf{k}',\mathbf{k}} = \left\langle \phi\left(\mathbf{k}'\right) \middle| V \sum_{n=0}^{\infty} \left(G_0^{(+)}(E_k)\, V\right)^n \middle| \phi\left(\mathbf{k}\right) \right\rangle. \tag{4.65}$$

[3]Note that the constant of proportionality between the scattering amplitude and the T-matrix depends on the choice of the normalization of the plane wave. We use $A = (2\pi)^{-3/2}$ and obtain $-2\pi^2\left(2\mu/\hbar^2\right)$. Some authors (like Austern in [Austern (1970)]) use $A = 1$. In this case it takes the value $-\mu/2\pi\hbar^2$.

Replacing $G_0^{(+)}(E_k)$ by its spectral representation (Eq. (4.16)), we obtain

$$T_{\mathbf{k}',\mathbf{k}} = \langle\phi(\mathbf{k}')|\,V\,|\phi(\mathbf{k})\rangle + \int d^3\mathbf{q}\,\frac{\langle\phi(\mathbf{k}')|\,V\,|\phi(\mathbf{q})\rangle\;\langle\phi(\mathbf{q})|\,V\,|\phi(\mathbf{k})\rangle}{E_k - E_q + i\epsilon}$$
$$+ \int d^3\mathbf{q}\,d^3\mathbf{q}'\,\frac{\langle\phi(\mathbf{k}')|\,V\,|\phi(\mathbf{q})\rangle\;\langle\phi(\mathbf{q})|\,V\,|\phi(\mathbf{q}')\rangle\;\langle\phi(\mathbf{q}')|\,V\,|\phi(\mathbf{k})\rangle}{(E_k - E_q + i\epsilon)\,(E_k - E_{q'} + i\epsilon)}$$
$$+ \cdots \text{ higher order terms.} \tag{4.66}$$

Eq. (4.65) (or equivalently Eq. (4.66)) is known as the *Born's series.* Neglecting in Eq.(4.65) terms beyond $n = 0$ (i.e. truncating the series of Eq. (4.66) after the first term) is the *first order Born approximation*, or simply the *Born approximation.* It will be discussed in detail in chapter 5.

Multiplying Eq. (4.53) (for the outgoing boundary condition) from the left by V and using Eq. (4.62), we get the Lippmann-Schwinger equation for the transition operator,

$$T = V + V\,G_0^{(+)}(E_k)\,T. \tag{4.67}$$

Using the same procedure for Eq. (4.54), we obtain the equivalent equation

$$T = V + V\,G^{(+)}(E_k)\,V. \tag{4.68}$$

If one writes Eqs. (4.67) and (4.68) in the momentum representation, both on-the-energy-shell and off-the-energy-shell matrix-elements must be considered. That is, the on-the-energy-shell matrix-elements $T_{\mathbf{k}',\mathbf{k}}$ (with $|\mathbf{k}'| = |\mathbf{k}|$), are given by the equation

$$T_{\mathbf{k}',\mathbf{k}} = V_{\mathbf{k}',\mathbf{k}} + \int d^3\mathbf{q}\;\frac{V_{\mathbf{k}',\mathbf{q}}\,T_{\mathbf{q},\mathbf{k}}}{E_k - E_q + i\epsilon},$$

with the vector $\mathbf{q}$ running over the full three-dimensional space (both $|\mathbf{q}| = |\mathbf{k}|$ and $|\mathbf{q}| \neq |\mathbf{k}|$).

The transition operator can also be given in terms of $\Omega^{(-)}$. Multiplying Eq. (4.54) (for the ingoing boundary condition, $^{(-)}$) from the left with V, we get

$$V\Omega^{(-)}(E) = V + V\,G^{(-)}(E)\,V. \tag{4.69}$$

Taking the hermitian conjugate of Eq. (4.69) and using the relations: $V^\dagger = V$ and $[G^{(-)}(E)]^\dagger = G^{(+)}(E)$ (Eq. (4.31)), the above equation becomes

$$[\Omega^{(-)}(E)]^\dagger\,V = V + V\,G^{(+)}(E)\,V. \tag{4.70}$$

Comparing Eqs. (4.70) and (4.68), we conclude that

$$T = V\,\Omega^{(+)} = \Omega^{(-)\dagger}\,V \tag{4.71}$$

and the scattering amplitude can be given by Eq. (4.61) or by

$$f(\theta) = -\,2\pi^2\left(\frac{2\mu}{\hbar^2}\right)\left\langle\psi^{(-)}\left(\mathbf{k}'\right)\middle|V\middle|\phi\left(\mathbf{k}\right)\right\rangle. \tag{4.72}$$

4.2.1 *The Optical Theorem*

We now show that the Optical Theorem can be derived from the Lippmann-Schwinger equation for the T-matrix . For this purpose, we employ the spectral representation of $G^{(+)}\left(E\right)$, using the completeness relation

$$\int\left|\psi^{(+)}\left(\mathbf{q}\right)\right\rangle\,d^3\mathbf{q}\,\left\langle\psi^{(+)}\left(\mathbf{q}\right)\right| + \sum_n\left|n\right\rangle\,\left\langle n\right| = 1,$$

where $|n\rangle$ stands for a bound state of the system[4] with energy $E_n = -B_n$. Taking matrix-elements of Eq. (4.68) between plane wave states and using the spectral representation of the full Green's operator, we find

$$T_{\mathbf{k}',\mathbf{k}}\left(E\right) = V_{\mathbf{k}',\mathbf{k}} + \sum_n\frac{V_{\mathbf{k}',n}\,V_{n,\mathbf{k}}}{E+B_n} + \int d^3\mathbf{q}\,\frac{\left\langle\phi\left(\mathbf{k}'\right)\right|V\left|\psi^{(+)}\left(\mathbf{q}\right)\right\rangle\,\left\langle\psi^{(+)}\left(\mathbf{q}\right)\right|V\left|\phi\left(\mathbf{k}\right)\right\rangle}{E-E_q+i\epsilon}.$$

Replacing:

$$\left\langle\phi\left(\mathbf{k}'\right)\middle|V\middle|\psi^{(+)}\left(\mathbf{q}\right)\right\rangle = T_{\mathbf{k}',\mathbf{q}}\left(E_q\right) \quad \text{and} \quad \left\langle\psi^{(+)}\left(\mathbf{q}\right)\middle|V\middle|\phi\left(\mathbf{k}\right)\right\rangle = T^*_{\mathbf{k},\mathbf{q}}\left(E_q\right),$$

we get[5]

$$T_{\mathbf{k}',\mathbf{k}}\left(E\right) = V_{\mathbf{k}',\mathbf{k}} + \sum_n\frac{V_{\mathbf{k}',n}\,V_{n,\mathbf{k}}}{E+B_n} + \int d^3\mathbf{q}\,\frac{T_{\mathbf{k}',\mathbf{q}}\left(E_q\right)\,T^*_{\mathbf{k},\mathbf{q}}\left(E_q\right)}{E-E_q+i\epsilon}. \tag{4.73}$$

The above relation is known as *Low's equation.*

We now set $\mathbf{k}' = \mathbf{k}$, $E = E_k$, and evaluate the difference between Eq. (4.73) and its complex conjugate. Using Eq. (4.32), we get

$$T_{\mathbf{k},\mathbf{k}}\left(E_k\right) - T^*_{\mathbf{k},\mathbf{k}}\left(E_k\right) = -2\pi i\int d^3\mathbf{q}\,T_{\mathbf{k},\mathbf{q}}\left(E_q\right)\,T^*_{\mathbf{k},\mathbf{q}}\left(E_q\right)\,\delta\left(E_k - E_q\right). \tag{4.74}$$

[4]Of course, if the potential is not attractive enough to have bound states the second term at the LHS of the above equation should be disregarded.

[5]Note that in general $T_{\mathbf{k}',\mathbf{k}}\left(E\right)$ can be on- or off-the-energy-shell.

Changing to spherical momentum coordinates, so that

$$d^3\mathbf{q} = q^2\, dq\, d\Omega_{\mathbf{q}} = \frac{\mu q}{\hbar^2}\, dE_q\; d\Omega_{\mathbf{q}}\,,$$

we obtain

$$\text{Im}\left\{T_{\mathbf{k},\,\mathbf{k}}\left(E_k\right)\right\} = -\pi\frac{\mu k}{\hbar^2}\int d\Omega_{\mathbf{q}}\; \left|T_{\mathbf{k},\,\mathbf{q}}\left(E_k\right)\right|^2. \tag{4.75}$$

Using Eq. (4.64), $T_{\mathbf{k},\,\mathbf{q}}$ is expressed in terms of the scattering amplitude and we get the Optical Theorem,

$$\frac{4\pi}{k}\;\text{Im}\Big\{f(0)\Big\} = \int d\Omega\;\Big|f\left(\theta\right)\Big|^2 = \sigma_{\text{el}}. \tag{4.76}$$

4.2.2 *The S-matrix*

To finish this section, we introduce the scattering operator, S. It is defined as

$$S = \Omega^{(-)\dagger}\;\Omega^{(+)}\,. \tag{4.77}$$

The set of on-the-energy-shell matrix-elements of the scattering operator is called the *S-matrix* and it plays a major role in Scattering Theory. The S-matrix elements can be written

$$S_{\mathbf{k}',\mathbf{k}} = \langle\phi\left(\mathbf{k}'\right)|\,\Omega^{(-)\dagger}\;\Omega^{(+)}\,|\phi\left(\mathbf{k}\right)\rangle \tag{4.78}$$

or, taking into account Eq. (4.52) and its hermitian conjugate,

$$S_{\mathbf{k}',\mathbf{k}} = \langle\psi^{(-)}\left(\mathbf{k}'\right)\,|\,\psi^{(+)}\left(\mathbf{k}\right)\rangle\,. \tag{4.79}$$

A very important property of the scattering operator is its unitarity. That is

$$S^\dagger S = SS^\dagger = \mathbf{1}\,, \tag{4.80}$$

or, in the momentum representation,

$$\left(S^\dagger S\right)_{\mathbf{k}',\mathbf{k}} = \left(SS^\dagger\right)_{\mathbf{k}',\mathbf{k}} = \delta\left(\mathbf{k}-\mathbf{k}'\right)\,. \tag{4.81}$$

To prove Eq. (4.80), one first uses Eq. (4.77) to express S and $S^\dagger$ in terms of the Møller operators. One then replaces $\Omega^{(\pm)\dagger}\,\Omega^{(\pm)}$ and $\Omega^{(\pm)}\,\Omega^{(\pm)\dagger}$ by the values given in Eqs. (4.56) and (4.57). The proof is completed with the help of the representation of $\Omega^{(\pm)}$ given in Eq. (4.55) and the orthogonality relation of Eq. (4.47).

The S-matrix is related with the on-the-energy-shell T-matrix through the expression

$$S_{\mathbf{k}',\mathbf{k}} = \delta\left(\mathbf{k}-\mathbf{k}'\right) - 2\pi i\,\delta\left(E_k - E_{k'}\right)\,T_{\mathbf{k}',\mathbf{k}}, \tag{4.82}$$

or in operator notation

$$S(E) = 1 - 2\pi i \, \delta \left(E - H_0\right) \, T(E) \, . \tag{4.83}$$

This property is proved below.

Using the Lippmann-Schwinger equation for $\Omega^{(-)\dagger}$ (the conjugate of Eq. (4.54)) and replacing $\Omega^{(+)} \left|\phi\left(\mathbf{k}\right)\right\rangle = \left|\psi^{(+)}\left(\mathbf{k}\right)\right\rangle$, Eq. (4.78) becomes

$$S_{\mathbf{k}',\mathbf{k}} = \left\langle \phi\left(\mathbf{k}'\right) \right| 1 + V\, G^{(-)\dagger} \left(E_{k'}\right) \left|\psi^{(+)}\left(\mathbf{k}\right)\right\rangle,$$

or, according to Eq. (4.31),

$$S_{\mathbf{k}',\mathbf{k}} = \left\langle \phi\left(\mathbf{k}'\right) \right| 1 + V\, G^{(+)} \left(E_{k'}\right) \left|\psi^{(+)}\left(\mathbf{k}\right)\right\rangle \equiv A + B. \tag{4.84}$$

Above,

$$A = \left\langle \phi\left(\mathbf{k}'\right) \middle| \psi^{(+)}\left(\mathbf{k}\right)\right\rangle = \left\langle \phi\left(\mathbf{k}'\right) \right| \Omega^{(+)}(E_k) \left|\phi\left(\mathbf{k}\right)\right\rangle \tag{4.85}$$

and

$$B = \left\langle \phi\left(\mathbf{k}'\right) \right| V\, G^{(+)} \left(E_{k'}\right) \left|\psi^{(+)}\left(\mathbf{k}\right)\right\rangle . \tag{4.86}$$

We now evaluate A with the help of Eq. (4.53) for $\Omega^{(+)}(E_k)$. We get

$$\begin{aligned} A &= \left\langle \phi\left(\mathbf{k}'\right) \right| 1 + G_0^{(+)} \left(E_k\right) \, V\, \Omega^{(+)} \left(E_k\right) \left|\phi\left(\mathbf{k}\right)\right\rangle \\ &= \left\langle \phi\left(\mathbf{k}'\right) \middle| \phi\left(\mathbf{k}\right)\right\rangle + \left\langle \phi\left(\mathbf{k}'\right) \right| G_0^{(+)} \left(E_k\right) \, V\, \Omega^{(+)} \left(E_k\right) \left|\phi\left(\mathbf{k}\right)\right\rangle . \end{aligned}$$

Using the orthogonality of the momentum states, applying the operator $G_0^{(+)}\left(E_k\right)$ on $\left\langle \phi\left(\mathbf{k}'\right) \right|$,

$$\left\langle \phi\left(\mathbf{k}'\right) \right| G_0^{(+)} \left(E_k\right) \equiv \left\langle \phi\left(\mathbf{k}'\right) \right| \frac{1}{E_k - H_0 + i\epsilon} = \frac{1}{E_k - E_{k'} + i\epsilon} \left\langle \phi\left(\mathbf{k}'\right) \right| ,$$

and replacing $V\Omega^{(+)}\left(E_k\right) = T\left(E_k\right)$, we obtain

$$\begin{aligned} A &= \left\langle \phi\left(\mathbf{k}'\right) \middle| \phi\left(\mathbf{k}\right)\right\rangle + \frac{1}{E_k - E_{k'} + i\epsilon} \left\langle \phi\left(\mathbf{k}'\right) \right| T \left|\phi\left(\mathbf{k}\right)\right\rangle \\ &= \delta(\mathbf{k}' - \mathbf{k}) + \frac{1}{E_k - E_{k'} + i\epsilon} \, T_{\mathbf{k}',\mathbf{k}} \, . \end{aligned} \tag{4.87}$$

To calculate the term B, we replace in Eq. (4.86),

$$G^{(+)}\left(E_{k'}\right) \left|\psi^{(+)}\left(\mathbf{k}\right)\right\rangle = \frac{1}{E_{k'} - H + i\epsilon} \left|\psi^{(+)}\left(\mathbf{k}\right)\right\rangle = \frac{1}{E_{k'} - E_k + i\epsilon} \left|\psi^{(+)}\left(\mathbf{k}\right)\right\rangle,$$

and get

$$B = \frac{1}{E_{k'} - E_k + i\epsilon} \left\langle \phi\left(\mathbf{k}'\right) \right| V \left|\psi^{(+)}\left(\mathbf{k}\right)\right\rangle = - \frac{1}{E_k - E_{k'} - i\epsilon} \, T_{\mathbf{k}',\mathbf{k}} \, . \tag{4.88}$$

Inserting Eqs. (4.87) and (4.88) into Eq. (4.84), and using the well known result involving Cauchy's principal value (Eq. (4.32))

$$\lim_{\epsilon\to 0}\left[\frac{1}{x\pm i\epsilon}\right]=\mathcal{P}\left(\frac{1}{x}\right)\mp i\pi\,\delta(x),$$

for $x=E_{k'}-E_k$ and $x=E_k-E_{k'}$, we obtain Eq. (4.82).

We point out that the unitarity of the scattering operator contains the Optical Theorem. Replacing Eq. (4.82) in Eq. (4.80) and carrying out some algebra, one reaches Eq. (4.75) and, therefore, the Optical Theorem.

4.3 The time-dependent picture

The time-dependent picture of the collision in terms of wave-packets describes a realistic scattering problem and leads to a more intuitive definition of the Scattering operator[6]. In this picture, the projectile is described by a wave-packet which evolves in time as dictated by the time-dependent Schrödinger equation with the full Hamiltonian, H. Therefore, this wave-packet can be written as a superposition of eigenstates of H. Since the asymptotic scattering state must contain an outgoing spherical wave, one should make the expansion in terms of $\psi^{(+)}$. That is

$$\Psi^{(+)}(\mathbf{k};\mathbf{r},t)=\int d^3\mathbf{k}'\;A(\mathbf{k}'-\mathbf{k})\,e^{-i\omega_{k'}\,t}\,\psi^{(+)}(\mathbf{k}';\mathbf{r})\,, \tag{4.89}$$

where the notation $\Psi^{(+)}$ is used to emphasize the outgoing nature of the wave-packet. The stationary states $\psi^{(+)}(\mathbf{k}';\mathbf{r})$ can be written as the sum of a plane wave $\phi\,(\mathbf{k}')$ with an outgoing spherical wave $\psi^{\rm sc}(\mathbf{k}';\mathbf{r})$ as,

$$\psi^{(+)}(\mathbf{k}';\mathbf{r})=\phi(\mathbf{k}';\mathbf{r})+\psi^{\rm sc}(\mathbf{k}';\mathbf{r}). \tag{4.90}$$

Using Eq. (4.90) in Eq. (4.89), $\Psi^{(+)}(\mathbf{k};\mathbf{r},t)$ takes the form

$$\Psi^{(+)}(\mathbf{k};\mathbf{r},t)=\Phi(\mathbf{k};\mathbf{r},t)+\Psi^{\rm sc}(\mathbf{k};\mathbf{r},t), \tag{4.91}$$

where $\Phi(\mathbf{k};\mathbf{r},t)$ corresponds to the incident wave-packet propagating freely and $\Psi^{\rm sc}(\mathbf{k};\mathbf{r},t)$ is an outgoing spherical wave-packet. In section 1.6.2, we have investigated the behavior of $\Psi^{\rm sc}(\mathbf{k};\mathbf{r},t)$ in the $t\to\pm\infty$ limits. We found that, far away from the scatterer, the different stationary states $\psi^{\rm sc}(\mathbf{k}';\mathbf{r})$ tend to interfere destructively so that $\Psi^{\rm sc}(\mathbf{k};\mathbf{r},t)$ vanishes. This, however, does not happen in the neighborhood of the spherical surface

[6] We assume in this section that the potential has a short range. Long range potentials, like those in the scattering of two charged particles, will be discussed in section 4.7.

where the integrand has stationary phase, given by the condition (see Eq. (1.95)),

$$r = v_0\, t. \tag{4.92}$$

Since r and the group velocity v_0 of the incident packet are positive definite quantities, this condition cannot be satisfied as $t \to -\infty$. Consequently, $\Psi^{\rm sc}(\mathbf{k};\mathbf{r},t)$ vanishes and we may write

$$\lim_{t\to-\infty} \Psi^{(+)}(\mathbf{k};\mathbf{r},t) = \Phi(\mathbf{k};\mathbf{r},t). \tag{4.93}$$

As expected, before the collision takes place, the scattering state corresponds to the incident free wave-packet. On the other hand, Eq. (4.92) can be satisfied in the $t \to +\infty$ limit. Taking the derivative of Eq. (4.92) with respect to the time, we find that the stationary phase surface moves away from the scatterer with the constant velocity $\dot{r} = v_0$. This corresponds to an outgoing spherical wave-packet, propagating with the same group velocity as the incident packet. Therefore, in the $t \to +\infty$ limit, $\Psi^{(+)}(\mathbf{k};\mathbf{r},t)$ is the sum of the incident wave-packet with an outgoing spherical packet.

Note that both in the $t \to -\infty$ and in the $t \to +\infty$ limits $\Psi^{(+)}(\mathbf{k};\mathbf{r},t)$ is given by wave-packets located far away from the scatterer. Therefore, in the initial and final stages of the collision $\Psi^{(+)}(\mathbf{k};\mathbf{r},t)$ propagates as a free particle. It is convenient to introduce the free states $\Phi^{\rm in}(\mathbf{k})$ and $\Phi^{\rm out}(\mathbf{k})$. The former and the latter are respectively specified by the behaviors in the remote past and in the distant future. That is,

$$\Phi^{\rm in}(\mathbf{k};\mathbf{r},t \to -\infty) = \Psi^{(+)}(\mathbf{k};\mathbf{r},t \to -\infty) \tag{4.94}$$

$$\Phi^{\rm out}(\mathbf{k};\mathbf{r},t \to \infty) = \Psi^{(+)}(\mathbf{k};\mathbf{r},t \to \infty). \tag{4.95}$$

Since $\Phi^{\rm in}$ and $\Phi^{\rm out}$ are always free states, their propagator is

$$U_0(t,t_0) = e^{-iH_0(t-t_0)/\hbar}.$$

Thus, they can easily be determined at any time.

Let us now consider the wave-packet $\Psi^{(-)}(\mathbf{k};\mathbf{r},t)$, obtained from stationary states with incoming wave boundary condition, $\psi^{(-)}(\mathbf{k}';\mathbf{r})$. It is given by

$$\Psi^{(-)}(\mathbf{k};\mathbf{r},t) = \int d^3\mathbf{k}'\; A(\mathbf{k}'-\mathbf{k})\, e^{-i\omega_{k'}\, t}\, \psi^{(-)}(\mathbf{k}';\mathbf{r}). \tag{4.96}$$

Although $\Psi^{(-)}(\mathbf{k};\mathbf{r},t)$ is not directly related to the scattering process, it plays an important role in the evaluation of the scattering cross section. Proceeding as in the previous case, we write $\psi^{(-)}(\mathbf{k}';\mathbf{r})$ as the sum

$$\psi^{(-)}(\mathbf{k}';\mathbf{r}) = \phi(\mathbf{k}';\mathbf{r}) + \bar{\psi}(\mathbf{k}';\mathbf{r})\,, \tag{4.97}$$

where $\bar{\psi}(\mathbf{k}';\mathbf{r})$ has the asymptotic behavior

$$\bar{\psi}(\mathbf{k}';\mathbf{r}\to\infty) \sim \frac{e^{-ik'r}}{r}.$$

Replacing $(+)\to(-)$ in Eq. (4.89) and using Eq. (4.97), we obtain

$$\Psi^{(-)}(\mathbf{k};\mathbf{r},t) = \Phi(\mathbf{k};\mathbf{r},t) + \bar{\Psi}(\mathbf{k};\mathbf{r},t). \tag{4.98}$$

To investigate the asymptotic behaviors of $\bar{\Psi}(\mathbf{k};\mathbf{r},t)$, we follow the same steps used for $\Psi^{\mathrm{sc}}(\mathbf{k};\mathbf{r},t)$ in section 1.6.2. The only difference is the replacement $k'r\to -k'r$ in the evaluation of the global phase (Eq. (1.92)) of the integrand. With this sign change, the stationary phase condition becomes

$$r = -v_0\, t.$$

Now it cannot be satisfied in the $t\to+\infty$ limit. The components $\bar{\psi}(\mathbf{k}';\mathbf{r})$ then interfere destructively everywhere and $\Psi'(\mathbf{k};\mathbf{r},t)$ vanishes. Therefore, we get the distant future asymptotic behavior

$$\lim_{t\to+\infty}\ \Psi^{(-)}(\mathbf{k};\mathbf{r},t) = \Phi(\mathbf{k};\mathbf{r},t)\,. \tag{4.99}$$

On the other hand, in the $t\to-\infty$ limit the stationary phase condition is satisfied on a spherical surface moving with velocity $\dot{r} = -v_0$. This corresponds to an incoming spherical wave-packet. Therefore, in the remote past $\Psi^{(-)}(\mathbf{k};\mathbf{r},t)$ is given by the superposition of a wave-packet propagating along the **k**-direction and an incoming spherical wave-packet. To summarize the above discussion, we present in figure 4.2 a schematic representation of the asymptotic behaviors of the wave-packets $\Psi^{(\pm)}(\mathbf{k};\mathbf{r},t)$.

The states $\Psi^{(\pm)}(\mathbf{k};\mathbf{r},t)$ can be obtained from the plane-wave-packet $\Phi(\mathbf{k};\mathbf{r},t)$ through the action of the operators $\Omega^{(\pm)}$. To prove this point, we replace $\psi^{(\pm)}(\mathbf{k})\to\Omega^{(\pm)}\phi(\mathbf{k})$ in Eqs. (4.89) and (4.96), and move $\Omega^{(\pm)}$ out of the integral. We get

$$|\Psi^{(\pm)}(\mathbf{k};t)\rangle = \Omega^{(\pm)}\ |\Phi(\mathbf{k};t)\rangle\,. \tag{4.100}$$

Combining Eq. (4.100) with the asymptotic limits of Eqs. (4.93) and (4.99), we obtain

$$\Omega^{(\pm)} = \lim_{t'\to\pm\infty}\left[e^{-iH\,(t'-t)/\hbar}\ e^{iH_0\,(t'-t)/\hbar}\right]. \tag{4.101}$$

Acting on the free wave-packet at any time t, the operator $\Omega^{(+)}$ propagates it as a free particle backwards in time to $t'\to-\infty$, where the free wave-packet coincides with $\Psi^{(+)}(\mathbf{k};\mathbf{r},t')$, and then brings it back to the time t with the full propagator. The operator $\Omega^{(-)}$ carries out the free evolution to the distant future, where the free wave-packet coincides with $\Psi^{(-)}(\mathbf{k};\mathbf{r},t')$, and then brings it back to t with the full propagator.

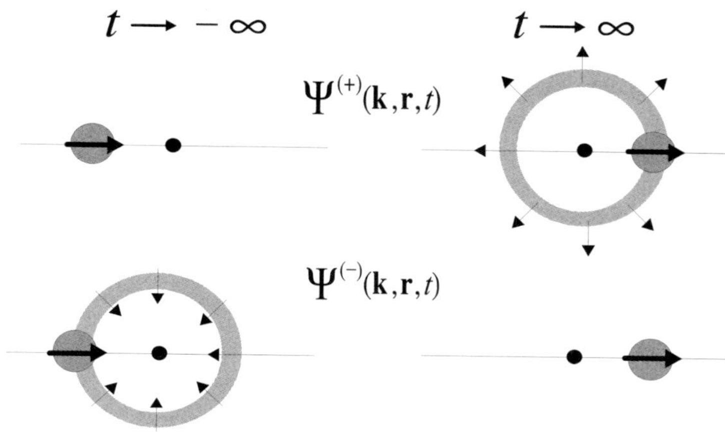

Fig. 4.2 Asymptotic behaviors of the states $\Psi^{(\pm)}(\mathbf{k};\mathbf{r},t)$ in the $t \to \pm\infty$ limits. The picture represents the regions filled by the wave packets (gray area) and the target is shown as a solid circle.

4.3.1 *Time-dependent definition of the scattering operator*

A scattering experiment begins with a beam of free wave-packets $\Phi^{\mathrm{in}}(\mathbf{k}_i)$. Since initially the wave-packets are far away from the target, they propagate freely, so that $\Psi^{(+)}(\mathbf{k}_i;\mathbf{r},t \to -\infty) = \Phi^{\mathrm{in}}(\mathbf{k}_i;\mathbf{r},t \to -\infty)$ (Eq. (4.94)). As the wave-packet gets into the reach of the target's potential, $\Psi^{(+)}(\mathbf{k}_i;\mathbf{r},t)$ deviates from $\Phi^{\mathrm{in}}(\mathbf{k}_i;\mathbf{r},t)$. After the collision, the wave-packet is again outside the range of the target's potential and it propagates as a free particle. Its asymptotic form is then given by $\Phi^{\mathrm{out}}(\mathbf{k}_i;\mathbf{r},t)$. Since the cross section is measured in this situation, all the relevant information about the scattering process must be contained in $\Phi^{\mathrm{out}}(\mathbf{k}_i;\mathbf{r},t)$. The scattering problem then reduces to finding the *scattering operator*, S, which transforms $\Phi^{\mathrm{in}}(\mathbf{k}_i;\mathbf{r},t)$ into $\Phi^{\mathrm{out}}(\mathbf{k}_i;\mathbf{r},t)$. That is[7],

$$\left|\Phi^{\mathrm{out}}(\mathbf{k}_i)\right\rangle = S\left|\Phi^{\mathrm{in}}(\mathbf{k}_i)\right\rangle . \tag{4.102}$$

When the cross section is measured by a detector located at some orientation θ the measurement process acts like a filter, projecting the state $|\Phi^{\mathrm{out}}(\mathbf{k}_i)\rangle$ onto some state $\left|\bar{\Phi}(\mathbf{k}_f)\right\rangle$, characteristic of the detector. It must be a wave-packet with dominant wave vector $\mathbf{k}_f$, where $\mathbf{k}_f$ is a vector with the same modulus as $\mathbf{k}_i$ and orientation θ with respect to $\mathbf{k}_i$. The shape of

[7] Of course, it is necessary to prove that this definition is consistent with that of Eq. (4.77). This will be done at the end of this section.

the packet depends on the energy resolution and detector opening. However, we have shown in chapter 1 that under typical conditions the cross section is independent of the detailed shape of the wave-packet. In this way, we can adopt the shape of the incident packet and replace $\left|\bar{\Phi}\left(\mathbf{k}_f\right)\right\rangle \simeq \left|\Phi\left(\mathbf{k}_f\right)\right\rangle .$ Therefore, the relevant information for the elastic scattering cross section are contained in the S-matrix elements,

$$S_{fi} \equiv \left\langle \Phi\left(\mathbf{k}_f\right)\right| S \left|\Phi(\mathbf{k}_i)\right\rangle = \left\langle \Phi\left(\mathbf{k}_f\right) \middle| \Phi^{\text{out}}(\mathbf{k}_i)\right\rangle . \tag{4.103}$$

Above, we use the simplified notation: $\left|\Phi^{\text{out}}(\mathbf{k}_i)\right\rangle = S\left|\Phi^{\text{in}}(\mathbf{k}_i)\right\rangle \rightarrow S\left|\Phi(\mathbf{k}_i)\right\rangle .$

In Eq. (4.103), $\left|\Phi\left(\mathbf{k}_f\right)\right\rangle$ and $\left|\Phi^{\text{out}}(\mathbf{k}_i)\right\rangle$ are asymptotic forms of exact states when the measurement takes place, i.e. $t\rightarrow +\infty$. However, since H is hermitian, the scalar product between these states - and hence S_{fi} - is conserved along the collision. Therefore, the S-matrix will have the same value if evaluated at an arbitrary time t_0. For this purpose, it is necessary to use the full propagator to take the asymptotic states from $t = \infty$ to $t = t_0$. According to Eqs. (4.95) and (4.99), this propagation yields

$$\left|\Phi^{\text{out}}(\mathbf{k}_i)\right\rangle \rightarrow \left|\Psi^{(+)}\left(\mathbf{k}_i\right)\right\rangle \quad \text{and} \quad \left|\Phi\left(\mathbf{k}_f\right)\right\rangle \rightarrow \left|\Psi^{(-)}\left(\mathbf{k}_f\right)\right\rangle .$$

Eq. (4.103) then becomes

$$S_{fi} = \left\langle \Psi^{(-)}\left(\mathbf{k}_f\right) \middle| \Psi^{(+)}\left(\mathbf{k}_i\right)\right\rangle \tag{4.104}$$

or, using Eq. (4.100),

$$S_{fi} = \left\langle \Phi\left(\mathbf{k}_f\right)\right| \Omega^{(-)\dagger}\, \Omega^{(+)} \left|\Phi\left(\mathbf{k}_i\right)\right\rangle . \tag{4.105}$$

Since Eq. (4.105) holds for any pair of states, we can write

$$S = \Omega^{(-)\dagger}\, \Omega^{(+)} ,$$

as given by Eq. (4.77).

The time-dependent and time-independent approaches to the S-matrix are equivalent. Even if one uses the time-dependent definition of the Møller operators (Eq. (4.101)) in Eq. (4.77), the S-matrix turns out to be time-independent. One gets

$$S = \lim_{\substack{t_1\rightarrow -\infty \\ t_2\rightarrow \infty}} \left[e^{iH_0 t_2/\hbar}\, e^{-iH(t_2-t_1)/\hbar}\, e^{-iH_0 t_1/\hbar} \right]. \tag{4.106}$$

4.3.2 *Energy conservation*

We now prove the important property:

$$[H_0, S]_- = 0\,. \tag{4.107}$$

This equation implies that the energy is conserved in the scattering process.

The starting point is the so-called *interwining relation*

$$H\Omega^{(\pm)} = \Omega^{(\pm)} H_0. \tag{4.108}$$

To prove this relation, we write $\Omega^{(\pm)}$ as in Eq. (4.101) and calculate the quantity

$$\begin{aligned} e^{iHt}\Omega^{(\pm)} &= e^{iHt} \lim_{t'\to\pm\infty} \left[e^{-\frac{i}{\hbar}H\,t'}\, e^{\frac{i}{\hbar}H_0\,t'} \right] \\ &= \lim_{t'\to\pm\infty} \left[e^{-\frac{i}{\hbar}H\,(t'-t)}\, e^{\frac{i}{\hbar}H_0\,t'} \right]. \end{aligned} \tag{4.109}$$

Calling $\tau = t' - t$, we can write

$$e^{iHt}\Omega^{(\pm)} = \lim_{\tau\to\pm\infty} \left[e^{-\frac{i}{\hbar}H\,\tau}\, e^{\frac{i}{\hbar}H_0 \tau} \right] e^{\frac{i}{\hbar}H_0\,t}\,,$$

or, according to Eq. (4.101),

$$e^{iHt}\Omega^{(\pm)} = \Omega^{(\pm)} e^{\frac{i}{\hbar}H_0\,t}\,.$$

Taking derivative of both sides with respect to t and setting $t = 0$, we have the proof of Eq. (4.108).

Let us now evaluate the commutator $[H_0, S]_-$. Using Eq. (4.77), we can write

$$[H_0, S]_- = H_0 S - \Omega^{(-)\dagger}\,\Omega^{(+)} H_0.$$

Applying the interwining relation, we can replace $\Omega^{(+)}H_0 = H\Omega^{(+)}$ and the above equation takes the form

$$[H_0, S]_- = H_0 S - \Omega^{(-)\dagger} H\,\Omega^{(+)}\,. \tag{4.110}$$

Since $H^\dagger = H$ and $H_0^\dagger = H_0$, we can write $\Omega^{(-)\dagger}H = (H\Omega^{(-)})^\dagger$ or, using again the interwining relation,

$$\Omega^{(-)\dagger}H = (H\Omega^{(-)})^\dagger = (\Omega^{(-)}H_0)^\dagger = H_0\Omega^{(-)\dagger}\,.$$

Inserting this result into Eq. (4.110), we obtain

$$[H_0, S]_- = H_0 S - H_0\Omega^{(-)\dagger}\Omega^{(+)} = H_0 S - H_0 S = 0\,.$$

4.3.3 *Time-reversal*

Although in actual physical system time develops towards increasing values, it is interesting to discuss states for which time flows backwards (see e.g. [Messiah (1999)]). Let us first consider classical solutions of the equations of motion for a standard Lagrangian, which depends on velocities quadratically. In this case, the Lagrangian has the time-reversal symmetry, so that for each solution $\{\mathbf{r}(t);\, \dot{\mathbf{r}}(t)\}$, there exists a time reversed solution $\{\mathbf{r}_{\mathbb{T}}(t);\, \dot{\mathbf{r}}_{\mathbb{T}}(t)\}$. These solutions are connected by the relations

$$\mathbf{r}_{\mathbb{T}}(t) = \mathbf{r}(-t) \quad \text{and} \quad \dot{\mathbf{r}}_{\mathbb{T}}(t) = -\dot{\mathbf{r}}(-t)\,. \tag{4.111}$$

Similar properties are found in Quantum Mechanics with hermitian Hamiltonian, which have time-reversal symmetry. Let us consider the Schrödinger equation

$$H\psi(\mathbf{r},t) = i\hbar\,\frac{\partial\psi(\mathbf{r},t)}{\partial t}.$$

We now take its hermitian conjugate and make the replacement $t \to -t$. For hermitian Hamiltonians, we recover the original equation. Therefore, this procedure gives rise to the new solution

$$\psi_{\mathbb{T}}(\mathbf{r},t) = \psi^*(\mathbf{r},-t)\,. \tag{4.112}$$

The transformation $\psi \to \psi_{\mathbb{T}}$ is performed by the time-reversal operator[8],

$$\mathbb{T}\,\psi(\mathbf{r},t) = \psi_{\mathbb{T}}(\mathbf{r},t) = \psi^*(\mathbf{r},-t). \tag{4.113}$$

The operator $\mathbb{T}$ is fully determined (except for a meaningless global phase) by its effects on the position and momentum operators,

$$\mathbb{T}\mathbf{r}\mathbb{T}^\dagger = \mathbf{r} \quad \text{and} \quad \mathbb{T}\mathbf{p}\mathbb{T}^\dagger = -\mathbf{p}\,. \tag{4.114}$$

Furthermore, since this operator expresses a symmetry of the Hamiltonian, it must be unitary or anti-unitary. Therefore, we can write

$$\mathbb{T}H\mathbb{T}^\dagger = H \quad \text{and} \quad \mathbb{T}\mathbb{T}^\dagger = \mathbb{T}^\dagger\mathbb{T} = \mathbf{1}. \tag{4.115}$$

Applying the time reversal transformation on the commutator $[r_i, p_i] = i\hbar$ and using Eqs. (4.114) and (4.115) we get

$$\mathbb{T}\,[r_i,p_i]\,\mathbb{T}^\dagger = \mathbb{T}\,(i\hbar)\,\mathbb{T}^\dagger = -i\hbar\,, \tag{4.116}$$

which indicates that $\mathbb{T}$ is an anti-unitary operator.

[8] This operator is discussed in standard Quantum Mechanics text books like [Messiah (1999)]. A nice discussion of this operator aiming at scattering theory can be found in [Taylor (2006)].

An important consequence of the above discussion is the time-reversal transformation of the time-development operators

$$\mathbb{T}\left[e^{-iHt/\hbar}\right]\mathbb{T}^{\dagger} = e^{iHt/\hbar} \tag{4.117}$$

$$\mathbb{T}\left[e^{-iH_0 t/\hbar}\right]\mathbb{T}^{\dagger} = e^{iH_0 t/\hbar}\,. \tag{4.118}$$

For stationary scattering states, Eq. (4.112) becomes

$$[\psi(\mathbf{r})]_{\mathbb{T}} = \psi^*(\mathbf{r})\,. \tag{4.119}$$

In particular, the time-reversal conjugate of $\phi(\mathbf{k};\mathbf{r})$ is

$$[\phi(\mathbf{k};\mathbf{r})]_{\mathbb{T}} = \left(\frac{1}{(2\pi)^{3/2}}e^{i\mathbf{k}\cdot\mathbf{r}}\right)^* = \phi(-\mathbf{k};\mathbf{r})\,. \tag{4.120}$$

We now consider the time-reversal transformation on the Møller operators and the scattering states. Using the representation of Eq. (4.101), the invariance of H and H_0, and the transformation law of the time-development operators associated with H and H_0 (Eqs. (4.117) and (4.118)), we get

$$\mathbb{T}\Omega^{(\pm)}\mathbb{T}^{\dagger} = \Omega^{(\mp)}. \tag{4.121}$$

The time-reversal conjugates of the scattering states are obtained from Eq. (4.52) with the help of the above equations

$$\begin{aligned}|\psi^{(\pm)}(\mathbf{k})\rangle_{\mathbb{T}} &= \mathbb{T}\,|\psi^{(\pm)}(\mathbf{k})\rangle = \mathbb{T}\Omega^{(\pm)}\,|\phi(\mathbf{k})\rangle \\ &= \mathbb{T}\Omega^{(\pm)}\mathbb{T}^{\dagger}\mathbb{T}\,|\phi(\mathbf{k})\rangle = \Omega^{(\mp)}\,|\phi(-\mathbf{k})\rangle = \left|\psi^{(\mp)}(-\mathbf{k})\right\rangle.\end{aligned} \tag{4.122}$$

Therefore, we can write

$$[\psi^{(\pm)}(\mathbf{k};\mathbf{r})]^* = \psi^{(\mp)}(-\mathbf{k};\mathbf{r}). \tag{4.123}$$

Eq. (4.123) can be used to derive the asymptotic form of $\psi^{(-)}(\mathbf{k};\mathbf{r})$. Since

$$\psi^{(-)}(\mathbf{k};\mathbf{r}) = [\psi^{(+)}(-\mathbf{k};\mathbf{r})]^*\,, \tag{4.124}$$

Eq. (2.3) yields

$$\psi^{(-)}(\mathbf{k};\mathbf{r}) \to A\left[e^{i\mathbf{k}\cdot\mathbf{r}} + f^*(\pi-\theta)\,\frac{e^{-ikr}}{r}\right]. \tag{4.125}$$

Note that the transformation $\mathbf{k}\to-\mathbf{k}$ leads to the changes $\theta\to\pi-\theta$ in the argument of the above scattering amplitude.

Using Eq. (4.122), in Eq. (4.79), we obtain the S-matrix in terms of outgoing waves and their time-reversal conjugates

$$S_{\mathbf{k}',\mathbf{k}} = \left\langle\,[\psi^{(+)}(-\mathbf{k}')]_{\mathbb{T}}\,\middle|\,\psi^{(+)}(\mathbf{k})\,\right\rangle. \tag{4.126}$$

4.4 Scattering from non-local separable potentials

In this section we discuss an exactly soluble problem: scattering from separable potentials.

When one evaluates matrix-elements in the coordinate representation, the operators in Quantum Mechanics fall into two categories: local and non-local. Local operators are those whose matrix-elements have the form

$$U(\mathbf{r},\mathbf{r}') \equiv \langle \mathbf{r}|\, U \,|\mathbf{r}'\rangle = U(\mathbf{r})\,\delta(\mathbf{r}-\mathbf{r}'). \tag{4.127}$$

Non-local operators are those which do not have this property. Most of the operators appearing in simple problems in quantum mechanics are local. In this category are the momentum and kinetic energy operators and the majority of the potentials describing interactions of physical systems.

However, there are situations where non-local potentials emerge naturally. One example is the Hartree-Fock approximation for systems of many interacting identical fermions. In this case, the single particle orbitals satisfy a Schrödinger equation where the potential is the sum of a local term and a non-local term. The latter arises from the exchange of identical particles. Another example is the potential describing the scattering of clusters of identical particles (see chapter 6). In this case, the wave function for the relative projectile-target motion satisfies also an equation with a non-local potential, arising from exchange. This can be seen explicitly in approximate treatments of the scattering problem, like the Resonating Group Method (for a detailed description of this method see, e.g., [Friedrich (1981)]).

In general, the matrix-elements of the potential can be written

$$V(\mathbf{r},\mathbf{r}') = V^{\mathrm{L}}(\mathbf{r})\,\delta(\mathbf{r}-\mathbf{r}') + V^{\mathrm{NL}}(\mathbf{r},\mathbf{r}'), \tag{4.128}$$

where V^{L} and V^{NL} stand respectively for the local and non-local parts of the potential. Accordingly, the stationary Schrödinger's equation takes the more complicated integro-differencial form

$$\left[-\frac{\hbar^2}{2\mu}\nabla^2_{\mathbf{r}} + V^{\mathrm{L}}(\mathbf{r})\right]\psi(\mathbf{r}) + \int d^3\mathbf{r}'\; V^{\mathrm{NL}}(\mathbf{r},\mathbf{r}')\,\psi(\mathbf{r}') = E\,\psi(\mathbf{r})\,. \tag{4.129}$$

Although this equation looks more complicated than those for local potentials, the Scattering problem can be easily solved in the case of a strictly non-local hermitian potential ($V^{\mathrm{L}}(r) = 0$) that can be approximated by the *separable form*

$$V(\mathbf{r},\mathbf{r}') = v(\mathbf{r})\,v(\mathbf{r}'), \tag{4.130}$$

with v representing a real function of the position vector. Using Dirac's notation for the operator V,

$$V = |v\rangle\,\langle v|, \tag{4.131}$$

and using Eq. (4.45), one gets the scattering amplitude

$$f(\theta) = -2\pi^2\left(\frac{2\mu}{\hbar^2}\right)\Big\langle\phi\,(\mathbf{k}')\,\Big|v\Big\rangle\,\Big\langle v|\psi^{(+)}\,(\mathbf{k})\,\Big\rangle. \tag{4.132}$$

The first scalar product at the RHS of Eq. (4.132),

$$\langle\phi\,(\mathbf{k}')\,|v\rangle = \frac{1}{(2\pi)^{3/2}}\int d^3\mathbf{r}\,e^{-i\mathbf{k}'\cdot\mathbf{r}}\,v(\mathbf{r})\,, \tag{4.133}$$

is simply the Fourier transform of the function $v(\mathbf{r})$. To obtain the scalar product $\langle v|\psi^{(+)}\,(\mathbf{k})\rangle$, we write the Lippmann-Schwinger equation for $|\psi^{(+)}\,(\mathbf{k})\rangle$ (Eq. (4.38)),

$$|\psi^{(+)}\,(\mathbf{k})\rangle = |\phi\,(\mathbf{k})\,\rangle \;+\; G_0^{(+)}(E)|v\rangle\,\langle v|\psi^{(+)}\,(\mathbf{k})\rangle\,, \tag{4.134}$$

and take scalar product of both sides with $\langle v|$. Eq. (4.134) becomes

$$\langle v|\psi^{(+)}\,(\mathbf{k})\rangle = \langle v|\phi\,(\mathbf{k})\rangle \;+\; \langle\,v|G_0^{(+)}(E)|v\rangle\,\langle v|\psi^{(+)}\,(\mathbf{k})\rangle.$$

Solving this equation for $\langle v|\psi^{(+)}\,(\mathbf{k})\rangle$ one obtains

$$\langle v|\psi^{(+)}\,(\mathbf{k})\rangle = \frac{\langle v|\phi\,(\mathbf{k})\rangle}{1-\langle\,v|\,G_0^{(+)}(E)\,|v\rangle}. \tag{4.135}$$

The expectation value $\langle\,v|\,G_0^{(+)}(E)\,|v\rangle$ appearing in the above equation can easily be calculated. Using the explicit form of the free particle's Green's function (Eq. (4.37)), one gets

$$\langle\,v|\,G_0^{(+)}(E)\,|v\rangle = -\frac{\mu}{2\pi\hbar^2}\int\!\!\int d^3\mathbf{r}\;d^3\mathbf{r}'\;v^*(\mathbf{r})\,\frac{e^{\,ik\,|\mathbf{r}-\mathbf{r}'|}}{|\mathbf{r}-\mathbf{r}'|}\,v(\mathbf{r}').$$

To evaluate the scattering amplitude, we Insert Eq. (4.135) into Eq. (4.132). One then obtains the closed expression,

$$f(\theta) = -2\pi^2\left(\frac{2\mu}{\hbar^2}\right)\frac{\langle\phi\,(\mathbf{k}')\,|v\rangle\,\langle v|\phi\,(\mathbf{k})\rangle}{1-\langle\,v|\;G_0^{(+)}(E)\;|v\rangle}, \tag{4.136}$$

and the cross section is given by the usual expression

$$\frac{d\sigma_{\mathrm{el}}(\theta)}{d\Omega} = |f(\theta)|^2\;.$$

It is straightforward to generalize the above procedure for non-local potentials that can be represented by a sum of n separable terms, namely,

$$V = \sum_{\lambda=1}^{n} |v_\lambda\rangle \, \langle v_\lambda| \, . \tag{4.137}$$

In this case, the scattering amplitude is

$$f(\theta) = -2\pi^2 \left(\frac{2\mu}{\hbar^2}\right) \sum_{\lambda=1}^{n} \langle \phi\,(\mathbf{k}')\,|v_\lambda\rangle \, \langle v_\lambda|\psi^{(+)}\,(\mathbf{k})\rangle. \tag{4.138}$$

The scalar products $\langle \phi\,(\mathbf{k}')\,|v_\lambda\rangle$ are trivial Fourier transforms of the functions $v_\lambda(\mathbf{r})$, as in Eq. (4.133), and the factors $\langle v_\lambda|\psi^{(+)}\,(\mathbf{k})\rangle$ can be obtained similarly to the previous case of separable potentials with a single term. One takes scalar products of the Lippmann-Schwinger equation by each of the factors $\langle v_\lambda|$. One gets the equations

$$\langle v_\lambda|\psi^{(+)}\,(\mathbf{k})\rangle = \langle v_\lambda|\phi\,(\mathbf{k})\rangle + \sum_{\lambda'=1}^{n} \langle\, v_\lambda|G^{(+)}(E)|v_{\lambda'}\rangle \, \langle v_{\lambda'}|\psi^{(+)}\,(\mathbf{k})\rangle \, , \quad \lambda = 1, ..., n \, .$$

This corresponds to a system of coupled linear equations with the general form

$$\begin{aligned} A_{11}X_1 + A_{12}X_2 + \ldots\ldots + A_{1n}X_n &= Y_1 \\ A_{21}X_1 + A_{22}X_2 + \ldots\ldots + A_{2n}X_n &= Y_2 \\ \ldots\ldots\ldots\ldots\ldots\ldots\ldots\ldots\ldots\ldots\ldots\ldots\ldots\ldots \\ A_{n1}X_1 + A_{n2}X_2 + \ldots\ldots + A_{nn}X_n &= Y_n, \end{aligned} \tag{4.139}$$

where

$$\begin{aligned} X_\lambda &= \langle v_\lambda|\psi^{(+)}\,(\mathbf{k})\rangle \, , \\ A_{\lambda\lambda'} &= \delta_{\lambda\lambda'} - \langle\, v_\lambda|G^{(+)}(E)|v_{\lambda'}\rangle \, , \\ Y_\lambda &= \langle v_\lambda|\phi\,(\mathbf{k})\rangle \, . \end{aligned}$$

This system can be easily be solved by standard methods. Using the solutions $\langle v_\lambda|\psi_{\mathbf{k}}^{(+)}\rangle = X_\lambda$ in Eq. (4.138) we get the scattering amplitude and then the cross-section.

Scattering from non-local separable potentials are encountered in several situations. One example in many-body scattering can be found in section 10.2.2.

4.5 Scattering from the sum of two potentials

In section 4.1 we wrote the Hamiltonian as $H = H_0 + V$, with H_0 being the kinetic energy operator, K, and V the total potential. We then showed that the exact solutions $|\psi^{(\pm)}(\mathbf{k})\rangle$ and the plane waves $|\phi(\mathbf{k})\rangle$ are related by Lippmann-Schwinger equations. In some situations, it is convenient to split the total potential into two parts,

$$V = V_1 + V_2, \tag{4.140}$$

and derive Lippmann-Schwinger equations involving the states distorted by the potential V_1 alone, which will be denoted by $|\chi^{(\pm)}(\mathbf{k})\rangle$. They are scattering states when one sets $V_2 = 0$. Thus, they satisfy the Schrödinger equation

$$(E - H_1)\,|\chi^{(\pm)}(\mathbf{k})\rangle = 0\,, \tag{4.141}$$

with

$$H_1 = K + V_1. \tag{4.142}$$

In the $r \to \infty$ limit, the distorted states have the asymptotic forms

$$\chi^{(+)}(\mathbf{k};\mathbf{r}) \to A\,\left(e^{i\mathbf{k}\cdot\mathbf{r}} + f_1(\theta)\,\frac{e^{ikr}}{r}\right), \tag{4.143}$$

$$\chi^{(-)}(\mathbf{k};\mathbf{r}) \to A\,\left(e^{i\mathbf{k}\cdot\mathbf{r}} + f_1^*(\pi-\theta)\,\frac{e^{-ikr}}{r}\right). \tag{4.144}$$

Above, $f_1(\theta)$ is the scattering amplitude associated with the potential V_1.

The distorted states satisfy the Lippmann-Schwinger equations

$$|\chi^{(\pm)}(\mathbf{k})\rangle = |\phi(\mathbf{k})\rangle + G_0^{(\pm)}(E_k)\,V_1\,|\chi^{(\pm)}(\mathbf{k})\rangle \tag{4.145}$$

and

$$|\chi^{(\pm)}(\mathbf{k})\rangle = \left[1 + G_1^{(\pm)}(E_k)\,V_1\right]\,|\phi(\mathbf{k})\rangle\,, \tag{4.146}$$

where

$$G_1^{(\pm)}(E) = \frac{1}{E - H_1 \pm i\epsilon}, \tag{4.147}$$

and they obey the completeness relation

$$\int |\chi^{(\pm)}(\mathbf{q})\rangle\, d^3q\, \langle\chi^{(\pm)}(\mathbf{q})| + \sum_n |\chi_n\rangle\,\langle\chi_n| = \mathbf{1}. \tag{4.148}$$

Above, $|\chi_n\rangle$ stands for the bound states that the potential V_1 may possibly have.

Applying the procedures of section 4.1 to the Schrödinger equation with the full potential,

$$(E - H)\,|\psi^{(\pm)}(\mathbf{k})\rangle \equiv (E - H_1 - V_2)\,\Big|\psi^{(\pm)}(\mathbf{k})\Big\rangle = 0,$$

one easily obtains the Lippmann-Schwinger equations for $|\psi^{(\pm)}(\mathbf{k})\rangle$ (see problem 2),

$$|\psi^{(\pm)}(\mathbf{k})\rangle = |\chi^{(\pm)}(\mathbf{k})\rangle + G_1^{(\pm)}(E_k)\,V_2\,|\psi^{(\pm)}(\mathbf{k})\rangle \tag{4.149}$$

and

$$|\psi^{(\pm)}(\mathbf{k})\rangle = [1 + G^{(\pm)}(E_k)\,V_2]\,|\chi^{(\pm)}(\mathbf{k})\rangle\,. \tag{4.150}$$

4.5.1 *The Gell-Mann Goldberger relations*

It is useful to introduce the T-matrix associated with the potential V_1 alone (setting $V_2 = 0$). It is given by Eq. (4.63), which in this case becomes

$$T^{(1)}_{\mathbf{k}',\mathbf{k}} = \langle\phi(\mathbf{k}')\,|V_1\,|\chi^{(+)}(\mathbf{k})\,\rangle. \tag{4.151}$$

This matrix is related to the T-matrix for the total potential, $T_{\mathbf{k}',\mathbf{k}}$, through the *Gell-Mann Goldberger relation*, also called the *two-potential formula*,

$$T_{\mathbf{k}',\mathbf{k}} = T^{(1)}_{\mathbf{k}',\mathbf{k}} + \langle\chi^{(-)}(\mathbf{k}')|\,V_2\,|\psi^{(+)}(\mathbf{k})\rangle\,. \tag{4.152}$$

This expression will be proved below.

As a first step, we rewrite the Lippmann-Schwinger equations involving $|\phi(\mathbf{k})\rangle$ and $|\psi^{(+)}(\mathbf{k})\rangle$,

$$|\psi^{(+)}(\mathbf{k})\rangle = |\phi(\mathbf{k})\rangle + G_0^{(+)}(E_k)\,[V_1 + V_2]\,|\psi^{(+)}(\mathbf{k})\rangle \tag{4.153}$$

and

$$|\psi^{(+)}(\mathbf{k})\rangle = \{1 + G^{(+)}(E_k)\,[V_1 + V_2]\}\,|\phi(\mathbf{k})\rangle\,. \tag{4.154}$$

Using the property[9] $G_0^{(-)\dagger} = G_0^{(+)}$, the hermitian conjugate of Eq. (4.145) can be written

$$\langle\phi(\mathbf{k}')| = \langle\chi^{(-)}(\mathbf{k}')| - \langle\chi^{(-)}(\mathbf{k}')|\,V_1\,G_0^{(+)}\,. \tag{4.155}$$

Inserting Eq. (4.155) into the definition of the T-matrix ,

$$T_{\mathbf{k}',\mathbf{k}} \equiv \Big\langle\phi(\mathbf{k}')\Big|\,(V_1 + V_2)\,\Big|\psi^{(+)}(\mathbf{k})\,\Big\rangle, \tag{4.156}$$

[9]Since in this proof we are dealing with matrix-elements on-the-energy-shell, $E_k = E_{k'}$, the energy labels become superfluous.

we get

$$T_{\mathbf{k}',\mathbf{k}} = \langle \chi^{(-)}(\mathbf{k}')| (V_1 + V_2) |\psi^{(+)}(\mathbf{k})\rangle - \langle \chi^{(-)}(\mathbf{k}')| V_1 \left\{ G_0^{(+)} (V_1 + V_2) \left|\psi^{(+)}(\mathbf{k})\right\rangle \right\}. \tag{4.157}$$

Now, we use Eq. (4.153) to write

$$G_0^{(+)} [V_1 + V_2] |\psi^{(+)}(\mathbf{k})\rangle = |\psi^{(+)}(\mathbf{k})\rangle - \left|\phi(\mathbf{k})\right\rangle, \tag{4.158}$$

and replace the term within curly brackets in Eq. (4.157) by the above result. We obtain

$$T_{\mathbf{k}',\mathbf{k}} = \langle \chi^{(-)}(\mathbf{k}') |V_1 + V_2 |\psi^{(+)}(\mathbf{k})\rangle - \langle \chi^{(-)}(\mathbf{k}') |V_1 |\psi^{(+)}(\mathbf{k})\rangle + \langle \chi^{(-)}(\mathbf{k}') |V_1 |\phi(\mathbf{k})\rangle,$$

or

$$T_{\mathbf{k}',\mathbf{k}} = \left\langle \chi^{(-)}(\mathbf{k}') \middle| V_1 \middle| \phi(\mathbf{k}) \right\rangle + \left\langle \chi^{(-)}(\mathbf{k}') \middle| V_2 \middle| \psi^{(+)}(\mathbf{k}) \right\rangle. \tag{4.159}$$

Now we re-write the first term at the RHS of Eq. (4.159) in terms of the Møller operator associated with H_1 and use Eq. (4.71). Eq. (4.159) then becomes

$$\left\langle \chi^{(-)}(\mathbf{k}') \middle| V_1 \middle| \phi(\mathbf{k}) \right\rangle = \left\langle \phi(\mathbf{k}') \middle| \Omega^{(-)^\dagger} V_1 \middle| \phi(\mathbf{k}) \right\rangle = T^{(1)}_{\mathbf{k}',\mathbf{k}}. \tag{4.160}$$

Inserting the above result into Eq. (4.159), we get the proof of the Gell-Mann Goldberger relation (Eq. (4.152)).

4.5.2 *The Distorted Wave series*

The Gell-Mann Goldberger relations can be written as power series. Applying recursively Eq. (4.149) on Eq. (4.152), we get

$$T_{\mathbf{k}',\mathbf{k}} = \left\langle \phi(\mathbf{k}') \middle| V_1 \middle| \chi^{(+)}(\mathbf{k}) \right\rangle + \left\langle \chi^{(-)}(\mathbf{k}') \middle| V_2 \sum_{m=0}^{\infty} \left(G_1^{(+)}(E_k)\, V_2 \right)^m \middle| \chi^{(+)}(\mathbf{k}) \right\rangle. \tag{4.161}$$

If one uses the spectral representation[10] of $G_1^{(+)}(E_k)$,

$$G_1^{(+)}(E_k) = \int d^3\mathbf{q}\, \frac{\left|\chi^{(+)}(\mathbf{q})\right\rangle \left\langle \chi^{(+)}(\mathbf{q})\right|}{E_k - E_q + i\epsilon}, \tag{4.162}$$

[10]For simplicity, we are neglecting the possibility that the potential V_1 have bound states. Bound states would lead to extra terms in the expansion, arising from the discrete summation of Eq. (4.148).

Eq. (4.161) becomes

$$T_{\mathbf{k}',\mathbf{k}} = \left\langle \phi\left(\mathbf{k}'\right) \middle| V_1 \middle| \chi^{(+)}\left(\mathbf{k}\right) \right\rangle + \langle \chi^{(-)}\left(\mathbf{k}'\right) | \, V_2 \, | \chi^{(+)}\left(\mathbf{k}\right) \rangle$$
$$+ \int d^3\mathbf{q} \, \frac{\langle \chi^{(-)}\left(\mathbf{k}'\right) | \, V_2 \, | \chi^{(+)}\left(\mathbf{q}\right) \rangle \, \langle \chi^{(+)}\left(\mathbf{q}\right) | \, V_2 \, | \chi^{(+)}\left(\mathbf{k}\right) \rangle}{E_k - E_q + i\epsilon}$$
$$+ \text{higher order terms.} \tag{4.163}$$

This is known as the *Distorted Wave series.* Truncating the expansion of Eq. (4.161) after $m = 0$, (i.e. keeping only the first line of Eq. (4.163)), we get the *Distorted Wave Born Approximation* (DWBA). This approximation will be considered in detail in chapter 5 and in chapter 11, in the cases of potential scattering and many-body scattering, respectively.

4.6 Partial-wave expansions

In this section we carry out partial-wave expansion of the Lippmann-Schwinger equations. As a starting point, we introduce the orthonormal sets of states $\{|\phi(Elm)\rangle\}$ and $\{|\psi(Elm)\rangle\}$, which are eigenstates of the operators L^2 and L_z, with eigenvalues $\hbar^2 l(l+1)$ and $\hbar m$, respectively. They satisfy the Schrödinger equations

$$(E - H_0) \, |\phi(Elm)\rangle = 0 \tag{4.164}$$
$$(E - H) \, |\psi(Elm)\rangle = 0 \tag{4.165}$$

and are normalized as to satisfy the relations

$$\langle \phi\left(E'l'm'\right) | \phi(Elm)\rangle = \delta_{ll'} \, \delta_{mm'} \, \delta(E - E') \tag{4.166}$$
$$\langle \psi\left(E'l'm'\right) | \, \psi\left(Elm\right)\rangle = \delta_{ll'} \, \delta_{mm'} \, \delta(E - E') \, . \tag{4.167}$$

In the coordinate representation, the wave functions $\phi(Elm; \mathbf{r})$ and $\psi(Elm; \mathbf{r})$ are products of Spherical Harmonics $Y_{lm}(\hat{\mathbf{r}})$ with appropriate radial wave functions. The completeness relation for the states $|\phi(Elm)\rangle$ has the form

$$\sum_{lm} \int |\phi(Elm)\rangle \, dE \, \langle \phi(Elm)| = \mathbf{1}. \tag{4.168}$$

The solutions of the Schrödinger equation with the Hamiltonian H, $\psi(Elm; \mathbf{r})$, depend on the particular potential of the problem while the free spherical waves $\phi(Elm; \mathbf{r})$ are completely determined (see chapter 2), except for an arbitrary global phase. They can be written as

$$\phi(Elm; \mathbf{r}) = A_l \, \frac{\hat{\jmath}_l(kr)}{kr} Y_{lm}(\hat{\mathbf{r}}), \tag{4.169}$$

where $k = \sqrt{2\mu E}/\hbar$ and A_l is a normalization constant. Using the orthonormality of the Spherical Harmonics and the radial integral of Ricatti-Bessel functions (Eq. (2.26)), Eq. (4.166) yields

$$|A_l|^2 = \frac{2\mu k}{\pi\hbar^2} \quad \rightarrow \quad A_l = i^l \left(\frac{2\mu k}{\pi\hbar^2}\right)^{1/2}, \tag{4.170}$$

where the arbitrary phase of A_l was chosen to be i^l. We can use an analogous procedure for the wave functions $\psi(Elm;\mathbf{r})$. We write

$$\psi(Elm;\mathbf{r}) = i^l \left(\frac{2\mu k}{\pi\hbar^2}\right)^{1/2} \frac{u_l(k,r)}{kr}\, Y_{lm}(\hat{\mathbf{r}}) \tag{4.171}$$

$$= i^l \left(\frac{2\mu k}{\pi\hbar^2}\right)^{1/2} \frac{\bar{u}_l(k,r)}{kr}\, Y_{lm}(\hat{\mathbf{r}})\; e^{i\delta_l}, \tag{4.172}$$

were $u_l(k,r)$ and $\bar{u}_l(k,r)$ are solutions of the radial equation with different choices of phase, as discussed in chapter 2. As will become clear, the inclusion of the arbitrary phase $e^{i\delta_l}$ leads to simpler expressions. Eq. (4.167) requires that the radial wave functions be normalized as,

$$\int_0^\infty u_l^*(k,r)\, u_l(k',r)\, dr = \frac{\pi}{2}\,\delta(k-k'). \tag{4.173}$$

Frequently, it is convenient to expand the plane waves and the scattering states in terms of the corresponding orthonormal angular momentum eigenstates, as

$$\phi(\mathbf{k};\mathbf{r}) = \sum_{l',m'} \int dE'\; C_{E'l'm'}(\mathbf{k})\; \phi(E'l'm';\mathbf{r}), \tag{4.174}$$

$$\psi^{(+)}(\mathbf{k};\mathbf{r}) = \sum_{l',m'} \int dE'\; C_{E'l'm'}(\mathbf{k})\; \psi(E'l'm';\mathbf{r}). \tag{4.175}$$

The expansion coefficients,

$$C_{E'l'm'}(\mathbf{k}) = \langle \phi(E'l'm') \,|\, \phi(\mathbf{k}) \rangle\,, \tag{4.176}$$

can be evaluated explicitly with the help of Eqs. (2.44a), (4.169), (4.170) and the orthogonality relations of the Ricatti-Bessel functions (Eq. (2.26)). One gets,

$$C_{E'l'm'}(\mathbf{k}) = \frac{\hbar}{\sqrt{\mu k}}\; Y^*_{l'm'}(\hat{\mathbf{k}})\; \delta(E-E'), \tag{4.177}$$

where $E = \hbar^2 k^2/2\mu$. The plane waves and the scattering states are then given by

$$\phi(\mathbf{k};\mathbf{r}) = \frac{\hbar}{\sqrt{\mu k}} \sum_{l,m} Y^*_{lm}(\hat{\mathbf{k}})\; \phi(Elm;\mathbf{r}) \tag{4.178a}$$

$$= \frac{1}{(2\pi)^{3/2}} \sum_{lm} 4\pi\, Y^*_{lm}(\hat{\mathbf{k}})\; Y_{lm}(\hat{\mathbf{r}})\; i^l\, \frac{\hat{j}_l(kr)}{kr} \tag{4.178b}$$

and

$$\psi^{(+)}(\mathbf{k},\mathbf{r}) = \frac{\hbar}{\sqrt{\mu k}} \sum_{l,m} Y^*_{lm}(\hat{\mathbf{k}})\ \psi(Elm;\mathbf{r}). \tag{4.179}$$

Using the explicit form of $\psi(Elm;\mathbf{r})$, Eq. (4.179) becomes (Eq. (2.57a) with $A = (2\pi)^{-3/2}$)

$$\psi^{(+)}(\mathbf{k},\mathbf{r}) = \frac{1}{(2\pi)^{3/2}} \sum_{lm} 4\pi\ Y^*_{lm}(\hat{\mathbf{k}})\ Y_{lm}(\hat{\mathbf{r}})\ i^l\ \frac{u_l(k,r)}{kr} \tag{4.180a}$$

$$= \frac{1}{(2\pi)^{3/2}} \sum_{l} (2l+1)\ P_l(\cos\theta)\ i^l\ \frac{u_l(k,r)}{kr}. \tag{4.180b}$$

The partial-wave expansion for $\psi^{(-)}(\mathbf{k},\mathbf{r})$ is easily obtained from the above equations, through the relation $\psi^{(-)}(\mathbf{k},\mathbf{r}) = (\psi^{(+)}(-\mathbf{k},\mathbf{r}))^*$ (see section 4.3.3). Considering the property $Y_{lm}(-\hat{\mathbf{k}}) = (-)^l\ Y_{lm}(\hat{\mathbf{k}})$, we get[11]

$$\psi^{(-)}(\mathbf{k},\mathbf{r}) = \frac{1}{(2\pi)^{3/2}} \sum_{lm} 4\pi\ Y^*_{lm}(\hat{\mathbf{k}})\ Y_{lm}(\hat{\mathbf{r}})\ i^l\ \frac{u^*_l(k,r)}{kr} \tag{4.181a}$$

$$= \frac{1}{(2\pi)^{3/2}} \sum_{l} (2l+1)\ P_l(\cos\theta)\ i^l\ \frac{u^*_l(k,r)}{kr}. \tag{4.181b}$$

To derive the partial-wave projected Lippmann-Schwinger equation, we carry out the partial-wave expansion of $G_0^{(\pm)}(E;\mathbf{r},\mathbf{r}')$ as[12]

$$G_0^{(\pm)}(E;\mathbf{r},\mathbf{r}') = \frac{1}{rr'} \sum_{lm} Y^*_{lm}(\hat{\mathbf{r}})\ g_{0,l}^{(\pm)}(r,r')\ Y_{lm}(\hat{\mathbf{r}}'). \tag{4.182}$$

Since we are supposing that the potential is spherically symmetric, the Hamiltonian is rotationally invariant. Therefore, the Green's function is diagonal in the angular momentum representation. We then insert this expansion and the expansions for the wave functions (Eqs. (2.44a) and (2.57a) with $A = 1/(2\pi)^{3/2}$) into the Lippmann-Schwinger equation (Eq. (4.40)). Using the linear independence and the orthonormality of the spherical harmonics, we get

$$u_l(k,r) = \hat{\jmath}_l(kr) + \int dr'\ g_{0,l}^{(+)}(r,r')\, V(r')\ u_l(k,r'). \tag{4.183}$$

In terms of radial functions with the phase chosen as in Eq. (2.41a), the above equation becomes

$$e^{i\delta_l}\ \bar{u}_l(k,r) = \hat{\jmath}_l(kr) + \int dr'\ g_{0,l}^{(+)}(r,r')\, V(r')\, e^{i\delta_l}\ \bar{u}_l(k,r'). \tag{4.184}$$

[11] We recall that $u^*_l(k,r) = e^{-i\delta_l}\ \bar{u}_l(k,r)$, where $\bar{u}_l(k,r)$ is the real radial wave function, which has the asymptotic behavior $\bar{u}_l(k,r) \to \sin(kr - l\pi/2 + \delta_l)$ (see chapter 2).

[12] The use of the factor $1/rr'$ in the partial-wave expansion of the Green's functions leads to a simpler form of the Lippmann-Schwinger equations for $u_l^{(\pm)}(k,r)$.

Equations involving the Green's function $g_{0,l}^{(-)}$ could be obtained using the partial-wave expansion of $\psi^{(-)}(\mathbf{k};\mathbf{r})$ (Eq. (4.181a). One gets,

$$u_l^*(k,r) = \hat{\jmath}_l(kr) + \int dr'\, g_{0,l}^{(-)}(r,r')\, V(r')\, u_l^*(k,r') \tag{4.185}$$

or

$$e^{-i\delta_l}\, \bar{u}_l(k,r) = \hat{\jmath}_l(kr) + \int dr'\, g_{0,l}^{(-)}(r,r')\, V(r')\, e^{-i\delta_l}\, \bar{u}_l(k,r'). \tag{4.186}$$

However, these equations are not lineraly independent of those involving $g_{0,l}^{(+)}$, since they are simply their complex conjugates.

Equivalent equations can be obtained with the full Green's function. Making the expansion

$$G^{(\pm)}(E;\mathbf{r},\mathbf{r}') = \frac{1}{rr'} \sum_{lm} Y_{lm}^*(\hat{\mathbf{r}})\; g_l^{(\pm)}(r,r')\; Y_{lm}(\hat{\mathbf{r}}') \tag{4.187}$$

and following the same procedures, we obtain

$$u_l(k,r) = \hat{\jmath}_l(kr) + \int dr'\, g_l^{(+)}(r,r')\, V(r')\, \hat{\jmath}_l(kr'). \tag{4.188}$$

and

$$e^{i\delta_l}\, \bar{u}_l(k,r) = \hat{\jmath}_l(kr) + \int dr'\, g_l^{(+)}(r,r')\, V(r')\, \hat{\jmath}_l(kr'). \tag{4.189}$$

Similar equations can be obtained in terms of $g_l^{(-)}$.

4.6.1 *Partial-wave projection of the S- and the T-matrices*

It is also convenient to carry out the partial-wave expansion of the T-matrix between states on the same energy-shell. For this purpose, we start from its definition (Eq. (4.63)) and use the partial-wave expansions of $\psi^{(+)}(\mathbf{k})$ (Eq. (4.179)) and $\phi(\mathbf{k}')$ (Eq. (4.178a). With the help of Eqs. (4.169) and (4.171), we get

$$T_{\mathbf{k}',\mathbf{k}} \equiv \left\langle \phi(\mathbf{k}') \middle| V \middle| \psi^{(+)}(\mathbf{k}) \right\rangle = \frac{\hbar^2}{\mu k} \sum_{lm} Y_{lm}(\hat{\mathbf{k}}') Y_{lm}^*(\hat{\mathbf{k}})\, T_l(E_k), \tag{4.190}$$

where we have introduced the partial-wave component of the T-matrix between states on the same energy-shell,

$$T_l(E_k) = \langle \phi(Elm)|\, V\, |\psi\,(Elm)\rangle \tag{4.191a}$$

$$= \frac{k}{\pi E_k} \int_0^\infty dr\, \hat{\jmath}_l(kr)\, V(r)\, u_l(k,r) \tag{4.191b}$$

$$= \frac{k}{\pi E_k}\, e^{i\delta_l} \int_0^\infty dr\, \hat{\jmath}_l(kr)\, V(r)\, \bar{u}_l(k,r)\,. \tag{4.191c}$$

Using the addition theorem for the Spherical Harmonics,

$$\sum_m Y_{lm}(\mathbf{k}')Y^*_{lm}(\hat{\mathbf{k}}) = \frac{2l+1}{4\pi}\, P_l(\cos\theta),$$

Eq. (4.190) becomes

$$T_{\mathbf{k}',\mathbf{k}} = \sum_l (2l+1)\, P_l(\cos\theta)\left[\frac{\hbar^2}{4\pi\mu k}\, T_l(E_k)\right]. \tag{4.192}$$

A similar procedure can be followed in the case of the S-matrix. First, we write the scattering operator in the representation of the $|\phi(Elm)\rangle$ states and use the completeness relation of Eq. (4.168). We get

$$\begin{aligned} S_{\mathbf{k}',\mathbf{k}} = &\sum_{l_1m_1,l_2m_2} \int\!\!\int dE_1\, dE_2\, \langle\phi\,(\mathbf{k}')\,|\,\phi\,(E_1l_1m_1)\rangle \\ &\times \langle\phi\,(E_1l_1m_1)\,|\,S\,|\phi(E_2l_2m_2)\rangle\, \langle\phi(E_2l_2m_2)|\,\phi\,(\mathbf{k})\rangle\,. \end{aligned} \tag{4.193}$$

Since we are dealing with spherically symmetric potentials, the Hamiltonians H_0 and H are rotationally invariant. This implies that the S-operator is diagonal in the angular momentum representation. This becomes clear when one writes $S = \Omega^{(-)\dagger}\,\Omega^{(+)}$, with $\Omega^{(\pm)}$ given by Eq. (4.101), and realizes that S commutes with any component of the angular momentum operator. Furthermore, $\langle\phi\,(E_1l_1m_1)|\; S\; |\phi(E_2l_2m_2)\rangle$ must be independent of m-values, since S also commutes with the operators $J_\pm = J_x \pm iJ_y$, which change $m \to m \pm 1$. Considering that the S-operator also commutes with the Hamiltonian H_0 (see Eq. (4.107)), we can write

$$\langle\phi\,(E_1l_1m_1)|\; S\; |\phi(E_2l_2m_2)\rangle = S_{l_1}(E_1)\; \delta_{l_1l_2}\; \delta_{m_1m_2}\; \delta(E_1-E_2)\,. \tag{4.194}$$

Now we use Eqs. (4.176) and (4.177) to evaluate $\langle\phi(\mathbf{k}')|\,\phi\,(E_1l_1m_1)\rangle$ and $\langle\phi(E_2l_2m_2)|\,\phi\,(\mathbf{k})\rangle$. Inserting the results into Eq. (4.193), we obtain

$$S_{\mathbf{k}',\mathbf{k}} = \delta(E_{k'}-E_k)\,\frac{\hbar^2}{\mu k}\sum_{lm} Y_{lm}(\hat{\mathbf{k}}')\; Y^*_{lm}(\hat{\mathbf{k}})\; S_l(E_k) \tag{4.195}$$

$$= \delta(E_{k'}-E_k)\,\frac{\hbar^2}{4\pi\mu k}\sum_l (2l+1)\, P_l(\cos\theta)\; S_l(E_k). \tag{4.196}$$

Above, we have adopted the notation: $l_1 = l_2 = l, m_1 = m_2 = m$, $E_q = \hbar^2q^2/2\mu$ and $E_k = \hbar^2k^2/2\mu$. Eq. (4.196) was obtained from to Eq. (4.195) through the Addition Theorem.

Now we can relate the partial-wave components of the T- and the S-operators. For this purpose we use in Eq. (4.82) the partial-wave expansion for the T-matrix (Eq. (4.192)). We obtain,

$$S_{\mathbf{k}',\mathbf{k}} = \delta(\mathbf{k}' - \mathbf{k}) - 2\pi i\,\delta(E_{k'} - E_k) \sum_l (2l+1)\,P_l(\cos\theta) \left[\frac{\hbar^2}{4\pi\mu k}\,T_l(E_k)\right]. \tag{4.197}$$

The next step is to express $\delta(\mathbf{k}' - \mathbf{k})$ in terms of $E_k, \Omega_{\mathbf{k}}, E_{k'}, \Omega_{\mathbf{k}'}$. We start from the relation

$$\int d\Omega_{\mathbf{k}}\,dk\,k^2\,\delta(\mathbf{k}' - \mathbf{k}) = \int d\Omega_{\mathbf{k}}\,dE_k\;\delta(E_{k'} - E_k)\,\delta(\Omega_{\mathbf{k}'} - \Omega_{\mathbf{k}}). \tag{4.198}$$

Writing

$$dE_k = \frac{\hbar^2}{\mu k}\,\left(k^2\,dk\right),$$

we get

$$\delta(\mathbf{k}' - \mathbf{k}) = \frac{\hbar^2}{\mu k}\,\delta(E_{k'} - E_k)\;\delta(\Omega_{\mathbf{k}'} - \Omega_{\mathbf{k}}). \tag{4.199}$$

We now use the completeness of the Spherical Harmonics to write

$$\delta(\Omega_{\mathbf{k}'} - \Omega_{\mathbf{k}}) = \sum_{lm} Y^*_{lm}(\Omega_{\mathbf{k}'})\,Y_{lm}(\Omega_{\mathbf{k}}) = \frac{1}{4\pi}\sum_l (2l+1)\,P_l(\cos\theta). \tag{4.200}$$

Using this relation in Eq. (4.199) and inserting the result into Eq. (4.197), we obtain

$$S_{\mathbf{k}',\mathbf{k}} = \delta(E_{k'} - E_k)\;\sum_l (2l+1)\,P_l(\cos\theta)\,\left[\frac{\hbar^2}{4\pi\mu k}\,(1 - 2\pi i T_l(E))\right]. \tag{4.201}$$

Comparing Eqs. (4.196) and Eq. (4.201), we get the desired expression

$$S_l(E) = 1 - 2\pi i\,T_l(E). \tag{4.202}$$

It is convenient to express also the scattering amplitude in terms of $T_l(E)$. Inserting Eq. (4.192) into Eq. (4.64), we get

$$\begin{aligned} f(\theta) &= -2\pi^2\left(\frac{2\mu}{\hbar^2}\right)\,\frac{\hbar^2}{4\pi\mu k}\sum_l (2l+1)\,P_l(\cos\theta)\,T_l(E) \\ &= -\frac{\pi}{k}\,\sum_l (2l+1)\,P_l(\cos\theta)\,T_l(E) \end{aligned}$$

or

$$f(\theta) = \frac{1}{2ik}\sum_l (2l+1)\,P_l(\cos\theta)\;\left[-2\pi i\,T_l(E)\right]. \tag{4.203}$$

Using Eq. (4.202), the scattering amplitude takes the form

$$f(\theta) = \frac{1}{2ik} \sum_l (2l+1)\, P_l(\cos\theta)\ [S_l(E) - 1]\,. \tag{4.204}$$

This equation was derived in chapter 2 (see Eq. (2.54b)) from the asymptotic behavior of the wave function.

4.6.2 *The partial-wave projected two-potential formula*

One can also perform partial-wave expansion of the Gell-Mann Goldberger relation. First, we write the angular momentum expansions of the distorted waves,

$$\chi^{(+)}(\mathbf{k},\mathbf{r}) = \frac{1}{(2\pi)^{3/2}} \sum_{lm} 4\pi\, Y^*_{lm}(\hat{\mathbf{k}})\, Y_{lm}(\hat{\mathbf{r}})\, i^l\, \frac{w_l(k,r)}{kr} \tag{4.205}$$

and

$$\chi^{(-)}(\mathbf{k},\mathbf{r}) = \frac{1}{(2\pi)^{3/2}} \sum_{lm} 4\pi\, Y^*_{lm}(\hat{\mathbf{k}})\, Y_{lm}(\hat{\mathbf{r}})\, i^l\, \frac{w^*_l(k,r)}{kr}. \tag{4.206}$$

Next, we use the expansion of $\psi^{(+)}(\mathbf{k},\mathbf{r})$ (Eq. (4.180a)) and expand the distorted and the total T-matrices as (Eq. (4.190))

$$\begin{aligned} T_{\mathbf{k}',\mathbf{k}} &= = \frac{\hbar^2}{\mu k} \sum_{lm} Y_{lm}(\hat{\mathbf{k}}')Y^*_{lm}(\hat{\mathbf{k}})\ T_l(E_k) \\ T^{(1)}_{\mathbf{k}',\mathbf{k}} &= = \frac{\hbar^2}{\mu k} \sum_{lm} Y_{lm}(\hat{\mathbf{k}}')Y^*_{lm}(\hat{\mathbf{k}})\ T^{(1)}_l(E_k), \end{aligned} \tag{4.207}$$

where

$$T^{(1)}_l(E_k) = \frac{k}{\pi E_k} \int_0^\infty dr\, \hat{\jmath}_l(kr)\, V_1(r)\, w_l(k,r). \tag{4.208}$$

Inserting these expansions into Eq. (4.152), we get the partial-wave projected Gell-Mann Goldberger relation,

$$T_l(E_k) = T^{(1)}_l(E_k) + \frac{k}{\pi E_k} \int_0^\infty dr\, w_l(k,r)\, V_2(r)\, u_l(k,r). \tag{4.209}$$

4.7 Long range potentials

The scattering from potentials that fall off slowly[13] in the $r \to \infty$ limit requires special treatment. This problem was considered in chapter 3, where

[13]Cases in which the condition $\lim_{r\to\infty} [r\, V(r)] = 0$ is not satisfied.

we discussed the scattering amplitude and the wave function in the cases of Coulomb and Coulomb plus short range potentials. We derived an analytical expression for the Coulomb scattering amplitude (Eq. (3.29)) and showed that the corresponding wave function is given by the partial-wave expansion (Eq. (3.102a))

$$\phi_{\mathrm{C}}^{(+)}(\mathbf{k};\mathbf{r}) = \frac{1}{(2\pi)^{3/2}} \sum_{lm} 4\pi\, Y_{lm}^{*}(\hat{\mathbf{k}})\, Y_{lm}(\hat{\mathbf{r}})\, i^{l}\, e^{i\sigma_l}\, \frac{F_l(\eta, kr)}{kr}. \tag{4.210}$$

Above, $F_l(\eta, kr)$ is the regular Coulomb wave function and σ_l is the Coulomb phase shift for the lth partial-wave (see e.g. [Abramowitz and Stegun (1972)]). Now we discuss Møller operators and the S- and the T-matrices in Coulomb scattering.

First, we obtain the partial-wave expansion for the wave function with incoming wave boundary condition, $\phi_{\mathrm{C}}^{(-)}(\mathbf{k};\mathbf{r})$. This can be done with the help of the time reversal property of $\phi_{\mathrm{C}}^{(+)}(\mathbf{k};\mathbf{r})$. Using Eq. (4.124), we obtain[14]

$$\phi_{\mathrm{C}}^{(-)}(\mathbf{k};\mathbf{r}) = \frac{1}{(2\pi)^{3/2}} \sum_{lm} 4\pi\, Y_{lm}^{*}(\hat{\mathbf{k}})\, Y_{lm}(\hat{\mathbf{r}})\, i^{l}\, e^{-i\sigma_l}\, \frac{F_l(\eta, kr)}{kr}. \tag{4.211}$$

This expression will be used later in this section.

Now we evaluate the on-shell matrix-elements of the transition operator[15]. Using Eqs. (3.29) and (4.64) we directly obtain

$$T_{\mathbf{k}',\mathbf{k}}^{\mathrm{C}} = \left(\frac{\hbar^2}{2\mu}\right) \frac{1}{2\pi^2} \left[\frac{\eta}{2k\sin^2(\theta/2)} e^{-i\eta\ln(\sin^2\theta/2)}\, e^{2i\sigma_0}\right]. \tag{4.212}$$

It is still possible to define Møller wave operators and the S-matrix as in the previous sections. The Møller operators are given by the condition

$$\Omega_{\mathrm{C}}^{(\pm)}\, |\phi(\mathbf{k})\rangle = |\phi_{\mathrm{C}}^{(\pm)}(\mathbf{k})\rangle\ . \tag{4.213}$$

We remark that the long range of the Coulomb field leads to serious mathematical problems involving the Møller operators [Dollard (1964); Alt and

[14]Although the partial-wave expansion for the Coulomb scattering amplitude is useless, since it does not converge, the expansions for $\phi_{\mathrm{C}}^{(\pm)}(\mathbf{k};\mathbf{r})$ are very Important. They are used in the description of scattering from Coulomb plus short-range potentials.

[15]In fact, owing to the infinite range of the Coulomb field, the transition matrix in momentum space has no on-shell limit. However, this difficulty can be avoided evaluating matrix-elements in the asymptotically Coulomb-modified free states [Nutt (1968); van Haeringen (1976)]. Although we do not state it explicitly, it should be understood that in this section the matrix-elements in the momentum representation are evaluated with this modification.

Sandhas (1966)]. In the time-dependent approach, where the incident particle is represented by a wave packet, $\Omega_C^{(\pm)}$ is no longer given by Eq. (4.101). In this case, the asymptotic equivalence of the Hamiltonians H and H_0 breaks down, since the Coulomb field distorts the incident packet even at asymptotic distances. However, this difficulty can be avoided with an appropriate modification of Eq. (4.101) [Dollard (1964)].

The Coulomb S-matrix is given in terms of the Møller operators as

$$S^c = \Omega_C^{(-)^\dagger}\, \Omega_C^{(+)} . \qquad (4.214)$$

According to Eq. (4.83), the Coulomb scattering and transition operators for a given collision energy are related by the equation

$$S^c(E) = 1 - 2\pi i\, \delta\,(E - H_0)\; T^c(E). \qquad (4.215)$$

Let us now consider the scattering from Coulomb plus short-range potentials. In this case the potential is given by the sum

$$V(r) = V_C(r) + \bar{V}(r)\,,$$

where $\bar{V}(r)$ vanishes or can be neglected for $r > \bar{R}$. In chapter 3 we have shown that the scattering wave function is given by the partial-wave expansion (Eq. (3.125a))

$$\psi^{(+)}(\mathbf{k};\mathbf{r}) = \frac{1}{(2\pi)^{3/2}} \sum_{lm} 4\pi\; Y_{lm}^*(\hat{\mathbf{k}})\; Y_{lm}(\hat{\mathbf{r}})\; i^l\; e^{i\sigma_l}\; \frac{u_l(k,r)}{kr}, \qquad (4.216)$$

where $u_l(k,r)$ is the solution of the radial equation for the lth partial-wave, with the asymptotic form (see Eq. (3.115))

$$u_l(k, r > \bar{R}) = \frac{i}{2}\; \left[H_l^{(-)}(\eta, kr) - \bar{S}_l\, H_l^{(+)}(\eta, kr)\right]\; .$$

Above, $H_l^{(-)}$ ($H_l^{(+)}$) is the Coulomb wave function with incoming (outgoing) wave boundary condition and $\bar{S}_l$ is the l-projected S-matrix associated with the short-range interaction. Similarly to the case of a pure Coulomb field, the wave function $\psi^{(-)}(\mathbf{k};\mathbf{r})$ can be derived from the time-reversal conjugate of $\psi^{(+)}(-\mathbf{k};\mathbf{r})$ and one obtains

$$\psi^{(-)}(\mathbf{k};\mathbf{r}) = \frac{1}{(2\pi)^{3/2}} \sum_{lm} 4\pi\; Y_{lm}^*(\hat{\mathbf{k}})\; Y_{lm}(\hat{\mathbf{r}})\; i^l\; e^{-i\sigma_l}\; \frac{u_l^*(k,r)}{kr}. \qquad (4.217)$$

It is convenient to write the Møller wave operators as

$$\Omega^{(\pm)} = \Omega_C^{(\pm)}\, \bar{\Omega}^{(\pm)}. \qquad (4.218)$$

In this way, $\bar{\Omega}^{(\pm)}$ is the operator which transforms Coulomb states $\left|\phi_{\mathrm{C}}^{(\pm)}(\mathbf{k})\right\rangle$ into solutions of the Schrödinger equation with the full potential. That is

$$\bar{\Omega}^{(\pm)} \left|\phi_{\mathrm{C}}^{(\pm)}(\mathbf{k})\right\rangle = \left|\psi^{(\pm)}(\mathbf{k})\right\rangle . \tag{4.219}$$

In the time-dependent approach, these operators are given by the trivial generalization of Eq. (4.101)

$$\bar{\Omega}^{(\pm)} = \lim_{t' \to \pm\infty} e^{-\frac{i}{\hbar} H \left(t'-t\right)} \, e^{\frac{i}{\hbar} H_{\mathrm{C}} \left(t'-t\right)} , \tag{4.220}$$

with $H_{\mathrm{C}} = K + V_{\mathrm{C}}$ and $H = H_{\mathrm{C}} + \bar{V}$. The scattering operator associated with the short-range interaction is defined as

$$\bar{S} = \bar{\Omega}^{(-)^{\dagger}} \, \bar{\Omega}^{(+)} . \tag{4.221}$$

Inserting Eq. (4.218) into Eq. (4.77), we get

$$S = \Omega_{\mathrm{C}}^{(-)^{\dagger}} \, \bar{S} \, \Omega_{\mathrm{C}}^{(+)} \tag{4.222}$$

It is useful to define a transition operator associated with the short-range interaction as

$$\bar{T} = \bar{V} \bar{\Omega}^{(+)} = \bar{\Omega}^{(-)^{\dagger}} \bar{V} . \tag{4.223}$$

It is related with the full transition matrix through the Gell-Mann-Goldberger relation. Setting in Eqs. (4.151) and (4.152)

$$V_1 = V_{\mathrm{C}}, \quad V_2 = \bar{V} , \quad \left|\chi^{(\pm)}(\mathbf{k})\right\rangle = \left|\phi_{\mathrm{C}}^{(\pm)}(\mathbf{k})\right\rangle ,$$

one gets

$$T_{\mathbf{k}',\mathbf{k}} = \left\langle \phi\left(\mathbf{k}'\right)\right| V_{\mathrm{C}} \left| \phi_{\mathrm{C}}^{(+)}(\mathbf{k})\right\rangle + \left\langle \phi_{\mathrm{C}}^{(-)}\left(\mathbf{k}'\right)\right| \bar{V} \left|\psi^{(+)}(\mathbf{k})\right\rangle \tag{4.224}$$

$$= T^{\mathrm{C}}_{\mathbf{k}',\mathbf{k}} + \left\langle \phi_{\mathrm{C}}^{(-)}\left(\mathbf{k}'\right)\right| \bar{T} \left|\phi_{\mathrm{C}}^{(+)}(\mathbf{k})\right\rangle , \tag{4.225}$$

or, simply

$$T = T^{\mathrm{C}} + \Omega_{\mathrm{C}}^{(-)^{\dagger}} \, \bar{T} \, \Omega_{\mathrm{C}}^{(+)} . \tag{4.226}$$

Inserting Eq. (4.226) into Eq. (4.83) and using Eq. (4.215), we get the relation,

$$S(E) = S^{\mathrm{C}}(E) - 2\pi i \, \delta\left(E - H_0\right) \Omega_{\mathrm{C}}^{(-)^{\dagger}} \, \bar{T} \, \Omega_{\mathrm{C}}^{(+)},$$

or, taking into account Eqs. (4.222) and (4.214),

$$\Omega_{\mathrm{C}}^{(-)^{\dagger}} \, \bar{S}(E) \, \Omega_{\mathrm{C}}^{(+)} = S^{\mathrm{C}}(E) - 2\pi i \, \delta\left(E - H_0\right) \Omega_{\mathrm{C}}^{(-)^{\dagger}} \, \bar{T} \, \Omega_{\mathrm{C}}^{(+)} \tag{4.227}$$

$$= \Omega_{\mathrm{C}}^{(-)^{\dagger}} \, \Omega_{\mathrm{C}}^{(+)} - 2\pi i \, \delta\left(E - H_0\right) \Omega_{\mathrm{C}}^{(-)^{\dagger}} \, \bar{T} \, \Omega_{\mathrm{C}}^{(+)} . \tag{4.228}$$

The above equation leads to a simple relation between the partial-wave components of $\bar{S}$ and $\bar{T}$. First, one takes matrix-elements in the momentum representation and let the operator $\delta(E - H_0)$ act on the bra. We get

$$\left\langle \phi_C^{(-)}(\mathbf{k}')\,|\bar{S}\,|\phi_C^{(+)}(\mathbf{k})\right\rangle = \langle \phi_C^{(-)}(\mathbf{k}')\,|\phi_C^{(+)}(\mathbf{k})\rangle + 2\pi i\,\delta(E - E_{k'})\left\langle \phi_C^{(-)}(\mathbf{k}')\,|\bar{T}\,|\phi_C^{(+)}(\mathbf{k})\right\rangle. \quad (4.229)$$

Now we carry out the partial-wave expansion of the T- and the S-matrices in the scattering from Coulomb plus short-range potentials. For this purpose, we follow the same procedures of section 4.6.1. In the Coulomb case, the orthonormal states of the energy and angular momentum basis are

$$\phi_C(Elm;\mathbf{r}) = i^l\left(\frac{2\mu k}{\pi\hbar^2}\right)^{1/2}\frac{F_l(\eta,kr)}{kr}\,Y_{lm}(\hat{\mathbf{r}}) \quad (4.230)$$

and the partial-wave components of the operators $\bar{S}$ and $\bar{T}$ are defined as

$$\bar{S}_l(E) = \langle\phi_C(Elm)|\;\bar{S}\;|\phi_C(Elm)\rangle \quad (4.231)$$

and

$$\bar{T}_l(E) = \langle\phi_C(Elm)|\;\bar{T}\;|\phi_C(Elm)\rangle = \langle\phi_C(Elm)|\;\bar{V}\;|\psi(Elm)\rangle \quad (4.232)$$

$$= \frac{k}{\pi E}\int dr\,F_l(\eta,kr)\,\bar{V}(r)\,u_l(k,r), \quad (4.233)$$

respectively. Proceeding as in section 4.6.1, with the replacement of Eq. (4.177) by its Coulomb analog,

$$\langle\phi_C(Elm)\,|\,\phi_C^{(\pm)}(\mathbf{k})\rangle = \frac{\hbar}{\sqrt{\mu k}}Y_{lm}^*(\hat{\mathbf{k}})\;e^{\pm i\sigma_l}\;\delta(E - E_k), \quad (4.234)$$

we obtain

$$\bar{S}_l(E) = 1 - 2\pi i\,\bar{T}_l. \quad (4.235)$$

4.8 Evaluation of partial-wave Green's functions

In this section we discuss the calculation of partial-wave projected Green's function. We begin with the case of a free particle, where the Green's functions can be determined in terms of Riccati-Bessel functions of different kinds. We then derive an expression for the Green's functions for an arbitrary potential. This expression involves the regular solution of the equation for the same potential and the solution with outgoing wave boundary condition.

4.8.1 *Free particle's Green's function*

Let us now evaluate the partial-wave projected Green's function for a free particle, $g_l^{(+)}(E;r,r')$. We start from the spectral representation of $G_0^{(+)}(E;\mathbf{r},\mathbf{r}')$ in the spherical basis[16],

$$\begin{aligned}G_0^{(+)}(E;\mathbf{r},\mathbf{r}') &= \sum_{lm}\int_0^\infty dE'\,\frac{\phi^*_{E'lm}(\mathbf{r})\,\phi_{E'lm}(\mathbf{r}')}{E-E'+i\epsilon}\\ &= \frac{1}{rr'}\sum_{lm}Y^*_{lm}(\hat{\mathbf{r}})\,Y_{lm}(\hat{\mathbf{r}}')\left[\frac{2\mu}{\pi\hbar^2}\int_0^\infty\frac{dE'}{k'}\,\frac{\hat{\jmath}_l(k'r)\,\hat{\jmath}_l(k'r')}{E-E'+i\epsilon}\right].\end{aligned}$$

Comparing the above equation with Eq. (4.182) we get

$$g_{0,l}^{(+)}(E;r,r') = \frac{2\mu}{\pi\hbar^2}\int_0^\infty\frac{dE'}{k'}\,\frac{\hat{\jmath}_l(k'r)\,\hat{\jmath}_l(k'r')}{E-E'+i\epsilon}\,. \tag{4.236}$$

Changing variable: $E'\to k'$, we replace $E'=\hbar^2k'^2/2\mu$ and $dE'=(\hbar^2k'/\mu)\,dk'$ and Eq. (4.236) becomes

$$g_{0,l}^{(+)}(E;r,r') = \frac{2}{\pi}\left(\frac{2\mu}{\hbar^2}\right)\int_0^\infty dk'\,\frac{\hat{\jmath}_l(k'r)\,\hat{\jmath}_l(k'r')}{k^2-k'^2+i\epsilon'}. \tag{4.237}$$

Considering the properties of Bessel functions (see, for example, appendix C of [Joachain (1983)]),

$$\begin{aligned}j_l(-z) &= (-)^l\,j_l(z)\to\hat{\jmath}_l(-z)=(-)^{l+1}\,\hat{\jmath}_l(z),\\ n_l(-z) &= (-)^{l+1}\,n_l(z)\to\hat{n}_l(-z)=(-)^l\,\hat{n}(z),\end{aligned} \tag{4.238}$$

we can put Eq. (4.237) in the form

$$g_{0,l}^{(+)}(E;r,r') = \frac{1}{\pi}\left(\frac{2\mu}{\hbar^2}\right)\int_{-\infty}^\infty dk'\,\frac{\hat{\jmath}_l(k'r)\,\hat{\jmath}_l(k'r')}{k^2-q^2+i\epsilon'}. \tag{4.239}$$

Writing

$$\frac{1}{k^2-k'^2+i\epsilon'} = -\frac{1}{2k'}\left[\frac{1}{k'-(k+i\varepsilon)}+\frac{1}{k'+(k+i\varepsilon)}\right],$$

where $\varepsilon=2k\epsilon'$ is also an infinitesimal quantity, and using this expression in Eq. (4.239) we obtain

$$g_{0,l}^{(+)}(E;r,r') = \frac{2\mu}{\hbar^2}\,I^{(+)}, \tag{4.240}$$

[16]For simplicity of notation, we do not write $\lim_{\epsilon\to0}$. We assume that this limit is implicitly taken.

with

$$I^{(+)} = -\frac{1}{2\pi}\int_{-\infty}^{\infty} dk'\, \frac{\hat{\jmath}_l(k'r)\,\hat{\jmath}_l(k'r')}{k'}\left[\frac{1}{k'-(k+i\varepsilon)} + \frac{1}{k'+(k+i\varepsilon)}\right]. \tag{4.241}$$

The integral $I^{(+)}$ can be evaluated on the complex plane. For this purpose, we consider separately the situations $r' > r$ and $r' < r$. In the former case, we write

$$\hat{\jmath}_l(k'r') = \frac{i}{2}\left(\hat{h}_l^{(-)}(k'r') - \hat{h}_l^{(+)}(k'r')\right) \tag{4.242}$$

and split $I^{(+)}$ into the terms

$$I^{(+)} = I_1^{(+)} + I_2^{(+)},$$

where

$$I_1^{(+)} = -\frac{i}{4\pi}\int_{-\infty}^{\infty} dk'\, \frac{\hat{\jmath}_l(k'r)\,\hat{h}_l^{(-)}(k'r')}{k'}\left[\frac{1}{k'-(k+i\varepsilon)} + \frac{1}{k'+(k+i\varepsilon)}\right]$$

$$I_2^{(+)} = \frac{i}{4\pi}\int_{-\infty}^{\infty} dk'\, \frac{\hat{\jmath}_l(k'r)\,\hat{h}_l^{(+)}(k'r')}{k'}\left[\frac{1}{k'-(k+i\varepsilon)} + \frac{1}{k'+(k+i\varepsilon)}\right].$$

The first integral is evaluated over the closed contour formed by the real axis and the lower half of a circle of infinite radius[17]. On the lower half-circle we can take the asymptotic form $\hat{h}_l^{(-)}(k'r') \to \exp\left[-i\left(\mathrm{Re}\,(k'r') - l\pi/2\right)\right] \times \exp\left[\mathrm{Im}\,(k'r')\right]$, with $\mathrm{Im}\,(k'r') < 0$. Since $|\mathrm{Im}\,(k'r')| \to \infty$, the integrand vanishes and the integral over the closed contour reduces to the contribution from the real axis. Applying Cauchy's theorem, the integral is given by the residue of the pole at $k' = -(k + i\varepsilon)$, enclosed by this contour. Therefore, we have

$$I_1^{(+)} = -2\pi i\left[-\frac{i}{4\pi}\left(\frac{\hat{\jmath}_l(-kr)\,\hat{h}_l^{(-)}(-kr')}{-k}\right)\right]. \tag{4.243}$$

Note that the minus sign on the RHS of the above equation arises from the clockwise sense followed along the contour. With the help of Eq. (4.238) and the property of the Ricatti-Haenkel functions

$$\hat{h}_l^{(\pm)}(-z) = (-)^l\hat{h}^{(\mp)}(z),$$

we get

$$I_1^{(+)} = -\frac{1}{2k}\left[\hat{\jmath}_l(kr)\,\hat{h}_l^{(+)}(kr')\right] \tag{4.244}$$

---

[17]The integrations on the complex plane here are analogous to the ones discussed in section 4.1.1.

The integral $I_2^{(+)}$ can be evaluated in a similar way. In this case, we close the contour with the upper half-circle, where the integrand vanishes, and follow it in the counterclockwise sense. The integral over the real axis is then given by the residue of the pole at $k' = k + i\varepsilon$. That is,

$$I_2^{(+)} = 2\pi i \left[\frac{i}{4\pi} \left(\frac{\hat{\jmath}_l(kr)\, \hat{h}_l^{(+)}(kr')}{k} \right) \right] = -\frac{1}{2k} \left[\hat{\jmath}_l(kr)\, \hat{h}_l^{(+)}(kr') \right] . \qquad (4.245)$$

Adding Eqs. (4.244) and (4.245) and inserting the result into Eq. (4.240) we obtain

$$g_{0,l}^{(+)}(E;r,r') = -\left(\frac{2\mu}{\hbar^2}\right) \frac{1}{k} \left[\hat{\jmath}_l(kr)\, \hat{h}_l^{(+)}(kr') \right] . \qquad (4.246)$$

For $r' < r$, we proceed in the same way, except that we use Eq. (4.242) for $\hat{\jmath}_l(kr)$, instead of $\hat{\jmath}_l(kr')$. It is clear that we obtain the same result, exchanging r and r'. Analogously, one can also derive the Green's functions with incoming wave boundary condition. These results can be summarized in the compact form

$$g_{0,l}^{(+)}(E;r,r') = -\left(\frac{2\mu}{\hbar^2}\right) \frac{1}{k} \left[\hat{\jmath}_l(kr_<)\, \hat{h}_l^{(+)}(kr_>) \right] , \qquad (4.247)$$

where $r_<$ and $r_>$ stand respectively for the smaller and the larger of the two radial distances r and r'. Using the properties $g_{0,l}^{(-)} = g_{0,l}^{(+)\,*}$ and $\hat{h}_l^{(-)} = \hat{h}_l^{(+)\,*}$, the above result can be generalized in the form

$$g_{0,l}^{(\pm)}(E;r,r') = -\left(\frac{2\mu}{\hbar^2}\right) \frac{1}{k} \left[\hat{\jmath}_l(kr_<)\, \hat{h}_l^{(\pm)}(kr_>) \right] .$$

Using the explicit form of $g_{0,l}^{(+)}(E;r,r')$ in Eq. (4.183) we can investigate the asymptotic form of $u_l(k,r)$. For large r, i.e. outside the range of the potential, we can write

$$\begin{aligned} u_l(k, r \to \infty) &= \hat{\jmath}_l(kr) \\ &\quad + \hat{h}_l^{(+)}(kr) \left[-\frac{1}{k} \left(\frac{2\mu}{\hbar^2}\right) \int_0^\infty dr'\, \hat{\jmath}_l(kr')\, V(r')\, u_l(k,r') \right] . \end{aligned} \qquad (4.248)$$

According to Eq. (4.191b), we can substitute the quantity within square brackets by $-\pi T_l(E)$. We get

$$u_l(k, r\to\infty) = \hat{\jmath}_l(kr) \; - \hat{h}_l^{(+)}(kr) \;\; \pi T_l(E).$$

Using Eq. (4.242) we can write

$$\begin{aligned} u_l(k, r\to\infty) &= \frac{i}{2} \left(\hat{h}_l^{(-)}(kr) - \hat{h}_l^{(+)}(kr) \right) \; - \hat{h}_l^{(+)}(kr)\, \pi T_l(E) \\ &= \frac{i}{2} \left[\hat{h}_l^{(-)}(k'r') - (1 - 2i\pi T_l(E))\, \hat{h}_l^{(+)}(kr) \right] , \end{aligned}$$

or, with the help of Eq. (4.202),

$$u_l(k, r \to \infty) = \frac{i}{2}\left[\hat{h}_l^{(-)}(k'r') - S_l(E)\,\hat{h}_l^{(+)}(kr)\right]. \tag{4.249}$$

This is the correct asymptotic form, given in Eq. (2.35) with $\gamma_l = i/2$.

4.8.2 *Green's functions for an arbitrary potential*

The result of Eq. (4.247) can be generalized to any potential, including combinations of Coulomb and short-range contributions. Let us consider the wave functions $\bar{u}_l(k,r)$, $\bar{\mathcal{W}}_l(k,r)$, $\mathcal{H}_l^{(+)}(k,r)$ and $\mathcal{H}_l^{(-)}(k,r)$ which satisfy an equation of the form

$$\mathbf{D}_r\, Z(k,r) = 0\,,$$

where $\mathbf{D}_r$ is the differential operator

$$\mathbf{D}_r = E + \frac{\hbar^2}{2\mu}\left(\frac{d^2}{dr^2} - \frac{l(l+1)}{r^2}\right) - V(r)$$

and $Z(k,r)$ stands for $\bar{u}_l(k,r)$, $\bar{\mathcal{W}}_l(k,r)$, $\mathcal{H}_l^{(+)}(k,r)$ or $\mathcal{H}_l^{(-)}(k,r)$. These functions are characterized by their asymptotic forms[18],

$$\bar{u}_l(k, r \to \infty) = \sin(kr - l\pi/2 + \delta_l)\,, \tag{4.250}$$

$$\bar{\mathcal{W}}_l(k, r \to \infty) = \cos(kr - l\pi/2 + \delta_l)\,, \tag{4.251}$$

$$\mathcal{H}_l^{(\pm)}(k, r \to \infty) = e^{\pm i(kr - l\pi/2 + \delta_l)}\,. \tag{4.252}$$

It can be easily checked that

$$\mathcal{H}_l^{(\pm)}(k,r) = \bar{\mathcal{W}}_l(k,r) \pm i\,\bar{u}_l(k,r)\,. \tag{4.253}$$

The partial-wave projected Green's function for the full Hamiltonian is then given by Eq. (4.247) with the replacements $\hat{\jmath}_l(kr_<) \to \bar{u}_l(kr_<)$ and $\hat{h}_l^{(+)}(kr_>) \to \mathcal{H}_l^{(+)}(kr_>)$. That is

$$g_l^{(\pm)}(E; r, r') = -\frac{2\mu}{\hbar^2 k}\left[\bar{u}_l(k, r_<)\,\mathcal{H}_l^{(\pm)}(k, r_>)\right], \tag{4.254}$$

or, according to Eq. (2.42),

$$g_l^{(\pm)}(E; r, r') = -\frac{2\mu}{\hbar^2 k}\, e^{-i\delta_l}\left[u_l(k, r_<)\,\mathcal{H}_l^{(\pm)}(k, r_>)\right]. \tag{4.255}$$

---

[18] If the potential includes a Coulomb part, the term $-\eta \ln 2kr + \sigma_l$ should be added to the argument of these functions.

In particular, the l-projected Green's function for pure Coulomb scattering is

$$g_{\mathrm{C},l}^{(\pm)}(E;r,r') = -\frac{2\mu}{\hbar^2 k}\left[F_l(\eta,kr_<)\,H_l^{(\pm)}(\eta,kr_>)\right], \tag{4.256}$$

where F_l and $H_l^{(\pm)}$ are the Coulomb functions discussed in chapter 3.

To prove Eq. (4.254), we must show that $g_l^{(\pm)}(E;r,r')$ satisfies the angular momentum projected version of Eq. (4.29), namely,

$$\mathbf{D}_r\, g_l^{(\pm)}(E;r,r') = \delta(r-r'), \tag{4.257}$$

and that it leads to a radial wave function with the right boundary condition. We first consider the Green's function with outgoing wave boundary condition, $g_l^{(+)}(E;r,r')$. We rewrite it in the form[19]

$$g_l^{(+)}(E;r,r') = -\left(\frac{2\mu}{\hbar^2}\right)\frac{1}{k}\left\{\bar{u}_l(r)\ \mathcal{H}^{(+)}(r')\ \left[1-\Theta(r-r')\right]\right.$$
$$\left.+\,\bar{u}_l(r')\ \mathcal{H}^{(+)}(r)\ \Theta(r-r')\right\} \tag{4.258}$$

and evaluate $d^2g^{(+)}(E;r,r')/dr^2$. We get

$$\frac{d^2g^{(+)}(E;r,r')}{dr^2} = -\frac{2\mu}{k\hbar^2}\left\{\bar{u}_l''(r)\,\mathcal{H}^{(+)}(r')\right.$$
$$\times\left[1-\Theta(r-r')\right] + \mathcal{H}^{(+)}(r)\,\bar{u}_l(r')\,\Theta(r-r')$$
$$-\,2\,\delta(r-r')\left[\bar{u}_l'(r)\,\mathcal{H}^{(+)}(r') - \mathcal{H}^{(+)}(r)\,\bar{u}_l(r')\right]$$
$$\left.-\,\delta'(r-r')\left[\bar{u}_l(r)\,\mathcal{H}^{(+)}(r') - \bar{u}_l(r')\,\mathcal{H}^{(+)}(r)\right]\right\}, \tag{4.259}$$

where primes stand for d/dr and double-primes for d^2/dr^2. Using Eq. (4.259) in Eq. (4.257) and considering that both $\bar{u}_l$ and $\mathcal{H}^{(+)}$ satisfy the Schrödinger equation, we obtain

$$\mathbf{D}_r\, g^{(+)}(E;r,r') = \frac{1}{k}\left\{2\,\delta(r-r')\ \left[\bar{u}_l'(r)\,\mathcal{H}^{(+)}(r') - \mathcal{H}^{(+)}(r)\,\bar{u}_l(r')\right]\right.$$
$$\left.+\,\delta'(r-r')\left[\bar{u}_l(r)\,\mathcal{H}^{(+)}(r') - \bar{u}_l(r')\,\mathcal{H}^{(+)}(r)\right]\right\}. \tag{4.260}$$

---

[19] Since there is no room for confusion, we use below the simplified notation $\mathcal{H}_l^{(\pm)}(k,r) \to \mathcal{H}_l^{(\pm)}(r)$, $\bar{u}_l(k,r) \to \bar{u}_l(r)$ and $\bar{\mathcal{W}}_l(k,r) \to \bar{\mathcal{W}}_l(r)$.

Using the property of the delta function,

$$\delta'(r-r')\,X(r,r') = -\,\delta(r-r')\,\frac{dX(r,r')}{dr},$$

Eq. (4.260) becomes

$$\mathbf{D}_r\, g^{(+)}(E;r,r') = \frac{1}{k}\left[\bar{u}'_l(r)\;\mathcal{H}^{(+)}(r) - \mathcal{H}^{(+)}(r)\;\bar{u}_l(r)\right]\delta(r-r')\,. \quad (4.261)$$

Since in Eq. (4.260) $[\bar{u}_l(r)\,\mathcal{H}^{(+)}(r') - \bar{u}_l(r')\,\mathcal{H}^{(+)}(r)]$ is a continuous function of r and r' multiplying $\delta(r-r')$, we have set $r'=r$. Using Eq. (4.253), we get

$$\mathbf{D}_r\, g^{(+)}(E;r,r') = \frac{1}{k}\left[\bar{u}'_l(r)\;\bar{\mathcal{W}}_l(r) - \bar{\mathcal{W}}'_l(r)\;\bar{u}_l(r')\right]\delta(r-r')\,. \quad (4.262)$$

The quantity within square brackets is the Wronskian $W\left\{\bar{u}_l,\bar{\mathcal{W}}_l\right\}$, which is given by Eq. (2.66) (with a different notation for the irregular solution: $\bar{\mathcal{W}}_l \to \bar{w}_l$). Thus, we can write,

$$W\left\{\bar{\mathcal{W}}_l,\bar{u}_l\right\} = k\,.$$

Inserting this result into Eq. (4.262) we obtain

$$\mathbf{D}_r\, g^{(+)}(E;r,r') = \delta(r-r')$$

and the first part of our proof is completed.

Now we must prove that Eq. (4.254) leads to the correct boundary condition. Let us first consider $g_l^{(+)}(E;r,r')$. We insert it in Eq. (4.188) and take the $r\to\infty$ limit. Owing to the finite range of the potential, the integrand vanishes for $r'>\bar{R}$. In this way, we can replace $r_> = r$, $r_< = r'$ and obtain the equation

$$\begin{aligned} u_l(k,r\to\infty) &= \hat{\jmath}_l(kr\to\infty) \\ &\quad - \left(\frac{2\mu}{\hbar^2}\right)\frac{1}{k}\mathcal{H}_l^{(+)}(k,r\to\infty)\int dr'\,\hat{\jmath}_l(kr')\;V(r')\,u_l(k,r'). \end{aligned}$$

Using Eqs. (2.42) and (4.191c), and replacing $\mathcal{H}_l^{(+)}(k,r\to\infty) = e^{i\delta_l}\,\hat{h}_l^{(+)}(kr)$, we get

$$u_l(k,r\to\infty) = \frac{i}{2}\left[\hat{h}_l^{(-)}(kr) - \hat{h}_l^{(+)}(kr)\right] - \pi\;T_l(E)\;\hat{h}_l^{(+)}(kr).$$

Using the relation $T_l(E) = [1 - S_l(E)]/2\pi i$ (Eq. (4.202)), we obtain

$$u_l(k, r \to \infty) = \frac{i}{2}\left[\hat{h}_l^{(-)}(kr) - \hat{h}_l^{(+)}(kr)\right] - \frac{1 - S_l(E)}{2i}\,\hat{h}_l^{(+)}(kr),$$

or

$$u_l(k, r \to \infty) = \frac{i}{2}\left[\hat{h}_l^{(-)}(kr) - S_l(E)\,\hat{h}_l^{(+)}(kr)\right], \tag{4.263}$$

Since $g_l^{(-)}(E; r, r') = g_l^{(+)\,*}(E; r, r')$, the above proof can immediately be extended to $g_l^{(-)}(E; r, r')$.

To finish this chapter, we point out that the partial-wave projected Lippmann-Schwinger equation in the case of Coulomb plus short range potentials is a straighforward generalization of Eq. (4.183). Performing partial-wave expansions in Eq. (4.149), with $V_0 = V_{\rm C}$ and $V' = \bar{V}$, we obtain

$$u_l(k, r) = F_l(kr) + \int dr'\, g_{{\rm C},l}^{(+)}(E; r, r')\,\bar{V}(r')\; u_l(k, r'). \tag{4.264}$$

Exercises

(1) Show that the solutions of the Lippmann-Schwinger equations:

$$|\psi^{(+)}(\mathbf{k})\rangle = |\phi(\mathbf{k})\rangle + G_0^{(+)}(E_k)\,V\,|\psi^{(+)}(\mathbf{k})\rangle\,;$$
$$|\psi^{(+)}(\mathbf{k})\rangle = (1 + G^{(+)}(E_k)\,V)\;|\phi(\mathbf{k})\rangle\,,$$

satisfy the Schrödinger equation

$$[E_k - H]\;|\psi^{(+)}(\mathbf{k})\rangle = 0,$$

and have the asymptotic ($|\mathbf{r}| \to \infty$) boundary condition

$$\psi^{(+)}(\mathbf{k}, \mathbf{r}) \to \frac{1}{(2\pi)^{3/2}}\left(e^{i\mathbf{k}\cdot\mathbf{r}} + f(\theta)\,\frac{e^{ikr}}{r}\right).$$

Above,

$$f(\theta) = -2\pi^2\left(\frac{2\mu}{\hbar^2}\right)\langle\phi(\mathbf{k}')|\;V\;\Omega^{(+)}\;|\phi(\mathbf{k})\rangle$$

and $\theta = \cos^{-1}\left(\hat{\mathbf{k}}\cdot\hat{\mathbf{k}}'\right)$.

(2) Starting from the Schrödinger equation for the Hamiltonian $H = K + V_1 + V_2$, derive Eqs. (4.149) and (4.150).

(3) Consider the scattering from the separable potential $V = |v\rangle\langle v|$, with matrix elements

$$V(\mathbf{r}, \mathbf{r}') = \lambda \exp\left(-\frac{r^2 + r'^2}{\alpha^2}\right).$$

a) Show that the scattering amplitude within the Born approximation is given by the expression

$$f^{\mathrm{B}}(k) = -\lambda \left(\frac{\mu\pi^2\alpha^6}{2\hbar^2}\right) e^{-\alpha^2k^2/2}.$$

b) Show that the exact scattering amplitude is isotropic and is given by[20]

$$f(k) = \frac{f^{\mathrm{B}}(k)}{A(\alpha, k)},$$

with

$$A(\alpha, k) = 1 + \lambda \left(\frac{\mu\pi\alpha^6}{\hbar^2}\right) \left\{ik\pi\left[1 - \frac{1}{2}\mathrm{erf}\left(-ik/\alpha\sqrt{2}\right) e^{-\alpha^2k^2/2}\right] + \frac{\sqrt{\pi}}{\alpha\sqrt{2}}\right\}.$$

(4) Consider the two-potential formula for the scattering from the potential

$$V(r) = V_{\mathrm{C}}(r) + V'(r),$$

where

$$V_{\mathrm{C}}(r) = \frac{q_{\mathrm{P}}q_{\mathrm{T}}}{4\pi\epsilon_0 r},$$

and $V'(r)$ is a spherically symmetric short-range potential. Show that the partial-wave components of the T-matrix can be written

$$T_l(E) = \frac{k}{\pi E_k} e^{i\sigma_l}\left[\int dr\, \hat{\jmath}_l(kr)\, V_{\mathrm{C}}(r)\, F_l(k,r) + \int dr\, F_l(k,r)\, V'(r)\, u_l(k,r)\right],$$

where $\hat{\jmath}_l(kr)$ is the Ricatti-Bessel function, $F_l(k,r)$ is the regular Coulomb wave function and $w_l(k,r)$ and $u_l(k,r)$ are the radial wave functions appearing in the partial-wave expansions of $\chi^{(+)}(\mathbf{k};\mathbf{r})$ and $\psi^{(+)}(\mathbf{k};\mathbf{r})$, respectively.

[20] *Hint*: Use Eq. (4.136) and employ the momentum representation for $\langle v|G_0^{(+)}|v\rangle$. The desired result can be obtained evaluating the on-shell and off-shell parts separately. For this purpose, use the mathematical properties of Error functions and Fresnel integrals, given on page 297 of [Abramowitz and Stegun (1972)].

(5) Prove that the wave function of Eq. (4.169) with the coefficient A_l of Eq. (4.170) satisfies the normalization condition of Eq. (4.166).

(6) Prove that the partial-wave projected free particle Green's functions are

$$g_l^{(\pm)}(E;r,r') = -\frac{2\mu}{\hbar^2 k}\left[\hat{\jmath}_l(kr_<)\,\hat{h}_l^{(\pm)}(kr_>)\right].$$

Then show that, if its off-shell part can be neglected, $g_l^{(+)}(E;r,r')$ takes the separable form

$$g_l^{(+)}(E;r,r') \simeq \left[g_l^{(+)}(E;r,r')\right]_{\rm on} = -\frac{2\mu}{\hbar^2 k}\,i\,\left[\hat{\jmath}_l(kr)\,\hat{\jmath}_l(kr')\right]\,,$$

where $\left[g_l^{(+)}(E;r,r')\right]_{\rm on}$ stands for the on-shell part of $g_l^{(+)}(E;r,r')$.

Chapter 5

Approximate Methods in Potential Scattering

The application of the scattering formalism developed in the previous sections relies heavily on the use of physically motivated approximation schemes. Of course a straightforward integration of the Schrödinger equation which describes the potential scattering with the appropriate boundary conditions is always an alternative. In many instances, however, the use of an approximate form of the scattering wave function that incorporates the basic physics is more convenient. Besides avoiding the numerical integration of the second-order differential equation, this approach may supply an invaluable insight of the physical problem. Other methods of approximation involve the Lippmann-Schwinger equation for the wave function or for the T-matrix. In this chapter we develop these approximation schemes for potential scattering.

The approximations discussed in this chapter can be classified in two categories. In the first are the approximations for weak potentials, based on perturbation theory. In the second are the approximations used in situations where the de Broglie wave length associated with the relative projectile-target motion is small as compared to the characteristic dimension of the system.

5.1 Perturbative approximations

5.1.1 *The Born series*

We begin with the Born approximation, which is valid when the interaction potential, V, is weak and/or the energy, E, is high (such that V/E is small). The starting point of this approximation scheme is the Lippmann-Schwinger (LS) equation for the T-matrix, already discussed at length in chapter 4,

$$T(E) = V + V\ G_0^{(+)}(E)\ T(E), \tag{5.1}$$

which can also be written in the equivalent form,

$$T(E) = V + V\ G^{(+)}(E)\ V. \tag{5.2}$$

The free Green function, $G_0^{(+)}(E)$, and the full Green function, $G^{+}(E)$, introduced in the previous chapter, are also related by an LS-type equation:

$$G^{(+)}(E) = G_0^{(+)}(E) + G_0^{(+)}(E)\ V\ G^{(+)}(E). \tag{5.3}$$

In general, the full Hamiltonian,

$$H = K + V, \tag{5.4}$$

where K is the kinetic energy operator[1] and V is the projectile-target interaction, has bound and unbound eigenstates. Their energies are respectively $E_n = -|E_n|$ and E, and they satisfies the Schrödinger equations

$$H\,|\Psi_n\rangle = -|E_n|\ |\Psi_n\rangle \quad \text{and} \quad H\,|\Psi^{(+)}(E)\rangle = E\,|\Psi^{(+)}(E)\rangle. \tag{5.5}$$

Therefore a spectral expansion of the full Green function can be readily written down,

$$G^{(+)}(E) = \sum_n \frac{|\Psi_n\rangle\ \langle\Psi_n|}{E + |E_n|} + \int_0^\infty dz\ \frac{|\Psi^{(+)}(z)\rangle\ \langle\Psi^{(+)}(z)|}{E - z + i\varepsilon}. \tag{5.6}$$

The LS equation for the transition operator (Eq. (5.1)) can be expressed as the Born series (see Eq. (4.65)),

$$T(E) = \sum_{m=0}^{\infty} V\ \left[(G_0^{(+)}(E)\,V\right]^m. \tag{5.7}$$

The above series converges if the kernel $\mathcal{K}(E) = G_0^{(+)}(E)V$ is of Hilbert-Schmidt type. That is, it satisfies the condition

$$\mathrm{Tr}\left[\mathcal{K}(z)\ \mathcal{K}^{\dagger}(z)\right] < \infty. \tag{5.8}$$

A more convenient criterion can be shown to exist if one uses a symmetrized version of the kernel [Schwinger (1961); Coester (1964); Scadron *et al.* (1964)], namely

$$\mathcal{K}(z) = V^{1/2}\ G_0^{(+)}(z)\ V^{1/2}. \tag{5.9}$$

Physically, one would expect that the kernel would be small in the high energy limit, since it is a product of the potential with the Green's function.

[1] To avoid confusion with the transition operator, we represent the kinetic energy operator by K, instead of the more frequent notation: T.

Thus, it goes as V/E. Accordingly, the kernel would be generally smaller than unity.

The first order approximation for the T-matrix is obtained truncating the series of Eq. (5.7) after $m = 0$. That is, setting

$$T^{(1)}(E) = V. \tag{5.10}$$

The matrix element of the transition operator between the plane waves states $\mathbf{k}$ and $\mathbf{k}'$ then becomes

$$T^{(1)}_{\mathbf{k}',\mathbf{k}} \equiv \langle\phi(\mathbf{k}')\,|T^{(1)}(E)|\,\phi(\mathbf{k})\rangle = \langle\phi(\mathbf{k}')\,|V|\,\phi(\mathbf{k})\rangle. \tag{5.11}$$

This procedure is called *first Born approximation*, *Plane Wave Born Approximation* (PWBA), or simply *Born approximation.*

Writing Eq. (5.11) in the coordinate representation and using the explicit form of the plane waves, the T-matrix becomes a function of the momentum transfer expressed in $\hbar$-units, $\mathbf{q} = \mathbf{k}' - \mathbf{k}$. That is

$$T^{(1)}_{\mathbf{k},\mathbf{k}'} \equiv T^{(1)}(\mathbf{q}) = \tilde{V}(\mathbf{q}), \tag{5.12}$$

with

$$\tilde{V}(\mathbf{q}) = \frac{1}{(2\pi)^3} \int d^3\mathbf{r}\ \ \mathrm{e}^{-i\mathbf{q}\cdot\mathbf{r}}\ V(\mathbf{r}). \tag{5.13}$$

Note that $\tilde{V}(\mathbf{q})$ is nothing but the Fourier transform of the potential with respect to the momentum transfer, divided by the factor $(2\pi)^{3/2}$. For spherically symmetrical potentials, the argument of the Fourier transform is the modulus of $\mathbf{q}$, which is related to the scattering angle, θ, by the expression

$$q = 2k \sin{(\theta/2)}. \tag{5.14}$$

For small scattering angles, we can approximate $q \simeq k\theta$. This means that the momentum imparted to the system in the collision corresponds to the incident wave number multiplied by the scattering angle.

The first Born approximation allows a very simple formula for the cross section. From the relation between the scattering amplitude and the plane wave matrix elements of the transition operator (Eq. (4.64)), we have

$$f(\theta) = -\frac{4\pi^2\mu}{\hbar^2}\ \tilde{V}(\mathbf{q}) \tag{5.15}$$

and the elastic scattering cross section is

$$\frac{d\sigma(\theta)}{d\Omega} \equiv \frac{d\sigma(\mathbf{q})}{d\Omega} = \left(\frac{4\pi^2\mu}{\hbar^2}\right)^2\ \left|\tilde{V}(\mathbf{q})\right|^2. \tag{5.16}$$

The above formula is a very important one as it allows an unambiguous way of relating the projectile-target interaction with measurements. Furthermore, it can be extended beyond the scope of potential scattering. In some situations, cross sections for collisions of many-body systems, can be approximated by a product of the cross section of Eq. (5.16) with a structure function, which contains information about the ground and excited states of the system. This approach has been extensively used in neutron physics to extract information about the structure of materials, and in high energy scattering to extract information about the internal structure of hadrons and other systems.

The first Born approximation is particularly simple for spherically symmetric potentials, where $V(r)$ can be moved out of the angular integration of Eq. (5.13). We then get

$$\tilde{V}(\mathbf{q}) \to \tilde{V}(q) = \frac{1}{2\pi^2} \int r^2 \, \frac{\sin(qr)}{qr} \, V(r) \, dr \tag{5.17}$$

and the scattering amplitude is given by the real function

$$f^{(1)}(\theta) = -\frac{2\mu}{\hbar^2} \int r^2 \, \frac{\sin(qr)}{qr} \, V(r) \, dr. \tag{5.18}$$

An interesting property of the Born approximation is that the elastic cross section at very forward angles is independent of the collision energy. In this situation $q\bar{R} \ll 1$, where $\bar{R}$ is the range of the potential, and we can approximate $\sin(qr)/qr \sim 1$. The cross section then takes the constant value

$$f^{(1)}(\theta = 0) = -\frac{2\mu}{\hbar^2} \int r^2 \, V(r) \, dr. \tag{5.19}$$

However, the Born approximation tends to be better at large momentum transfers and in some situations the Born series diverges at $q \simeq 0$.

One shortcoming of the first order Born approximation is that it violates unitarity, since the optical theorem is not obeyed. This can be easily checked in the case of an Hermitian potential with spherical symmetry. Inspecting Eq. (5.19), one immediately concludes that $\mathrm{Im}\,\{f^{(1)}(0)\} = 0$. Thus, the optical theorem leads to a vanishing total elastic cross section, which is not correct. From the practical point of view, however, this does not belittle the importance of this approximation. To guarantee unitarity one has to go to higher order terms in the Born series, where the T-matrix elements becomes complex. The second order term in the series is $V\, G_0^{(+)}(E)\, V$. Its imaginary part, which can be evaluated with the help of Eq. (4.32), is just $-\pi V\, \delta(E-K)\, V$. This term would render the optical theorem applicable (see exercise 2).

5.1.1.1 *Applications of the first Born approximation*

Now we illustrate the use of the first Born approximation with a few examples. We first consider the scattering of a particle of mass μ off the Yukawa potential

$$V(r) = \gamma \, \frac{e^{-r/b}}{r}. \tag{5.20}$$

Above, γ is the strength parameter[2] and b gives the range of the interaction. Inserting this potential into Eq. (5.17) and performing the integration, we obtain

$$T^{(1)}(q) = \tilde{V}(q) = \frac{\gamma b^2}{2\pi^2} \left[\frac{1}{1 + q^2 b^2} \right] \tag{5.21}$$

and the corresponding scattering amplitude is (see Eq. (4.64))

$$f^{(1)}(\theta) = -\frac{2\mu\gamma b^2}{\hbar^2 \left[1 + q^2 b^2\right]}. \tag{5.22}$$

Writing the momentum transfer in terms of the scattering angle and evaluating the cross section, we get

$$\frac{d\sigma^{(1)}(\theta)}{d\Omega} = \frac{\sigma_0}{\left[1 + q^2 b^2\right]^2} = \frac{\sigma_0}{\left[1 + 4k^2 b^2 \sin^2(\theta/2)\right]^2}, \tag{5.23}$$

where

$$\sigma_0 = \left(\frac{2\mu\gamma b^2}{\hbar^2} \right) \tag{5.24}$$

is the cross section at $\theta = 0$.

The results for the Yukawa potential can be used to derive the first Born approximation for Coulomb scattering. In a collision of particles with charges Q_P and Q_T, the Coulomb interaction corresponds to the Yukawa potential of Eq. (5.20) with the particular strength

$$\gamma = \gamma_C = \frac{Q_P \, Q_T}{4\pi\epsilon_0}, \tag{5.25}$$

and in the infinite-range limit. That is,

$$V_C(r) = \lim_{b \to \infty} \left[\gamma_C \, \frac{e^{-r/b}}{r} \right] = \frac{Q_P \, Q_T}{4\pi\epsilon_0 \, r}. \tag{5.26}$$

Replacing $\gamma \to \gamma_C$ in Eqs. (5.21) to (5.24), and letting $b \to \infty$, we obtain

$$T_C^{(1)}(q) = \frac{Q_P \, Q_T}{(2\pi)^3 \epsilon_0 \, q^2} \tag{5.27}$$

[2] Note that γ has the dimension of energy times length.

and

$$f_{\rm C}^{(1)}(\theta) = -\left(\frac{{\rm Q_P\,Q_T}}{4\pi\epsilon_0}\right)\frac{2\mu}{\hbar^2 q^2} = -\left(\frac{{\rm Q_P\,Q_T}}{4\pi\epsilon_0}\right)\frac{\mu}{2\hbar^2 k^2\sin^2(\theta/2)}. \qquad (5.28)$$

The above equation takes a more familiar form if we replace $\hbar^2 k^2/2\mu = E$ and introduce the half distance of closest approach in a head-on collision (see chapter 3),

$$a = \left(\frac{{\rm Q_P Q_T}}{4\pi\epsilon_0}\right)\frac{1}{2E}. \qquad (5.29)$$

We get

$$f_{\rm C}^{(1)}(\theta) = -\frac{a}{2\sin^2(\theta/2)} \qquad (5.30)$$

and the corresponding cross section is

$$\frac{d\sigma_{\rm C}^{(1)}(\theta)}{d\Omega} = \frac{a^2}{4\sin^4(\theta/2)}. \qquad (5.31)$$

We have here a very curious situation. The Born approximation for the Coulomb cross section is identical to the cross section of Eq. (3.35), which is exact. However, this does not mean that the Born approximation is reliable for coulomb Scattering. Comparing Eqs. (5.30) and (3.29), we conclude that the phase of $f_{\rm C}^{(1)}(\theta)$ is completely wrong, although its modulus is correct. Since the cross section depends only on the modulus, the first Born approximation leads to the correct result. However, other predictions of the Born approximation are wrong. Using this approximation for angular momentum projected quantities, for example, leads to completely wrong results.

We now discuss another application of the Born approximation. We evaluate phase shifts in the simple case of scattering from an attractive square well (see section 2.8). The starting point is the approximation of Eq. (2.72), which we reproduce below,

$$\sin\delta_l \simeq -\frac{1}{k}\left(\frac{2\mu}{\hbar^2}\right)\int dr\; V(r)\,\hat{\jmath}_l^2(kr).$$

It was obtained replacing the exact radial wave function by its free particle counterpart in the integrand of Eq. (2.71). This is essentially the first Born approximation. We used the above equation to calculate s-wave phase shifts for a few square well potentials. As in section 2.8, the strength of a well with depth V_0 and range $\bar{R}$ is measured by the quantity

$$K_0\bar{R} \equiv \frac{\sqrt{2\mu V_0}}{\hbar}\,\bar{R}. \qquad (5.32)$$

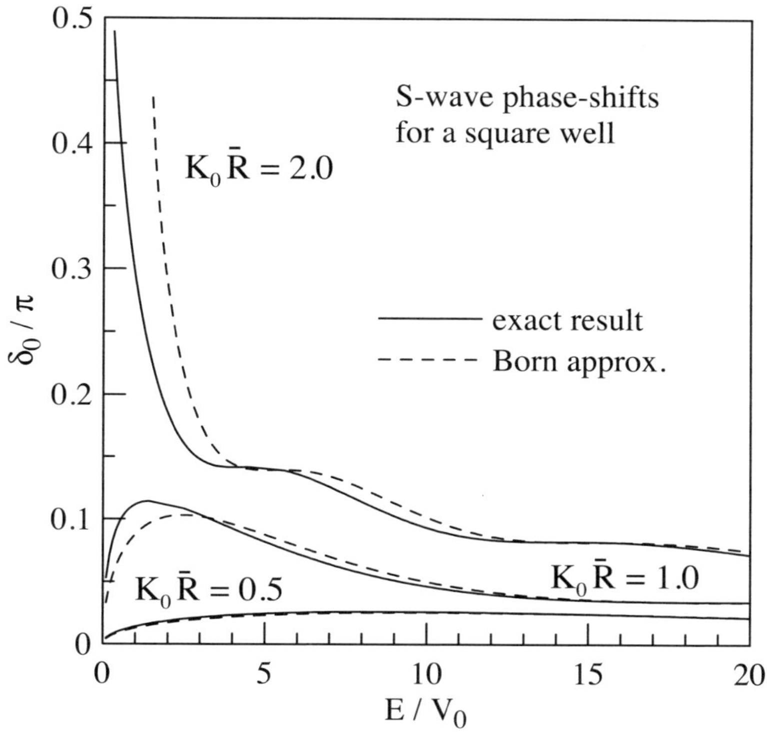

Fig. 5.1 Comparison between exact phase shifts and phase shifts obtained with the first Born approximation. The figure shows s-wave phase shifts normalized with respect to π and the results are for square well potentials of different strengths.

In that section it was shown that the value of the phase shift in the $E \to 0$ limit is given by Levinson Theorem,

$$\delta_0(E \to 0) = n_0 \pi, \tag{5.33}$$

where n_0 is the number of bound states of the potential. Potentials with $K_0\bar{R} < \pi/2$ have no bound states, while potentials with $\pi/2 < K_0\bar{R} < 3\pi/2$ have a single one (see section 2.8 for details).

In figure 5.1, we compare exact and approximate s-wave phase shifts for square well potentials with $K_0\bar{R} = 0.5, 1.0$ and 2.0. The first two have no bound states while the third is strong enough to bind one state. The phase shifts are normalized with respect to π and they are plotted against the collision energy divided by V_0. We see that the Born approximation works quite well for the two weaker potentials and also for the third one at large collision energies. On the other hand, for the strongest potential at $E/V_0 < 5$ the Born approximation is very poor.

Our last example is the first Born approximation in collisions of an electron with an atom with atomic number Z. The electron-atom interaction potential is

$$V(\mathbf{r}) = V_{\mathrm{e-Z}}(\mathbf{r}) + V_{\mathrm{e-e}}(\mathbf{r}), \tag{5.34}$$

with

$$V_{\mathrm{e-Z}}(\mathbf{r}) = -\frac{Ze^2}{4\pi\epsilon_0\, r} \tag{5.35}$$

and

$$V_{e-e}(\mathbf{r}) = \int d^3\mathbf{r}'\, \frac{e^2\, \rho(\mathbf{r}')}{4\pi\epsilon_0\, |\mathbf{r} - \mathbf{r}'|}. \tag{5.36}$$

The potentials $V_{\mathrm{e-Z}}$ and V_{e-e} represent respectively the attractive electron-nucleus interaction and the repulsion arising from the interaction of the incident electron with the electronic charge distribution of the atom. The density ρ is normalized by the condition

$$\int d^3\mathbf{r}'\, \rho(\mathbf{r}') = Z. \tag{5.37}$$

Before we discuss the Born approximation in this example, we point out that we are making two simplifying assumptions. The first is that the collision can be treated as a problem of potential scattering. In a proper treatment of the collision, the degrees of freedom of the atomic electrons should be taken into account. The density $\rho(\mathbf{r}')$ should be given by the operator

$$\hat{\rho}(\mathbf{r}') = \sum_{i=1}^{Z} \delta\left(\mathbf{r}' - \mathbf{r}_i\right), \tag{5.38}$$

rather than by a simple function of $\mathbf{r}'$. This operator is defined in the space of the intrinsic many-body wave functions of the atomic electrons, $\varphi_\alpha\left(\mathbf{r}_1, ..., \mathbf{r}_Z\right)$. In our *static* approximation of *frozen density*, we only consider the expectation value of this operator in the ground state of the atom. That is,

$$\rho(\mathbf{r}') = \left\langle \varphi_0 \left| \hat{\rho}(\mathbf{r}') \right| \varphi_0 \right\rangle. \tag{5.39}$$

Usually, this is not a good approximation. However, it is suitable for the qualitative discussion of this section[3].

The second approximation adopted here is the neglect of all effects of the Pauli Principle. Since the projectile and the electrons of the target atom are identical fermions, the wave function of the system should be fully anti-symmetrized for the exchange of any pair of electrons, including the projectile. This renders the problem much more complicated[4].

The first Born approximation can be easily obtained. Since in this approximation the scattering amplitude is linear with respect to $V_{\mathrm{e-Z}}$ and V_{e-e}, we can write

$$T^{(1)}(\mathbf{q}) = T^{(1)}_{\mathrm{e-Z}}(\mathbf{q}) + T^{(1)}_{\mathrm{e-e}}(\mathbf{q}). \tag{5.40}$$

The term $T^{(1)}_{\mathrm{e-Z}}$, associated with $V_{\mathrm{e-Z}}$, can readily be written. It is given by Eq. (5.27) for $\mathsf{Q}_{\mathrm{P}} = -e$ and $\mathsf{Q}_{\mathrm{T}} = Ze$. That is

$$T^{(1)}_{\mathrm{e-Z}}(q) = -\frac{Ze^2}{(2\pi)^3\,\epsilon_0\,q^2}. \tag{5.41}$$

The term $T^{(1)}_{\mathrm{e-e}}$ is given by,

$$\begin{aligned} T^{(1)}_{\mathrm{e-e}}(q) = \tilde{V}_{\mathrm{e-e}}(q) &= \frac{1}{(2\pi)^3}\int d^3\mathbf{r}\ \mathrm{e}^{-i\mathbf{q}\cdot\mathbf{r}}\ V_{\mathrm{e-e}}(\mathbf{r}) \\ &= \frac{1}{(2\pi)^3}\int d^3\mathbf{r}\ \mathrm{e}^{-i\mathbf{q}\cdot\mathbf{r}}\int d^3\mathbf{r}'\ \frac{e^2\ \rho(\mathbf{r}')}{4\pi\epsilon_0\ |\mathbf{r}-\mathbf{r}'|}. \end{aligned} \tag{5.42}$$

introducing the variable $\mathbf{x} = \mathbf{r} - \mathbf{r}'$ and inverting the order of integration, we get

$$T^{(1)}_{\mathrm{e-e}}(q) = \left[\int d^3\mathbf{r}'\ \mathrm{e}^{-i\mathbf{q}\cdot\mathbf{r}'}\rho(\mathbf{r}')\right] \times \left[\frac{1}{(2\pi)^3}\int d^3\mathbf{x}\ \mathrm{e}^{-i\mathbf{q}\cdot\mathbf{x}}\frac{e^2}{4\pi\epsilon_0 x}\right]. \tag{5.43}$$

Above, the integral within the second set of square brackets is the Fourier transform of the Coulomb potential for the charges $Q_{\mathrm{P}} = Q_{\mathrm{T}} = -e$. It is then given by Eq. (5.27). Introducing the Fourier transform of the charge distribution[5]

$$F(q) = \int d^3\mathbf{r}'\ \mathrm{e}^{-i\mathbf{q}\cdot\mathbf{r}'}\rho(\mathbf{r}'), \tag{5.44}$$

---

[3]Proper treatments of the many-body scattering problem will be considered in the second part of this book, in chapters 9 and 11.

[4]For a detailed discussion of this problem in collisions of identical particles and/or clusters of identical particles see chapter 6.

[5]We are using loosely the term Fourier transforms for the integrals $\tilde{V}(q)$ and $F(q)$. In fact they correspond to Fourier transforms divided by the constant factor $(2\pi)^{3/2}$.

and using Eq. (5.27), we get

$$T^{(1)}_{e-e}(q) = -\frac{e^2}{(2\pi)^3\epsilon_0\; q^2} \times F(q). \tag{5.45}$$

Combining the results of Eqs. (5.41) and (5.45) and using the relation between the T-matrix and $f(\theta)$, we obtain the first Born approximation for the scattering amplitude

$$f^{(1)}(q) = \frac{2Ze^2\mu}{4\pi\epsilon_0\hbar^2\; q^2}\left[1 - \frac{F(q)}{Z}\right]. \tag{5.46}$$

Since the mass of the incident electron, m_e, is negligible when compared to the atomic mass, we approximate $\mu \simeq m_e$. The cross section then can be written

$$\frac{d\sigma^{(1)}}{d\Omega} = \left(\frac{2Ze^2 m_e}{4\pi\epsilon_0\hbar^2\; q^2}\right)^2 \left|1 - \frac{F(q)}{Z}\right|^2. \tag{5.47}$$

The above equation can be expressed in terms of the scattering angle as

$$\frac{d\sigma^{(1)}}{d\Omega} = \left(\frac{Ze^2}{4\pi\epsilon_0 E}\right)^2 \left|1 - \frac{F(2k\sin(\theta/2))}{Z}\right|^2 \frac{1}{\sin^4(\theta/2)}. \tag{5.48}$$

Eq. (5.48) is very useful, since it provides a relation between the Fourier transform of the density of atomic electrons and an observable cross section.

5.1.2 *The Distorted Wave Born series*

In some situations the effects of the projectile-target interaction are too strong to be handled by Born approximation. This is, for example, the case of charged collision partners, where the long range Coulomb interaction cannot be treated by perturbation theory. The interaction is then written as the sum of the Coulomb interaction of two point charges plus a short-range term. The latter accounts for the charge distribution inside the projectile and the target (deviations from the point charge picture). In nucleus-nucleus collisions, the short-range part of the potential contains also the contribution from nuclear forces. Usually, these forces are are too strong to be treated by perturbation methods. This will be illustrated with an example later in this section. On the other hand, the DWBA is very useful in many-body scattering, where exact treatments of the collision may be very complicated. In typical situations, the projectile-target interaction is written as the sum of a dominant potential, which is diagonal in channel space, with weaker channel-coupling terms. These terms can then be treated as perturbations in the space of the waves distorted by the dominant potential. In such cases, the DWBA is valid and turns out to be a very useful approximation. Collisions of this kind are discussed in part II of this book (see chapters 9, 10 and 11). Although the DWBA is not very useful for applications in potential scattering, it is important to discuss it in this simple context. Once it is well understood, a generalization to many-body scattering becomes straightforward. Below, we recall the main results of sections 4.5 and 4.7, and illustrate the method with an application.

Let us consider the scattering off the potential

$$V(r) = V_1(r) + V_2(r), \tag{5.49}$$

where V_1 is the dominant part of the interaction and V_2 is a weaker potential, which can be handled by perturbation theory. The starting point for the DWBA is the Gell-Mann Goldberger relation (Eq. (4.152)),

$$T_{\mathbf{k}',\mathbf{k}} = T^{(1)}_{\mathbf{k}',\mathbf{k}} + \langle \chi^{(-)}(\mathbf{k}')| \, V_2 \, |\psi^{(+)}(\mathbf{k})\rangle \,, \tag{5.50}$$

where $T^{(1)}_{\mathbf{k}',\mathbf{k}}$ is the T-matrix for the potential V_1 alone (with $V_2 = 0$) and $\psi^{(+)}(\mathbf{k})$ is the exact solution of the scattering problem with incident wave vector $\mathbf{k}$. In Eq. (5.50), $\chi^{(-)}(\mathbf{k}')$ is the wave distorted by the strong potential $V_1(r)$ (with $V_2(r)$ switched off) with incident wave vector $\mathbf{k}'$ and ingoing asymptotic behavior. That is, it satisfies the equation

$$\Big[K + V_1\Big] \Big|\chi^{(-)}(\mathbf{k}')\Big\rangle = E_{k'} \Big|\chi^{(-)}(\mathbf{k}')\Big\rangle \tag{5.51}$$

and has the asymptotic form (see Eq. (4.144)),

$$\chi^{(-)}(\mathbf{k};\mathbf{r}) \to A \left(e^{i\mathbf{k}\cdot\mathbf{r}} + f_1^*(\pi - \theta)\, \frac{e^{-ikr}}{r} \right). \tag{5.52}$$

We recall that Eq. (5.50) is an exact result. In chapter 4, we have used this equation in the case of a Coulomb plus short-range interaction and obtained Eq. (4.225). The same idea was implicitly used in chapter 3, when we wrote $f(\theta)$ as the sum of the Coulomb amplitude with a correction arising from the short-range part of the potential.

Eq. (5.50) can be put in operator form with the help of the Møller wave operators associated with the V_1 and the total potential V, $\Omega_1^{(\pm)}$ and $\Omega^{(\pm)}$, respectively. These operators, which were defined in chapter 4, have the properties

$$\Omega_1^{(\pm)} \Big|\phi(\mathbf{k})\Big\rangle = |\chi^{(\pm)}(\mathbf{k})\rangle \tag{5.53}$$

$$\Omega^{(\pm)} \Big|\phi(\mathbf{k})\Big\rangle = |\psi^{(\pm)}(\mathbf{k})\rangle \,. \tag{5.54}$$

The distorted wave $|\chi^{(+)}(\mathbf{k})\rangle$, appearing in Eq. (5.53), is the solution of Eq. (5.51) with the outgoing wave boundary condition of Eq. (4.143). We obtain

$$T(E) = T^{(1)}(E) + \Omega_1^{(-)\dagger} V_2 \, \Omega^{(+)}. \tag{5.55}$$

The correction to the T-matrix arising from $V_2(r)$ then is

$$\Delta T(E) \equiv T(E) - T^{(1)}(E) = \Omega_1^{(-)\dagger} V_2 \, \Omega^{(+)}. \tag{5.56}$$

As we shall see in the following, the kernel involving the short range potential V_2 appears in a conspicuous way. As long as this kernel is of the Hilbert-Schmidt type, $\Delta T(E)$ can be expressed as a series. To obtain this series, we use the LS equation for the Møller operator, $\Omega^{(+)}$, namely

$$\Omega^{(+)} = \Omega_1^{(+)} + G_1^{(+)} \, V_2 \, \Omega^{(+)}, \tag{5.57}$$

where $G_1^{(+)}$ is the exact Green's function for the dominant potential, V_1. The distorted Born series then follows,

$$\Delta T(E) = \sum_{m=0}^{\infty} \Omega_1^{(-)\dagger} \, V_2 \left[G_1^{(+)} \, V_2 \right]^m \, \Omega_1^{(+)}. \tag{5.58}$$

The above series is called the Distorted Wave Born Series. The first term of this series,

$$\Delta T^{\mathrm{DWBA}}(E) = \Omega_1^{(-)\dagger} V_2 \, \Omega_1^{(+)}, \tag{5.59}$$

is called the Distorted Wave Born Approximations (DWBA). In the momentum representation, it is given by

$$\Delta T^{\mathrm{DWBA}}_{\mathbf{k}',\mathbf{k}} = \left\langle \chi^{(-)} \left(\mathbf{k}' \right) \right| V_2 \left| \chi^{(+)} \left(\mathbf{k} \right) \right\rangle . \tag{5.60}$$

Inserting this result into Eq. (4.152), we obtain

$$T^{\mathrm{DWBA}}_{\mathbf{k}',\mathbf{k}} = T^{(1)}_{\mathbf{k}',\mathbf{k}} + \left\langle \chi^{(-)} \left(\mathbf{k}' \right) \right| V_2 \left| \chi^{(+)} \left(\mathbf{k} \right) \right\rangle . \tag{5.61}$$

5.1.2.1 *Partial-wave projections*

For practical purposes, it is convenient to derive the partial-wave projected DWBA. The first step is to use the partial-wave expansions of the distorted waves[6]. Following the procedures of section 4.6, we write

$$\chi^{(+)}(\mathbf{k};\mathbf{r}) = \frac{1}{(2\pi)^{3/2}} \sum_{lm} 4\pi \; Y^*_{lm}(\hat{\mathbf{k}}) \; Y_{lm}(\hat{\mathbf{r}}) \; i^l \; e^{i\sigma_l} \; \frac{w_l(k,r)}{kr}, \tag{5.62}$$

[6]As frequently the dominant potential includes a Coulomb term, we are giving general expressions valid for long range potentials

and[7],

$$\chi^{(-)}(\mathbf{k};\mathbf{r}) \equiv \left(\chi^{(+)}(-\mathbf{k};\mathbf{r})\right)^* = \frac{1}{(2\pi)^{3/2}} \sum_{lm} 4\pi\, Y^*_{lm}(\hat{\mathbf{k}})\, Y_{lm}(\hat{\mathbf{r}})\, i^l\, e^{-i\sigma_l}\, \frac{w^*_l(k,r)}{kr}. \quad (5.63)$$

Above, $w_l(k,r)$ are radial wave functions associated with the potential V_1, which satisfy the equation

$$\left[E + \frac{\hbar^2}{2\mu}\left(\frac{d^2}{dr^2} - \frac{l\,(l+1)}{r^2}\right) - V_1(r)\right] w_l(k,r) = 0. \quad (5.64)$$

The angular momentum projected T-matrices are given by (see section 4.6)

$$T_l(E) = \left\langle \phi(Elm) \left| V^{(1)} + V^{(2)} \right| \psi(Elm) \right\rangle, \quad (5.65)$$

$$T^{(1)}_l(E) = \left\langle \phi(Elm) \left| V^{(1)} \right| \chi(Elm) \right\rangle, \quad (5.66)$$

where $\phi(Elm;\mathbf{r})$, $\chi(Elm;\mathbf{r})$ and $\psi(Elm;\mathbf{r})$ are eigenfunction of the energy and angular momentum operators, normalized as in Eq. (4.167). They are given by[8]

$$\phi(Elm;\mathbf{r}) = i^l \left(\frac{2\mu k}{\pi\hbar^2}\right)^{1/2} \frac{\hat{\jmath}_l(kr)}{kr} Y_{lm}(\hat{\mathbf{r}}) \quad (5.67)$$

$$\chi(Elm;\mathbf{r}) = i^l \left(\frac{2\mu k}{\pi\hbar^2}\right)^{1/2} \frac{w_l(k,r)}{kr} Y_{lm}(\hat{\mathbf{r}}) \quad (5.68)$$

$$\psi(Elm;\mathbf{r}) = i^l \left(\frac{2\mu k}{\pi\hbar^2}\right)^{1/2} \frac{u_l(k,r)}{kr} Y_{lm}(\hat{\mathbf{r}}). \quad (5.69)$$

Above, $\hat{\jmath}_l(kr)$ is the Ricatti-Bessel function (see section 2.1), $w_l(k,r)$ is the solution of Eq. (5.64) and $u_l(k,r)$ is the radial wave function for the full interaction, $V_1 + V_2$.

Using in Eq. (5.61) the partial wave expansions of Eqs. (5.67) to (5.69), one gets

$$T_l(E) = T^{(1)}_l(E) + \Delta T_l(E), \quad (5.70)$$

[7] We use the time-reversal property: $\chi^{(-)}(\mathbf{k};\mathbf{r}) = \left[\chi^{(-)}(-\mathbf{k};\mathbf{r})\right]^*$.

[8] If V_1 is just the point-charge Coulomb potential, we should replace $w_l(k,r)$ by the regular Coulomb function (see chapter 3), $F_l(\eta,kr)$, throughout this section.

with

$$T_l^{(1)}(E) = \frac{k}{\pi E} \int dr\, \hat{\jmath}_l(kr)\, V_2(r)\, u_l(k,r) \tag{5.71}$$

$$\Delta T_l(E) = \frac{k}{\pi E} \int dr\, w_l(k,r)\, V_2(r)\, u_l(k,r)\,. \tag{5.72}$$

These radial integrals are analogous to the one of Eq. (4.233).

The DWBA consists of replacing in Eq. (5.72) $u_l(k,r) \rightarrow w_l(k,r)$. That is

$$\Delta T_l^{\text{DWBA}} = \frac{k}{\pi E} \int dr\, w_l^2(k,r)\, V_2(r). \tag{5.73}$$

5.1.2.2 *An illustrative application of the DWBA*

The DWBA is an extremely useful approximation in problems of many-body scattering, where intrinsic degrees of freedom of the collision partners play very important roles. The intrinsic states of the system must then be taken explicitly into account. This can be done through the use of the coupled channel method, which will be extensively discussed in chapter 9. This method associates a radial wave function to each intrinsic state of the system (channel), and these wave functions are the solutions of a set of coupled differential equations. Since the complexity of the calculation grows rapidly with the number of channels, the calculations include only the channels which are strongly coupled to the entrance channel. The cross sections for the remaining channels are usually accounted for by approximate methods like the DWBA.

The use of the DWBA in problems of potential scattering does not present great advantages. The reason is that it usually requires numerical evaluation of radial distorted waves and the computational effort involved is similar to the one needed for the direct calculation of the exact wave functions. Nevertheless, a discussion of applications of this approximation in potential scattering is helpful to familiarize the reader with the DWBA, aiming at the more complex examples discussed in part 2 of this book.

As an example, we employ the DWBA to evaluate phase shifts in the scattering of two alpha-particles. This collision was discussed in section 3.4.1, to illustrate the scattering from a Coulomb plus short-range potential. The potential is written as

$$V(r) = V_{\text{C}}'(r) + V_{\text{N}}(r). \tag{5.74}$$

Above, V_{N} is the attractive nuclear potential, which we approximate by the gaussian

$$V_{\text{N}}(r) = -\,V_0\, e^{-r^2/\bar{R}^2}\,, \tag{5.75}$$

and V'_{C} is the modified Coulomb interaction (see problem 2 of chapter 3)

$$\begin{aligned} V'_{\text{C}}(r) &= \frac{1}{4\pi\varepsilon_0}\,\frac{4e^2}{r}, \quad \text{for } r > R_{\text{C}} \simeq R_{\text{P}} + R_{\text{T}} = 2R_\alpha, \\ &= \frac{1}{4\pi\varepsilon_0}\,\frac{4e^2}{2R_{\text{C}}}\left(3 - \frac{r^2}{R_{\text{C}}^2}\right), \quad \text{for } r < R_{\text{C}}. \end{aligned}$$

This interaction takes into account deviations from the usual potential between two point charges,

$$V_{\text{C}}(r) = \frac{1}{4\pi\varepsilon_0}\,\frac{4e^2}{r}, \tag{5.76}$$

at projectile-target distances where the proton densities of the two alpha-particles overlap.

The radial equation involves the effective l-dependent interaction, which can be put in the form

$$V_l(r) = V_{\text{C}}(r) + V_{\text{cent}}(r) + \bar{V}(r), \tag{5.77}$$

where $V_{\text{cent}}(r)$ is the usual centrifugal term,

$$V_{\text{cent}}(r) = \frac{\hbar^2}{2\mu r^2}\; l(l+1). \tag{5.78}$$

The the short-range part of the interaction can be written,

$$\bar{V}(r) = V_{\text{N}}(r) + \Delta V_{\text{C}}(r). \tag{5.79}$$

It is the sum of the nuclear term with the correction to the point charge Coulomb potential,

$$\Delta V_{\text{C}}(r) = V'_{\text{C}}(r) - V_{\text{C}}(r). \tag{5.80}$$

According to the notation of sections 4.5 and 4.7, the dominant part of the interaction is

$$V_1(r) = V_{\text{C}}(r) + V_{\text{cent}}(r) \tag{5.81}$$

and the perturbation is

$$V_2(r) = \bar{V}(r). \tag{5.82}$$

We have evaluated the phase shifts associated with the short-range interaction, $\bar{\delta}_l$, using the potential of section 3.4.1, with the same parameters ($V_0 = 60$ MeV and $\bar{R} = 4.5$ fm). We compared the phase shifts obtained from the asymptotic behavior of the exact radial wave function (see chapter 2) with the phase shifts calculated with the DWBA. For the latter, we first evaluate the ΔT_l^{DWBA} using Eq. (5.73). Since the dominant potential is just the point charge Coulomb interaction, the distorted waves are the regular Coulomb functions, $F_l(\eta, kr)$, which are real (see chapter 3). Eq. (5.73) then becomes

$$\Delta T_l^{\text{DWBA}} = \frac{k}{\pi E} \int dr\, F_l^2(\eta, kr)\, V_2(r). \tag{5.83}$$

Next, we use Eq. (4.235) to get the corresponding S-matrix,

$$\bar{S}_l^{\text{DWBA}} = 1 - 2\pi i\, \Delta T_l^{\text{DWBA}}. \tag{5.84}$$

The phase-shits are then given by the standard relation

$$\bar{S}_l^{\text{DWBA}} = e^{2i\bar{\delta}_l^{\text{DWBA}}}. \tag{5.85}$$

In figure 5.2 we show phase shifts for the lowest partial-waves[9], $l = 4, 6, 8, 10$. Although the adopted nuclear potential is not strong enough to have bound states, it is too strong for the DWBA. Comparing exact results (solid lines) with DWBA results (dashed lines), we conclude that the approximation is poor for all partial-waves, except at very low collision energies. The reason for the inaccuracy is clear in panel (b), where the effective potentials $V_l(r)$ (solid lines) are compared to the corresponding results when the nuclear potential is switched off (dashed lines). The comparison is made for the partial-waves $l = 0$ and $l = 10$. At small separations, the modulus of $\bar{V}_{l=0}(r)$ is over one order of magnitude larger than $V_{\text{C}}(r)$. Even at $l = 10$, where the centrifugal term is very large, the perturbation below $r = 8$ fm is of the same order of the dominant potential. For this reason the DWBA approximation does not work satisfactorily.

To emphasize this point, we have repeated this study using a weak nuclear potential. We used a Gaussian with depth $V_0 = 1$ MeV, instead of the previous value of 60 MeV. Although this is not a realistic value for this parameter, it helps understanding the physics involved. The results are presented in figure 5.3. Now the DWBA reproduces the exact phase shift very well, except in the case of $l = 0$. Again, it is clearly justified by the behavior of the potentials, exhibited in panel (b) of the same figure.

[9]Note that there are no odd partial-waves. This is because the projectile and target are identical particles with an even number of fermions (see section 6.3.3).

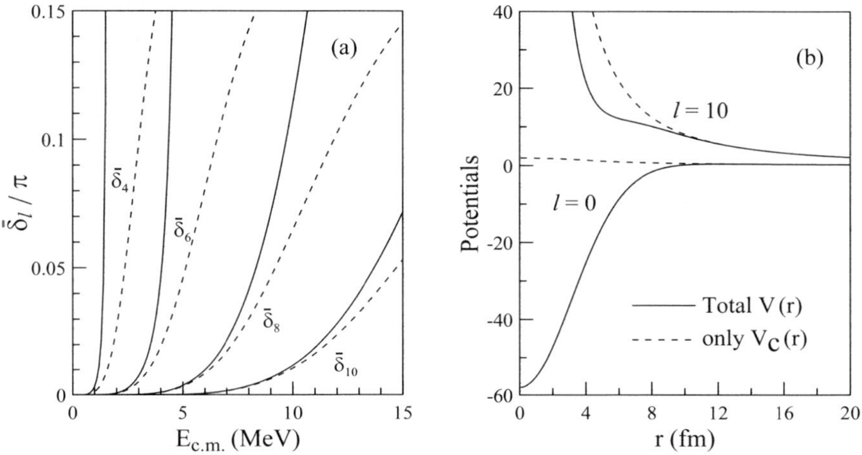

Fig. 5.2 (a) Comparison between exact (solid lines) and DWBA (dashed lines) phase shifts for a few partial-waves in α-α scattering. The phase shifts are normalized with respect to π; (b) Effective potentials in the radial equation at a few partial-waves, for the same collision.

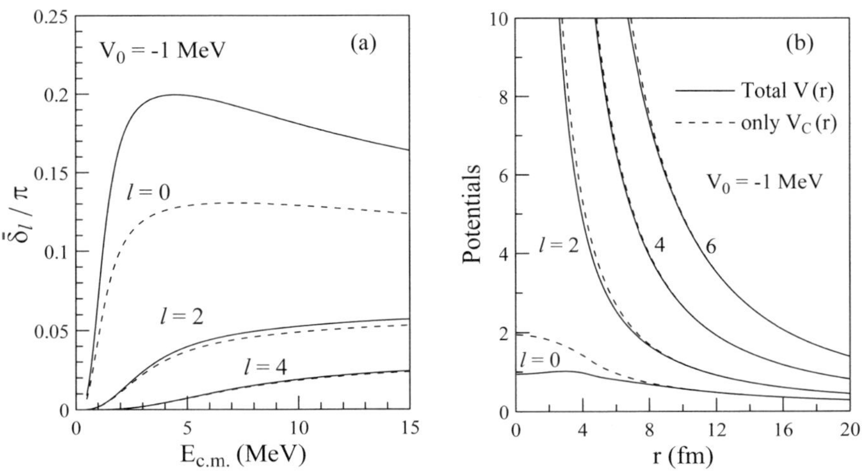

Fig. 5.3 Same as in figure 5.2, except that now the nuclear potential is very weak. For details, see the text.

Now the potential represented by the solid and the dashed lines are much closer, except for $l = 0$. Accordingly, the DWBA for this partial wave is considerably poorer than for the other ones[10]

Before finishing this section, we remark that in the present example the accuracy of the DWBA does not improve as the collision energy increases, as in the Born approximation. In applications where the dominant potential is repulsive, the projectile has to overcome a potential barrier to reach the region where the short-range interaction is important. For low collision energies, the classical turning point is beyond the reach of $\bar{V}(r)$. Therefore, the influence of the short range interaction on the collision is very weak. In this way, the DWBA is more successful in this energy range.

5.1.2.3 *Other perturbative approximations*

In this section we discuss an interesting representation of the T-matrix which is quite useful for discussing convergence issues, as well as developing approximations. This representation is derived using operator identities.

The Lippmann-Schwinger equation for the T-matrix is[11],

$$T = V + VG_0^{(+)}T, \tag{5.86}$$

where the T-matrix elements is

$$T_{\mathbf{k}',\mathbf{k}} \equiv \langle\, \phi(\mathbf{k}')|T|\phi(\mathbf{k})\,\rangle = \langle\, \phi(\mathbf{k}')|V|\psi^{(+)}(\mathbf{k})\,\rangle\,. \tag{5.87}$$

Above, the exact scattering state $|\psi^{(+)}(\mathbf{k})\rangle$ is calculated with the full action of the interaction V. If we now scale this interaction $V \to \lambda V$, then the LS equation becomes

$$T(\lambda) = \lambda V + \lambda V\, G_0^{(+)}\, T(\lambda). \tag{5.88}$$

We have used the notation $T(\lambda)$ to represent the transition operator associated with the scaled potential, λV. Then, using operator manipulations, one derive an expression for the derivative of $T(\lambda)$ with respect the scaling parameter, which ranges in values between 0 and 1. The result is

$$\frac{dT(\lambda)}{d\lambda} = V + VG_0\, T(\lambda) + \lambda V G_0^{(+)}\, \frac{dT(\lambda)}{d\lambda} \tag{5.89}$$

---

[10] We point out that the dashed lines represent the potentials with the nuclear part switched off. In this way, for $l = 0$ it corresponds to V_C'. Thus, it is not V_1 (point charge Coulomb interaction), which diverges at $r = 0$. Therefore, the deviation between V_C and the full potential at small distances is still much larger than the figure indicates.

[11] To simplify the notation, in this section we will omit the energy dependence of the T-matrix and of the Green's functions.

The above equation can be formally solved for $dT(\lambda)/d\lambda$ to give

$$\frac{dT(\lambda)}{d\lambda} = \Big[1 - \lambda V G_0^{(+)}\Big]^{-1} \Big[V + V G_0 T(\lambda)\Big]. \tag{5.90}$$

This result can be simplified with the help of Eq. (4.53). For this purpose, we take the conjugate of Eq. (4.53) for the potential λV. Namely,

$$\Omega_\lambda^{(-)^\dagger} = 1 + \Omega_\lambda^{(-)^\dagger} \lambda V\, G_0^{(-)^\dagger} \tag{5.91}$$

$$= 1 + \Omega_\lambda^{(-)^\dagger} \lambda V\, G_0^{(+)}. \tag{5.92}$$

Reorganizing the terms, the above equation becomes

$$\Omega_\lambda^{(-)^\dagger} \left[1 - \lambda V\, G_0^{(+)}\right] = 1 \quad \rightarrow \quad \left[1 - \lambda V\, G_0^{(+)}\right]^{-1} = \Omega_\lambda^{(-)^\dagger}. \tag{5.93}$$

Inserting this result into Eq. (5.90), we obtain

$$\frac{dT(\lambda)}{d\lambda} = \Omega_\lambda^{(-)^\dagger} \Big[V + V\, G_0\, T(\lambda)\Big]. \tag{5.94}$$

Using Eq. (5.88) to replace the term within square brackets by $T(\lambda)/\lambda$, the above equation becomes

$$\frac{dT(\lambda)}{d\lambda} = \frac{1}{\lambda}\, \Omega_\lambda^{(-)^\dagger} T(\lambda). \tag{5.95}$$

Now we derive an equation for the matrix elements of the transition operator. Taking matrix elements of Eq. (5.95). We get

$$\frac{dT_{\mathbf{k}',\mathbf{k}}(\lambda)}{d\lambda} = \frac{1}{\lambda} \left\langle \phi(\mathbf{k}') \left| \Omega_\lambda^{(-)^\dagger} T(\lambda) \right| \phi(\mathbf{k}) \right\rangle. \tag{5.96}$$

Using Eq. (4.62), Eq. (4.52) and its conjugate, Eq. (5.96) becomes

$$\frac{dT_{\mathbf{k}',\mathbf{k}}(\lambda)}{d\lambda} = \left\langle \psi_\lambda^{(-)}(\mathbf{k}') \left| V \right| \psi_\lambda^{(+)}(\mathbf{k}) \right\rangle. \tag{5.97}$$

Above, we used the notations $\psi_\lambda^{(\pm)}$ to stress the fact that they are solutions of the Schrödinger equation with the potential λV. Integrating the above equation over the scaling parameter from 0 to 1 yields the following interesting symmetrical representation of the T-matrix elements,

$$T_{\mathbf{k}',\mathbf{k}} = \int_0^1 d\lambda \left\langle \psi_\lambda^{(-)}(\mathbf{k}') \left| V \right| \psi_\lambda^{(+)}(\mathbf{k}) \right\rangle. \tag{5.98}$$

The interesting feature of the above equation is that the integrand involves a matrix element of the potential with the full scattering states in the scaled potential, both in the incident and outgoing waves. This symmetry, not present in the original form of the T-matrix elements, can be exploited to discuss convergence of the Born series, and to develop the high energy, small angle scattering, eikonal approximation, so that it is applicable in a wider range of scattering angles.

We venture to write the identity,

$$T = \int_0^1 d\lambda \, \Omega_\lambda^{(-)\dagger} \, V \, \Omega_\lambda^{(+)}. \tag{5.99}$$

It is easy to derive the usual Born approximations using the symmetrical representation of the T-matrix above. One just expands the Møller operators using their respective LS equations (Eqs. (4.1.4)),

$$\begin{aligned} \Omega_\lambda^{(+)} &= 1 + G_0^{(+)} \, (\lambda V) + \cdots \\ \Omega_\lambda^{(-)\dagger} &= 1 + \, (\lambda V) \, G_0^{(+)} \, + \cdots, \end{aligned}$$

in Eq. (5.99). Keeping terms up to first order in λ, one obtains

$$T \simeq V \int_0^1 d\lambda \, + 2 \, V \, G_0^{(+)} \, V \int_0^1 d\lambda \, \lambda = V \, + \, V \, G_0^{(+)} \, V, \tag{5.100}$$

which corresponds to the Lippmann-Schwinger series for the transition operator truncated above second order in the potential. That is, the second order Born approximation. Higher-order terms in the Born series can be derived in a similar way.

The extension of the symmetrical form of the T-matrix to the case of Coulomb plus short-range potentials, is easy. The long-range potential is left untouched, while the short-range potential is scaled by λ. In this way, the operators $\Omega_\lambda^{(\pm)}$ are associated with the potential $V_\lambda = V_{\mathrm{C}} + \lambda \bar{V}$. In this case, the correction to the T-matrix arising from the short- range potential is given by

$$\Delta T = \int_0^1 d\lambda \, \Omega_\lambda^{(-)\dagger} \, \bar{V} \, \Omega_\lambda^{(+)}. \tag{5.101}$$

The usual distorted wave Born series can be obtained from the LS for $\Omega_\lambda^{(\pm)}$ in terms of the Coulomb Green's function,

$$\begin{aligned} \Omega_\lambda^{(\pm)} &= \Omega_{\mathrm{C}}^{(\pm)} + G_{\mathrm{C}}^{(\pm)} \, \lambda \bar{V} \, \Omega_\lambda^{(\pm)} && (5.102) \\ &= \left[1 + G_{\mathrm{C}}^{(\pm)} \, \lambda \bar{V} + \left(G_{\mathrm{C}}^{(\pm)} \, \lambda \bar{V} \right)^2 + \cdots \right] \, \Omega_{\mathrm{C}}^{(+)}, && (5.103) \end{aligned}$$

where $G_{\mathrm{C}}^{(\pm)}$ and $\Omega_{\mathrm{C}}^{(+)}$ are respectively the Green's functions and the Møller operators associated with the long-range potential. The DWBA series is obtained through the replacement of the above expansion in Eq. (5.101) and truncating in increasing powers of λ. We leave this as an exercise.

5.2 Semiclassical approximations

In scattering theory, the term *semiclassical* is used for several approximations which may be based on different assumptions. However, they all assume that the de Broglie wavelength associated with the projectile-target separation degree of freedom is small compared with some characteristic length of the interaction. These approximations are particularly useful in many-body scattering, where sometimes the collision degrees of freedom are treated by classical mechanics. We discuss below the main semiclassical approximations in the context of potential scattering.

5.2.1 *The eikonal approximation*

This approximation is very useful and its generalization to many-body scattering, developed by Glauber (see Sec. 11.2.3), can be applied to many physical problems. It relies on the assumption that at high enough energies the trajectory of the scattered particle is basically a straight line and there is very small momentum transfer along the collision. This means that the momentum distribution of the scattering wave function is sharply peaked at the incident momentum. For an incident wave along the z-axis, with wave vector $\mathbf{k} = k\hat{\mathbf{z}}$, the eikonal approximation can be used if $\Delta k_z \ll k_z$ and $\Delta k_\perp \ll k_z$, where Δk_z and $\Delta k_\perp$ denote respectively the widths of the momentum distributions along the longitudinal and transversal directions. These conditions are met for high energies ($E \gg |V| \to k_z \gg \Delta k_z$) and small scattering angles ($\tan\theta = k_\perp / k_z \ll 1$).

To describe the eikonal approximation, we use the cylindrical coordinates, $\mathbf{r} \equiv \{z, b, \varphi\}$. We can write $\mathbf{r} = z\,\hat{\mathbf{z}}+\mathbf{b}$, where z extends from $-\infty$ to $+\infty$, and the impact parameter vector $\mathbf{b}$ subtends the azimuthal angle, φ, within the range $(0, 2\pi)$. Accordingly, the volume element is

$$d^3\mathbf{r} = b\,db\,dz\,d\varphi. \tag{5.104}$$

The first step to derive the eikonal approximation is to cast the scattering wave function in the factorized form[12]

$$\psi^{(+)}(\mathbf{k};\mathbf{r}) = \frac{e^{ikz}}{(2\pi)^{3/2}}\;\hat{\psi}^{(+)}(\mathbf{b}, z)\,. \tag{5.105}$$

The first factor at the RSH of the above equation is a momentum eigenstate with eigenvalue $\hbar k\,\hat{\mathbf{z}}$, whereas $\hat{\psi}^{(+)}$ contains the dispersion in momentum

[12] To simplify the notation, we do not include the incident wave vector in the argument of $\hat{\psi}^{(+)}$.

space resulting from the collision. Inserting the factorized wave function into the Schrödinger equation and reorganizing the terms, we get the equation

$$\left[2ik\,\frac{\partial}{\partial z} - \frac{2\mu}{\hbar^2}\,V(\mathbf{b},z)\right]\hat{\psi}^{(+)}(\mathbf{b},z) = \nabla^2\,\hat{\psi}^{(+)}(\mathbf{b},z)\,. \tag{5.106}$$

Since the eikonal wave function is assumed to be a sharp wave packet in momentum space, with $\hat{\psi}^{(+)}$ giving the dispersion around $k\hat{\mathbf{z}}$, we can assume that the relevant Fourier components of $\hat{\psi}^{(+)}$ have negligible momentum, as compared with its incident value. This leads to the relation: $\nabla^2\hat{\psi}^{(+)} \ll 2k\,\partial\hat{\psi}^{(+)}/\partial z$, which implies that the term at the right hand side of Eq. (5.106) is very small. The eikonal approximation consists of neglecting this term. In this way, the eikonal approximated wave function, $\hat{\psi}_{\mathrm{e}}^{(+)}$, satisfies the equation,

$$\left[2ik\,\frac{\partial}{\partial z} - \frac{2\mu}{\hbar^2}\,V(\mathbf{b},z)\right]\hat{\psi}_{\mathrm{e}}^{(+)}(\mathbf{b},z) = 0. \tag{5.107}$$

This equation can be immediately integrated and the result is

$$\hat{\psi}_{\mathrm{e}}^{(+)}(\mathbf{b},z) = \exp\left[-i\frac{k}{2E_k}\int_{-\infty}^{z} dz'\;V(\mathbf{b},z')\right], \tag{5.108}$$

or

$$\hat{\psi}_{\mathrm{e}}^{(+)}(\mathbf{b},z) = \exp\left[-\frac{i}{\hbar v}\int_{-\infty}^{z} dz'\;V(\mathbf{b},z')\right]. \tag{5.109}$$

Above, v is the velocity along the z-axis, which within the eikonal approximation has a constant value. Accordingly, the scattering wave function of Eq. (5.105) becomes

$$\psi_{\mathrm{e}}^{(+)}(\mathbf{k};\mathbf{b},z) = \frac{e^{ikz}}{(2\pi)^{3/2}}\;\exp\left[-i\frac{k}{2E_k}\int_{-\infty}^{z} dz'\;V(\mathbf{b},z')\right]. \tag{5.110}$$

A similar formula can be found for the ingoing solution, by merely performing the time reversal operation on the above outgoing solution. We get[13]

$$\left(\psi_{\mathrm{e}}^{(-)}(\mathbf{k};\mathbf{b},z)\right)^* = \frac{e^{-ikz}}{(2\pi)^{3/2}}\;\exp\left[-i\frac{k}{2E_k}\int_{z}^{\infty} dz'\;V(\mathbf{b},z')\right]. \tag{5.111}$$

The scattering amplitude within the eikonal approximation is obtained replacing in Eq. (4.45): $\psi^{(+)}(\mathbf{k};\mathbf{r}) \to \psi_{\mathrm{e}}^{(+)}(\mathbf{k};\mathbf{b},z)$. One gets

$$f_{\mathrm{e}}(\mathbf{q}) = -\frac{\mu}{2\pi\hbar^2}\int d^2\mathbf{b}\;e^{-i\mathbf{q}\cdot\mathbf{b}}\int_{-\infty}^{\infty} dz\,V(\mathbf{b},z)\,\hat{\psi}_{\mathrm{e}}^{(+)}(\mathbf{b},z)\,, \tag{5.112}$$

[13]Since we are usually interested in the complex conjugate of the ingoing solution, we write it directly.

where $\mathbf{q} = \mathbf{k}' - \mathbf{k}$ is the momentum transfer (in $\hbar$ units). Note that we have used the approximation, $\mathbf{q}\cdot\mathbf{r} = zq_z + \mathbf{q}_\perp\cdot\mathbf{r} \simeq \mathbf{q}_\perp\cdot\mathbf{r} = \mathbf{q}\cdot\mathbf{b}$, which corresponds to neglecting the longitudinal momentum transfer. This is justified within the conditions for applying the eikonal approximation, namely high energy and small scattering angles.

Using in Eq. (5.112) the explicit form of $\hat{\psi}_{\rm e}^{(+)}(\mathbf{b}, z)$ (Eq. (5.109)), one gets

$$f_{\rm e}(\mathbf{q}) = -\frac{\mu}{2\pi\hbar^2}\int d^2\mathbf{b}\, e^{-i\mathbf{q}\cdot\mathbf{b}}\int_{-\infty}^{\infty} dz\, V(\mathbf{b}, z)\, \exp\left[-i\frac{k}{2E_k}\int_{-\infty}^{z} dz'\, V(\mathbf{b}, z')\right]. \tag{5.113}$$

Now we change variable:

$$z \to w = -i\frac{k}{2E_k}\int_{-\infty}^{z} dz'\; V(\mathbf{b}, z'). \tag{5.114}$$

With this transformation, the differential and the integration limits become

$$dz = i\,\frac{2E_k}{kV(\mathbf{b}, z)}\; dw \tag{5.115}$$

$$z = -\infty \to w = 0 \tag{5.116}$$

$$z = \infty \to w = i\,\chi(\mathbf{b}). \tag{5.117}$$

Above, we have introduced the eikonal phase[14]

$$\chi(\mathbf{b}) = -\frac{k}{2E_k}\int_{-\infty}^{\infty} dz\;V(\mathbf{b}, z) \equiv -\frac{1}{\hbar v}\int_{-\infty}^{\infty} dz\; V(\mathbf{b}, z). \tag{5.118}$$

Eq. (5.113) then takes the form,

$$f_{\rm e}(\mathbf{q}) = -\frac{ik}{2\pi}\int d^2\mathbf{b}\; e^{-i\mathbf{q}\cdot\mathbf{b}}\int_0^{i\,\chi(\mathbf{b})} dw\; e^{w} \tag{5.119}$$

or, performing the integration over w and using Eq. (5.117),

$$f_{\rm e}(\mathbf{q}) = -\frac{ik}{2\pi}\int d^2\mathbf{b}\; e^{-i\mathbf{q}\cdot\mathbf{b}}\left[e^{i\,\chi(\mathbf{b})} - 1\right]. \tag{5.120}$$

It is convenient to express the above integral over $\mathbf{b}$ in terms of its modulus, b, and azimuthal angle $\varphi_{\mathbf{b}}$. We then have,

$$d^2\mathbf{b} = b\,db\,d\varphi_{\mathbf{b}} \quad \text{and} \quad -\mathbf{q}\cdot\mathbf{b} = bq\,\cos(\varphi - \varphi_{\mathbf{b}}), \tag{5.121}$$

[14] In this section, we use the symbol χ to denote the eikonal phase. Is should not be confused with the distorted waves $\chi^{(\pm)}$ appearing in other sections of this book.

where φ is the azimuthal angle associated with $-\mathbf{q}$. Note that φ is also the angle between the scattering plane and the plane x-z. Next, we change variable: $\varphi \to \phi = \varphi - \varphi_{\mathbf{q}}$, and the above equation takes the form[15]

$$f_{\mathrm{e}}(\mathbf{q}) = -\frac{ik}{2\pi} \int b\, db \int_0^{2\pi} d\phi\, e^{iqb\cos\phi} \left[e^{i\,\chi(\mathbf{b})} - 1\right]. \tag{5.122}$$

The eikonal scattering amplitude takes a simpler form when the eikonal phase has axial symmetry. In this case, the term within square brackets becomes independent of ϕ and one gets

$$f_{\mathrm{e}}(\mathbf{q}) = -ik \int b\, db \left[e^{i\,\chi(b)} - 1\right] \left(\frac{1}{2\pi} \int_0^{2\pi} d\phi\; e^{iqb\cos\phi}\right). \tag{5.123}$$

One identifies within the round brackets the cylindrical Bessel function of order zero,

$$J_0(qb) = \frac{1}{2\pi} \int_0^{2\pi} e^{qb\cos\phi}\, d\phi. \tag{5.124}$$

Thus, the scattering amplitude can be written

$$f_{\mathrm{e}}(\mathbf{q}) \to f_{\mathrm{e}}(q) = -ik \int b\, db \left[e^{i\,\chi(b)} - 1\right] J_0(qb). \tag{5.125}$$

At forward angles, one can use the relation [Abramowitz and Stegun (1972)]

$$P_l(\cos\theta) = \left(\frac{\theta}{\sin\theta}\right)^{1/2} J_0\left(\left[l + {}^1/_2\right]\theta\right). \tag{5.126}$$

For $\theta \approx 0$, $\sin\theta \simeq \theta$, $q = 2k\sin(\theta/2) \simeq k\theta$, $l + 1/2 \simeq kb$, and thus $P_l(\cos\theta) \simeq J_0(qb)$. Replacing

$$\int_0^{\infty} b\; db \;\to\; \frac{1}{k^2} \sum_{l=0}^{\infty} \left(l + {}^1/_2\right),$$

Eq. (5.125) becomes,

$$f_{\mathrm{e}}(\theta) = \frac{1}{2ik} \sum_{l=0}^{\infty} (2l+1) \left[e^{i\,\chi(b_l)} - 1\right] P_l(\cos\theta).$$

Comparing the above equation with the partial-wave expansion of Eq. (2.54a), we conclude that within the eikonal approximation the phase shift and the S-matrix are given by

$$\delta_l = \frac{1}{2}\,\chi(b_l), \quad S_l = e^{i\,\chi(b_l)}, \quad \text{with } b_l = \frac{l + {}^1/_2}{k}. \tag{5.127}$$

[15] In principle the change of variable leads to the new integration limits $\{-\varphi, 2\pi - \varphi\}$. If the integrand does not change under the transformation $\phi \to \phi + 2\pi$, the result of the integration remains unchanged. We assume that this condition is met and do not change the integration limits.

It is interesting to observe that Eq. (5.107), which determines the eikonal wave function, is quite similar to the time-dependent Schrödinger equation. Since $z = vt$, and v is taken to be a constant, the derivative with respect to z can be written as a derivative with respect to t: $v\,\partial/\partial z \to \partial/\partial t$. Inserting the factorized wave function of Eq. (5.105) in the time-dependent Schrödinger equation in the xy-plane ($\mathbf{b} \equiv x\,\hat{\mathbf{x}} + y\,\hat{\mathbf{y}}$), we get

$$\left[-\frac{\hbar^2}{2\mu}\,\nabla^2 + V(\mathbf{b}, vt)\right]\,\hat{\psi}^{(+)}(\mathbf{b}, t) = i\hbar\,\frac{\partial}{\partial t}\hat{\psi}^{(+)}(\mathbf{b}, t)\,. \tag{5.128}$$

In the eikonal approximation it is further assumed that the action of the kinetic energy operator on $\hat{\psi}^{(+)}$ can be neglected in the above equation. Solving the time-dependent equation exactly amounts to solving for the scattering wave function exactly and would take us well beyond the simple eikonal approximation since the kinetic energy operator is fully taken into into account. To exhibit the exact solution we first write

$$-\frac{\hbar^2}{2\mu}\nabla^2 = H_0 \tag{5.129}$$

and use the interaction representation for $\hat{\psi}^{(+)}(\mathbf{b}, t)$,

$$\hat{\psi}^{(+)}(\mathbf{b}, t) = e^{-\frac{i}{\hbar}H_0 t}\,\hat{\phi}^{(+)}(\mathbf{b}, t)\,. \tag{5.130}$$

The equation that governs the time evolution of $\hat{\phi}^{(+)}$ is then just

$$i\hbar\frac{\partial}{\partial t}\hat{\phi}^{(+)}(\mathbf{b}, t) = V(\mathbf{b}, vt)\,\hat{\phi}^{(+)}(\mathbf{b}, t)\,. \tag{5.131}$$

The above equation solves immediately for $\hat{\phi}^{(+)}$ as in Eq. (5.107). Accordingly, the exact solution which permits calculation of corrections to the eikonal approximation can be written down as

$$\hat{\psi}^{(+)}(\mathbf{k}; \mathbf{b}, t) = \frac{e^{ikvt}}{(2\pi)^{3/2}}\,e^{-\frac{i}{\hbar}H_0 t}\,\exp\left[-\frac{i}{\hbar}\int_{-\infty}^{\infty} dt\;V(\mathbf{b}, vt)\right]. \tag{5.132}$$

We observe that H_0 and $V(\mathbf{b}, vt)$ are two non-commuting operators. To calculate the effect of H_0, one resorts to perturbation through an expansion of the exponential $\exp\left[-iH_0 t/\hbar\right]$ in powers of H_0, namely

$$e^{-\frac{i}{\hbar}H_0 t} = 1 - \frac{i}{\hbar}H_0 t - \frac{1}{2\hbar^2}\,H_0^2 t^2 + \cdots \tag{5.133}$$

The operation of H_0 on the eikonal phase involves derivatives of the potential with respect to b and z. We leave as an exercise the calculation of the first such correction.

The eikonal approximation can be applied to many problems in atomic and nuclear physics. One example is the nuclear scattering of neutrons at forward angles. In this case, the interaction potential is

$$V_{\mathrm{N}}(r) = -\frac{V_0 + iW_0}{1 + \exp\left[\left(r - R_0\right)/a\right]}$$

and the eikonal phase is given by the integral

$$\chi_{\mathrm{C}}(b) = -\int_{-\infty}^{+\infty} dz\; V_{\mathrm{N}}\left(\sqrt{b^2 + z^2}\right). \tag{5.134}$$

The use of the eikonal approximation in Coulomb scattering presents difficulties. They arise from the infinite range of the Coulomb field. Writing

$$V_{\mathrm{C}}(\mathbf{b}, z) = \frac{Q_{\mathrm{P}}\, Q_{\mathrm{T}}}{4\pi\epsilon_0\, \sqrt{z^2 + b^2}}, \tag{5.135}$$

the eikonal phase takes the form

$$\chi_{\mathrm{C}}(b) = -\eta \int_{-\infty}^{+\infty} dz\; \frac{1}{\sqrt{z^2 + b^2}}, \tag{5.136}$$

where $\eta = Q_{\mathrm{P}}\, Q_{\mathrm{T}}/4\pi\epsilon_0\, \hbar v$ is the usual Sommerfeld parameter (see chapter 3). The problem is that this integral diverges. Thus, the eikonal approximation cannot be used in the scattering of two point charges. To circumvent this, Glauber [Glauber (1959)], resorted to a physically motivated modification of the Coulomb interaction, namely screening. The actual potential felt by the projectile is not the one of Eq. (5.135), but rather

$$V_{\mathrm{C}}(\mathbf{b}, z) = \frac{Q_{\mathrm{P}}\, Q_{\mathrm{T}}}{4\pi\epsilon_0\, \sqrt{z^2 + b^2}}\; F(r). \tag{5.137}$$

The screening factor, F, is taken as the simple exponential

$$F(r) \equiv F\left(\sqrt{z^2 + b^2}\right) = \exp\left[-\frac{\sqrt{z^2 + b^2}}{\alpha}\right], \tag{5.138}$$

with α being orders of magnitudes larger than the size of the system. In the case of nuclei, where Glauber [Glauber (1959)] employed his analysis, the charge of the nucleus is not the number of protons Z times e, as it is shielded by the atomic electrons surrounding the nucleus. Such a shielding is referred to as atomic screening and α is normally of the order of the radius of the atom (about four orders of magnitude larger than the nuclear radius).

For angles not too close to zero[16], the screening modified Coulomb eikonal can be evaluated in closed form. In this case, it is given by [Glauber (1959)]

$$\chi_{\mathrm{C}}(b) = 2\eta \left[\ln\left(\frac{b}{2\alpha} \right) - \gamma \right], \tag{5.139}$$

where γ is Euler's constant, $\gamma = 0.57721$. It is important to note that the factor $\ln(b/2\alpha)$ in the above expression comes about even if the screening is treated through a mere cutting off the z-integral through the condition,

$$F(r) = \Theta(\alpha - r), \tag{5.140}$$

where $\Theta(x)$ is the step function. Thus the exponential form brings in the Euler constant, but does not change the impact parameter dependence of the case with the sharp-cutoff screening above. This result has the inconvenient feature of depending on the particular value of the screening parameter, α. The same happens with the scattering amplitude. However, this dependence is limited to the phase of $f(\theta)$ and the eikonal approximation leads correctly to the Rutherford cross section [Glauber (1959)].

The importance of the eikonal approximation resides in the simple linear dependence of the phase of the wave function on the interaction potential. In particular, the two-potentials scattering problem within the eikonal theory is realized through two distinct contributions to the eikonal phase. In typical nuclear collisions, the interaction is the sum of the Coulomb interaction with the nuclear potential and the eikonal phase can be written

$$\chi(b) = \chi_{\mathrm{C}}(b) + \chi_{\mathrm{N}}(b). \tag{5.141}$$

5.2.1.1 *Improved versions of the eikonal approximation*

The eikonal approximation assumes that the momentum transfer is very small as compared with the incident momentum, so that the projectile's trajectory is close to a straight line. This condition may not be satisfied in low energy nuclear physics. This is frequently the case in heavy ion collisions, where the Coulomb field is very strong and the trajectories deviate from straight lines, even for large impact parameters.

As an example of the use of the eikonal approximation in heavy ion collisions, we discuss the elastic scattering of ^{12}C projectiles off a ^{16}O target,

[16]By *angles too close to zero*, we mean $\Theta \ll 1/k\alpha$ or, equivalently, $b \ll \alpha$.

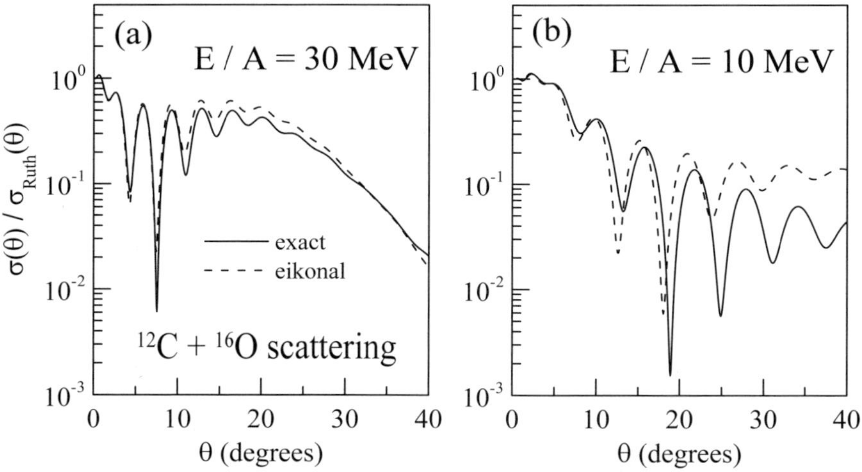

Fig. 5.4 Elastic cross sections for the $^{12}C + {}^{16}O$ collision at $E/A = 30$ MeV (panel (a)) and $E/A = 10$ MeV (panel (b)). The solid lines and the dashed lines represent respectively results obtained with full quantum mechanics and with the eikonal approximation. The results are normalized with respect to the Rutherford cross sections.

at the collisions energies[17] $E/A = 30$ and 10 MeV. In figure 5.4 we show the eikonal cross sections calculated by Aguiar, Zardi and Vituri [Aguiar *et al.* (1997)], normalized with respect to the Rutherford cross section. The calculations were performed with the nuclear potential,

$$V(r) = \frac{V_{0,r}}{1 + \exp\left[(r - R_r)/a_r\right]} + i\,\frac{V_{0,i}}{1 + \exp\left[(r - R_i)/a_i\right]},$$

with the parameters: $V_{0,r} = -63.7$ MeV, $R_r = 5.1$ fm, $a_r = 0.63$ fm and $V_{0,i} = -27.2$ MeV, $R_i = 5.1$ fm, $a_i = 0.69$ fm. Further detail of the calculations can be found in [Aguiar *et al.* (1997)]. For comparison, the exact cross sections are also shown. At $E/A = 30$ MeV, the results of the eikonal approximation are not far from the exact ones. They agree very well below 10°, the agreement gets worse between 10° and 30° and then improves between 30° and 40°. The situation for $E/A = 10$ MeV is much worse. The agreement is reasonable below 10° but deteriorates rapidly at higher angles. At $\theta \geq 30^o$, the cross section of the eikonal approximation is nearly one order of magnitude higher than the results of full quantum

[17] It is usual in Nuclear Physics to express collision energies in MeV per nucleon (E/A). To obtain the collision energy in the laboratory frame one should then multiply by the projectile's mass number.

mechanics. In such cases, this approximation cannot be applied. There are, however, simple methods to improve the eikonal approximation. Some of these methods are discussed below.

Wallace [Wallace (1973)] wrote the scattering amplitude of Eq. (5.125) as

$$f(q) = -ik \int_0^\infty b\, db \left[\exp\left(i\chi^{\mathrm{W}}(b)\right) - 1 \right] J_0(qb), \tag{5.142}$$

but leaving $\chi^{\mathrm{W}}(b)$ as a function to be determined. It was then written as the power series of $1/k$,

$$\chi^{\mathrm{W}}(b) = \chi_0(b) + \frac{\chi_1(b)}{k} + \frac{\chi_2(b)}{k^2} + \cdots, \tag{5.143}$$

where the lowest order term, $\chi_0(b)$, is the eikonal phase. Closed (but complicated) expressions were found for the two subsequent corrections, $\chi_1(b)/k$ and $\chi_2(b)/k^2$.

Other approaches introduce corrections to the trajectory. Instead of evaluating the eikonal phase along the straight line determined by the impact parameter and the incident velocity, they use a straight line determined by the values of the distance and the velocity at closest approach. That is, the radial turning point is corrected, taking into account the potentials, and the velocity is modified as to conserve angular momentum. There are two possibilities: (*i*) the corrections only take into account the dominant Coulomb potential and (*ii*) they include also the effects of the nuclear potential. If the nuclear potential has an imaginary part, the corrected turning point will be complex. We follow here the approach of [Aguiar *et al.* (1997)], where details of the method and references to earlier works can be found.

The Coulomb and nuclear contributions to the eikonal are separated as in Eq. (5.141), but they modified the nuclear part following the prescription,

$$\chi_{\mathrm{N}}(b) \rightarrow \tilde{\chi}_{\mathrm{N}}(b) = -\frac{1}{\hbar v_{\mathrm{ca}}} \int_{-\infty}^{\infty} dz\, V_{\mathrm{N}}(d_{\mathrm{ca}}, z). \tag{5.144}$$

Above, $\tilde{\chi}_{\mathrm{N}}(b)$ is the nuclear eikonal evaluated along a modified straight line trajectory determined by the corrected turning point, d_{ca}, and the velocity at closest approach, v_{ca}. These quantities are determined by the classical trajectory generated by the Hamiltonian with the full interaction potential.

In figure 5.5, we show the elastic cross sections of the previous figure at the energy $E/A = 10$ MeV, where direct application of the eikonal approximation lead to poor results. Now we evaluate the scattering amplitude

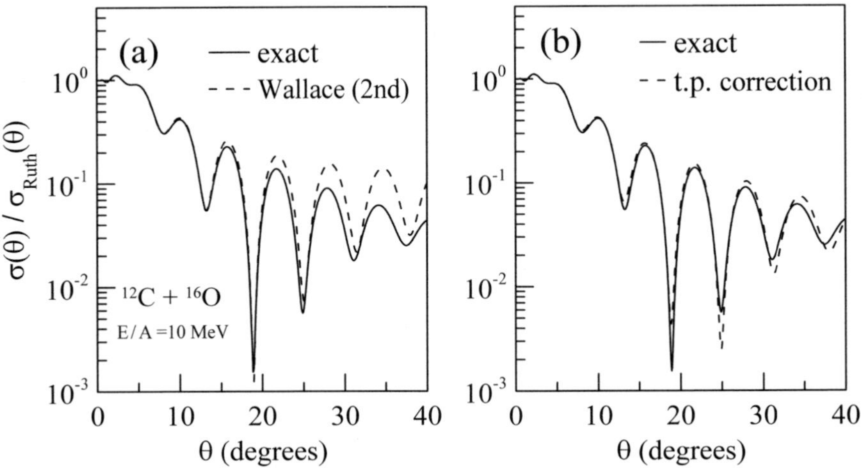

Fig. 5.5 Improved eikonal approximations for $^{12}C + ^{16}O$ elastic scattering at $E/A = 10$ MeV. As the previous figure, the results are normalized with respect to the Rutherford cross section. The solid lines represent exact results whereas the dashed lines correspond to the improved eikonal approximations of Wallace (panel (a)) and of turning point correction (panel (b)).

using improved forms of the nuclear eikonal. On panel (a), the dashed line represents the results of the expansion of Eq. (5.143), taken to second order. The agreement with the exact results is clearly better than in the case of the original eikonal approximation. On panel (b) of the same figure, one finds the results of the eikonal with turning point correction (Eq. (5.144)). The agreement with the exact results is excellent. We remark that in this calculation it is important to include nuclear effects in the turning point correction. The situation is simpler in collisions of heavier nuclei, where the Coulomb potential is dominant and the nuclear effects on the trajectory are less pronounced [Aguiar *et al.* (1997)].

5.2.1.2 *Extension to many-body systems*

In applications to charged particle scattering and collision with many-body target systems at high energies, Glauber [Glauber (1959)] developed the theory that carries his name. He employed, among other things, the screening-modified Coulomb eikonal discussed earlier in this section. The importance of the eikonal approximation resides in the simple linear dependence of the phase of the wave function on the interaction potential. One example is the above discussed two-potential scattering problem, where the eikonal phase is given by $\chi(b) = \chi_C(b) + \chi_N(b)$. The Coulomb phase becomes less important at higher energies and angles not close to zero. In this case, $\chi_N(b)$ becomes dominated by the absorption due to the imaginary part[18] of $V_N(r)$. The resulting angular distribution is a typical Fraunhofer diffraction (see section 7.5.1). Here one sees a decaying rather regular oscillatory shape with an angular period directly related to the radius of the system. Through an analysis of the elastic cross section, one can extract the radius of the nucleus and compare it with that extracted from electron scattering data, using the first-order Born approximation. Glauber explored this linear dependence of the eikonal on the potentials to relate the scattering amplitude directly to the density distribution of charge or matter. In a sense, this makes the eikonal approximation a powerful tool to study the structure of many-body targets which goes far beyond the simple perturbative Born approximation discussed in previous sections. In another sense, one may say that the eikonal/Glauber approximation is a re-summation of the Born series. The demonstration of this is rather complicated as one needs the high energy limit of the free Green's function.

We now consider the scattering of protons from nuclei at intermediate energies (a few MeV's bombarding energy). The potential $V(r)$ is the Optical Potential (OP) describing the average interaction of the proton with the nucleons in the target nucleus (see chapter 10). The form of this OP is given by the Lax interaction [Lax (1951)] and is

$$V(r) = \bar{t}_{NN}(E)\ \rho(r), \tag{5.145}$$

where $\bar{t}_{NN}(E)$ is the nucleon-nucleon T-matrix element taken at zero momentum transfer[19] and $\rho(r)$ is the density of the target nucleus. Since $\bar{t}_{NN}(E)$ is complex, the potential $V(r)$ is complex too. The imaginary part of the Lax optical potential represents processes where a nucleon inside the target is knocked out. This imaginary part can be calculated using the

[18]The modifications of scattering theory to include non-hermitian potentials are described in chapter 7 and the meaning of complex potentials is discussed in chapter 10.
[19]We adopt the notation $\bar{t}_{NN}(E)$ to indicate a different normalization of the free states. In Eq. (5.145) the normalization factor for the plane waves is $A = 1$, while we use throughout this book $A = (2\pi)^{3/2}$. With the normalization of [Lax (1951)], the relation between the zero momentum transfer T-matrix and the forward scattering amplitude becomes

$$f(\theta = 0) = -\frac{1}{2\pi}\left(\frac{\mu}{\hbar^2}\right)\bar{t}_{NN},$$

instead of our Eq. (4.64).

optical theorem for the NN system[20],

$$\mathrm{Im}\{V(r)\} = \mathrm{Im}\{\bar{t}_{\mathrm{NN}}(E)\}\ \rho(r) = -\frac{E_k}{k}\ \sigma_{\mathrm{NN}}(E)\ \rho(r), \tag{5.146}$$

where $\sigma_{\mathrm{NN}}(E)$ is the total NN cross section known and tabulated. It is customary to write for the real part of the interaction

$$\mathrm{Re}\{V(r)\} = \alpha(E) \times \mathrm{Im}\{V(r)\}.$$

The values of $\alpha(E)$ are also known and tabulated. With this optical potential, the eikonal/Glauber phase $\chi(b)$ can be evaluated as an integral over the density profile of the target nucleus. Before we present this, we consider the limiting case of a sharp matter distribution for the density,

$$\rho(r) = \rho_0\ \Theta(R - r), \tag{5.147}$$

where $\Theta(x)$ is the step function. Then the eikonal phase becomes,

$$\begin{aligned}\chi(b) &= -\frac{k}{2E_k}\bar{t}_{\mathrm{NN}}(E)\ \rho_0 \int_{-\infty}^{\infty} dz\ \Theta(R - \sqrt{b^2 + z^2}) \\ &= -\frac{k}{E_k}\bar{t}_{\mathrm{NN}}(E)\ \rho_0 \sqrt{R^2 - b^2}\ \ \Theta(R - b),\end{aligned} \tag{5.148}$$

and the corresponding scattering amplitude is

$$f(q) = ik \int_0^R b\, db \left[1 - \exp\left(-i\frac{k}{E_k}\ \bar{t}_{\mathrm{NN}}(E)\, \rho_0\ \sqrt{R^2 - b^2}\right)\right]\ J_0(qb). \tag{5.149}$$

The above expression can be evaluated using the near/far decomposition to be described later.

In general, the density function has a smooth variation with r, and the eikonal phase is calculated as

$$\chi(b) = -\frac{k}{2E_k}\ \bar{t}_{\mathrm{NN}}(E) \int_{-\infty}^{\infty} dz\ \rho(\sqrt{z^2 + b^2}). \tag{5.150}$$

It is known that for light nuclei such as ^{12}C, the density is approximately a Gaussian, while for heavy targets such as ^{208}Pb, the density has a Fermi shape, namely,

$$\rho(r) = \frac{\rho_0}{1 + \exp\left[(r - R)/a\right]}. \tag{5.151}$$

[20]Note that the constant factor $-E_k/k$ below corresponds to waves normalized with $A = 1$ (see previous footnote).

The Gaussian case can be worked out analytically, but the Fermi-shape case can only be performed numerically. Approximate evaluation of the eikonal/Glauber phase for the latter, however, can be performed by extending the r-integral into the complex r-plane. Note, that the Fermi function has poles when

$$\exp\left[\frac{r-R}{a}\right] = -1 \quad \rightarrow \quad r = R + i\, m\pi a. \tag{5.152}$$

That is, at

$$r = R + i\, m\pi a, \quad m = 1, 3, \cdots . \tag{5.153}$$

Using Cauchy residue theorem the r-integral, or equivalently, the z-integral, can be evaluated by taking into account the first one or two poles. By taking into account only the first pole, the form factor $F(q)$ can be written as

$$F(q) = \frac{8\pi^2 \rho_0 a}{q}\, e^{-\pi a q}\, [\pi a \sin qR - R\cos qR] \tag{5.154}$$

and the eikonal phase can be analyzed in a similar way.

An important property of the eikonal/Glauber phase is its additivity. If the interaction is composed of several pieces as

$$V(r) = \sum_j V_j(r), \tag{5.155}$$

then the eikonal phase is

$$\chi(b) = \sum_j \chi_j(b). \tag{5.156}$$

This implies that

$$e^{i\,\chi(b)} = \prod_j e^{i\,\chi_j(b)} \tag{5.157}$$

and the scattering amplitude becomes

$$f(q) = 2\pi \int_0^\infty b\, db \left[1 - \prod_j e^{i\,\chi_j(b)}\right] J_0(qb). \tag{5.158}$$

The above representation of the amplitude is very useful in application to scattering involving many-body systems.

5.2.2 *The WKB approximation*

The Wentzel-Kramers-Brillouin (WKB) approximation is a well known method that has been applied to several problems of quantum mechanics. It relies on the close connection between the canonical formulation of classical mechanics through the action-angle canonical variables and the short wave length condition encountered in some of the problems in quantum mechanics. The latter condition is that the collision energy is high enough for the potential to have a small variation over a radial distance corresponding to one de Broglie wave length. In this section we discuss the WKB approximation in scattering problems.

We start with the simple case of a collision in one dimension described by the coordinate x, varying in the range $\{-\infty, \infty\}$. A particle of mass μ and energy E approaches a potential $V(x)$ from the right, as represented in figure 5.6. In collisions with a repulsive potential, as the one represented in the figure, classical physics restricts the motion to the classically allowed region, where the collision energy is larger than the repulsive potential. This region is delimited by the coordinate x_0, given by the condition $V(x_0) = E$. This coordinate is the turning point, where an incident classical particle would be reflected back. However, quantum mechanics gives assess also to the classically forbidden region ($x < x_0$).

For slowly varying potentials, the local wave number,

$$k(x) = \frac{1}{\hbar} \sqrt{2\mu \, [E - V(x)]}, \tag{5.159}$$

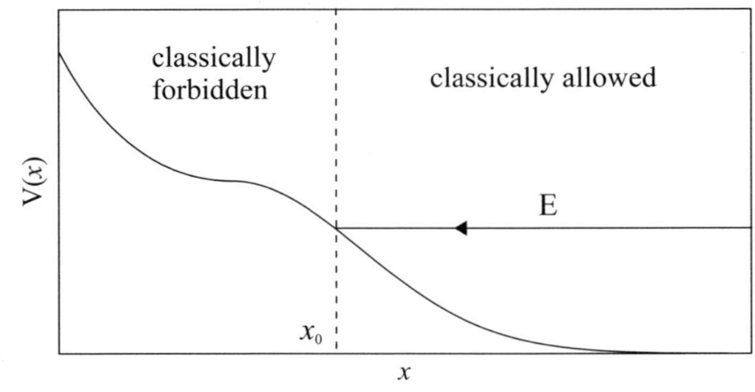

Fig. 5.6 Schematic representation of a one-dimensional collision of a particle with energy E. The classical turning point is represented by x_0.

is expected to suffer a small change over one de Broglie wave length, $\lambda\!\!\!^{-}$. That is, calling the variation $\Delta k = |k(x + \lambda\!\!\!^{-}/2) - k(x - \lambda\!\!\!^{-}/2)|$, one should have[21]

$$\left|\frac{\Delta k}{k}\right| \simeq \left|\frac{k'\,\lambda\!\!\!^{-}}{k}\right| = \left|\frac{k'}{k^2}\right| \ll 1. \tag{5.160}$$

It is clear that the above condition cannot be satisfied in the vicinity of the turning point, where the wave number vanishes. This is the source of considerable complications that can only be handled by the connection formulae discussed later in this section.

The WKB ansatz is the following expression for the wave function,

$$\psi(x) = A(x)\; e^{i\,\mathcal{S}(x)/\hbar}. \tag{5.161}$$

Above, $A(x)$ and $\mathcal{S}(x)$ are real functions of x, giving respectively the modulus and the phase of the wave function. It will be clear latter that $\mathcal{S}(x)$ corresponds to the classical action. In the trivial case of a constant potential, which is simply a change in the energy scale, one gets the free particle wave function of the previous chapters. The wave function is given by Eq. (5.161) with $A(x) = (2\pi)^{-3/2}$ and $\mathcal{S}(x) = k_0\, x$, with the notation $k_0 \equiv k(x \to \infty)$. This implies that $A'(x) = 0$ and $\mathcal{S}''(x) = 0$. It is then reasonable to assume that if the potential is slowly varying (but not constant), $A'(x)$ and $\mathcal{S}''(x)$ are small. This is the basis of the WKB approximation.

We now insert the ansatz of Eq. (5.161) into the Schrödinger equation,

$$\left[\frac{d^2}{dx^2} + k^2(x)\right]\; A(x)\, e^{i\,\mathcal{S}(x)/\hbar} = 0. \tag{5.162}$$

Evaluating the second derivative and separating real and imaginary parts, one gets,

$$\hbar^2\, A''(x) + \hbar^2\, k^2(x)\, A(x) = A(x)\; \left[\mathcal{S}'(x)\right]^2 \tag{5.163}$$

and

$$2\, A'(x)\, \mathcal{S}'(x) + A(x)\, \mathcal{S}''(x) = 0. \tag{5.164}$$

The lowest order approximation consists of neglecting the term in $A''(x)$ in Eq. (5.163). This equation then reduces to[22]

$$\left[\mathcal{S}'(x)\right]^2 = \hbar^2\, k^2(x), \tag{5.166}$$

[21]Primes stand for derivatives with respect to x.

[22]Dividing both sides of Eq. (5.166) by 2μ, one identifies the equation satisfied by Hamilton's characteristic function in classical mechanics [Goldstein (1980)],

$$H\left(\partial\mathcal{S}(x)/dx, x\right) = E \tag{5.165}$$

and its solution is $\mathcal{S}(x) = \int_{x_0}^{x} dx'\, \hbar k(x')$.

or

$$\mathcal{S}'(x) = \pm\hbar\, k(x). \tag{5.167}$$

The two linearly independent solutions of this equation are immediately obtained as[23]

$$\mathcal{S}(x) = \pm\hbar \int_{x_0}^{x} dx'\; k(x'). \tag{5.168}$$

To obtain the WKB approximation, one uses Eq. (5.167) and its derivative, $\mathcal{S}''(x) = \pm\hbar k'(x)$, in Eq. (5.164). We get

$$\frac{A'(x)}{A(x)} = -\frac{1}{2}\,\frac{k'(x)}{k(x)}, \tag{5.169}$$

which has the solution

$$A(x) = \frac{A_0}{\sqrt{k(x)}}. \tag{5.170}$$

The linearly independent WKB wave functions then are

$$\psi_{\pm}^{\mathrm{WKB}}(x) = \frac{A_0}{\sqrt{k(x)}}\; \exp\left[\pm i \int_{x_0}^{x} dx'\; k(x')\right]. \tag{5.171}$$

In the classically allowed region ($x > x_0$), the local wave number is real and the WKB solution is an oscillatory function of x. The general solution can then be written as a linear combination of $\psi_{+}^{\mathrm{WKB}}(x)$ and $\psi_{-}^{\mathrm{WKB}}(x)$. It is convenient to write the combinations in terms of two real constants: a normalization (A) and a phase (α), as

$$\psi^{\mathrm{WKB}}(x > x_0) = \frac{A}{\sqrt{k(x)}}\; \sin\left[\int_{x_0}^{x} dx'\; k(x') + \alpha\right]. \tag{5.172}$$

In the classically forbidden region ($x < x_0$), the local wave number is imaginary and the wave function has a completely different behavior. Writing $k(x) = i\gamma(x)$, with

$$\gamma(x) = \frac{1}{\hbar}\,\sqrt{2\mu\,[V(r) - E]}, \tag{5.173}$$

the two linearly independent solutions of Eq. (5.171) becomes

$$\psi_{\pm}^{\mathrm{WKB}}(x < x_0) = \frac{A_0}{\sqrt{\gamma(x)}}\; \exp\left[\pm \int_{x}^{x_0} dx'\; \gamma(x')\right]. \tag{5.174}$$

[23]The lower limit of the integral gives the arbitrary constant in the solution of the differential equation. We use the turning point, x_0.

In opposition to what happens in the allowed region, these wave functions do not oscillate as $x \to -\infty$. They have the asymptotic limits

$$\psi_-^{\text{WKB}}(x \to -\infty) = 0, \quad \psi_+^{\text{WKB}}(x \to -\infty) = \infty. \tag{5.175}$$

The divergent solution $\psi_+(x)$ should then be discarded[24]. Therefore, the physically acceptable WKB wave function in the forbidden region is

$$\psi^{\text{WKB}}(x < x_0) = \frac{B}{\sqrt{\gamma(x)}} \exp\left[- \int_x^{x_0} dx' \, \gamma(x') \right]. \tag{5.176}$$

5.2.2.1 *Connection formulae - one turning point*

The problem now is to determine the wave function in the classically allowed region that corresponds to the convergent solution of Eq. (5.176). That is, finding the correct values of the constants A and α of Eq. (5.172). This is a difficult task. The usual procedure of matching the wave functions and their derivatives at the border of the two regions cannot be followed here because Eqs. (5.172) and (5.176) are not valid near the turning point. In this case, it is necessary to resort to connection formulae. There are different methods to derive them. One is to do the analytical continuation of the variable x onto the complex plane and then to go from the forbidden to the allowed regions along a path that remains far enough from the turning point for the WKB approximation to be valid (see [Berry and Mount (1972)] for a review). The other possibility is to make a linear expansion of the potential in the neighborhood of the turning point. This leads to an equation that can be solved[25]. One then looks at the behavior of the two linearly independent approximate solutions at some point x_f far inside the forbidden region and discard the divergent one. Inspecting the convergent solution at a coordinate $x_a \gg x_0$ where the approximate wave function reaches its asymptotic form, one finds the particular combination of ψ_+^{WKB} and ψ_-^{WKB} that corresponds to it. In this way, the constants A and α are determined. We remark that the application of this method requires that the coordinates x_f and x_a be close enough of the turning point for the first order expansion of the potential to remain valid. The details of this connection method can be found in several papers and books [Langer (1937); Fröman and Fröman (1965); Berry and Mount (1972); Brink (1985); Merzbacher (1998)]. Such studies lead to the conclusion that $\alpha = \pi/4$. The constant

[24]If below x_0 there is another classically allowed region, as in the case of a potential barrier, ψ_-^{WKB} cannot be discarded (see section 5.2.2.5).

[25]This is a particular case of the method of comparison equations, which will be used later to derive the uniform approximation.

A is arbitrary, since it is just a normalization. This procedure leads to the following connection formulae for the convergent and divergent WKB solutions in the forbidden region:

Forbidden region ($x < x_0$) Allowed region ($x > x_0$)

$$\frac{1}{2\sqrt{\gamma(x)}} \exp\left[-\int_x^{x_0} dx' \; \gamma(x')\right] \Longleftrightarrow \frac{1}{\sqrt{k(x)}} \sin\left[\int_{x_0}^{x} dx' \; k(x') + \frac{\pi}{4}\right]. \tag{5.177}$$

$$\frac{1}{\sqrt{\gamma(x)}} \exp\left[\int_x^{x_0} dx' \; \gamma(x')\right] \Longleftrightarrow \frac{1}{\sqrt{k(x)}} \cos\left[\int_{x_0}^{x} dx' \; k(x') + \frac{\pi}{4}\right]. \tag{5.178}$$

The right hand side of Eq. (5.177) is then the physically acceptable solution in the classically allowed region.

5.2.2.2 *The radial equation*

Now we consider the WKB approximation in actual scattering problems. It can be used in the radial equation, which is of fundamental importance to evaluate phase shifts and the scattering cross section. The differences with respect to the previously discussed one-dimensional problems are that here the radial equation contains an effective potential (including the centrifugal term) and that the variable r is limited to the domain $0 \leq r < \infty$. This limitation corresponds to having an infinite repulsive potential at $r < 0$. The radial equation can be handled similarly to the one-dimensional equation discussed above. It can be put in the form

$$\left[\frac{d^2}{dr^2} + k_l^2(r)\right] \; u_l(k_0, r) = 0, \tag{5.179}$$

with the l-dependent local wave number, $k_l(r)$, defined as

$$k_l(r) = \sqrt{\frac{2\mu}{\hbar^2}\left[E - V(r) - \frac{\hbar^2\lambda^2}{2\mu r^2}\right]}. \tag{5.180}$$

Above and throughout the remaining part of this chapter we use the Langer modification[26]: $l(l+1) \to (l+1/2)^2$, and adopt the notation $\lambda = l + 1/2$. One then writes

$$u_l^{\mathrm{WKB}}(k_0, r) = A_l(r) \; e^{i\,\mathcal{S}_l(r)/\hbar} \tag{5.181}$$

[26]Owing to the singularity at $r = 0$ this modification is necessary for the WKB approximation to be valid at low partial waves [Langer (1934, 1937)].

and insert the above expression into the radial equation. Adopting the same approximations as before and using the connection formulae, we get the WKB approximation for the radial wave function in the classically allowed region,

$$u_l^{\text{WKB}}(k_0, r > r_l) = \frac{A_0}{\sqrt{k_l(r)}} \sin\left[\int_{r_l}^{r} dr'\ k_l(r') + \frac{\pi}{4}\right]. \tag{5.182}$$

Above, r_l is the turning point at the lth partial-wave, determined by the condition $k_l(r_l) = 0$. It is convenient to introduce the notation

$$I_l(r) = \int_{r_l}^{r} dr'\ k_l(r') \tag{5.183}$$

and write

$$u_l^{\text{WKB}}(k_0, r > r_l) = \frac{A_0}{\sqrt{k_l(r)}} \sin\left[I_l(r) + \frac{\pi}{4}\right]. \tag{5.184}$$

We now evaluate WKB phase shifts for a short-range potential, which vanishes for $r > \bar{R}$. For this purpose, we find the WKB wave function at large radial distances ($r \gg \bar{R}$) and compare with the asymptotic expression of the exact radial wave function. Through this comparison, one gets the phase shift within the WKB approximation. To illustrate this procedure, we consider the trivial case of a free particle, where all phase shifts vanish. For $V(r) = 0$ the local wave wave number is

$$k_l^{(0)}(r) = \sqrt{\frac{2\mu}{\hbar^2}\left[E - \frac{\hbar^2\lambda^2}{2\mu r^2}\right]} = k_0\sqrt{1 - \frac{\lambda^2}{k_0^2 r^2}}. \tag{5.185}$$

In this case the integral within the square brackets of Eq. (5.182) should be evaluated between the turning point $r_l^{(0)} = \lambda/k_0$ and an asymptotic upper limit, $r \to \infty$. Changing variable: $r \to \rho = k_0\, r$ and using results from standard integral tables one gets,

$$I_l^{(0)}(r) \equiv \int_{r_l^{(0)}}^{r} dr'\ k_l^{(0)}(r') \equiv \int_{\lambda}^{\rho} d\rho'\ \frac{\sqrt{\rho'^2 - \lambda^2}}{\rho'} = \rho - \lambda\,\pi/2 = k_0\, r - \lambda\,\pi/2. \tag{5.186}$$

Inserting this result into Eq. (5.182), we get the asymptotic form of the WKB wave function

$$\begin{aligned} u_l^{\text{WKB}}(k_0, r \to \infty) &\sim \sin\left[k_0\, r - \lambda\,\pi/2 + \pi/4\right] \\ &= \sin\left[k_0\, r - l\,\pi/2\right]. \end{aligned} \tag{5.187}$$

Above, we have dropped the normalization constant $A_0/\sqrt{k_0}$, which is irrelevant for the determination of phase shifts. This asymptotic behavior is the same as the one obtained in chapter 2. For free particles, the exact regular solutions of the radial wave equation are the Ricatti-Bessel functions, $\hat{j}_l(kr)$, which have the same asymptotic form as that of Eq. (5.187) (see Eq. (2.22a)). Therefore, all phase shifts are equal to zero, as expected for a free particle.

Now we consider the phase shifts for a short-range potential, that vanishes for $r > \bar{R}$. We begin with the identity

$$I_l(r) = \left[I_l(r) - I_l^{(0)}(r)\right] + I_l^{(0)}(r) \tag{5.188}$$

and use Eq. (5.186) for the term outside the square brackets. The result is then inserted in Eq. (5.184). We get the asymptotic expression

$$u_l^{\text{WKB}}(k_0, r \to \infty) \sim \sin\left[k_0\, r - l\,\pi/2 + \left[I_l(r) - I_l^{(0)}(r)\right]\right). \tag{5.189}$$

Comparing this result with the asymptotic form of the exact radial wave function of Eq. (2.39a),

$$u_l(k, r \to \infty) \sim \sin\left(kr - l\pi/2 + \delta_l\right), \tag{5.190}$$

we conclude that the WKB approximation for the phase shift is

$$\begin{aligned} \delta_l^{\text{WKB}} &= \lim_{r\to\infty} \left\{I_l(r) - I_l^{(0)}(r)\right\} \\ &= \int_{r_l}^{\infty} dr'\ k_l(r') - \int_{r_l^{(0)}}^{\infty} dr'\ k_l^{(0)}(r'). \end{aligned} \tag{5.191}$$

Since the integrands $k_l(r')$ and $k_l^{(0)}(r')$ become identical for $r' > \bar{R}$, the WKB phase shift reduces to the expression

$$\delta_l^{\text{WKB}} = \int_{r_l}^{\bar{R}} dr'\ k_l(r') - \int_{r_l^{(0)}}^{\bar{R}} dr'\ k_l^{(0)}(r'). \tag{5.192}$$

Phase shifts for long range potentials can be derived through similar procedures. We first consider the point charge Coulomb potential

$$V(r) \equiv V_{\text{C}}(r) = \frac{1}{4\pi\,\varepsilon_0}\,\frac{\text{Q}_\text{P}\,\text{Q}_\text{T}}{r}. \tag{5.193}$$

The local wave number now is

$$k_l^{(\text{C})}(r) = \sqrt{\frac{2\mu}{\hbar^2}\left[E - \frac{\text{Q}_\text{P}\,\text{Q}_\text{T}}{4\pi\,\varepsilon_0\, r} - \frac{\hbar^2\,\lambda^2}{2\mu\, r^2}\right]} = k_0\sqrt{1 - \frac{2\eta}{k_0 r} - \frac{\lambda^2}{k_0^2 r^2}}, \tag{5.194}$$

and the turning point becomes (see Eq. (3.9)),

$$r_l^{(\mathrm{C})} = \frac{\eta + \sqrt{\eta^2 + \lambda^2}}{k_0}, \tag{5.195}$$

where η is the Sommerfeld parameter, defined in Eq. (3.3). The asymptotic form of the WKB wave function can be written

$$u_l^{\mathrm{WKB}}(k_0, r \to \infty) \sim \sin\left[I_l^{(\mathrm{C})} + \pi/4\right], \tag{5.196}$$

where $I_l^{(\mathrm{C})}(r)$ is the integral

$$I_l^{(\mathrm{C})}(r) = \int_{r_l^{(\mathrm{C})}}^{r} dr'\ k_l^{(\mathrm{C})}(r'). \tag{5.197}$$

Using the explicit form of $k_l^{(\mathrm{C})}$ (Eq. (5.194)) and changing $r \to \rho = k_0\, r$, the above equation becomes

$$I_l^{(\mathrm{C})}(r) = \int_{\rho_l^{(\mathrm{C})}}^{\rho} d\rho'\ \frac{\sqrt{\rho'^{\,2} - 2\eta\rho' + \lambda^2}}{\rho'}, \tag{5.198}$$

with $\rho_l^{(\mathrm{C})} = \eta + \sqrt{\eta^2 + \lambda^2}$. This integral can also be evaluated with the help of standard integral tables. After a lengthy calculation, one obtains

$$\begin{aligned} I_l^{(\mathrm{C})}(\rho/k_0) &= k_l^{(\mathrm{C})}\, r - \eta\, \ln\left[k_l^{(\mathrm{C})}\, r + k_0\, r - \eta\right] + \frac{1}{2}\,\eta\, \ln\left[\eta^2 + \lambda^2\right] \\ &\quad + \lambda \sin^{-1}\left[\frac{\eta k_0 r + \lambda^2}{k_0 r\left(\eta^2 + \lambda^2\right)^{1/2}}\right] - \lambda\frac{\pi}{2}. \end{aligned} \tag{5.199}$$

At asymptotic distances we can approximate,

$$k_l^{(\mathrm{C})}\, r \simeq k_0\, r\left(1 - \frac{\eta}{k_0\, r}\right) = k_0\, r - \eta. \tag{5.200}$$

Using this approximation[27] in Eq. (5.199) we obtain

$$\begin{aligned} I_l^{(\mathrm{C})}(r) &= k_0\, r - \eta\, \ln 2k_0\, r - \lambda\, \pi/2 \\ &\quad - \eta + \frac{1}{2}\,\eta\, \ln\left[\eta^2 + \lambda^2\right] + \lambda \sin^{-1}\left[\frac{\eta}{\sqrt{\eta^2 + \lambda^2}}\right]. \end{aligned} \tag{5.201}$$

Finally, we insert this result into Eq. (5.196) and get

$$u_l^{\mathrm{WKB}}(k_0, r \to \infty) \sim \sin\left[k_0\, r - \eta\, \ln 2k_0\, r - l\,\pi/2 + \Delta_l^{(\mathrm{C})}\right], \tag{5.202}$$

[27] Note that the correction $-\eta$ inside the argument of the logarithm vanishes when proper expansion of $\ln\left[k_l^{(\mathrm{C})}\, r + k_0\, r - \eta\right]$ is carried out.

where $\Delta_l^{(\mathrm{C})}$ is the phase,

$$\Delta_l^{(\mathrm{C})} = \frac{1}{2}\,\eta\,\ln\left[\eta^2+\lambda^2\right] + \lambda\sin^{-1}\left[\frac{\eta}{\sqrt{\eta^2+\lambda^2}}\right] - \eta. \tag{5.203}$$

Using the identity

$$\sin^{-1}\left(\eta/\sqrt{\eta^2+\lambda^2}\right) = \tan^{-1}\left(\eta/\lambda\right),$$

we put the above equation in the form

$$\Delta_l^{(\mathrm{C})} = -\eta + \eta\,\ln\sqrt{\eta^2+\lambda^2} + \lambda\tan^{-1}\eta/\lambda. \tag{5.204}$$

Comparing Eq. (5.202) with the asymptotic form of the radial wave function in Coulomb scattering (Eq. (3.44)),

$$F_l(\eta, k_0\,r\to\infty) \to \sin\left[k_0\,r - \frac{l\pi}{2} - \eta\ln 2k_0\,r + \sigma_l\right], \tag{5.205}$$

we find the WKB approximation for the Coulomb phase shift,

$$\sigma_l^{\mathrm{WKB}} = \Delta_l^{(\mathrm{C})} = -\eta + \eta\,\ln\sqrt{\eta^2+\lambda^2} + \lambda\tan^{-1}\eta/\lambda. \tag{5.206}$$

Note that this corresponds to the large l and/or large η approximation of section 3.3.1. Since in this limit $\lambda \simeq l$, Eqs. (5.206) and (3.87) are equivalent.

We now evaluate WKB phase shifts in the scattering from the Coulomb plus short-range potential,

$$V(r) = V_{\mathrm{C}}(r) + \bar{V}(r), \tag{5.207}$$

where $\bar{V}(r)$ vanishes for $r > \bar{R}$. The local wave number is

$$k_l(r) = \sqrt{\frac{2\mu}{\hbar^2}\left[E - \frac{\mathrm{Q_P\,Q_T}}{4\pi\,\varepsilon_0\,r} - \bar{V}(r) - \frac{\hbar^2\,\lambda^2}{2\mu\,r^2}\right]} \tag{5.208}$$

and the turning point r_l is determined by the condition $k_l(r_l) = 0$. The asymptotic form of the WKB wave function can be written

$$u_l^{\mathrm{WKB}}(k_0, r\to\infty) \sim \sin\left(I_l(r) + \pi/4\right), \tag{5.209}$$

with

$$I_l(r) = \int_{r_l}^{r} dr'\;k_l(r'). \tag{5.210}$$

The procedure to determine the WKB phase shifts is similar to the one employed in the case of a potential with short range. We add and subtract $I_l^{(C)}(r)$ to the argument of the sinus function in Eq. (5.209). That is,

$$u_l^{\text{WKB}}(k_0, r \to \infty) \sim \sin\left(\left[I_l(r) - I_l^{(C)}(r)\right] + I_l^{(C)}(r) + \pi/4\right). \tag{5.211}$$

Using Eqs. (5.201) and (5.206), the above equation can be put in the form

$$u_l^{\text{WKB}}(k_0, r \to \infty) \sim \sin\Big(k_0 r - l\,\pi/2 - \eta \ln 2k_0 r + \sigma_l^{\text{WKB}} + \left[I_l(r) - I_l^{(C)}(r)\right]\Big). \tag{5.212}$$

Comparing Eq. (5.212) with the asymptotic wave function in Coulomb plus short-range scattering (Eq. (3.127)), we obtain the WKB approximation for the phase shift of the short-range potential in the presence of the Coulomb interaction,

$$\bar{\delta}_l^{\text{WKB}} = I_l(\infty) - I_l^{(C)}(\infty) = \int_{r_l}^{\infty} dr'\; k_l(r') - \int_{r_l^{(C)}}^{\infty} dr'\; k_l^{(C)}(r'). \tag{5.213}$$

Since the two integrands in the above equation become identical beyond the reach of the short-range potential, we can write

$$\bar{\delta}_l^{\text{WKB}} = \int_{r_l}^{\bar{R}} dr'\; k_l(r') - \int_{r_l^{(C)}}^{\bar{R}} dr'\; k_l^{(C)}(r'). \tag{5.214}$$

5.2.2.3 *WKB deflection function and time delay*

In the following we discuss the classical action, namely the WKB phase, and indicate how it is related to the classical deflection function and the time delay. Although we restrict the discussion to the case of a short-range potential, the extension to long range potentials is straightforward. Let us first consider the derivative $d\delta_l^{\text{WKB}}/dl$ (note that $d/dl = d/d\lambda$). Deriving both sides of Eq. (5.191), we get

$$\frac{d\delta_l^{\text{WKB}}}{dl} = \lim_{r\to\infty}\left\{\frac{dI_l(r)}{dl}\right\} - \lim_{r\to\infty}\left\{\frac{dI_l^{(0)}(r)}{dl}\right\}. \tag{5.215}$$

The derivative of the second term on the RHS of the above equation can be immediately evaluated. Using the explicit form of $I_l^{(0)}(r)$ (Eq. (5.186)), we find

$$\lim_{r\to\infty}\left\{\frac{dI_l^{(0)}(r)}{dl}\right\} = -\frac{\pi}{2}. \tag{5.216}$$

The derivative of the first term is[28]

$$\lim_{r\to\infty}\left\{\frac{dI_l(r)}{dl}\right\} = \int_{r_l}^{\infty} dr\,\frac{dk_l(r)}{dl}. \tag{5.217}$$

Using the explicit form of the local wave number (Eq. (5.180)) and evaluating the derivative, we get

$$\lim_{r\to\infty}\left\{\frac{dI_l(r)}{dl}\right\} = -\int_{r_l}^{\infty} dr\,\frac{\hbar\,\lambda}{r^2\,\sqrt{2\mu\,(E - V(r) - \hbar^2\lambda^2/r^2)]}}. \tag{5.218}$$

Adopting the notations $\hbar\lambda = L$ and $V_L(r) = V(r) + \hbar^2\lambda^2/r^2$ and inserting Eqs. (5.216) and (5.218) in Eq. (5.215), we obtain the derivative of the WKB phase shift

$$\frac{d\delta_l^{\text{WKB}}}{dl} = \frac{\pi}{2} - \int_{r_l}^{\infty} dr\,\frac{L}{r^2\,\sqrt{2\mu\,(E - V_L(r))}}. \tag{5.219}$$

Comparing the above equation with the expression for the classical deflection angle, $\Theta(L)$ (Eq. (1.43)), we obtain the important result,

$$\Theta(L) = 2\left[\frac{d\delta_l^{\text{WKB}})}{dl}\right]. \tag{5.220}$$

We will come back to this relation later in this section, when we discuss the semiclassical approximation for the scattering amplitude.

We now consider the derivative of the WKB phase shift with respect to the collision energy. First, we point out that the phase shifts depend on the collision energy, although we are not representing this dependence explicitly. This derivative can be written as a sum of two terms, as in Eq. (5.215). That is

$$\frac{d\delta_l^{\text{WKB}}}{dE} = \int_{r_l}^{\infty} dr\,\frac{dk_l(r)}{dE} - \int_{r_l^{(0)}}^{\infty} dr\,\frac{dk_l^{(0)}(r)}{dE}. \tag{5.221}$$

Using the explicit forms of $k_l(r)$ and $k_l^{(0)}(r)$, we get the derivatives

$$\frac{dk_l(r)}{dE} = \frac{\mu}{\hbar^2\,k_l(r)} = \frac{1}{\hbar\,v_l(r)} \tag{5.222}$$

and

$$\frac{dk_l^{(0)}(r)}{dE} = \frac{\mu}{\hbar^2\,k_l^{(0)}(r)} = \frac{1}{\hbar\,v_l^{(0)}(r)}. \tag{5.223}$$

[28] Note that this derivative does not have contributions from the integration limits. The upper limit does not contribute because it is independent of l. The contribution from the lower limit, $k_l(r_l)\,dr_l/dl$, also vanishes because r_l is the turning point.

Above, we have introduced the local velocities with and without interaction, $v_l(r) = \hbar\, k_l(r)/\mu$ and $v_l^{(0)}(r) = \hbar\, k_l^{(0)}(r)/\mu$, respectively. Using the fact that the two integrands of Eq. (5.221) are the same for radial distances larger than the range of the potential, the upper limits can be replaced by $\bar{R}$. Multiplying both sides of this equation by $2\hbar$, it takes the form

$$2\,\hbar\,\frac{d\delta_l^{\mathrm{WKB}}}{dE} = \tau - \tau^{(0)}, \tag{5.224}$$

with

$$\tau = 2\int_{r_l}^{\bar{R}} \frac{dr}{v_l(r)} \quad \text{and} \quad \tau^{(0)} = 2\int_{r_l^{(0)}}^{\bar{R}} \frac{dr}{v_l^{(0)}(r)}. \tag{5.225}$$

The integrands appearing above have a simpler interpretation: $dr/v_l(r)$ is the time spent by the system to traverse the distance dr and $dr/v_l^{(0)}(r)$ is the corresponding time when the potential is switched of. Therefore, the first integral represents the time spent by the projectile to go from $\bar{R}$ to closest approach and then back to $\bar{R}$, as it re-separates. The second integral has the same meaning but with the potential switched off. Therefore, the difference

$$\Delta_t = 2\hbar\left[\frac{d\delta_l^{\mathrm{WKB}}}{dE}\right] \tag{5.226}$$

is the time-delay caused by the interaction. We point out that an equivalent relation has been derived in chapter 2 (Eq. (2.112)), in the case of a resonance.

5.2.2.4 *Eikonal vs. WKB*

It is interesting to find the conditions under which the WKB approximation reduces to the eikonal. As it has been stressed, the intermediate energy eikonal limit is valid when the interaction is much smaller than the energy and the trajectory is close to a straight line. That is,

$$\mathbf{v} \simeq v_0\,\hat{\mathbf{z}} = \frac{\hbar\,k_0}{\mu}\,\hat{\mathbf{z}} \quad \text{and} \quad \mathbf{r} \simeq \mathbf{b} + z\,\hat{\mathbf{z}} = \mathbf{b} + v_0 t\,\hat{\mathbf{z}}, \tag{5.227}$$

where $\mathbf{b}$ is the constant vector associated with the impact parameter and $\hat{\mathbf{z}}$ is the unit vector along the z-axis. In this case the momentum transfer is very small and the radial component of the wave vector, $k_l(r)$, can be approximated to first order in V/E. The integrand of $I_l(r)$ in Eq. (5.191) then becomes

$$k_l(r') \simeq k_l^{(0)}(r') - \frac{\mu\,V(r')}{\hbar^2\,k_l^{(0)}(r')}. \tag{5.228}$$

Using this expansion in Eq. (5.191) and neglecting the effect of the interaction on the turning point, the WKB phase shifts become

$$\delta_l^{\rm WKB} = -\frac{\mu}{\hbar^2} \int_{r_l^{(0)}}^{\infty} dr' \, \frac{V(r')}{k_l^{(0)}(r')}. \tag{5.229}$$

We now change the integration variable $r' \to z'$ according to the relation $r' = \sqrt{b^2 + z'^2}$. Differentiating this relation, we find that $dr' = z' \, dz'/r'$. The new lower limit of the integral is $z' = 0$ and the upper limit remains infinity. Eq. (5.229) then becomes

$$\delta_l^{\rm WKB} = -\frac{\mu}{2\hbar^2} \int_{-\infty}^{\infty} dz \, \frac{z}{r \, k_l^{(0)}(r)} V(z, b). \tag{5.230}$$

Above, we have used the time-reversal invariance of the potential to replace

$$\int_0^\infty dz \to \frac{1}{2} \int_{-\infty}^{\infty} dz.$$

Since $k_l^{(0)}(r)$ is the radial component of the wave vector, we can write $k_l^{(0)}(r) = \mathbf{k} \cdot \mathbf{r}/r$ and, owing to the straight line approximation for the trajetory, $\mathbf{k} = k_0 \, \hat{\mathbf{z}}$. Thus, the denominator in the integrand of Eq. (5.230) becomes

$$r \, k_l^{(0)}(r) = r \, \frac{k_0 \, \hat{\mathbf{z}} \cdot \mathbf{r}}{r} = k_0 \, z.$$

Inserting this result into Eq. (5.230), we get

$$\delta_l^{\rm WKB} = -\frac{\mu}{2\hbar^2 \, k_0} \int_{-\infty}^{\infty} dz \, V(z, b) = \frac{1}{2} \left[-\frac{1}{\hbar \, v_0} \int_{-\infty}^{\infty} dz \, V(z, b) \right], \tag{5.231}$$

or

$$\delta_l^{\rm WKB} = \frac{1}{2} \left[-\frac{k_0}{2E} \int_{-\infty}^{\infty} dz \, V(z, b) \right]. \tag{5.232}$$

Comparing this expression with Eq. (5.118), we obtain the high energy limit of the WKB phase shift,

$$\delta_l^{\rm WKB} \to \frac{\chi(b)}{2} = \delta^{\rm eikonal}(b), \quad \text{with} \quad b = \frac{l + {}^1\!/_2}{k_0}. \tag{5.233}$$

As will be clear later in this section, the integration over impact parameter of Eq. (5.125) corresponds to a semiclassical approximation for the partial-wave series giving the scattering amplitude and $\chi(b)/2$ plays the role of a phase shift. Corrections to the above can be made, but we feel that a better strategy is simply to use these approximations as they are constructed. One now knows that the WKB contains the eikonal as a limiting case. The major hurdle with the WKB are the singular turning points, which makes the numerical evaluation of the phase shift a bit subtle.

5.2.2.5 *Scattering from potential barriers*

So far we have discussed the WKB approximation in the simple case of a single turning point. The situation is more difficult when there are two or more turning points, or when there is none. Additional complications arise when the two turning points are close together. This occurs when the collision energy approaches the height of the barrier, V_B, or it is slightly above it. These points will be addressed later in this chapter.

Let us first consider the situation of two well separated turning points, represented in figure 5.8. In this case, the projectile approaches the barrier from the right, with energy $E < V_B$. The external and the internal turning points are denoted by r_2 and r_1, respectively. Regions I ($r > r_2$) and III ($r < r_1$) are classically allowed whereas region II ($r_2 > r > r_1$) is classically forbidden. At r-values not too close to the turning points we can use the WKB approximation to evaluate the wave functions $u_l^{\mathrm{I}}(k_0, r)$, $u_l^{\mathrm{II}}(k_0, r)$ and $u_l^{\mathrm{III}}(k_0, r)$. Since the incident wave propagates from the right to the left, the wave function in region III has only the transmitted component, which propagates to the left. Therefore the wave function in this region is well determined (except for an irrelevant normalization factor) and we can write[29]

$$u_l^{\mathrm{III}}(k_0, r) = \frac{A}{\sqrt{k_l(r)}} \; \exp\left[-i\left(\int_r^{r_1} dr'\, k_l(r') + \pi/4\right)\right]. \tag{5.234}$$

The corresponding wave functions in regions II and I have the general forms

$$u_l^{\mathrm{II}}(k_0, r) = \frac{1}{\sqrt{\gamma_l(r)}} \left\{ B_- \; \exp\left[-\int_{r_1}^{r} dr'\, \gamma_l(r')\right] \right. \\ \left. + B_+ \; \exp\left[\int_{r_1}^{r} dr'\, \gamma_l(r')\right] \right\} \tag{5.235}$$

and

$$u_l^{\mathrm{I}}(k_0, r) = \frac{1}{\sqrt{k_l(r)}} \left\{ C_- \; \exp\left[-i\left(\int_{r_2}^{r} dr'\, k_l(r') + \pi/4\right)\right] \right. \\ \left. + C_+ \; \exp\left[i\left(\int_{r_2}^{r} dr'\, k_l(r') + \pi/4\right)\right] \right\}. \tag{5.236}$$

[29]The factor $\pi/4$ was introduced in the argument of the exponential to make the calculation simpler. This is just a particular choice of phase for the normalization constant A.

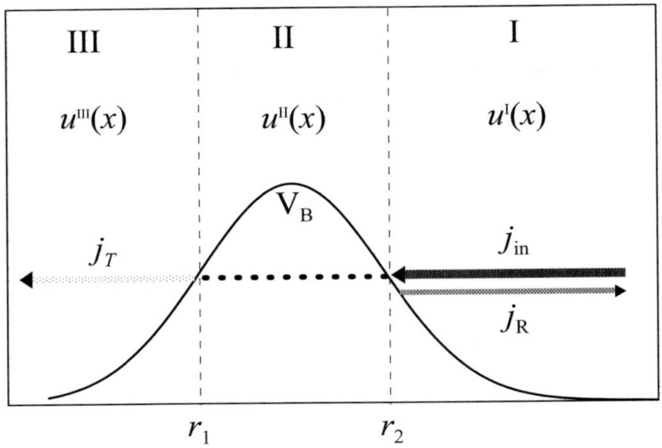

Fig. 5.7 Schematic representation of a collision with two well separated turning points, r_1 and r_2. When the incident current j_{in} reaches the barrier, it is partially transmitted (j_{T}) and partially reflected (j_{R}).

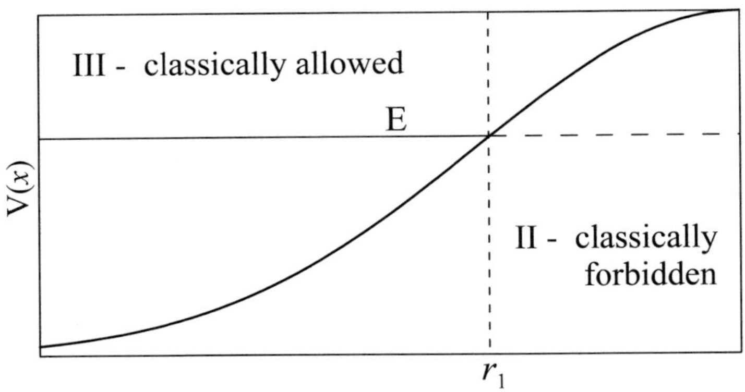

Fig. 5.8 Schematic representation of the inner turning point in the transmission through a potential barrier.

To solve the scattering problem, we should express the coefficients $B_\pm$ and $C_\pm$ in terms of A. For this purpose, we need connection formulae. Eqs. (5.177) and (5.178) (with the replacement $x_0 \to r_2$) relate the increasing and decreasing solutions in the forbidden region with the waves propagating to the left and to the right in region I. However, we do not know the wave function there. We know it in region III. Therefore, we need connection formulae relating WKB wave functions in regions III and II.

These regions are delimited by the inner turning point, r_1, as represented in figure 5.8. The desired formulae connect oscillating functions in region III with the increasing and the decreasing solutions of region II. The derivation of these formulae is analogous to that of Eqs. (5.177) and (5.178) and the results are

Allowed region III ($r < r_1$) Forbidden region II ($r > r_1$)

$$\frac{1}{\sqrt{k(r)}} \sin\left[\int_r^{r_1} dr'\ k(r') + \frac{\pi}{4}\right] \iff \frac{1}{2\sqrt{\gamma(r)}} \exp\left[-\int_{r_1}^{r} dr'\ \gamma(r')\right] \tag{5.237}$$

$$\frac{1}{\sqrt{k(r)}} \cos\left[\int_r^{r_1} dr'\ k(r') + \frac{\pi}{4}\right] \iff \frac{1}{\sqrt{\gamma(r)}} \exp\left[\int_{r_1}^{r} dr'\ \gamma(r')\right]. \tag{5.238}$$

The transmitted wave of Eq. (5.234) can be written as a linear combination of the oscillating functions at the left hand sides of the above two equations. Accordingly, the wave function in region II is a linear combination of the right hand sides of the corresponding equations, with the same coefficients. In this way, we get $B_\pm$. Using the connection formulae between regions II and I (Eqs. (5.177) and (5.178)), we find that the coefficients $C_\pm$ are given by

$$C_- = iA\left[e^{\Phi} + e^{-\Phi}/4\right] \quad \text{and} \quad C_+ = -iA\left[e^{\Phi} - e^{-\Phi}/4\right], \tag{5.239}$$

with Φ standing for the integral

$$\Phi = \int_{r_1}^{r_2} dr\ \gamma_l(r). \tag{5.240}$$

We can now check the influence of the inner turning point on the WKB S-matrix. For this purpose, we look at the asymptotic form of Eq. (5.236), using the compact notation of Eq. (5.183) (with the replacement $r_l \to r_2$). We get

$$u_l^{\rm I}(k_0, r \to \infty) = \frac{C_-}{\sqrt{k_l(r)}} \left\{ \exp\left[-i\left(I_l(r) + \frac{\pi}{4}\right)\right] + \frac{C_+}{C_-} \exp\left[i\left(I_l(r) + \frac{\pi}{4}\right)\right] \right\}. \tag{5.241}$$

Now, we add and subtract $I_l^{(0)}(r)$ to $I_l(r)$ and use Eq. (5.189). The above equation then becomes

$$u_l^{\rm I}(k_0, r \to \infty) = \frac{C_-}{\sqrt{k_l(r)}} e^{-i\delta_l^{\rm WKB}} \left\{ \exp\left[-i\left(k_0 r - l\pi/2\right)\right] + \left(\frac{C_+}{C_-}\right) e^{2i\delta_l^{\rm WKB}} \exp\left[i\left(k_0 r - l\pi/2\right)\right] \right\}. \tag{5.242}$$

Dropping irrelevant multiplicative constants, the radial wave function can be put in the form

$$u^{\rm I}(k_0, r \to \infty) \sim \left[e^{-i(k_0\, r - l\pi/2)} \;-\; S_l^{\rm WKB}\; e^{i(k_0\, r - l\pi/2)}\right], \tag{5.243}$$

where

$$S_l^{\rm WKB} \equiv |S_l^{\rm WKB}|\; e^{2i\delta_l^{\rm WKB}} \tag{5.244}$$

is a complex quantity with modulus

$$|S_l^{\rm WKB}| = \frac{C_+}{C_-} = \frac{e^{\Phi} - e^{-\Phi}/4}{e^{\Phi} + e^{-\Phi}/4}. \tag{5.245}$$

Since C_+/C_- is real, we conclude that the inner classically accessible region (region III in figure 5.7) does not affect the phase-shift. However it modifies the modulus of the S-matrix, making it smaller than one. The physical interpretation of this result is that the current transmitted throughout the barrier is absorbed. That is, the projectile-target system is excited so that it stays no longer in the elastic channel[30].

Now we evaluate the reflection and transmission coefficients of the incident wave. The reflection coefficient, $\mathcal{R}$, is defined in terms of the current densities in regions I (see figure 5.8) as

$$\mathcal{R} = \left|\frac{j_R}{j_{in}}\right|^2. \tag{5.246}$$

Above, j_{in} and j_R are respectively associated with the asymptotic components of the $u_l^{(\rm I)}(k_0, r)$ propagating to the left and to the right. We adopt the notation:

$$u_l^{(\rm I)}(k_0, r) = u_-(r) + u_+(r), \tag{5.247}$$

with

$$u_-(r) \sim \exp\left[-i\left(I_l(r) + \frac{\pi}{4}\right)\right] \tag{5.248}$$

and

$$u_+(r) \sim \frac{C_+}{C_-}\; \exp\left[i\left(I_l(r) + \frac{\pi}{4}\right)\right]. \tag{5.249}$$

Above, we have dropped constant factors which multiplies simultaneously u_- and u_+. The radial currents j_{in} and j_R are then evaluated by the well known expression

$$j(r) = \frac{\hbar}{2\mu i}\left[u^*(r)\,\frac{du(r)}{dr} - u(r)\,\frac{du^*(r)}{dr}\right], \tag{5.250}$$

[30] We come back to this point later in this section.

with the replacements: $u \to u_-$ and $u \to u_+$, respectively. The current densities can be easily evaluated and the reflection coefficient of Eq. (5.246) is

$$\mathcal{R} = \left(\frac{e^{\Phi} - e^{-\Phi}/4}{e^{\Phi} + e^{-\Phi}/4} \right)^2 = |S_l^{\text{WKB}}|^2 . \tag{5.251}$$

We now turn to the transmission coefficient. Here there is a complication. The transmitted current density depends on the radial distance chosen for its evaluation. On the other hand, if we consider that the transmitted current is fully absorbed in region III, we can write

$$\mathcal{T} = 1 - \mathcal{R} = \left(\frac{1}{e^{\Phi} + e^{-\Phi}/4} \right)^2 . \tag{5.252}$$

Frequently, there is a third turning point inside the potential barrier. It is due to the contribution from the centrifugal potential, which is infinitely repulsive at $r = 0$. In the schematic representation of figure 5.9 (b) it is denoted by r_0. If the potential is real, the transmitted wave is reflected back at r_0 and eventually reemerges. In this way, the unitary character of the S-matrix can be restored. However, in the scattering of complex systems, like molecules, atoms and nuclei, the collision frequently involves internal degrees of freedom that are strongly excited at small radial distances. Therefore, the probability that the transmitted current reemerges in the elastic channel is very small. As far as elastic scattering is concerned, this corresponds to absorption. In potential scattering, this effect is mocked up by a negative imaginary potential with short range (see section 7). Thus, neglecting reflections at this inner turning point in the WKB approximation is well justified in collisions subject to strong absorption.

When the energy is well below the barrier top, $\Phi \gg 1$ and the transmission coefficient of Eq. (5.252) can be approximated by

$$\mathcal{T} \simeq e^{-2\,\Phi}. \tag{5.253}$$

This result is frequently found in the literature. It should be stressed that this expression is only valid for energies much below the barrier height. At $E = V_B$, for example, the transmission coefficient of Eq. (5.253) is equal to 1, and that of Eq. (5.252) is equal to 16/25. These results are quite different from the value of 1/2 usually associated with full quantum mechanics.

The predictions of the WKB approximation in the scattering from a potential barrier are reasonable for low energies, where there are two well separated turning points. As the energy approaches V_B, the turning points get close and the connection formulae used in the above derivations break

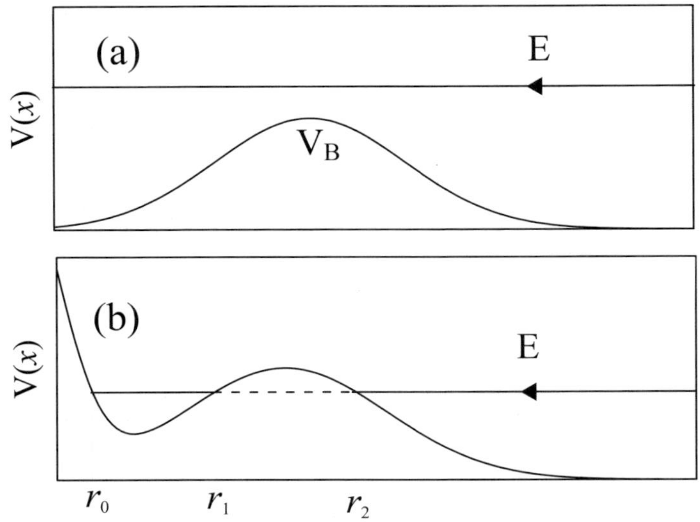

Fig. 5.9 Representations of a collision with $E > V_{\rm B}$, without turning points (panel (a)), and a collision with $E < V_{\rm B}$ with three turning points (panel (b)).

down. The WKB results for energies just above the barrier are also wrong. In this situation, the transmission coefficient of Eq. (5.253) is equal to one, meaning that there is no reflection. In such cases it is necessary to go beyond the WKB approximation and take into account terms that have been neglected. The origin of the problem is that the WKB wave functions of Eq. (5.171) are not solutions of the original Schrödinger equation,

$$\left[-\frac{\hbar^2}{2\mu}\frac{d^2}{dr^2} + V_l(r)\right] u_l(r) = E\, u_l(r), \tag{5.254}$$

or, equivalently,

$$\left[\frac{d^2}{dr^2} + k_l^2(r)\right] u_l(r) = 0. \tag{5.255}$$

This can be checked inserting the WKB wave function of Eq. (5.171) into the above equation. One easily finds that the right hand side does not vanish. We get instead

$$\left[\frac{d^2}{dr^2} + k_l^2(r)\right] u_l^{\rm WKB}(r) = -\left[\frac{3k'(r)}{4\,k^2(r)} - \frac{k''(r)}{2\,k(r)}\right] u_l^{\rm WKB}(r). \tag{5.256}$$

This means that the WKB wave functions are solutions of the modified Schrödinger equation

$$\left[-\frac{\hbar^2}{2m}\,\frac{d^2}{dr^2}+V_{\rm WKB}(r)\right]\,u_l^{\rm WKB}(r)=E\,u_l^{\rm WKB}(r). \qquad (5.257)$$

Above,

$$V_{\rm WKB}(r)=V_l(r)+V_q(r) \qquad (5.258)$$

and $V_q(r)$ is the so called *quantum potential* [Langer (1937)],

$$V_q(r)=-\frac{\hbar^2}{2m}\left[\frac{3k'(r)}{4\,k^2(r)}-\frac{k''(r)}{2\,k(r)}\right]. \qquad (5.259)$$

When the conditions for the validity of the WKB approximation are met, $|V_q(r)|\ll|V_l(r)|$ and the quantum potential can be neglected. Otherwise, it is necessary to take it into account, using the effective potential $V_{\rm WKB}(r)$ instead of $V_l(r)$.

The shortcomings of the WKB approximation can be handled in different ways. For energies near the top of the barrier the first derivative of the potential vanishes and the linear expansion used for an isolate turning point cannot be used. However, it is possible to use the method of comparison equation for a parabolic potential and obtain reasonable results [Berry and Mount (1972)]. Equivalent results can be derived through the analytical continuation into the complex plane [Kemble (1935); Fröman and Fröman (1965); Berry and Mount (1972)]. These methods lead to the much better expression for the transmission coefficient,

$$\mathcal{T}=\frac{1}{1+e^{2\Phi}}\,, \qquad (5.260)$$

which was first derived by Kemble [Kemble (1935)]. This result can be extended to energies above the barrier, with the integral defining Φ (Eq. (5.240)) being evaluated between two complex turning points.

Eq. (5.260) becomes particularly simple if the potential barrier can be approximated by the parabola

$$V_l(r)\simeq V_B-\frac{1}{2}\,m\omega^2\,(r-R_B)^2\,. \qquad (5.261)$$

The transmission coefficient then becomes

$$\mathcal{T}=\frac{1}{1+\exp\left[2\pi\,(V_B-E)\,/\hbar\omega\right]}. \qquad (5.262)$$

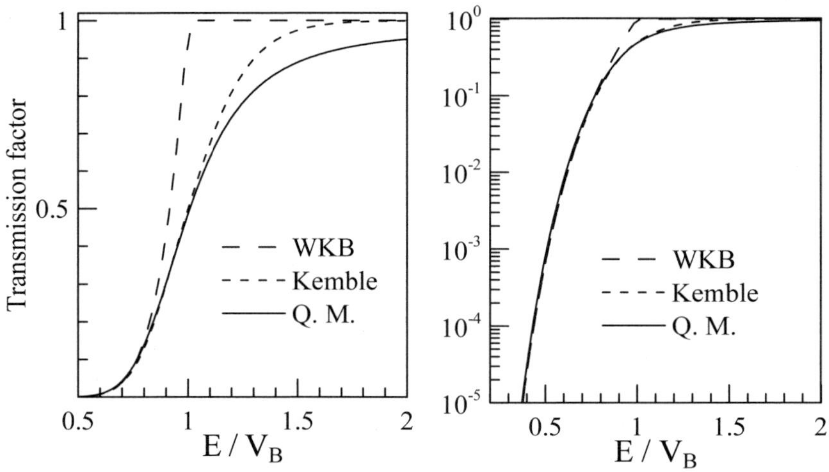

Fig. 5.10 Comparison of semiclassical transmission factors with exact quantum mechanics results (solid lines). The long dashed lines were obtained with Eq. (5.253) and the short-dashed lines with the Kemble's formula (Eq. (5.260)). This example corresponds to collisions of ^{6}Li projectiles on a ^{12}C target. The same results are presented in linear (left panel) and logarithmic (right panel) scales.

This result, which is exact for a parabolic potential, is known as the Hill-Wheeler transmission coefficient [Hill and Wheeler (1953)]. For other barriers one can expand the potential to second order around the top and obtain the height, radius and curvature parameters, denoted respectively by V_B, R_B and $\hbar\omega$. This approximation is reasonable at collision energies near V_B but it may be inaccurate much below the barrier, where the shape of the barrier tends to be very different from its parabolic approximation. In this energy range, it is better to use Kemble's formula (Eq. 5.260)).

As an illustration, in figure 5.10 we compare the transmission factor obtained by full Quantum Mechanics with the results of the versions of the WKB approximations discussed above. The example is the $^{6}\text{Li} + {}^{12}\text{C}$ collision and the nuclear potential used in the exact calculation includes a strong absorptive term (see section 7). Thus, the contributions from the innermost turning point is not expected to be relevant. It is clear that the Kemble formula is a much better approximations then the standard WKB.

In collisions where short-range absorption is not too important and the potential barrier has the behavior of figure 5.9 (b), the contribution from reflections at the innermost turning point r_0 cannot be neglected. This problem has been discussed by several authors (see e.g. [Berry and Mount (1972); Brink and Takigawa (1977)]). The WKB wave function still has the general form of Eq. (5.243) but the strength of the emergent wave is modified as $S_l \to \tilde{S}_l = S_l + \Delta S_l$, with ΔS_l arising from the reflection at r_0. In this way, the internal turning point r_0 changes both the modulus and the phase of the S-matrix.

To close this section, we point out that the WKB approximation does not account for quantum reflection above the effective barrier. There are however other approximations to handle this problem. Bremmner [Bremmer (1951)] approximated the barrier by a series of steps and evaluated the reflection coefficient for this system. He then lets the width of the steps go to zero, with the number os steps going to infinity. In this way, he obtains a reflected wave. Other possibility is to use the WKB wave functions as distorted waves in a DWBA calculation, in which the quantum potential of Eq. (5.259) plays the role of a perturbation. This method has been used by Maitra and Heller [Maitra and Heller (1996)] for a collision in one dimension.

5.2.3 *The semiclassical scattering amplitude*

In section 5.2.2 we presented a semiclassical method to calculate the phase shift in scattering problems. Of course the final aim of all of this is to find a mean to evaluate the partial-wave sum leading to the scattering amplitude, which gives the cross section. In the present section we present the relevant theory for this endeavor. The basic idea is to relate an infinite sum of terms to an infinite sum of integrals that converge much faster. Usually, the first term is strongly dominant and it is alone a reasonable approximation to the scattering amplitude. Performing the integration itself would be much easier than to evaluate the infinite sum of terms.

There is an important theorem due to Poisson that turns out to be very useful for the calculation of the scattering amplitude. Suppose that we want to evaluate the infinite sum of functions $a_l(\theta)$,

$$f(\theta) = \sum_{l=0}^{\infty} a_l(\theta). \tag{5.263}$$

Poisson's theorem states that[31]

$$f(\theta) = \sum_{l=0}^{\infty} a_l(\theta) = \sum_{m=-\infty}^{\infty} (-)^m \int_0^{\infty} d\lambda\, e^{2i\pi m\lambda}\, a(\lambda, \theta), \tag{5.264}$$

[31] A simple proof of this theorem can be found in [Brink (1985)].

where $a(\lambda, \theta) = a_l(\theta)$, with $\lambda = l + 1/2$. If the $m = 0$ term dominates, then the sum above can be replaced by a single one-dimensional integral.

The scattering amplitude,

$$f(\theta) = \frac{1}{2ik} \sum_{l=0}^{\infty} (2l+1)\ [S_l - 1]\ P_l(\cos\theta), \tag{5.265}$$

has the general form of Eq. (5.263), with

$$a_l(\theta) = \frac{1}{2ik}\ (2l+1)\ [S_l - 1]\ P_l(\cos\theta). \tag{5.266}$$

Then, according to Poisson's theorem, it can be written as

$$f(\theta) = \frac{1}{ik} \sum_{m=-\infty}^{\infty} (-)^m \int_0^{\infty} d\lambda\ e^{2i\pi m\lambda}\ \lambda\,[S(\lambda) - 1]\ P_\lambda(\cos\theta), \tag{5.267}$$

where $S(\lambda) = S_l$. The dominant term ($m = 0$) is simply

$$f(\theta) = \frac{1}{ik} \int_0^{\infty} d\lambda\ \lambda\,[S(\lambda) - 1]\ P_\lambda(\cos\theta). \tag{5.268}$$

Excluding the angle $\theta = 0$, which is of no practical interest, we can drop the contribution from the term -1 within the square brackets in Eqs. (5.265) and in the subsequent equations. In this way, Eq. (5.268) becomes[32]

$$f(\theta) = \frac{1}{ik} \int_0^{\infty} d\lambda\ \lambda\ |S(\lambda)|\, e^{2i\delta(\lambda)}\ P_\lambda(\cos\theta). \tag{5.269}$$

The integral in the above equation can be best evaluated if we use the Legendre polynomial with the continuous order λ,

$$P_l(\cos\theta) \quad \longrightarrow \quad P(\lambda, \cos\theta) \simeq \left(\frac{2}{\pi\lambda\sin\theta}\right)^{1/2} \cos\,(\lambda\theta - \pi/4), \tag{5.270}$$

and decompose it into its two branches,

$$P(\lambda, \cos\theta) = \left(\frac{1}{2\pi\lambda\sin\theta}\right)^{1/2} \left[e^{i(\lambda\theta - \pi/4)} + e^{-i(\lambda\theta - \pi/4)}\right]. \tag{5.271}$$

[32]Using the addition theorem and the completeness relation for spherical harmonics, we can write

$$\sum_{l=0}^{\infty} (2l+1)\ P_l(\cos\theta) = \sum_{l=0}^{\infty} \sum_{m=-l}^{l} 4\pi\, Y^*_{lm}(\hat{\mathbf{k}}')\, Y_{lm}(\hat{\mathbf{k}}) = \delta\,(\theta)\,.$$

This shows that the contribution from the term -1 within the square brackets is proportional to a delta function at the forward direction.

Eq. (5.270) is valid asymptotically for large values of λ and/or values of θ not close to zero or to π. With this decomposition, the asymptotic form of the θ-standing wave of Eq. (5.270) is rewritten as a sum of two running waves.

Inserting the above form of the Legendre function into Eq. (5.269), the scattering amplitude splits into two distinct terms,

$$f(\theta) = f^{(+)}(\theta) + f^{(-)}(\theta), \tag{5.272}$$

given by

$$f^{(+)}(\theta) = \frac{1}{ik}\left(\frac{1}{2\pi\sin\theta}\right)^{1/2}\int_0^\infty d\lambda\,\lambda^{1/2}\,|S(\lambda)|\;e^{i[2\delta(\lambda)+\lambda\theta-\pi/4]} \tag{5.273}$$

and

$$f^{(-)}(\theta) = \frac{1}{ik}\left(\frac{1}{2\pi\sin\theta}\right)^{1/2}\int_0^\infty d\lambda\,\lambda^{1/2}\,|S(\lambda)|\;e^{i[2\delta(\lambda)-\lambda\theta+\pi/4]}. \tag{5.274}$$

Above and throughout this section, we write separately the modulus and the phase of the S-matrix[33], as we have done in Eq. (5.244). Note that the only difference between the amplitudes $f^{(+)}$ and $f^{(-)}$ is the sign of $(\lambda\theta - \pi/4)$ in the exponential factors. These different signs correspond to opposite projections of the orbital angular momentum on the direction perpendicular to the scattering plane. In $f^{(+)}(\theta)$, it is parallel to the product $\mathbf{k} \times \mathbf{k}'$ whereas in $f^{(-)}(\theta)$ it is anti-parallel. The branches $f^{(+)}(\theta)$ and $f^{(-)}(\theta)$ are called, respectively, the far-side and the near-side amplitudes, for reasons that will become clear latter. We label these by the aforementioned signs of $(\lambda\theta - \pi/4)$ in the exponentials of the running θ-waves.

In nuclear scattering applications, the phase shift is complex owing to the absorption, which simulates the virtual excitation of internal degrees of freedom of the collision partners (inelastic scattering, transfer of particle, fusion etc.). Instead of using a complex phase shift, we save the notation δ_l for its real part. Its imaginary part is then contained in $|S_l|$, which becomes less than one. In some nuclear collisions the influence of the imaginary part of the potential may be so strong that the elastic cross section is fully dominated by diffractive effects. This situation will be discussed in details in section 7.5.

[33]Since we are dealing with hermitian hamiltonians, this modulus is equal to one and this factor could be omitted. However, keeping this factor, we can use the results of this section in chapter 7, which deals with complex potentials.

We now evaluate the λ integrals in Eqs. (5.273) and (5.274). For this purpose we resort to the stationary phase approximation (SPA). This approximation can be used in integrals of the type

$$I = \int_a^b dx\ F(x)\, e^{i\varphi(x)} \tag{5.275}$$

when two conditions are satisfied. The first is that the exponential factor oscillates rapidly compared with the rate of variation of $F(x)$, so that there are strong cancellations. The second is that there is a region within the range of integration where $\varphi(x)$ is slowly varying. This condition is satisfied in the vicinity of a point where the derivative of the exponent with respect to x vanishes. Thus, the *stationary point* is given by the condition

$$\frac{d\varphi(x)}{dx} = 0. \tag{5.276}$$

In principle, there may be several stationary points, x_j ($j = 1, ..., N$), within the interval $\{a, b\}$. One can then expand the phase $\varphi(x)$ around the stationary points as

$$\varphi(x) \simeq \varphi(x_j) + \frac{\varphi''(x_j)}{2}\,(x - x_j)^2\,. \tag{5.277}$$

The procedure described below can be used when there is a single stationary point or when there are several well separated stationary points. Two neighboring stationary points are *well separated* when their contributions come from regions that do not overlap. In this case, the result of the integration will be given by the sum of the independent contributions from the stationary points. If there are no other stationary point outside the interval $\{a, b\}$, the integration limits can be extended to $\{-\infty, \infty\}$ and the integration can be carried out analytically. The problem reduces to a sum of Gaussian integrals and the result is

$$\int_{-\infty}^{\infty} dx\, F(x)\, e^{i\varphi(x)} \simeq \sum_{j=1}^{N} \left[\frac{2i\pi}{\varphi''(x_j)}\right]^{1/2} F(x_j)\, e^{i\varphi(x_j)}. \tag{5.278}$$

We can use the SPA to evaluate the integral over λ that gives the near-side and far-side amplitudes, $f^{(-)}(\theta)$ and $f^{(+)}(\theta)$. In these cases, the functions $F(x)$ and $\varphi(x)$ are

$$F(x) \to F(\lambda) = \frac{1}{ik}\left(\frac{1}{2\pi\sin\theta}\right)^{1/2} \lambda^{1/2}\,|S(\lambda)| \tag{5.279}$$

$$\varphi(x) \to \varphi^{(\pm)}(\lambda) = 2\delta(\lambda) \pm \lambda\theta \mp \pi/4, \tag{5.280}$$

where $\varphi^{(\pm)}$ are phases associated with $f^{(\pm)}(\theta)$. In principle, the amplitudes $f^{(+)}$ and $f^{(-)}$ can have several stationary points, which we denote by λ_{j_+} and λ_{j_-}, respectively. They are given by the equations

$$\left[\frac{d\varphi^{(\pm)}(\lambda)}{d\lambda}\right]_{\lambda=\lambda_{j\pm}} = 0. \tag{5.281}$$

Using the explicit values of $\varphi^{(\pm)}(\lambda)$, the above condition yields

$$2\left[\frac{d\delta(\lambda)}{d\lambda}\right]_{\lambda=\lambda_{j_+}} = -\,\theta \tag{5.282}$$

and

$$2\left[\frac{d\delta(\lambda)}{d\lambda}\right]_{\lambda=\lambda_{j_-}} = \theta. \tag{5.283}$$

The above equations have a simple interpretation. According to Eq. (5.220), twice the derivative of the phase shift with respect to the angular momentum is the classical deflection function. That is $2\,d\delta(\lambda)/d\lambda = \Theta(\lambda)$. Therefore, the stationary phase condition for the near and far amplitudes are

$$f^{(-)}(\theta) \;\rightarrow\; \Theta(\lambda_{j_-}) = \theta \tag{5.284}$$

$$f^{(+)}(\theta) \;\rightarrow\; \Theta(\lambda_{j_+}) = -\,\theta. \tag{5.285}$$

With the above results, the far-side and near-side amplitudes can be easily calculated using the SPA. One gets respectively,

$$f^{(+)}(\theta) = \frac{1}{ik}\sum_{j_+}\left(\frac{\lambda_{j_+}}{2\pi\sin\theta}\right)^{1/2}\left[\frac{2i\pi}{\Theta'(\lambda_{j_+})}\right]^{1/2} e^{i\left[2\delta(\lambda_{j_+})+\lambda_{j_+}\theta-\pi/4\right]} \tag{5.286}$$

$$f^{(-)}(\theta) = \frac{1}{ik}\sum_{j_-}\left(\frac{\lambda_{j_-}}{2\pi\sin\theta}\right)^{1/2}\left[\frac{2i\pi}{\Theta'(\lambda_{j_-})}\right]^{1/2} e^{i\left[2\delta(\lambda_{j_-})-\lambda_{j_-}\theta+\pi/4\right]}. \tag{5.287}$$

The above discussion can be easily extended to any term in the Poisson series. In the case of the mth term, the far-side amplitude becomes

$$f^{(+)}(\theta) = \frac{e^{i\pi m}}{ik}\left(\frac{1}{2\pi\sin\theta}\right)^{1/2}\int_{-\infty}^{\infty} d\lambda\,\lambda^{1/2}\,e^{i\,[2\delta(\lambda)+\lambda\theta+2\pi m\lambda-\pi/4]} \tag{5.288}$$

and the stationary point condition is

$$\Theta(\lambda_{j_+,m}) = -\theta - 2\pi m. \tag{5.289}$$

Similarly, for the near-side amplitude one gets the stationary point condition

$$\Theta(\lambda_{j_-,m}) = \theta - 2\pi m. \qquad (5.290)$$

We remark that in typical nuclear physics applications, the interaction responsible for scattering is absorptive (complex with negative imaginary part) and all the $m \neq 0$ terms, which follow a trajectory going deep inside the strong absorption region, are damped out.

The physical interpretation of the stationary phase conditions is now clear. The far-side amplitude is sensitive to negative-angle deflection, usually encountered for attractive potentials, while the near-side amplitude receives its main contribution from positive-angle deflection, arising from repulsive interactions. The Poisson contribution to the stationary point, namely -$2\pi m$, qualifies the number of times (m) the scattered particle encircles the target before it escapes to infinity. For repulsive interactions one can only have $m = 0$. However, for attractive interactions, the $m \neq 0$ contributions may be of importance. This situation is schematically represented in figure 5.11, which exhibits stationary points for $m = 0$ (figures 5.11(a) and (b)) and $m = 1$ (figure 5.11(c)).

To illustrate the above procedures, we consider the scattering from a repulsive potential. In this case, the deflection function is always positive and the far-side scattering amplitude has no stationary point. We then have to consider only the near-side contribution, given by Eq. (5.287). Dropping the lower index j of the stationary point and using the relation

$$\frac{d}{d\lambda} = \frac{db}{d\lambda}\frac{d}{db} = \frac{1}{k}\frac{d}{db},$$

the derivative of the deflection function appearing in the denominator of Eq. (5.287) takes the form

$$\Theta'(\lambda_-) \equiv \left[\frac{d\Theta(\lambda)}{d\lambda}\right]_{\lambda=\lambda^-} = \frac{1}{k}\left[\frac{d\Theta(b)}{db}\right], \qquad (5.291)$$

where $b = \lambda_-/k$ is the impact parameter associated with the stationary value of the angular momentum. Calculating the other quantities appearing Eq. (5.287), and noticing that the constant phase $i\pi/4$ gives rise to $i^{1/2}$, which combines with the other $i^{1/2}$ in the pre-exponential factor of the amplitude to cancel the i in the denominator, we obtain

$$f^{(-)}(\theta) = \left[\frac{b}{\sin\theta}\frac{db}{d\Theta}\right]^{1/2} e^{i\,[2\,\delta(\lambda_-) - \theta\,\lambda_-]}\,. \qquad (5.292)$$

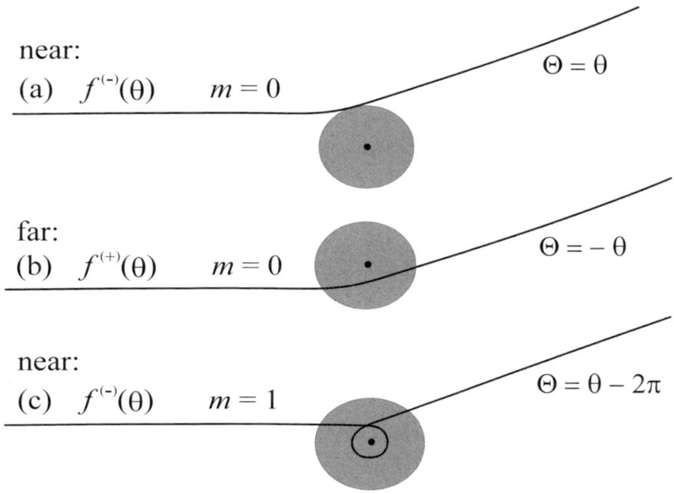

Fig. 5.11 Trajectories corresponding to the stationary points in the $m = 0$ and $m = 1$ near-side amplitudes, and in the $m = 0$ far-side amplitude. The relation between the classical deflection function, Θ, and the scattering angle, θ, is indicated for each trajectory.

It is clear that only one solution of the stationary phase point will be real and accordingly the near-side amplitude will contain just one term. The cross section then reduces to

$$\frac{d\sigma(\theta)}{d\Omega} = |f^{(-)}(\theta)|^2 = \frac{b}{\sin\theta}\left|\frac{db}{d\Theta}\right|. \tag{5.293}$$

This cross section is identical to its classical counterpart, given in Eq. (1.46). It is thus clear that when there is only one stationary point the scattering amplitude may be written as the square root of the classical cross section multiplied by a phase factor, which depends on the nature of the interation, and the semiclassical cross section coincides with the classical one. In the particular case of the Coulomb interaction, one gets the Rutherford cross section. This point is discussed in more detail in section 7.5.2.

In cases involving several solutions of the stationary point condition, the above result is just a sum of terms. We thus write for the far-side amplitude

$$f^{(+)}(\theta) = \sum_{j_+} e^{i\alpha_{j_+}(\theta)} \sqrt{\frac{d\sigma_{cl,j_+}}{d\Omega}}, \tag{5.294}$$

where j_+ stands for the jth stationary point of $f^{(+)}(\theta)$. For the near-side amplitude we have

$$f^{(-)}(\theta) = \sum_{j_-} e^{i\alpha_{j_-}(\theta)} \sqrt{\frac{d\sigma_{cl,j_-}}{d\Omega}} \,, \tag{5.295}$$

with an analogous meaning for j_-. Above, $\sigma_{cl,j\pm}$ are the classical cross sections for the impact parameters $b_{j_\pm} = \lambda_{j\pm}/k$ and $\alpha_{j\pm}(\theta)$ are the phases

$$\alpha_{j_+}(\theta) = 2\,\delta(\lambda_{j_+}) + \lambda_{j_+}\theta - \pi/2 + \Delta \tag{5.296}$$

and

$$\alpha_{j_-}(\theta) = 2\,\delta(\lambda_{j_-}) - \lambda_{j_-}\,\theta + \Delta. \tag{5.297}$$

Note that the stationary points $\lambda_{j\pm}$ are themselves functions of the scattering angle θ, as indicated in Eqs. (5.284) and (5.285). The constant phase, Δ can be either 0 or $\pi/2$ depending on the sign of $d\Theta/db$. One has to keep track of this phase when more than one stationary phase point contributes to these amplitudes.

The above results can be summarized in the expression for the cross section

$$\frac{d\sigma(\theta)}{d\Omega} = \left| \sum_{j\pm} e^{i\,\alpha_{j\pm}(\theta)} \sqrt{\frac{d\sigma_{cl,j\pm}}{d\Omega}} \right|^2 . \tag{5.298}$$

This is the principal result of the semiclassical scattering theory. It shows clearly that the quantum amplitude can be represented as a sum of contributions whose amplitudes are the square roots of the classical cross sections and whose phases are completely determined from knowledge of the phase shift at the stationary points. Accordingly, interference in the quantum cross section arises if there are more than one stationary point contribution. Otherwise, within the many approximations made, the quantum and the classical cross sections are identical !

Let us discuss in details the simple case of two stationary points, say λ_1 and λ_2. For the moment, they can be both associated with the same branch of the scattering amplitude, or with different ones. The cross section of Eq. (5.298) becomes

$$\frac{d\sigma(\theta)}{d\Omega} = \frac{d\sigma_{cl,1}}{d\Omega} + \frac{d\sigma_{cl,2}}{d\Omega} + 2\,\sqrt{\frac{d\sigma_{cl,1}}{d\Omega}\,\frac{d\sigma_{cl,2}}{d\Omega}} \times \cos\left(\alpha_{\rm rel}\right), \tag{5.299}$$

where $\alpha_{\rm rel}$ is the relative phase

$$\alpha_{\rm rel} = \alpha_2(\theta) - \alpha_1(\theta). \tag{5.300}$$

Let us now consider separately the cases of two stationary points of different branches (near-far interference) and that of two stationary points of the same branch (near-near or far-far). In the former case one should use Eq. (5.296) for one point, say α_2, and Eq. (5.297) for the other. The relative phase then becomes

$$\alpha_{\rm rel} = 2\left[\delta(\lambda_2) - \delta(\lambda_1)\right] - \pi/2 + (\lambda_2 + \lambda_1)\,\theta \qquad (5.301)$$

The period of the angular oscillation corresponds to a variation of the relative phase of 2π. Neglecting the variation of the phase shifts, the period is

$$\Delta\theta = \frac{2\pi}{\lambda_2 + \lambda_1}. \qquad (5.302)$$

In the case of near-near or far-far interference, the same equation should be used for α_1 and α_2. One gets

$$\Delta\theta = \frac{2\pi}{|\lambda_2 - \lambda_1|}. \qquad (5.303)$$

Since stationary points of the same branch of the scattering amplitude are usually at angular momenta of the same order, the above period may be very large.

We conclude that cross sections influenced by near-far interference oscillate much faster than those dominated by near-near or far-far interference. More details of this problem will be given in a latter sub-section, in an illustration of scattering from a Lennard-Jones potential. In this case, there are one stationary point in the near-side amplitude and two in the far-side one. The cross section then exhibits both far-far and near-far interferences.

5.2.3.1 *Caustics in the semiclassical cross section*

We have seen that the semiclassical structure of the scattering amplitude is dictated by the classical deflection function. If the deflection function has a monotonic dependence on the impact parameter, as in Coulomb scattering, there is only one term in the sum of Eq. (5.298) and the resulting cross section coincides with its classical counterpart. On the other hand, if the deflection function exhibits maxima or minima, the stationary point condition may have more than one solution. The resulting semiclassical amplitude is then a sum of as many terms as the number of solutions of the stationary phase condition. These terms are given by products of the square root of the classical cross section multiplied by a phase. Owing to these phases, the resulting quantum cross section exhibits interference patterns.

Realistic deflection functions in atomic and nuclear scattering may shows several singular points, called caustics. The first one of these caustics is the rainbow, which takes place when the deflection function has a maximum or a minimùm with respect to the impact parameter or, equivalently, to the semiclassical angular momentum. At the rainbow angle, the derivative of the deflection function vanishes and the classical cross section diverges (see Eq. (1.46)). The classical cross section also diverges when the deflection function crosses the impact parameter axis for a non-zero value of b, say b_{G}. Here, the divergence is due to the vanishing of $\sin\theta$ in Eq. (1.46). This is called the forward glory scattering[34]. If after the crossing at b_{G} the deflection dives to $-\infty$, the singularity is referred to as orbiting, and it represents the classical analogue of resonances. In classical mechanics this situation occurs for pair of values of the impact parameter and collision energy, $\{b_{\mathrm{o}}, E_{\mathrm{o}}\}$, for which the particle reaches the top of the effective potential barrier (potential + centrifugal term) with vanishing radial kinetic energy. In this case the particle remains for a long time with nearly the same radial coordinate. It describes orbits around the target, before it emerges. Examples of a deflection function exhibiting rainbow, glory and orbiting were given in figure 1.6 of chapter 1. We discuss below these phenomena in more details.

Rainbow scattering

The derivation of the semiclassical cross section of Eq. (5.298) was based on the assumption that the stationary points were sufficiently far appart. Calling λ_j and λ_{j+1} two neighboring stationary points of $f^{(+)}$ or $f^{(-)}$, the adopted gaussian approximations around the stationary points require that they do not interfere. This condition can be expressed by the equations

$$\frac{1}{2}\,\Theta'(j)\ |\lambda_j - \lambda_{j+1}| \gg; \quad \frac{1}{2}\,\Theta'(j+1)\ |\lambda_j - \lambda_{j+1}| \gg 1. \tag{5.304}$$

Rainbow scattering occurs when the deflection function has an extremum (maximum or minimum), Θ_r, and the observation angle is close to $|\Theta_r|$. In this case one branch of the scattering amplitude (near-side or far-side) has two stationary points, say λ_1 and λ_2 (with $\lambda_2 > \lambda_1$), associated

[34]There is a similar divergence when the potential is strong enough for the deflection function to reach $\pm\pi$. In this case, the divergence is called backward glory.

with that observation angle, and these points are close together. As the observation angle approaches the rainbow angle, $\theta_r \equiv |\Theta_r|$, we have

$$\theta \to \theta_r \implies \lambda_1 \to \lambda_r \leftarrow \lambda_2 \text{ and } \Theta'(\lambda_1) \to \Theta'(\lambda_r) \leftarrow \Theta'(\lambda_2), \tag{5.305}$$

where λ_r is the rainbow angular momentum, defined by the condition: $\Theta(\lambda_r) = \Theta_r$. We will frequently use the notation $\Theta_i \equiv \Theta(\lambda_i)$ and a similar notation for the first and second derivatives of the deflection function.

Although $\Theta'(\lambda_1) = \Theta'(\lambda_2) = 0$, the derivative of the deflection function does not exactly vanish at λ_r. However, assuming that the deflection function is not singular in the rainbow region, $\Theta'(\lambda)$ must be very close to zero for any λ in the range $\{\lambda_1, \lambda_2\}$. Therefore, it is a good approximation to assume that λ_r is nearly an stationary point for observation angles close to θ_r, and expand the phase of the integrand around λ_r. There is however one difficulty: in rainbow scattering the second order term, corresponding to the derivative of the deflection function, vanishes at λ_r[35]. One should then use a different approximation. The simplest way to handle this situation is to go to next order in the expansion, including the third order term. Airy used this procedure to study rainbow scattering of light in 1839 [Airy (1839)]. The expansion of the phase in the integrand of Eqs. (5.273) and (5.274) becomes (the signs (+) and (-) should be used in the cases of rainbow in $f^{(+)}$ and $f^{(-)}$, respectively)

$$\varphi^{(\pm)}(\lambda, \theta) \simeq \alpha_0^{(\pm)} + \alpha_1^{(\pm)} (\lambda - \lambda_r) + \frac{\alpha_3^{(\pm)}}{3!} (\lambda - \lambda_r)^3 . \tag{5.306}$$

Using the explicit form of the phase,

$$\varphi^{(\pm)}(\lambda, \theta) = 2\delta(\lambda) \pm \lambda\theta \mp \pi/4, \tag{5.307}$$

we get the expansion coefficients[36]

$$\alpha_0^{(\pm)} = \varphi_r^{(\pm)} \tag{5.308}$$

$$\alpha_1^{(\pm)} = \left[\frac{d\varphi^{(\pm)}}{d\lambda}\right]_{\lambda=\lambda_r} = \Theta_r \pm \theta \tag{5.309}$$

$$\alpha_3^{(\pm)} = \left[\frac{d^3\varphi^{(\pm)}}{d\lambda^3}\right]_{\lambda=\lambda_r} = \Theta''_r, \tag{5.310}$$

where

$$\varphi_r^{(\pm)} \equiv \varphi^{(\pm)}(\lambda_r) = 2\,\delta(\lambda_r) \pm \lambda_r\,\theta \mp \pi/4. \tag{5.311}$$

[35] In this case, Eqs. (5.287) and (5.286) lead to a divergent cross section.

[36] We remark that we did not include the second order term in the expansion because it is equal to the first derivative of the deflection function, which vanishes at λ_r.

Note that $\varphi_r^{(\pm)}$ and $\alpha_1^{(\pm)}$ are functions of θ and λ_r, whereas $\alpha_3^{(\pm)}$ depends only on λ_r. To keep the notation simple, we avoid indicating these dependences explicitly.

The integrals giving the amplitudes $f^{(\pm)}(\theta)$ are then evaluated with the approximate phase of Eq. (5.306). Using Eqs. (5.306) to (5.310) in Eqs. (5.273) and (5.274) and moving out of the integral the slowly varying factors. One obtains

$$f^{(\pm)}(\theta) = \frac{1}{ik}\left(\frac{\lambda_r}{2\pi\sin\theta}\right)^{1/2} |S(\lambda_r)|\, e^{i\varphi_r^{(\pm)}}$$
$$\times \int_{-\infty}^{\infty} d\lambda\, \exp\left[i\,\alpha_1^{(\pm)}\,(\lambda-\lambda_r) + i\,\frac{\alpha_3}{6}\,(\lambda-\lambda_r)^3\right]. \qquad (5.312)$$

Note that we have extended to $-\infty$ the lower limit of the integral of Eq. (5.312). This is based on the assumption that the relevant contributions to the integral comes from the vicinity of λ_r, which is much larger than 0.

Now we change variable: $\lambda \to t$, according to the equation

$$\frac{\Theta_r''}{6}\,(\lambda-\lambda_r)^3 \to \frac{t^3}{3}. \qquad (5.313)$$

This implies that

$$\lambda - \lambda_r = (\Theta_r''/2)^{-1/3}\, t \quad \text{and} \quad d\lambda = (\Theta_r''/2)^{-1/3}\, dt. \qquad (5.314)$$

Using this transformation, Eq. (5.312) becomes

$$f^{(\pm)}(\theta) = \frac{1}{ik}\left(\frac{\lambda_r}{2\pi\sin\theta}\right)^{1/2} (\Theta_r''/2)^{-1/3}$$
$$\times |S(\lambda_r)|\, e^{i\,\varphi_r^{(\pm)}} \int_{-\infty}^{\infty} dt\, \exp\left[i\left(yt + \frac{t^3}{3}\right)\right], \qquad (5.315)$$

with $\varphi_r^{(\pm)}$ given by Eq. (5.311) and

$$y = \left(\frac{\Theta_r''}{2}\right)^{-1/3} (\Theta_r \pm \theta)\,. \qquad (5.316)$$

The integral of Eq. (5.315) is proportional to the regular Airy function, defined as

$$Ai(y) = \frac{1}{2\pi}\int_{-\infty}^{\infty} dt\, \exp\left[i\left(yt + \frac{t^3}{3}\right)\right]. \qquad (5.317)$$

Thus, Eq. (5.315) can be cast in the form

$$f^{(\pm)}(\theta) = \frac{1}{ik}\left(\frac{2\pi\lambda_r}{\sin\theta}\right)^{1/2} (\Theta_r''/2)^{-1/3}\, |S(\lambda_r)|\, e^{i\,\varphi_r^{(\pm)}}\, Ai(y). \qquad (5.318)$$

The Airy function is the regular solution of the well known equation of Mathematical Physics,

$$\frac{d^2\psi(y)}{dy^2} - y\,\psi(y) = 0. \tag{5.319}$$

This equation has the general form of a Schrödinger equation in the representation of the generalized coordinate y. Multiplying both sides of Eq. (5.319) by $\left(-\hbar^2/2\mu\right)$, it can be put in the form

$$-\frac{\hbar^2}{2\mu}\frac{d^2\psi(y)}{dy^2} + V(y)\,\psi(y) = E\,\psi(y), \tag{5.320}$$

where $V(y)$ is the linear energy-dependent potential,

$$V(y) = E + \frac{\hbar^2}{2\mu}\,y, \tag{5.321}$$

corresponding to a constant force. This potential has a single turning point, at $y = 0$. Thus, $y = 0$ is the border between the classically allowed ($y < 0$) and the classically forbidden ($y > 0$) regions of the configuration space. This becomes clear by looking at the asymptotic form of the Airy function for large positive y,

$$Ai(y) \to \frac{1}{2y^{1/4}\sqrt{\pi}}\,\exp\left(-\frac{2}{3}\,y^{3/2}\right) \tag{5.322}$$

and for large negative y

$$Ai(-y) \to \frac{1}{y^{1/4}\sqrt{\pi}}\,\sin\left(\frac{2}{3}\,y^{3/2} + \frac{\pi}{4}\right). \tag{5.323}$$

It oscillates for $y < 0$ and goes asymptotically to zero as $y \to \infty$. These behaviors are illustrated in figure 5.12, where we show the regular Airy function.

The cross section for rainbow scattering is given by the square of the modulus of Eq. (5.318),

$$\frac{d\sigma_r(\theta)}{d\Omega} = \frac{2\pi\,\lambda_r}{k^2\sin\theta}\left(\frac{2}{\Theta_r''}\right)^{2/3}|S(\lambda_r)|^2\,\left[Ai\,(y)\right]^2, \tag{5.324}$$

where the argument of the Airy function is (see Eq. (5.316))

$$y = \left(\frac{\Theta_r''}{2}\right)^{-1/3}(\Theta_r \pm \theta)\,.$$

The signs + and - should be used for rainbow in $f^{(+)}$ and $f^{(-)}$, respectively.

The period of the angular oscillations of the cross section is given by the period of the square of the Airy function. For large negative values

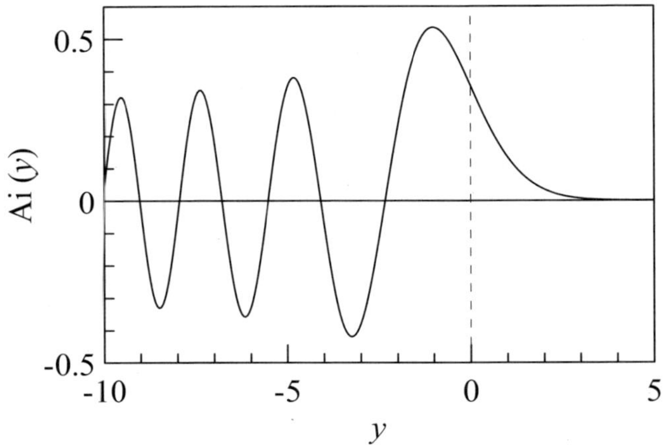

Fig. 5.12 The regular Airy function $Ai(y)$.

of the argument, one can use the asymptotic form given by Eq. (5.323). An estimate of the period can be obtained as follows. Suppose that a maximum of the sinus square function occurs at a given value of y and the next maximum is at $y + \Delta y$. Calling $\xi(y + \Delta y)$ and $\xi(y)$ the arguments of the Airy function for $y+\Delta y$ and y, the period Δy is given by the condition[37]

$$\Delta\xi \equiv \xi(y + \Delta y) - \xi(y) = \left(\frac{2}{3}(y + \Delta y)^{3/2} + \pi/4\right) - \left(\frac{2}{3}y^{3/2} + \pi/4\right) = \pi. \tag{5.325}$$

For large negative values of y, $\Delta y \ll y$ and one can expand the left hand side of the above equation to first order around y. In this way, one obtains

$$\Delta y = \frac{\pi}{\sqrt{|y|}}. \tag{5.326}$$

Therefore, the period decreases slowly as $|y|$ increases. The angular period can be easily obtained from Eq. (5.316) as

$$\Delta\theta = \left|\frac{\Theta_r''}{2\,(\Theta_r - \theta)}\right|^{1/2}. \tag{5.327}$$

The uniform approximation

The Airy approximation is very good for angles close to θ_r. However, as the observation angle deviates from θ_r, the linear term in the expansion of

[37]We used π instead of 2π because $\sin^2\alpha$ behaves as $\cos 2\alpha$.

Eq. (5.306) takes large values, leading to strong cancellations. Therefore, the main contributions to the integral no longer comes from the vicinity of λ_r. The Airy approximation to the cross section then breaks down. On the other hand, far away from the rainbow angle the expansions around the stationary points do not overlap and the cross section is given by Eq. (5.298). However, neither of these expressions can be used at intermediate angles. It would be convenient to have a semiclassical approximation that works in the whole angular range, having Eqs. (5.324) and Eq. (5.298) as limiting cases. The uniform approximation has these characteristic. It consists of making a transformation $\lambda \rightarrow t$ similar to the one used in Airy's approximation, but without expanding around λ_r.

Before introducing the uniform approximation, we re-derive Airy's approximation in a more convenient way. First, we write Eq. (5.273) as

$$f^{(+)}(\theta) = \frac{1}{ik} \left(\frac{1}{2\pi \sin\theta} \right)^{1/2} I^{(+)}(\theta) \tag{5.328}$$

with

$$I^{(+)}(\theta) = \int_{-\infty}^{\infty} d\lambda \; F(\lambda) \; e^{i\varphi(\lambda)}, \tag{5.329}$$

$$F(\lambda) = \lambda^{1/2} \quad \text{and} \quad \varphi(\lambda) = 2\delta(\lambda) + \lambda\theta - \frac{\pi}{4}. \tag{5.330}$$

For simplicity, we are considering a rainbow in the far-side amplitude and omitting the superscripts $(+)$. We consider here the scattering from a real potential, so that we can set $|S(\lambda)| = 1$. To evaluate the integral of Eq. (5.329), we change from λ to a new variable t. In this way, λ becomes itself a function of t and the functions of λ are mapped onto functions of t. Using a tilde to represents the functions of the new variable we have

$$\lambda \quad \rightarrow \quad \tilde{\lambda}(t) \tag{5.331}$$

$$F(\lambda) \quad \rightarrow \quad \tilde{F}(t) = F\left(\tilde{\lambda}(t)\right) \tag{5.332}$$

$$\varphi(\lambda) \quad \rightarrow \quad \tilde{\varphi}(t) = \varphi\left(\tilde{\lambda}(t)\right). \tag{5.333}$$

Once $\tilde{\lambda}(t)$ is specified, $\tilde{F}(t)$, $\tilde{\varphi}(t)$ and $d\tilde{\lambda}/dt$ are determined and $I^{(+)}(\theta)$ can be put in the form

$$I^{(+)}(\theta) = \int_{-\infty}^{\infty} dt \; \mathcal{F}(t) \; e^{i\,\tilde{\varphi}(t)}, \tag{5.334}$$

with

$$\mathcal{F}(t) = \tilde{F}(t) \, \frac{d\tilde{\lambda}}{dt}. \tag{5.335}$$

Airy's approximation can be obtained through the above procedure, using an approximate form of the phase $\varphi(\lambda)$. It is expanded to third order around λ_r, as

$$\varphi(\lambda) \simeq \alpha_0 + \alpha_1 (\lambda - \lambda_r) + \frac{\alpha_3}{6} (\lambda - \lambda_r)^3. \tag{5.336}$$

One then looks for the transformation $\tilde{\lambda}(t)$ which maps the cubic term of the above equation as

$$\frac{\alpha_3}{6} (\lambda - \lambda_r)^3 \to \frac{t^3}{3}. \tag{5.337}$$

One immediately gets

$$\tilde{\lambda}(t) = \lambda_r + \left(\frac{2}{\alpha_3}\right)^{1/3} t\ . \tag{5.338}$$

With this result, $\tilde{F}(t)$, $\tilde{\varphi}(t)$ and $d\lambda/dt$ can easily be determined and the integration can be evaluated in terms of Airy functions. Replacing the slowly varying function $\mathcal{F}$ by its value corresponding to the rainbow angular momentum, one gets the cross section of Eq. (5.324). The problem with this approximation is that the transformation is based on an expansion of the phase around λ_r which is only valid in the neighborhood of the rainbow angle.

The uniform approximation follows the general procedure described above but does not use the expansion around λ_r to find the transformation $\tilde{\lambda}(t)$. In the present case, the transformation is determined by two criteria. The first is that it keeps the cubic form of $\tilde{\varphi}(t)$,

$$\tilde{\varphi}(t) = \beta + \gamma\, t + \frac{t^3}{3}. \tag{5.339}$$

The second is that it establishes a one-to-one correspondence of the points of stationary phase in the λ- and in the t-spaces. Using Eq. (5.339) in the stationary phase condition

$$\frac{d\tilde{\varphi}(t)}{dt} = 0, \tag{5.340}$$

one gets the stationary points

$$t_1 = -i\sqrt{\gamma} \quad \text{and} \quad t_2 = i\sqrt{\gamma}. \tag{5.341}$$

The one-two-one correspondence requires that

$$\tilde{\lambda}(t_1) = \lambda_1 \quad \text{and} \quad \tilde{\lambda}(t_2) = \lambda_2. \tag{5.342}$$

Before searching for a transformation which satisfies the above requirements, let us determine the coefficients β and γ of Eq. (5.339). For this purpose, we evaluate the phase $\tilde{\varphi}$ at the points of stationary phase. Using Eqs. (5.339) and (5.341), we can write

$$\tilde{\varphi}(t_1) = \beta - \frac{2i}{3}\,\gamma^{3/2} \quad \text{and} \quad \tilde{\varphi}(t_2) = \beta + \frac{2i}{3}\,\gamma^{3/2}. \tag{5.343}$$

Taking into account Eqs. (5.333) and (5.342), the above relations become

$$\varphi(\lambda_1) = \beta - \frac{2i}{3}\,\gamma^{3/2} \quad \text{and} \quad \varphi(\lambda_2) = \beta + i\,\frac{2}{3}\,\gamma^{3/2}. \tag{5.344}$$

Combining these two equations, we find the coefficients

$$\beta = \frac{1}{2}\,\left[\varphi(\lambda_2) + \varphi(\lambda_1)\right] \tag{5.345}$$

and

$$\gamma = e^{-i\pi/3}\left(\frac{3}{4}\,\left[\varphi(\lambda_2) - \varphi(\lambda_1)\right]\right)^{2/3}, \tag{5.346}$$

with the phase $\varphi(\lambda)$ given by Eq. (5.330). We remark that there are three possible roots in the above equation. However, since γ is the argument of the Airy function one should choose the root that makes it real. That is

$$\gamma = \pm\,|\gamma|\,, \quad \text{with} \quad |\gamma| = \left|\,\left(\frac{3}{4}\left[\varphi(\lambda_2) - \varphi(\lambda_1)\right]\right)^{2/3}\right|, \tag{5.347}$$

where the signs + and the - should be used respectively for $\theta > \theta_r$ and $\theta < \theta_r$.

The remaining task to calculate the integral of Eq. (5.334) is to determine the function $\mathcal{F}(t)$. Let us look for its values at the stationary points t_1 and t_2. We know that

$$\tilde{F}(t_1) = F(\tilde{\lambda}(t_1)) = F(\lambda_1) = \sqrt{\lambda_1} \tag{5.348}$$

$$\tilde{F}(t_2) = F(\tilde{\lambda}(t_2)) = F(\lambda_2) = \sqrt{\lambda_2}, \tag{5.349}$$

but we still need $d\lambda/dt$ at these points. It can be obtained taking the second derivative with respect to t of the relation

$$\varphi(\lambda) = \tilde{\varphi}(t), \tag{5.350}$$

with $\varphi(\lambda)$ and $\tilde{\varphi}(t)$ given respectively by Eqs. (5.330) and (5.339). The first derivative gives

$$\left[\Theta(\lambda) + \theta\right]\,\frac{d\lambda}{dt} = \gamma + t^2 \tag{5.351}$$

and the second is

$$\Theta'(\lambda)\,\left(\frac{d\lambda}{dt}\right)^2 + [\Theta(\lambda)+\theta]\,\frac{d^2\lambda}{dt^2} = 2\,t. \tag{5.352}$$

At the stationary point t_1, the sum $[\Theta(\lambda_1)+\theta]$ vanishes and the same happens at the other stationary point. The above equation then leads to

$$\left[\frac{d\lambda}{dt}\right]_{\lambda=\lambda_1} = \left(\frac{2t_1}{\Theta_1'}\right)^{1/2} = \left(\frac{-2i\sqrt{\gamma}}{\Theta_1'}\right)^{1/2} \tag{5.353}$$

and

$$\left[\frac{d\lambda}{dt}\right]_{\lambda=\lambda_2} = \left(\frac{2t_2}{\Theta_2'}\right)^{1/2} = \left(\frac{2i\sqrt{\gamma}}{\Theta_2'}\right)^{1/2}. \tag{5.354}$$

Using the above equations and Eq. (5.349), we get $\mathcal{F}(t)$ at the stationary points,

$$\mathcal{F}_1 \equiv \mathcal{F}(t_1) = \sqrt{\lambda_1}\,\left(\frac{-2i\sqrt{\gamma}}{\Theta_1'}\right)^{1/2} \tag{5.355}$$

and

$$\mathcal{F}_2 \equiv \mathcal{F}(t_2) = \sqrt{\lambda_2}\,\left(\frac{2i\sqrt{\gamma}}{\Theta_2'}\right)^{1/2}. \tag{5.356}$$

The problem now is to chose a transformation such that $\mathcal{F}(t)$ passes by the points $\{t_1,\mathcal{F}_1\}$ and $\{t_2,\mathcal{F}_2\}$ on the $t\times\mathcal{F}$ plane. The simplest solution is the straight line

$$\mathcal{F}(t) = a + b\,t \tag{5.357}$$

with

$$a = \frac{\mathcal{F}_2+\mathcal{F}_1}{2}. \tag{5.358}$$

and

$$b = \frac{\mathcal{F}_2-\mathcal{F}_1}{2i\sqrt{\gamma}}. \tag{5.359}$$

We are now ready to evaluate the integral $I^{(+)}(\theta)$. Inserting Eq. (5.357) into Eq. (5.334), we get

$$I^{(+)}(\theta) = e^{i\beta}\left[a\int_{-\infty}^{\infty} dt\ \exp\left[i\left(\gamma t+\frac{t^3}{3}\right)\right] + b\int_{-\infty}^{\infty} dt\ t\ \exp\left[i\left(\gamma t+\frac{t^3}{3}\right)\right]\right]. \tag{5.360}$$

The above expression can be written in terms of the Airy function and its derivative,

$$Ai(z) = \frac{1}{2\pi} \int_{-\infty}^{\infty} dt \, \exp\left[i\left(zt + \frac{t^3}{3}\right)\right] \tag{5.361}$$

$$Ai'(z) = \frac{i}{2\pi} \int_{-\infty}^{\infty} dt \, t \, \exp\left[i\left(zt + \frac{t^3}{3}\right)\right]. \tag{5.362}$$

Adopting the notation $\gamma = z$, we write

$$I^{(+)}(\theta) = 2\pi \, e^{i\beta} \left[a \, Ai(z) \; - \; ib \, Ai'(z)\right], \tag{5.363}$$

and using this result in Eq. (5.328), we get

$$f^{(+)}(\theta) = \frac{e^{i\beta}}{ik} \left(\frac{2\pi}{\sin\theta}\right)^{1/2} \left[a \, Ai(z) \; - \; ib \, Ai'(z)\right]. \tag{5.364}$$

The argument of the Airy's function and of its derivative, $z = \gamma$, is given by Eq. (5.346). The phase β and the coefficients a and b are respectively given by Eqs. (5.345), (5.358) and (5.359). Finally, we can write the cross section for rainbow scattering within the uniform approximation:

$$\frac{d\sigma(\theta)}{d\theta} \equiv |f^{(+)}(\theta)|^2 = \frac{2\pi}{k^2 \sin\theta} \left|a \, Ai(z) \; - \; ib \, Ai'(z)\right|^2. \tag{5.365}$$

Although the linear expansion of the function $\mathcal{F}(t)$ is usually good enough, this expansion can be extended to higher orders. The resulting cross section is then expressed in terms of Airy's function and derivatives of several orders. This approximation is widely discussed in the literature. A detailed presentation can be found in [Chester *et al.* (1957)].

Examples of rainbow scattering

We now give a few examples of the Airy and the uniform approximations in rainbow scattering. We begin with the simple case of a purely attractive potential, where only the far-side amplitude has stationary phase points. For simplicity, we consider the attractive gaussian potential

$$V(r) = -V_0 \, e^{-\,r^2/\bar{R}^2}, \tag{5.366}$$

where V_0 is a positive constant giving the strength of the potential and $\bar{R}$ is a parameter associated with its range. It can be easily checked that the deflection function depends exclusively on the dimensionless parameters[38]

$$\epsilon = \frac{E}{V_0} \quad \text{and} \quad \xi = k\,\bar{R}. \tag{5.367}$$

[38]This results from the fact that the equations of motion are completely determined by these two parameters.

Similarly, the cross section normalized with respect to its geometric value, $\pi\bar{R}^2$, is completely determined by ϵ and ξ.

In figure 5.13, we show the classical deflection function for an attractive gaussian potential with $\epsilon = 1$ and $\xi = 50$. As expected, the deflection function is always negative. It shows a rainbow minimum at the angular momentum $\lambda_r = 51.2$ with value $\Theta_r = -37.9^o$. The corresponding cross sections are shown in figure 5.14. They are normalized with respect to $\pi\bar{R}^2$ and multiplied by $\sin\theta$, to eliminate the divergence at the origin. This divergence arises from the asymptotic form of the Legendre Polynomial (Eq. (5.270)), which is not valid at small angles. The results of an exact quantum mechanical calculation (open circles) are compared with the ones obtained with the Airy (dashed line) and the uniform (solid line) approximations. To get the exact cross section, we solved the radial equation for all relevant partial-waves and carried out the partial-wave summation (see chapter 2). Below the rainbow angle, the cross sections oscillate with a large period, owing to the interference of the contributions from the two stationary points of the far-side amplitude (far-far interference). One sees that the results of the Airy approximation are very accurate around the rainbow maximum and above. However, the agreement with the exact results progressively deteriorates as the observation angle decreases. This is not surprising, since it is well known that Airy's approximation breaks down far away from the rainbow angle. The figure also shows the results of the uniform approximation (solid line). They are extremely accurate in the whole angular range shown in the picture. They can hardly be distinguished from the exact results. This indicates that the linear approximation for $\mathcal{F}(t)$ is quite good. The use of a better description of this function (see e.g. [Chester *et al.* (1957)]) is not necessary, at least in the present case.

Our next example is the scattering from a Lennard-Jones potential. This potential, frequently used to describe molecular forces, is given by

$$V(r) = -V_0 \left[\left(\frac{\bar{R}}{r} \right)^{12} - 2 \left(\frac{\bar{R}}{r} \right)^{6} \right] . \tag{5.368}$$

It is attractive at large distances and strongly repulsive for $r \ll \bar{R}$. We consider the example discussed by Berry in [Berry (1966)], where $k\bar{R} = 346$ and $E/V_0 = 4.7$.

The classical deflection function is shown in figure 5.15. It is positive at small angular momenta and negative at large ones. This behavior expresses the fact that the repulsive nature of the potential is predominant

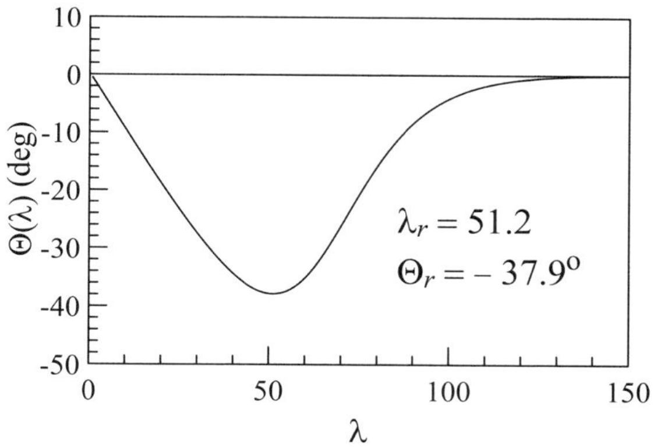

Fig. 5.13 Classical deflection function for an attractive gaussian potential. The results are for $k\bar{R} = 50$ and $E/V_0 = 1.0$. For details see the text.

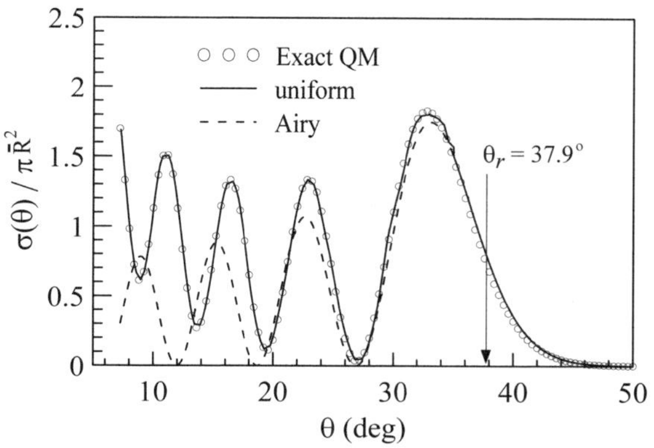

Fig. 5.14 Cross sections for the same potential of the previous figure, normalized with respect to the geometric cross section.

in collisions with small impact parameters whereas the opposite situation occurs when the impact parameter is large. Here we have a more complicated situation: both the near-side and the far-side amplitudes have stationary points. The deflection function exhibits a rainbow at the angular momentum $\lambda_r = 405$, with value $\Theta_r = -25.6^o$. The stationary point of the near-side amplitude (λ_{1-}) and the two stationary points of the far-side amplitude (λ_{1+} and λ_{2+}) for an arbitrary observation angle θ ($< |\Theta_r|$) are indicated by solid circles.

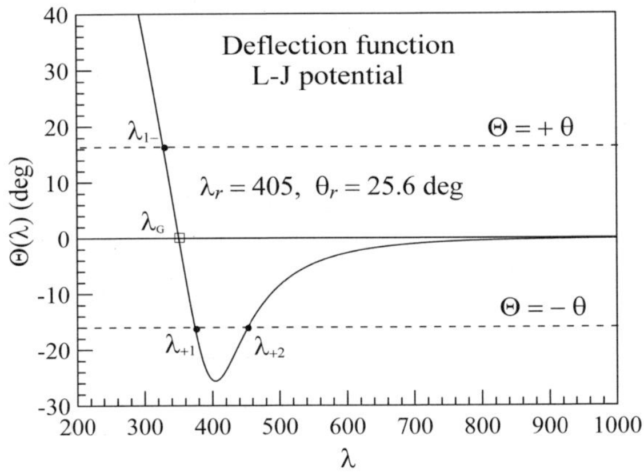

Fig. 5.15 Classical deflection function for the Lennard-Jones potential. The results are for $k\bar{R} = 346$ and $E/V_0 = 4.7$. The points of stationary phase for an arbitrary angle $\theta < \theta_r$ are represented by solid circles and the glory angular momentum, which is discussed later in this section, corresponds to the open square. For details see the text.

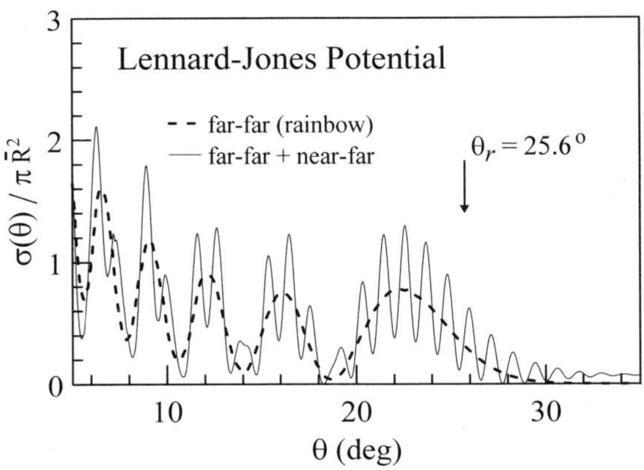

Fig. 5.16 Cross sections for the same potential of the previous figure, normalized with respect to the geometric cross section.

The corresponding elastic scattering cross section is shown in figure 5.16. As in the previous example, the cross sections are normalized with respect to their geometrical values, and multiplied by $\sin\theta$. The contribution from the angular momenta within the rainbow region was evaluated by the uniform approximation, which was shown to be very accurate. The dashed line was obtained neglecting the contribution from the stationary point of $f^{(-)}$. It exhibits the same qualitative behavior of figure 5.14, falling off above θ_r and oscillating with large period in the illuminated region ($\theta < \theta_r$). The solid line includes also the contribution from the near-side amplitude. Now there are high frequency oscillations, superimposed to the slow oscillations of the rainbow cross section (dashed line). The high frequency oscillations corresponds to near-far interference while the low frequency oscillations are due to far-far interference. The periods of these oscillations are roughly estimated in Eqs. (5.302) and (5.303).

Rainbow scattering has been experimentally observed in several situations in molecular, atomic and nuclear physics. In collisions of heavy nuclei, the classical deflection function may present a maximum and also a minimum, at a lower angular momentum. This gives rise to two rainbows, which are referred to as *Coulomb rainbow* (larger λ) and *nuclear rainbow* (smaller λ). Since in such collisions the absorption plays an important role, it is necessary to include an imaginary term in the potential (see chapter 7). If the absorption is too strong, it damps completely the nuclear rainbow. However, in collisions of lighter systems it can be observed. In the present book, we do not consider in detail the modifications in rainbow scattering arising from the absorptive part of the potential. T his problem is presented in detail in [Hussein and McVoy (1984); Brink (1985)].

Rainbow patterns can be seen more clearly in scattering data for atomic and molecular systems. This is illustrated in figure 5.17, which shows the angular distribution in the $H^+ - Kr$ elastic scattering. The similarity with figure 5.14 is very clear. A richer example is the data of Buck and coworkers [Buck *et al.* (1973)] for Li-Hg scattering, shown in figure 5.18. In this case the rapid oscillations on top of the rainbow pattern are resolved. These oscillations resemble the ones of figure 5.16 , which arise from near-far interference. The main difference is the smaller amplitude of the oscillations for the actual system. This suggests that the near-side amplitude here is weaker than the one in our schematic example.

Digression on meteorological rainbow according to geometrical optics[39]

It is of value to have a feeling about the actual colorful rainbow seen in the sky after rain. The mathematics is simple enough to warrant detailed

[39]A nice discussion of the meteorological rainbow can be found in Ref. [Nussenzveig (1977)].

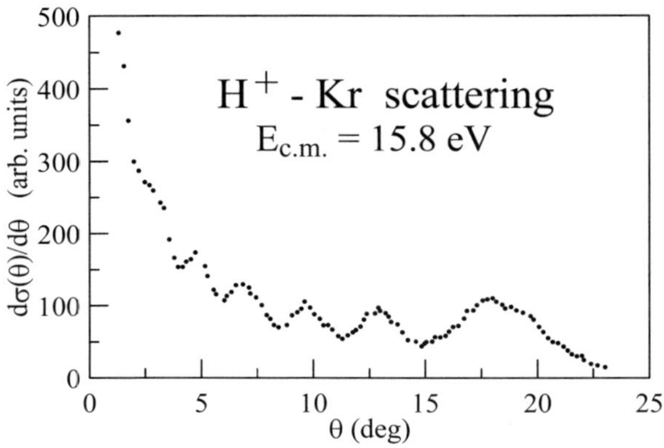

Fig. 5.17 Angular distribution in $H^+ - Kr$ elastic scattering. The data are from [Weise *et al.* (1971)] (see also [Buck (1974)]).

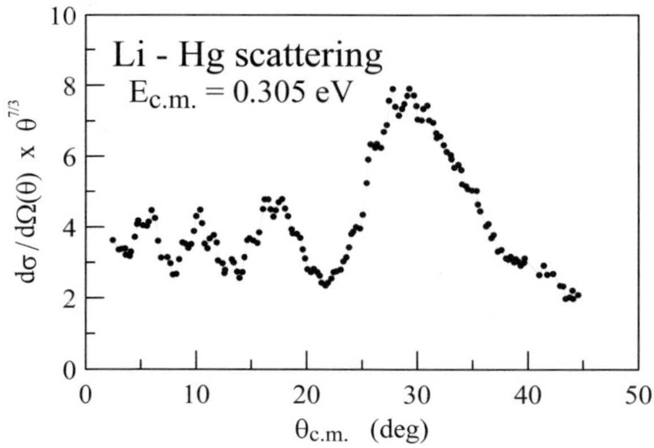

Fig. 5.18 Angular distribution in Li-Hg elastic scattering. The data (solid circles) are from [Buck *et al.* (1973)]. The solid line is simply to guide the eyes. Note that the cross section is multiplied by the factor $\theta^{7/3}$, to remove the steep increase of the cross section at small angles.

development. When light rays pass from the air with an index of refraction of $N_\mathrm{a} = 1$ to the water inside the drop, with $N_\mathrm{w} = 4/3$, internal reflection occurs. In the case of the primary rainbow, there is one total reflection at the internal surface of the drop, as represented in figure 5.19. In the case of secondary rainbow, the ray suffers two total reflections inside the drop before it emerges. Similarly, rainbows of any order k are defined. We discuss below the primary rainbow in detail.

A ray contributing to the primary rainbow is depicted in figure 5.19. The incident ray (denoted by "**1**" in the figure) reaches the drop at A, with incident angle α_i. It is refracted to an angle α_r, propagates within the drop (**2**) and suffers total reflection at B. It then propagates (**3**) until reaching the inner surface of the drop at C, to suffer another refraction. After that, the ray propagates freely through the air (**4**). The light ray is represented by the line (**1**) $\rightarrow$ (**2**) $\rightarrow$ (**3**) $\rightarrow$ (**4**). At each of the points A, B and C, it suffers a clockwise rotation. The rotation angles (see figure 5.19) are respectively

$$\Delta\theta_{\rm A} = \alpha_i - \alpha_r, \quad \Delta\theta_{\rm B} = \pi - 2\alpha_r \quad \text{and} \quad \Delta\theta_{\rm C} = \alpha_i - \alpha_r. \tag{5.369}$$

Thus, the total deflection angle in primary rainbow is

$$\Theta = 2\left(\alpha_i - \alpha_r\right) + \pi - 2\alpha_r = \pi + 2\alpha_i - 4\alpha_r. \tag{5.370}$$

The deflection angle has an implicit dependence on the impact parameter, b. This becomes clear if we write the incidence angle as

$$\sin\alpha_i = b/a, \tag{5.371}$$

with a standing for the radius of the drop. The refraction angle can also be expressed in terms of b. First, one uses Snell's law and write

$$\sin\alpha_r = \frac{\sin\alpha_i}{n}, \tag{5.372}$$

with the shorthand notation $n = N_{\rm w}/N_{\rm a}$. Using Eqs. (5.371) and (5.372) in Eq. (5.370), we obtain

$$\Theta(b) = \pi + 2\sin^{-1}\left(b/a\right) - 4\sin^{-1}\left(b/na\right). \tag{5.373}$$

The deflection function of Eq. (5.373) is represented in figure 5.20, as a function of the impact parameter, normalized with respect to the radius of the drop, $x = b/a$. It begins with the value of 180° at $x = 0$. It then decreases as x increases, until it reaches the rainbow minimum $\Theta_r \simeq$ 138°, at $x_r \simeq 0.85$. These results can be obtained analytically. Deriving Eq. (5.373) with respect to b one obtains

$$\frac{d\Theta^{(1)}(b)}{db} = \frac{2}{\sqrt{a^2 - b^2}} - \frac{4}{\sqrt{n^2a^2 - b^2}}, \tag{5.374}$$

or, using the dimensionless variable $x = b/a$,

$$\frac{d\Theta^{(1)}(x)}{db} = \frac{2}{a}\left[\frac{1}{\sqrt{1-x^2}} - \frac{2}{\sqrt{n^2 - x^2}}\right]. \tag{5.375}$$

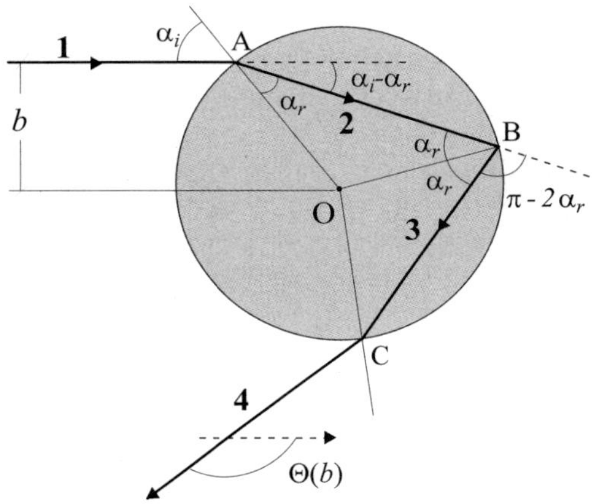

Fig. 5.19 Schematic representation of a light ray reaching a raindrop, in the the situation giving rise to the primary rainbow. The details are given in the text.

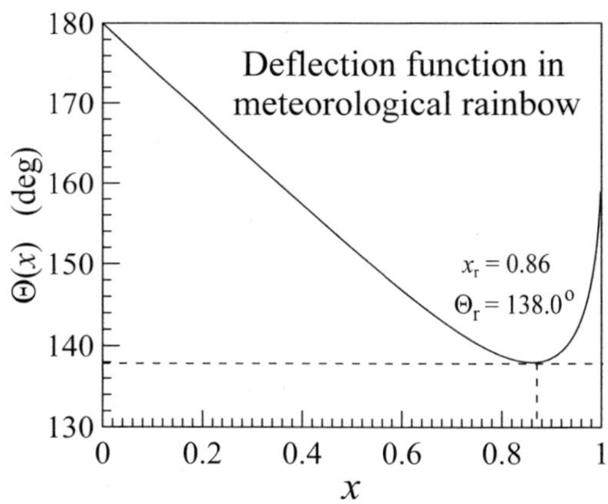

Fig. 5.20 Deflection function of the primary rainbow, as a function of $x = b/a$. For detail, see the text.

The second derivative is

$$\frac{d^2\Theta^{(1)}(b)}{db^2} = \frac{2x}{a^2}\left[\frac{1}{(1-x^2)^{3/2}} - \frac{2}{(n^2-x^2)^{3/2}}\right]. \tag{5.376}$$

The primary rainbow condition, $[d\Theta^{(1)}(x)/dx]_{x=x_r} = 0$, then yields

$$x_r = \sqrt{\frac{4-n^2}{3}}. \tag{5.377}$$

In the case of water droplets, $n = 4/3$ and one gets $x_r = 0.86$. The corresponding value of the rainbow angle, is $\Theta_r = 138.0^o$, as qualitatively found in figure 5.19. It can be easily checked that the second derivative at $x = x_r$ is positive, as required for a minimum.

As in rainbow scattering of massive particles, the classical cross section diverges at the rainbow angle. Although its quantum mechanical counterpart remains finite, it presents a pronounced maximum at the rainbow angle. The same happens in the scattering of light. Therefore, a brighter region is observed in the vicinity of the rainbow angle. In fact, when the rainbow phenomenon is observed the observer has the sun at his back and the rainbow region is seen at the supplement of the rainbow angle. That is at $\theta_r = 180^o - \Theta_r = 42^o$. This is the angle with the horizon, at which the primary rainbow is seen in the sky. Since the index of refraction has a weak dependence on the light wavelength, the rainbow angle is slightly different for each color of the spectrum. For red light, $\lambda = 6563\ \mathring{A}$, $n = 1.3318$. At the other extremum of the spectrum, the wavelength of the violet light is $\lambda = 4047\ \mathring{A}$ and $n = 1.3435$. The rainbow condition for the red light is met at $\theta_r = 42.3^o$, while for the violet it is $\theta_r = 40.6^o$. The difference is 1.7^o. This is the angular width of the main bow. The colors start with the violet at 40.6^o and end with the red at 42.3^o. The band of colors between the violet and red constitute the color band of the primary rainbow. The angle difference of 1.7^o is in fact in error. The correct value is about 2.2^o. The reason is the fact that the angular diameter of the sun is just about 0.5^o, and the incident rays will arrive at the water drop not quite parallel by this amount.

The above discussion can be extended to drops of other fluids and to higher order rainbows. The order corresponds to the number of total reflections at the inner surface of the drop before another refraction takes place and the ray leaves the drop. At each total reflection the direction of the ray is rotated clockwise, by the angle $\pi - 2\alpha_r$. Thus, a general expression for the deflection function in a rainbow of order k can readily be obtained, as

$$\Theta^{(k)} = k\,(\pi - 2\alpha_r) + 2\,(\alpha_i - \alpha_r) = k\pi + 2\alpha_i - 2\,(k+1)\,\alpha_r. \qquad (5.378)$$

Taking into account Snell's law and using the dimensionless variable x, we get

$$\Theta^{(k)}(x) = k\pi \ + \ 2\sin^{-1} x \ - \ 2\,(k+1)\ \sin^{-1}(x/n)\,. \qquad (5.379)$$

The condition for a rainbow of order k, $\left[d\Theta^{(k)}(x)/dx\right]_{x=x_r^{(k)}} = 0$, yields

$$x_r^{(k)} = \sqrt{\frac{(k+1)^2 - n^2}{(k+1)^2 - 1}}, \tag{5.380}$$

and inserting this value into Eq. (5.378) one obtains the corresponding value of the rainbow angle. In the case of secondary rainbow ($k = 2$), one gets for the red light $\theta_r = 36.2^\circ$. The correct secondary rainbow angle is about 51°. For the violet light, one gets $\theta_r = 47.31^\circ$ and the correct value is about 61.1°. The angular width of the secondary rainbow is 10°, and the order of colors is inverted when compared to the primary rainbow: The first bow is the red and the last one is the blue.

The third rainbow, $k = 3$, can be easily calculated, and the rainbow angle comes out at 139°! This means that the bow will be behind the observer as a ring surrounding the sun.The order of colors in the third rainbow is the same as that of the primary rainbow: starts with the blue and ends with the red. However, the sun will supply a bright background. For this reason, it has been very difficult to observe the third rainbow.

Glory scattering

In this case the singularity resides in the $\sin\theta$ which appears in the denominator of the classical cross section of Eq. (1.45). This can happen at $\theta = 0$ or at $\theta = \pi$. These phenomena are known respectively as *forward glory* and *backward glory*. As in the cases of the other previously discussed caustics, the quantum mechanical cross section for glory scattering remains finite. The same happens with the properly derived semiclassical cross section, which we discuss below. Our discussion is restricted to the main features of glory scattering. A detailed study of the problem, with derivation of more accurate expressions, can be found in [Berry (1969)].

The starting point of the semiclassical approximation for glory scattering is to replace the approximate form the Legendre Polynomial of Eq. (5.270), which breaks down at forward and backward angles, by other expressions that are valid in these angular regions. The appropriate expression in each case is given below:

forward Glory ($\theta \approx 0$):

$$P_l(\cos\theta) \to \left(\frac{\theta}{\sin\theta}\right)^{1/2} J_0(\lambda\theta), \tag{5.381}$$

backward Glory ($\vartheta \equiv \pi - \theta \approx 0$):

$$P_l(\cos\theta) \to e^{-i\lambda\pi} \left(\frac{\vartheta}{\sin\vartheta}\right)^{1/2} J_0\left(\lambda\vartheta\right), \tag{5.382}$$

with $\lambda = l + 1/2$. We discuss below the case of forward glory. The results for backward glory can be derived similarly.

To evaluate the scattering amplitude in forward glory scattering, we insert Eq. (5.381) into Eq. (5.269) and obtain

$$f_{\rm G}(\theta) = \frac{1}{ik} \left(\frac{\theta}{\sin\theta}\right)^{1/2} \int_0^\infty d\lambda\ \lambda\, e^{2i\,\delta(\lambda)}\, J_0(\lambda\theta). \tag{5.383}$$

We have assumed that $|S_\lambda| = 1$, which is correct for any real potential[40]. Now we use the integral representation of the Bessel function [Abramowitz and Stegun (1972)],

$$J_0(\lambda\theta) = \frac{1}{2\pi} \int_0^{2\pi} d\gamma\ e^{-i\,\lambda\theta\ \cos\gamma}\ , \tag{5.384}$$

in Eq. (5.383). We get

$$f_{\rm G}(\theta) = \frac{1}{2\pi ik} \left(\frac{\theta}{\sin\theta}\right)^{1/2} \int_0^{2\pi} d\gamma\ I_{\rm G}(\theta,\gamma), \tag{5.385}$$

with

$$I_{\rm G}(\theta,\gamma) = \int_0^\infty d\lambda\ \lambda\ e^{2i\delta(\lambda) - i\,\lambda\theta\ \cos\gamma}. \tag{5.386}$$

The above integral can be evaluated using the stationary phase approximation. The stationary point, λ_s, is determined by requiring that the phase

$$\varphi(\lambda) \equiv 2\delta(\lambda) - \lambda\theta\ \cos\gamma \tag{5.387}$$

be stationary. That is,

$$\left[\frac{d\varphi(\lambda)}{d\lambda}\right]_{\lambda=\lambda_s} = 0 \quad \to \quad 2\left[\frac{d\delta(\lambda)}{d\lambda}\right]_{\lambda=\lambda_s} \equiv \Theta(\lambda_s) = \theta\ \cos\gamma. \tag{5.388}$$

The exponential phase in the integrand of Eq. (5.386) can then expanded to second order around λ_s, as

$$\varphi(\lambda) \simeq \varphi(\lambda_s) + \frac{1}{2}\,\Theta'(\lambda_s)\,(\lambda - \lambda_s)^2. \tag{5.389}$$

Since we are interested in small scattering angles, the stationary phase condition of Eq. (5.388) can be approximated as $\Theta(\lambda_s) = 0$. Therefore, it

[40] We postpone the discussion of complex potentials to chapters 7 and 10.

is approximately satisfied at the glory angular momentum, $\lambda_{\rm G}$, where the deflection function vanishes.

Using Eq. (5.389), with the replacements: $\lambda_s \to \lambda_{\rm G}$, $\Theta'(\lambda_s) \to \Theta'(\lambda_{\rm G}) \equiv \Theta'_{\rm G}$, $\delta(\lambda_s) \to \delta(\lambda_{\rm G}) \equiv \delta_{\rm G}$, and replacing the multiplicative factor in the integrand $\lambda \to \lambda_{\rm G}$, Eq. (5.386) becomes

$$I_{\rm G}(\theta,\gamma) = \lambda_{\rm G}\, e^{2i\varphi_{\rm G}} \int_0^\infty d\lambda\ e^{i\,\Theta'_{\rm G}\,(\lambda-\lambda_{\rm G})^2/2}. \tag{5.390}$$

Since all the relevant contributions to the integral comes from angular momenta in the neighborhood of $\lambda_{\rm G}$, which is much larger then zero, the lower limit of the integral can be extended to $-\infty$. In this way, the gaussian integral can easily be calculated and one gets

$$I_{\rm G}(\theta,\gamma) = \lambda_{\rm G}\, e^{2i\delta_{\rm G}} \left(\frac{2\pi i\nu}{|\Theta'_{\rm G}|}\right)^{1/2} e^{-i\,\lambda_{\rm G}\theta\,\cos\gamma}. \tag{5.391}$$

Above, we have used the explicit form of $\varphi_{\rm G}$,

$$2\varphi_{\rm G} = 2\delta_{\rm G} - \theta\lambda_{\rm G}\,\cos\gamma. \tag{5.392}$$

The quantity ν accounts for the sign of $\Theta'_{\rm G}$, which has been replaced by its modulus. The values $\nu = 1$ and $\nu = -1$ should be used for positive and negative $\Theta'_{\rm G}$, respectively. Using the above expressions in Eq. (5.385), the glory scattering amplitude becomes

$$f_{\rm G}(\theta) = e^{i\alpha}\left(\frac{\sin\theta}{\theta}\right)^{1/2}\left(\frac{2\pi\,\lambda_{\rm G}^2}{k^2\,|\Theta'_{\rm G}|}\right)^{1/2}\frac{1}{2\pi}\int_0^{2\pi} d\gamma\ e^{-i\,\lambda_{\rm G}\theta\,\cos\gamma}, \tag{5.393}$$

or

$$f_{\rm G}(\theta) = e^{i\alpha}\left(\frac{\sin\theta}{\theta}\right)^{1/2}\left(\frac{2\pi\,\lambda_{\rm G}^2}{k^2\,|\Theta'_{\rm G}|}\right)^{1/2} J_0\left(\lambda_{\rm G}\theta\right). \tag{5.394}$$

Above, α stands for the constant phase

$$\alpha = 2\delta_{\rm G} - \frac{\pi}{2}\left(1-\nu/2\right). \tag{5.395}$$

The cross section for glory scattering can be immediately written as

$$\frac{d\sigma_{\rm G}(\theta)}{d\Omega} \equiv |f_{\rm G}(\theta)|^2 = \left[\frac{2\pi\,\lambda_{\rm G}^2}{k^2\,|\Theta'_{\rm G}|}\right]\left(\frac{\sin\theta}{\theta}\right) J_0^2\left(\lambda_{\rm G}\theta\right). \tag{5.396}$$

As an illustration of forward glory, we consider the scattering from the Lennard-Jones potential discussed in the previous section. In this case, the deflection function is the one showed in figure 5.16. The system exhibits both rainbow and forward glory. The glory angular momentum is $\lambda_{\rm G} = 351$ (represented in the figure by an open square). The corresponding cross

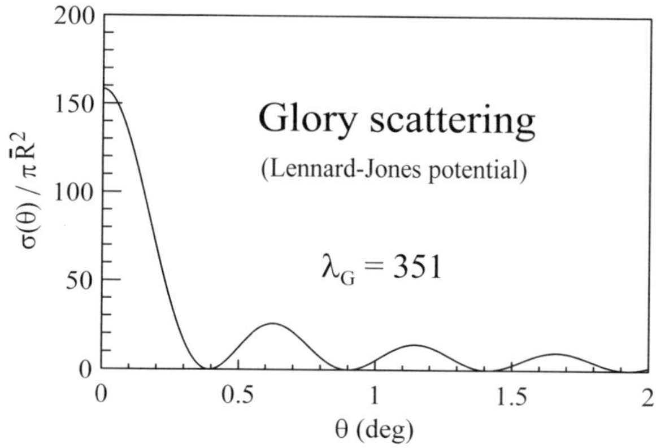

Fig. 5.21 Semiclassical cross section for forward glory scattering, normalized with respect to the geometrical cross seciton. The results are for the same Lennard-Jones potential of figures 5.15 and 5.16.

section at forward angles is shown in figure 5.21. In the present situation, it is not necessary to multiply the cross section by $\sin\theta$, since the approximate expression for the Legendre polynomials is valid at small angles. At the very forward angles shown in the figure ($\theta < 2^{\circ}$), the factor $\theta/\sin\theta \simeq 1$ and the cross section oscillates as the square of the Bessel function. The period (in degrees) then is $\Delta\theta^{\circ} = 180^{\circ}/\lambda_{\rm G}$. In the present case, $\lambda_{\rm G} \simeq 350$, so that $\Delta\theta \simeq 0.5^{\circ}$. It is also interesting to estimate the forward concentration of the cross section evaluating the ratio of the cross sections at first maximum to that at the second maximum, $\alpha_{\rm G}$. Denoting by σ_1 the cross section at $\theta = 0$ and by σ_2 the cross section at the subsequent maximum, according to Eq. (5.396) this ratio is

$$\alpha_{\rm G} \equiv \frac{\sigma_1}{\sigma_2} \simeq \left[\frac{J_0(0)}{J_0(x_2)}\right]^2. \tag{5.397}$$

The second maximum of $J_0(x)$ occurs at $x_2 = 3.8$ and the ratio in the above equation is

$$\alpha_{\rm G} = 6.2. \tag{5.398}$$

Note that this ratio is a characteristic of glory scattering. Contrasting with the period of the oscillation, which depends on $\lambda_{\rm G}$, $\alpha_{\rm G}$ does not depend on the collision energy or any detail of the system.

The scattering amplitude for backward glory can be derived through similar procedures and the results are very similar. The backward glory scattering amplitude and the corresponding cross section are:

$$f_{\mathrm{bG}}(\theta) = e^{i\alpha_{\mathrm{bG}}} \left(\frac{\sin\vartheta}{\vartheta}\right)^{1/2} \left(\frac{2\pi\,\lambda_{\mathrm{bG}}^2}{k^2\,|\Theta'_{\mathrm{bG}}|}\right)^{1/2} J_0\left(\lambda_{\mathrm{bG}}\vartheta\right), \tag{5.399}$$

with

$$\alpha_{\mathrm{bG}} = 2\delta_{\mathrm{bG}} - \frac{\pi}{2}\left(1 - \frac{\nu}{2} - 2\pi\lambda_{\mathrm{bG}}\right), \tag{5.400}$$

and

$$\frac{d\sigma_{\mathrm{bG}}(\phi)}{d\Omega} = \left[\frac{2\pi\,\lambda_{\mathrm{bG}}^2}{k^2\,|\Theta'_{\mathrm{bG}}|}\right] \left(\frac{\sin\vartheta}{\vartheta}\right) J_0^2\left(\lambda_{\mathrm{bG}}\vartheta\right). \tag{5.401}$$

It may be difficult to discern backward glory from other physically related oscillations, but scientists have found means to observe the glory energy oscillations by looking at cases involving charged particles and relied on the Coulomb-short range interference effects. Such attempts were made in pion and heavy ion scattering in nuclear physics.

Glory in the shadow of rainbow

It may happen that the extrema in the deflection function occur at angles very close to zero degrees. In such a case, one would be dealing with both forward glory and rainbow at the same value of the impact parameter. The rainbow supplies the Airy function while the glory supplies the Bessel function. Such a situation has been discussed by several authors (see e.g. [Ueda *et al.* (1999)]), and one would be seing a glory illuminated dark side of the rainbow. The calculation of the amplitude for this case is straightforward and would result in an Airy function evaluated at the rainbow angle, multiplying the forward glory zero order Bessel function as a function of angle, as

$$f(\theta) = \frac{2\pi\,\lambda_r}{ik} \left(\frac{\theta}{\sin\theta}\right)^{1/2} e^{2i\delta_r} \left(\frac{2}{\Theta''_r}\right)^{1/3} Ai\left[\left(\frac{2}{\Theta''_r}\right)^{1/3}\theta_r\right] J_0\left(\lambda_r\theta\right). \tag{5.402}$$

Above, λ_r is the rainbow angular momentum and δ_r and Θ''_r are respectively the phase shift and the second derivative of the deflection function, evaluated at λ_r.

Accordingly, one sees damped glory oscillations, or illuminated rainbow shadow! Such cases are encountered in nuclear scattering, where another factor contributes to the damping: the absorption due to the coupling of the elastic channel to the non-elastic ones (see part II of this book). It would certainly be present in atomic-molecular scattering as well, since the excitation of the rho-vibrational states in the molecule also leads to absorption.

Orbiting

Orbiting arises for particular combinations of collision energy and angular momentum, $\{E_o, \lambda_o\}$, as the ones represented in figure 1.5. The pair $\{E_o, \lambda_o\}$ is determined by two conditions. The first is that the λ-dependent effective potential[41]

$$V(\lambda, r) = V(r) + \frac{\hbar^2}{2\mu}\frac{\lambda^2}{r^2} \tag{5.403}$$

have a maximum at some radial distance, $R(\lambda_o)$, namely

$$\left[\frac{V(\lambda, r)}{dr}\right]_{r=R(\lambda_o)} = 0. \tag{5.404}$$

The second condition is the collision energy be equal to the height of the barrier of the effective potential at the corresponding angular momentum, $B(\lambda_o)$. That is,

$$E_o = B(\lambda_o) \equiv V\left(\lambda_o, R(\lambda_o)\right). \tag{5.405}$$

When these conditions are satisfied the incident particle sits at the constant radial distance $R(\lambda_o)$. This means that the angular variable goes to $-\infty$, as the projectile rotates around the target, describing a circular orbit. Then, the deflection function exhibits the behavior illustrated in figure 1.6. Near the orbiting angular momentum, the deflection function will vary logarithmically and can be parametrized as [Ford and Wheeler (1959)]

$$\Theta(\lambda) = \alpha + c \;\ln\left[\frac{\lambda - \lambda_o}{\lambda_o}\right] \qquad \text{for } \lambda > \lambda_o \tag{5.406}$$

$$\Theta(\lambda) = \beta + 2\,c \;\ln\left[\frac{\lambda_o - \lambda}{\lambda_o}\right] \qquad \text{for } \lambda < \lambda_o. \tag{5.407}$$

The constants α, β and c depend on the nature of the effective potential. The corresponding phase shift can be calculated by integrating the above forms of the deflection functions.

At a given scattering angle θ, the classical cross section will receive an infinite number of contributions. For Θ in the interval $\{0, -\pi\}$ it will receive

[41]This is a generalization of the l-dependent effective potential of chapter 2 for the continuous angular momentum $\lambda = l + 1/2$.

a contribution from $\lambda > \lambda_o$ and one from $\lambda < \lambda_o$, and the same happens in each of the intervals: $\{-\pi, -2\pi\}, ..., \{-n\pi, -(n+1)\pi\}, ...$. The dominant contribution coming from the first interval $0 > \Theta = -\theta > -\pi$, is [Friedman and Goebel (1977)],

$$\frac{d\sigma_o(\theta)}{d\Omega} = \frac{\lambda_o}{ck \sin\theta} \left[\exp\left(-\frac{\theta+\alpha}{c} \right) + \frac{1}{2} \exp\left(-\frac{\theta+\beta}{2c} \right) \right]. \tag{5.408}$$

Quantum mechanically, there will be interferences among the different contributions. Full QM treatment of orbiting is complicated as the phenomenon involves quasi-bound configurations (resonances). A few words about this are in order. As mentioned, the orbiting condition implies the particle sits at the top of the effective barrier at the angular momentum λ_o, $B(\lambda_o)$. Thus one can expand the potential around $R(\lambda_o)$ as

$$V(\lambda_o, r) \simeq B(\lambda_o) - \frac{1}{2}\mu\omega^2 \left[r - R(\lambda_o) \right]^2 . \tag{5.409}$$

To establish an analogy with an inverted harmonic oscillator, it is convenient to write the coefficient of the quadratic term as $\mu\omega^2$ and treat ω as an adjustable parameter. Inserting the above expansion into the Schrödinger equation and changing to the dimensionless variables

$$x = \sqrt{\frac{\mu\omega}{\hbar}} \left[r - R(\lambda_o) \right], \quad \varepsilon = \frac{E_o - B(\lambda_o)}{\hbar\omega}, \tag{5.410}$$

we get

$$\left[\frac{d^2}{dx^2} + x^2 + 2\varepsilon \right] \psi(x) = 0. \tag{5.411}$$

The solution of the above equation is a decaying one, as the potential is an inverted harmonic oscillator. The energies of the system are complex attesting to this fact.

Exercises

(1) Show that, within the Born approximation, the scattering amplitude for the Yukawa potential $V(r) = \gamma \exp[-r/b]/r$ is given by Eq. (5.22).

(2) Consider the total elastic cross section, σ_{el}, in the scattering off a spherical potential, $V(r)$. An approximation for this cross section can be obtained using the Optical Theorem and evaluating $\mathrm{Im}\{f(0)\}$ by the Born series, truncated after the non-vanishing term of lowest order.

a) Prove that the first order Born approximation leads to $\sigma_{el} = 0$.

b) Show that using the second order Born approximation one gets[42]

$$\sigma_{el} = 64 \, \frac{\pi^5 \mu^2}{\hbar^4} \, \tilde{V}^2(q),$$

where $\tilde{V}(q)$ is the integral of Eq. (5.17).

(3) Calculate the eikonal phase for the potential $V(r) = g/r^2$.

(4) Consider the eikonal approximation in a collision dominated by strong absorption, where the projectile-target interaction can be approximated by a black disk of radius R_0. That is,

$$\begin{aligned} V(r) &= -\, i\, W_0, \quad \text{for } r \equiv \sqrt{b^2 + z^2} \le R_0 \\ &= 0, \qquad\quad\; \text{for } r \equiv \sqrt{b^2 + z^2} > R_0, \end{aligned}$$

where W_0 is positive and infinitely large (this corresponds to Fraunhofer diffraction, discussed in section 7.5.1).

a) Derive an expression for the scattering amplitude within the eikonal approximation.

b) Estimate the period of the angular oscillations[43] in the cross section for a $E_{\text{lab.}} = 300$ MeV proton beam, incident upon a ^{12}C target. Take for the radius of the disk: $R_0 = 4$ fm.

(5) Calculate the quantum potential of Eq. (5.259) for particle of mass m, angular momentum $l = 0$ and energy E, scattered off the parabolic potential

$$V(r) = V_B + \frac{1}{2} m\omega^2 \, (r - R_B)^2 \, .$$

(6) Show that the deflection function in the collision of a projectile of charge Q_P with a target of charge Q_P, uniformly distribution over a sphere of radius R_c, exhibits a positive rainbow. In this case, the projectile-target interaction is given by the potential

$$V_c(r) = \frac{V_0}{r}, \text{ for } r \ge R_c; \quad V_c(r) = \frac{V_0}{2R_c} \left(3 - \frac{r^2}{R_c^2} \right), \text{ for } r < R_c \, ,$$

[42] *Hint*: Calculate the contributions from the on-shell and from the off-shell parts of the Green's function separately and show that the off-shell contribution leads to a real scattering amplitude.

[43] *Hint*: Use the properties of Bessel functions (see section 7.5.1) :

$$\frac{1}{x} \frac{d}{dx} \left[x \, J_1(x) \right] = J_0(x), \quad J_m(x) \simeq \sqrt{\frac{2}{mx}} \, \cos\left(x - mp/2 - \pi/4 \right) .$$

with $V_0 = \mathrm{Q_P\,Q_P}/4\pi\varepsilon_0$. Evaluate the deflection function at energies $E > E_c$, $E = E_c$ and $E < E_c$, where E_c is value of the potential at the origin, $E_c = 3V_0/2R_c$. With this deflection function, calculate the near-side amplitude and show that it exhibits Airy shape[44] for $E > E_c$.

(7) A near far decomposition of the eikonal scattering amplitude can be carried out using in Eq. (5.125) the approximate expression

$$J_0(x) \simeq \frac{1}{\sqrt{2\pi x}} \left[e^{i(x-\pi/4)} + e^{-i(x-\pi/4)}\right].$$

(This equation can be derived from Eqs. (5.126) and (5.271).)

a) Calculate the momentum transfer function $Q(b) = d\,\chi(b)/db$ for the nuclear scattering at high energies, where the Coulomb contribution can be neglected, even at angles as small as a few degrees. Assume that the potential can be approximated by the attractive gaussian

$$V(r) = -\,V_0\; e^{-r^2/\alpha^2},$$

where V_0 and α are positive constants.

b) Show that in this case the near-side amplitude is negligible and there is a nuclear rainbow at a negative angle. Calculate this angle.

(8) It is known that the atmosphere of the Saturn moon Titan is mostly methane CH_4, or, more popularly, natural gas. Its index of refraction is $N_m = 1.286$ (to be compared to water: $N_w = 1.33$). The temperature on Titan is 179°C below zero. Nevertheless, methane remains a liquid even in this frost. Show that the primary rainbow angle in this case is 49°, compared to 42.5^o for water. The order of colors would be the same in both cases. Since Titan has an orange sky, there is a background orange color behind the rainbow colors, making the observation of the blue - red primary bow more difficult for an astronaut in Titan.

[44] *Hint*: look at Hussein et al., Am. J. Phys. **52**, 650 (1984).

Chapter 6

Spin and Identical Particles

In the preceding chapters we dealt with spherically symmetric potentials and distinguishable spinless particles. In such cases, the only particular orientation in space is that of the incident beam, and the exchange symmetry of the wave function can be ignored. In the present chapter we discuss how the collisions are influenced by the spin of the collision partners and the modifications required when the projectile and the target are identical particles.

6.1 Collisions of particles with spin

In collisions of discernible spinless particles and rotationally invariant interactions, the scattering amplitude has axial symmetry with respect to the orientation of the collision plane. In this way, taking the z-axis along the incident beam, one eliminates $m \neq 0$ components in partial-wave expansions (see section 2.3). The situation is more complicated when the projectile or/and the target has/have internal degrees of freedom and the intrinsic state does not behave as a scalar with respect to rotations. This occurs when the system has intrinsic spin or any other kind of angular momentum, e.g. that associated with the rotation of a deformed target[1]. For simplicity[2], we consider collisions of a spinless partner with another with spin s. We assume that the incident beam is polarized so that the spin component along the z-axis is equal to ν, and that the experimental setup is capable of distinguishing among the possible final spin projections, ν'. Other situations will be considered at the end of this section.

[1] Here, we loosely use the term *spin* for intrinsic angular momentum of any kind.

[2] For a general treatment of the scattering of particles with spin, we refer to [Satchler (1983)] and [Newton (2002)].

The spin states, $|s\nu\rangle$, have the properties

$$S^2\,|s\nu\rangle = \hbar^2 s(s+1)\,|s\nu\rangle \tag{6.1}$$

$$S_z\,|s\nu\rangle = \hbar\nu\,|s\nu\rangle\,. \tag{6.2}$$

In the time-independent approach, the incident wave is

$$\phi(\nu\mathbf{k};\mathbf{r}) = \frac{1}{(2\pi)^{3/2}}\,e^{i\mathbf{k}\cdot\mathbf{r}}\,|s\nu\rangle \tag{6.3}$$

and the scattering state satisfies the Schrödinger equation

$$\Big[E - H\Big]\,|\psi^{(+)}(\nu\mathbf{k})\rangle = 0. \tag{6.4}$$

It satisfies also the Lippmann-Schwinger equation,

$$|\psi^{(+)}(\nu\mathbf{k})\rangle = |\phi(\nu\mathbf{k})\rangle + G_0^{(+)}\,V\,|\psi^{(+)}(\nu\mathbf{k})\rangle\,. \tag{6.5}$$

Above, $H = K + V$ is the system's Hamiltonian, with K representing the kinetic energy operator and V the potential, which may act on the collision and spin coordinates, and $G_0^{(+)} = [E-K]^{-1}$ is the free particle's Green operator.

Usually, the scattering wave function is written as

$$\psi^{(+)}(\nu\mathbf{k};\mathbf{r}) = \phi(\nu\mathbf{k};\mathbf{r}) \,+\, \psi^{\mathrm{sc}}(\nu\mathbf{k};\mathbf{r})\,. \tag{6.6}$$

In Eq. (6.6),

$$\psi^{\mathrm{sc}}(\nu\mathbf{k};\mathbf{r}) = \sum_{\nu'=-s}^{+s}\,\psi^{\mathrm{sc}}_{\nu'}(\nu\mathbf{k};\mathbf{r})\,|s\nu'\rangle \tag{6.7}$$

is the scattered wave, which has the asymptotic behavior

$$\psi^{\mathrm{sc}}(\nu\mathbf{k};\mathbf{r}\to\infty) = \frac{1}{(2\pi)^{3/2}}\sum_{\nu'=-s}^{+s}\,|s\nu'\rangle\,\left[f_{\nu',\nu}(\Omega)\,\frac{e^{ikr}}{r}\right]. \tag{6.8}$$

In this way, $\psi^{(+)}(\nu\mathbf{k};\mathbf{r})$ has the asymptotic form

$$\psi^{(+)}(\nu\mathbf{k};\mathbf{r}) \to \frac{1}{(2\pi)^{3/2}}\sum_{\nu'=-s}^{+s}\,|s\nu'\rangle\,\left[\delta_{\nu\nu'}\,e^{i\mathbf{k}\cdot\mathbf{r}} + f_{\nu',\nu}(\Omega)\,\frac{e^{ikr}}{r}\right] \tag{6.9}$$

The distribution of the scattering amplitude over the spin components, ν', is determined by the spin-dependence of the projectile-target interaction. Clearly, when the potential V is spin-indedependent, we have

$$f_{\nu',\nu}(\Omega) = f_{\nu,\nu}(\Omega)\, \delta_{\nu\nu'}.$$

with Ω standing for the angular coordinates, θ, φ. Assuming that the potential is spherically symmetric[3] and independent of spin, the scattering amplitude is independent of φ and conserves the z-component of the spin. That is,

$$f_{\nu,\nu'}(\theta,\varphi) \equiv f(\theta)\, \delta_{\nu\nu'}.$$

If the potential contains a Coulomb term, modifications along the lines of chapter 3 are necessary. Is this case, the asymptotic form of Eq. (6.9) becomes

$$\psi^{(+)}(\nu\mathbf{k};\mathbf{r}) \to \phi_{\mathrm{C}}\left(\mathbf{k};\mathbf{r}\right)\, |s\nu) + \frac{1}{(2\pi)^{3/2}} \sum_{\nu'=-s}^{+s} \bar{f}_{\nu',\nu}(\theta)\, \frac{e^{i\Theta(r)}}{r}\, |s\nu')\,, \tag{6.10}$$

where $\phi_{\mathrm{C}}(\mathbf{k};\mathbf{r})$ is the Coulomb scattering wave function of section 3.2, $\bar{f}_{\nu',\nu}(\theta)$ is the partial scattering amplitude arising from the short-range part of the interaction, and $\Theta(r)$ is the asymptotic phase distorted by the Coulomb field,

$$\Theta(r) = kr - \eta \ln 2kr. \tag{6.11}$$

The scattering amplitude is then given by

$$f_{\nu',\nu}(\theta) = f_{\mathrm{C}}(\theta)\, \delta_{\nu\nu'} + \bar{f}_{\nu',\nu}(\theta), \tag{6.12}$$

where $f_{\mathrm{C}}(\theta)$ is the Coulomb amplitude of Eq. (3.29), and the elastic cross section for a final state with spin component ν' is

$$\frac{d\sigma_{\nu',\nu}(\theta)}{d\Omega} = \left|f_{\nu',\nu}(\theta)\right|^2. \tag{6.13}$$

For unpolarized beams and inclusive experiments, where the final component of the spin along the z-axis is not determined, the cross section should be averaged over the initial spin orientations and summed over the final spin states. That is

$$\frac{d\sigma(\theta)}{d\Omega} = \frac{1}{(2s+1)} \sum_{\nu'\nu} \left|f_{\nu',\nu}(\theta)\right|^2. \tag{6.14}$$

[3]We make this assumption throughout this chapter.

6.1.1 *Partial-wave expansions*

The aim of the partial-wave expansion method is to reduce the three-dimensional Schrödinger equation (Eq. (6.4)) to an angular momentum dependent equation on a single variable, the radial coordinate r. The reduced equation is then solved for each value of the angular momentum and its contribution to the scattering amplitude is evaluated.

To generalize the partial-wave expansion method to systems with spin, we re-introduce it using a more general approach. The first step is to single out the radial coordinate and look for an appropriate basis for the vector space associated with the remaining degrees of freedom, which we formally represent by ζ. In the case of spinless particles, discussed in chapter 2, ζ stands for the orientation variables $\{\hat{\mathbf{r}} \equiv \theta, \varphi\}$. We call $\mathcal{Y}_n(\zeta)$ the states of this basis, with the label n standing for the set of quantum numbers $\{\alpha_n^{(1)}, \alpha_n^{(2)}, \alpha_n^{(3)}, \cdots\}$, which specify the state. The wave funtion is then expanded as

$$\psi^{(+)}(\nu\mathbf{k}; r, \zeta) = \sum_{n'} \mathcal{Y}_{n'}(\zeta)\, X_{n'}(r)\,. \tag{6.15}$$

Using this expansion in the Schrödinger equation and taking scalar product[4] with $(\mathcal{Y}_n|$, one gets

$$\sum_{n'} [E\, \delta_{nn'} - H_{nn'}]\, X_{n'}(r) = 0\,, \tag{6.16}$$

with

$$H_{n,n'} = (\mathcal{Y}_n|\, H\, |\mathcal{Y}_{n'})\,. \tag{6.17}$$

Eq. (6.16) is a matrix equation in the vector space spanned by $|\mathcal{Y}_n)$, coupling the r-dependent components of the vector $\psi^{(+)}$, $X_{n'}(r)$, and their derivatives. The essential point of the method is to take advantage of the symmetries of the system, looking for a basis in which the Hamiltonian is diagonal. This aim is achieved if each quantum number of the set $\{\alpha_n^{(1)}, \alpha_n^{(2)}, \alpha_n^{(3)}, \cdots\}$ is a constant of motion[5]. The equations for $X_n(r)$ are then completely decoupled and can be solved as described in section 2.2. In the case of spinless particles, the rotational invariance of H leads to

[4] We use round brackets to represent scalar producs in the space of the coordinates ζ.

[5] In some situations, e.g. multi-channel scattering (see chapter 9), one cannot find a basis which diagonalizes the Hamiltonian. One should then look for a basis in which the set $\left\{\alpha_n^{(1)}, \alpha_n^{(2)}, \alpha_n^{(3)}, \cdots\right\}$ contains as many constants of motion as possible. The Hamiltonian matrix then breaks into idependent blocks, leading to sets of coupled differential equations of smaller dimensions.

orbital angular momentum conservation, so that the set of spherical harmonics is the appropriate expansion basis. In the Y_{lm} representation, the Hamiltonian matrix is diagonal and independent of m. That is,

$$H_{n,n'} \equiv H_{lm,l'm'} = \delta_{ll'}\,\delta_{mm'}\left\{-\frac{\hbar^2}{2\mu}\left[\frac{1}{r^2}\frac{\partial}{\partial r}\left(r^2\frac{\partial}{\partial r}\right)-\frac{l(l+1)}{r^2}\right]+V(r)\right\}. \tag{6.18}$$

Inserting these matrix-elements into Eq. (6.16), with $\psi^{(+)}(\nu\mathbf{k};\mathbf{r})$ expanded as in Eq. (2.8), and taking scalar product with $(Y_{lm}|$, we have obtained the radial equation for spinless systems (Eqs. (2.6) and (2.9)).

The partial-wave expansion method for particles with spin is more complicated. If the potential is spin-dependent, the rotational invariance of H may not guarantee conservation of the orbital angular momentum and the spherical harmonics may no longer be a convenient expansion basis. We clarify this point with the example of a potential which contains a spin-orbit term, as in the expression

$$\begin{aligned} V(r) &= V_0(r) + V_{\mathrm{so}}(r)\,\mathbf{L}\cdot\mathbf{s} \\ &\equiv V_0(r) + \frac{1}{2}\,V_{\mathrm{so}}(r)\,\left[\mathbf{J}^2 - \mathbf{L}^2 - \mathbf{s}^2\right], \end{aligned} \tag{6.19}$$

where $\mathbf{J} = \mathbf{L} + \mathbf{s}$ is the total angular momentum operator. In this case, the method requires a basis for the vector space associated with the orientation and spin coordinates (i.e., $\{\zeta \equiv \hat{\mathbf{r}}, \nu\}$) which diagonalizes the Hamiltonian. Since $[L_z, H] \neq 0$, the quantum number m is not conserved. Therefore, a trivial generalization like $|\mathcal{Y}_{lm\nu}) = |s\nu)\,|Y_{lm})$ would not be satisfactory. However, rotational invariance guarantees conservation of any component of the total angular momentum, $\mathbf{J}$. In this case, the appropriate basis is the set of *spin-angle wave functions*[6] defined as[7]

$$\mathcal{Y}^{lJM}(\zeta) = i^l \sum_{m'\nu'} \langle lm's\nu'\,|JM\rangle\; Y_{lm'}(\hat{\mathbf{r}})\;|s\nu'\rangle\,, \tag{6.20}$$

---

[6]We write the quantum numbers of the constants of motion of the spin-angle basis as superscripts. This choice will prove convenient for the extension of the procedures of this chapter to coupled-channel problems (see chapter 9). For simplicity of notation, we do not include the quantum number s, because it has a fixed value.

[7]It is convenient to define the spin-angle wave functions with the arbitrary phase i^l. With this phase, the wave function has convenient time-reversal properties [Huby (1954); Lane (1962); Thomas (1958); Satchler (1983)].

where $\langle lms\nu | JM \rangle$ are the usual Clebsch-Gordan coefficients [Edmonds (1974)]. Since the above Clebsch-Gordan coefficients vanish if $m' + \nu' \neq M$, we can write

$$\mathcal{Y}^{lJM}(\zeta) = i^l \sum_{\nu'} \langle l\,(M-\nu')\; s\nu' \,|JM\rangle\; Y_{l(M-\nu')}(\hat{\mathbf{r}})\; |s\nu'\rangle\,. \tag{6.21}$$

Using properties of the Clebsch-Gordan coefficients and the orthogonality of angular momentum eigestates (in the coordinate and in the spin spaces), it can easily be checked that

$$\left(\mathcal{Y}^{lJM} \,\middle|\, \mathcal{Y}^{l'J'M'}\right) = \delta_{ll'}\,\delta_{JJ'}\,\delta_{MM'}. \tag{6.22}$$

The rotational invariance of the Hamiltonian guarantees that the H is diagonal in the total angular momentum representation. On the other hand, the commutators $[\mathbf{L}^2, H]$ and $[\mathbf{s}^2, H]$ depend on the nature of the potential. In most cases they vanish[8], so that H is diagonal in the basis of spin-angle wave functions. One then obtains a set of fully decoupled radial equations. Otherwise, the partial-wave method leads to a set of differential equations for each invariant subspace, in which radial wave functions with the same J and different l are coupled. This situation is analogous to the coupled-channel equations in many-body scattering, which will be discussed in detail in chapter 9. We consider below the simpler situation where the orbital angular momentum is conserved. In this case, the set of spin-angle wave functions diagonalizes the Hamiltonian and we can write

$$\left(\mathcal{Y}^{lJM}\right| H \left|\mathcal{Y}^{l'J'M'}\right) = \left\{ -\frac{\hbar^2}{2\mu}\left[\frac{1}{r^2}\frac{\partial}{\partial r}\left(r^2\frac{\partial}{\partial r}\right) - \frac{l(l+1)}{r^2}\right] + V_l^J(r)\right\} \times \delta_{ll'}\,\delta_{JJ'}\,\delta_{MM'}. \tag{6.23}$$

The incident wave can be expanded in the spin-angle basis as[9]

$$\phi(\nu\mathbf{k};\mathbf{r}) = \frac{1}{(2\pi)^{3/2}} \sum_{lJM} C_{JM}(\nu l\mathbf{k}) \left|\mathcal{Y}^{lJM}\right) \frac{\hat{\jmath}_l(kr)}{kr}. \tag{6.24}$$

To determine the coefficients $C_{JM}(\nu l\mathbf{k})$, we first use Bauer's expansion (Eq. (2.43a)) in Eq. (6.3). We get

$$\phi(\nu\mathbf{k};\mathbf{r}) = \frac{1}{(2\pi)^{3/2}} \sum_{lm'} 4\pi\, Y^*_{lm'}(\hat{\mathbf{k}}) \left[i^l\, Y_{lm'}(\hat{\mathbf{r}})\; |s\nu\rangle\right] \frac{\hat{\jmath}_l(kr)}{kr}\,. \tag{6.25}$$

[8] There are, however, important exceptions, like the tensor force in the nucleon-nucleon interactions.

[9] This equation corresponds to Eq. (6.15), with the replacement $X_n(r) \equiv X_{lJM}(r) \rightarrow (2\pi)^{-3/2}\; C_{lJM}\; \hat{\jmath}_l(kr)/kr$.

Using the inverse of Eq. (6.21),

$$i^l\, Y_{lm'}(\hat{\mathbf{r}})\;|s\nu\rangle = \sum_{JM} \langle JM\,|lm's\nu\rangle\; \mathcal{Y}^{lJM}(\zeta)\,, \tag{6.26}$$

Eq. (6.25) takes the form

$$\phi(\nu\mathbf{k};\mathbf{r}) = \frac{1}{(2\pi)^{3/2}} \sum_{lm'JM} 4\pi\, Y^*_{lm'}(\hat{\mathbf{k}})\,\langle JM\,|lm's\nu\rangle\, \mathcal{Y}^{lJM}(\zeta)\, \frac{\hat{\jmath}_l(kr)}{kr}\,.$$

Since the non-vanishing Clebsch-Gordan coefficients have $M = m' + \nu$, we can replace $m' \to M - \nu$ and the above equation becomes

$$\begin{aligned}\phi(\nu\mathbf{k};\mathbf{r}) &= \frac{1}{(2\pi)^{3/2}} \sum_{lJM} \Big[4\pi\,\langle JM\,|l\,(M-\nu)\;s\nu\rangle \\ &\quad \times Y^*_{l\,(M-\nu)}(\hat{\mathbf{k}})\Big]\, \mathcal{Y}^{lJM}(\zeta)\, \frac{\hat{\jmath}_l(kr)}{kr}\,. \end{aligned} \tag{6.27}$$

Comparing Eqs. (6.24) and (6.27), we get

$$C_{JM}(\nu l\mathbf{k}) = 4\pi\,\langle JM\,|l(M-\nu)\,s\nu\rangle\;\, Y^*_{l\,(M-\nu)}(\hat{\mathbf{k}}). \tag{6.28}$$

We now evaluate Eq. (6.25) in the $r \to \infty$ limit and express it in terms of incoming and outgoing waves, as we did in section 2.3. Using in Eq. (6.25) the asymptotic form of the Ricatti-Bessel function (Eq. (2.22a)),

$$\frac{\hat{\jmath}_l(kr)}{kr} \to \frac{\sin{(kr - l\pi/2)}}{kr} = \frac{e^{-ikr}}{r}\left(\frac{-i^l}{2ik}\right) + \frac{e^{ikr}}{r}\left(\frac{i^{-l}}{2ik}\right), \tag{6.29}$$

we get

$$\phi(\nu\mathbf{k};\mathbf{r}) \to \frac{1}{(2\pi)^{3/2}} \sum_{\nu'} |s\nu'\rangle \left[\frac{e^{-ikr}}{r}\, Z^{\nu'}_- + \frac{e^{ikr}}{r}\, Z^{\nu'}_+\right], \tag{6.30}$$

with

$$Z^{\nu'}_- = -\,\delta_{\nu\nu'}\,\frac{1}{2ik} \sum_{lm} 4\pi\, Y^*_{lm}(\hat{\mathbf{k}})\, Y_{lm}(\hat{\mathbf{r}})\, i^{2l} \tag{6.31}$$

and

$$Z^{\nu'}_+ = \delta_{\nu\nu'}\,\frac{1}{2ik} \sum_{lm} 4\pi\, Y^*_{lm}(\hat{\mathbf{k}})\, Y_{lm}(\hat{\mathbf{r}})\,. \tag{6.32}$$

If the potential contains a Coulomb term, changes along the lines of chapter 3 are required. In first place, Eq. (6.3) becomes

$$\phi(\nu\mathbf{k};\mathbf{r}) = \phi_{\rm C}(\mathbf{k};\mathbf{r})\,|s\nu\rangle\,. \tag{6.33}$$

Using the partial-wave expansion of $\phi_{\rm C}(\mathbf{k};\mathbf{r})$ (Eq. (3.102a)) instead of Bauer's expansion, Eq. (6.24) takes the form

$$\phi(\nu\mathbf{k};\mathbf{r}) = \frac{1}{(2\pi)^{3/2}}\sum C_{JM}(\nu l\mathbf{k})\,\big|\mathcal{Y}^{lJM}\big)\,\frac{F_l(\eta,r)}{kr}, \tag{6.34}$$

with the coefficients $C_{JM}(\nu l\mathbf{k})$ given by

$$C_{JM}(\nu l\mathbf{k}) = 4\pi\,\langle JM\,|l(M-\nu)\,s\nu\rangle\;Y^{*}_{l\,(M-\nu)}(\hat{\mathbf{k}})\;e^{i\sigma_l} \tag{6.35}$$

Above, $F_l(\eta,r)$ is the regular Coulomb function and σ_l is the Coulomb phase shift at the lth partial-wave.

Evaluating Eq. (6.34) in the $r\to\infty$ limit (see Eq. (3.103)) we obtain

$$\phi(\nu\mathbf{k};\mathbf{r}) \to \frac{1}{(2\pi)^{3/2}}\sum_{\nu'}|s\nu'\rangle\left[\frac{e^{-i\Theta(r)}}{r}\,Z_-^{\nu'} + \frac{e^{i\Theta(r)}}{r}\,Z_+^{\nu'}\right], \tag{6.36}$$

with $Z_-^{\nu'}$ and Θ given respectively by Eqs. (6.31) and (6.11), and

$$Z_+^{\nu'} = \delta_{\nu\nu'}\,\frac{1}{2ik}\sum_{lm}4\pi\,Y^{*}_{lm}(\hat{\mathbf{k}})\,Y_{lm}(\hat{\mathbf{r}})\,e^{2i\sigma_l}. \tag{6.37}$$

We now consider the wave function in the case of scattering from a short range potential. First, we expand it in the spin-angle basis as

$$\psi^{(+)}(\nu\mathbf{k};\mathbf{r}) = \frac{1}{(2\pi)^{3/2}}\sum_{lJM}C_{JM}(\nu l\mathbf{k})\,\big|\mathcal{Y}^{lJM}\big)\,\frac{u_l^J(k,r)}{kr}. \tag{6.38}$$

Inserting the expansion into the Schrödinger equation, taking scalar product with $\big(\mathcal{Y}^{lJM}\big|$ and using the orthogonality of the spin-angle wave functions, we obtain the radial equation

$$\left[-\frac{\hbar^2}{2\mu}\left(\frac{d^2}{dr^2}-\frac{l(l+1)}{r^2}\right)+V_l^J(r)\right]\,u_l^J(k,r) = E\,u_l^J(k,r), \tag{6.39}$$

where[10]

$$V_l^J(r) = \big(\mathcal{Y}^{lJM}\big|\;V\;\big|\mathcal{Y}^{lJM}\big)\,. \tag{6.40}$$

[10]For simplicity, we assume that the interaction is local.

For the potential of Eq. (6.19), for example, these matrix-elements are

$$V_l^J(r) = V_0(r) + \frac{1}{2}\, V_{\rm so}(r)\, [J(J+1) - l(l+1) - s(s+1)]\,.$$

Beyond the range of the interaction, $\bar{R}$, the radial wave functions can be written (Eq. (2.35) with $\gamma = i/2$)

$$u_l^J(k, r > \bar{R}) = \frac{i}{2}\left[\hat{h}_l^{(-)}(kr) - S_l^J\, \hat{h}_l^{(+)}(kr)\right] \tag{6.41}$$

and in the $r \to \infty$ limit they become

$$u_l^J(k, r \to \infty) = e^{-ikr}\left(-\frac{i^l}{2i}\right) + e^{ikr}\left(\frac{i^{-l}}{2i}\, S_l^J\right). \tag{6.42}$$

The coefficients $C_{JM}(\nu l\mathbf{k})$ in the partial-wave expansion of $\psi^{(+)}$ (Eq. (6.38)) are still given by Eq. (6.28). That is, they are the same as those in the expansion of the incident wave. This statement is justified as follows. First, we remark that the difference between $\psi^{(+)}(\nu\mathbf{k};\mathbf{r})$ and $\phi(\nu\mathbf{k};\mathbf{r})$ is the scattered wave $\psi^{\rm sc}(\nu\mathbf{k};\mathbf{r})$, which has the asymptotic behavior of an outgoing wave (Eq. (6.8)). In this way, $\psi^{(+)}(\nu\mathbf{k};\mathbf{r})$ and $\phi(\nu\mathbf{k};\mathbf{r})$ have identical incoming contents. The same happens for the radial wave functions u_l^J and $\hat{\jmath}_l$. This becomes clear if the Ricatti-Bessel function is written as

$$\hat{\jmath}_l(kr) = \frac{\hat{h}_l^{(+)}(kr) - \hat{h}_l^{(-)}(kr)}{2i} = \frac{i}{2}\left[\hat{h}_l^{(-)}(kr) - \hat{h}_l^{(+)}(kr)\right]. \tag{6.43}$$

Comparing Eqs. (6.41) and (6.43), we conclude that these radial wave functions have the same incoming part (the term in $\hat{h}_l^{(-)}$) and their outgoing parts (the term in $\hat{h}_l^{(+)}$) differ only by the factor S_l^J. The equality of the incoming parts,

$$\begin{aligned}
\left[\psi^{(+)}(\nu\mathbf{k};\mathbf{r}\to\infty)\right]_{in} &= \left[\phi(\nu\mathbf{k};\mathbf{r}\to\infty)\right]_{in}\\
\left[u_l^J(k, r\to\infty)\right]_{in} &= \left[\hat{\jmath}_l(kr\to\infty)\right]_{in},
\end{aligned}$$

implies that the partial-wave expansions of $\psi^{(+)}$ and ϕ have the same coefficients.

The scattering wave function can then be written

$$\begin{aligned}
\psi^{(+)}(\nu\mathbf{k};\mathbf{r}) = \frac{1}{(2\pi)^{3/2}} &\sum_{lJM} 4\pi\, \langle JM\,|l(M-\nu)\,s\nu\rangle\\
&\times Y^*_{l(M-\nu)}(\hat{\mathbf{k}})\; \mathcal{Y}^{lJM}(\zeta)\, \frac{u_l^J(k,r)}{kr}.
\end{aligned} \tag{6.44}$$

Using Eq. (6.21) and carrying out some algebra, the above equation can be put in the form

$$\psi^{(+)}(\nu\mathbf{k};\mathbf{r}) = \frac{1}{(2\pi)^{3/2}} \sum_{\nu'} |s\nu'\rangle \left[\sum_{Jlm} 4\pi\, Y^{*}_{lm}(\hat{\mathbf{k}})\, Y_{l(m+\nu-\nu')}(\hat{\mathbf{r}}) \right.$$
$$\times\, i^{l}\, \langle l\,(m+\nu-\nu')s\nu'\,|J(m+\nu)\rangle$$
$$\left. \times\, \langle J(m+\nu)\,|lms\nu\rangle\, \frac{u^{J}_{l}(k,r)}{kr} \right]. \tag{6.45}$$

Eq. (6.45) can be considerably simplified if one chooses the z-axis along the incident beam. In this case we can replace $Y^{*}_{lm}(\hat{\mathbf{k}}) \rightarrow Y^{*}_{lm}(\hat{\mathbf{z}}) = \delta_{m0}\sqrt{(2l+1)/4\pi}$ and obtain

$$\psi^{(+)}(\nu\mathbf{k};\mathbf{r}) = \frac{1}{(2\pi)^{3/2}} \sum_{\nu'} |s\nu'\rangle \left[\sum_{Jl} \sqrt{4\pi\,(2l+1)}\, Y_{l(\nu-\nu')}(\hat{\mathbf{r}})\, i^{l} \right.$$
$$\left. \times\, \langle l\,(\nu-\nu')s\nu'\,|J\nu\rangle\, \langle J\nu\,|l0s\nu\rangle\, \frac{u^{J}_{l}(k,r)}{kr} \right]. \tag{6.46}$$

If the potential contains a Coulomb term, the coefficients of the partial-wave expansion of $\psi^{(+)}$ are given by Eq. (6.35). Using the same arguments as above, we conclude that they are the same as those in the expansion of $\phi_{\mathrm{C}}(\mathbf{k};\mathbf{r})\,|s\nu\rangle$. Therefore, we can write

$$\psi^{(+)}(\nu\mathbf{k};\mathbf{r}) = \frac{1}{(2\pi)^{3/2}} \sum_{\nu'} |s\nu'\rangle \left[\sum_{Jl} \sqrt{4\pi\,(2l+1)}\, Y_{l(\nu-\nu')}(\hat{\mathbf{r}})\, i^{l} \right.$$
$$\left. \times\, \langle l\,(\nu-\nu')s\nu'\,|J\nu\rangle\, \langle J\nu\,|l0s\nu\rangle\, e^{i\sigma_l}\, \frac{u^{J}_{l}(k,r)}{kr} \right]. \tag{6.47}$$

Now, however, the asymptotic radial wave functions are expressed in terms of Coulomb functions (see chapter 3)

$$u^{J}_{l}(k, r > \bar{R}) = \frac{i}{2}\left[\hat{H}^{(-)}_{l}(kr) - \bar{S}^{J}_{l}\, \hat{H}^{(+)}_{l}(kr)\right] \tag{6.48}$$

and in the $r \rightarrow \infty$ limit it becomes

$$u^{J}_{l}(k, r \rightarrow \infty) = e^{-i\Theta(r)}\left(-\frac{i^{l}}{2i}\right) + e^{i\Theta(r)}\left(\frac{i^{-l}}{2i}\,\bar{S}^{J}_{l}\right). \tag{6.49}$$

with $\Theta(r)$ given by Eq. (6.11).

6.1.1.1 *Partial-wave projected scattering amplitudes and cross sections*

To find the scattering amplitude, we compare the asymptotic form of Eq. (6.45) with the scattering boundary condition of Eq. (6.9). Using Eq. (6.42) in Eq. (6.45), we obtain

$$\psi^{(+)}(\nu\mathbf{k};\mathbf{r}) \to \frac{1}{(2\pi)^{3/2}} \sum_{\nu'} |s\nu'\rangle \left[\frac{e^{-ikr}}{r}\, Z_-^{\nu'} + \frac{e^{ikr}}{r}\, \bar{Z}_+^{\nu'}\right], \tag{6.50}$$

with the amplitude $Z_-^{\nu'}$ given by Eq. (6.31) and $\bar{Z}_+^{\nu'}$ given by

$$\begin{aligned}\bar{Z}_+^{\nu'} &= \frac{1}{2ik} \sum_{Jlm} 4\pi\, Y_{lm}^*(\hat{\mathbf{k}})\, Y_{l(m+\nu-\nu')}(\hat{\mathbf{r}})\, \langle J(m+\nu)\,|lms\nu\rangle \\ &\quad \times S_l^J\, \langle l\,(m+\nu-\nu')\; s\nu'\,|J(m+\nu)\rangle\,. \end{aligned}\tag{6.51}$$

The partial-wave expansion of the Eq. (6.9) can be easily obtained if we consider Eqs. (6.6) and (6.8). According to these equations, the incoming part of $\psi^{(+)}$ is identical to that of ϕ. On the other hand, the outgoing part of $\psi^{(+)}$ is the sum of the outgoing part of ϕ with the contribution of ψ^{sc}, given in Eq. (6.8). We get

$$\psi^{(+)}(\nu\mathbf{k};\mathbf{r}) \to \frac{1}{(2\pi)^{3/2}} \sum_{\nu'} |s\nu'\rangle \left[\frac{e^{-ikr}}{r}\, Z_-^{\nu'} + \frac{e^{ikr}}{r}\, \tilde{Z}_+^{\nu'}\right], \tag{6.52}$$

with $Z_-^{\nu'}$ still given by Eq. (6.31) and the outgoing amplitude

$$\tilde{Z}_+^{\nu'} = Z_+^{\nu'} + f_{\nu',\nu}(\Omega)\,. \tag{6.53}$$

Setting $\tilde{Z}_+^{\nu'} = \bar{Z}_+^{\nu'}$, and defining $\hat{\mathbf{k}}'$ as the vector parallel to $\mathbf{r}$ with modulus k, we obtain

$$\begin{aligned}f_{\nu',\nu}(\hat{\mathbf{k}}',\hat{\mathbf{k}}) &= \frac{1}{2ik} \sum_{Jlm} 4\pi\, Y_{lm}^*(\hat{\mathbf{k}})\; Y_{l(m+\nu-\nu')}(\hat{\mathbf{k}}') \Big[\langle J(m+\nu)\,|lms\nu\rangle \\ &\quad \times S_l^J\, \langle l\,(m+\nu-\nu')\; s\nu'\,|J(m+\nu)\rangle - \delta_{\nu'\nu}\Big]. \end{aligned}\tag{6.54}$$

A simpler expression for the scattering amplitude is obtained choosing the z-axis along the incident beam. Substituting $Y_{lm}^*(\hat{\mathbf{k}}) \to \delta_{m0}\sqrt{(2l+1)/4\pi}$, the above equation becomes

$$\begin{aligned}f_{\nu',\nu}(\hat{\mathbf{k}}',\hat{\mathbf{k}}) &= \frac{1}{2ik} \sum_{lJ} \sqrt{4\pi\,(2l+1)}\; Y_{l(\nu-\nu')}(\hat{\mathbf{k}}') \\ &\quad \times \left[\langle l\,(\nu-\nu')\; s\nu'\,|J\nu\rangle\; S_l^J\, \langle J\nu\,|l\,0\,s\nu\rangle - \delta_{\nu'\nu}\right]. \end{aligned}\tag{6.55}$$

It is convenient to use the orthogonality property of the Clebsch-Gordan coefficients [Edmonds (1974)],

$$\sum_{JM} \langle j_1\, m_1\, j_2\, m_2 | JM \rangle \ \langle JM | j_1\, m_1'\, j_2\, m_2' \rangle = \delta_{m_1' m_1}\, \delta_{m_2' m_2} \tag{6.56}$$

to write

$$\begin{aligned} \delta_{\nu'\nu} &= \sum_{JM} \langle l\,(\nu-\nu')\; s\,\nu' \,| J\, M \rangle \ \langle JM\, | l\, 0\, s\nu \rangle \\ &= \sum_{J} \langle l\,(\nu-\nu')\; s\nu' \,| J\nu \rangle \ \langle J\nu\, | l\, 0\, s\nu \rangle\,. \end{aligned}$$

Note that the sum of the first line of the above equation reduces to the sum of the second line because the Clebsch-Gordan coefficients with $M \neq \nu$ vanish. Using this result, Eq. (6.56) becomes

$$\begin{aligned} f_{\nu',\nu}(\hat{\mathbf{k}}', \hat{\mathbf{k}}) = \frac{1}{2ik} \sum_{lJ} \sqrt{4\pi\,(2l+1)}\, Y_{l(\nu-\nu')}(\hat{\mathbf{k}}') \\ \times \langle l\,(\nu-\nu')\; s\nu' \,| J\nu \rangle \ \langle J\nu\, | l\, 0\, s\nu \rangle \left[S_l^J - 1 \right]. \end{aligned} \tag{6.57}$$

For long range potentials, one takes the $r \to \infty$ limit in Eq. (6.47) and in the partial-wave expansion of Eq. (6.10), and the comparison leads to the partial amplitude $\bar{f}_{\nu',\nu}(\hat{\mathbf{k}}', \hat{\mathbf{k}})$. Following the same procedures as above, we obtain

$$\begin{aligned} \bar{f}_{\nu',\nu}(\hat{\mathbf{k}}', \hat{\mathbf{k}}) = \frac{1}{2ik} \sum_{lJ} \sqrt{4\pi\,(2l+1)}\, Y_{l(\nu-\nu')}(\hat{\mathbf{r}}) \\ \times \langle l\,(\nu-\nu')\; s\nu' \,| J\nu \rangle \ \langle J\nu\, | l\, 0\, s\nu \rangle \ e^{2i\sigma_l} \left[\bar{S}_l^J - 1 \right]. \end{aligned} \tag{6.58}$$

This expression is analogous to Eq. (6.57), except for the inclusion of the Coulomb phase factor, $\exp(2i\sigma_l)$. Above, we use the notation $\bar{S}_l^J$ (instead of S_l^J) to stress that it is associated with the partial amplitude $\bar{f}_{\nu',\nu}$. The bar indicates association with the short-range part of the interaction. The full scattering amplitude is then obtained adding the Coulomb amplitude, according to Eq. (6.12).

Eq. (6.54) is the analog of Eq. (2.53a) when the system's spin is different from zero. This can be seen more clearly if we introduce the S-matrix elements averaged over the total angular momentum J, for the transition $\nu \to \nu'$,

$$\left\langle S_{ls}^{\nu,\nu'} \right\rangle = \sum_J \langle J(m+\nu)\, | lms\nu \rangle \ \ S_{ls}^J \ \langle l\,(m+\nu-\nu')\; s\nu' \,| J(m+\nu) \rangle\,. \tag{6.59}$$

Using the properties of Clebsch-Gordan coefficients, it is straightforward to show that Eq. (6.54) can be written as

$$f_{\nu,\nu'}(\hat{\mathbf{k}},\hat{\mathbf{k}}') = \frac{1}{2ik}\sum_{lm} 4\pi\, Y^*_{lm}(\hat{\mathbf{k}})\, Y_{l(m-\Delta\nu)}(\hat{\mathbf{k}}')\, \left[\left\langle S^{\nu,\nu'}_{ls}\right\rangle - 1\right], \qquad (6.60)$$

where $\Delta\nu = \nu' - \nu$. In the case of spin zero ($s = 0$, $\nu = \nu' = 0$) the above equation reduces to Eq. (2.53a).

In practical situations, we must solve numerically the radial equation for $l = 0, 1, ..., l_{\max}$. For the partial-wave expansion to converge, $l_{\max}$ should satisfy the usual condition $l_{\max} \gg k\bar{R}$, where $\bar{R}$ is the range of the potential. However, for each partial wave the radial equation must be solved several times, once for each of the possible J-values. That is, $2s + 1$ times for partial-waves where $l \geq s$ or $2l + 1$ times for partial-waves where $l < s$. In this way, the procedure to obtain the scattering amplitude is very similar to the one described in chapter 2, except that, for the same potential range, the radial equation must be solved a larger number of times.

In a more general situation, both the projectile and the target may have spin and the algebra involved becomes more complicated. However, a generalization of the present section to this case is straightforward. For problems of this kind, we refer to [Satchler (1983)].

6.1.1.2 *Partial-wave projected T-matrix, S-matrix and Lippmann-Schwinger equations*

First, we derive Lippmann-Schwinger equations for the radial wave functions of systems with spin. They are similar to the ones for spinless systems. We start from Eq. (6.5) in the coordinate representation[11],

$$\psi^{(+)}(\nu\mathbf{k};\mathbf{r}) = \phi(\nu\mathbf{k};\mathbf{r}) + \int d^3\mathbf{r}'\, G_0^{(+)}(E;\mathbf{r},\mathbf{r}')\, V(\mathbf{r}')\, \psi^{(+)}(\nu\mathbf{k};\mathbf{r}'). \qquad (6.61)$$

Using the partial-wave expansions for $\phi(\nu\mathbf{k};\mathbf{r})$ (Eq. (6.24)) and $\psi^{(+)}(\nu\mathbf{k};\mathbf{r})$ (Eq. (6.38)) in Eq. (6.61), we obtain[12]

$$u_l^J(k,r) = \hat{\jmath}_l(kr) + \int dr'\, g^{(+)}_{0,l}(E;r,r')\, V_l^J(r')\, u_l^J(k,r'). \qquad (6.62)$$

[11]For simplicity, we consider the case of a rotationally invariant local potential.
[12]We are assuming that the potential is diagonal in the representation of orbital angular momentum. The general situation where this condition is not satisfied is included in the discussion of multi-channel scattering (chapter 9).

Above, $V_l^J(r')$ is given by Eq. (6.40) and the angular momentum projected components of the free Green's function are

$$g_{0,l}^{(+)}(E;r,r') = \int d\zeta\, d\zeta'\, \left(\mathcal{Y}^{lJM}(\zeta)\right)^* \; G_0^{(+)}(E;\mathbf{r},\mathbf{r}')\; \mathcal{Y}^{lJM}(\zeta'). \qquad (6.63)$$

When the potential contains a Coulomb part, the Born series does not converge and the above expressions are not useful. However the results of section 4.5 may be applied. The potential can be split into two parts,

$$V = V_1 + V_2, \qquad (6.64)$$

with the long-range Coulomb interaction included in V_1. The Lippman-Schwinger equation can then be written in terms of the exact solutions for the potential V_1, $\chi^{(+)}(\nu\mathbf{k};\mathbf{r})$, which satisfies Eq. (4.141). Proceeding as in the case of spinless particles (see chapter 4), we get

$$\psi^{(+)}(\nu\mathbf{k};\mathbf{r}) = \chi^{(+)}(\nu\mathbf{k};\mathbf{r}) + \int d^3\mathbf{r}'\, G_1^{(+)}(E;\mathbf{r},\mathbf{r}')\, V_2(\mathbf{r}')\, \psi^{(+)}(\nu\mathbf{k};\mathbf{r}'), \quad (6.65)$$

where

$$G_1^{(+)}(E;\mathbf{r},\mathbf{r}') = \left\langle \mathbf{r} \left| \frac{1}{E - K - V_1 + i\epsilon} \right| \mathbf{r}' \right\rangle. \qquad (6.66)$$

It is straightforward to carry out angular momentum expansions in Eq. (6.65) and derive the projected Lippmann-Schwinger equation involving distorted waves,

$$u_l^J(k,r) = w_l^J(k,r) \; + \; \int dr'\, g_{1,l}^{(+)}(E;r,r')\; V_{2,l}^J(r')\; u_l^J(k,r'). \qquad (6.67)$$

Above, $g_{1,l}^{(+)}(E;r,r')$ and $V_{2,l}^J(r')$ are given by Eqs. (6.63) and (6.40), replacing respectively $G_0^{(+)}(E;\mathbf{r},\mathbf{r}') \to G_1^{(+)}(E;\mathbf{r},\mathbf{r}')$ and $V(\mathbf{r}') \to V_2(\mathbf{r}')$, and $w_l^J(k,r)$ is the radial wave function distorted by the potential V_1 (solution of the radial equation setting $V_2(r) = 0$). In the particular case where $V_1 = V_C$, the distorted wave is equal to the Coulomb wave function, $F_l(\eta, kr)$.

Now we derive the angular momentum projection of the T- and the S-matrices. For this purpose we introduce the normalized energy and angular momentum eigenfunctions,

$$\phi\,(ElJM;\zeta) = \left(\frac{2\mu k}{\pi\hbar^2}\right)^{1/2} \; \mathcal{Y}^{lJM}(\zeta)\; \frac{\hat{\jmath}(kr)}{kr} \qquad (6.68)$$

$$\psi\,(ElJM;\zeta) = \left(\frac{2\mu k}{\pi\hbar^2}\right)^{1/2} \; \mathcal{Y}^{lJM}(\zeta)\; \frac{u_l^J(k,r)}{kr}. \qquad (6.69)$$

One can easily prove that they satisfy the orthonormality relations (see section 4.6.1)

$$\left\langle \phi\,(ElJM) \,\middle|\, \phi\,(E'l'J'M')\right\rangle = \delta_{ll'}\,\delta_{JJ'}\,\delta_{MM'}\,\delta(E-E'), \tag{6.70}$$

$$\left\langle \phi\,(ElJM) \,\middle|\, \phi\,(E'l'J'M')\right\rangle = \delta_{ll'}\,\delta_{JJ'}\,\delta_{MM'}\,\delta(E-E'). \tag{6.71}$$

The partial wave projected T-matrix is then defined as

$$\begin{aligned} T_l^J(E) &\equiv \left\langle \phi(ElJM)\right| T \left|\phi(ElJM)\right\rangle \\ &= \left\langle \phi(ElJM)\right| V \left|\psi(ElJM)\right\rangle. \end{aligned}$$

The above matrix-elements can be evaluated in the coordinate representation, using the explicit forms of $\phi\,(ElJM;\zeta)$ and $\psi\,(ElJM;\zeta)$ (Eqs. (6.68) and (6.69)). We get

$$T_l^J(E) = \frac{k}{\pi E}\int dr\; \hat{\jmath}_l(kr)\, V_l^J(r)\, u_l^J(k,r). \tag{6.72}$$

To express the T-matrix in momentum space,

$$T_{\nu'\mathbf{k}';\nu\mathbf{k}} = \left\langle \phi(\nu'\mathbf{k}')\right| V \left|\psi^{(+)}(\nu\mathbf{k})\right\rangle, \tag{6.73}$$

in terms of $T_l^J(E)$, we re-write the angular momentum expansions of $\phi(\nu'\mathbf{k}',\zeta)$ and $\psi^{(+)}(\nu\mathbf{k};\zeta)$ in terms of $\mathcal{Y}_{lJM}(\zeta)$. Using Eqs. (6.68) and (6.69) in Eqs. (6.27) and (6.44), we get the expansion

$$\begin{aligned} T_{\nu'\mathbf{k}';\nu\mathbf{k}} = \frac{\hbar^2}{\mu k} &\sum_{Jlm} Y_{l(m+\nu-\nu')}(\hat{\mathbf{k}}')\; Y^*_{lm}(\hat{\mathbf{k}})\; \langle lms\nu|\; J(m+\nu)\rangle \\ &\times \langle J(m+\nu)|l(m+\nu-\nu')s\nu\rangle\;\; T_l^J(E). \end{aligned} \tag{6.74}$$

This expression may be simplified if one takes the z-axis along the incident beam. One can then replace: $Y_{lm}(\hat{\mathbf{k}}) \to Y_{lm}(\hat{\mathbf{z}}) = \delta_{m0}\sqrt{(2l+1)/4\pi}$, and get

$$\begin{aligned} T_{\nu'\mathbf{k}';\nu\mathbf{k}} = \frac{\hbar^2}{\mu k} &\sum_{Jl}\sqrt{\frac{2l+1}{4\pi}}\; Y_{l(\nu-\nu')}(\hat{\mathbf{k}}')\;\; \langle l0s\nu|\; J\nu\rangle \\ &\times \langle J\nu|l(\nu-\nu')s\nu\rangle\;\; T_l^J(E). \end{aligned} \tag{6.75}$$

A relation between the angular momentum projected T- and S-matrices can easily be derived. This can be achieved multiplying the above equation by the factor $-4\pi^2\mu/\hbar^2$ and comparing the result with the scattering amplitude of Eq. (6.57). According to the general relation between the T-matrix and the scattering amplitude (Eq. (4.64)), they should be identical. In this way, we get

$$S_l^J(E) = 1 - 2\pi i\; T_l^J(E). \tag{6.76}$$

6.2 The scattering of a spin $1/2$ particle

Nucleons and electrons are spin-$1/2$ particles and their scattering and spin polarization are quite important. The formalism developed above becomes quite simple in the case of a spin-$1/2$ particle colliding with a particle with spin-zero. In this case, it is convenient to write the wave function and the scattering amplitude as vectors and matrices in the two-dimensional space of spin. This procedure becomes particularly useful when the initial state is not polarized along the beam axis and the incident state is described in terms of the density matrix.

In the collision of one spin-$1/2$ particle with one spin-zero particle, the spin quantum numbers take the values, $s = 1/2$ and $\nu = \pm 1/2$. We then use the short hand notations $|+\rangle$ and $|-\rangle$ to represent the spin-up and spin-down states, respectively. The scattering amplitude is a 2x2 matrix in the spinor space, which can be written as

$$f(\theta,\varphi) = \sum_{\nu,\nu'=\pm} |\nu'\rangle\ f_{\nu',\nu}(\theta,\varphi)\ \langle\nu|\,. \tag{6.77}$$

The matrix-elements $f_{\nu',\nu}$ are given by Eq. (6.57), for $s = 1/2$ and $\nu, \nu' = \pm 1/2$. The amplitudes f_{++} and f_{+-} are

$$\begin{aligned} f_{++}(\hat{\mathbf{k}}') = \frac{1}{2ik} \sum_l \sqrt{4\pi\,(2l+1)}\ Y_{l0}(\hat{\mathbf{k}}') \\ \times \Big\{ \left|\langle l\,0\,{}^1\!/_2\,{}^1\!/_2\,|(l+{}^1\!/_2)\,{}^1\!/_2\rangle\right|^2\ [S_l^+ - 1] \\ + \left|\langle l\,0\,{}^1\!/_2\,{}^1\!/_2\,|(l-{}^1\!/_2)\,{}^1\!/_2\rangle\right|^2\ [S_l^- - 1] \Big\} \end{aligned} \tag{6.78}$$

and

$$\begin{aligned} f_{-+}(\hat{\mathbf{k}}') = \frac{1}{2ik} \sum_l \sqrt{4\pi\,(2l+1)}\ Y_{l1}(\hat{\mathbf{k}}') \\ \times \Big\{ \langle l\,1\,{}^1\!/_2\,(-{}^1\!/_2)\ |(l+{}^1\!/_2)\,{}^1\!/_2\rangle \langle (l+{}^1\!/_2)\,{}^1\!/_2|l\,0\,{}^1\!/_2\,{}^1\!/_2\rangle\ [S_l^+ - 1] \\ + \langle l\,1\,{}^1\!/_2\ (-{}^1\!/_2)\ |(l-{}^1\!/_2)\,{}^1\!/_2\rangle \langle (l-{}^1\!/_2)\,{}^1\!/_2|l\,0\,{}^1\!/_2\,{}^1\!/_2\rangle\ [S_l^- - 1] \Big\}. \end{aligned} \tag{6.79}$$

Above, we used the short-hand notation: $S_l^{\,l\,\pm\,1/2} = S_l^{\pm}$. These expressions are greatly simplified if one uses the explicit values of the Clebsch-Gordan coefficients ([Edmonds (1974)], table 5.2, p.89). The non-spin-flip amplitudes become

$$f_{++}(\hat{\mathbf{k}}') = \sum_l \left[\frac{(l+1)\,\left[S_l^+ - 1\right] + l\,\left[S_l^- - 1\right]}{2ik} \right] P_l(\cos\theta) \tag{6.80}$$

$$f_{--}(\hat{\mathbf{k}}') = f_{++}(\hat{\mathbf{k}}') \tag{6.81}$$

and the spin-flip amplitudes can be put in the form

$$f_{-+}(\hat{\mathbf{k}}') = \sum_l - \left[\frac{S_l^+ - S_l^-}{2ik} \right] \frac{dP_l(\cos\theta)}{d\cos\theta} \sin\theta\, e^{i\varphi} \tag{6.82}$$

$$f_{+-}(\hat{\mathbf{k}}') = \sum_l \left[\frac{S_l^+ - S_l^-}{2ik} \right] \frac{dP_l(\cos\theta)}{d\cos\theta} \sin\theta\, e^{-i\varphi}. \tag{6.83}$$

To get the above results we have used the relations [Jackson (1975)]

$$Y_{l1}(\theta,\phi) = -\sqrt{\frac{2l+1}{4\pi\, l(l+1)}} \left[\frac{dP_l(\cos\theta)}{d\cos\theta} \right] \sin\theta\, e^{i\varphi} \tag{6.84}$$

and

$$Y_{l(-1)}(\theta,\varphi) = -Y_{l1}^*(\theta,\varphi). \tag{6.85}$$

It is convenient to express the matrix elements of the scattering amplitude in terms of two scalar functions of θ:

$$g(\theta) = \sum_l \left[\frac{(l+1)\,\left[S_l^+ - 1\right] + l\,\left[S_l^- - 1\right]}{2ik} \right] P_l(\cos\theta) \tag{6.86}$$

and

$$h(\theta) = \sum_l \left[\frac{S_l^+ - S_l^-}{2ik} \right] \sin\theta\, \frac{dP_l(\cos\theta)}{d\cos\theta}. \tag{6.87}$$

In this way, we get

$$f_{++}(\theta) = f_{--}(\theta) = g(\theta) \tag{6.88}$$

$$f_{+-}(\theta) = h(\theta)\, e^{-i\,\varphi(\theta)} \tag{6.89}$$

$$f_{-+}(\theta) = -\,h(\theta)\, e^{i\,\varphi(\theta)} \tag{6.90}$$

or, in matrix form

$$f_{2\times 2}(\theta) \equiv \begin{pmatrix} f_{++}(\theta) & f_{+-}(\theta) \\ f_{-+}(\theta) & f_{--}(\theta) \end{pmatrix} = \begin{pmatrix} g(\theta) & h(\theta)\, e^{-\,i\varphi} \\ -\,h(\theta)\, e^{i\,\varphi} & g(\theta) \end{pmatrix}. \tag{6.91}$$

The scattering amplitude matrix can be expressed as a linear combination of the Pauli matrices and the identity,

$$\mathbb{I}_{2\text{x}2} = \begin{pmatrix} 1 & 0 \\ 0 & 1 \end{pmatrix}, \quad \sigma_x = \begin{pmatrix} 0 & 1 \\ 1 & 0 \end{pmatrix}, \quad \sigma_y = \begin{pmatrix} 0 & -i \\ i & 0 \end{pmatrix} \quad \text{and} \quad \sigma_z = \begin{pmatrix} 1 & 0 \\ 0 & -1 \end{pmatrix}.$$

These four matrices form a complete set for the space of complex 2x2 matrices. It can be easily checked that Eq. (6.91) can be written as (see problem 4)

$$f_{2\text{x}2}(\theta) = g(\theta)\,\mathbb{I}_{2\text{x}2} + i\,h(\theta)\;\hat{\mathbf{n}}\cdot\boldsymbol{\sigma}, \tag{6.92}$$

where $\boldsymbol{\sigma} \equiv \sigma_x\,\hat{\mathbf{x}} + \sigma_y\,\hat{\mathbf{y}} + \sigma_z\,\hat{\mathbf{z}}$ and $\hat{\mathbf{n}} = \mathbf{k}\times\mathbf{k}'/|\mathbf{k}\times\mathbf{k}'|$ is the unit vector perpendicular to the scattering plane. Taking the incident beam along the z-axis, $\mathbf{k} = \hat{\mathbf{z}}$ and we can write

$$\hat{\mathbf{n}} = \frac{\hat{\mathbf{z}}\times\mathbf{k}'}{|\hat{\mathbf{z}}\times\mathbf{k}'|} = \frac{-\sin\theta\sin\varphi\;\hat{\mathbf{x}} + \sin\theta\cos\varphi\;\hat{\mathbf{y}}}{\sin\theta} = -\sin\varphi\;\hat{\mathbf{x}} + \cos\varphi\;\hat{\mathbf{y}}. \tag{6.93}$$

The spin of the scattered wave, $|\chi^{\text{sc}}\rangle$, is obtained applying the 2x2 scattering amplitude matrix on the spin of the incident wave, $|\chi^{\text{in}}\rangle$. That is

$$|\chi^{\text{sc}}\rangle = f(\mathbf{k}')\;|\chi^{\text{in}}\rangle\,. \tag{6.94}$$

The cross section is then given by

$$\frac{d\sigma(\mathbf{k}')}{d\Omega} = \langle\chi^{\text{sc}}|\chi^{\text{sc}}\rangle\,. \tag{6.95}$$

For incident beams polarized with spin-up and spin-down the corresponding cross sections are respectively

$$\frac{d\sigma_+(\mathbf{k}')}{d\Omega} = |f_{++}(\mathbf{k}')|^2 + |f_{-+}(\mathbf{k}')|^2 \tag{6.96}$$

$$\frac{d\sigma_-(\mathbf{k}')}{d\Omega} = |f_{--}(\mathbf{k}')|^2 + |f_{+-}(\mathbf{k}')|^2\,. \tag{6.97}$$

For the arbitrary polarization,

$$|\chi^{\text{in}}\rangle = \begin{pmatrix} \alpha \\ \beta \end{pmatrix}, \tag{6.98}$$

the cross section is

$$\begin{aligned}\frac{d\sigma(\mathbf{k}')}{d\Omega} &= \Big|\alpha f_{++}(\mathbf{k}') + \beta f_{-+}(\mathbf{k}')\Big|^2 + \Big|\alpha f_{+-}(\mathbf{k}') + \beta f_{--}(\mathbf{k}')\Big|^2 \\ &\quad + 2\,\text{Re}\Big\{\,(\alpha^* f^*_{++}(\mathbf{k}') + \beta^* f^*_{-+}(\mathbf{k}'))\;\;(\alpha f_{+-}(\mathbf{k}') + \beta f_{--}(\mathbf{k}'))\,\Big\}.\end{aligned} \tag{6.99}$$

The above discussion assumes that the incident beam is in a pure state in the spin space. However in most experiments the incident beam is in a mixed state. The only available information is the spin polarization of the incident beam. For a pure state $|\chi\rangle$, the components of the polarization $\mathbf{P} = P_x\,\hat{\mathbf{x}} + P_y\,\hat{\mathbf{y}} + P_z\,\hat{\mathbf{z}}$ are the expectation values of the Pauli matrices,

$$P_i = \langle\chi|\,\sigma_i\,|\chi\rangle\,, \tag{6.100}$$

with $i = x, y, z$. On the other hand, mixed states are represented by density matrices, ρ, in the spin space. In this case, the components of the polarization are given by

$$P_i = \operatorname{tr}\{\sigma_i\,\rho\}\,. \tag{6.101}$$

It can be easily checked that an incident beam in a mixed state with polarization $\mathbf{P}^{\rm in}$ is described by the density matrix[13]

$$\rho^{\rm in} = \frac{1}{2}\left[\mathbb{I}_{2\times 2} + \boldsymbol{\sigma}\cdot\mathbf{P}^{\rm in}\right] = \frac{1}{2}\left[\mathbb{I}_{2\times 2} + \sum_{i=x,y,z} P_i^{\rm in}\,\sigma_i\right]. \tag{6.102}$$

We have seen that in the scattering process a pure incident spin state, $\chi^{\rm in}$, is transformed into an scattered spin state. The transformation is performed through the action of the matrix f. That is,

$$\left|\chi^{\rm in}\right\rangle \longrightarrow \left|\chi^{\rm sc}\right\rangle = f\left|\chi^{\rm in}\right\rangle. \tag{6.103}$$

Accordingly, an operator O acting on the spin space is transformed as $O \to fOf^\dagger$. Thus, the density matrix describing the incident wave transforms into the scattered density matrix

$$\rho^{\rm in} \longrightarrow \rho^{\rm sc} = f\rho^{\rm in}f^\dagger \tag{6.104}$$

and the cross section is given by

$$\frac{d\sigma(\mathbf{k}')}{d\Omega} = \operatorname{tr}\{\rho^{\rm sc}\} = \operatorname{tr}\left\{f\rho^{\rm in}f^\dagger\right\}. \tag{6.105}$$

In experiments with polarized beams described by an initial density matrix, $\rho^{\rm in}$, the cross section should be expressed in terms of the known polarization of the incident beam, $\mathbf{P}^{\rm in}$. For this purpose, we first use Eq. (6.102) and (6.92) in Eq. (6.104), to get

$$\rho^{\rm sc} = \frac{1}{2}\left[g(\theta)\,\mathbb{I}_{2\times 2} + ih(\theta)\,\hat{\mathbf{n}}\cdot\boldsymbol{\sigma}\right]\left[\mathbb{I}_{2\times 2} + \boldsymbol{\sigma}\cdot\mathbf{P}^{\rm in}\right]\left[g^*(\theta)\,\mathbb{I}_{2\times 2} - i\,h^*(\theta)\,\hat{\mathbf{n}}\cdot\boldsymbol{\sigma}\right]. \tag{6.106}$$

[13]It is an immediate consequence of the facts that $\operatorname{tr}\mathbb{I}_{2\times 2} = 2$ and the Pauli matrices are traceless.

The above expression can be considerably simplified using the property of Pauli matrices,

$$(\boldsymbol{\sigma} \cdot \mathbf{A})\ (\boldsymbol{\sigma} \cdot \mathbf{B}) = (\mathbf{A} \cdot \mathbf{B})\ \mathbb{I}_{2\times 2} + i\,\boldsymbol{\sigma} \cdot (\mathbf{A} \times \mathbf{B})\,, \tag{6.107}$$

where $\mathbf{A}$ and $\mathbf{B}$ are vectors and $\mathbb{I}_{2\times 2}$ is the identity matrix in 2 dimensions. Thus, one gets

$$\begin{aligned}\rho^{\mathrm{sc}} = \frac{1}{2}\ \Big[&|g(\theta)|^2 + |h(\theta)|^2 \\ &+ 2\,\mathrm{Im}\,\{h^*(\theta)\, g(\theta\}\ \hat{\mathbf{n}} \cdot \mathbf{P}^{\mathrm{in}}\Big]\, \mathbb{I}_{2\times 2} + \frac{\mathbf{1}}{\mathbf{2}}\ \boldsymbol{\sigma} \cdot \boldsymbol{\Delta}.\end{aligned} \tag{6.108}$$

Above, we have introduced the angle-dependent vector, $\boldsymbol{\Delta}$,

$$\begin{aligned}\boldsymbol{\Delta} = 2\,\mathrm{Im}\,\{g(\theta)\, h^*(\theta)\}\ \hat{\mathbf{n}} + \left(|g(\theta)|^2 - |h(\theta)|^2\right)\ \mathbf{P}^{\mathrm{in}} \\ + 2\ |h(\theta)|^2\ \left(\hat{\mathbf{n}} \cdot \mathbf{P}^{\mathrm{in}}\right)\ \hat{\mathbf{n}} - 2\,\mathrm{Re}\,\{g(\theta)\, h^*(\theta)\}\, \hat{\mathbf{n}} \times \mathbf{P}^{\mathrm{in}}.\end{aligned} \tag{6.109}$$

Since the traces of the Pauli matrices are zero, the cross section for a initial polarization $\mathbf{P}^{\mathrm{in}}$ is given by,

$$\left[\frac{d\sigma(\hat{\mathbf{k}}')}{d\Omega}\right]_{\mathbf{P}^{\mathrm{in}}} = \mathrm{tr}\,\{\rho^{\mathrm{sc}}\} = |g(\theta)|^2 + |h(\theta)|^2 + 2\,\mathrm{Im}\,\{g(\theta)\, h^*(\theta)\}\ \hat{\mathbf{n}} \cdot \mathbf{P}^{\mathrm{in}}, \tag{6.110}$$

or,

$$\left[\frac{d\sigma(\hat{\mathbf{k}}')}{d\Omega}\right]_{\mathbf{P}^{\mathrm{in}}} = \left(\frac{d\sigma(\hat{\mathbf{k}}')}{d\Omega}\right)\ \left[1 + \mathbf{P}(\theta) \cdot \mathbf{P}^{\mathrm{in}}\right]. \tag{6.111}$$

Above, we have introduced the polarization vector,

$$\mathbf{P}(\theta) = \left[\frac{2\,\mathrm{Im}\,\{g(\theta) h^*(\theta)\}}{|g(\theta)|^2 + |h(\theta)|^2}\right]\ \hat{\mathbf{n}} \tag{6.112}$$

and the unpolarized cross section,

$$\frac{d\sigma(\hat{\mathbf{k}}')}{d\Omega} = |g(\theta)|^2 + |h(\theta)|^2\,. \tag{6.113}$$

Note that the polarization vector is independent of the polarization of the incident beam.

The final spin polarization,

$$\mathbf{P}^{\mathrm{sc}}(\theta) = \frac{\mathrm{tr}\,\{\boldsymbol{\sigma} \rho^{\mathrm{sc}}(\theta)\}}{\mathrm{tr}\,\{\rho^{\mathrm{sc}}(\theta)\}}, \tag{6.114}$$

then becomes

$$\mathbf{P}^{\rm sc}(\theta) = \frac{\mathbf{P}(\theta) + \mathbf{P}^{\rm in} - \mathbf{Q}(\theta) \times \mathbf{P}^{\rm in}}{1 + \mathbf{P}(\theta) \cdot \mathbf{P}^{\rm in}} + \frac{2\, |g(\theta)|^2 \left(\hat{\mathbf{n}} \cdot \mathbf{P}^{\rm in}\right) \hat{\mathbf{n}} - |h(\theta)|^2\, \mathbf{P}^{\rm in}}{\left(1 + \mathbf{P}(\theta) \cdot \mathbf{P}^{\rm in}\right) \left(|g(\theta)|^2 + |h(\theta)|^2\right)}, \tag{6.115}$$

where $\mathbf{Q}(\theta)$ is called the spin rotation vector,

$$\mathbf{Q}(\theta) = \left[\frac{2\,\mathrm{Re}\,\{g(\theta)\, h^*(\theta)\}}{|g(\theta)|^2 + |h(\theta)|^2}\right] \hat{\mathbf{n}}. \tag{6.116}$$

The important feature of Eq. (6.115) is that even if the incident beam is unpolarized ($\mathbf{P}^{\rm in} = 0$), the final polarization is not. Thus one can produce polarized particles by merely scattering them. The interaction responsible for this final polarization is the spin-orbit term. Thus by measuring the final polarization of unpolarized spin-$1/2$ beam of particles one can learn about the nature of the spin-orbit interaction.

To summarize, for unpolarized spin-$1/2$ particles, the cross section and final polarization of the scattered beam are given by,

$$\frac{d\sigma(\hat{\mathbf{k}}')}{d\Omega} = |g(\theta)|^2 + |h(\theta)|^2.$$

and

$$\mathbf{P}^{\rm sc}(\theta) = \mathbf{P}(\theta) = \left[\frac{2\,\mathrm{Im}\,\{g(\theta) h^*(\theta)\}}{|g(\theta)|^2 + |h(\theta)|^2}\right] \hat{\mathbf{n}}.$$

The final polarization is perpendicular to the scattering plane. Any component of the polarization which lies on the scattering plane is an indication of violation of parity. Such component would be present if the interaction were to contain a term of the type $\boldsymbol{\sigma} \cdot \mathbf{p}$, where $\mathbf{p}$ is the linear momentum vector.

How to measure the polarization? If the initial polarization is not zero, then the cross section is given by Eq. (6.110). Taking $\mathbf{P}^{\rm in}$ to be pointing at an azimuthal angle of φ from $\hat{\mathbf{n}}$, namely, $\mathbf{P}^{\rm in} \cdot \hat{\mathbf{n}} = P^{\rm in} \cos\varphi$, then by flipping the direction of $\hat{\mathbf{n}}$ we can measure the cross section for scattering to the right and to the left. The difference is directly related to the magnitude of initial polarization. We take $\mathbf{P}^{\rm in}$ to be parallel to $\hat{\mathbf{n}}$, namely $\varphi = 0$. This defines, say, the scattering to the right. Then the scattering to the left is defined by $\varphi = \pi$. The right-left asymmetry, called the analyzing power, $A(\theta)$, is then given by

$$A(\theta) = \frac{[d\sigma/d\Omega]_{\mathbf{P}_R^{\rm in}} - [d\sigma/d\Omega]_{\mathbf{P}_L^{\rm in}}}{[d\sigma/d\Omega]_{\mathbf{P}_R^{\rm in}} + [d\sigma/d\Omega]_{\mathbf{P}_L^{\rm in}}} = \frac{2\,|\mathrm{Im}\,\{g(\theta)\,h^*(\theta)\}|}{|g(\theta)|^2 + |h(\theta)|^2}\,P^{\rm in}, \tag{6.117}$$

with L and R standing for scattering to the left and to the right, respectively. The above type of measurement requires a rather complicated experimental set up with magnetic field used to fix the value and direction of $\mathbf{P}^{\rm in}$. An easier way to measure the strength of the spin-orbit interaction is through a double scattering experiment. The initial polarization is taken to be zero. After the first scattering, the cross section is given by,

$$\frac{d\sigma(\mathbf{k}')}{d\Omega} = |g(\theta)|^2 + |h(\theta)|^2\,. \tag{6.118}$$

The scattered beam is polarized according to Eq. (6.2). This first target is thus a polarizer. The beam, now with an 'initial' polarization vector $\mathbf{P}(\theta)$ along $\hat{\mathbf{n}}$, is then scattered from another target in the same scattering plane, both to the right and to the left. This second target acts as an analyzer. According to Eq. (6.117), we get, for the analyzing power, or right-left asymmetry,

$$A_2(\theta) = \left[\frac{2\,|\mathrm{Im}\,\{g(\theta)\,h^*(\theta)\}|}{|g(\theta)|^2 + |h(\theta)|^2}\right]^2 \tag{6.119}$$

From the first scattering, the unpolarized cross section $d\sigma/d\Omega = |g(\theta)|^2 + |h(\theta)|^2$ is determined. For all practical purposes the term $|g|^2$ dominates over $|h|^2$, and thus one can safely say that the first scattering determines $|g|^2$. After the second scattering, the asymmtery is found proportional to $\left[\mathrm{Im}\,\{g(\theta)h^*(\theta)\}\,/|g|^2\right]^2$. In this way, the absolute value of the term $\mathrm{Im}\,\{g(\theta)h^*(\theta)\}$ is determined. The above double scattering experiment with a spin unpolarized beam determines only the absolute value of g. The determination of the phase requires a more complicated set up. The spin rotation vector $\mathbf{Q}(\theta)$ of Eq. (6.116) is a quantity which can furnish this information. The determination of the spin rotation is more involved as it requires the triple experiment set up.

It is instructive to calculate the amplitudes h and g at very low energies. In this case, we have,

$$g = \frac{1}{2ik}\left[S_0^{1/2} - 1\right] \tag{6.120}$$

$$h = \frac{1}{2ik}\left[S_1^{3/2} - S_1^{1/2}\right], \tag{6.121}$$

and the spin polarization and rotation projected along the normal $\hat{\mathbf{n}}$,

$$\hat{\mathbf{n}} \cdot \mathbf{P}(\theta) = 2\sin\theta \left[\frac{\mathrm{Im}\left\{ \left(S_0^{1/2} - 1\right) \left(S_1^{3/2} - S_1^{1/2}\right)^* \right\}}{|S_0^{1/2} - 1|^2} \right] \tag{6.122}$$

$$\hat{\mathbf{n}} \cdot \mathbf{Q}(\theta) = 2\sin\theta \left[\frac{\mathrm{Re}\left\{ \left(S_0^{1/2} - 1\right) \left(S_1^{3/2} - S_1^{1/2}\right)^* \right\}}{|S_0^{1/2} - 1|^2} \right] \tag{6.123}$$

6.2.1 *The eikonal approximation in collisions of a spin-1/2 particle*

At high energies, the eikonal approximation is applicable (see section 5.2.1). Here, we use the eikonal approximation in collisions of a spin-1/2 projectile with a spin-zero target. We consider the case of a spherically symmetric potential containing a spin-orbit term, with the form

$$V(\mathbf{r}) = V_0(r) + V_1(r)\,\boldsymbol{\sigma}\cdot\mathbf{l} \equiv V_0(r) + V_1(r)\,\boldsymbol{\sigma}\cdot(\mathbf{r}\times\mathbf{k})\,. \tag{6.124}$$

In high energy collisions at forward angles, $\mathbf{k} \simeq k\,\hat{\mathbf{z}}$. One can then approximate $\mathbf{r}\times\mathbf{k} \simeq \mathbf{b}\times\mathbf{k}$ and the eikonal phase of Eq. (5.118) becomes

$$\chi(b) = \frac{k}{2E}\int_{-\infty}^{\infty} dz \left[V_0\left(\sqrt{z^2+b^2}\right) + \boldsymbol{\sigma}\cdot(\mathbf{b}\times\mathbf{k})\; V_1\left(\sqrt{z^2+b^2}\right)\right], \tag{6.125}$$

or

$$\chi(b) = \Delta_0(b) + \Delta_1(b)\;\; \boldsymbol{\sigma}\cdot(\mathbf{b}\times\mathbf{k}) \tag{6.126}$$

with

$$\Delta_i(b) = -\frac{k}{2E}\int_{-\infty}^{\infty} dz\; V_i\left(\sqrt{z^2+b^2}\right). \tag{6.127}$$

The S-matrix within the eikonal approximation then can be written

$$S(b) = e^{i\,\Delta_0(b) + i\,\Delta_1(b)\,\boldsymbol{\sigma}\cdot(\mathbf{b}\times\mathbf{k})} = e^{i\,\Delta_0(b)}\;\; e^{i\,\Delta_1(b)\,\boldsymbol{\sigma}\cdot(\mathbf{b}\times\mathbf{k})}. \tag{6.128}$$

The functions $\Delta_0(b)$ and $\Delta_1(b)$ can easily be calculated using numerical methods. Thus, the evaluation of the first factor on the RHS of the above equation is straightforward. Let us then discuss the second factor. It has the general form $\exp\left[ia\,\boldsymbol{\sigma}\cdot\hat{\mathbf{A}}\right]$, where a is a scalar function and $\mathbf{A}$ is a vector. In our case they are given by

$$\hat{\mathbf{A}} = \frac{\mathbf{b}\times\mathbf{k}}{|\mathbf{b}\times\mathbf{k}|} \quad \text{and} \quad a = |\mathbf{b}\times\mathbf{k}|\;\Delta_1(b). \tag{6.129}$$

Since at forward angles $\mathbf{k} \simeq k\,\hat{\mathbf{z}}$, $\mathbf{k}$ is nearly perpendicular to $\mathbf{b}$ and thus $|\mathbf{b} \times \mathbf{k}| \simeq kb$. In this way, we can approximate

$$a = kb\,\Delta_1(b). \tag{6.130}$$

To evaluate $\exp\left[ia\,\boldsymbol{\sigma}\cdot\hat{\mathbf{A}}\right]$, we use the series expansion of the exponential, writing the even and the odd terms separately. That is

$$e^{ia\,\boldsymbol{\sigma}\cdot\hat{\mathbf{A}}} = \sum_{n=0}^{\infty} \frac{(ia)^n}{n!}\left(\boldsymbol{\sigma}\cdot\hat{\mathbf{A}}\right)^n = \sum_{m=0}^{\infty} \frac{(ia)^{2m}}{(2m)!}\left(\boldsymbol{\sigma}\cdot\hat{\mathbf{A}}\right)^{2m} + \sum_{m=0}^{\infty} \frac{(ia)^{2m+1}}{(2m+1)!}\left(\boldsymbol{\sigma}\cdot\hat{\mathbf{A}}\right)^{2m+1}, \tag{6.131}$$

or

$$e^{ia\,\boldsymbol{\sigma}\cdot\hat{\mathbf{A}}} = \sum_{m=0}^{\infty} (-)^m \frac{a^{2m}}{(2m)!}\left(\boldsymbol{\sigma}\cdot\hat{\mathbf{A}}\right)^{2m} + i\,\boldsymbol{\sigma}\cdot\hat{\mathbf{A}} \sum_{m=0}^{\infty} (-)^m \frac{a^{2m+1}}{(2m+1)!}\left(\boldsymbol{\sigma}\cdot\hat{\mathbf{A}}\right)^{2m}. \tag{6.132}$$

According to Eq. (6.107), we can replace

$$\left(\boldsymbol{\sigma}\cdot\hat{\mathbf{A}}\right)^{2m} = \mathbb{I}_{2\text{x}2}. \tag{6.133}$$

Using this result in Eq. (6.132), identifying the cosinus (even terms) and sinus (odd terms) series, and replacing a by its explicit form (Eq. (6.130)), we get

$$e^{ia\,\boldsymbol{\sigma}\cdot\hat{\mathbf{A}}} = \cos\left(kb\,\Delta_1(b)\right)\,\mathbb{I}_{2\text{x}2} + i\,\boldsymbol{\sigma}\cdot\hat{\mathbf{A}}\;\sin\left(kb\,\Delta_1(b)\right). \tag{6.134}$$

Using Eq. (6.134), the eikonal S-matrix becomes

$$S(b) = e^{i\Xi(b)} = e^{i\Delta_0(b)}\,\left[\cos\left(kb\,\Delta_1(b)\right)\,\mathbb{I}_{2\text{x}2} + i\,\boldsymbol{\sigma}\cdot\hat{\mathbf{A}}\;\sin\left(kb\,\Delta_1(b)\right)\right]. \tag{6.135}$$

The scattering amplitude is obtained inserting the above S-matrix into Eq. (5.120). We get

$$f_{\text{e}}(\hat{\mathbf{k}}') \equiv f_{\text{e}}(\theta,\varphi) = \frac{ik}{2\pi}\int d\mathbf{b}\; e^{i\mathbf{q}\cdot\mathbf{b}}\left[\left[e^{i\Delta_0(b)}\cos\left(kb\,\Delta_1(b)\right) - 1\right]\mathbb{I}_{2\text{x}2} + i\,e^{i\Delta_0(b)}\sin\left(kb\,\Delta_1(b)\right)\,\boldsymbol{\sigma}\cdot\hat{\mathbf{A}}\right]. \tag{6.136}$$

The above expression can be written as a sum of two terms,

$$f_{\mathrm{e}}(\theta,\varphi) = f_{\mathrm{nsf}}(\theta) + f_{\mathrm{sf}}(\theta,\varphi), \tag{6.137}$$

corresponding respectively to the non-spin-flip and spin-flip, amplitudes. The first one is given by

$$f_{\mathrm{nsf}}(\theta) = \mathbb{I}_{2\mathrm{x}2}\left[ik\int b\,db\,\left[e^{i\Delta_0(b)}\,\cos\big(kb\,\Delta_1(b)\big) - 1\right]\right.$$
$$\left.\times\left(\frac{1}{2\pi}\int_0^{2\pi} e^{ikb\cos\phi}\,d\varphi_{\mathbf{b}}\right)\right]. \tag{6.138}$$

Above, $\varphi_{\mathbf{b}}$ is the azimuthal angle associated with the variable $\mathbf{b}$, whereas ϕ is the angle between the vectors $\mathbf{q}$ and $\mathbf{b}$. With the neglect of longitudinal momentum transfer, both $\mathbf{q}$ and $\mathbf{b}$ are vectors parallel to the x-y plane. Thus, the angle between them is simply the difference $\phi = \varphi_{\mathbf{b}} - \varphi$, with φ being the azimuthal angle describing $\mathbf{q}$. Note that φ is also the azimuthal angle determining the scattering plane. This plane contains the z-axis and form an angle φ with the plane x-z. To evaluate the integral within round brackets, we change variable $\varphi_{\mathbf{b}} \to \phi = \varphi_{\mathbf{b}} - \varphi$. This transformation leaves the differential unchanged and the integration limits become $\{-\varphi, 2\pi - \varphi\}$. However, this change does not affect the result of the integration[14]. In this way, the integral over ϕ reduces to the integral representation of the Bessel function of order zero (Eq. (5.124)) and the non-spin-flip scattering amplitude becomes

$$f_{\mathrm{nsp}}(\theta) = \mathbb{I}_{2\mathrm{x}2}\left[ik\int b\,db\,\left[e^{i\Delta_0(b)}\,\cos\big(kb\,\Delta_1(b)\big) - 1\right]\,J_0(qb)\right], \tag{6.140}$$

where the momentum transfer can be expressed in terms of the scattering angle through the relation $q = 2k\sin(\theta/2)$.

We now turn to the spin-flip amplitude,

$$f_{\mathrm{sf}}(\theta,\varphi) = -\frac{k}{2\pi}\,\boldsymbol{\sigma}\cdot\left[\int b\,db\,e^{i\Delta_0(b)}\,\sin\big(kb\,\Delta_1(b)\big)\int \hat{\mathbf{A}}\,e^{iqb\cos\phi}\,d\varphi_{\mathbf{b}}\right]. \tag{6.141}$$

[14]Since $\cos(-\phi) = \cos(2\pi - \phi)$,

$$\frac{1}{2\pi}\int_{-\varphi}^{2\pi-\varphi} e^{ikb\cos\phi}\,d\phi = \frac{1}{2\pi}\int_0^{2\pi} e^{ikb\cos\phi}\,d\phi. \tag{6.139}$$

To evaluate the integral over $\varphi_{\mathbf{b}}$, we must determine $\hat{\mathbf{A}}$. First, we point out that the vectors $\mathbf{b}$ and $\mathbf{k}$ are orthogonal and thus we can write

$$\hat{\mathbf{A}} \equiv \frac{\mathbf{b} \times \mathbf{k}}{|\mathbf{b} \times \mathbf{k}|} = \hat{\mathbf{b}} \times \hat{\mathbf{k}}. \tag{6.142}$$

The unit vectors $\hat{\mathbf{k}}$ and $\hat{\mathbf{b}}$ are: $\hat{\mathbf{k}} = \hat{\mathbf{z}}$ (incident beam parallel to the z-axis) and $\hat{\mathbf{b}} = \cos\varphi_{\mathbf{b}}\ \hat{\mathbf{x}} + \sin\varphi_{\mathbf{b}}\ \hat{\mathbf{y}}$. The unit vector $\hat{\mathbf{A}}$ then is

$$\hat{\mathbf{A}} \equiv \hat{\mathbf{b}} \times \hat{\mathbf{k}} = (\cos\varphi_{\mathbf{b}}\ \hat{\mathbf{x}} + \sin\varphi_{\mathbf{b}}\ \hat{\mathbf{y}}) \times \hat{\mathbf{z}} = -\cos\varphi_{\mathbf{b}}\ \hat{\mathbf{y}} + \sin\varphi_{\mathbf{b}}\ \hat{\mathbf{x}}. \tag{6.143}$$

Using this result, the integral over $\varphi_{\mathbf{b}}$ of Eq. (6.141) becomes

$$\begin{aligned}\int_0^{2\pi} \hat{\mathbf{A}}\, e^{iqb\cos\phi}\ d\varphi_{\mathbf{b}} = -\hat{\mathbf{y}} &\int_0^{2\pi} e^{iqb\cos\phi}\ \cos\varphi_{\mathbf{b}}\ d\varphi_{\mathbf{b}} \\ +\hat{\mathbf{x}} &\int_0^{2\pi} e^{iqb\cos\phi}\ \sin\varphi_{\mathbf{b}}\ d\varphi_{\mathbf{b}}.\end{aligned} \tag{6.144}$$

Changing the integration variable to $\phi = \varphi_{\mathbf{b}} - \varphi$ and replacing

$$\begin{aligned}\cos\varphi_{\mathbf{b}} &= \cos(\phi + \varphi) = \cos\phi\ \cos\varphi - \sin\phi\ \sin\varphi \\ \sin\varphi_{\mathbf{b}} &= \sin(\phi + \varphi) = \sin\phi\ \cos\varphi + \cos\phi\ \sin\varphi,\end{aligned}$$

Eq. (6.144) takes the form

$$\begin{aligned}\int_0^{2\pi} \hat{\mathbf{A}}\, e^{iqb\cos\phi}\ d\varphi_{\mathbf{b}} = (-\cos\varphi\ \hat{\mathbf{y}} + \sin\varphi\ \hat{\mathbf{x}}) &\int_0^{2\pi} e^{iqb\cos\phi}\ \cos\phi\ d\phi \\ +(\sin\varphi\ \hat{\mathbf{y}} + \cos\varphi\ \hat{\mathbf{x}}) &\int_0^{2\pi} e^{iqb\cos\phi}\ \sin\phi\ d\phi.\end{aligned} \tag{6.145}$$

Again, we are keeping the original integration limits because the modifications induced by the variable change do not change the integral (see footnote on the previous page). It can easily be checked that

$$\int_0^{2\pi} e^{iqb\cos\phi}\ \sin\phi\ d\phi = 0. \tag{6.146}$$

and

$$\begin{aligned}\int_0^{2\pi} e^{iqb\cos\phi}\ \cos\phi\ d\phi &= -i\,\frac{d}{d(qb)}\left(\int_0^{2\pi} e^{iqb\cos\phi}\ d\phi\right) \\ &= -2\pi i\,\frac{d}{d(qb)} J_0(qb) = 2\pi i\, J_1(qb).\end{aligned} \tag{6.147}$$

The above result was obtained with the help of the property of Bessel functions: $dJ_0(x)/dx = -J_1(x)$. Inserting the integrals of Eqs. (6.146) and (6.147) in Eq. (6.145) we get the result of the integration over $\varphi_{\mathbf{b}}$:

$$\int_0^{2\pi} \hat{\mathbf{A}}\, e^{iqb\cos\phi}\, d\varphi_{\mathbf{b}} = 2\pi i\, [\sin\varphi\, \hat{\mathbf{x}} - \cos\varphi\, \hat{\mathbf{y}}]\, J_1(qb). \tag{6.148}$$

The vector within square brackets in Eq. (6.148) is orthogonal to the scattering plane. Comparing it with Eq. (6.93), we conclude that

$$\sin\varphi\, \hat{\mathbf{x}} - \cos\varphi\, \hat{\mathbf{y}} = -\hat{\mathbf{n}}. \tag{6.149}$$

Thus, we can write

$$\int_0^{2\pi} \hat{\mathbf{A}}\, e^{iqb\cos\phi}\, d\varphi_{\mathbf{b}} = -2\pi i\, J_1(qb)\, \hat{\mathbf{n}}. \tag{6.150}$$

Inserting this result in Eq. (6.141) we get the spin-flip amplitude

$$f_{\text{sf}}(\theta,\varphi) = i\,\boldsymbol{\sigma}\cdot\hat{\mathbf{n}}\, \left[k\int b\,db\; e^{i\Delta_0(b)}\, \sin\left(kb\,\Delta_1(b)\right)\, J_1(qb)\right]. \tag{6.151}$$

Adding Eqs. (6.140) and (6.151), the scattering amplitude takes the general form of Eq. (6.92), with

$$g(\theta) = ik\int b\,db\, \left[e^{i\Delta_0(b)}\, \cos\left(kb\,\Delta_1(b)\right) - 1\right]\, J_0(qb) \tag{6.152}$$

and

$$h(\theta,\varphi) = k\int b\,db\; e^{i\Delta_0(b)}\, \sin\left(kb\,\Delta_1(b)\right)\, J_1(qb). \tag{6.153}$$

6.3 Identical particles

If the projectile and the target are identical, Quantum Mechanics imposes a few changes in the results of the previous chapters. When a particle reaches a detector, the experiment cannot tell if this particle is the projectile or the target. On the other hand, momentum conservation guarantees that whenever one collision partner emerges at the orientation $\hat{\mathbf{r}} \equiv \{\theta,\varphi\}$ (in the center of mass frame), the other emerges at the opposite orientation, i.e., $-\hat{\mathbf{r}} \equiv \{\pi-\theta, \varphi+\pi\}$. Therefore, the cross sections for these orientations will be mixed in some way. From the point of view of classical mechanics, the cross section, which we denote σ_{inc}, is an incoherent sum of the cross sections for the corresponding scattering of discernible particles at these orientations[15]. That is[16]

$$\sigma_{\text{inc}}(\theta, \varphi) = \sigma(\theta, \varphi) + \sigma(\pi - \theta, \varphi + \pi). \tag{6.154}$$

Note that the total elastic scattering cross section obtained by angular integration of the above equation is twice that for the collision of discernible particle under the same conditions. This result is not surprising since σ_{inc} contains equal contributions from the projectile and from the target particles. The Quantum Mechanical description of such collisions is more complicated, as discussed below.

In Quantum mechanics, the total wave function for pairs of identical particles with integer spins, i.e., two bosons, must be symmetric with respect to the exchange of these particles. On the other hand, in the case of particles with half-integer spin, i.e., two fermions, the wave function must be anti-symmetric with respect to exchange. The wave function of a particle with spin can be expressed in terms of the set of coordinates $\{q \equiv \mathbf{r}, \mu\}$, where $\mathbf{r}$ is the position vector and μ is the spin component along the z-axis. In collisions of identical particles, the wave function must be an eigenstate of the exchange operator P_{PT} (subscripts p and t stand for *projectile* and *target*). This operator can be written

$$P_{\text{PT}} = P_{\text{PT}}^{\mathbf{r}} \, P_{\text{PT}}^{s} \,,$$

where $P_{\text{PT}}^{\mathbf{r}}$ exchanges $\mathbf{r}_{\text{P}} \leftrightarrows \mathbf{r}_{\text{T}}$ and P_{PT}^{s} exchanges $\mu_{\text{P}} \leftrightarrows \mu_{\text{T}}$. That is

$$P_{\text{PT}} \, \psi_\lambda(q_{\text{P}}, q_{\text{T}}) = \psi_\lambda(q_{\text{T}}, q_{\text{P}}) = \lambda \, \psi_\lambda(q_{\text{P}}, q_{\text{T}}),$$

with $\lambda = 1$ for boson-boson collisions and $\lambda = -1$ for fermion-fermion collisions. Symmetric $(\psi_+(q_{\text{P}}, q_{\text{T}}))$ or anti-symmetric $(\psi_-(q_{\text{P}}, q_{\text{T}}))$ wave functions can be generated with the help of the projector $\mathcal{A}_\lambda$, given by

$$\mathcal{A}_\lambda = \frac{1}{2} \left[1 + \lambda \, P_{\text{PT}}\right]. \tag{6.155}$$

That is,

$$\psi_+(q_{\text{P}}, q_{\text{T}}) = N \, \mathcal{A}_+ \, \psi(q_{\text{P}}, q_{\text{T}}) = \frac{1}{\sqrt{2}} \left[\psi(q_{\text{P}}, q_{\text{T}}) + \psi(q_{\text{T}}, q_{\text{P}})\right] \tag{6.156}$$

$$\psi_-(q_{\text{P}}, q_{\text{T}}) = N \, \mathcal{A}_- \, \psi(q_{\text{P}}, q_{\text{T}}) = \frac{1}{\sqrt{2}} \left[\psi(q_{\text{P}}, q_{\text{T}}) - \psi(q_{\text{T}}, q_{\text{P}})\right], \tag{6.157}$$

where $N = \sqrt{2}$ is an appropriate normalization factor.

[15]In our discussion, the cross-section $d\sigma_{\text{inc}}(\theta, \varphi)/d\Omega$ is only classical in the sense that the contributions from the target and the projectile to the cross section are summed incoherently. The cross sections $d\sigma(\theta, \varphi)/d\Omega$ and $d\sigma(\pi - \theta, \pi + \varphi)/d\Omega$ are evaluated by Quantum Mechanics.

[16]Henceforth, we adopt the compact notation: $\frac{d\sigma(\theta)}{d\Omega} \to \sigma(\theta)$.

6.3.1 *Collisions of identical spinless particles and particles with aligned spins*

We consider here the simple situation were the wave function in the spin space is symmetric with respect to projectile-target exchange. This condition is trivially satisfied when the identical particles have spin $s = 0$. A general discussion of the $s \neq 0$ case, which is more complicated, will be postponed to the next section. However, a simple situation is also encountered when both the projectile and the target are polarized with their spins aligned. The total spin is then $S = 2s$, and the spin wave function is symmetric with respect to exchange (see section 6.3.2). In this case, the spin degree of freedom can be ignored and the eigenvalue λ is given by the exchange symmetry in the coordinate space. The wave function should then be written

$$\psi_{\pm}(\mathbf{r}_{\rm P},\mathbf{r}_{\rm T}) = \frac{1}{\sqrt{2}}\left[\psi(\mathbf{r}_{\rm P},\mathbf{r}_{\rm T}) \pm \psi(\mathbf{r}_{\rm T},\mathbf{r}_{\rm P})\right], \tag{6.158}$$

with ' $+$ ' for bosons and ' $-$ ' for fermions.

The usual treatment of the scattering problem replaces the position vectors $\mathbf{r}_{\rm P}$ and $\mathbf{r}_{\rm T}$ by center of mass ($\mathbf{R}_{\rm c.m.}$) and relative ($\mathbf{r}$) coordinates, which are given by the transformations (see section 1.3.1),

$$\mathbf{r}_{\rm P};\, \mathbf{r}_{\rm T} \to \mathbf{R}_{\rm c.m.} = \frac{\mathbf{r}_{\rm P}+\mathbf{r}_{\rm T}}{2}; \quad \mathbf{r} = \mathbf{r}_{\rm P} - \mathbf{r}_{\rm T}, \tag{6.159}$$

Therefore, we should discuss the effects of exchange on these variables. Inspection of Eq. (6.159) indicates that they transform as

$$P^{\dagger}_{\rm PT}\,\mathbf{R}_{\rm c.m.}\,P_{\rm PT} = \mathbf{R}_{\rm c.m.} \tag{6.160}$$

$$P^{\dagger}_{\rm PT}\,\mathbf{r}\,P_{\rm PT} = -\,\mathbf{r}\,. \tag{6.161}$$

In this way, we have

$$P_{\rm PT} = \Pi\,.$$

Above, Π is the parity operator, which carries out space inversion of the vector $\mathbf{r}$. Eq. (6.158) then yields

$$\psi_{\pm}(\mathbf{r}) = \frac{1}{\sqrt{2}}\left[\psi(\mathbf{r}) \pm \psi(-\mathbf{r})\right]. \tag{6.162}$$

Therefore, the exchange symmetry requires that the wave function have even parity, $\pi = +$, in the case of bosons, or odd parity, $\pi = -$, in the case of fermions.

Eq. (6.162) leads to a modification of the scattering boundary condition of Eq. (2.3). Writing

$$\psi^{(+)}_{\pm}(\mathbf{r}) = \phi_{\pm}(\mathbf{r}) + \psi^{\rm sc}_{\pm}(\mathbf{r})\,,$$

the asymptotic forms of the incident plane wave and the scattered wave are respectively[17]

$$\phi_{\pm}(\mathbf{r}) \to \frac{1}{(2\pi)^{3/2}} \left[\frac{e^{i\mathbf{k}\cdot\mathbf{r}} \pm e^{-i\mathbf{k}\cdot\mathbf{r}}}{\sqrt{2}}\right] \tag{6.163}$$

and

$$\psi_{\pm}^{\mathrm{sc}}(\mathbf{r}) \to \frac{1}{(2\pi)^{3/2}} \left[\frac{f(\theta) \pm f(\pi-\theta)}{\sqrt{2}}\right] \frac{e^{ikr}}{r}. \tag{6.164}$$

Above, we have assumed that the potential is spherically symmetric, so that the wave function has axial symmetry and we can write $f(\hat{\mathbf{r}}) \equiv f(\theta,\varphi) = f(\theta)$ and $f(-\hat{\mathbf{r}}) = f(\pi-\theta)$. We now consider the implications of these modifications on the cross section. Inspecting Eq. (6.163), we see that there are two incident currents. One approaching from the left, arising from $\exp(i\mathbf{k}\cdot\mathbf{r})$, and one approaching from the right, arising from $\exp(-i\mathbf{k}\cdot\mathbf{r})$. The incident flux appearing in Eq. (1.68) should have contributions from both. Since the opposite signs of the currents are compensated by the opposite orientation of the normals, the two contributions are identical. In this way one gets a factor 2 that cancels the factor 1/2 arising from the $1/\sqrt{2}$ normalization of $\phi_{\pm}(\mathbf{r})$. Therefore, the incident flux is unchanged with respect to the value obtained in chapter 1. We now consider the flux scattered onto the detector. Comparing to section 1.5, one should make the replacement

$$|f(\theta)|^2 \to \left|\frac{f(\theta) \pm f(\pi-\theta)}{\sqrt{2}}\right|^2 = \frac{1}{2}\left|f(\theta) \pm f(\pi-\theta)\right|^2,$$

where f is the scattering amplitude for discernible particles under the same conditions. Since the factor 1/2 is cancelled by the factor 2 arising from the indiscernibility of projectile and target, discussed in the beginning of the present section, we get the cross section

$$\sigma_{\pm}(\theta) = |f_{\pm}(\theta)|^2 = |f(\theta) \pm f(\pi-\theta)|^2. \tag{6.165}$$

Above, we have introduced the notation

$$f_{\pm}(\theta) = f(\theta) \pm f(\pi-\theta). \tag{6.166}$$

Evaluating Eq. (6.165) and comparing with Eq. (6.154), we can write

$$\sigma_{\pm}(\theta) = \sigma_{\mathrm{inc}}(\theta) + \sigma_{\mathrm{int}}(\theta), \tag{6.167}$$

where $\sigma_{\mathrm{int}}(\theta)$ is the interference term

$$\sigma_{\mathrm{int}}(\theta) = \pm 2\,\mathrm{Re}\Big\{f^*(\theta)\, f(\pi-\theta)\Big\}. \tag{6.168}$$

[17]Henceforth, we are adopting the normalization constant $A = (2\pi)^{-3/2}$.

This term, which has no classical analog, plays a very important role. It has a strong spin-dependence, whereas $\sigma_{\rm inc}(\theta)$ is independent of the spin. This point will be discussed in the next section.

Direct inspection of Eqs. (6.154), (6.167) and (6.168), indicates that the cross sections $\sigma_{\rm inc}(\theta)$, $\sigma_{\rm int}(\theta)$ and $\sigma_{\pm}(\theta)$ are symmetric with respect to $\theta = \pi/2$. This feature is connected with an important property of the partial-wave expansion of $f_{\pm}(\theta)$. For simplicity, we discuss the case of scattering from a short-range spherically symmetric potential.

Inserting Eq. (2.57b) into Eq. (6.166) and using the property of Legendre Polynomials

$$P_l(\cos(\pi-\theta)) = P_l(-\cos\theta) = (-)^l P_l(\cos\theta),$$

we get

$$f_{\pm}(\theta) = \frac{1}{2ik}\sum_l \Big[(2l+1)\,P_l(\cos\theta)\,(S_l-1)\Big] \times \Big[1 \pm (-)^l\Big]. \tag{6.169}$$

This result is analogous to Eq. (2.57b), except for the factor $\left[1 \pm (-)^l\right]$ within the partial-wave summation. In boson-boson (fermion-fermion) collisions, this factor doubles the contribution from even-waves (odd-waves) and eliminates those from odd-waves (even-waves). This is the justification for excluding odd-waves in the calculations for identical spinless bosons shown in figures 3.2 and 3.5.

6.3.1.1 *Coulomb scattering*

Let us apply the results of the previous section to the case of Coulomb scattering. We first consider in detail collisions of identical spinless bosons. In this case $\sigma_+(\theta)$ is known as the Mott cross section and for this reason we use the notation $\sigma_{\rm M}(\theta)$. We must evaluate

$$\sigma_{\rm M}(\theta) = \left|f_{\rm R}(\theta) + f_{\rm R}(\pi-\theta)\right|^2, \tag{6.170}$$

using the analytical expression for the Rutherford scattering amplitude, $f_{\rm R}$ (Eq. (3.29)). Writing the cross section as in Eq. (6.167) and normalizing it with respect to the Rutherford cross section at $\theta = 90^{\rm o}$, we get[18]

$$\bar\sigma_{\rm M}(\theta) \equiv \frac{\sigma_{\rm M}(\theta)}{\sigma_{\rm R}(\pi/2)} = \bar\sigma_{\rm inc}(\theta) + \bar\sigma_{\rm int}(\theta), \tag{6.171}$$

with

$$\bar\sigma_{\rm inc}(\theta) \equiv \frac{\sigma_{\rm inc}(\theta)}{\sigma_{\rm R}(90^{\rm o})} = \frac{1}{4}\left[\frac{1}{\sin^4(\theta/2)} + \frac{1}{\cos^4(\theta/2)}\right] \tag{6.172}$$

[18]We use the notation $\bar\sigma(\theta)$ for the cross section normalized with respect to $\sigma_{\rm R}(90^{\rm o})$.

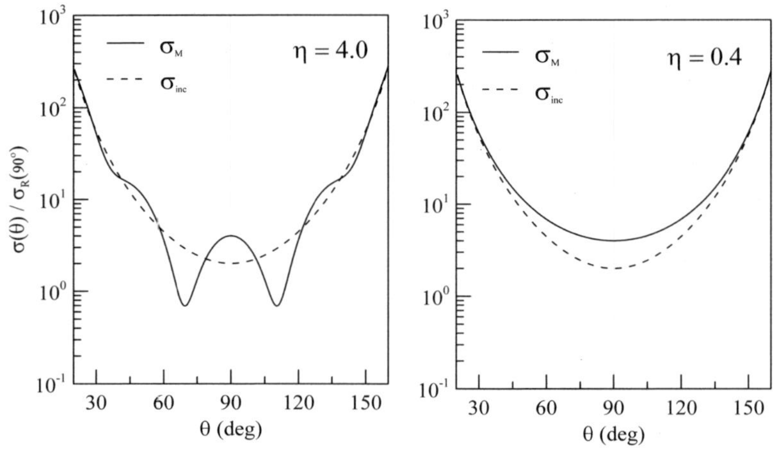

Fig. 6.1 Comparison between the Mott (solid lines) and the incoherent (dashed lines) cross sections, for two values of the Sommerfeld parameter. The cross sections are normalized with respect to the Rutherford cross section at $\theta = 90°$.

and

$$\bar{\sigma}_{\rm int}(\theta) \equiv \frac{\sigma_{\rm int}(\theta)}{\sigma_{\rm R}(90^{\rm o})} = \frac{1}{4}\left[2\,\frac{\cos\left[2\eta\ln\left(\tan(\theta/2)\right)\right]}{\sin^2(\theta/2)\,\cos^2(\theta/2)}\right].$$

Note that $\bar{\sigma}_{\rm inc}(\theta)$ is energy-independent, whereas $\bar{\sigma}_{\rm int}(\theta)$ depends on the collision energy through the Sommerfeld parameter. This energy-dependence is carried over to $\bar{\sigma}_{\rm M}(\theta)$. This is illustrated in figure 6.1. The normalized cross sections are shown for the values of the Sommerfeld parameter $\eta = 4.0$ and $\eta = 0.4$. Inspecting this figure, one notices two interesting points. The first is that the values of the normalized cross sections at $\theta = 90^{\rm o}$, are independent of η. They are: $\bar{\sigma}_{\rm M}(90^{\rm o}) = 4$ and $\bar{\sigma}_{\rm inc}(90^{\rm o}) = 2$. The second is that the shape of $\bar{\sigma}_{\rm M}(\theta)$ depends on the collision energy. For $\eta = 4.0$ it has a maximum at $\theta = 90^{\rm o}$ while for $\eta = 0.4$ it has a minimum. This behavior can be understood if we separately analyze the two terms contributing to $\bar{\sigma}_{\rm M}$ in Eq. (6.171). The incoherent cross section (dashed lines in figure 6.1) has always a minimum at $\theta = 90^{\rm o}$. On the other hand, at this angle the amplitudes $f(\theta)$ and $f(180^{\rm o} - \theta)$ interfere constructively, so that the interference term has a maximum. The shape of $\bar{\sigma}_{\rm M}(\theta)$ depends on which of these trends dominates. If the interference term is slowly varying, which happens for small η values, the incoherent part dominates and $\bar{\sigma}_{\rm M}(\theta)$ has a minimum at $90^{\rm o}$. The opposite situation occurs when η is large. This condition can be quantitatively stated if we evaluate the second derivative

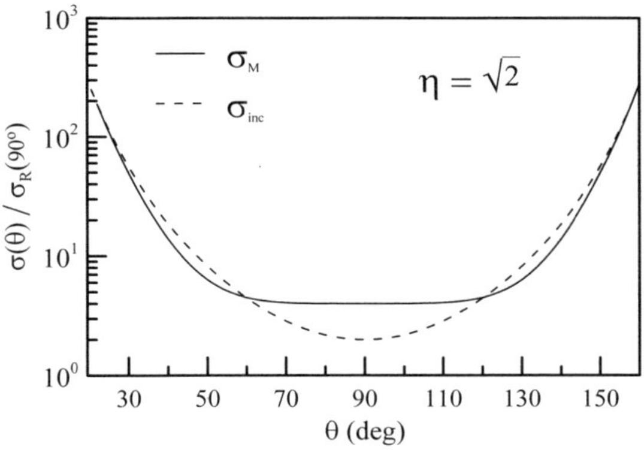

Fig. 6.2 Same as figure 6.1 but for the transitional value $\eta = \sqrt{2}$.

of $\bar{\sigma}_{\rm M}(\theta)$ with respect to θ, at $\theta = 90°$. We find[19]

$$\left[\frac{d^2\bar{\sigma}_{\rm M}(\theta)}{d\theta^2}\right]_{\theta=\pi/2} = 8\,\left(2-\eta^2\right)\,.$$

The above equation indicates that the transition from a minimum to a maximum occurs at $\eta = \sqrt{2}$. This situation is illustrated in figure 6.2. Note that the cross section in this case becomes remarkably flat over a broad angular interval around $\theta = 90°$.

A better insight of the influence of the interference term on the Mott cross section can be achieved if one plots $\sigma_{\rm M}$ normalized with respect to $\sigma_{\rm inc}(\theta)$, instead of to $\sigma_{\rm R}(90°)$. This is done in figure 6.3. In this case, there is a maximum at $\theta = 90°$ independently of the η-value. Thus, a maximum of $\sigma/\sigma_{\rm inc}$ ratio at $\theta = 90°$ could be taken as a signature for collisions of identical charged bosons.

We now extend the results of this section to collisions of identical fermions with aligned polarizations. In this case, the incoherent cross section remains the same but $\sigma_{\rm int}$ changes sign. This change leads to an important consequence: the cross section associated with the antisymmetric wave function, which we here denote by $\sigma_{\rm A}(\theta)$, has always a minimum at $\theta = 90°$. It can be readily checked that the cross section at this minimum is zero. This situation is illustrated in figure 6.4, for the same values of the Sommerfeld parameter of the previous two figures.

[19]This lengthy derivative was evaluated with the help of an algebraic computer program.

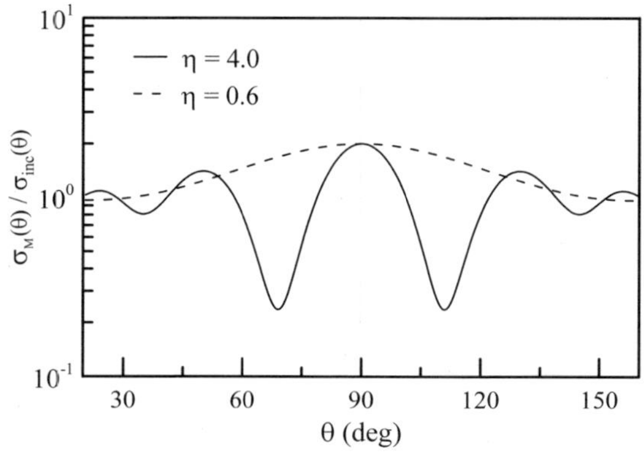

Fig. 6.3 The ratio between the Mott and the incoherent cross sections for the same collision energies as in figure 6.1.

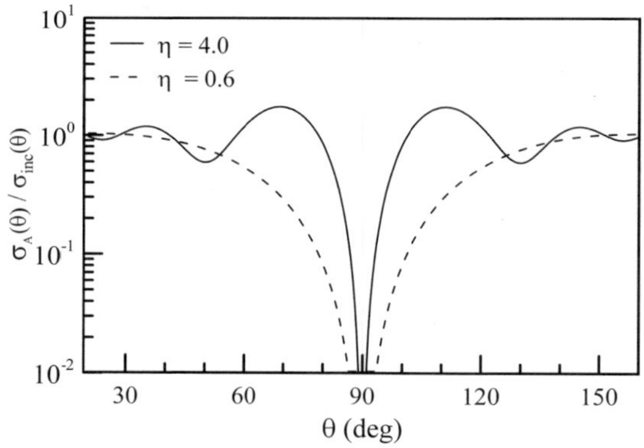

Fig. 6.4 The ratio between the Mott and the incoherent cross sections for a fermion-fermion collision with aligned spins, for two values of the Sommerfeld parameter.

6.3.2 *Collisions of identical particles with spin*

If the identical particles have spin, the exchange operator is no longer equivalent to the parity (space reflection) operator. In this case we have

$$P_{\rm PT} = \Pi \, P^s_{\rm PT}. \tag{6.173}$$

An important consequence of this equation is that there are always two ways to get symmetric or anti-symmetric wave functions. Since the eigenvalue λ is the product of the parity, π, with the eigenvalue of the exchange operator in spin space, λ_s, we may have:

boson-boson collisions

$$\lambda = + \quad \Rightarrow \left\{ \begin{array}{c} \pi = +, \quad \lambda_s = + \\ \text{or} \\ \pi = -, \quad \lambda_s = - \end{array} \right\}$$

fermion-fermion collisions

$$\lambda = - \quad \Rightarrow \left\{ \begin{array}{c} \pi = +, \quad \lambda_s = - \\ \text{or} \\ \pi = -, \quad \lambda_s = + \end{array} \right\} .$$

In this way, the parity of the wave function in the position space, which determines the angular distribution, depends on the symmetry of the wave function in the spin space. We discuss this point in detail below.

Let us consider the collision of two identical particles with spin s and call μ_{P} and μ_{T} the spin projections of the projectile and the target, respectively. The exchange symmetry in spin space is determined by the quantum number S, associated with the total intrinsic angular momentum operator, $\mathbf{S}^2 = (\mathbf{s}_{\text{P}} + \mathbf{s}_{\text{T}})^2$. The eigenstates of the operators $\mathbf{S}^2$ and S_z, $|\chi_{SM}\rangle$, are linear combinations of products $|\chi_{s\nu}\rangle_{\text{P}} |\chi_{s\nu'}\rangle_{\text{T}}$, where $|\chi_{s\nu}\rangle_{\text{P}}$ and $|\chi_{s\nu'}\rangle_{\text{T}}$ are respectively eigenstates of the pairs of operators $\left\{\mathbf{s}^2, s_z\right\}_{\text{P}}$ and $\left\{\mathbf{s}^2, s_z\right\}_{\text{T}}$, with Clebsch-Gordan coefficients. That is,

$$|\chi_{SM}\rangle = \sum_{\nu\nu'} \langle s\nu s\nu'|\, SM\rangle \; |\chi_{s\nu}\rangle_{\text{P}} |\chi_{s\nu'}\rangle_{\text{T}} . \tag{6.174}$$

Since the Clebsch-Gordan Coefficients have the property (see, e.g., [Edmonds (1974)])

$$\langle s\nu' s\nu|\, SM\rangle = (-1)^{2s+S} \langle s\nu s\nu'|\, SM\rangle \tag{6.175}$$

and projectile-target exchange in spin space amounts to exchanging the dummy indices ν and ν', the eigenvalues λ_s are

$$\lambda_s = (-1)^{2s+S} . \tag{6.176}$$

Therefore, we have

$$\begin{aligned} &\textit{bosons}\,(2s = \textit{even}) : \\ &\quad\Rightarrow S = \textit{even} \rightarrow \lambda_s = +, \quad \pi = + \\ &\quad\Rightarrow S = \textit{odd} \;\; \rightarrow \lambda_s = -, \quad \pi = - \end{aligned} \tag{6.177}$$

$$\begin{aligned} &\textit{fermions } (2s = odd): && (6.178)\\ &\qquad \Rightarrow S = even \rightarrow \lambda_s = -, \quad \pi = + \\ &\qquad \Rightarrow S = odd \;\; \rightarrow \lambda_s = +, \quad \pi = - . \end{aligned}$$

In general, one cannot control the eigenvalue S in an experiment. One can polarize both the projectile and the target but this usually leads to admixtures of even and odd values of S. There is however one exception. When the polarization is such that $\nu = \nu' = \pm s$, the only non-vanishing Clebsch-Gordan coefficient has $M = \pm 2s$. Therefore the total spin quantum number is $S = 2s$ and, according to Eqs. (6.177) and (6.178), $\lambda_s = +1$. In this way, one must have $\pi = +$ for bosons and $\pi = -$ for fermions. This simple case was considered in section (6.3.1).

If the system is not polarized as above, proper averages must be taken. Let us first consider the inclusive cross section in the scattering of unpolarized identical particles. In this case, the initial spins are randomly oriented and the cross section is summed over final spins. In this way, each of the N possible spin state $|\chi_{SM}\rangle$, with $S = 0, .., 2s$ and $M = -S, .., S$, contributes to the cross section with the same weight, $1/N$. As we have seen, such contributions are independent of M and depends on S exclusively through the parity. That is, states with eigenvalue $\lambda_s = +1$ have $\pi = \lambda$ while those with $\lambda_s = -1$ have $\pi = -\lambda$. Calling respectively N_+ and N_- the number of states with eigenvalues $\lambda_s = +1$ and $\lambda_s = -1$ and using Eqs. (6.167) and (6.168), the cross section is given by the average

$$\langle \sigma_\lambda(\theta) \rangle = \sigma_{\rm inc}(\theta) + \lambda \left(\frac{N_+ - N_-}{N} \right) \sigma_{\rm int}(\theta) \,. \qquad (6.179)$$

Since above the exchange symmetry, λ, appears explicitly, the interference cross section should be that of Eq. (6.168) with positive sign.

The total number of states can easily be calculated. It is given by the sum of the arithmetic series

$$N = 1 + 3 + 5 + ... + (2S_{\rm max} + 1) = 1 + 3 + 5 + ... + (4s + 1) = (2s + 1)^2 \,. \qquad (6.180)$$

N_+ and N_- are also sums of arithmetic series. That is

$$\begin{aligned} N_+ &= 1 + 5 + ... + (4s + 1); \qquad \text{for } \lambda = +1 \\ &= 3 + 7 + ... + (4s + 1)\,; \qquad \text{for } \lambda = -1 \end{aligned}$$

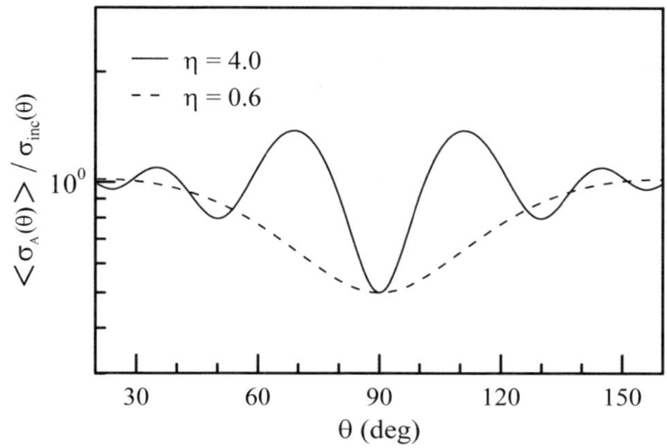

Fig. 6.5 Unpolarized proton-proton cross sections, normalized as in figure 6.3.

and

$$\begin{aligned} N_- &= 3 + 7 + \dots + (4s - 1); \quad \text{for } \lambda = +1 \\ &= 1 + 5 + \dots + (4s - 1); \quad \text{for } \lambda = -1 \end{aligned}$$

Evaluating the series for N_+ and N_-, we obtain

$$\frac{N_+}{N} = \frac{s+1}{2s+1}; \quad \frac{N_-}{N} = \frac{s}{2s+1}, \tag{6.181}$$

and the cross section of Eq. (6.179) becomes

$$\langle \sigma_\lambda(\theta) \rangle = \sigma_{\text{inc}}(\theta) + \lambda \, \frac{\sigma_{\text{int}}(\theta)}{2s+1} \, . \tag{6.182}$$

As an illustration, we give in figure 6.5 the cross section for the collision of unpolarized protons for the same values of the Sommerfeld parameter as those in figure 6.4. The results are normalized with respect to the incoherent cross section. In this case, there are two possible values of the total spin. The singlet, with $S = 0$, and the triplet, with $S = 1$. The former, with $\lambda_s = -1$ and positive parity, contributes with the weight $1/4$ (see Eq. (6.181)). The latter, with $\lambda_s = 1$ and negative parity, has weight $3/4$. Since the cross section is dominated by the triplet, which has larger weight, it has a minimum at $\theta = \pi/2$. However, this fermionic signature is much less pronounced then in the case of aligned polarization, where only the triplet contributes. Note that $\langle \sigma_-(\theta) \rangle / \bar{\sigma}(\theta)$ remains finite at $\pi/2$, owing to the contribution from the singlet.

6.3.3 *Clusters of identical fermions*

In typical atomic and nuclear collisions the projectile and/or the target are composite particles. Usually, they are clusters of fermions[20]. In this way, the Pauli Principle requires that the wave function be anti-symmetric with respect to the exchange of any two identical fermions, either in the same cluster or in different ones. If the projectile contains $N_{\rm P}$ fermions with coordinates $\mathbf{r}_1, \mathbf{r}_2, ..., \mathbf{r}_{\rm N_P}$ and the target contains $N_{\rm T}$ particles with position vectors $\mathbf{r}_{\rm N_P+1}, \mathbf{r}_{\rm N_P+2}, ..., \mathbf{r}_{\rm N_P+N_T}$, the system's wave function can be written

$$\Psi(\mathbf{r}_1, ..., \mathbf{r}_{\rm N_P}; \mathbf{r}_{\rm N_P+1}, ..., \mathbf{r}_{\rm N}) = \mathcal{A}\left\{\Phi_{\rm P}(\mathbf{r}_1, ..., \mathbf{r}_{\rm N_P})\ \Phi_{\rm T}\left(\mathbf{r}_{\rm N_P+1}, ..., \mathbf{r}_N\right)\right\}. \tag{6.183}$$

Above, $\Phi_{\rm P}$ and $\Phi_{\rm T}$ are respectively the anti-symmetrized intrinsic wave functions of the projectile and the target, and $N = N_{\rm P}+N_{\rm T}$. $\mathcal{A}$ is the many-body anti-symmetrization operator, which accounts for the $N_{\rm P}!N_{\rm T}!/N!$ different ways of permuting fermions between the two clusters. It can written

$$\mathcal{A} = \sum (-)^{p_\alpha}\, P_\alpha\,, \tag{6.184}$$

where the operator P_α exchanges fermion coordinates until the permutation α is reached. The sign $(-)^{p_\alpha}$ is positive for permutations obtained through an even number of fermion pair exchanges, and negative otherwise.

Solving the Schrödinger equation for the many-body scattering problem is, in most cases, a hopeless task. However, the effects of Pauli Principle can be included in approximations like Wheeler's Resonating Group Method (RGM) [Wheeler (1937); Friedrich (1981)], or the essentially equivalent Generator Coordinate Method (GCM) [Hill and Wheeler (1953); Griffin and Wheeler (1957)]. These methods avoid the complications of a Coupled-Channel calculation, approximating the intrinsic wave functions of the projectile and the target by the Slater determinants corresponding to their ground states when they are far apart. In this way, the intrinsic states of the projectile and the target are kept frozen along the collision.

To derive the RGM equation, one introduces the coordinates of the projectile's and target's centers of mass,

$$\mathbf{r}_{\rm P} = \frac{1}{N_{\rm P}}\left(\mathbf{r}_1 + ... + \mathbf{r}_{\rm N_P}\right); \quad \mathbf{r}_{\rm T} = \frac{1}{N_{\rm T}}\left(\mathbf{r}_{\rm N_P+1} + ... + \mathbf{r}_{\rm N}\right), \tag{6.185}$$

[20]In the case of nuclei, there are two different fermions: protons and neutrons. However, with the isotopic spin formalism they are treated as different states of the same particle - the nucleon.

and the intrinsic coordinates[21]

$$\xi_i = \mathbf{r}_i - \mathbf{r}_{\rm P}\,; \quad i = 1, .., N_{\rm P}$$
$$\xi_j = \mathbf{r}_j - \mathbf{r}_{\rm T}\,; \quad j = N_{\rm P} + 1, .., N\,, \tag{6.186}$$

and then adopts the ansatz

$$\Psi(\mathbf{r}_1, ..., \mathbf{r}_{\rm N}) = \Psi_{\rm CM}(\mathbf{R}_{\rm c.m.}) \cdot \mathcal{A}\left\{\psi(\mathbf{r})\, \Phi_{\rm P}(\xi_1, ..., \xi_{\rm N_P})\, \Phi_{\rm T}\left(\xi_{\rm N_P+1}, ..., \xi_{\rm N}\right)\right\}. \tag{6.187}$$

Above, $\mathbf{R}_{\rm c.m.}$ and $\mathbf{r}$ are respectively the center of mass and the projectile-target separation vectors, and $\psi(\mathbf{r})$ is the relative wave function, which contains the relevant information for the scattering cross section. This wave function is determined by the variational condition

$$\delta\left[\frac{\langle\Psi|\, H\, |\Psi\rangle}{\langle\Psi\, |\Psi\rangle}\right] = 0, \tag{6.188}$$

where H is the many-body collision Hamiltonian

$$H = \left[-\frac{\hbar^2}{2\mu}\sum_{i=1}^{N}\nabla^2_{\mathbf{r}_i}\right] + \left[\sum_{i>j=1}^{N} v(\mathbf{r}_i - \mathbf{r}_j)\right] - T_{\rm CM}. \tag{6.189}$$

It should be understood that the variations in Eq. (6.188), are taken through changes of the relative wave function of Eq. (6.187).

For the implementation of the RGM, one rewrites the ansatz of Eq. (6.187) in the form,

$$\Psi = \int d^3\mathbf{r}''\, \psi(\mathbf{r}'')\, \Phi_{\mathbf{r}''}, \tag{6.190}$$

with

$$\Phi_{\mathbf{r}''} = \Psi_{\rm CM}(\mathbf{R}_{\rm c.m.}) \cdot \mathcal{A}\left\{\delta(\mathbf{r} - \mathbf{r}'')\, \Phi_{\rm P}(\xi_1, ..., \xi_{\rm N_P})\, \Phi_{\rm T}\left(\xi_{\rm N_P+1}, ..., \xi_{\rm N}\right)\right\}, \tag{6.191}$$

and inserts it into the variational principle. Using the property $\mathcal{A}^2 = \mathcal{A}$, one obtains the the RGM equation (for details, see [Friedrich (1981)]),

$$\left[-\frac{\hbar^2}{2\mu}\nabla^2_{\mathbf{r}} + V_{\rm D}(\mathbf{r})\right]\psi(\mathbf{r}) + \int d^3\mathbf{r}'\, V_{\rm E}(\mathbf{r}, \mathbf{r}')\, \psi(\mathbf{r}') = E\, \psi(\mathbf{r}). \tag{6.192}$$

[21]Note that these intrinsic coordinates are not linearly independent as they are related by the constraints $\sum_P \xi_i = \sum_T \xi_j = 0$. This feature must be taken into account in the evaluation of the RGM potentials which will be introduced below.

Above, $V_{\rm D}(\mathbf{r})$ is the double-folding potential

$$V_{\rm D}(\mathbf{r})\,\delta(\mathbf{r}-\mathbf{r}') = \int d\xi_{\rm P}\, d\xi_{\rm T}\, d^3\mathbf{r}'' \left\{\Phi_{\rm P}(\xi_{\rm P})\,\Phi_{\rm T}\,(\xi_{\rm T})\,\delta(\mathbf{r}-\mathbf{r}'')\right\} \times \mathcal{V}(\mathbf{r}'',\xi_{\rm P},\xi_{\rm T}) \left\{\Phi_{\rm P}(\xi_{\rm P})\,\Phi_{\rm T}\,(\xi_{\rm T})\,\delta(\mathbf{r}'-\mathbf{r}'')\right\}, \quad (6.193)$$

where

$$\mathcal{V}(\mathbf{r},\,\xi_{\rm P},\xi_{\rm T}) = \sum_{i=1}^{N_{\rm P}} \sum_{j=N_{\rm P}+1}^{N} v(\mathbf{r}_i - \mathbf{r}_j),$$

and $V_{\rm E}(\mathbf{r},\mathbf{r}')$ is the non-local energy-dependent potential[22]

$$V_{\rm E}(\mathbf{r},\mathbf{r}') = \int d\xi_{\rm P}\, d\xi_{\rm T}\, d^3\mathbf{r}'' \left\{\,\Phi_{\rm P}(\xi_{\rm P})\,\Phi_{\rm T}\,(\xi_{\rm T})\,\delta(\mathbf{r}-\mathbf{r}'')\,\right\} \times (H-E)\;(\mathcal{A}-1)\left\{\Phi_{\rm P}(\xi_{\rm P})\,\Phi_{\rm T}\,(\xi_{\rm T})\,\delta(\mathbf{r}'-\mathbf{r}'')\right\}, \quad (6.194)$$

arising from exchange.

Although the direct potential can be calculated without major difficulties, the exchange part is very complicated. The difficulty arises from the fact that as $\mathcal{A}$ exchange particles from the projectile with those from the target, the relative coordinate in the delta functions becomes different combinations of the single particle coordinates. To handle this problem, specific technics have been developed (see, for example, [Friedrich (1981)] and references therein). In typical situations, the exchange potential has a short range. However, when the projectile and the target are identical clusters of fermions, the exchange potential contains a local term. It arises from the permutation $P_{\rm tot}$, where all projectile's fermions are exchanged with those of the target. In this case, it is convenient to include this term in $V_{\rm D}(\mathbf{r})$, replacing in Eq. (6.194) $(\mathcal{A}-1) \to (\mathcal{A}-1-P_{\rm tot})$. This leads to the replacement

$$\psi(\mathbf{r}) \to \frac{1}{\sqrt{2}}\left(\psi(\mathbf{r}) + \lambda\psi(-\mathbf{r})\right),$$

where λ gives the symmetry for the total projectile-target exchange ($\lambda = +1$ for even $N_{\rm P}$ and $\lambda = -1$ for odd $N_{\rm P}$). The relevance of $V_{\rm E}(\mathbf{r},\mathbf{r}')$ is illustrated in figure 6.6, in an example of Nuclear Physics: the scattering of two alpha particles. We compare the results of the RGM, which includes both the local and the non-local potentials, with those obtained approximating the alpha particles by identical bosons. In the latter case, only the projectile-target exchange, corresponding to the action of $P_{\rm tot}$, is taken into account.

[22]We use the short hand notations $\xi_{\rm P} \equiv \{\xi_1, ..., \xi_{\rm N_P}\}$ and $\xi_{\rm T} \equiv \{\xi_{\rm N_P+1}, ..., \xi_{\rm N}\}$.

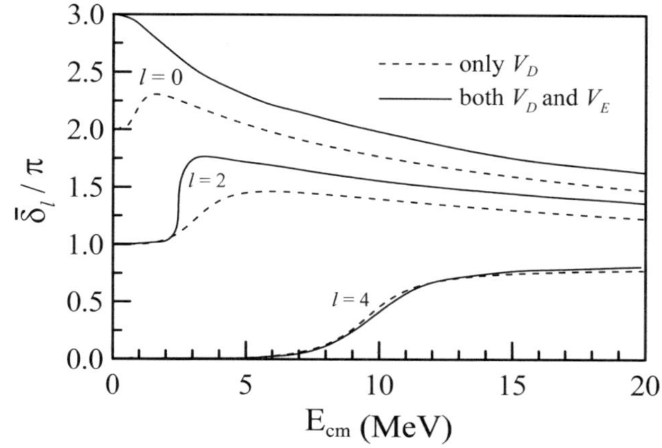

Fig. 6.6 Phase shift in α-α scattering, treated as a collision of identical clusters of fermions (solid lines) and as a collision of identical structureless bosons (dashed lines). The results are normalized with respect to π and plotted according to the generalized Levinson's theorem (see footnote 8 of chapter 3).

The figure shows the phase shifts arising from the short-range term in the potential, $\bar{\delta}_l$ (see chapter 3), for the partial waves $l = 0, 2$ and 4, and collision energies in the range $0 - 20$ MeV. The RGM results, for a Volkov's V1 [Volkov (1965)] nucleon-nucleon interaction[23] and Harmonic Oscillators orbitals, are taken from [Brink and Canto (1977)]). One notices that the non-local potential can be neglected for $l = 4$ (or higher partial-waves), but it is very important for $l = 0$ and 2. Collisions at the lowest partial-waves ($l = 0$ and 2) are dominated by small projectile-target separations, where the single particle orbitals in the two clusters becomes very similar. In this way, the Pauli Principle is violated, unless the wave function if fully anti-symmetrized. Since the exchange potential, which takes care of this effect, is neglected in the *boson-boson* approximation (dashed lines), the agreement with the RGM calculation becomes very poor.

Exercises

(1) Consider the collision of a spin-$1/2$ projectile with a spin-zero target, interacting through the potential $V(r) = V_0(r) + V_{so}(r)\,\mathbf{L}\cdot\mathbf{s}$, with $\mathbf{L}$ and

[23] The Volkov potential is a sum of Gaussians with parametes adjusted to fit experimental nuclear data. For details see [Volkov (1965)].

s representing respectively the orbital and the spin angular momenta, measured in $\hbar$ units. Show that the matrix elements of the interaction are given by (using the short-hand notation: $|\mathcal{Y}^{lJM}\rangle = |lJM\rangle$)

$$\langle lJM|\, V\, |l'J'M'\rangle = \left[V_0(r) + \frac{l}{2}\, V_{\rm so}(r)\right]\, \delta_{ll'}\, \delta_{JJ'}\, \delta_{MM'},$$

for total angular momentum $J = l + {}^1\!/_2$, and

$$\langle lJM|\, V\, |l'J'M'\rangle = \left[V_0(r) - \frac{l+1}{2}\, V_{\rm so}(r)\right]\, \delta_{ll'}\, \delta_{JJ'}\, \delta_{MM'},$$

for $J = l - {}^1\!/_2$.

(2) Show that, in the collisions of the previous problem,

$$\Lambda_l^+ = \frac{1}{2l+1}\, \left[l + 1 + \boldsymbol{\sigma}\cdot\mathbf{L}\right]$$

and

$$\Lambda_l^- = \frac{1}{2l+1}\, \left[l - \boldsymbol{\sigma}\cdot\mathbf{L}\right]$$

are projection operators onto eigenstates of total angular momentum, with eigenvalues $J = l \pm {}^1\!/_2$.

(3) Prove that in collisions of an unpolarized beam of spin=${}^1\!/_2$ projectiles with a spin-zero target, the polarization vector is perpendicular to the scattering plane.

(4) Show that in collisions of one spin-${}^1\!/_2$ particle with a spin-zero one, the scattering amplitude in the two-dimensional spinor space can be written as the matrix

$$f_{2\text{x}2}(\theta) = g(\theta)\, \mathbb{I}_{2\text{x}2} + i\, h(\theta)\; \boldsymbol{\sigma}\cdot\hat{\mathbf{n}},$$

with the amplitudes $g(\theta)$ and $h(\theta)$given by Eqs. (6.86) and (6.87).

(5) Use the optical theorem to show that the total cross section for collisions of one neutral spin-${}^1\!/_2$ projectile with a spin-zero target can be written as,

$$\sigma_T(E) = \frac{4\pi}{k^2} \sum_l \left[(l+1)\, \sin^2\delta_l^+ + l\, \sin^2\delta_l^-\right],$$

where the $\delta_l^{\pm}$ are the phase shift at the lth partial-wave for $J = l \pm {}^1\!/_2$.

(6) Consider a small perturbation to the point charge Coulomb potential, arising from the dipole polarizability of the target,

$$V_d(r) = -\frac{C}{r^4}.$$

Assume that this interaction is very weak, so that the scattering amplitude is dominated by its near-side component, and that the semiclassical approximation is applicable.

a) Neglecting the influence of V_d on the classical trajectory and on the modulus of the cross section, show that the scattering amplitude can be written in the general form of Eq. (5.295),

$$f(\theta) = \sqrt{\frac{d\sigma_{\mathrm{R}}(\theta)}{d\Omega}}\, e^{i\alpha(\theta)}, \quad \text{with} \quad \alpha(\theta) \simeq \alpha_{\mathrm{C}}(\theta) + \alpha_d(\theta).$$

Above, $\alpha_{\mathrm{C}}(\theta)$ is a phase arising purely from the Coulomb field: $\alpha_{\mathrm{C}}(\theta) \equiv 2\sigma(\lambda) - \lambda\,\theta - \pi/2 \simeq 2\sigma_0 - \eta \ln\left[\sin(\theta/2)\right] + \phi_0$, where λ is the angular momentum associated with θ through the Coulomb trajectory, $\sigma(\lambda)$ is the corresponding Coulomb phase-shift and ϕ_0 is an irrelevant constant phase. The term $\alpha_d(\theta)$ is the correction due to the potential V_d, given by

$$\alpha_d(\theta) = \frac{\eta\, C}{4Ea^4} \tan^4\frac{\theta}{2}\left[\left(\cot^2\frac{\theta}{2} + 3\right)\left(\frac{\pi-\theta}{2}\right)\tan\frac{\theta}{2} - 3\right],$$

where a is one-half of the distance of closest approach in a head-on collision[24].

b) Assuming that the projectile and the target are identical particles, evaluate the Mott cross section and determine the shifts in the maxima and minima owing to the presence of V_d[25].

---

[24] *Hint*: Evaluate the phase-shift by the WKB approximation, expanding the local wave number around $V_d = 0$. In this way, the contributions from the V_{C} and V_d become additive. Next, use the approximate expressions for $\sigma(\lambda)$ and σ_0 of chapters 3 and 5.

[25] *Hint*: From the interference term,

$$\cos\left[2\left(\alpha_{\mathrm{C}}(\theta) + \alpha_d(\theta) - \alpha_{\mathrm{C}}\left(\pi-\theta\right) - \alpha_d\left(\pi-\theta\right)\right)\right] = \cos\Delta(\theta) \tag{6.195}$$

find the period P_θ, as,

$$\Delta(\theta + P_\theta) - \Delta(\theta) = 2\pi \tag{6.196}$$

which can be reduced to,

$$P_\theta = \frac{2\pi}{d\Delta(\theta)/d\theta}. \tag{6.197}$$

Chapter 7

Scattering from Complex Potentials

So far we have discussed the scattering of structureless particles off real potentials. In most applications in the physical world, one deals with collisions of molecular, atomic or nuclear system, where the process of scattering is very complicated. Such situations require the introduction of new concepts, which will be expanded upon in the second part of this book. A first step in this direction is the generalization of the interaction potential into a complex one. This potential enables us to describe some features of the elastic scattering of composite particles. The imaginary part of this interaction mocks up the loss of flux into non-elastic processes. A fuller description of the complex potential (usually called the Optical Potential) using a microscopic treatment of the collision process will be developed later. In the following, we simply take the interaction to be complex and seek the consequences in so far as the elastic scattering observables are concerned. We also generalize the formal scattering theory (T-matrix, S-matrix, etc.) to this new frontier.

7.1 The absorption cross section

Since the Hamiltonian is usually Hermitian, the total current,

$$\mathbf{j} = \frac{\hbar}{2\mu i}\Big[\psi^*(\mathbf{r})\,\nabla\psi(\mathbf{r}) - \psi(\mathbf{r})\,\nabla\psi^*(\mathbf{r})\Big], \tag{7.1}$$

satisfies the continuity equation, which for a stationary state is

$$\nabla\cdot\mathbf{j} = 0\,.$$

However, this is not valid for complex potentials. In this case, the Hamiltonian has the form [Feshbach (1962)]

$$H = -\frac{\hbar^2}{2\mu}\nabla^2 + V(r), \quad \text{with} \quad V(r) = U(r) + i\,W(r), \tag{7.2}$$

where $U(r)$ is real and $W(r)$ is a short range negative[1] function of r. The term $i\,W$ takes care of the flux lost to excited channels. A peculiar aspect of H which is brought to light is that, naively, one may imagine that its energy spectrum will be complex. However, the energy eigenvalue is implicitly that of the exact many-channel problem (see chapter 9) with a real interaction. The imaginary complex potential comes about through the process of reducing the coupled channel equations to a single effective equation for the elastic channel, with a complex potential (see part II). Nevertheless, the energy has the same real value as the one appearing in the original set of coupled equations.

In the case of complex potentials, the continuity equation must be modified. For this purpose, we write the Schrödinger equation and its complex conjugate,

$$\left[-\frac{\hbar^2}{2\mu}\nabla^2 + U(r) + i\,W(r)\right]\ \psi\,(\mathbf{r}) = E\ \psi\,(\mathbf{r}) \tag{7.3}$$

$$\left[-\frac{\hbar^2}{2\mu}\nabla^2 + U(r) - i\,W(r)\right]\ \psi^*\,(\mathbf{r}) = E\ \psi^*\,(\mathbf{r})\,, \tag{7.4}$$

and evaluate $\psi^*\,(\mathbf{r}) \times$ Eq. (7.3) - $\psi\,(\mathbf{r}) \times$ Eq. (7.4). The resulting equation can be written

$$\frac{\hbar}{2\mu i}\ \left[\psi^*\,(\mathbf{r})\ \nabla^2\,\psi\,(\mathbf{r})\ -\ \psi\,(\mathbf{r})\ \nabla^2\,\psi^*\,(\mathbf{r})\right] = \frac{2}{\hbar}\,W(r)\ |\psi\,(\mathbf{r})|^2\,, \tag{7.5}$$

or, identifying the LHS of Eq. (7.5) as the divergence of the total current,

$$\nabla\cdot\mathbf{j} = \frac{2}{\hbar}\ W(\mathbf{r})\ \ |\psi\,(\mathbf{r})|^2\,. \tag{7.6}$$

Eq. (7.6) can be put in an integral form. For this purpose, we integrate it over the volume of a large sphere, which encloses the region of non-vanishing absorption ($W(\mathbf{r}) \neq 0$). We get

$$\int d^3\mathbf{r}\ \nabla\cdot\mathbf{j} = \frac{2}{\hbar}\int d^3\mathbf{r}\ W(r)\quad |\psi\,(\mathbf{r})|^2\,,$$

or, using the divergence theorem,

$$\int d\mathbf{s}\cdot\mathbf{j} = \frac{2}{\hbar}\int d^3\mathbf{r}\ W(r)\quad |\psi\,(\mathbf{r})|^2\,. \tag{7.7}$$

[1]If the optical potential is derived from coupled-channel equations (see part II of this book), it is non-local and state-dependent. In such situations it may be replaced by a local effective potential which leads to the same elastic wave function. This potential may have a positive imaginary part within some $r-$range. However, the overall effect of this potential must be that of a sink of flux.

This result can be used to obtain the absorption cross section in a scattering problem. In this case, the wave function, $\psi^{(+)}$, has scattering boundary conditions and

$$N_a = -\int d\mathbf{s} \cdot \mathbf{j}$$

is the number of absorbed particles per unit time. The absorption cross section is then obtained dividing N_a by the incident flux $J = |A|^2 v$ (Eq. (1.68)). We get

$$\sigma_a = \left(\frac{1}{|A|^2}\right) \frac{2}{\hbar v} \int d^3\mathbf{r} \; [-W(r)] \; \left|\psi^{(+)}(\mathbf{r})\right|^2 . \tag{7.8}$$

Note that the wave function $\psi^{(+)}$ contains the full effect of W. Writing v in terms of E and k and using a more compact notation, Eq. (7.8) takes the form

$$\sigma_a = \frac{1}{|A|^2} \times \frac{k}{E} \left\langle \psi^{(+)} \right| \Big[-W(r) \Big] \left| \psi^{(+)} \right\rangle . \tag{7.9}$$

Some authors (e.g. [Satchler *et al.* (1987)]) normalizes $\psi^{(+)}_{\mathbf{k}}$ such that it has the asymptotic behavior (Eq. (1.61) with $A = 1$)

$$\psi^{(+)}_{\mathbf{k}}(\mathbf{r}) \to e^{i\mathbf{k}\cdot\mathbf{r}} + f(\Omega) \frac{e^{ikr}}{r} . \tag{7.10}$$

In this case, Eq. (7.9) takes the simpler form

$$\sigma_a = \frac{k}{E} \left\langle \psi^{(+)} \middle| W \middle| \psi^{(+)} \right\rangle . \tag{7.11}$$

It should be stressed that the above equation is not valid for the normalization constant $A = (2\pi)^{-3/2}$, adopted in the present text. The absorption cross section is then given by Eq. (7.9).

The absorption cross section is a very inclusive piece of information. It gives the total flux removed from the elastic channel, without specifying how it is distributed over the open channels. Eq. (7.9) becomes particularly useful when the absorptive potential is a sum of terms[2], $W = W_1 + W_2 + ..$, each one associated with the coupling with one specific channel. In this case, the cross section for the corresponding channel, i, can be estimated by Eq. (7.9), with the same wave function but with the replacement $W \to W_i$.

For practical purposes, σ_a can be expressed in terms of radial integrals. Using in Eq. (7.9) the partial-wave expansion of the wave function

[2]The derivation of these potentials is given in chapter 9

(Eq. (2.57a)) and the orthogonality relation for the Legendre Polynomials (Eq. (2.60)), we obtain the result

$$\sigma_a = \frac{\pi}{k^2} \sum_l (2l+1)\, \mathcal{T}_l \,, \tag{7.12}$$

where $\mathcal{T}_l$ is the absorption coefficient at the partial-wave l,

$$\mathcal{T}_l = \frac{4k}{E} \int_0^\infty dr \, \Big[- W(r) \Big] \, \Big| u_l(k,r) \Big|^2 . \tag{7.13}$$

It can be easily shown that the absorption coefficients are related to the $l-$components of the S-matrix as

$$\mathcal{T}_l = 1 - |S_l|^2 \,. \tag{7.14}$$

For this purpose, we evaluate (see section (1.5))

$$\sigma_a \equiv \frac{N_a}{J} = -\frac{1}{v} \int d\mathbf{s} \cdot \mathbf{j} = -\frac{r^2}{v} \int d\Omega \; j_r \,, \tag{7.15}$$

where j_r is the radial component of the total current,

$$\begin{aligned} j_r &= \frac{\hbar}{2\mu i} \left[\left(\psi^{(+)}_{\mathbf{k}_i}\right)^* \frac{\partial}{\partial_r} \left(\psi^{(+)}_{\mathbf{k}_i}\right) - \psi^{(+)}_{\mathbf{k}_i} \frac{\partial}{\partial_r} \left(\psi^{(+)}_{\mathbf{k}_i}\right)^* \right] \\ &= \frac{\hbar}{\mu} \, \mathrm{Im} \left\{ \left(\psi^{(+)}_{\mathbf{k}_i}\right)^* \frac{\partial}{\partial_r} \left(\psi^{(+)}_{\mathbf{k}_i}\right) \right\} . \end{aligned} \tag{7.16}$$

To calculate j_r, we recall Eq. (2.50);

$$\psi^{(+)}_{\mathbf{k}} = X_- \, \frac{e^{-\,ikr}}{r} + X_+ \, \frac{e^{\,ikr}}{r} . \tag{7.17}$$

Using the expressions for $X_\pm$ (Eqs. (2.48) and (2.47)) with the coefficients C_l of Eq. (2.45), we can write

$$X_- = -\sum_l \frac{2l+1}{2ik} \, (-)^l \; P_l(\cos\theta) \tag{7.18}$$

$$X_+ = \sum_l \frac{2l+1}{2ik} \, S_l \; P_l(\cos\theta) \,. \tag{7.19}$$

Inserting Eq. (7.17) into Eq. (7.16), we obtain

$$j_r = \frac{v}{r^2} \, \Big(|X_+|^2 - |X_-|^2 \Big) \,.$$

Using this result in Eq. (7.15), we get

$$\sigma_a = \int d\Omega \; |X_-|^2 - \int d\Omega \; |X_+|^2 \,.$$

Above, the first integral represents the flux of the incident spherical wave while the second integral gives the corresponding emergent flux. The explicit form of $X_\pm$ (Eqs. (7.18) and (7.19)) together with the orthogonality relation for the Legendre Polynomials (Eq. (2.60)) lead to the result

$$\sigma_a = \frac{\pi}{k^2} \sum_l (2l+1) \left[1 - |S_l|^2\right] . \tag{7.20}$$

Note that for real potentials $|S_l| = 1$ and σ_a vanishes. Comparing Eqs. (7.20) and (7.12), we have the proof of Eq. (7.14).

An interesting limit, discussed further in chapter 5, is the Born approximation. In this case, the wave function $\psi^{(+)}$ is replaced by a plane wave and thus the absorption cross section of Eq. (7.9) becomes simply

$$\sigma_a = \frac{k}{E} \int d^3\mathbf{r} \; [-W(r)] . \tag{7.21}$$

If we further take $W(r)$ to be a step function, namely

$$W(r) = -W_0 \;\; \Theta(\bar{R} - r),$$

where $\bar{R}$ is the absorption range and W_0 (a positive quantity) is the strength of the absorptive potential, we get the simple expression

$$\sigma_a = \frac{4\pi k}{3E} \; W_0 \bar{R}^3 = \left(\frac{8\pi W_0 \, \bar{R}^3}{3\hbar}\right) \frac{1}{v} . \tag{7.22}$$

Thus, the above approximation gives an absorption cross section which goes inversely with the velocity of the absorbed particle. This behavior is in fact observed in very low energy neutron scattering from nuclei as well as in atoms, where the phenomenon is refered to as *Wigner's law.* We should warn the reader that in these systems the Born approximation is not valid and the above result should be considered as a mere coincidence.

To close this section we show the generalization of the Optical Theorem for complex potentials. The essential difference is that for them $|S_l| < 1$. In this case, we get

$$\frac{4\pi}{k} \mathrm{Im}\left\{f(0)\right\} = \frac{2\pi}{k^2} \sum_l (2l+1) \left[1 - \mathrm{Re}\left\{S_l\right\}\right]. \tag{7.23}$$

It is straightforward to check that the same result is obtained summing Eqs. (7.20) and (2.61a). Therefore, the Optical Theorem for complex potentials becomes

$$\sigma_{\mathrm{el}} + \sigma_a = \frac{4\pi}{k} \mathrm{Im}\left\{f(0)\right\} . \tag{7.24}$$

Special care is required when use is made of Eq. (7.24) for charged particle scattering. In this case the Coulomb interaction renders σ_{el} infinite. In such situations, one relies on the two-potential formula for the scattering amplitude (see section 4.5),

$$f(\theta) = f_{\rm C}(\theta) + \bar{f}(\theta)\,, \tag{7.25}$$

where $f_{\rm C}(\theta)$ is the Coulomb scattering amplitude (Eq. (3.29)) and $\bar{f}(\theta)$ is the correction owing to the short range complex potential. Carrying out the partial-wave expansion for $\bar{f}(\theta)$, we get (see section 3.4)

$$\bar{f}(\theta) = \frac{1}{2ik}\sum_{l}(2l+1)\,e^{2i\sigma_l}\left[e^{2i\,\bar{\delta}_l} - 1\right]. \tag{7.26}$$

Above, σ_l are the Coulomb phase shift and $\bar{\delta}_l$ is the complex phase shift arising from the short range complex potential. The cross section $d\sigma/d\Omega$ can be related to the Coulomb cross section $d\sigma_{\rm C}/d\Omega$ as

$$\frac{d\sigma}{d\Omega} - \frac{d\sigma_{\rm C}}{d\Omega} = \left|\bar{f}(\theta)\right|^2 + 2{\rm Re}\left\{f_{\rm C}^*(\theta)\,\bar{f}(\theta)\right\}. \tag{7.27}$$

Following Holdeman and Thaler [Holdeman and Thaler (1965)], a modified Optical Theorem can be obtained by integrating the above equation in the range $\theta_0 < \theta < \pi$, with θ_0 chosen to be very small. This procedure, known as the *Sum-of-Differences Method* (SOD), results in an expression which is convenient for the extraction of the σ_a. The SOD form of the Optical Theorem can be written as

$$\begin{aligned}\sigma_a &+ 2\pi\int_{\theta_0}^{\pi} d\theta\,\sin\theta\left(\frac{d\sigma}{d\Omega} - \frac{d\sigma_{\rm C}}{d\Omega}\right)\\ &= \frac{4\pi}{k}\left|\bar{f}(\theta_0)\right| + \sin\left[\arg\left(\bar{f}(\theta_0)\right) - 2\sigma_0 + 2\eta\ln\left(\sin\theta_0/2\right)\right].\end{aligned} \tag{7.28}$$

Eq. (7.28) is the appropriate form of the Optical Theorem used in the analysis of the elastic scattering of charged particles in the presence of absorption.

7.2 Lippmann-Schwinger equations with complex potentials

If the potential is complex, an extension of the dimension of the Hilbert space is required. Since $H = H_0 + V$ and $H^\dagger = H_0 + V^\dagger$ are different, one needs to introduce bi-orthogonal sets of scattering wave functions for each boundary condition.

In chapter 4 we have shown that the relevant quantities in Scattering Theory are contained in the state $|\psi_{\mathbf{k}}^{(+)}\rangle$ or, alternatively, in $|\psi_{\mathbf{k}}^{(-)}\rangle$, which is obtained from the $|\psi_{-\mathbf{k}}^{(+)}\rangle$ through the time-reversal operation (see section (4.3.3)). They are both solutions of the equation

$$\Big[E_k - H\Big] \Big|\psi_{\mathbf{k}}^{(\pm)}\Big\rangle = 0, \tag{7.29}$$

with different boundary conditions. When $H \neq H^\dagger$, the situation is more complicated. The physical state

$$\Big|\psi_{\mathbf{k}}^{(-)}\Big\rangle = \mathbb{T}\,\Big|\psi_{-\mathbf{k}}^{(+)}\Big\rangle, \tag{7.30}$$

where $\mathbb{T}$ is the time-reversal operator, is no longer a solution of Eq. (7.29) but satisfies the equation

$$\Big[E_k - H^\dagger\Big] \Big|\psi_{\mathbf{k}}^{(-)}\Big\rangle = 0\,. \tag{7.31}$$

It may look unreasonable that the physical state $\psi_{\mathbf{k}}^{(-)}$ satisfies a Schrödinger equation with the Hamiltonian $H^\dagger$, which acts as a source. However, one should have in mind that calculations of the S- and the T-matrix elements involve $\left(\psi_{\mathbf{k}}^{(-)}(\mathbf{r})\right)^*$, which satisfies an equation with H.

Following the same procedures as in chapter 4 but starting from Eqs. (7.29) (for $|\psi_{\mathbf{k}}^{(+)}\rangle$) and (7.31) (for $|\psi_{\mathbf{k}}^{(-)}\rangle$), one obtains the Lippmann-Schwinger equations

$$|\psi_{\mathbf{k}}^{(+)}\rangle = |\phi_{\mathbf{k}}\rangle + G_0^{(+)}(E_k)\,V\,|\psi_{\mathbf{k}}^{(+)}\rangle = |\phi_{\mathbf{k}}\rangle + G^{(+)}(E_k)V\,|\phi_{\mathbf{k}}\rangle \tag{7.32}$$

$$|\psi_{\mathbf{k}}^{(-)}\rangle = |\phi_{\mathbf{k}}\rangle + G_0^{(-)}(E_k)\,V^\dagger\,|\psi_{\mathbf{k}}^{(-)}\rangle = |\phi_{\mathbf{k}}\rangle + G^{(-)}(E_k)V^\dagger\,|\phi_{\mathbf{k}}\rangle\,, \tag{7.33}$$

where $G_0^{(\pm)}$ are the free particle Green's functions of chapter 4 and $G^{(\pm)}$ are given by

$$G^{(+)}(E_k) = \frac{1}{E_k - H + i\epsilon}, \tag{7.34}$$

$$G^{(-)}(E_k) = \frac{1}{E_k - H^\dagger - i\epsilon}\,. \tag{7.35}$$

Note that the property (see Eq. (4.31)) $G^{(\pm)}(E_k) = \left[G^{(\mp)}(E_k)\right]^\dagger$ is still valid.

For complex potentials, the states $|\psi_{\mathbf{k}}^{(+)}\rangle$ no longer form an orthogonal set. That is, $\langle\psi_{\mathbf{k}'}^{(+)}|\psi_{\mathbf{k}}^{(+)}\rangle \neq \delta(\mathbf{k}-\mathbf{k}')$. The same occurs with the set of states

$|\psi_{\mathbf{k}}^{(-)}\rangle$. However, one can introduce appropriate *dual* states $|\tilde{\psi}_{\mathbf{k}}^{(+)}\rangle$ and $|\tilde{\psi}_{\mathbf{k}'}^{(-)}\rangle$ which do satisfy the orthonormality conditions

$$\langle\tilde{\psi}_{\mathbf{k}}^{(+)}|\psi_{\mathbf{k}'}^{(+)}\rangle = \delta(\mathbf{k}-\mathbf{k}'), \tag{7.36}$$

$$\langle\psi_{\mathbf{k}'}^{(-)}|\tilde{\psi}_{\mathbf{k}}^{(-)}\rangle = \delta(\mathbf{k}-\mathbf{k}'). \tag{7.37}$$

The completeness relations should be modified accordingly, taking the form

$$\int |\psi_{\mathbf{k}}^{(+)}\rangle\, d^3\mathbf{k}\, \left\langle\tilde{\psi}_{\mathbf{k}}^{(+)}\right| + P_{\rm B} = \mathbf{1}, \tag{7.38}$$

$$\int \left|\tilde{\psi}_{\mathbf{k}}^{(-)}\right\rangle d^3\mathbf{k}\, \langle\psi_{\mathbf{k}}^{(-)}| + P_{\rm B} = \mathbf{1}, \tag{7.39}$$

with the term $P_{\rm B}$ accounting for the bound states of the potential.

The dual states are defined by their Lippmann-Schwinger equations[3],

$$|\tilde{\psi}_{\mathbf{k}}^{(+)}\rangle = |\phi_{\mathbf{k}}\rangle + G_0^{(+)}(E_k)\, V^\dagger\, |\tilde{\psi}_{\mathbf{k}}^{(+)}\rangle \tag{7.40}$$

$$= |\phi_{\mathbf{k}}\rangle + \tilde{G}^{(+)}(E_k)V^\dagger\, |\phi_{\mathbf{k}}\rangle, \tag{7.41}$$

and

$$|\tilde{\psi}_{\mathbf{k}}^{(-)}\rangle = |\phi_{\mathbf{k}}\rangle + G_0^{(-)}(E_k)\, V\, |\tilde{\psi}_{\mathbf{k}}^{(-)}\rangle \tag{7.42}$$

$$= |\phi_{\mathbf{k}}\rangle + \tilde{G}^{(-)}(E_k)V\;|\phi_{\mathbf{k}}\rangle, \tag{7.43}$$

where the dual full Green's functions $\tilde{G}^{(\pm)}$ are

$$\tilde{G}^{(+)}(E_k) = \frac{1}{E_k - H^\dagger + i\epsilon}, \tag{7.44}$$

$$\tilde{G}^{(-)}(E_k) = \frac{1}{E_k - H - i\epsilon}. \tag{7.45}$$

Multiplying Eqs. (7.41) and (7.43) from the left by the operators $(E - H_0 + i\epsilon)$ and $(E - H_0 + i\epsilon)$, respectively, and rearranging terms, we obtain the Schrödinger equations for the dual states

$$\left(E_k - H^\dagger\right)\, |\tilde{\psi}_{\mathbf{k}}^{(+)}\rangle = 0\,,$$

$$(E_k - H)\, |\tilde{\psi}_{\mathbf{k}}^{(-)}\rangle = 0\,.$$

The physical scattering wave functions $\psi_{\mathbf{k}}^{(\pm)}$ and their dual partners $\tilde{\psi}_{\mathbf{k}}^{(\pm)}$ can now be used to generate the spectral representations of Eqs. (7.34) and (7.35)

$$G^{(+)}(E_k) = \int \frac{|\psi_{\mathbf{q}}^{(+)}\rangle\, d^3\mathbf{q}\, \langle\tilde{\psi}_{\mathbf{q}}^{(+)}|}{E_k - E_q + i\epsilon} \tag{7.46}$$

[3]The proof that the dual states $|\tilde{\psi}_{\mathbf{k}}^{(+)}\rangle$ and $|\tilde{\psi}_{\mathbf{k}'}^{(-)}\rangle$, defined by Eqs. (7.41) and (7.43), are respectively orthogonal to $|\psi_{\mathbf{k}'}^{(+)}\rangle$ and $|\psi_{\mathbf{k}'}^{(-)}\rangle$ is analogous to that given in section 4.1 for real potentials.

and

$$G^{(-)}(E_k) = \int \frac{|\tilde{\psi}_{\mathbf{q}}^{(-)}\rangle \; d^3\mathbf{q} \; \langle \psi_{\mathbf{q}}^{(-)}|}{E_k - E_q - i\epsilon}. \tag{7.47}$$

The spectral representations of $\tilde{G}^{(\pm)}$ can be obtained analogously.

In chapter 4, we used the explicit form of the free Green's function (Eq. (4.37)) in the Lippmann-Schwinger equation for $\psi_{\mathbf{k}}^{(+)}(\mathbf{r})$ to get the asymptotic form of the scattering function. The asymptotic form of $\psi_{\mathbf{k}}^{(-)}(\mathbf{r})$ was then obtained from the property $\left(\psi_{\mathbf{k}}^{(-)}(\mathbf{r})\right)^* = \psi_{-\mathbf{k}}^{(+)}(\mathbf{r})$. Analogous procedures can be used to derive the asymptotic form of $\psi_{\mathbf{k}}^{(\pm)}(\mathbf{r})$. The results are summarized below

$$\psi_{\mathbf{k}}^{(+)}(\mathbf{r}) \to A \; \left(e^{i\mathbf{k}\cdot\mathbf{r}} + f(\theta) \; \frac{e^{ikr}}{r}\right), \tag{7.48}$$

$$\psi_{\mathbf{k}}^{(-)}(\mathbf{r}) \to A \; \left(e^{i\mathbf{k}\cdot\mathbf{r}} + f^*(\pi - \theta) \; \frac{e^{-ikr}}{r}\right), \tag{7.49}$$

$$\tilde{\psi}_{\mathbf{k}}^{(+)}(\mathbf{r}) \to A \; \left(e^{i\mathbf{k}\cdot\mathbf{r}} + \tilde{f}(\theta) \; \frac{e^{ikr}}{r}\right), \tag{7.50}$$

$$\tilde{\psi}_{\mathbf{k}}^{(-)}(\mathbf{r}) \to A \; \left(e^{i\mathbf{k}\cdot\mathbf{r}} + \tilde{f}^*(\pi - \theta) \; \frac{e^{-ikr}}{r}\right), \tag{7.51}$$

with

$$f(\theta) = -2\pi^2 \left(\frac{2\mu}{\hbar^2}\right) \left\langle \phi_{\mathbf{k}'} \middle| V \middle| \psi_{\mathbf{k}}^{(+)} \right\rangle \tag{7.52}$$

$$\tilde{f}(\theta) = -2\pi^2 \left(\frac{2\mu}{\hbar^2}\right) \left\langle \phi_{\mathbf{k}'} \middle| V^\dagger \middle| \tilde{\psi}_{\mathbf{k}}^{(+)} \right\rangle. \tag{7.53}$$

We have demonstrated above that the complex nature of the potential leads to a deviation of the scalar products $\langle \psi_{\mathbf{k}}^{(+)} | \psi_{\mathbf{k}'}^{(+)} \rangle$ from the Dirac's delta function. Now we explicitly evaluate this deviation. We first relate $\psi_{\mathbf{k}}^{(+)}$ to $\tilde{\psi}_{\mathbf{k}}^{(+)}$ through a Lippmann-Schwinger like equation. Using the relation

$$H^\dagger = H - V + V^\dagger$$

in Eq. (7.29) we can write

$$\left(E_k - H^\dagger\right) |\psi_{\mathbf{k}}^{(+)}\rangle = \left(V - V^\dagger\right) |\psi_{\mathbf{k}}^{(+)}\rangle. \tag{7.54}$$

This equation can be formally solved as

$$|\psi_{\mathbf{k}}^{(+)}\rangle = |\tilde{\psi}_{\mathbf{k}}^{(+)}\rangle + \tilde{G}^{(+)}(E_k) \left(V - V^\dagger\right) |\psi_{\mathbf{k}}^{(+)}\rangle. \tag{7.55}$$

Formal manipulation of this equation together with the orthogonality relation of Eq. (7.36) leads to

$$\langle \psi_{\mathbf{k}'}^{(+)} | \psi_{\mathbf{k}}^{(+)} \rangle = \delta(\mathbf{k} - \mathbf{k}') + \frac{\langle \psi_{\mathbf{k}'}^{(+)} | \left(V - V^{\dagger}\right) | \psi_{\mathbf{k}}^{(+)} \rangle}{E_k - E_{k'} + i\epsilon}. \tag{7.56}$$

A similar relation can be obtained for the scalar product $\langle \psi_{\mathbf{k}'}^{(-)} | \psi_{\mathbf{k}}^{(-)} \rangle$, following the same procedures as above. The result is

$$\langle \psi_{\mathbf{k}'}^{(-)} | \psi_{\mathbf{k}}^{(-)} \rangle = \delta(\mathbf{k} - \mathbf{k}') + \frac{\langle \psi_{\mathbf{k}'}^{(-)} | \left(V^{\dagger} - V\right) | \psi_{\mathbf{k}}^{(-)} \rangle}{E_k - E_{k'} - i\epsilon}. \tag{7.57}$$

The matrix-elements $\langle \psi_{\mathbf{k}'}^{(-)} | \left(V^{\dagger} - V\right) | \psi_{\mathbf{k}}^{(-)} \rangle$ can be reduced to those appearing in Eq. (7.56). For this purpose we use Eq. (7.30) written in the form

$$|\psi_{\mathbf{k}}^{(-)}\rangle = \mathbb{T}\, |\psi_{-\mathbf{k}}^{(+)}\rangle = \mathbb{T}\mathbb{R}\, |\psi_{\mathbf{k}}^{(+)}\rangle$$

where $\mathbb{R}$ is the space reflection operator. We obtain

$$\begin{aligned}\langle \psi_{\mathbf{k}'}^{(-)} | \left(V^{\dagger} - V\right) | \psi_{\mathbf{k}}^{(-)} \rangle &= \langle \psi_{\mathbf{k}'}^{(+)} | \mathbb{R}^{\dagger}\mathbb{T}^{\dagger} \left(V^{\dagger} - V\right) \mathbb{T}\mathbb{R} | \psi_{\mathbf{k}}^{(+)} \rangle \\ &= \langle \psi_{\mathbf{k}'}^{(+)} | \mathbb{R}^{\dagger} \left(V - V^{\dagger}\right) \mathbb{R} | \psi_{\mathbf{k}}^{(+)} \rangle \\ &= \langle \psi_{\mathbf{k}'}^{(+)} | \left(V - V^{\dagger}\right) | \psi_{\mathbf{k}}^{(+)} \rangle .\end{aligned} \tag{7.58}$$

The last step was carried out as a consequence of the assumed space reflection invariance of the potential. Therefore, the only difference between Eqs. (7.56) and (7.57) is the sign of $i\epsilon$ in the denominator of the second term. Accordingly,

$$\begin{aligned}\langle \psi_{\mathbf{k}'}^{(+)} | \psi_{\mathbf{k}}^{(+)} \rangle - \langle \psi_{\mathbf{k}'}^{(-)} | \psi_{\mathbf{k}}^{(-)} \rangle &= \left[\frac{1}{E_k - E_{k'} + i\epsilon} - \frac{1}{E_k - E_{k'} - i\epsilon} \right] \\ &\quad \times \langle \psi_{\mathbf{k}'}^{(+)} | \left(V - V^{\dagger}\right) | \psi_{\mathbf{k}}^{(+)} \rangle .\end{aligned}$$

or

$$\begin{aligned}\langle \psi_{\mathbf{k}'}^{(+)} | \psi_{\mathbf{k}}^{(+)} \rangle - \langle \psi_{\mathbf{k}'}^{(-)} | \psi_{\mathbf{k}}^{(-)} \rangle &= - \left[|A|^2\, 4\pi \left(\frac{E_k}{k} \right) \delta\left(E_k - E_{k'}\right) \right] \\ &\quad \times \frac{1}{|A|^2} \frac{k}{E_k} \langle \psi_{\mathbf{k}'}^{(+)} | W | \psi_{\mathbf{k}}^{(+)} \rangle ,\end{aligned}$$

where $W = -\mathrm{Im}\{V\}$. Since we are assuming the normalization of Eqs. (7.36) and (7.37), we should set $|A|^2 = (2\pi)^{-3}$ within the square brackets of the above equation. It then becomes

$$\begin{aligned}\langle \psi_{\mathbf{k}'}^{(+)} | \psi_{\mathbf{k}}^{(+)} \rangle - \langle \psi_{\mathbf{k}'}^{(-)} | \psi_{\mathbf{k}}^{(-)} \rangle &= - \left[\frac{1}{2\pi^2} \left(\frac{E_k}{k} \right) \delta\left(E_k - E_{k'}\right) \right] \\ &\quad \times \frac{1}{|A|^2} \frac{k}{E_k} \langle \psi_{\mathbf{k}'}^{(+)} | W | \psi_{\mathbf{k}}^{(+)} \rangle .\end{aligned} \tag{7.59}$$

Note that, with the constraint imposed by the delta function, the quantity

$$\Sigma^a_{\mathbf{k}',\mathbf{k}} \equiv \frac{1}{|A|^2}\,\frac{k}{E_k}\left\langle \psi^{(+)}_{\mathbf{k}'}\right| W \left|\psi^{(+)}_{\mathbf{k}}\right\rangle \tag{7.60}$$

is an angle-dependent version of the absorption cross section of Eq. (7.9). The angle refered to here is $\theta = \cos^{-1}\left(\hat{\mathbf{k}}'\cdot\hat{\mathbf{k}}\right)$.

We can define Møller wave operators in the dual space similarly to $\Omega^{(\pm)}$ (see section 4.1). That is, through their action on free states,

$$\Omega^{(\pm)}\left|\phi_{\mathbf{k}}\right\rangle = \left|\psi^{(\pm)}_{\mathbf{k}}\right\rangle \tag{7.61a}$$

$$\tilde{\Omega}^{(\pm)}\left|\phi_{\mathbf{k}}\right\rangle = \left|\tilde{\psi}^{(\pm)}_{\mathbf{k}}\right\rangle. \tag{7.61b}$$

From Eqs. (7.32), (7.33), (7.41) and (7.43), we obtain Lippmann-Schwinger equations

$$\Omega^{(+)}(E_k) = 1 + G_0^{(+)}(E_k)\,V\,\Omega^{(+)}(E_k) = 1 + G^{(+)}(E_k)V \tag{7.62}$$

$$\Omega^{(-)}(E_k) = 1 + G_0^{(-)}(E_k)\,V^\dagger\,\Omega^{(-)}(E_k) = 1 + G_0^{(-)}(E_k)\,V^\dagger \tag{7.63}$$

$$\tilde{\Omega}^{(+)}(E_k) = 1 + G_0^{(+)}(E_k)\,V^\dagger\,\tilde{\Omega}^{(+)}(E_k) = 1 + \tilde{G}^{(+)}(E_k)V^\dagger \tag{7.64}$$

$$\tilde{\Omega}^{(-)}(E_k) = 1 + G_0^{(-)}(E_k)\,V\,\tilde{\Omega}^{(-)}(E_k) = 1 + \tilde{G}^{(-)}(E_k)V\,. \tag{7.65}$$

The time-dependent representations of these operators are straightforward extensions of Eq. (4.101), as follows

$$\Omega^{(+)} = \lim_{t'\to\infty} e^{-\frac{i}{\hbar}H\,(t'-t)}\,e^{\frac{i}{\hbar}H_0\,(t'-t)}\,, \tag{7.66}$$

$$\Omega^{(-)} = \lim_{t'\to-\infty} e^{-\frac{i}{\hbar}H^\dagger\,(t'-t)}\,e^{\frac{i}{\hbar}H_0\,(t'-t)}\,, \tag{7.67}$$

$$\tilde{\Omega}^{(+)} = \lim_{t'\to\infty} e^{-\frac{i}{\hbar}H^\dagger(t'-t)}\,e^{\frac{i}{\hbar}H_0\,(t'-t)}\,, \tag{7.68}$$

$$\tilde{\Omega}^{(-)} = \lim_{t'\to-\infty} e^{-\frac{i}{\hbar}H\,(t'-t)}\,e^{\frac{i}{\hbar}H_0\,(t'-t)}\,. \tag{7.69}$$

7.3 The scattering and the transition operators

In the scattering from a complex potential, the definition of the transition operator, its Lippmann-Schwinger equation and its relation with the scattering amplitude remain the same. That is

$$T = V\,\Omega^{(+)}\,, \tag{7.70}$$

$$T = V + VG_0^{(+)}T = V + VG^{(+)}V \tag{7.71}$$

$$f(\theta) = -2\pi^2\left(\frac{2\mu}{\hbar^2}\right)T_{\mathbf{k}',\mathbf{k}}\,, \tag{7.72}$$

with $\theta = \cos^{-1}\left(\hat{\mathbf{k}}'\cdot\hat{\mathbf{k}}\right)$.

The S-matrix is also given as before (see Eq. (4.79))

$$S_{\mathbf{k}',\mathbf{k}} = \langle \psi_{\mathbf{k}'}^{(-)} \mid \psi_{\mathbf{k}}^{(+)} \rangle \,. \tag{7.73}$$

It is easy to prove (problem 1) that the inverse of the S-matrix is given by [Feshbach (1985)]

$$\left(S^{-1}\right)_{\mathbf{k}',\mathbf{k}} = \langle \tilde{\psi}_{\mathbf{k}'}^{(+)} \mid \tilde{\psi}_{\mathbf{k}}^{(-)} \rangle. \tag{7.74}$$

The fact that the non-hermiticity of the Hamiltonian requires no special care in the evaluation of the scattering amplitude may suggest that a detailed discussion of the dual states is superfluous. However, these states play an important role in some equations involving the T-matrix. An important case is the Distorted-Wave series of Eq. (4.161). Let us consider the two-potential problem (see section 4.5) when V_1 is complex. The Gell-Mann Goldberger formula is still given by

$$T_{\mathbf{k}',\mathbf{k}} = \left\langle \phi_{\mathbf{k}'} \right| V_1 \left| \chi_{\mathbf{k}}^{(+)} \right\rangle + \left\langle \chi_{\mathbf{k}'}^{(-)} \right| V_2 \sum_{m=0}^{\infty} \left(G^{(+)}(E_k)\, V_2 \right)^m \left| \chi_{\mathbf{k}}^{(+)} \right\rangle.$$

However, to write the expansion terms explicitly, we must use the spectral representation of $G^{(+)}$. Since Eq. (4.161) is no longer valid, it must be replaced by Eq. (7.46), which involves the dual states. In this way, Eq. (4.161) becomes

$$\begin{aligned} T_{\mathbf{k}',\mathbf{k}} = &\left\langle \chi_{\mathbf{k}'}^{(-)} \right| V_1 \left| \phi_{\mathbf{k}} \right\rangle + \left\langle \chi_{\mathbf{k}'}^{(-)} \right| V_2 \left| \chi_{\mathbf{k}}^{(+)} \right\rangle \\ &+ \int d^3\mathbf{q} \, \frac{\left\langle \chi_{\mathbf{k}'}^{(-)} \right| V_2 \left| \chi_{\mathbf{q}}^{(+)} \right\rangle \left\langle \tilde{\chi}_{\mathbf{q}}^{(+)} \right| V_2 \left| \chi_{\mathbf{k}}^{(+)} \right\rangle}{E_k - E_q \pm i\epsilon} \\ &+ \int d^3\mathbf{q}\, d^3\mathbf{q}' \, \frac{\left\langle \chi_{\mathbf{k}'}^{(-)} \right| V_2 \left| \chi_{\mathbf{q}}^{(+)} \right\rangle \left\langle \tilde{\chi}_{\mathbf{q}}^{(+)} \right| V_2 \left| \chi_{\mathbf{q}'}^{(+)} \right\rangle \left\langle \tilde{\chi}_{\mathbf{q}'}^{(+)} \right| V_2 \left| \chi_{\mathbf{k}}^{(+)} \right\rangle}{(E_k - E_q \pm i\epsilon)\,(E_k - E_{q'} \pm i\epsilon')} + \cdots . \end{aligned} \tag{7.75}$$

Note that the DWBA, which consists of keeping only the first line of the above equation, only involves the physical states $\left|\chi_{\mathbf{k}}^{(+)}\right\rangle$ and $\left\langle\chi_{\mathbf{k}'}^{(-)}\right|$. However, the dual states $\left\langle\tilde{\chi}_{\mathbf{q}}^{(+)}\right|$ become essential to evaluate higher order terms in the Distorted Wave series. This is the case of a second order DWBA, which includes the second line of Eq. (7.75).

Further applications of the dual states are encountered in the Low's equation and in the generalized Optical Theorem, discussed below.

7.3.1 *Low's equation and the generalized Optical Theorem*

We recall from chapter 4 the Low equation (Eq. (4.73)),

$$T_{\mathbf{k}',\mathbf{k}} = V_{\mathbf{k}',\mathbf{k}} + \int d^3\mathbf{q}\, \frac{T_{\mathbf{k}',\mathbf{q}}\, T^*_{\mathbf{k},\mathbf{q}}}{E_k - E_q + i\epsilon} + \sum_n \frac{V_{\mathbf{k}',n}\, V_{n,\mathbf{k}}}{E + B_n}\,, \tag{7.76}$$

In the case of complex potentials, some modifications are needed. First, we introduce the operator $\tilde{T}$, defined as

$$\tilde{T} = \left(\tilde{\Omega}^{(+)}\right)^\dagger V. \tag{7.77}$$

Note that $\tilde{T}$ can be considered a generalization of $T^\dagger$, since it reduces to the latter when the imaginary part of V vanishes. Following the same steps as in chapter 4 and neglecting bound states, Low's equation becomes

$$T_{\mathbf{k}',\mathbf{k}} = V_{\mathbf{k}',\mathbf{k}} + \int d^3\mathbf{q}\, \frac{T_{\mathbf{k}',\mathbf{q}}\, \tilde{T}_{\mathbf{k},\mathbf{q}}}{E_k - E_q + i\epsilon}\,, \tag{7.78}$$

where

$$\tilde{T}_{\mathbf{q},\mathbf{k}} = \tilde{T}^\dagger_{\mathbf{q},\mathbf{k}} = \langle \tilde{\psi}^{(+)}_{\mathbf{q}} |\, V\, | \phi_{\mathbf{k}} \rangle\,. \tag{7.79}$$

Now we write the analog of Low equation for $\tilde{T}$. The only difference with respect to Eq. (7.76) is the replacement $i\epsilon \to -i\epsilon$. Thus

$$\tilde{T}_{\mathbf{k}',\mathbf{k}} = V_{\mathbf{k}',\mathbf{k}} + \int d^3\mathbf{q}\, \frac{T_{\mathbf{k}',\mathbf{q}}\, \tilde{T}_{\mathbf{k},\mathbf{q}}}{E_k - E_q - i\epsilon}\,. \tag{7.80}$$

Subtracting Eq. (7.80) from Eq. (7.78), we obtain

$$T_{\mathbf{k}',\mathbf{k}} - \tilde{T}_{\mathbf{k}',\mathbf{k}} = -2\pi i \int d^3\mathbf{q}\; \delta(E_k - E_q)\, T_{\mathbf{k}',\mathbf{q}}\, \tilde{T}_{\mathbf{k},\mathbf{q}}\,. \tag{7.81}$$

Rewriting Eq. (7.79) in the form

$$\tilde{T}_{\mathbf{q},\mathbf{k}} = \langle \tilde{\psi}^{(+)}_{\mathbf{q}} |\, V^\dagger\, | \phi_{\mathbf{k}} \rangle + \langle \tilde{\psi}^{(+)}_{\mathbf{q}} | \left(V - V^\dagger\right) | \phi_{\mathbf{k}} \rangle$$

and using

$$\langle \tilde{\psi}^{(+)}_{\mathbf{q}} | = \langle \psi^{(+)}_{\mathbf{q}} | + \langle \tilde{\psi}^{(+)}_{\mathbf{q}} | \left(V - V^\dagger\right) \frac{1}{E_q - H^\dagger - i\epsilon}\,, \tag{7.82}$$

which is derived similarly to Eq. (7.55), we get

$$\tilde{T}_{\mathbf{q},\mathbf{k}} = T^\dagger_{\mathbf{q},\mathbf{k}} + \langle \tilde{\psi}^{(+)}_{\mathbf{q}} | \left(V - V^\dagger\right) | \psi^{(-)}_{\mathbf{k}} \rangle\,.$$

Inserting this result into Eq. (7.81), we obtain

$$T_{\mathbf{k}',\mathbf{k}} - T^\dagger_{\mathbf{k}',\mathbf{k}} = I_{\mathbf{k}',\mathbf{k}} - 2\pi i \int d^3\mathbf{q}\; \delta(E_k - E_q)\, T_{\mathbf{k},\mathbf{q}} T^\dagger_{\mathbf{q},\mathbf{k}'}\,, \tag{7.83}$$

where $I_{\mathbf{k}',\mathbf{k}}$ stands for

$$I_{\mathbf{k}',\mathbf{k}} = \langle\tilde{\psi}^{(+)}_{\mathbf{k}'}|\left(V - V^{\dagger}\right)|\psi^{(-)}_{\mathbf{k}}\rangle - 2\pi i \int d^3\mathbf{q}\, \delta(E_k - E_q)\, T_{\mathbf{k},\mathbf{q}}\, \langle\tilde{\psi}^{(+)}_{\mathbf{q}}|\left(V - V^{\dagger}\right)|\psi^{(-)}_{\mathbf{k}}\rangle\,,$$

or

$$I_{\mathbf{k}',\mathbf{k}} = \int d^3\mathbf{q}\, \left[\delta(\mathbf{k}-\mathbf{q}) - 2\pi i\, \delta(E_k - E_q)\, T_{\mathbf{k},\mathbf{q}}\right] \langle\tilde{\psi}^{(+)}_{\mathbf{q}}|\left(V - V^{\dagger}\right)|\psi^{(-)}_{\mathbf{k}}\rangle\,. \tag{7.84}$$

Now we replace (see Eqs. (4.82) and (4.79))

$$\delta(\mathbf{k}-\mathbf{q}) - 2\pi i\, \delta(E_k - E_q)\, T_{\mathbf{k},\mathbf{q}} = S_{\mathbf{k},\mathbf{q}} = \langle\psi^{(-)}_{\mathbf{k}}|\psi^{(+)}_{\mathbf{q}}\rangle$$

and move $\langle\psi^{(-)}_{\mathbf{k}}|$ out of the integral over $\mathbf{q}$. Eq. (7.84) then becomes

$$I_{\mathbf{k}',\mathbf{k}} = \langle\psi^{(-)}_{\mathbf{k}}|\left[\int |\psi^{(+)}_{\mathbf{q}}\rangle\, d^3\mathbf{q}\, \langle\tilde{\psi}^{(+)}_{\mathbf{q}}|\right]\left(V - V^{\dagger}\right)|\psi^{(-)}_{\mathbf{k}}\rangle, \tag{7.85}$$

or (see Eq. (7.38)),

$$I_{\mathbf{k}',\mathbf{k}} = \langle\psi^{(-)}_{\mathbf{k}}|\left(V - V^{\dagger}\right)|\psi^{(-)}_{\mathbf{k}}\rangle. \tag{7.86}$$

Next, we take into account Eq. (7.58) to write $I_{\mathbf{k}',\mathbf{k}}$ as

$$I_{\mathbf{k}',\mathbf{k}} = -\langle\psi^{(+)}_{\mathbf{k}}|\left(V - V^{\dagger}\right)|\psi^{(+)}_{\mathbf{k}}\rangle, \tag{7.87}$$

and insert this result into Eq. (7.83). We obtain

$$T_{\mathbf{k}',\mathbf{k}} - T^{\dagger}_{\mathbf{k}',\mathbf{k}} = -\langle\psi^{(+)}_{\mathbf{k}}|\left(V - V^{\dagger}\right)|\psi^{(+)}_{\mathbf{k}}\rangle - 2\pi i \int d^3\mathbf{q}\, \delta(E_k - E_q)\, T_{\mathbf{k},\mathbf{q}} T^{\dagger}_{\mathbf{q},\mathbf{k}'}, \tag{7.88}$$

or, using Eq. (7.60)

$$T_{\mathbf{k}',\mathbf{k}} - T^{\dagger}_{\mathbf{k}',\mathbf{k}} = -2\pi i \int d^3\mathbf{q}\, \delta(E_k - E_q)\, T_{\mathbf{k},\mathbf{q}} T^{\dagger}_{\mathbf{q},\mathbf{k}'} + 2i\frac{E}{k}\,\Sigma^{a}_{\mathbf{k}',\mathbf{k}}\,. \tag{7.89}$$

For $\mathbf{k}' = \mathbf{k}$, the above equation can be cast in the form of the generalized optical theorem, Eq. (7.24), since

$$\Sigma^{a}_{\mathbf{k},\mathbf{k}} = \sigma_a\,.$$

When the Hamiltonian has an absorptive term, the S-matrix is not unitary. Eq. (7.89) can be used to construct the unitarity defect operator, defined as

$$\Delta = \mathbf{1} - SS^{\dagger}\,, \tag{7.90}$$

where $\mathbf{1}$ stands for the identity operator. Employing the relation between the S- and the T-matrices (Eq. (4.82)), which remains valid for complex potentials, and using Eq. (7.89), one easily finds

$$\Delta_{\mathbf{k}'\mathbf{k}} = -4\pi\,\delta\left(E_{k'} - E_k\right)\frac{E_k}{k}\;\Sigma^{a}_{\mathbf{k}'\mathbf{k}}. \tag{7.91}$$

7.4 Partial-wave expansions

In the following we derive the quantities that are modified by absorption (complexity of the potential). In particular, we check how the partial-wave expansions are affected. We mention at the outset that, formally, the complex nature of the interaction requires the introduction of the scattering wave functions discussed in section 7.2. However, in so far as the calculation of the T and S operators and their matrix elements are concerned, only the physical scattering states $|\psi_{\mathbf{k}}^{(+)}\rangle$ and $|\psi_{\mathbf{k}}^{(-)}\rangle$, defined in Eqs. (7.32) and (7.33), enter. It can easily be checked that the procedures of section 2.3 can still be followed, so that the partial-wave expansion of $\psi_{\mathbf{k}}^{(+)}(\mathbf{r})$ is given by Eq. (2.57a) with $A = (2\pi)^{-3/2}$. That is,

$$\psi_{\mathbf{k}}^{(+)}(\mathbf{r}) = \frac{1}{(2\pi)^{3/2}} \sum_{lm} 4\pi\, Y_{lm}^{*}(\hat{\mathbf{k}})\, Y_{lm}(\hat{\mathbf{r}})\, i^{l}\, \frac{u_l(k,r)}{kr}\,. \tag{7.92}$$

However, the phase shifts contained in the radial wave function are now complex. The partial-wave expansion of the ingoing solution, which is obtained through the time-reversal operation, $\psi_{\mathbf{k}}^{(-)}(\mathbf{r}) = \left(\psi_{-\mathbf{k}}^{(+)}(\mathbf{r})\right)^{*}$ (see section 4.6), is

$$\psi_{\mathbf{k}}^{(-)}(\mathbf{r}) = \frac{1}{(2\pi)^{3/2}} \sum_{lm} 4\pi\, Y_{lm}^{*}(\hat{\mathbf{k}})\, Y_{lm}(\hat{\mathbf{r}})\, i^{l}\, \frac{u_l^{*}(k,r)}{kr} \tag{7.93}$$

The corresponding dual wave functions can be expanded as

$$\tilde{\psi}_{\mathbf{k}}^{(+)}(\mathbf{r}) = \frac{1}{(2\pi)^{3/2}} \sum_{lm} 4\pi\, Y_{lm}^{*}(\hat{\mathbf{k}})\, Y_{lm}(\hat{\mathbf{r}})\, i^{l}\, \frac{\tilde{u}_l(k,r)}{kr} \tag{7.94}$$

$$\tilde{\psi}_{\mathbf{k}}^{(-)}(\mathbf{r}) = \frac{1}{(2\pi)^{3/2}} \sum_{lm} 4\pi\, Y_{lm}^{*}(\hat{\mathbf{k}})\, Y_{lm}(\hat{\mathbf{r}})\, i^{l}\, \frac{\tilde{u}_l^{*}(k,r)}{kr}\,. \tag{7.95}$$

The wave functions $u_l(k,r)$ and $\tilde{u}_l(k,r)$ are respectively the solutions of the radial equations with the Hamiltonians H and $H^{\dagger}$, with the boundary condition of an outgoing scattered wave. At asymptotic distances they are

$$u_l(k, r\to\infty) = \frac{i}{2}\left[e^{-i(kr-l\pi/2)} - S_l\, e^{i(kr-l\pi/2)}\right] \tag{7.96}$$

$$\tilde{u}_l(k, r\to\infty) = \frac{i}{2}\left[e^{-i(kr-l\pi/2)} - \tilde{S}_l\, e^{i(kr-l\pi/2)}\right]. \tag{7.97}$$

Above, S_l is the partial-wave component of the S-matrix in the scattering from the potential[4] $V = U - iW$ while $\tilde{S}_l$ is that in the scattering from the potential $V^\dagger = U + iW$. Since $-iW$ is a sink of flux, one must have $|S_l| \equiv |e^{2i\delta_l}| < 1$. On the other hand, $+iW$ acts as a source and therefore $|\tilde{S}_l| \equiv |e^{2i\tilde{\delta}_l}| > 1$. The orthonormality relation involving u_l and $\tilde{u}_l$ is a straightforward extension of Eq. (4.173) to complex potentials. Inserting the partial-wave expansions of Eqs. (7.92) and (7.94) into Eq. (7.36), we obtain

$$\int_0^\infty dr\; \tilde{u}_l^*(k,r)\, u_l(k',r) = \frac{\pi}{2}\,\delta(k-k')\,. \tag{7.98}$$

The operators S and $\tilde{S}$ can also be expressed in terms of their partial-wave components. Proceeding as in chapter 4, we write

$$S_{\mathbf{k}',\mathbf{k}} = \delta(E_{k'} - E_k)\,\frac{\hbar^2}{\mu k}\sum_{lm} Y_{lm}(\hat{\mathbf{k}}')Y_{lm}^*(\hat{\mathbf{k}})\,S_l(E_k) \tag{7.99}$$

and

$$\tilde{S}_{\mathbf{k}',\mathbf{k}} = \delta(E_{k'} - E_k)\,\frac{\hbar^2}{\mu k}\sum_{lm} Y_{lm}(\hat{\mathbf{k}}')Y_{lm}^*(\hat{\mathbf{k}})\,\tilde{S}_l(E_k)\,. \tag{7.100}$$

It is straightforward to show that (see exercise 6.2)

$$\tilde{S}_l = \frac{1}{S_l^*}. \tag{7.101}$$

Thus, the phase of $\tilde{S}_l$ is identical to that of S_l while its modulus is

$$\left|\tilde{S}_l\right| = \frac{1}{|S_l|}. \tag{7.102}$$

The wave-functions $\tilde{u}_l(k,r)$ and $u_l(k,r)$ are connected through the expression

$$\tilde{u}_l(k,r) = u_l^*(k,r)\;\tilde{S}_l. \tag{7.103}$$

The validity of Eq. (7.103) at asymptotic distances can be trivially confirmed. Taking the conjugate of Eq. (7.96), replacing $\delta_l^* = \tilde{\delta}_l$ and comparing with Eq. (7.97) we get the above result. The generalization to any r requires more work. Using the partial-wave expansions of Eqs. (7.92)

[4]From this point, we use a different notation for the imaginary potential. To emphasize its absorptive nature, we write $V = U - iW$, where W is a positive function. In section 7.1 we wrote (see Eq. (7.2)) $V = U + iW$ with W standing for a negative function. Both notations are commonly found in the literature.

and (7.93) in Eq. (7.73) and comparing the result with the partial-wave expansion of the S-matrix (Eq. (7.99)) we get

$$\int_0^\infty dr\ u_l(k,r)\, u_l(k',r) = \frac{\pi}{2}\ \ S_l\ \delta(k-k'). \tag{7.104}$$

In the derivation of the above equation, the relation

$$\delta(E_k - E_{k'}) = \frac{\mu}{k\hbar^2}\,\delta(k-k')$$

was used. To prove Eq. (7.103), we must show that it satisfies the orthonormality relation that defines the dual states. That is,

$$\int_0^\infty dr\ \left[u_l^*(k,r)\ \tilde{S}_l\right]^*\ u_l(k',r) = \frac{\pi}{2}\,\delta(k-k')\,.$$

Calling

$$\begin{aligned} X &= \int_0^\infty dr\ \left[u_l^*(k,r)\ \tilde{S}_l\right]^*\ u_l(k',r) \\ &= \frac{1}{S_l}\int_0^\infty dr\ u_l(k,r)\ u_l(k',r) \end{aligned}$$

and replacing the above radial integral by the result of Eq. (7.104), we get the desired proof.

7.5 Diffractive scattering

In this section we use the semiclassical approximation for the scattering amplitude (see section 5.2.3) to discuss the diffractive nature of the scattering from complex potentials. We start with the semiclassical expression of the scattering amplitude of Eq. (5.268),

$$f(\theta) = \frac{1}{ik}\int_0^\infty d\lambda\,\lambda\ \left[S(\lambda)-1\right]\ P_\lambda(\cos\theta), \tag{7.105}$$

and the cross section

$$\frac{d\sigma(\theta)}{d\Omega} = |f(\theta)|^2\,.$$

The above integral can easily be evaluated in the limiting case of strong absorption where the modulus of the l-components of the S-matrix can be approximated by the *sharp cut-off* absorption,

$$\begin{aligned} \left|S(\lambda)\right| &= 0 \quad \text{for } \lambda < \bar{\Lambda}, \\ &= 1 \quad \text{for } \lambda \geq \bar{\Lambda}. \end{aligned}$$

The cut-off angular momentum is given by $\bar{\Lambda} = k\bar{R}$, with $\bar{R}$ being the characteristic radius of the scatterer. There are two types of known diffraction scenarios borrowed from Physical Optics:

(1) *Fraunhofer diffraction* - characterized by pure black-disk absorption (no refraction associated with a real phase shift). It corresponds to the diffraction of a beam of parallel rays of light emitted by a source at an infinite distance from the disc.
(2) *Fresnel diffraction* - characterized by black-disc absorption but with real Coulomb phase shift, which renders the source of the rays to be at a finite distance from the disk. It corresponds to the diffraction of light from a source located near the scatterer.

Both cases are based on the guiding optical properties of Babinet principle, which states that the scattering from a disc is equivalent to that from an aperture with the same radius. In the following we describe the scattering in these limiting cases, which represent the qualitative effects of absorptive interactions (more details can be find in [Hussein and McVoy (1984); Nussenzveig (1992); Frahn (1985)]).

7.5.1 *Fraunhofer diffraction*

In this case one neglects any refractive effect, setting $\delta(\lambda) = 0$. The S-matrix can then be written

$$S(\lambda) \equiv \left|S(\lambda)\right| e^{2i\delta(\lambda)} \simeq \left|S(\lambda)\right|. \tag{7.106}$$

Simple analytical results can be obtained if one adopts the sharp cut-off approximation,

$$\begin{aligned} S(\lambda) &= 0, \quad \text{for } \lambda < \bar{\Lambda}, \\ &= 1, \quad \text{for } \lambda \geq \bar{\Lambda}. \end{aligned} \tag{7.107}$$

Above, $\bar{\Lambda}$ represents the angular momentum for a grazing collision, which is given in terms of the radius of the scatterer, $\bar{R}$, and wave number, k, by the relation, $\bar{\Lambda} = k\bar{R}$.

In typical situations Franhofer diffraction occurs at small scattering angles. Thus, one can adopt the asymptotic (large λ) approximation of Eq. (5.381),

$$P_\lambda(\cos\theta) \to \left(\frac{\theta}{\sin\theta}\right)^{1/2} J_0\left(\lambda\theta\right). \tag{7.108}$$

Using Eqs. (7.107) and (7.108), the scattering amplitude of Eq. (7.105) becomes

$$f(\theta) = \frac{1}{ik}\left(\frac{\theta}{\sin\theta}\right)^{1/2} \int_0^\infty d\lambda\ \left(|S(\lambda)| - 1\right)\ \left[\lambda\, J_0(\lambda\theta)\right], \qquad (7.109)$$

or, changing variable: $\lambda \to x = \lambda\theta$,

$$f(\theta) = \frac{1}{ik\theta^2}\left(\frac{\theta}{\sin\theta}\right)^{1/2} \int_0^\infty dx\ \left(|S(x/\theta)| - 1\right)\ \left[x\, J_0(x)\right]. \qquad (7.110)$$

The above integral can be evaluated by parts. Using the result[5]

$$\frac{d}{dx}\left[x J_1(x)\right] = x J_0(x), \qquad (7.111)$$

where J_0 and J_1 are respectively the cylindrical Bessel functions of order 0 and 1, and having in mind that

$$\left[x\, J_1(x)\ \left(|S(x/\theta)| - 1\right)\right]_0^\infty = 0,$$

we get

$$f(\theta) = \frac{i}{k\theta^2}\left(\frac{\theta}{\sin\theta}\right)^{1/2} \int_0^\infty dx\ \left[x\, J_1(x)\right]\ \frac{d}{dx}|S(x/\theta)|. \qquad (7.112)$$

Within the sharp cut-off approximation for the S-matrix, we can write[6],

$$|S(x/\theta)| = \Theta(x - \bar{X}) \ \longrightarrow\ \frac{d}{dx}\left(|S(x/\theta)| - 1\right) = \delta(x - \bar{X}), \qquad (7.113)$$

and $f(\theta)$ becomes

$$f(\theta) = \frac{i\bar{\Lambda}}{k}\left(\frac{\theta}{\sin\theta}\right)^{1/2} \left[\frac{J_1(\bar{\Lambda}\theta)}{\theta}\right] = i\bar{R}\left(\frac{\theta}{\sin\theta}\right)^{1/2} \left[\frac{J_1(\bar{\Lambda}\theta)}{\theta}\right]. \qquad (7.114)$$

The cross section for Fraunhofer scattering, given by $\left|f(\theta)\right|^2$, is

$$\frac{d\sigma(\theta)}{d\Omega} = \bar{R}^2\left(\frac{\theta}{\sin\theta}\right)\left[\frac{J_1(\bar{\Lambda}\theta)}{\theta}\right]^2. \qquad (7.115)$$

For small angles, $\sin\theta/\theta \approx 1$ and the cross section reduces to

$$\frac{d\sigma(\theta)}{d\Omega} = \bar{R}^2\left[\frac{J_1(\bar{\Lambda}\theta)}{\theta}\right]^2. \qquad (7.116)$$

[5]This can easily be obtained using the ascending series for the Bessel functions [Abramowitz and Stegun (1972)].

[6]In this sub-section, we use the standard notations Θ and δ for the step-function and the Dirac delta function, respectively. They should not be confused with the classical deflection function and the phase shifts, which are represented by these symbols in other sections of this book.

For large values of $\bar{\Lambda}$ and not too small θ, $x = \bar{\Lambda}\theta \gg 1$ and one can use the asymptotic expression for cylindrical Bessel functions [Watson (1966)]

$$J_m(x) \simeq \sqrt{\frac{2}{\pi x}}\, \cos\left(x - m\pi/2 - \pi/4\right). \tag{7.117}$$

The scattering amplitude then becomes

$$f(\theta) = \frac{i\bar{R}}{\theta}\left(\frac{1}{\bar{\Lambda}\pi \sin\theta}\right)^{1/2}\left[\sin(\bar{\Lambda}\theta) - \cos(\bar{\Lambda}\theta)\right] \tag{7.118}$$

and the corresponding cross section is

$$\frac{d\sigma(\theta)}{d\Omega} = \bar{R}^2\left[\frac{1 - \sin\left(2\bar{\Lambda}\theta\right)}{\pi\,\bar{\Lambda}\theta^2 \sin\theta}\right]. \tag{7.119}$$

In figure 7.1 we show the cross sections for Fraunhofer scattering given by Eq. (7.115) (solid line) and by the approximate expression of Eq. (7.119) (dashed line). The cross sections are shown both in linear (top panel) and logarithmic (bottom panel) scales. The two curves are quite close for angles $\theta \gtrsim \theta_0 = 5^{\circ}$. This corresponds to $x \gtrsim \theta_0\,\bar{\Lambda} \times \pi/180 \simeq 4.4$, where the approximation of Eq. (7.117) is accurate. At angles below θ_0, the two curves are very different. The approximate curve is out of phase and diverges at $\theta = 0$.

To discuss the period of oscillation, it is convenient to re-write Eq. (7.116) as

$$\frac{d\sigma(\theta)}{d\Omega} = \left(\bar{\Lambda}R\right)^2\left[\frac{J_1(x)}{x}\right]^2. \tag{7.120}$$

The period corresponds to the difference between the x-values where $J_1(x)/x$ reaches the first two consecutive maxima. The first is clearly at $x_1 = 0$. The second, determined through numerical calculation of the $J_1(x)/x$ ratio, is $x_2 = 5.1$. Thus, we can write, $\Delta x = x_2 - x_1 = 5.1$. The corresponding angular period is

$$\Delta\theta^{\mathrm{o}} = \frac{x_2}{\bar{\Lambda}} \times \frac{180}{\pi} = 5.8^{o}. \tag{7.121}$$

It is interesting to check the concentration of the Fraunhofer cross section at the forward direction. Similarly to the case of forward glory, we can measure this concentration through the ratio of the cross sections at the first two consecutive maxima. The *forward concentration parameter* for Fraunhofer scattering, α_{F}, then is

$$\alpha_{\mathrm{F}} = \lim_{x_1 \to 0}\left[\frac{x_2\, J_1(x_1)}{x_1\, J_1(x_2)}\right]^2 = 57. \tag{7.122}$$

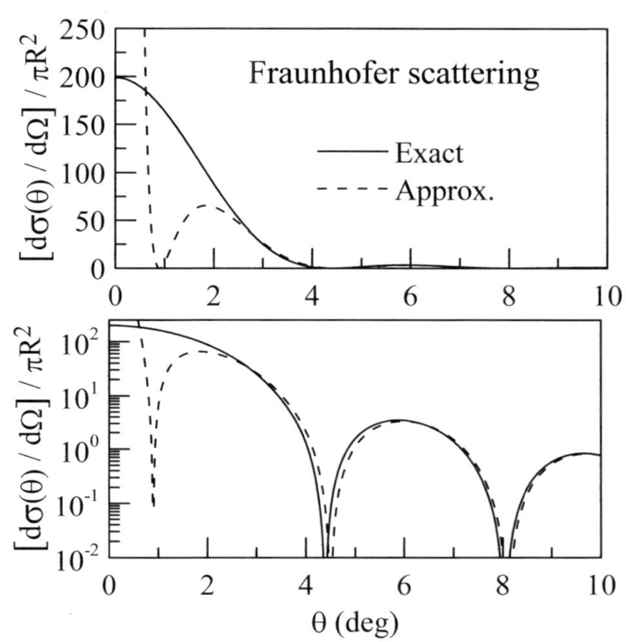

Fig. 7.1 Fraunhofer scattering cross sections normalized with respect to the geometrical cross section, shown in linear (top panel) and logarithmic (bottom panel) scales. The semiclassical calculations were made with a sharp cut-off S-matrix, with $\bar{\Lambda} = 50$. The solid and the dashed lines correspond respectively to the results obtained with the semiclassical expression of Eq. (7.115) and with its approximate version (Eq. (7.119)).

As in the case of forward glory, the above parameter is a general characteristic of Fraunhofer scattering. It does not depend on the collision energy or any property of the system. The only requirement is that the modulus of the S-matrix can be approximated by a sharp cut-off.

The forward angle concentration in Fraunhofer scattering is much more pronounced than that in glory scattering [Nussenzveig (1992)]. The concentration parameter for the former is almost one order of magnitude larger than the one for the latter ($\alpha_g = 6.2$). This may be a practical mean to distinguish the nature of observed cross sections. Frauhofer scattering and glory arise from very different physical processes. The former is a diffractive effect associated with the imaginary part of the optical potential, whereas the latter is a refractive effect, related with the real part of the potential. However, the period of oscillation for the two processes can be rather similar. Furthermore, it depends on some angular momentum that may be model dependent. In this way, the ratio of the cross sections at the first two consecutive maxima gives a better indication of the physics involved.

The sharp-cutoff limit is seldom really attained, except in light scattering from water droplets [Nussenzveig (1992)]. Most physical systems have diffused absorption surfaces. In such cases, the modulus of the partial-wave projected S-matrix can be expressed in terms of two parameters: the cut-off parameter, $\bar{\Lambda}$, and a diffuseness parameter, Δ. In the *smooth cut-off* model (SCO), $|S(\lambda)|$ is approximated by some simple function of λ, depending on these two parameters. As we will show, the diffuseness of the absorptive potential brings in more damping in the angular oscillation. We discuss below a simple implementation of the smooth cut-off model.

As a starting point, we rewrite Eq. (7.112) as

$$f(\theta) = \frac{i}{k\theta} \left(\frac{\theta}{\sin\theta} \right)^{1/2} \int_{-\infty}^{\infty} d\lambda \; [\lambda \, J_1(\lambda\theta)] \; \frac{d}{d\lambda} |S(\lambda)|. \tag{7.123}$$

Note that in the above equation we have extended the lower limit of the integral to $-\infty$. This limit originally was zero. Since $\Delta \ll \bar{\Lambda}$, the integrand at $\lambda \leq 0$ is vanishingly small and thus the modification of the lower limit has no practical consequence.

In the smooth cut-off model, the derivative of $|S(\lambda)|$ is no longer a delta function. However, it is sharply peaked at $\lambda = \bar{\Lambda}$ and the width of the peak is measured by the parameter Δ ($\Delta \ll \bar{\Lambda}$). We use the notation

$$D_\Delta(\lambda - \bar{\Lambda}) = \frac{d}{d\lambda} \big| S(\lambda) \big|, \tag{7.124}$$

and point out that this function has the limit

$$\lim_{\Delta \to 0} \left[D_\Delta(\lambda - \bar{\Lambda}) \right] = \delta(\lambda - \bar{\Lambda}). \tag{7.125}$$

Now we use the asymptotic form[7] of the cylindrical Bessel functions (Eq. (7.117)) for $m = 1$,

$$J_1(x) \simeq \sqrt{\frac{2}{\pi x}} \; \cos\left(x - \pi/2 - \pi/4\right) \equiv \sqrt{\frac{2}{\pi x}} \; \sin\left(x - \pi/4\right). \tag{7.126}$$

[7]In principle, the use of the asymptotic form of the Bessel function would imply that the resulting scattering amplitude is not valid at small scattering angles. However, it is not so. The same results can be obtained writing the Legendre polynomials in terms of traveling waves components, $Q_l^{(\pm)}(x)$, like in Fuller's near-side decomposition [Fuller (1975a); Brink (1985)], instead of $J_1(x)$. The asymptotic forms would then enter only in the evaluation of the integrals in the complex plane over an asymptotic contour, which appear later in this section.

Inserting this result into Eq. (7.123) and neglecting the variation of $\sqrt{\lambda}$ within the range of $D_\Delta\left(\lambda-\bar{\Lambda}\right)$, we get

$$f_\Delta(\theta)=\frac{1}{k\theta}\left(\frac{\bar{\Lambda}}{2\pi\sin\theta}\right)^{1/2}\left[I_\Delta^{(+)}(\theta)\,e^{i\left(\bar{\Lambda}\theta-\pi/4\right)}-I_\Delta^{(-)}(\theta)\,e^{-i\left(\bar{\Lambda}\theta-\pi/4\right)}\right]. \tag{7.127}$$

Above, we have introduced the diffraction damping functions

$$\begin{aligned} I_\Delta^{(\pm)}(\theta)&=\int_{-\infty}^{\infty}d\lambda\;D_\Delta\left(\lambda-\bar{\Lambda}\right)\;e^{\pm i\left(\lambda-\bar{\Lambda}\right)\,\theta}\\ &=\int_{-\infty}^{\infty}dz\;D_\Delta\left(z\right)\;e^{\pm i\,z\theta}. \end{aligned} \tag{7.128}$$

Note that we have changed variable $\lambda\to z=\lambda-\bar{\Lambda}$.

Closed expressions for the scattering amplitude can be obtained if one adopts convenient parametrizations for the modulus of the S-matrix. We consider here the choice

$$\left|S(\lambda)\right|=\frac{1}{1+\exp\left[\left(\bar{\Lambda}-\lambda\right)/\Delta\right]}, \tag{7.129}$$

which reduces to the sharp cut-off approximation in the $\Delta\to 0$ limit. In this case, one immediately obtains

$$D_\Delta(z)=\frac{e^{z/\Delta}}{\Delta\left[1+e^{z/\Delta}\right]^2}, \tag{7.130}$$

and the integrals of Eq. (7.128) become

$$I_\Delta^{(\pm)}(\theta)=\frac{1}{\Delta}\int_{-\infty}^{\infty}dz\;\frac{e^{z/\Delta}}{\left[1+e^{z/\Delta}\right]^2}\;e^{\pm i\,z\theta}. \tag{7.131}$$

These integrals can be calculated taking closed contours in the complex plane. The integrands have double poles at the points

$$z_n=i\pi\;(2n+1)\;\Delta. \tag{7.132}$$

The evaluation of the diffraction damping functions can be performed using the residue theorem for double poles and taking only the leading pole ($n=0$) into account. One gets

$$I_\Delta^{(+)}(\theta)=I_\Delta^{(-)}(\theta)\equiv I_\Delta(\theta)=\frac{\pi\theta\Delta}{\sin\left(\pi\theta\Delta\right)}. \tag{7.133}$$

Now one inserts this result into Eq. (7.127) and move $I_\Delta(\theta)$ out of the square brackets. One is left with $2i\sin(\bar{\Lambda}\theta-\pi/4)$, which can be expressed in terms

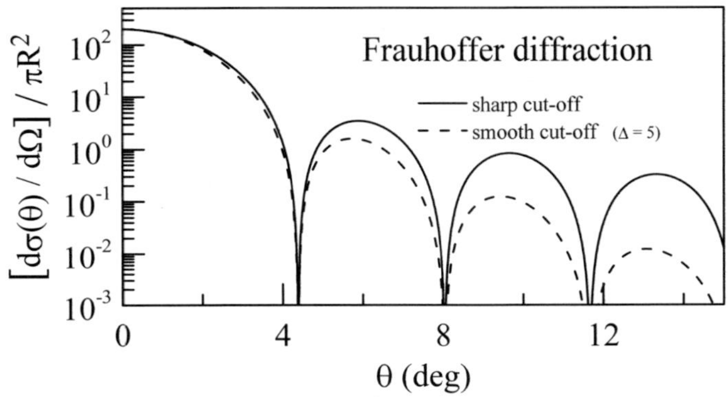

Fig. 7.2 Comparison between the semiclassical cross sections obtained with the sharp-cutt-off (solid line) and the smooth-cut-off (dashed line) approximations. In the latter, we used $\Delta = 5$. As in the previous figure, the cross sections are normalized with respect to the geometric value.

of $J_1(\bar{\Lambda}\theta)$ (see Eq. (7.126)). In this way, the smooth cut-off scattering amplitude can be put in the form

$$f_{\Delta}(\theta) = f_{\rm sco}(\theta) \times I_{\Delta}(\theta), \tag{7.134}$$

and the cross section becomes

$$\frac{d\sigma_{\Delta}(\theta)}{d\Omega} = \frac{d\sigma_{\rm sco}(\theta)}{d\Omega} \times I_{\Delta}^2(\theta). \tag{7.135}$$

In the above equations, $f_{\rm sco}(\theta)$ is the scattering amplitude within the sharp cut-off approximation, given in Eq. (7.114), and $d\sigma_{\rm sco}(\theta)/d\Omega$ is the corresponding cross section (Eq. (7.115)).

An example of the damping produced by smooth cut-off absorption is given in figure 7.2. The system is the same of figure 7.1 and the sharp cut-off cross section is still represented by a solid line. The dashed line corresponds to smooth cut-off absorption, with $|S(\lambda)|$ given by Eq. (7.129), with $\Delta = 5$ units of angular momentum. This corresponds to 10% of the grazing angular momentum, $\bar{\Lambda} = 50$. One notices that the smooth cut-off absorption leads to a stronger concentration of the cross section in the forward region. In this case, the forward concentration parameter of Eq. (7.122) is considerably larger then for sharp cut-off ($\alpha_{\rm F} = 57$), reaching the value

$$\alpha_{\rm F}(\Delta = 5) = 125. \tag{7.136}$$

7.5.2 *Fresnel diffraction*

We now consider the effect of a strong Coulomb field on the diffraction pattern. It is expected to introduce refractive effects in the near-side (Coulomb repulsion) or in the far-side (Coulomb attraction) scattering amplitude. We concentrate here on the Fresnel diffraction associated with the near-side amplitude and to simplify the notation we omit the superscript $(-)$ in the near-side branch of the scattering amplitude.

The starting point is the semiclassical expression for the near-side amplitude of Eq. (5.274), using the the sharp cut-off approximation for the modulus of the S-matrix and Coulomb phase shifts for its phase,

$$\begin{aligned} S(\lambda) &= 0, \qquad \text{for } \lambda \leq \bar{\Lambda} \\ &= e^{2i\sigma(\lambda)}, \quad \text{for } \lambda > \bar{\Lambda}. \end{aligned} \tag{7.137}$$

In this case, the scattering amplitude can be written as

$$f(\theta) = \frac{e^{-i\pi/4}}{k} \left(\frac{1}{2\pi \sin\theta}\right)^{1/2} \int_{\bar{\Lambda}}^{\infty} d\lambda\, \lambda^{1/2}\, e^{i\,[2\sigma(\lambda) - \lambda\theta]}. \tag{7.138}$$

Before we evaluate the above integral, it is convenient to consider the pure Coulomb amplitude (no absorption), which is given by Eq. (5.274) with the S-matrix

$$S(\lambda) = e^{2i\sigma(\lambda)}. \tag{7.139}$$

That is[8]

$$f_C(\theta) = \frac{e^{-i\pi/4}}{k} \left(\frac{1}{2\pi \sin\theta}\right)^{1/2} \int_0^{\infty} d\lambda\ \lambda^{1/2}\ e^{i\,[2\sigma(\lambda) - \lambda\theta]}. \tag{7.140}$$

Note that the only difference between the above semiclassical Coulomb amplitude and the sharp cut-off scattering amplitude of Eq. (7.138) is the lower limit of integration. Above, this limit is zero whereas in Eq. (7.138) it is the grazing angular momentum $\bar{\Lambda}$.

The integrals of Eqs. (7.138) and (7.140) can be evaluated by the stationary phase approximation. The stationary phase occurs at the angular momentum λ_0 where the phase of the integrand,

$$\varphi(\lambda) \equiv 2\sigma(\lambda) - \lambda\theta, \tag{7.141}$$

[8] Although the Coulomb amplitude cannot be calculated by a truncated summation of the partial-wave series, the integral introduced in the semiclassical approximation can be evaluated with the upper limit going to infinity. As it will be shown below, the semiclassical approximation to the Coulomb scattering amplitude leads to the correct cross section.

satisfies the condition

$$\left[\frac{d\varphi(\lambda)}{d\lambda}\right]_{\lambda=\lambda_0} = \left[\frac{d}{d\lambda}\left(2\sigma(\lambda) - \lambda\theta\right)\right]_{\lambda=\lambda_0} = 0. \tag{7.142}$$

Since $d[2\sigma(\lambda)]/d\lambda$ is the classical deflection function for the Coulomb potential, this condition can be written

$$\Theta(\lambda_0) = \theta. \tag{7.143}$$

Using the above equation in the relation between the Coulomb deflection function and the angular momentum (Eq. (3.11c)), one gets an analytical expression for λ_0 in terms of the observation angle, θ, and the Sommerfeld parameter, η,

$$\lambda_0 = \frac{\eta}{\tan\theta/2}. \tag{7.144}$$

Next, we expand $\varphi(\lambda)$ around λ_0 and keep terms up to second order. Owing to the stationary phase condition, the term of first order vanishes and the truncated expansion gives

$$\varphi(\lambda) \simeq \varphi_0 + \frac{\Theta_0'}{2}\left(\lambda - \lambda_0\right)^2. \tag{7.145}$$

Above, we used the short-hand notations: $\varphi_0 \equiv \varphi(\lambda_0)$ and $\Theta_0' \equiv \Theta'(\lambda_0)$.

In typical situations, $\lambda_0 \gg 0$, except for very backward angles, which are not important in the study of Fresnel diffraction. The lower limit of the integrals can then be extended to $-\infty$, without affecting the values of the scattering amplitude. Adopting this procedure and approximating $\lambda^{1/2} \simeq \lambda_0^{1/2}$, Eq. (7.138) becomes

$$f_{\mathrm{C}}(\theta) = \frac{e^{i\varphi_0 - i\pi/4}}{k}\left(\frac{\lambda_0}{2\pi\sin\theta}\right)^{1/2}\int_{-\infty}^{\infty} d\lambda\; e^{i\Theta_0'(\lambda-\lambda_0)^2/2}. \tag{7.146}$$

Using the explicit form of λ_0 (Eq. (7.144)) and the relation $\sin\theta = 2\sin\theta/2\,\cos\theta/2$, the scattering amplitude takes the form

$$f_{\mathrm{C}}(\theta) = \frac{e^{i\varphi_0 - i\pi/4}}{2k\,\sin\theta/2}\left(\frac{\eta}{\pi}\right)^{1/2}\int_{-\infty}^{\infty} d\lambda\; e^{i\Theta_0'(\lambda-\lambda_0)^2/2}. \tag{7.147}$$

To evaluate the above integral, we change to a new variable t such that

$$i\,\frac{\Theta_0'}{2}\left(\lambda - \lambda_0\right)^2 \longrightarrow -t^2. \tag{7.148}$$

This is achieved by the transformation[9]

$$t = e^{i\pi/4}\left(\frac{-\Theta_0'}{2}\right)^{1/2}\left(\lambda - \lambda_0\right), \tag{7.149}$$

[9]It is convenient to keep the minus sign together with Θ_0' to cancel the minus sign contained in the derivative of the Coulomb deflection function.

which leads to the relation

$$d\lambda = e^{-i\pi/4}\left(\frac{2}{-\Theta_0'}\right)^{1/2} dt. \tag{7.150}$$

Inserting Eqs. (7.149) and (7.150) into Eq. (7.147) and using the value of the gaussian integral,

$$\int_{-\infty}^{\infty} dt\; e^{-t^2} = \sqrt{\pi}, \tag{7.151}$$

we get

$$f_{\rm C}(\theta) = \frac{e^{i\varphi_0 - i\pi/2}}{2k\,\sin\theta/2}\left(\frac{2\eta}{-\Theta_0'}\right)^{1/2}. \tag{7.152}$$

The derivative of the deflection function appearing above, which can be obtained through direct differentiation of Eq. (3.11c), is

$$\Theta_0' = -\frac{2}{\eta}\,\sin^2(\theta/2)\,. \tag{7.153}$$

Inserting this result into Eq. (7.162), one obtains the semiclassical approximation for the Coulomb amplitude

$$f_{\rm C}(\theta) = \frac{a}{2\,\sin^2\theta/2}\; e^{i[2\sigma(\lambda_0) - \lambda_0\theta - \pi/2]}\,, \tag{7.154}$$

where $a = \eta/k$ is the half-distance of closest approach in a head-on collision, discussed in chapter 3. The semiclassical Coulomb cross section then is

$$\frac{d\sigma_{\rm C}(\theta)}{d\Omega} = \frac{a^2}{4}\left[\frac{1}{\sin^4\theta/2}\right]. \tag{7.155}$$

Note that the semiclassical approximation leads to the correct cross section of Eq. (3.35). The phase of the semiclassical scattering amplitude is similar to the exact one, differing only by a constant. To show this point, we first use the WKB approximation for the Coulomb phase-shift (Eq. (5.206)),

$$\begin{aligned}\sigma_l \simeq \sigma^{\rm WKB}(\lambda) &= -\eta + \eta\,\ln\sqrt{\eta^2+\lambda^2} + \lambda\tan^{-1}\eta/\lambda \\ &\equiv \eta\,[\ln\eta - 1] - \eta\,\ln\left[\frac{\eta}{\sqrt{\eta^2+\lambda^2}}\right] + \lambda\tan^{-1}\eta/\lambda.\end{aligned}$$

According to Eqs. (3.11a) and (3.11c), we can write

$$\sigma(\lambda) \simeq \eta\,[\ln\eta - 1] - \eta\,\ln[\sin(\theta/2)] + \lambda\,\frac{\theta}{2}.$$

With the above approximation, the phase of the scattering amplitude of Eq. (7.154) becomes

$$\alpha(\theta) = 2\sigma(\lambda_0) - \lambda_0\theta - \pi/2 = 2\left(\eta\left[\ln\eta - 1\right]\right) - \eta\ln\left[\sin^2(\theta/2)\right] + \pi/2. \tag{7.156}$$

Now we use the approximation of Eq. (3.88),

$$\sigma_0 = \frac{1}{2}\tan^{-1}\eta + \eta\left[\ln(\eta) - 1\right],$$

to write

$$\alpha(\theta) = 2\sigma_0 - \eta\ln\left[\sin^2(\theta/2)\right] + \pi - \phi_0, \tag{7.157}$$

with

$$\phi_0 = \frac{\pi}{2} + \tan^{-1}\eta. \tag{7.158}$$

Comparing Eq. (7.157) with the exact Coulomb scattering amplitude (Eq. (3.29)),

$$\begin{aligned} f_{\mathrm{C}}(\theta) &= -\frac{a}{2\sin^2(\theta/2)}e^{-i\eta\ln(\sin^2\theta/2)}\,e^{2i\sigma_0} \\ &\equiv |f_{\mathrm{C}}(\theta)| \times \exp\left[i\left(2\sigma_0 - \eta\ln\left[\sin^2\theta/2\right] + \pi\right)\right], \end{aligned}$$

we conclude that the phase of the semiclassical scattering amplitude differs from the exact one only by the constant ϕ_0.

Now we come back to the scattering amplitude of Eq. (7.138), which corresponds to the case of sharp cut-off absorption (Eq. (7.137)). We can follow the same steps as in the case of pure Coulomb scattering but keeping in mind that the lower limit of integration is the cut-off angular momentum, $\bar{\Lambda}$, instead of $-\infty$. Eq. (7.147) then becomes

$$f(\theta) = \frac{e^{i\varphi_0 - i\pi/4}}{2k\sin\theta/2}\left(\frac{\eta}{\pi}\right)^{1/2}\int_{\bar{\Lambda}}^{\infty} d\lambda\, e^{i\Theta_0'(\lambda-\lambda_0)^2/2}. \tag{7.159}$$

Accordingly, the integral in terms of the new variable t corresponds to Eq. (7.152) with a different lower integration limit, T. That is,

$$f(\theta) = \frac{e^{i\varphi_0 - i\pi/2}}{2k\sin\theta/2}\left(\frac{2\eta}{-\Theta_0'}\right)^{1/2}\left[\frac{1}{\sqrt{\pi}}\int_T^{\infty} dt\, e^{-t^2}\right]. \tag{7.160}$$

This limit is related with $\bar{\Lambda}$ through Eq. (7.149), as

$$T = e^{i\pi/4}\left(\frac{-\Theta_0'}{2}\right)^{1/2}\left(\bar{\Lambda} - \lambda_0\right). \tag{7.161}$$

Note that taking the $T \to -\infty$ limit in Eq. (7.160) one gets the scattering amplitude for pure Coulomb scattering. In this case, the quantity within square brackets becomes equal to one and Eq. (7.160) reduces to Eq. (7.152). Thus, Eq. (7.160) can be written as

$$f(\theta) = f_C(\theta) \left[\frac{1}{\sqrt{\pi}} \int_T^\infty dt\, e^{-t^2} \right]. \tag{7.162}$$

It is convenient to express the lower integration limit in terms of the observation angle, θ, and the grazing angle, $\bar{\Theta}$, which corresponds to the deflection function for the grazing angular momentum, $\bar{\Lambda}$. Using Eqs. (3.11c) and (7.153), and standard trigonometric relations, we get

$$T = e^{i\pi/4} \sqrt{\eta} \left[\frac{\sin\left(\frac{\theta - \bar{\Theta}}{2}\right)}{\sin \frac{\bar{\Theta}}{2}} \right]. \tag{7.163}$$

Usually, the cross section for Fresnel diffraction is expressed in terms of the Fresnel integrals $C(W)$ and $S(W)$, which are real functions of a real argument. They are defined as [Abramowitz and Stegun (1972)]

$$C(W) = \int_0^W dw \cos\left(\frac{\pi}{2} w^2\right) \tag{7.164}$$

and

$$S(W) = \int_0^W dw \sin\left(\frac{\pi}{2} w^2\right). \tag{7.165}$$

It is clear from the above definitions that these functions are antisymmetric:

$$C(-W) = -C(W) \quad \text{and} \quad S(-W) = -S(W). \tag{7.166}$$

The Fresnel integrals are illustrated in figure 7.3. They vanish at $W = 0$ and converge to the limit 0.5 as $W \to \infty$, oscillating around this limit with decreasing amplitudes.

To express the scattering amplitude in terms of Fresnel integrals, we first rewrite Eq. (7.162) as[10]

$$\begin{aligned} \frac{f(\theta)}{f_C(\theta)} &= \left[\frac{1}{\sqrt{\pi}} \int_0^\infty dt\, e^{-t^2} - \frac{1}{\sqrt{\pi}} \int_0^T dt\, e^{-t^2} \right] \\ &= \frac{1}{2} \left[1 - \frac{2}{\sqrt{\pi}} \int_0^T dt\, e^{-t^2} \right]. \end{aligned} \tag{7.167}$$

[10] To get the second line of the equation below, we used the value of the Gaussian integral, given in Eq. (7.151).

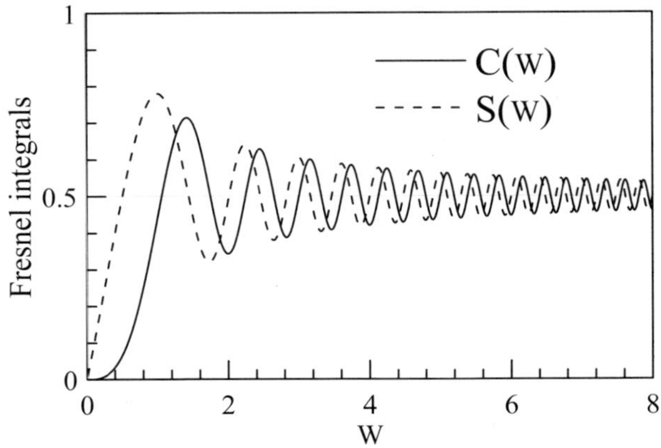

Fig. 7.3 The Fresnel integrals $C(W)$ and $S(W)$.

The next step is to introduce a new integration variable, w, such that

$$-t^2 \to -i\,\frac{\pi}{2}\,w^2. \tag{7.168}$$

This is achieved by the transformation

$$w = e^{-i\pi/4}\sqrt{\frac{2}{\pi}}\,t. \tag{7.169}$$

Replacing in the above equation $t \to T$, with T given by Eq. (7.163), we obtain the upper integration limit for the new variable:

$$W = \sqrt{\frac{2\eta}{\pi}}\left[\frac{\sin\left(\frac{\theta-\bar{\Theta}}{2}\right)}{\sin\frac{\bar{\Theta}}{2}}\right]. \tag{7.170}$$

From Eq. (7.169), we have

$$dt = e^{i\pi/4}\sqrt{\frac{\pi}{2}}\,dw \tag{7.171}$$

and using this result in Eq. (7.167), we obtain

$$\frac{f(\theta)}{f_{\rm C}(\theta)} = \frac{1}{2}\left[1-(1+i)\int_0^W dw\;e^{-i\pi w^2/2}\right]. \tag{7.172}$$

The above result can be expressed in terms of Fresnel integrals (Eqs. (7.164) and (7.165)), as

$$\frac{f(\theta)}{f_{\rm C}(\theta)} = \frac{1}{2}\left[1-(1+i)\,\left(C(W)-i\,S(W)\right)\right]. \tag{7.173}$$

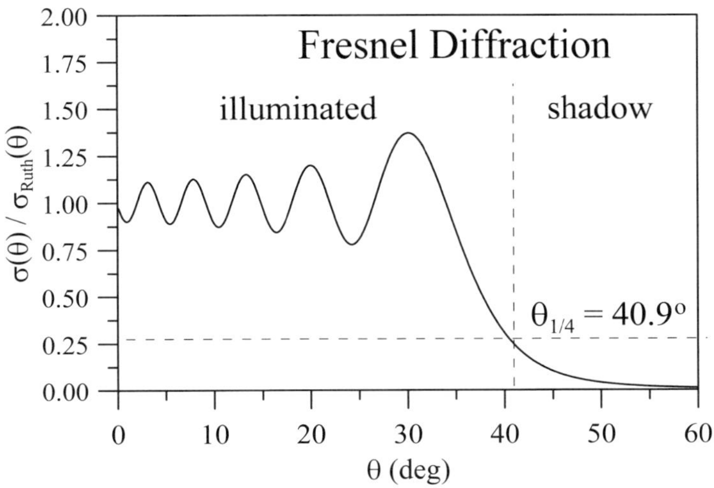

Fig. 7.4 Fresnel diffraction within the sharp cut-off approximation. For details see the text.

Taking the square modulus of the above equation on gets the ratio of the cross section for Fresnel diffraction to the Rutherford cross section. The result can be put in the form

$$\frac{\sigma(\theta)}{\sigma_{\text{Ruth}}(\theta)} = \frac{1}{2}\left[\left(\frac{1}{2} - C(W)\right)^2 + \left(\frac{1}{2} - S(W)\right)^2\right], \tag{7.174}$$

with W given by Eq. (7.170).

Fresnel diffraction within the sharp cut-off approximation is illustrated in figure 7.4. The example is the nuclear $^{16}O + {}^{208}Pb$ collision, discussed by Frahn in [Frahn (1966)]. The grazing angle is $\bar{\Theta} = 40.9^\circ$ and the Sommerfeld parameter is $\eta = 32.05$. The grazing angular momentum, determined through the Rutherford trajectory, is $\bar{\Lambda} = 86$. Frahn [Frahn (1966)] has shown that the results obtained with the sharp cut-off approximation are in qualitative agreement with the data.

The ratio of the diffractive cross section to the Rutherford one, given by Eq. (7.174), has a very interesting property. At the grazing angle $\theta = \bar{\Theta}$, the argument of $C(W)$ and $S(W)$ is $W = 0$. Thus, the Fresnel integrals vanish and the ratio of the two cross sections takes the simple value of $1/4$. In this way, one can estimate the grazing angle as the scattering angle at which the ratio $\sigma/\sigma_{\text{Ruth}}$ has the value of $1/4$ (see excercise 1, chapter 3).

The angular dependence of $\sigma/\sigma_{\text{Ruth}}$ in Fresnel diffraction is similar to the one in rainbow scattering: oscillatory behavior in the illuminate region, where $\theta < \bar{\Theta}$ ($W < 0$), and decaying in the shadow region, where $\theta > \bar{\Theta}$ ($W > 0$). However, the physics involved in each of these processes is entirely different. Fresnel diffraction is dictated by the imaginary part of the potential whereas rainbow scattering is determined by its real part. In the case of Fresnel diffraction, the shadow region is an immediate consequence of the asymptotic value of the Fresnel integrals, illustrated in figure 7.3. For large positive values of the argument, these functions tend to the limit 1/2. Therefore, each of the two terms within the square brackets of Eq. (7.174) tends to zero asymptotically and σ/σ_{Ruth} vanishes. The situation is quite different in the illuminated region, where the argument of the Fresnel integrals is negative. Using the antisymmetric properties of Eq. (7.166), for $W < 0$ we can write

$$\frac{\sigma(\theta)}{\sigma_{\text{Ruth}}(\theta)} = \frac{1}{2}\left[\left(\frac{1}{2} + C\left(|W|\right)\right)^2 + \left(\frac{1}{2} + S\left(|W|\right)\right)^2\right], \tag{7.175}$$

For large $|W|$, both $C(|W|)$ and $S(|W|)$ oscillate around the value 1/2, with decreasing amplitude as $|W|$ grows. Therefore the ratio of Eq. (7.175) oscillates around the value 1, with decreasing amplitude for smaller angles, well below $\bar{\Theta}$. This behavior can be observed in the illuminated side of figure 7.4.

More quantitative information about the Fresnel diffraction cross section within the sharp cut-off approximation can be obtained with the help of the asymptotic forms of the Fresnel integrals. Here, we only mention the main results [Almeida and Hussein (1985)]. The first one is the value of the maximum of $\sigma/\sigma_{\text{Ruth}}$, reached at the first maximum below the grazing angle. One finds

$$\frac{\sigma(\theta)}{\sigma_{\text{Ruth}}(\theta)} = 1.4. \tag{7.176}$$

The second is the approximate expression for the period of oscillation in the illuminated region,

$$\Delta\theta = \left[\frac{2\pi\eta}{\bar{\Lambda}^2 + \eta^2}\right] \frac{1}{\sin\left(\bar{\Theta} - \theta\right)}. \tag{7.177}$$

The sharp cut-off approximation is a poor description of realistic strong absorption situation. Owing to the step function approximation for the modulus of the S-matrix as a function of angular momentum, the Poisson series converges slowly and terms with $m \neq 0$ should be taken into account in Eq. (5.264).

This shortcoming is eliminated in Frahn's strong absorption model (see e.g. [Frahn (1966); Brink (1985)]). In this model, the modulus of the S-matrix is still expressed in terms of the grazing angular momentum. However, $|S(\lambda)|$ does not jump abruptly from zero to 1, as λ reaches $\bar{\Lambda}$. It varies gradually between these limits over a finite range of angular momentum values, characterized by a *diffuseness* parameter, Δ. The smooth cut-off model for Fresnel diffraction is similar to the one described in the previous section, in the case of Fraunhofer diffraction. Frequently, it is also important to include a small correction in the phase of the S-matrix, arising from the short-range part of the real potential. In [Frahn (1966)], an accurate description of the experimental data for $^{16}O + {}^{208}Pb$ elastic scattering was obtained in this way.

It looks surprising that collisions dominated by the Coulomb field plus strong absorption are analogous to Fresnel diffraction in optics. While in the latter the distance between the source and the scatterer (or the distance between the scatterer and the observation point) is comparable to its size, the situation in atomic and nuclear collisions is completely different. In such cases, the projectile and the target have microscopic dimensions whereas the distance between the source of the incident beam and the target is measured in a macroscopic scale. This suggests that atomic and nuclear collisions could be analogous to Fraunhofer diffraction but not to Fresnel diffraction. The reason for the observed Fresnel pattern is that the Coulomb field deflects the incident particles, producing a focalization effect. This is illustrated in figure 7.5, in the case of a repulsive Coulomb field. The figure shows two classical trajectories with grazing angular momentum. In three dimensions, the emergent grazing trajectories describe a cone and its vertex corresponds to a virtual source, placed close to the scatterer.

7.6 Realistic treatments of elastic scattering with absorptive potentials

In the previous sections we have developed the formal scattering theory of absorptive potentials. In the present section we treat the physical effects inflicted by such potentials on scattering observables. In particular we consider the elastic scattering of neutral and charged particles subject to both real and imaginary interactions, and discuss the qualitative changes

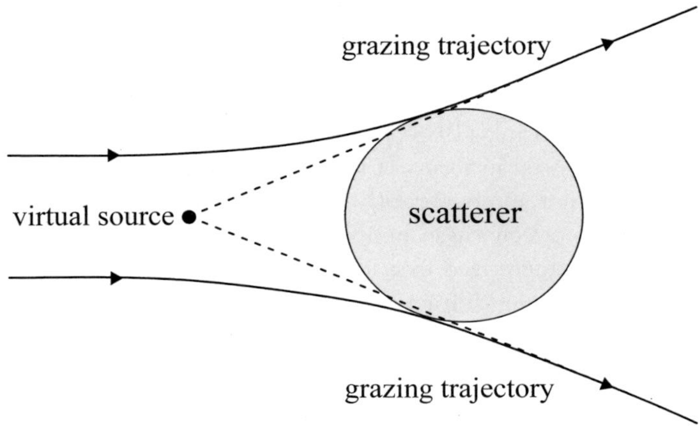

Fig. 7.5 Illustration of the Coulomb focalization effect leading to Fresnel diffraction, which occur, for example, in heavy ion scattering.

that the latter brings on the angular distribution. To be specific, we take the interaction to be,

$$V(r) = U(r) + iW(r) \tag{7.178}$$

The effect of the imaginary part of the potential is to dampen the wave function in the inner region and to supply a sub-unitary S-matrix element, as we discussed before. Besides, it affects the reflection properties in the sense that the real part of the phase shift is modified. This modification in the scattering, can best be seen when analyzing the near- and far-side amplitudes. Although, these amplitudes were introduced and analyzed in chapter 5 using the asymptotic form of the Legendre polynomial, they can in fact be calculated exactly. This can be done through the decomposition of the Legendre polynomial into running waves, as

$$P_l(\cos\theta) = Q_l^{(+)}(\theta) + Q_l^{(-)}(\theta). \tag{7.179}$$

The running waves are defined as [Fuller (1975a)]

$$Q_l^{(\pm)}(\theta) = \frac{1}{2}\left[P_l(\cos\theta) \pm \frac{2i}{\pi}\, Q_l(\cos\theta)\right], \tag{7.180}$$

where $P_l(x)$ is the Legendre polynomial of order l, $Q_l(x)$ is the Legendre function of second kind of the same order and $x = \cos\theta$. For large values of l and $\cos\theta$ not too close to ± 1, the running waves can be approximated by their asymptotic forms

$$Q_l^{(\pm)}(\theta) \to \left(\frac{1}{2\pi l \sin\theta}\right)^{1/2} \exp\left[\pm i\left((l+1/2)\theta - \pi/4\right)\right]. \tag{7.181}$$

These approximations were used to derive the semiclassical scattering amplitudes of chapter 5. With the above, the near-side (-) and far-side (+) amplitudes can be calculated exactly, once the radial Schrödinger equation is integrated and the phase shift is extracted as explained in chapter 2. They are then expressed as,

$$f^{(\pm)}(\theta) = \frac{1}{2ik} \sum_{l}^{\infty} (2l+1) \left[|S_l| \, e^{2i\delta_l} - 1 \right] Q_l^{(\pm)}(\theta). \tag{7.182}$$

We observe the important property,

$$Q_l^{(-)}(-\theta) = Q_l^{(+)}(\theta) \tag{7.183}$$

which results in the identity,

$$f^{(-)}(-\theta) = f^{(+)}(\theta). \tag{7.184}$$

We remind the reader that both the near-side, $f^{(-)}(\theta)$, and the far-side, $f^{(+)}(\theta)$, amplitudes can be written as an infinite Poisson sum of integrals (see section 5.2.3),

$$f^{(\pm)}(\theta) = \sum_{-\infty}^{\infty} f_m^{(\pm)}(\theta). \tag{7.185}$$

The presence of absorption will damp most terms in the above sum except, at most, the $m = 0$ and $m = -1$ terms. We then approximate,

$$f^{(\pm)}(\theta) \simeq f_0^{(\pm)}(\theta) + f_1^{(\pm)}(\theta), \tag{7.186}$$

where the above two terms are integrals in $\lambda = l + 1/2$.

The interference pattern in the calculated angular distribution can best be appreciated with the actual calculation of the near-side and the far-side contributions. Thus, two auxiliary cross sections are introduced,

$$\frac{d\sigma^{(-)}(\theta)}{d\Omega} = |f^{(-)}(\theta)|^2 \tag{7.187}$$

$$\frac{d\sigma^{(+)}(\theta)}{d\Omega} = |f^{(+)}(\theta)|^2. \tag{7.188}$$

In collisions of charged particles, the Coulomb contribution to the far-side component of the scattering amplitude tends to be small. This arises from the fact that the integral over angular momentum giving the semiclassical Coulomb far-side amplitude has no stationary points. Thus, in the scattering from Coulomb plus short-range potentials, it should be a reasonable approximation to include the Coulomb contribution fully in the near-side amplitude. However, this is only true for small angles. In fact, Fuller derived explicit expressions for the near- and far-side Coulomb amplitudes [Fuller (1975b)], which are

$$\frac{f_{\mathrm{C}}^{(-)}(\theta)}{f_{\mathrm{C}}(\theta)} = \frac{1}{1-e^{-2\pi\eta}} - iC(\eta,\theta) \tag{7.189}$$

$$\frac{f_{\mathrm{C}}^{(+)}(\theta)}{f_{\mathrm{C}}(\theta)} = \frac{-e^{-2\pi\eta}}{1-e^{-2\pi\eta}} + iC(\eta,\theta), \tag{7.190}$$

with

$$C(\eta,\theta) = \frac{\left[\sin^2(\theta/2)\right]^{1+i\eta}}{2\pi} S(\eta,\theta). \tag{7.191}$$

Above, the function $S(\eta,\theta)$ is given in terms of an infinite series involving the ψ function [Abramowitz and Stegun (1972)]. It is clear that the magnitude of $f_{\mathrm{C}}^{(+)}$ is a factor $e^{-2\pi\eta}$ smaller than the near-side Coulomb amplitude, $f_{\mathrm{C}}^{(-)}$. This justifies ignoring the former in the near/far calculation in the forward angle region. However, at backward angles the far-side Coulomb amplitude is significant, as it smoothes out the conspicuous, albeit spurious oscillations in the total far-side amplitude.

With the above preliminaries we can identify four types of interference patterns; the near-near (NN), arising from two stationary points of $f^{(-)}$, the far-far (FF), arising from two stationary points of $f^{(+)}$, the near-far (NF), interference of one stationary point of $f^{(+)}$ with one of $f^{(-)}$, and the interference between f_0 and f_1, which we call the Poisson interference pattern (PP). All realistic angular distributions of systems with absorption, can be analyzed within the general scheme of the NN, FF, NF, PP, interference patterns. This can be demonstrated through the realistic calculation to follow.

Let us numerically calculate the elastic scattering of a typical nuclear system, ^{4}He + ^{90}Zr, studied by Put and Paans [Put and Paans (1974)]. We remind that the helium nucleus, i.e. the alpha particle, has 2 protons and 2 neutrons, while Zirconium-90 has 40 protons and 50 neutrons. A phenomenological potential was used by these authors to account for the elastic scattering angular distribution data at several laboratory energies. One of these energies is $E_{\mathrm{lab}} = 79.5$ MeV, which we will discuss below as it exhibits a remnant of a rainbow. The phenomenological nuclear potential in question is $V(r) = U(r) + i\,W(r)$ with $U(r) = U_{\mathrm{N}}(r) + U_{\mathrm{C}}(r)$. The short-range components of the potential, $U_{\mathrm{N}}(r)$ and $W(r)$, are parametrized by the Woods-Saxon shapes

$$U_{\rm N}(r)=\frac{U_0}{1+\exp\left[\left(r-R_r\right)/a_r\right]},\quad W(r)=\frac{W_0}{1+\exp\left[\left(r-R_i\right)/a_i\right]},\tag{7.192}$$

with [Put and Paans (1974)] $U_0=-140.0$ MeV, $W_0=-18.7$ MeV, $R_r=5.50$ fm, $R_i=7.03$ fm, $a_r=0.815$ fm and $a_i=0.575$. The above short-range nuclear potential, the optical potential, is added to the long-range Coulomb potential given by,

$$U_{\rm C}(r)=\frac{1}{4\pi\epsilon_0}\,\frac{Z_1Z_2e^2}{r},\qquad \text{for } r\geq R_{\rm C}\tag{7.193}$$

$$=\frac{1}{4\pi\epsilon_0}\,\frac{Z_1Z_2e^2}{2R_{\rm C}}\left[3-\frac{r^2}{R_{\rm C}^2}\right],\qquad \text{for } r<R_{\rm C},\tag{7.194}$$

with $R_{\rm C}=5.50$ fm and the atomic numbers $Z_1=2$ and $Z_2=40$. Thus the full real potential is,

$$V(r)=U_{\rm N}(r)+U_{\rm C}(r).\tag{7.195}$$

For the purpose of understanding the structure of the data and the calculation we first evaluate the real effective potential defined by,

$$U_{\rm eff}(r,l)=U_{\rm N}(r)+U_{\rm C}(r)+\frac{\hbar^2}{2\mu}\,\frac{l(l+1)}{r^2}.\tag{7.196}$$

This potential is illustrated in figure 7.6, for a few values of the orbital angular momentum, l. Clearly $U_{\rm eff}(r,l)$, shows maxima, $U_{\rm eff}(r_m,l_m)$, at all values of $l\leq 26$. At the energies $E_{\rm c.m.}=U_{\rm eff}(r_m,l_m)$, the colliding system exhibits orbiting, as discussed in Chapter 5 (Eqs. (5.404) and (5.405)). The maximum of the $l=26$ effective potential is very close to an inflection point, indicating that above $l=26$ the effective potential grows monotonically as r decreases. Thus, $E_c=V(r_c,l=26)$ is approximately the transitional energy above which orbiting ceases to exist, and the scattering would be dominated by rainbow.

Now, we compare the theoretical cross section with experimental results. We consider the data of [Put and Paans (1974)], which were taken at several laboratory α particle energies. In figure 7.7, we exhibit the case at $E_{\rm lab}=79.6$ MeV or, equivalently, a center of mass energy of $E_{\rm c.m.}=76.1$ MeV, which coincidentally is very close to the transitional energy, E_c. The agreement between the experimental and the theoretical results is quite reasonable. One sees in the data the NF interference at small angles, followed by damped rainbow oscillations with the first Airy minimum quite conspicuous at 85 degrees. It corresponds to the nuclear rainbow, associated with

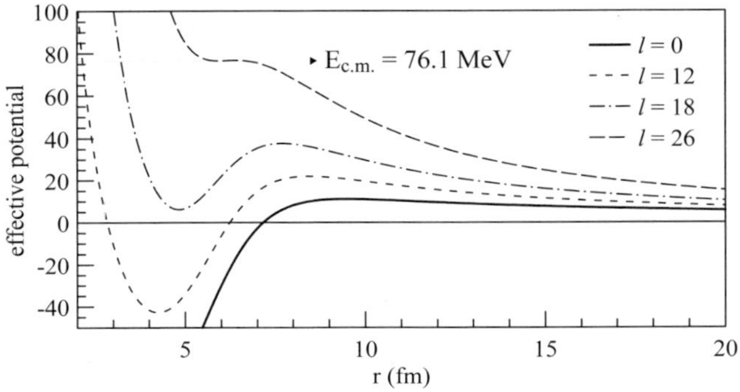

Fig. 7.6 Radial dependence of the effective potential of Eq. (7.196), for the partial-waves $l = 0$, 12, 18, and 26. The collision energy of the detailed discussion in the text is indicated by a horizontal line.

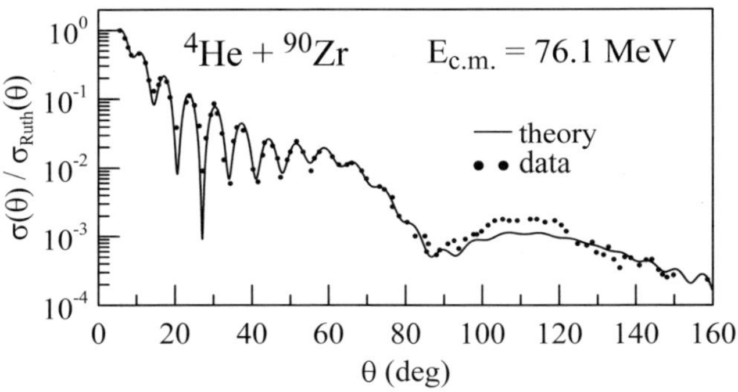

Fig. 7.7 The differential cross divided by the Rutherford cross section at $E_{\mathrm{c.m.}} = 76.1$ MeV. The data points are from [Put and Paans (1974)] and the solid line corresponds to a quantum mechanical calculation with the complex potential discussed in the text.

negative values of the deflection function. As discussed above, the energy here is very close to the transitional energy, and the rainbow is not quite formed as a well behaved minimum in the deflection function. In fact one would expect an almost orbiting case. To demonstrate this we exhibit the classical deflection function for this case in the top panel of figure 7.8. At $l = 26$ one finds an onset of orbiting, though not a singular one ($\theta \to -\infty$). The sharp very deep rainbow/orbiting situation above makes the discussion of the interference patterns of the angular distribution at $E_{\mathrm{c.m.}} = 76.1$ MeV rather complicated.

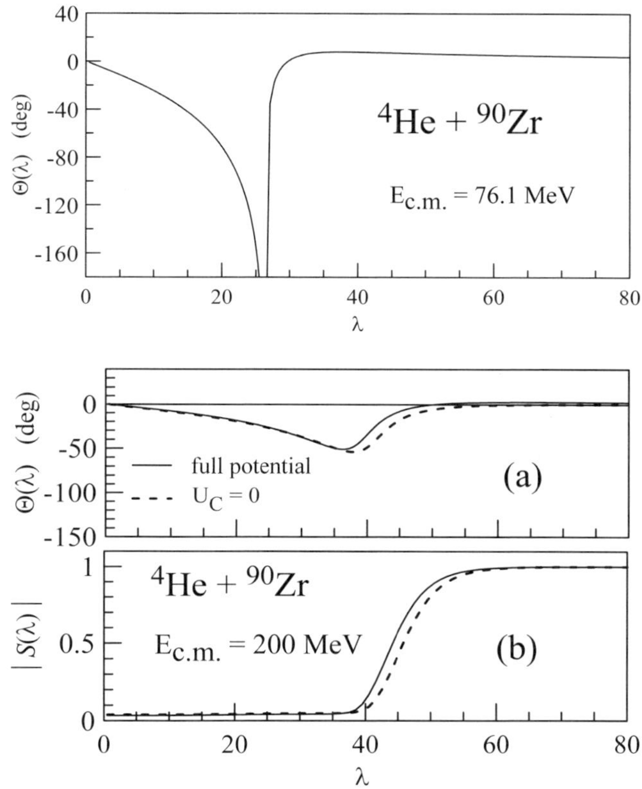

Fig. 7.8 Deflection functions for the ^{4}He + ^{90}Zr system at $E_{\text{c.m.}} = 76.1$ MeV (top panel) and $E_{\text{c.m.}} = 200$ MeV (panel (a)). The modulus of the S-matrix at $E_{\text{c.m.}} = 200$ MeV is shown in panel (b). The dashed lines in the calculations at $E_{\text{c.m.}} = 200$ MeV were obtained with the Coulomb field switched off.

To avoid this problem, we consider the higher energy of $E_{\text{c.m.}} = 200$ MeV, at which, no orbiting would be present, but just a mild rainbow. This is illustrated in the middle panel of figure 7.8 (indicated by (a)). One observes a rainbow at $\Theta(\lambda) \simeq 51$ degrees for $\lambda_r \simeq 36$. In the bottom panel of the figure (indicated by (b)), we show the absolute values of $|S(\lambda)|$, using the same angular momentum scale on the horizontal axis. Inspecting the λ-dependence of the modulus of the S-matrix, we conclude that the contribution from the stationary points below λ_r should be strongly attenuated by absorption. For comparison, we show also the 200 MeV deflection function and S-matrix in the absence of the Coulomb field (dashed lines).

We now present a detailed discussion of the structure of the angular distribution at $E_{\text{c.m.}} = 200$ MeV, considering four different scenarios:

(1) Scattering with the full potential containing repulsive Coulomb, short-range attractive nuclear and absorptive components.
(2) Scattering with the absorptive potential set equal to zero, $W(r) = 0$.
(3) The short-range nuclear component of the potential set to zero, $V_N(r) = 0$.
(4) The Coulomb potential is turned off to simulate neutral particle scattering.

The differential cross sections for the situations listed above are shown in figure 7.9. The curves in the top panel take into account the Coulomb field whereas in the bottom panel the Coulomb potential is set equal to zero. The angular distribution for the full potential shows very sharp NF interference, followed by the first Airy minimum at $\theta \simeq 25^o$, which cannot be seen clearly because this angular region is still affected by the NF oscillations. It then presents the usual drop of the far-side dominated cross section at angles in the shadow region of the nuclear rainbow. Again, an important point to emphasize is the absence of more Airy minima, as compared to the case of atomic scattering with the Lennard-Jones potential considered in chapter 5. This is a consequence of both absorption, which *hides* the inner part of the deflection function, and, when absorption is absent, the smallness of the inner contribution compared to the outer one due to the smaller pre-exponential factor alluded to above, in the former.

The elastic scattering of neutral particles can be discussed by setting the Coulomb interaction equal to zero. Angular distributions for $U_{\text{C}} = 0$ are shown in the lower panel of figure 7.9. The meaning of each curve is indicated inside the figure. Here the case with no absorption comes out similar to the atomic rainbow scattering considered in chapter 5 (see figures 5.14 and 5.15), while the case with both Coulomb and nuclear interactions are turned off, $U_{\text{C}} = U_{\text{N}} = 0$, resembles very much the Fraunhofer diffraction of neutral particles in the smooth-cutoff limit considered in the previous section, and shown as the dashed curve in figure 7.2. Therefore the angular distribution exhibits typical rainbow features in the far-side scattering, while the near side shows the effect of repulsion due to absorption. There are no NN and strong FF interferences, nor NF interference when the nuclear attraction is turned off.

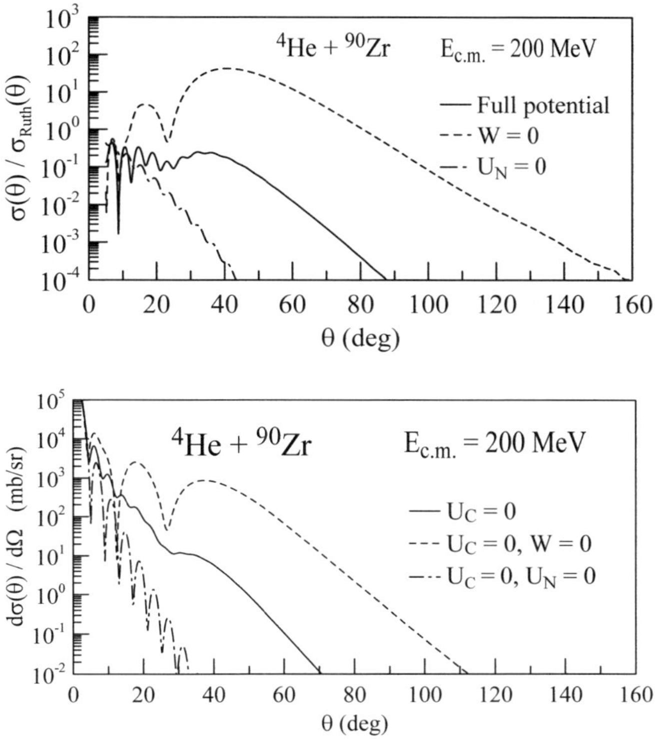

Fig. 7.9 Angular distributions in ^{4}He + ^{90}Zr scattering at $E_{\text{c.m.}} = 200$ MeV for the different scenarios listed in the text. On the top panel the Coulomb field is taken into account whereas on the bottom panel it is switched off. The meaning of each curve is indicated in legend of the figures.

The structure of the angular distribution can be better analyzed if one plots separately the auxiliary cross sections $d\sigma^{(-)}/d\Omega$ and $d\sigma^{(+)}/d\Omega$, of Eqs. (7.187) and (7.188). This is done in figure 7.10. Here, all calculations include both $U_{\text{N}}(r)$ and $W(r)$. The results in the top panel take into account the effects of the Coulomb field, whereas those in the bottom panel do not. In the calculations including the Coulomb field (top panel), the near-side cross section dominates at very forward angles, but drops rapidly as θ increases. It reaches a minimum at 40 degrees and then rises at back angles. The far-side cross section, which is small at very forward angles, grows rapidly with θ, until it approaches the rainbow region. The two cross sections cross at $\theta \simeq 10$ degrees. Accordingly, the total cross section exhibits very strong NF interference in the neighborhood of $\theta = 10$ degrees. Above

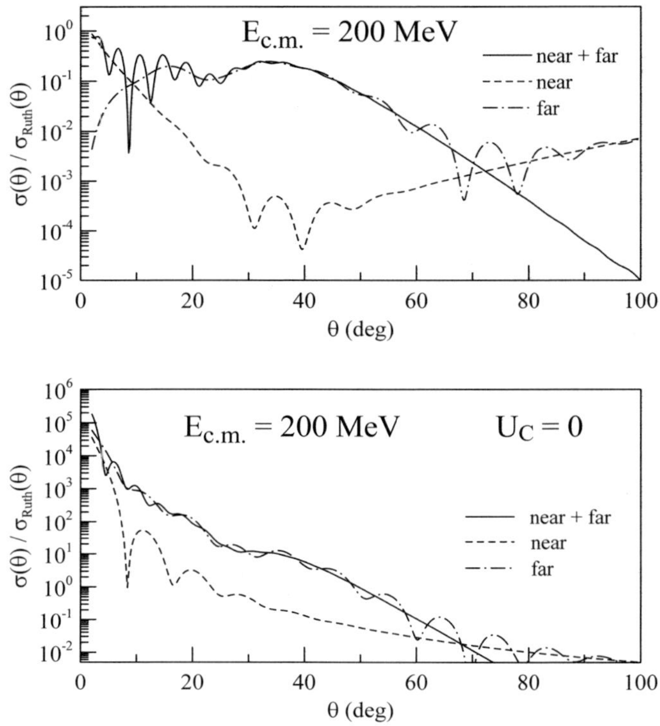

Fig. 7.10 The near/far decomposition of the differential cross section divided by the Rutherford cross section. The energy and the system are the same as in the previous figure and the meaning of each curve is indicated in the legend. On the top panel the Coulomb field is taken into account whereas on the bottom panel it is switched off.

this region, the far-side cross section exhibits a typical rainbow pattern damped by absorption and the asymmetrical deflection function. Superimposed on this are the spurious oscillations which would be smoothed out had we included the far-side Coulomb amplitude, as mentioned above and discussed in details by Fuller [Fuller (1975b)]. Thus, the total cross section, contains the NF interference in the forward angle region, and the partial rainbow. There is no FF interference, owing to the absence of orbiting. Studying the cross section in the region where the far-side is dominant, would supply much needed information about the attractive short range nuclear interaction which mostly contributes to the negative branch of the deflection function. An important activity in nuclear scattering research is how to devise a potential that can result in easily identifiable optical scattering pattern. Since W repels, as well as absorbs, one can fiddle with the potential to increase the repulsion and enhance the near-side, while damping the far-side, to make the angular distribution of the desired form. In particular, in many instances one would like to get as close as possible to the Fraunhofer pattern, characterized by the almost equality of $|f^{(-)}(\theta)|$ and $|f^{(+)}(\theta)|$. We leave it as an exercise to find the conditions on the potentials $U_{\rm N}(r), W(r)$ and $U_{\rm C}(r)$, that would lead to perfect equality of these amplitudes,

$$|f^{(-)}(\theta)| = |f^{(+)}(\theta)|. \tag{7.197}$$

A corresponding Near-Far study of the neutral case is shown in the lower panel of figure 7.10. This is simulated setting the Coulomb potential equal to zero. In this case, the cross section is completely dominated by the far-side contribution, arising from nuclear attraction. Once again, one sees the spurious oscillations in the far-side amplitude which would be gone if the far-side Coulomb amplitude were added in the calculation. Further, one sees a practically constant near-side cross section.

Finally, in figure 7.11, we show the absorption cross section, $\sigma_{\rm a}$, for the system as a function of both $E_{\rm c.m.}$ (panel (a)) and $1/E_{\rm c.m.}$ (panel (b)). The full curves represent the absorption cross section for the actual charged system, while the dashed curves are the absorption cross section when the charges are turned off. The charged particle results exhibit clearly the effect of barrier penetration at lower energies, and a practically geometrical behavior at higher energies. In fact the dependence of $\sigma_{\rm a}$ on $1/E_{\rm c.m.}$ is practically linear at higher energies. To understand the result, we remind that the absorption cross section is just,

$$\sigma_{\rm a} = \frac{\pi}{k^2}\sum_{l=0}^{\infty}(2l+1)\ [1-|S_l(E)|^2] = \frac{\pi}{k^2}\sum_{0}^{\infty}(2l+1)\ T_l(E), \tag{7.198}$$

where $T_l(E) = 1-|S_l(E)|^2$ is the transmission coefficient. For the charged particle scattering considered here, the Coulomb barrier has an important röle in absorption. To see this, we consider the parabolic approximation for the barrier, discussed in chapter 5, and consider the effect of W as merely to capture the flux that passes the barrier into the region of attraction. Then, taking the real potential in the barrier region to be approximated by Eq. (5.261), as $V(r) = V_{\rm B} - 1/2\mu\omega^2\ (r-R_{\rm B})^2$, we get for the transmission coefficient the Hill-Wheeler formula (Eq. (5.262),

$$T_l(E) = \frac{1}{1+\exp\left\{2\pi\left[E-\hbar^2\ l(l+1)/2\mu R_{\rm B}^2 - V_{\rm B}\right]/\hbar\omega\right\}} \tag{7.199}$$

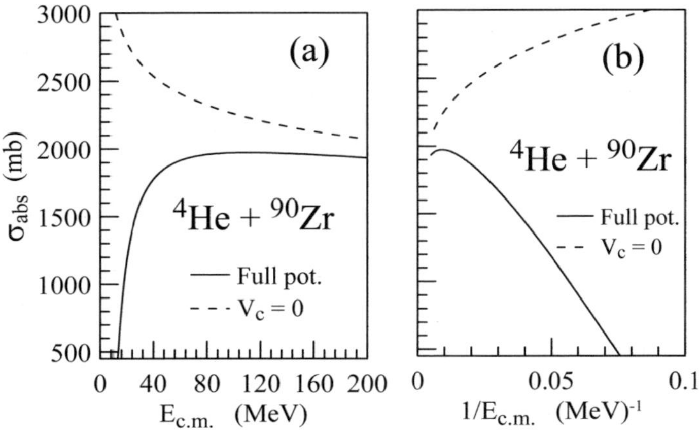

Fig. 7.11 The absorption cross section in ^{4}He + ^{90}Zr scattering. The cross section is plotted both against $E_{\text{c.m.}}$ (panel (a)) and against $1/E_{\text{c.m.}}$ (panel (b)). The solid lines correspond to calculations including the Coulomb potential whereas the dashed lines were obtained with $V_C = 0$.

The calculation of the absorption cross section is made doable by approximating the l-sum by an integral over λ, $l \to \lambda = l + 1/2, l(l+1) \to \lambda^2$ (see section 5.2.3)

$$\sigma_a = \frac{2\pi}{k^2} \int_0^\infty \frac{\lambda \, d\lambda}{1 + \exp\left\{2\pi \left[E - \hbar^2\lambda^2/2\mu R_B^2 - V_B\right]/\hbar\omega\right\}}. \tag{7.200}$$

The above can be evaluated to give what is known as the Wong formula,

$$\sigma_a = R_B^2 \, \frac{\hbar\omega}{2E} \, \ln\left\{1 + \exp\left[\frac{2\pi(E - V_B)}{\hbar\omega}\right]\right\}. \tag{7.201}$$

The above formula for absorption of charged particles has the two desired limits; exponential decrease at energies below the barrier height, and a geometrical behavior at higher energies,

(1) Low energies, $E \ll V_B$, $\ln(1+x) \to x$,

$$\sigma_a = R_b^2 \, \frac{\hbar\omega}{2E} \, \exp\left[\frac{-2\pi|E - V_b|}{\hbar\omega}\right] \tag{7.202}$$

(2) Higher energies, $E \gg V_B$, $\ln[1 + e^y] \to y$,

$$\sigma_a = \pi R_B^2 \, \left[1 - \frac{V_B}{E}\right]. \tag{7.203}$$

Clearly the Wong formula captures most of the salient features of $\sigma_{\rm a}$; exponential drop with decreasing center of mass energy, and a linear dependence on $1/E_{\rm c.m.}$ at higher energies, just as the exact calculation of figure 7.11 (b).

At very low energies, the parabolic approximation is not appropriate and one has to evaluate the absorption of charged particles using the $l = 0$ tunneling of the Coulomb barrier. The exact details of the complex nuclear potential are not important, except for the radius at which it starts acting, R. Once the flux reaches R, absorption sets in and the particle is captured. Then through the use of the WKB tunneling expression of a barrier composed of a $1/r$ potential and a square well at R, we find for the absorption cross section,

$$\sigma_{\rm a} \simeq \frac{\pi}{k^2}\, T_0(E) = \frac{\pi}{k^2}\, \exp\left[-2\int_R^{2a} dr\,\sqrt{\frac{2\mu}{\hbar^2}\left(E - \frac{Z_1Z_2e^2}{4\pi\epsilon_0 r}\right)}\right], \quad (7.204)$$

where $a = Z_1Z_2e^2/4\pi\epsilon_0 E$, is half the distance of closest approach in a head-on collision, introduced in chapter 3. The above can be evaluated in closed form to give,

$$\sigma_{\rm a} = \frac{\pi\hbar^2}{2\mu E}\, \exp\left\{-2\eta\left[\pi - \sqrt{\frac{R}{2\mu}}\left(1 - \frac{R}{2a}\right) - \sin^{-1}\left(\sqrt{R/2a}\right)\right]\right\}, \quad (7.205)$$

where $\eta = ka$ is the Sommerfeld parameter (see chapter 3). Since in most cases $a \gg R$, one may approximate the above by,

$$\sigma_{\rm a} = \frac{\pi\hbar^2}{2\mu E}\; e^{-2\pi\eta}. \quad (7.206)$$

The factor $e^{-2\pi\eta}$ is called the Gamow factor and it plays an important rôle in astrophysically relevant charged particle reactions, where the energies involved are extremely low. We discuss this problem further in section 12.5.

The absorption cross section in the neutral particle case, $U_{\rm C} = 0$, shown as the dashed curves in figure 7.11, increases with $1/E$, and, at very low energies, shows the well known $1/\sqrt{E}$ ($1/k$ or $1/v$) behavior (see e.g. [Blatt and Weisskopf (1991)]). This behavior is very important in application to thermal neutron capture reactions, the main process used in nuclear technologies. Of course the actual cross section here is full of resonances, called compound nuclear resonances to be discussed in chapter 10. The average cross section, where the resonances are averaged out, can be calculated using complex potential scattering. It is this cross section which shows the $1/k$ behavior. In atomic and molecular systems, the low energy scattering is described by a real potential. However, one does encounter cases where a complex interaction is required. One such case is the elastic scattering of excited molecules encountered in cold gases physics. The absorption in this case is caused by transitions to the ground state of the molecule. The usual effective range description which is employed here, would then require dealing with complex scattering lengths and effective range. We refer the reader to [Balakrishnan *et al.* (1997)] for a more detailed discussion. Similar considerations are used in the description of the scattering of hydrogen from anti-hydrogen [Froelich *et al.* (2000)]. The discussion of [Balakrishnan *et al.* (1997)] and [Froelich *et al.* (2000)] relies on the low-energy form of the scattering amplitude discussed in section 2.11.1, namely,

$$f = \frac{1}{-\frac{1}{a} - ik} = -\frac{a}{1 + ika}. \tag{7.207}$$

The absorption cross section can be obtained from the generalized optical theorem relating the total cross section, σ_{t}, with the imaginary part of the forward scattering amplitude. In the low energy limit, only the s-wave component is relevant and we can write

$$\sigma_{\mathrm{t}} = \sigma_{\mathrm{el}} + \sigma_{\mathrm{a}} = \frac{4\pi}{k} \operatorname{Im}\{f\}. \tag{7.208}$$

The integrated elastic cross section is then given by $\sigma_{\mathrm{el}} = 4\pi \left|f\right|^2$. Using this result in the above equation, we get the absorption cross section as

$$\sigma_{\mathrm{a}} = \frac{4\pi}{k} \left(\operatorname{Im}\{f\} - k\left|f\right|^2\right). \tag{7.209}$$

In the case of scattering from complex potentials, the scattering length is taken to be complex and we can write

$$a = a_r + i\, a_i. \tag{7.210}$$

Using the above equation in Eq. (7.207), we obtain

$$\operatorname{Im} f = \frac{k\,\left|a\right|^2 - a_i}{1 + k^2\,\left|a\right|^2}, \qquad \left|f\right|^2 = \frac{\left|a\right|^2}{1 + k^2\,\left|a\right|^2} \tag{7.211}$$

and inserting these results into Eq. (7.209) we get

$$\sigma_{\mathrm{a}} = \frac{4\pi}{k} \left[\frac{-\,a_i}{1 + k^2\,\left|a\right|^2}\right]. \tag{7.212}$$

At low energies, the term in k^2 in the denominator can be neglected and the cross section reduces to

$$\sigma_{\mathrm{a}} = (-4\pi\, a_i)\ \frac{1}{k}. \tag{7.213}$$

Clearly a_i must be negative! The above equation demonstrates the famous $1/k$ law.

The $1/\sqrt{E}$, or $1/k$, low-energy behavior of σ_{a} for neutral particles, is easily seen by considering the $l = 0$ transmission and considering the limit of a complex square well potential with radius R and the complex strength $V_0 = U_0\,(1 + i\,\chi)$, with χ standing for the ratio of its imaginary to its real part. Using Eq. (2.84) for $\tan\delta_0$, where now δ_0 is complex, the absorption cross section, can be written as,

$$\sigma_{\mathrm{a}} = \frac{\pi}{k^2}\left[1 - \left|e^{2i\delta_0}\right|^2\right] = \frac{\pi}{k^2}\left[1 - \left|\frac{(1 - i\tan\delta_0)^2}{(1 + \tan^2\delta_0)}\right|^2\right], \tag{7.214}$$

where the complex $\tan\delta_0$ is, from Eq. (2.84),

$$\tan\delta_0 = \frac{\frac{k}{K}\tan(KR) - \tan(kR)}{1 + \frac{k}{K}\tan(KR)\tan(kR)}, \tag{7.215}$$

with, $K = \sqrt{2\mu/\hbar^2(E + V_0(1 + i\chi))}$, and $k = \sqrt{2\mu E/\hbar^2}$. It is an easy matter to show that the absorption cross section behaves as $1/k$ at very low energy. Further, the above formula for σ_{a} exhibits oscillations as a function of the potential radius, R. This behavior is well known in nuclear physics, and these resonances are referred to as shape resonances. They occur at specific values of R, R_n. The obtention of R_n is left as an exercise (exercise 6).

Exercises

(1) Prove the relation $\left(S^{-1}\right)_{\mathbf{k}',\mathbf{k}} = \langle\tilde{\psi}^{(+)}_{\mathbf{k}'} \,|\, \tilde{\psi}^{(-)}_{\mathbf{k}}\rangle\,.$

(2) Show that the partial-wave expansion of S^{-1} is

$$\left(S^{-1}\right)_{\mathbf{k}',\mathbf{k}} = \delta(E_{k'} - E_k)\ \frac{\hbar^2}{\mu k}\ \sum_{lm} Y_{lm}(\hat{\mathbf{k}}')Y^*_{lm}(\hat{\mathbf{k}})\frac{e^{-2i\delta_l}}{|S_l|}$$

and then go through the details of the derivation of Eqs. (7.101) and (7.102).

(3) Consider the case of pure diffractive scattering within the sharp cut-off approximation for the S-matrix,

$$S_l \equiv |S_l|\, e^{2i\delta_l} \quad \text{with} \quad \delta_l = 0 \quad \text{and} \quad |S_l| = \Theta(l - \bar{L}).$$

Above, $\Theta(l - \bar{L})$ is the step function at the cut-off angular momentum, $\bar{L} = k\bar{R}$, with $\bar{R}$ representing the radius of the scatterer.

Show that evaluating the partial-wave series for the scattering amplitude one gets,[11]

$$f(\theta) = \frac{\bar{L}}{2ik}\left[\frac{P_{\bar{L}}(\cos\theta) - P_{\bar{L}-1}(\cos\theta)}{1 - \cos\theta}\right].$$

(4) Calculate the scattering amplitude and the cross section at forward angles, in the case of sharp cut-off absorption. Assume that the cut-off angular momentum is large ($\bar{L} \gg 1$) and that the scattering angle is small enough for the use of the approximation: $P_l(\cos\theta) \simeq J_0(\lambda\theta)$, with $\lambda = l + 1/2 \simeq l$. In this case, the angular momentum can be handled as a continuous variable. Show that the result agrees with the one obtained semiclassically[12] (considering only the leading term in the Poisson series):

$$f(\theta) = i\,\bar{R}\,\frac{J_1(\bar{L}\theta)}{\theta} \quad \text{and} \quad \frac{d\sigma(\theta)}{d\Omega} \simeq \bar{R}^2\left[\frac{J_1(\bar{L}\theta}{\theta}\right]^2.$$

(5) Evaluate the total cross section

$$\sigma_{\rm t} = \sigma_a + \int \frac{d\sigma(\theta)}{d\Omega}\, d\Omega$$

in the case of sharp cut-off absorption. Then show that the optical theorem, in its version generalized for complex potentials (Eq. (7.24)) is valid.

---

[11] *Hint*: Do not use the semiclassical approximation for the sum over partial-waves. Use the modified expression for the scattering amplitude arising from the recursion relation of Legendre polynomials, given by Eq. (2.55).

[12] *Hint*: approximate

$$P_{\bar{L}}(\cos\theta) - P_{\bar{L}-1}(\cos\theta) \simeq J_0(\bar{L}\theta) - J_0\left((\bar{L}-1)\theta\right),$$

and use the fist order expansion

$$J_0\left(x + \Delta x\right) \simeq J_0(x) + \left[\frac{dJ_0(x)}{dx}\right]_x \Delta x, \quad \text{and the relation} \quad \frac{dJ_0(x)}{dx} = -J_1(x).$$

(6) Calculate the positions of the shape resonances using the expression of Eq. (7.214). Use the approximation $\chi \ll 1$ and thus write $K = K_0 + i\kappa$. With this you should be able to find the condition of maxima in $\sigma_{\rm a}$. This then results in the values of the R-resonances,

$$R_n = \left(n + \frac{1}{2}\right)\frac{\pi}{K_0}$$

Chapter 8

Additional Topics

In this chapter, we complement our study of potential scattering with the discussion of four topics which have been left out of the previous chapters. The first is the analytical continuation of the S-matrix in the complex l-, E- and k-planes. We show that the poles of the S-matrix in the complex planes are directly related to resonances and bound states of the system. The second topic is the problem of potential scattering in one-dimensional and two-dimensional spaces. The third topic is the inverse scattering problem. That is: what can we learn about the potential looking at scattering data. Finally, the fourth is a brief discussion of the Calogero's equation for the scattering of neutral particles.

8.1 Analytical properties of the S-matrix

8.1.1 *The Jost function*

We begin this section by introducing the Jost function, which plays a central role in the analytical continuation of the S-matrix.

We have seen that different normalizations of the radial equation can be used. This is because the relevant information for the scattering cross section is contained in its logarithmic derivative, which is independent of the normalization. As a solution of a second order differential equation, the radial wave function requires two conditions to be fully specified. Usually, these conditions are the regularity of u_l and its asymptotic form[1]. That is

$$u_l(k, r \to 0) = 0 \tag{8.1}$$

[1]In fact, the numerical integration of the radial equation starts from the origin. Thus, it requires the values of $u_l(k,r)$ and $u_l'(k,r)$ at $r=0$. One sets $u_l(k,0)=0$ and takes an arbitrary value for $u_l'(k,0)$. As discussed in section 2.2, the choice of $u_l'(k,0)$ only affects the normalization of the radial wave function. Eventually the wave function is renormalized to have the asymptotic form of Eq. (8.2). In this way, the radial wave wave function is specified by Eqs. (8.1) and (8.2).

and (Eq. (2.35) with $\gamma = i/2$)

$$u_l(k,R) = \frac{i}{2}\left[\hat{h}_l^{(-)}(kR) - S_l(k)\,\hat{h}_l^{(+)}(kR)\right]. \tag{8.2}$$

Above, R is the matching radius, which can take any value beyond the potential range $\left(R \geq \bar{R}\right)$. To stress the energy-dependence of the S-matrix, we have used the notation $S_l(k)$, where $k = \sqrt{2\mu E}/\hbar$. The square brackets in Eq. (8.2) corresponds to the superposition of an incoming spherical wave with an outgoing spherical wave. The amplitude of the former is 1 while that of the latter is $S_l(k)$. For real potentials the strengths of the two waves are the same and the multiplication by $S_l(k)$ only shifts the phase of the outgoing wave. The resulting radial wave functions satisfy the orthonormality relation (see Eq. (4.173))

$$\int_0^\infty u_l^*(k,r)\,u_l(k',r)\,dr = \frac{\pi}{2}\,\delta(k-k').$$

According to the above procedure, the scattering wave function is determined from its values at two different points: $r = 0$ and $r = R$.

In some situations it is more convenient to determine the scattering wave function by its behavior at the origin[2], exclusively. A convenient choice is to normalize the radial wave function so that it converges to the Ricatti-Bessel function in the $r \to 0$ limit. That is,

$$\tilde{u}_l(k,0) = \hat{\jmath}_l(0) = 0 \quad \text{and} \quad u_l'(k,0) = \left[\frac{d\hat{\jmath}_l(kr)}{dr}\right]_{r=0}. \tag{8.3}$$

Above, we use the notation $\tilde{u}_l$ to stress the fact that the normalization of this solution differs from those of u_l and of $\bar{u}_l$ (see chapter 2). Outside the potential range, $\tilde{u}_l$ must be a linear combination of $\hat{h}_l^{(-)}$ and $\hat{h}_l^{(+)}$. Assuming that the potential is real, the coefficients must have the same modulus so that we can write

$$\tilde{u}_l(k, r\to\infty) = \frac{i}{2}\left[\mathcal{J}_l(k)\,\hat{h}_l^{(-)}(kr) - \mathcal{J}_l^*(k)\,\hat{h}_l^{(+)}(kr)\right]. \tag{8.4}$$

The k-dependent coefficient $\mathcal{J}_l(k)$ is called the *Jost function* and it is uniquely determined by the conditions of Eq. (8.3). To find an expression for $\mathcal{J}_l(k)$, we first look for an integral equation for $\tilde{u}_l(k,r)$. It can be checked by direct substitution in the radial equation (exercise 1) that $\tilde{u}_l(k,r)$ satisfies the Volterra type integral equation

$$\tilde{u}_l(k,r) = \hat{\jmath}_l(kr) + \int_0^\infty dr'\,\tilde{g}_l(r,r')\,V(r')\,\tilde{u}_l(k,r'), \tag{8.5}$$

[2]See, e.g., the variable phase method [Calogero (1967)].

with the Green's function

$$\tilde{g}_l(r,r') = \frac{2\mu}{\hbar^2 k}\left[\hat{\jmath}_l(kr)\,\hat{n}_l(kr') - \hat{\jmath}_l(kr')\,\hat{n}_l(kr)\right]\,\Theta(r-r'), \tag{8.6}$$

or, equivalently,

$$\tilde{g}_l(r,r') = \frac{\mu i}{\hbar^2 k}\left[\hat{h}_l^{(-)}(kr)\,\hat{h}_l^{(+)}(kr') - \hat{h}_l^{(+)}(kr)\,\hat{h}_l^{(-)}(kr')\right]\,\Theta(r-r'). \tag{8.7}$$

Above, $\Theta(r-r')$ is the usual step function,

$$\begin{aligned}\Theta(r-r') &= 1 \quad \text{for } r \geq r', \\ &= 0 \quad \text{for } r < r'.\end{aligned}$$

Replacing in Eq. (8.5) $\hat{\jmath}_l(kr) = (i/2)\left[\hat{h}_l^{(-)}(kr) - \hat{h}_l^{(+)}(kr)\right]$, taking the $r \to \infty$ limit and comparing with Eq. (8.4), we get

$$\mathcal{J}_l(k) = 1 + \int_0^\infty d(kr)\,\hat{h}_l^{(+)}(kr)\,\frac{V(r)}{E}\,\tilde{u}_l(k,r). \tag{8.8}$$

It can be shown that for physical values of k, i.e. real and non-negative, this integral always converges. Since in most relevant cases the effects of the potential for very large r-values are negligible (with the exception of the Coulomb potential) one can approximate $V(r > \bar{R}) = 0$. In this case, the integrand is always bound and the proof of convergence is immediate. Throughout this section we will restrict our discussion to finite-range potentials, where this approximation holds.

8.1.2 *Analytical continuation in the complex k- and E-planes*

Although only real and non-negative values of k are meaningful to the scattering problem, the radial equation can be generalized as to include complex values of the wave number (analytical continuation). The usefulness of such solutions will be discussed in the present section.

As a starting point, we should show that the analytical continuation discussed here is uniquely defined. For this purpose, we begin with the introduction of some basic concepts and properties of functions of a complex variable. The first one is the concept of an analytical function. A function $f(z)$ of a complex variable z is said to be analytical at some point z_0 if it has a uniquely defined derivative at this point and also at all points in some neighborhood of z_0. The function is analytical on a domain $\mathcal{D}$ if it is analytical at all points on this domain. If the analyticity domain is the full complex z-plane, $f(z)$ is called an *entire function.*

It is of fundamental importance that the analytical continuation of the radial wave funtion be an analytical function of the variable k. To discuss this point, we introduce the following theorem: *if a function $f(z)$ is analytical on the domain $\mathcal{D}$, the function $f^*(z^*)$ is analytical on the domain $\mathcal{D}^*$.* The analyticity of $f(z)$ guarantees the existence of the derivative $df/dz = a + bi$, where a and b are real numbers. If z' is some point on $\mathcal{D}^*$, we have

$$\frac{df^*(z'^*)}{dz'} = \left[\frac{df(z'^*)}{dz'^*}\right]^* . \tag{8.9}$$

Calling in Eq. (8.9) $z'^* = z$, z is a point on $\mathcal{D}$ so that we can write

$$\frac{df^*(z'^*)}{dz'} = \left[\frac{df(z)}{dz}\right]^* = (a+bi)^* = a - bi.$$

Therefore, the derivative $df^*(z'^*)/dz'$ does exist and $f^*(z'^*)$ is an analytical function on the domain $\mathcal{D}^*$. Owing to this theorem it is convenient to rewrite Eq. (8.4) in the form

$$\tilde{u}_l(k, r > \bar{R}) = \frac{i}{2}\left[\mathcal{J}_l(k)\,\hat{h}_l^{(-)}(kr) - \mathcal{J}_l^*(k^*)\,\hat{h}_l^{(+)}(kr)\right] . \tag{8.10}$$

On the physical region, k is real and Eq. (8.10) is identical to Eq. (8.4). However, the former has an important advantage for complex k-values. When $\mathcal{J}_l(k)$ is an entire function of k, as is the case for real finite-range potentials, the same holds for $\mathcal{J}_l^*(k^*)$.

We now discuss the unicity of the analytical continuation of the radial wave function. For this purpose, we use an important theorem involving complex functions[3]: *If a function $f(z)$ is analytical on a domain $\mathcal{D}$ and vanishes on a continuous line C contained in this domain, then it vanishes on the whole domain.* This theorem has two important consequences. The first one is that the values of an analytical function on a curve C determine the function on any domain which contains this curve. To prove this statement, one makes the hypothesis that two functions $f_1(z)$ and $f_2(z)$ are analytical on a domain $\mathcal{D}$ containing a curve C and coincide on this curve. If one define $f(z) = f_1(z) - f_2(z)$, this function is analytical on $\mathcal{D}$ and vanishes on C. Therefore, $f(z)$ vanishes on the whole domain and

[3]For detailed proofs of this theorem and the subsequent ones, see, for example, [Churchill (1974)] (section 65 and chapter 12).

$f_1(z) = f_2(z)$. This property ensures that $\mathcal{J}_l(k)$ has a unique analytical continuation in the complex plane and so does the radial function $\tilde{u}_l(k,r)$.

The second important consequence of the theorem is the Schwartz reflection principle. It states that: *if an analytical function $f(z)$ is real on the real axis then $f(z) = f^*(z^*)$*. In this case, the function $g(z) = f(z) - f^*(z^*)$ vanishes over the real axis. Therefore it vanishes on the whole analyticity domain and we can write $f(z) = f^*(z^*)$. To apply this relation to the Jost function, we use the property that $\mathcal{J}_l(k)$ is real on the imaginary axis[4]. Next, we introduce a new variable $\kappa = -ik$, which is real for imaginary values of k. Replacing $k = i\kappa$ in the Jost function we get a new function $\mathcal{I}_l(\kappa)$, such that $\mathcal{I}_l(\kappa) = \mathcal{J}_l(k)$. The function $\mathcal{I}_l(\kappa)$ now satisfies the condition for application of the Schwartz reflection principle and we can write

$$\mathcal{I}_l(\kappa) = \mathcal{I}_l^*(\kappa^*).$$

Changing back to the variable k, we get back the function $\mathcal{J}_l$ and the above equation becomes

$$\mathcal{J}_l(-k) = \mathcal{J}_l^*(k^*). \tag{8.11}$$

We are interested in the analytical continuation in the complex k-plane of the radial wave functions $\hat{h}_l^{(+)}(kr)$, $\hat{h}_l^{(-)}(kr)$ and $\tilde{u}_l(k,r)$, and also of the Jost function $\mathcal{J}_l(k)$. It can be easily shown that these radial wave functions are all entire functions. This can be done through the series expansion solutions of the radial equations (see, e.g. [Taylor (2006)]). However, the analyticity of the Jost function must be considered more carefully. For this purpose we use its integral form, given in Eq. (8.8). The convergence of the integral for complex k-values depends on the radial dependence of the potential. For complex k-values the Riccatti-Haenkel functions have the asymptotic values

$$\hat{h}_l^{(+)}(kr \to \infty) = e^{-r\,\mathrm{Im}\,k}\, e^{ir\,\mathrm{Re}\,k}, \tag{8.12}$$

$$\hat{h}_l^{(-)}(kr \to \infty) = e^{r\,\mathrm{Im}\,k}\, e^{-ir\,\mathrm{Re}\,k}. \tag{8.13}$$

Considering the asymptotic form of $\tilde{u}_l$, given in Eq. (8.4), one concludes that the integrand of Eq. (8.8) at large r-values contains terms with orders of magnitude $V(r)\,e^{-2\,r\,\mathrm{Im}\,k}$ and $V(r)$. At the lower half of the complex k-plane, $\mathrm{Im}k < 0$ and $V(r)\,e^{-2\,r\,\mathrm{Im}k}$ diverges if the asymptotic fall off of the potential is not fast enough. In this case, $\mathcal{J}_l(k)$ is not an analytical function of k at the lower half plane. This is the case of potentials falling off as $1/r^n$. In the case of an exponential fall off, namely $V(r \to \infty) = V_0\,\exp(-\gamma r)$,

[4]This property is proved in text books like [Taylor (2006)].

the analytical continuation of $\mathcal{J}_l(k)$ can be performed in the portion of the complex plane with $\mathrm{Im}\,k > -\gamma/2$. The analytical continuation can be extended to the whole complex k-plane if the potential has a finite range. Since in any relevant case, with the exception of the Coulomb potential[5], the potential can be neglected beyond some value $\bar{R}$, $\mathcal{J}_l(k)$ can be approximated by an entire function.

Using Eq. (8.11), Eq. (8.10) can be put in the form

$$\tilde{u}_l(k, r > \bar{R}) = \frac{i}{2}\left[\mathcal{J}_l(k)\,\hat{h}_l^{(-)}(kr) - \mathcal{J}_l(-k)\,\hat{h}_l^{(+)}(kr)\right]. \tag{8.14}$$

Comparing Eqs. (2.35) and (8.14), we obtain the relations

$$u_l(k,r) = \frac{\tilde{u}_l(k,r)}{\mathcal{J}_l(k)}, \tag{8.15}$$

$$S_l(k) = \frac{\mathcal{J}_l(-k)}{\mathcal{J}_l(k)} \tag{8.16}$$

and, recalling that $S_l(k) = \exp\left(2i\delta_l(k)\right)$, we get

$$\delta_l(k) = -\arg\left(\mathcal{J}_l(k)\right). \tag{8.17}$$

Since we are dealing with finite-range potentials, $S_l(k)$ is an analytical function of k on the whole complex k-plane, except at the zeroes of the Jost function, where $S_l(k)$ has poles. The poles of the S-matrix play a very important role, as will be discussed in the next subsections.

8.1.3 *Bound states*

For a bound state with angular momentum l, the radial equation has negative energy, which corresponds to an imaginary wave number. In this case, the wave function outside the range of the potential has the asymptotic form

$$\varphi_l(k, r > \bar{R}) = C\,e^{-\kappa r}, \tag{8.18}$$

where $\kappa \equiv \mathrm{Im}\,k = \sqrt{-2\mu E/\hbar^2}$. Taking the asymptotic form of Eq. (8.14) and considering Eqs. (8.12) and (8.13) for an imaginary wave number, we see that the wave function $\tilde{u}_l$ will have the same asymptotic form as φ_l (Eq. (8.18)) when the coefficient $\mathcal{J}_l(k)$ vanishes. Therefore, bound states of the projectile-target system correspond to zeroes of the Jost function in the upper part of the imaginary axis in the complex k-plane. Eq. (8.16)

[5] For the extension of the present section to the Coulomb potential, we refer to Newton [Newton (2002)].

indicates that these zeroes give rise to poles in the S-matrix. Thus, bound states correspond to poles of $S_l(k)$ on the imaginary k-axis.

One frequently discusses the location of poles of the S-matrix in the complex E-plane. If $k = \sqrt{2\mu/\hbar^2}\,E^{1/2}$ is considered as a function of a complex variable E, it has two branches. Consequently, the domain of this function is a Riemann surface with two sheets. The first and the second sheets lead to k-values on the upper and on the lower halves of the complex k-plane, respectively. The bound state associated with a pole of the S-matrix at $k = k_0 = i\kappa$ corresponds to a pole on the negative part of the real E-axis, at the location $E_0 = \left(\hbar^2/2\mu\right)\,k_0^2 = -\hbar^2\kappa^2/2\mu$.

An interesting situation occurs when the Jost function has a pole at the origin. For $l \neq 0$ the resulting wave function can be properly normalized so that it corresponds to a physically acceptable state. However, the opposite occurs for $l = 0$. Such unphysical states are called *virtual states*.

8.1.3.1 *Coulomb bound states and the Hydrogen-like spectrum*

In the Coulomb case, bound states correspond to poles of the S-matrix of Eq. (3.51),

$$S_l^{(\mathrm{C})} = \frac{\Gamma(1+l+i\eta)}{\Gamma(1+l-i\eta)}, \tag{8.19}$$

located on the negative part of the real axis of the complex E-plane. They occur at energies where the denominator vanishes or the numerator diverges. For negative energies, the Sommerfeld parameter is imaginary and thus the arguments of the gamma functions in the numerator and denominator are real. Since $\Gamma(x)$ does not have real zeroes, one is left with the second possibility. Figure 8.1 shows the gamma function for real values of x. It has a pole at $x = 0$, which leads to a pole in the S-matrix at the energy associated with the Sommerfeld parameter satisfying the relation

$$1 + l + i\eta = 0. \tag{8.20}$$

Using the notation E_n for the energy of the bound state with principal quantum number n and writing the Sommerfeld parameter in the form[6]

$$\eta = \frac{q_{\mathrm{P}}\, q_{\mathrm{T}}}{4\pi\epsilon_0\, \hbar v} = \sqrt{\frac{E_{\mathcal{R}}}{E_n}}, \tag{8.21}$$

[6]Note that the sign of the square root or, equivalently, the sign of the imaginary velocity, has to be such that the imaginary Sommerfeld parameter has a positive sign. This comes from the condition of Eq. (8.20).

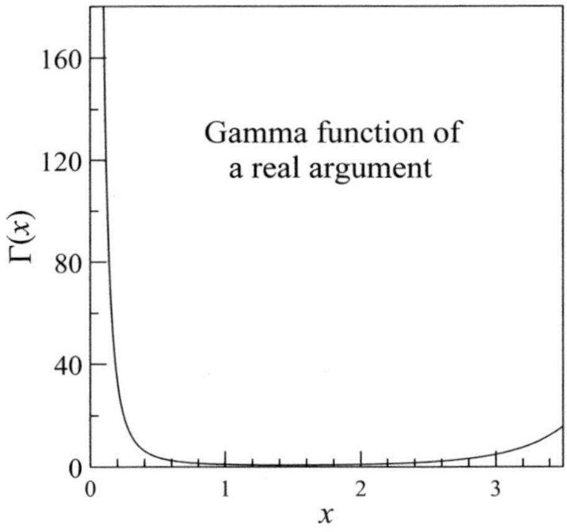

Fig. 8.1 Gamma function for real arguments.

where $E_{\mathcal{R}}$ is the Rydberg energy

$$E_{\mathcal{R}} = \frac{\mu\, q_{\rm P}^2\, q_{\rm T}^2}{2\hbar^2\,(4\pi\epsilon_0)^2}, \tag{8.22}$$

we get

$$E_n = -\frac{E_{\mathcal{R}}}{n^2}. \tag{8.23}$$

Above, $n = l + 1$ is the Coulomb principal quantum number, which takes the values $n = 1, 2, 3, \cdots$, for $l = 0, 1, 2, \cdots$.

The radius of the bound system in the state with principal quantum number n, R_n, is calculated easily using the virial theorem (see, e.g. [Merzbacher (1998)]), and equating $-2E_n$ with $Z_1 Z_2 e^2/R_n$,

$$R_n = \left[\frac{q_{\rm P}\, q_{\rm T}}{8\pi\epsilon_0 E_{\mathcal{R}}}\right] n^2 = \left[\frac{4\pi\epsilon_0\,\hbar^2}{\mu\, q_{\rm P} q_{\rm T}}\right] n^2. \tag{8.24}$$

Clearly, bound states occur when the Coulomb interaction is attractive. One important example is the Hydrogen atom, which corresponds to poles of the S-matrix in $e + p$ scattering. In this case, we have

$$q_{\rm P}\, q_{\rm T} = -\,e^2 \quad \text{and} \quad \mu = \frac{m_e m_p}{m_e + m_p}, \tag{8.25}$$

with m_e and m_p standing for the masses of the electron and the proton, respectively. Since $m_p \gg m_e$, $\mu \simeq m_e$. Inserting these values into Eqs. (8.22) to (8.24), we get the well known result for the spectrum of the Hydrogen atom,

$$E_n = -\frac{13.605698}{n^2}\ \text{eV}. \tag{8.26}$$

Accordingly, the corresponding radii of Hydrogen bound states (ground and excited), R_n, are,

$$R_n = a_0\, n^2, \tag{8.27}$$

where

$$a_0 = 0.529177249 \times 10^{-8}\,\text{cm} \tag{8.28}$$

is the Bohr radius, which corresponds to the radius of the ground state electron ($n = 1$) in the Hydrogen atom.

Another example is the positronium, which is the bound system of the electron and its anti-particle, the positron. Ignoring the annihilation effects, the binding energy of the positronium is similar to Hydrogen, except for the value of the reduced mass, which in the former is $\mu = m_e/2$. In this case, the energies and radii of the bound states are

$$E_n = -\frac{E_{\mathcal{R}}}{2\,n^2} \quad \text{and} \quad R_n = 2a_0\,n^2. \tag{8.29}$$

For the ground state we obtain

$$E_0 = -6.802849\ \text{eV} \quad \text{and} \quad R_0 = 1.058354498 \times 10^{-8}\,\text{cm}. \tag{8.30}$$

In contrast to the Hydrogen atom, the Positronium is an unstable bound state against the annihilation into multiple gamma rays. In fact the life time of the para-positronium, in which the spins of the electron and the positron are anti-parallel, is $t_{\text{p}} = 1.25 \times 10^{-10}$ s, and the dominant decay mode is the emission of two photons, while the ortho-positronium, with parallel spins of the electron and the positron, has a life time of $t_{\text{p}} = 1.4 \times 10^{-7}$ s, with the dominant decay mode being the emission of three photons. Other decay channels involve other even numbers (4, 6, etc.) of γ's for the para-positronium and other odd numbers (5, 7, etc.) of γ's for the ortho-positronium. The contributions of these other decay channels are, however, orders of magnitude smaller than the dominant two and three γ's. For more information about the current understanding of this exotic system we refer to [Rich (1981)].

8.1.4 *Resonances*

We now consider poles of the S-matrix on the lower half of the complex k-plane. It can be shown (see [Roman (1965)] section 4.4e) that all zeroes of the Jost function on the upper half of the complex k-plane are located on the imaginary axis. As we have seen, these zeroes correspond to the bound states of the projectile-target system. However, the Jost function may have zeroes on the lower half of the complex k-plane and accordingly the S-matrix will have poles at these points. Eq. (8.11) implies that these zeroes must occur in pairs $k_{\pm} = \pm a - bi$, where a and b are positive real numbers. Although the zeroes at $k_{-} = -a - bi$ are irrelevant for scattering theory, those at $k_{+} = a - bi$ will have a very important role when they are close to the real k-axis. To see this point, we expand $\mathcal{J}_l(k)$ around k_+ and take the expansion to first order. We get

$$\mathcal{J}_l(k) \simeq \left[\frac{d\mathcal{J}_l(k)}{dk}\right]_{k=a-bi} [(k-a)+bi] \,. \tag{8.31}$$

If k_+ is close to the real-axis, we can use Eq. (8.31) in Eq. (8.17) for k close to a and get

$$\delta_l(k) = -\arg\{\mathcal{J}_l(k)\} \simeq -\arg\left\{\left[\frac{d\mathcal{J}_l(k)}{dk}\right]_{k=a-bi}\right\} - \tan^{-1}\left(\frac{b}{k-a}\right) . \tag{8.32}$$

Eq. (8.32) shows that the phase shift is the sum of a constant "background" term

$$\delta_{\rm bg} = -\arg\left\{\left[\frac{d\mathcal{J}_l(k)}{dk}\right]_{k=a-bi}\right\} \tag{8.33}$$

with a resonant term

$$\delta_{\rm res} = -\tan^{-1}\left(\frac{b}{k-a}\right) .$$

The resonant term goes through $\pi/2$ as the wave number reach the value of the real part of the pole, a. As k moves away from the resonance the first order expansion no longer holds. It should be pointed out that the typical resonant behavior discussed in section 2.9 is observed when the background phase shift is very small. This always happens when the resonance energy is much lower than the barrier in the effective potential, resulting from the sum of the potential with the centrifugal term. In this case, the first order expansion of the resonant phase shift leads to a maximum in the cross section, as represented by the Breit-Wigner shape. However, when the background part is relevant, the pole only gives rise to a rapid change of the cross section as a function of the collision energy. This change may even be a sharp decrease, coming back to the original value after the resonance.

The discussion of resonances is usually carried out in terms of the collision energy. In this case, the poles of the S-matrix in the lower half of the complex k-plane are mapped into the second sheet of the Riemann surface of the complex E-plane. In particular, a pole close to the physical axis ($\mathrm{Re}\, k > 0$) corresponds to a pole on the lower half of the complex E-plane with positive real part and negative imaginary part. It is convenient to write it in the form

$$\bar{E} = E_r - i\,\frac{\Gamma}{2}\,. \tag{8.34}$$

In this way, near the resonance energy, the Jost function can be approximated by the lowest order expansion

$$\mathcal{J}_l(E) \simeq C\left[(E - E_r) + i\frac{\Gamma}{2}\right]. \tag{8.35}$$

Proceeding similarly to the expansion in the complex k-plane, we get the phase shift

$$\delta_{\mathrm{res}} = \tan^{-1}\left(\frac{\Gamma/2}{E_r - E}\right). \tag{8.36}$$

If the background phase shifts for all partial-waves are negligible and the resonance is isolated, the integrated elastic cross section, given by Eq. (2.61b), reduces to[7]

$$\sigma_{\mathrm{el}} \simeq \frac{4\pi}{k^2}\,(2l+1)\,\sin^2\delta_{\mathrm{res}}\,. \tag{8.37}$$

From Eq. (8.36), we have

$$\sin\delta_{\mathrm{res}} = \frac{\tan\delta_{\mathrm{res}}}{\sqrt{1+\tan^2\delta_{\mathrm{res}}}} = \frac{\Gamma/2}{\sqrt{(E-E_r)^2 + \Gamma^2/4}}$$

and Eq. (8.37) takes the form

$$\sigma_{el} \simeq \frac{\pi}{k^2}\,(2l+1)\left[\frac{\Gamma^2}{(E-E_r)^2 + \Gamma^2/4}\right]. \tag{8.38}$$

This is the Breit-Wigner formula, which reduces to Eq. (2.105) in the case of S-waves in real potentials.

[7] Note that $\delta_{\mathrm{res}}, \Gamma$ and E_r are functions of l. To keep the notation simple we do not indicate it explicitly.

It is interesting to study the behavior of a pole of the S-matrix in the scattering from an attractive potential as a function of its strength V_0. Let us assume that the potential is strong enough, to bind a state with energy $E = -E_0(V_0)$, which corresponds to a pole of the S-matrix on the upper part of the imaginary k-axis, or, equivalently, on the negative part of the real-axis of the first sheet of the Riemann surface in the complex E-plane. As V_0 decreases, so does the binding energy. At some critical value, the bound state becomes a sharp resonance, and the trajectory of the pole goes into the lower half of the complex k-plane. At this point, it is split into two branches, symmetrically located with respect to the imaginary k-axis. In the complex E-plane, at this critical potential depth, the pole moves to the second sheet of the Riemann surface and the trajectory is also split into two branches. The situation is illustrated in figure 8.2a, in the k-plane, and in figure 8.2b in the complex E-plane. In the latter case, the two sheets of the Riemann surface are superimposed. As discussed above, the poles in the left half of the complex k-plane, which correspond to the ones on the upper half of the E-plane are not relevant for the scattering cross section. These *pseudo-resonances* correspond to non-physical states, with wave functions growing in time.

The above discussion is of academic interest since in a real collision the depth of the attractive potential is fixed. However, the effective potential, resulting from the sum of a fixed attractive potential with the centrifugal term, has its depth decreased as the angular momentum increases and this situation is similar to the one represented in figure 8.2. Let us consider a square well deep enough to bind at least one s-state. It should satisfy the condition (see section 2.8)

$$K_0\bar{R} > \pi/2 \quad \Longrightarrow \quad V_0 > \pi^2\hbar^2/8\mu\bar{R}^2.$$

We take the square-well of section 2.10, with $K_0\bar{R} = 10$. Figure 8.3 shows the poles of the S-matrix in the complex E-plane associated with the lowest energy state of this well, as a function of the angular momentum. The energy scales are normalized with respect to V_0. For $l = 0, 1, ..., 6$ the states are bound, so that the poles are located on the real-axis, at the energy eigenvalues. Above $l = 6$, the effective potential has no bound states and the poles become resonances. They are located at $\varepsilon = \varepsilon_r - i\,\gamma/2$, where $\varepsilon_r = E_r/V_0$ and $\gamma = \Gamma/V_0$. The calculated bound state energies are given in Table 2.1. The resonance energies and the corresponding widths are given in Table 2.2.

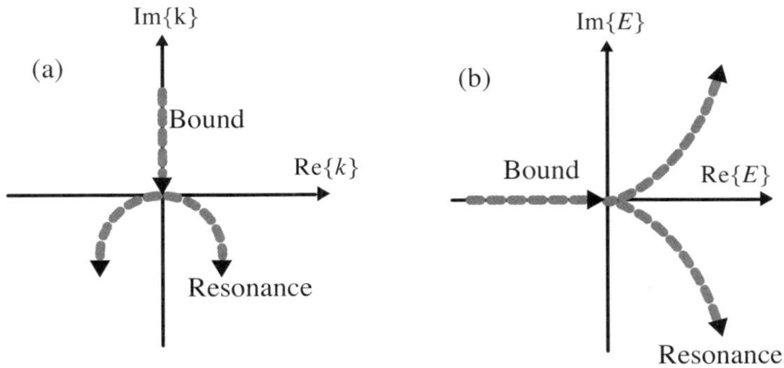

Fig. 8.2 Poles of the S-matrix in the complex k-plane (panel (a)) and in the complex E-plane (panel (b)), for an attractive square well. As the depth of the potential decreases, the poles follow the trajectories indicated in the figure. For details see the text.

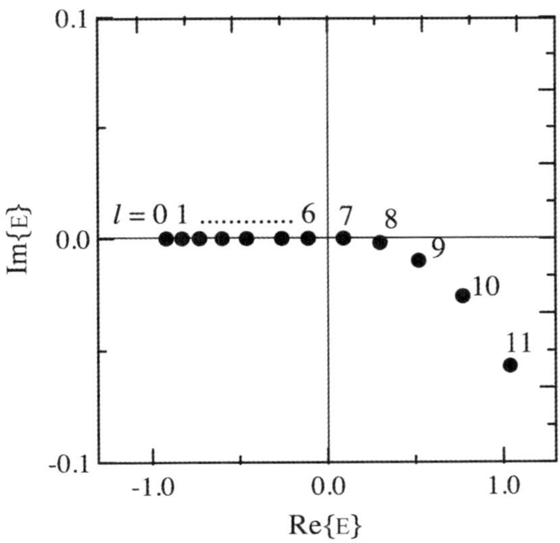

Fig. 8.3 Poles of the S-matrix in the complex E-plane for $l = 0, 1, \ldots.., 11$..

8.1.5 *Analytical continuation into the complex l-plane*

Inspecting the differential equation for the radial wave function (Eq. (2.9)), it becomes clear that there are no mathematical difficulties to extend it to complex l-values. It is true that radial solutions with complex, or even non-integer, angular momenta would lead to meaningless scattering states, owing to the angular part of the wave function. Nevertheless, such solutions are a useful tool for the calculation of the scattering amplitude. There is, however, one restriction on the values of the complex variable l. For $\text{Re}\{l\} < -1/2$, the usual boundary condition for a regular solution may be satisfied by solutions which becomes divergent for positive l-values. Therefore, the domain of l must be the semi-plane with $\text{Re}\{l\} > -1/2$. For a finite range potential, the Jost function, now denoted $\mathcal{J}(l, E)$, is an analytical function of l on this semi-plane and of E on the whole complex-plane.

8.1.5.1 *Regge poles and Regge trajectories*

One can now consider zeroes of $\mathcal{J}(l, E)$ in the complex l-plane, for real values of the collision energy. These zeroes give rise to poles of the S-matrix, $S(l, E)$, in the complex l-plane, which are known as *Regge poles.* The position of these poles appear as a function of E. As the energy is continuously varied, each Regge pole describe a continuous line in the complex l-plane. These lines are called *Regge trajectories.* If the potential has bound states with angular momentum $l_1,\ l_{2,..},l_n$ (real and integer values) with energies[8] $E_1, E_{2,..}, E_n$ (real and negative values), $l_1,\ l_{2,..},l_n$ are clearly points on the same Regge trajectory. As the energy becomes positive, the Regge trajectory moves away from the real axis and is given by a function $\tilde{l}(E) = \tilde{L}(E) + i\tilde{\Delta}(E)$. It is easy to show that the Regge trajectory has the following properties (see, for example, [Joachain (1983)]).

- $\tilde{\Delta}(E)$ is always positive.
- If for a given energy E_m, $\tilde{L}(E_m) = m$, where m is an integer, and $\tilde{\Delta}(E)$ is small, the phase shift δ_m has a resonance at this energy.

As an example, we consider the attractive square-well potential of the previous section. According to the above discussion, it leads to Regge poles at the real and integer values $l = 0, 1, ...$and 6, at the negative energies given in Table 2.1. The Regge trajectory passes also on the points $\tilde{l}(E) = 7 + i\tilde{\Delta}(\varepsilon_7)$, $8 + i\tilde{\Delta}(\varepsilon_8)$, ..., $11 + i\tilde{\Delta}(\varepsilon_{11})$, where $\varepsilon_7, ..., \varepsilon_{11}$ are the resonance energies given in Table 2.2. This situation is represented in figure 8.4.

[8]In this example, we are considering the Regge trajectory for the lowest energy states. If there are two or more bound states for a partial wave l, the state E_l referred to in the text is that with lowest energy.

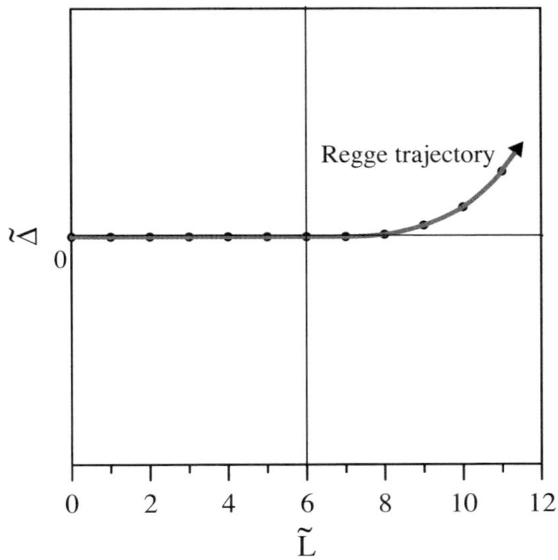

Fig. 8.4 Illustration of the Regge trajectory of the lowest energy state. The potential is the square-well of section 2.10.

8.2 Quantum scattering in 1D and 2D

The quantum mechanical scattering problem in one (1D) and two (2D) space dimensions have been addressed in several publications [Lapidus (1992); Adhikari (1986); Adhikari and Hussein (2008)]. In these references the whole machinery of quantum scattering theory was developed, especially in [Adhikari (1986)]. The critical issue in the 2D scattering is the inherent cylindrical symmetry, which renders the free partial-wave radial wave function to be the ordinary Bessel (regular at the origin) and Neumann functions (irregular), in contradistinction to the 3D case where these functions become their spherical counterpart (spherical Bessel and Neumann functions). Further, the scattering amplitude is found to be similar to the 3D, namely an infinite sum over angular momentum, with the distinct feature of having the s-wave contribution having half the weight when compared to all other terms in the series. The scattering amplitude has the dimension of $(\text{length})^{1/2}$, which necessarily makes the equivalent to the 3D differential cross section, a *differential scattering length*. 2D scattering has application in surface physics, besides being an interesting topic to discuss with students of scattering theory at the graduate level.

One topic not addressed in [Lapidus (1992); Adhikari (1986)] is the semiclassical limit of the 2D scattering. This is an important subject both from the pedagogical view, as it shows the connection between quantum and classical mechanics in a particularly transparent form and the working of the angle vs. angular momentum Heisenberg uncertainty relation, and for potential application to topics such as small magnetic particles where vortex -antivortex scattering, an important issue here [Komineas and Papanicolau (2008)], is usually dealt with using rudiments of classical scattering at the computational level. The purpose of this section is to develop the semi-classical version of the quantum scattering in 2D.

8.2.1 *Brief account of quantum scattering in 1D*

The scattering of particles in one space dimension is mostly of pure academic/pedagogical interest. It supplies ways of understanding how the dimensionality of the system manifests itself in the scattering observables. This will become clear when we turn to scattering in 2D. In this subsection we discuss the scattering in 1D using the formalism of scattering theory developed in chapter 4 (see also [Barlette *et al.* (2001, 2000); Eberly (1965); Formanek (1976); Kamal (1984)]). Of course all elementary text book on quantum mechanics do deal with particle motion in 1D as it passes by a potential. But, what is the Lippmann-Schwinger equation and the corresponding S- and T-matrices in this case? We answer these questions in this subsection. There is not really any practical use of this, except for the intelectual exercise and to lay the grounds for the following discussion of scattering in 2D, which is of great use in surface physics, among other things. For completeness we supply a self-contained discussion of this topic. First we note that the wave function describing the particle motion in one dimension is naturally decomposed into an incoming and an outgoing wave (motion to the left and to the right). The Schrödinger equation for the scattering of a particle with mass μ from a potential $V(x)$ is

$$\left[-\frac{\hbar^2}{2\mu}\frac{d^2}{dx^2} + V(x)\right]\psi^{(+)}(k;x) = E\,\psi^{(+)}(k;x). \tag{8.39}$$

The potential is taken to be centrally symmetric and of finite range. It is a function of $|x|$, namely $V(x) = V(-x)$. The solutions of the above equation satisfy the following condition,

$$\int_{-\infty}^{\infty} dx\,\left[\psi^{(+)}(k;x)\right]^*\,\psi^{(+)}(k';x) = \delta(k-k'). \tag{8.40}$$

Clearly, the above 1D Schrödinger equation can be cast into an integral Lipmmann-Schwinger equation, just as in the 3D case. The free solutions are

$$\phi(k;x) = \frac{e^{ikx}}{\sqrt{2\pi}} \tag{8.41}$$

and the free Green's function, being the solution of

$$\left[E_k + \frac{\hbar^2}{2\mu}\frac{d^2}{dx^2}\right] G_0^{(+)}(E_k;x,x') = \delta(x-x'), \tag{8.42}$$

is [Morse and Feshbach (1953)],

$$G_0^{(+)}(E_k;x,x') = -\left(\frac{2\mu}{\hbar^2}\right)\frac{i}{2k}\,e^{ik\,|x-x'|}. \tag{8.43}$$

Accordingly the LS equation, which in operator form is the same as in 3D (or any dimension for that matter), is

$$|\psi^{(+)}(k)\rangle = |\phi(k)\rangle + G_0^{(+)}(E_k)\,V\;|\psi^{(+)}(k)\rangle\,. \tag{8.44}$$

Taking representations in the coordinate space, it becomes

$$\psi^{(+)}(k;x) = \phi(k;x) + \int_{-\infty}^{\infty} G_0^{(+)}(E_k;x,x')\,V(x')\,\psi_k^{(+)}(k;x')\,dx'. \tag{8.45}$$

The scattering amplitude is obtained as usual by looking for the asymptotic behavior of $\psi^{(+)}(k;x)$. However, in the present case there are two asymptotic limits; one forward ($x \to \infty$) and one backward($x \to -\infty$). From Eq. (8.43), we get

$$G_0^{(+)}(E_k;x,x') \to -\left(\frac{2\mu}{\hbar^2}\right)\frac{i}{2k}\;e^{ikx}\,e^{-ikx'}, \quad \text{for } x \to \infty, \tag{8.46}$$

$$G_0^{(+)}(E_k,x,x') \to -\left(\frac{2\mu}{\hbar^2}\right)\frac{i}{2k}\;e^{-ikx}\,e^{ikx'}, \quad \text{for } x \to -\infty. \tag{8.47}$$

Using these results in the LS equation, one gets their $x \to +\infty$ limits,

$$\psi_+^{(+)}(k;x) \equiv \psi^{(+)}(k;x \to +\infty) = \frac{1}{\sqrt{2\pi}}\;\left[e^{ikx} + f_+(k)\;e^{ikx}\right], \tag{8.48}$$

$$\psi_-^{(+)}(k;x) \equiv \psi^{(+)}(k;x \to -\infty) = \frac{1}{\sqrt{2\pi}}\;\left[e^{ikx} + f_-(k)\;e^{-ikx}\right]. \tag{8.49}$$

The amplitudes $f_+(k)$ are

$$f_+(k) = -\left(\frac{2\mu}{\hbar^2}\right)\frac{i\pi}{k}\int dx'\,\frac{e^{-ikx'}}{\sqrt{2\pi}}\,V(x')\,\psi^{(+)}(k;x') \tag{8.50}$$

$$\equiv -\left(\frac{2\mu}{\hbar^2}\right)\frac{i\pi}{k}\;\langle\phi(k)|\,V\,|\psi^{(+)}(k)\rangle\,. \tag{8.51}$$

and

$$f_-(k) = -\left(\frac{2\mu}{\hbar^2}\right)\frac{i\pi}{k}\int dx'\,\frac{e^{ikx'}}{\sqrt{2\pi}}V(x')\,\psi^{(+)}(k;x') \qquad (8.52)$$

$$\equiv -\left(\frac{2\mu}{\hbar^2}\right)\frac{i\pi}{k}\,\langle\phi(-k)|\,V\,|\psi^{(+)}(k)\rangle\,. \qquad (8.53)$$

The above equations suggest the notation, $f(k')$, with $k' = +k$. In this way, we have: $f_+ = f(k)$ and $f_- = f(-k)$. One can now introduce the 1D version of the transition operator, which has the matrix elements,

$$T_{k',k} = \big\langle\phi(k')\big|\,V\,\big|\psi^{(+)}(k)\big\rangle. \qquad (8.54)$$

The scattering amplitude is expressed in terms of the T-matrix by the relations

$$f(k') = -\left(\frac{2\mu}{\hbar^2}\right)\frac{i\pi}{k}T_{k',k} \quad\rightarrow\quad T_{k',k} = i\left(\frac{\hbar^2 k}{2\pi\mu}\right)f(k'). \qquad (8.55)$$

Using Eqs. (8.44) and (8.54), it is straighforward to show that the T-matrix satisfies the same LS equation as in 3D scattering, namely

$$T = V + V\,G_0^{(+)}(E)\,T. \qquad (8.56)$$

The scattering amplitudes can be directly related with the usual transmission and reflection coefficients,

$$\mathcal{T} = \frac{|j_{\rm T}|}{|j_{\rm in}|} \quad\text{and}\quad \mathcal{R} = \frac{|j_{\rm R}|}{|j_{\rm in}|}. \qquad (8.57)$$

Above,

$$j_{\rm in} = \frac{\hbar}{2\mu i}\left[\phi^*(k;x)\frac{d\phi(k;x)}{dx} - \phi(k;x)\frac{d\phi^*(k;x)}{dx}\right] \qquad (8.58)$$

is the incident current and $j_{\rm T}$ and $j_{\rm R}$ are respectively the transmitted and the reflected currents. They are defined as in the above equation but replacing $\phi(k;x)$ by asymptotic forms of the scattering wave functions (Eqs. (8.48) and (8.49)). The current $j_{\rm T}$ is evaluated with $\psi_+^{(+)}$ whereas $j_{\rm R}$ is derived from the reflected part of $\psi_-^{(+)}$ (second term within the square bracked of Eq. (8.49)). Using Eqs. (8.41), (8.48) and (8.49)), we obtain

$$\mathcal{T} = |1 + f(+k)|^2 \quad\text{and}\quad \mathcal{R} = |f(-k)|^2\,. \qquad (8.59)$$

Given the scattering amplitude, the forward ($\sigma_{\rm F}^{\rm 1D}$) and backward ($\sigma_{\rm B}^{\rm 1D}$) one-dimensional *cross sections* can be calculated as usual,

$$\sigma_{\rm F}^{\rm 1D} = |f(+k)|^2 \quad\text{and}\quad \sigma_{\rm B}^{\rm 1D} = |f(-k)|^2\,. \qquad (8.60)$$

Summing the cross sections in both forward and backward directions, one obtains the total cross section,

$$\sigma_{\text{total}}^{\text{1D}} = |f(+k)|^2 + |f(-k)|^2 = \sum_{k'=+k} |f(k')|^2 . \tag{8.61}$$

Note that these cross sections do not have the dimension of an area. They are dimensionless quantities which represent probabilities.

A modified version of the optical theorem is valid for collisions in 1D. The starting point to derive this theorem is the one-dimensional version of Eq. (4.74), for $k' = k$, namely

$$T_{k,k}(E_k) - T_{k,k}^*(E_k) = -2\pi i \int dq\, T_{k,q}(E_q)\, T_{k,q}^*(E_q)\, \delta(E_k - E_q), \tag{8.62}$$

or

$$\text{Im}\{T_{k,k}(E_k)\} = -\pi \int dq\, |T_{k,q}(E_q)|^2\, \delta(E_k - E_q). \tag{8.63}$$

Using standard properties of delta functions we can write

$$\delta(E_q - E_k) = \delta\left(\frac{\hbar^2}{2\mu}(k+q)(k-q)\right) = \frac{\mu}{\hbar^2 k}\left[\delta(q+k) + \delta(q-k)\right] \tag{8.64}$$

and inserting this result into Eq. (8.63) we get

$$\text{Im}\{T_{k,k}(E_k)\} = -\frac{\pi\mu}{\hbar^2 k} \sum_{k'=\pm k} |T_{k',k}(E_k)|^2 . \tag{8.65}$$

Now we use Eqs. (8.55) and (8.61) to obtain

$$-\left(\frac{4\pi\mu}{\hbar^2 k}\right) \text{Im}\{T_{k,k}(E_k)\} = \sigma_{\text{total}}^{\text{1D}}. \tag{8.66}$$

Finally, expressing the T-matrix in terms of the scattering amplitude (Eq. (8.55)), the above equation becomes

$$\sigma_{\text{total}}^{\text{1D}} = -2\,\text{Re}\{f(k)\}. \tag{8.67}$$

This is the version of the optical theorem for 1D scattering.

Although the usual form of the optical theorem in 3D involves the factor $4\pi/k$ and $\text{Im}\{f(\theta = 0)\}$ (which here corresponds to $\text{Im}\{f(k)\}$), in 1D the multiplicative factor is just 2. If $f(k)$ is redefined as $f(k) \to i\, f(k)$, the optical theorem becomes,

$$\sigma_{\text{total}}^{\text{1D}} = 2\,\text{Im}\{f(k)\}, \tag{8.68}$$

to be compared to its 3D counterpart

$$\sigma_{\text{total}}^{\text{3D}} = \frac{4\pi}{k}\text{Im}\{f(\theta = 0)\}. \tag{8.69}$$

To understand the result, we mention that the factor 4π is nothing but the total solid angle, while the factor 2 in 1D accounts for the two possibilities: scattering forward and scattering backward. The factor $1/k$ in 3D is to guarantee that the dimension of σ is that of an area, while in 1D, both cross section and the scattering amplitude are dimensionless.

Another way to get the optical theorem in 1D is to use the current conservation,

$$j_{\rm in} = |\, j_{\rm T}\,| + |\, j_{\rm R}\,| \,. \tag{8.70}$$

Dividing both sides of the above equation by $j_{\rm in}$ and using Eqs. (8.57) to (8.59), we get

$$1 = \mathcal{T} + \mathcal{R} = |1 + f(k)|^2 + |f(-k)|^2 \,, \tag{8.71}$$

or

$$|f(k)|^2 + |f(-k)|^2 = -f(k) - f^*(k) = -2\,\mathrm{Re}\,\{f(k)\}\,. \tag{8.72}$$

Since the left hand side of the above equation is equal to $\sigma^{\rm 1D}_{\rm total}$ (see Eq. (8.61)), we get the 1D version of the optical theorem, given in Eq. (8.67).

8.2.2 *Brief account of quantum scattering in 2D*

The starting point of the scattering problem is to find positive energy solutions of the non-relativistic time-independent Schrödinger equation

$$\left[-\frac{\hbar^2}{2\mu}\nabla^2 + V(r)\right]\,\psi^{(+)}(\mathbf{k};\mathbf{r}) = E\,\psi^{(+)}(\mathbf{k};\mathbf{r}), \tag{8.73}$$

with appropriate scattering boundary conditions. In the present situation, $\mathbf{k}$ and $\mathbf{r}$ are vectors in a two-dimensional space and $V(r)$ is a central potential. In polar coordenates, $\{r,\theta\}$, the Laplacian is

$$\nabla^2 = \frac{1}{r}\frac{\partial}{\partial r}\left(r\frac{\partial}{\partial r}\right) + \frac{1}{r^2}\left(\frac{\partial^2}{\partial \theta^2}\right). \tag{8.74}$$

Eq. (8.73) can then be put in the form

$$\left[\frac{1}{r}\frac{\partial}{\partial r}\left(r\frac{\partial}{\partial r}\right) + \frac{1}{r^2}\,\frac{\partial^2}{\partial^2 \theta} \;+\; k^2 - U(r)\right]\,\psi^{(+)}(\mathbf{k};\mathbf{r}) = 0, \tag{8.75}$$

or

$$\left[r\,\frac{\partial}{\partial r}\left(r\frac{\partial}{\partial r}\right) \;+\; r^2\left(k^2 - U(r)\right)\right]\,\psi^{(+)}(\mathbf{k};\mathbf{r}) = -\,\frac{\partial^2}{\partial\theta^2}\,\psi^{(+)}(\mathbf{k};\mathbf{r}). \tag{8.76}$$

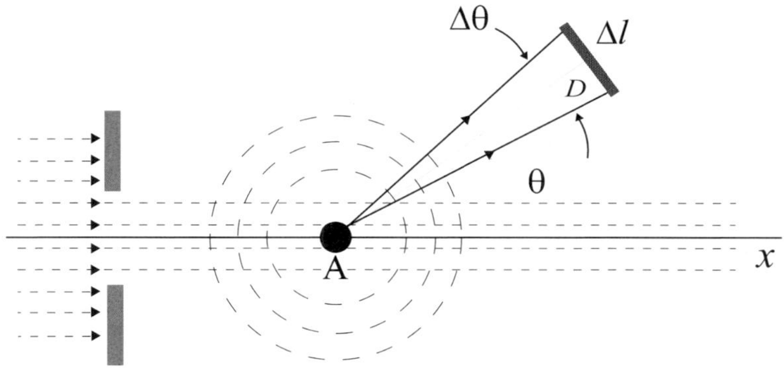

Fig. 8.5 Schematic representation of a scattering experiment in two dimensions.

Above, we used the notations

$$k^2 = \frac{2\mu E}{\hbar^2} \quad \text{and} \quad U(r) = \frac{2\mu V(r)}{\hbar^2}. \tag{8.77}$$

One wants to find a solution of Eq. (8.76) with the scattering boundary condition

$$\psi^{(+)}(\mathbf{k}; \mathbf{r} \to \infty) = \frac{1}{2\pi} \left[e^{ikx} + f(\theta) \frac{e^{i\mathbf{k}\,\mathbf{r}}}{\sqrt{r}} \right], \tag{8.78}$$

where $\mathbf{r} \to \infty$ means that $|\mathbf{r}| \to \infty$. We adopt the usual normalization of $1/\sqrt{2\pi}$ for each degree of freedom. The factor $1/\sqrt{r}$, which is the analog of $1/r$ in 3D scattering, guarantees that the flux of the radial current through any arc subtending an angle $\Delta\theta$ is constant. Eq. (8.78), represents a wave propagating on the x-y plane. The first and the second terms within the square bracket correspond respectively to an incident plane wave with wave vector parallel to the x-axis and modulus k, and an outgoing circular wave centered at the scatterer.

The cross section in 2D scattering is analogous to that in 3D. Figure 8.5 is a schematic representation of the problem. It shows the incident wave with wave vector along the x-axis, the scatterer (A) and the scattered circular wave. The figure also shows a detector placed at a distance D from the scatterer, at the orientation θ with respect to the x-axis. It covers an angular range of $\Delta\theta$ and is represented in the figure by a segment with length Δl, corresponding to its projection onto the tangential direction. The number of particles reaching the detector per unit time, $N(\theta, \Delta\theta)$, is given by the flux of the radial scattered current through Δl, and the differential scattering cross section, $d\sigma^{2D}(\theta)/d\theta$ is defined as

$$\frac{d\sigma^{2D}(\theta)}{d\theta} = \frac{N(\theta, \Delta\theta)}{j_{\rm in}\,\Delta\theta}. \tag{8.79}$$

Above, $j_{\rm in}$ is the modulus of the incident current, which is still given by Eq. (8.58), and here its value is

$$j_{\rm in} = \frac{v}{(2\pi)^2}, \quad \text{with} \quad v = \frac{\hbar k}{\mu}. \tag{8.80}$$

The classical cross section can be easily obtained from the deflection function, as in the case of scattering in three dimensions (see section 1.4.1). An observation angle θ is associated with one or more impact parameters through the classical trajectory. Let us assume that there are N values of the impact parameter leading to the angle θ. That is[9]

$$\Theta(b_1) = \Theta(b_2) = \cdots = \Theta(b_N) = \pm\theta. \tag{8.81}$$

The number of particles emerging within an angular interval $\Delta\theta$ around the angle θ is then the product of the incident current with the sum of the infinitesimal lengths Δb_j around each impact parameter that satisfies the above equation,

$$N(\theta, \Delta\theta) = j_{\rm in} \sum_{j=1}^{N} \Delta b_j. \tag{8.82}$$

The infinitesimal impact parameter intervals, Δb_j, are given by the relation

$$\Delta b_j = \frac{\Delta\theta}{|d\Theta(b)/db|_{b=b_j}} \tag{8.83}$$

Using Eq. (8.83) in Eq. (8.82), inserting the result into Eq. (8.79), and identifying the observation angle with the deflection function, we get

$$\frac{d\sigma^{2D}_{\rm cl}(\theta)}{d\theta} = \sum_{j=1}^{N} \frac{d\sigma^{2D}_{{\rm cl},j}(\theta)}{d\theta} = \sum_{j=1}^{N} \left|\frac{d\Theta(b)}{db}\right|^{-1}_{b=b_j}. \tag{8.84}$$

This result is similar to the one obtained in three dimensions. The difference is that it lacks the factor $b/\sin\theta$ of the classical 3D cross section. This difference is quite natural. The absence of the factor b results from the fact that here the cross section is a length whereas in 3D it is an area. On

[9]The sign $\pm$ comes from the fact that the observation angle is always positive. Experiments cannot distinguish particle coming from the near side of the scatterer from particles coming from its far side.

the other hand, the denominator $\sin\theta$ appearing in the expression of the 3D classical cross section arises from $d\Omega = \sin\,\theta\, d\theta\, d\phi$. The classical cross section in 2D does not contain this denominator because in this case the angular differential is simply $d\theta$.

We now turn to the quantum mechanical cross section. The radial scattered current is

$$j_r = \frac{\hbar}{2\mu i}\left[\psi^*_{\rm sc}(k;r\,\theta)\;\frac{d\psi_{\rm sc}(k;r\,\theta)}{dr} \;-\; \psi_{\rm sc}(k;r\,\theta)\;\frac{d\psi^*_{\rm sc}(k;r\,\theta)}{dr}\right], \tag{8.85}$$

where (see Eq. (8.78)),

$$\psi_{\rm sc}(k;r) = \frac{1}{2\pi}\; f(\theta)\;\frac{e^{ikr}}{\sqrt{r}}. \tag{8.86}$$

At the detector location ($r = D$), one gets

$$j_r = |f(\theta)|^2\;\frac{v}{(2\pi)^2\,D}. \tag{8.87}$$

The number of particles reaching the detector corresponds to the flux os this current through the detector, which is the product of the radial current with the arc $\Delta l = D\,\Delta\Theta$. That is,

$$N(\theta,\Delta\theta) = \frac{v\,\Delta\theta}{(2\pi)^2}\;|f(\theta)|^2\,. \tag{8.88}$$

Inserting Eqs. (8.80) and (8.88) into Eq. (8.79), we obtain

$$\frac{d\sigma^{\rm 2D}(\theta)}{d\theta} = |f(\theta)|^2\,, \tag{8.89}$$

as in the cases of 1D and 3D scattering.

To evaluate the cross section one then needs to calculate the scattering amplitude. For this purpose it is necessary to solve Eq. (8.76). This can be done using the separation of variables method. One looks for separable solutions of the form,

$$\psi^{(+)}(\mathbf{k};\mathbf{r}) = \Xi(\theta)\,R(k;r). \tag{8.90}$$

Inserting this factorized form into Eq. (8.76) and multiplying from the left by $1/\Xi_m(\theta)\,R_m(r)$, one obtains

$$\frac{1}{R(k;r)}\left[r\,\frac{d}{r}\left(r\,\frac{d}{dr}\right) \;+\; r^2\left(k^2 - U(r)\right)\right]\,R(k;r) = -\frac{1}{\Xi(\theta)}\;\frac{d^2}{d\theta^2}\,\Xi(\theta). \tag{8.91}$$

Since the two sides of this equation depend on different variables, the equality implies that each side is equal to a constant. We call this constant m^2

and use the notations $\Xi_m(\theta), R_m(r)$ and $\psi_m^{(+)}$ for the solutions associated with a given value of m. The angular equation then becomes

$$\frac{d^2\Xi_m(\theta)}{d\theta^2} + m^2\,\Xi_m(\theta) = 0, \tag{8.92}$$

and the solution is[10]

$$\Xi_m(\theta) = \frac{1}{\sqrt{\pi}}\cos(m\,\theta). \tag{8.93}$$

The symmetry $\Xi_m(\theta) = \Xi_m(\theta + 2\pi)$ requires that the constant m be an integer. With the normalization of Eq. (8.93), the angular wave functions satisfy the orthonormality relation

$$\int d\theta\,\Xi_m(\theta)\,\Xi_{m'}(\theta) = \delta_{mm'}. \tag{8.94}$$

Note that Eq. (8.92) corresponds to the eigenvalue equation for the square of the angular momentum operator associated with rotations in the x-y plane,

$$L_z = -i\hbar\,\frac{\partial}{\partial\theta}, \tag{8.95}$$

since it can be cast in the form

$$L_z^2\,\Xi_m(\theta) = m^2\,\Xi_m(\theta). \tag{8.96}$$

Therefore, the constant m corresponds to the angular momentum of the two-dimensional system, in $\hbar$ units.

Setting the right hand side of Eq. (8.91) equal to m^2 and reorganizing the terms, we get the radial equation

$$\frac{1}{r}\frac{d}{dr}\left(r\,\frac{dR_m(k,r)}{dr}\right) + \left[k^2 - U(r) - \frac{m^2}{r^2}\right] R_m(k,r) = 0. \tag{8.97}$$

For radial separations larger than the range of the potential, $\bar{R}$, $U(r)$ vanishes and Eq. (8.97) reduces to

$$\frac{1}{r}\frac{d}{dr}\left(r\,\frac{dR_m(k,r)}{dr}\right) + \left[k^2 - \frac{m^2}{r^2}\right] R_m(k,r) = 0, \tag{8.98}$$

[10] Note that we have discarded the other linearly independent solution,

$$\Xi_m(\theta) = \frac{1}{\sqrt{\pi}}\sin(m\,\theta),$$

because of the reflection symmetry around the x-axis. This symmetry is the analog of the axial symmetry in 3D collisions, when the beam is parallel to the z-axis. Axial symmetry requires that the wave function be independent of φ. Here this corresponds to the requirement $\Xi_m(\theta) = \Xi_m(-\theta)$.

or, using the notation $\rho = kr$ and evaluating the derivatives, the above equation takes the form

$$\rho^2 \, \frac{d^2 R_m(\rho)}{d\rho^2} + \rho \, \frac{dR_m(\rho)}{d\rho} + \left[\rho^2 - m^2\right] \, R_m(\rho) = 0. \tag{8.99}$$

This is the Bessel equation [Jackson (1975); Abramowitz and Stegun (1972)]. Its regular solution is the cylindrical Bessel function, $J_m(kr)$, and its irregular solution is the Neumann function, $N_m(kr)$.

Outside the range of the potential, the radial wave function must be a linear combination of $J_m(kr)$ and $N_m(kr)$. That is,

$$R_m(k, r > \bar{R}) = C_m \, J_m(kr) + D_m \, N_m(kr). \tag{8.100}$$

At asymptotic distances, the Bessel and Neumann functions have the behaviors

$$J_m(kr \to \infty) \to \sqrt{\frac{2}{\pi kr}} \, \cos\left[kr - \frac{\pi}{2} \, (m + 1/2)\right] \tag{8.101}$$

$$N_m(kr \to \infty) \to \sqrt{\frac{2}{\pi kr}} \, \sin\left[kr - \frac{\pi}{2} \, (m + 1/2)\right]. \tag{8.102}$$

and the wave function $R_m(k, r)$ can be cast in the form

$$R_m(k, r \to \infty) = \frac{A_m}{\sqrt{kr}} \, \cos\left[kr - \frac{\pi}{2} \, (m + 1/2) + \delta_m\right]. \tag{8.103}$$

Above, δ_m is the phase shift for the wave propagating on the x-y plane with angular momentum $\hbar m$. It is the analog of δ_l in 3D scattering.

To find the scattering amplitude and σ^{2D}, we proceed similarly to chapter 2. We expand the wave function $\psi^{(+)}(\mathbf{k}, \mathbf{r})$ as

$$\psi^{(+)}(\mathbf{k}, \mathbf{r}) = \sum_m A_m \, \Xi_m(\theta) \, \frac{u_m(k; r)}{\sqrt{kr}}. \tag{8.104}$$

This expansion leads to the simpler radial equation

$$\frac{d^2 u_m(k, r)}{dr^2} + \left[k^2 - U(r) - \frac{(m^2 - 1/4)}{r^2}\right] \, u_m(k, r) = 0. \tag{8.105}$$

The above equation is quite similar to the corresponding one in 3D (see chapter 2), except for the important fact that the centrifugal barrier in the later contains the usual $l(l + 1)$ while in the former it is $m^2 - 1/4$. This implies that for s-wave scattering in 2D, one still has an angular momentum contribution, which brings into the radial equation the effective potential

$U_{m=0}(r) = -1/4r^2$. In this case the effective potential is *centripetal*! This requires that the $m = 0$ case be treated with special care.

Usually, Eq. (8.105) is integrated numerically. One starts from the origin, using the boundary condition condition $u_m(k, 0) = 0$. The derivative of the wave function at the origin is arbitrary, since it does not affect the phase-shitf (see chapter 2). The phase shift is then determined matching the numerically obtained logarithmic derivative with the value derived from its asymptotic form,

$$u_m(k, r \to \infty) = B_m \cos\left[kr - \frac{\pi}{2}\,(m + 1/2) + \delta_m\right]. \tag{8.106}$$

The next step is to express the scattering amplitude as a sum over angular momenta, involving the phase shifts [Adhikari and Hussein (2008)]. This is done through the comparison of the asymptotic forms of Eqs. (8.104) and (8.78), with the help of Eq. (8.106). To obtain the asymptotic form of the incident plane wave, we take the Jacobi-Anger expansion [Colton and Kress (1998); Cuyt *et al.* (2008); Morse and Feshbach (1953)],

$$e^{ikx} \equiv e^{ikr\cos\theta} = \sum_{m=-\infty}^{\infty} i^m\, J_m(kr)\, e^{im\theta}, \tag{8.107}$$

and use the asymptotic form of the Bessel functions. The Jacobi-Anger expansion is the 2D analog of Bauer's formula. The next steps are very similar to the ones in the case of 3D. Sinus and cosinus are written in terms of incident (e^{-ikr}) and emergent (e^{ikr}) circular waves, and the factors multiplying these waves obtained from Eqs. (8.104) and (8.78) are set to be identical. One then obtains the desired expansion (see exercise 2),

$$f(\theta) = \frac{1}{\sqrt{2\pi ik}} \sum_{m=-\infty}^{\infty} e^{im\theta}\,\left[e^{2i\delta_m} - 1\right]. \tag{8.108}$$

If we consider that the dependence of Eq. (8.105) on m is through m^2, we conclude that $\delta_{-m} = \delta_m$. Therefore, we can evaluate together the contributions from m and $-m$, getting: $\left(e^{im\theta} + e^{-im\theta}\right)\left[e^{2i\delta_m} - 1\right]$, and exclude negative values of m from the summation. In this way, we can put Eq. (8.108) in the form

$$f(\theta) = \frac{2}{\sqrt{2\pi ik}} \sum_{m=0}^{\infty} \cos(m\theta)\left[e^{2i\delta_m} - 1\right]. \tag{8.109}$$

The cross section is then given by Eq. (8.89),

$$\frac{d\sigma^{2\mathrm{D}}(\theta)}{d\theta} = |f(\theta)|^2. \tag{8.110}$$

Note that here the scattering amplitude has the dimension of $1/\sqrt{k}$ and thus $\sqrt{\text{length}}$. Therefore the quantity $|f(\theta)|^2$, which in the 3D case is the differential cross section, becomes here a measure of a *differential scattering length.* In order not to confuse this nomenclature with the usual zero-energy scattering length, of much use these days in the cold atom research [Newton (2002)], we shall call the 2D cross section, the *scattering width.* The scattering length itself is just $a = -f_{k\to 0}(0)$ where, according to [Lapidus (1992)], the $m = 0$ form of the amplitude is

$$f_{k\to 0}(0) = \sqrt{\frac{2i}{\pi}} \left[\frac{1}{k^{\frac{1}{2}}\cot\delta - ik^{\frac{1}{2}}}\right]. \tag{8.111}$$

We now relate the scattering amplitude with the T-matrix elements. The starting point is the Lippmann-Schwinger equation, which has the same general form of 1D and 3D scattering. In the present case, the free particle Green's function in its spectral representation is

$$G_0^{(+)}(E;\mathbf{r},\mathbf{r}') = \frac{1}{(2\pi)^2} \int d^2\mathbf{k}' \, \frac{e^{-i\mathbf{k}'\cdot\mathbf{r}} \, e^{i\mathbf{k}'\cdot\mathbf{r}'}}{E - \hbar^2 k'^2/2\mu + i\epsilon} \tag{8.112}$$

Carrying out the integration in polar coordinates, one obtains [Morse and Feshbach (1953)]

$$G_0^{(+)}(E;\mathbf{r},\mathbf{r}') = -\left(\frac{2\mu}{\hbar^2}\right) \frac{i}{4} \, H_0^{(1)}\left(k\,|\mathbf{r}-\mathbf{r}'|\right), \tag{8.113}$$

where

$$H_0^{(1)}(\rho) = J_0(\rho) + i\,N_0(\rho) \tag{8.114}$$

is the usual Haenkel function for $m = 0$, with outgoing boundary condition. As $|\mathbf{r}| \to \infty$ the above Green's function takes the form

$$\begin{aligned} G_0^{(+)}(E;\mathbf{r}\to\infty,\mathbf{r}') &= -\left(\frac{2\mu}{\hbar^2}\right) \frac{e^{i\pi/4}}{4} \sqrt{\frac{2}{\pi k r}} \, e^{-i\mathbf{k}'\cdot\mathbf{r}'} \, e^{ikr} \\ &= \frac{1}{2\pi}\left[-\left(\frac{2\mu}{\hbar^2}\right)\sqrt{\frac{2\pi^3}{k}} \, e^{i\pi/4} \, \frac{e^{-i\mathbf{k}'\cdot\mathbf{r}'}}{2\pi}\right] \frac{e^{ikr}}{\sqrt{r}}. \end{aligned} \tag{8.115}$$

Using the above Green's function in the Lippmann-Schwinger equation in the coordinate representation, we get

$$\begin{aligned} \psi^{(+)}(\mathbf{k},\mathbf{r}) &= \frac{1}{2\pi}\left[e^{ikx} + \left(-\frac{2\mu}{\hbar^2}\right)\sqrt{\frac{2\pi^3}{k}} \, e^{i\pi/4} \, \langle\phi(\mathbf{k}')\,|V|\,\psi^{(+)}(\mathbf{k})\rangle \, \frac{e^{ikr}}{\sqrt{r}}\right] \\ &= \frac{1}{2\pi}\left[e^{ikx} + \left(-\frac{2\mu}{\hbar^2}\sqrt{\frac{2\pi^3}{k}} \, e^{i\pi/4} \, T_{\mathbf{k}',\mathbf{k}}\right) \frac{e^{ikr}}{\sqrt{r}}\right]. \end{aligned} \tag{8.116}$$

Comparing the above equation with Eq. (8.78), we get the relation

$$f(\theta) = -\frac{2\mu}{\hbar^2}\sqrt{\frac{2\pi^3}{k}} \, e^{i\pi/4} \, T_{\mathbf{k}',\mathbf{k}}. \tag{8.117}$$

8.2.3 *Semiclassical scattering in 2D*

The validity of using concepts of classical mechanics to discuss features of quantum scattering hinges on the short wave length limit, namely when the particle's de Broglie local wave length is short compared to the distance over which is the potential changes appreciably. Put a bit differently, and in terms of the classical scattering quantities, such as the impact parameter, b, and the deflection function, $\Theta(b)$, advanced by N. Bohr [Bohr (1948)] and further discussed in [Hussein *et al.* (1984)], the validity criterion for a collision with energy E states the following,

$$\frac{\delta\theta_{\text{opt}}}{\Theta(b)} \ll 1 \tag{8.118}$$

for all b. In the above $\delta\theta_{\text{opt}}$ is the optimal angular dispersion of the scattering particle around a classical path. This dispersion arises from diffraction (related to de Broglie wave length), uncertainty in the value of b (not an observable quantity), and the inherent uncertainty in the beam energy.

The starting point of the semiclassical development in 2D [Adhikari and Hussein (2008)] is to use the Poisson summation formula [Morse and Feshbach (1953); Brink (1985)] to evaluate the scattering amplitude of Eq. (8.109). The procedures are similar to the ones adopted in section 5.2.3. The scattering amplitude has the general form

$$f(\theta) = \sum_{m=0}^{\infty} a_m(\theta), \tag{8.119}$$

with

$$a_m(\theta) = \frac{2}{\sqrt{2i\pi k}} \cos(m\theta) \left[e^{2i\delta_m} - 1\right]. \tag{8.120}$$

As in section 5.2.3, we drop the unity within the square brackets because it only produces a delta-function at exactly $\theta = 0$. Therefore, we can write

$$f(\theta) = f^{(+)}(\theta) + f^{(-)}(\theta), \tag{8.121}$$

with

$$f^{(\pm)}(\theta) = \sum_{m=0}^{\infty} a_m^{(\pm)}(\theta), \tag{8.122}$$

and

$$a_m^{(\pm)}(\theta) = \frac{1}{\sqrt{2i\pi k}} e^{i(2\delta_m \pm m\theta)}. \tag{8.123}$$

Using Poisson's formula (Eq. (5.264)), m (which corresponds to l in 3D scattering) is treated as a continuum variable, λ, and $f^{(\pm)}(\theta)$ can be written

$$f^{(\pm)}(\theta) = \frac{1}{\sqrt{2\pi ik}} \sum_{n=-\infty}^{\infty} e^{-i\pi n} \int_0^{\infty} d\lambda \; e^{i\,[2\delta(\lambda)+2\pi n\lambda\pm\lambda\theta]}. \tag{8.124}$$

The above integrals can be evaluated when the stationary phase approximation is applicable. This is the case of the semiclassical regime, where the phase shift, being an action divided over $\hbar$, is very large. The stationary points of the amplitudes $f^{(+)}$ and $f^{(-)}$, which we denote as λ_{j_+} and λ_{j_-}, are values of the integration variable where the derivative of the phase

$$\varphi^{(\pm)}(x) = 2\delta(\lambda) + 2\pi n\lambda \pm \lambda\theta \tag{8.125}$$

vanishes. That is

$$\left[\frac{d\varphi^{(\pm)}(x)}{dx}\right]_{\lambda_{j_\pm}} = 0. \tag{8.126}$$

Using the explicit value of the phase, we find that the stationary points are the angular momenta for which the deflection function is related with the observation angle as

$$\Theta(\lambda_{j_\pm}) = \mp\theta - 2\pi n. \tag{8.127}$$

Clearly, the last term in the above equation represents the number of times the particle circles around the scatterer.

We discuss below the situation where the Poisson formula can be approximated by its leading term, corresponding to $n = 0$. The inclusion of other terms is straightforward and can be done through the same procedures of section 5.2.3. In this case, the scattering amplitude reduces to

$$f^{(\pm)}(\theta) = \frac{1}{\sqrt{2\pi ik}} \int_0^{\infty} d\lambda \; e^{i\,[2\delta(\lambda)\pm\lambda\theta]}. \tag{8.128}$$

Using the stationary phase approximation to evaluate the integrals (see section 5.2.3) one obtains

$$f(\theta) = \sum_{\lambda_{j_\pm}} \frac{1}{\sqrt{k\,\Theta'(\lambda_{j_\pm})}} \; e^{i\left[2\delta(\lambda_{j_\pm}) \pm \,\theta\lambda_{j_\pm}\right]}. \tag{8.129}$$

The above scattering amplitude is given by a sum of the contributions from all stationary points (λ_{j_-} for $f^{(-)}(\theta)$ and λ_{j_+} for $f^{(+)}(\theta)$), which are assumed to be independent. It can be directly related with the classical cross section. Since the angular momenta $\lambda_{j\pm}$ correspond to the impact parameters $b_{j\pm} = \lambda_{j\pm}/k$, one can write

$$k\,\Theta'(\lambda_{j\pm}) \equiv k\,\frac{d\Theta}{d\lambda_{j\pm}} = \frac{d\Theta}{db_{j\pm}} \tag{8.130}$$

and

$$\frac{1}{\sqrt{k\Theta'(\lambda_{j\pm})}} = \left|\frac{d\Theta}{db}\right|^{-1/2}_{b=b_{j\pm}} e^{i\Delta}. \tag{8.131}$$

The phase Δ appears when we replace $d\Theta/db \to |d\Theta/db|$. It takes care of the sign of the derivative. It is 0 for positive values of the derivative and $\pi/2$ for negative values. Considering the expression for the classical 2D cross section (Eq. (8.84)), one can write the semiclassical cross section as

$$\frac{d\sigma^{2\mathrm{D}}(\theta)}{d\theta} = \left|\sum_{j\pm} e^{i\,\alpha_{j\pm}(\theta)}\,\sqrt{\frac{d\sigma^{2\mathrm{D}}_{cl,j\pm}}{d\theta}}\right|^2. \tag{8.132}$$

where

$$\alpha_{j\pm}(\theta) = 2\,\delta(\lambda_{j\pm}) \pm \lambda_{j\pm}\,\theta + \Delta. \tag{8.133}$$

Eq. (8.132) is the 2D analog of Eq. (5.298).

8.2.3.1 *Caustics in semiclassical 2D scattering*

We have seen in the previous section that the semi-classical 2D cross section, the scattering width, can be expressed as a sum of classical cross sections plus interference terms, which represents the quantum mechanical origin of our derivation. In this sub-section we discuss in more detail the situation where there are contributions of only two stationary points. In this case the inverse of $\Theta(b)$, or equivalently $\Theta(\lambda)$, is a double valued function. That is, a given deflection angle corresponds to two incident impact parameters b_1 and b_2, associated respectively with the angular momenta λ_1 and λ_2. To deal with a concrete situation, we further assume that $\Theta(b) < 0$, and accordingly only $f^{(+)}$ contributes (see Eq. (8.127). Under these circumstances, the 2D cross section of Eq. (8.132) can be written as

$$\frac{d\sigma^{2\mathrm{D}}(\theta)}{d\theta} = \left|\frac{d\Theta(b)}{db}\right|^{-1}_{b=b_1} + \left|\frac{d\Theta(b)}{db}\right|^{-1}_{b=b_2} + \left(\left|\frac{d\Theta(b)}{db}\right|^{-1}_{b=b_1}\left|\frac{d\Theta(b)}{db}\right|^{-1}_{b=b_2}\cos(\alpha_2-\alpha_1)\right)^{1/2}. \tag{8.134}$$

The above cross section oscillates as the argument of the cosine[11],

$$\alpha_2 - \alpha_1 = 2\,[\delta(\lambda_2) - \delta(\lambda_1)] + \frac{\pi}{2} - \theta\,(\lambda_2 - \lambda_1)\,, \tag{8.135}$$

changes with θ. The angular period of oscillation (angular distance between two adjacent maxima in the 2D scattering cross section) can be immediately obtained from the above equation as,

$$\Delta\theta = \frac{2\pi}{|\lambda_2 - \lambda_1|}. \tag{8.136}$$

When λ_1 appoaches λ_2, the above expression breaks down. They converge to the rainbow angular momentum, λ_r, for which the deflection function reaches a maximum of a minimum. To simplify our discussion, we consider a potential that is attractive for any value of the projectile-target separation. In this case, the deflection function is negative for any angular momentum. We assume further that it has the qualitative behavior of figure 5.13, with a single minimum. This minimum is associated with the rainbow angle, θ_r, through the relation $\Theta_r = -\theta_r$. It is clear that $\Theta'(\lambda_r) \equiv \Theta'_r$ vanishes and therefore the semiclassical scattering amplitude of Eq. (8.129) diverges. The reason for this divergence (see section section 5.2.3) is that the second order expansion of the phase is not valid in the neighborhood of λ_r. In such situations one has to resort to other treatments, as the Airy's and the uniform approximations, discussed in detail in section 5.2.3. In Airy's approximation, the phase $\varphi(\lambda)$ is expanded to third order around λ_r, as

$$\begin{aligned}\varphi(\lambda) &\simeq \varphi(\lambda_r) + \varphi'(\lambda_r)\,(\lambda - \lambda_r) + \frac{1}{6}\,\varphi'''(\lambda_r)\,(\lambda - \lambda_r)\\ &= \varphi(\lambda_r) + \frac{1}{2}\,(\theta + \Theta_r)\,(\lambda - \lambda_r) + \frac{1}{6}\,\Theta''_r\,(\lambda - \lambda_r)^3\,. \end{aligned} \tag{8.137}$$

Inserting this expansion into Eq. (8.128), we get

$$f^{(+)}(\theta) = \frac{e^{i\,\varphi_r}}{\sqrt{2\pi i k}} \int_{-\infty}^{\infty} d\lambda\; e^{i\,(\theta+\Theta_r)\,(\lambda-\lambda_r)+i\,(\lambda-\lambda_r)^3/6}. \tag{8.138}$$

As in section 5.2.3, we change variable: $\lambda \to t$ so that

$$\frac{1}{6}\,\Theta''_r\,(\lambda - \lambda_r)^3 = \frac{1}{3}\,t^3. \tag{8.139}$$

[11]The $\pi/2$ term comes from the fact that the derivatives of the deflection function at the two stationary points have opposite signs. Thus, one will have $\Delta = 0$ and the other $\Delta = \pi/2$.

This transformation leads to

$$\lambda - \lambda_r = \left(\frac{2}{\Theta_r''}\right)^{1/3} t \quad \text{and} \quad d\lambda = \left(\frac{2}{\Theta_r''}\right)^{1/3} dt. \tag{8.140}$$

With this transformation, the scattering amplitude can be written as

$$f^{(+)}(\theta) = \sqrt{\frac{2\pi}{k}} \left(\frac{2}{\Theta_r''}\right)^{1/3} e^{i\varphi_r - i\pi/4} \left[\frac{1}{2\pi}\int_{-\infty}^{\infty} dt\, e^{iyt+it^3/3}\right], \tag{8.141}$$

with

$$y = \left(\frac{2}{\Theta_r''}\right)^{1/3} (\theta + \Theta_r). \tag{8.142}$$

The quantity within the square brackets is the Airy function [Abramowitz and Stegun (1972)], $Ai(y)$. Thus, one can write

$$f^{(+)}(\theta) = \sqrt{\frac{2\pi}{k}} \left(\frac{2}{\Theta_r''}\right)^{1/3} e^{i\varphi_r - i\pi/4}\, Ai(y) \tag{8.143}$$

and the corresponding 2D cross section is

$$\frac{d\sigma^{2\mathrm{D}}(\theta)}{d\theta} = \frac{2\pi}{k} \left(\frac{2}{\Theta_r''}\right)^{2/3} Ai^2(y). \tag{8.144}$$

This cross section has the same angular dependence as the corresponding rainbow cross section in 3 dimensions (Eq. (5.324)). Only their strengths are different (including the dimensions).

Airy's approximation is quite accurate in the dark side of the rainbow, and also in the illuminated region at angles close to θ_r. As illustrated in section 5.2.3 for 3D scattering, it only shows appreciable deviations with respect to the quantum mechanical cross section at small scattering angles, well below the first minimum under the rainbow angle. A better treatment of the problem, accurate in the whole angular region, is the uniform approximation. This approximation can be derived similarly to the case of 3D scattering.

The scattering in 3D exhibits more interesting optical effects. Besides the rainbow, one has the effect of glory scattering, both near $\theta = 0$ (forward glory) and near $\theta = \pi$ (backward glory). The presence of $\sin\theta$ in the denominator of the classical cross section is responsible for these caustics. In 2D scattering the classical cross section does not contain this denominator and therefore does not exhibits this kind of caustic.

Another effect seen in the 3D case is the phenomenon of orbiting. It can also be seen in 2D scattering. The orbiting condition is that at some angular momentum λ_o and energy E_o, corresponding effective potential

$$U_{\text{eff}}(r) = U(r) + \frac{\lambda^2}{k^2 r^2} \tag{8.145}$$

has a maximum at a certain radial distance r_o and value of this maximum is equal to the collision energy. That is

$$\left[\frac{dU_{\text{eff}}(r)}{dr}\right]_{r=r_o} = 0 \quad \text{and} \quad U_{\text{eff}}(r_o) = E_o. \tag{8.146}$$

8.3 The inverse scattering problem

Given a set of scattering data, can one obtain the interaction by inversion? The answer depends on several factors. Here we shall dwell on the use of the methods developed in this chapter to obtain some information about the potential responsible for the scattering. There are several sophisticated techniques to determine the interaction associated with a given data set in potential scattering. This is a very complicated problem, which may not have a unique solution. Thus, we only discuss the inverse scattering problem in a few simple situations. For a detailed presentation of the problem, we refer to [Newton (2002)].

The simplest situation is when the first-order Born approximation is a valid description of the physics. This happens for weak interactions and/or high energy. Assuming a spherically symmetric interaction, the scattering amplitude within the first Born approximation is given by Eq. (5.18), which can be written as

$$qf^{(1)}(q) = -\frac{2\mu}{\hbar^2} \int \sin(qr)\, rV(r)\, dr, \tag{8.147}$$

where

$$q = 2k \sin(\theta/2) \tag{8.148}$$

is the momentum transfer (divided by $\hbar$). Multiplying both sides of the above equation by $\sin(qr')$, integrating over q and using the orthogonality relation[12]

$$\int_0^\infty \sin(qr)\, \sin(qr')\, dq = \frac{\pi}{2}\, \delta(r - r'), \tag{8.149}$$

[12] This expression can easily be derived using the exponential form of the sinus functions and the well know relation: $\int_{-\infty}^{\infty} \exp\left[iq(r - r')\right] dq = 2\pi\, \delta(r - r')$.

we obtain

$$rV(r) = -\frac{\hbar^2}{\pi\mu}\int_0^\infty q f^{(1)}(q)\ \sin(qr)\ dq. \tag{8.150}$$

Thus, the knowledge of $f^{(1)}(q)$ is sufficient to determine the potential. The data, however, supply $|f|^2$. Thus there is an ambiguity of sign[13].

In typical situations, the first Born approximation is not valid and Eq. (8.150) cannot be used. It is then necessary to resort to more complicated procedures [Newton (2002)]. One possibility is to obtain information on the interaction through indirect procedures. For example, one can parametrize the interaction potential and fit the parameters as to reproduce scattering data. In such cases, the success of the method is strongly dependent on the details of the parametrization.

Usually it is easier to investigate the interaction through comparisons of theoretical phase shifts and phase shifts obtained by the partial-wave analysis of experimental angular distributions. Such studies are then performed in two steps. First, the angular distribution given by the partial-wave expansion for the scattering amplitude is compared with the corresponding experimental values, and the phase shift are treated as fitting parameters. Of course, from the quantum mechanical point of view, the scattering amplitude is an infinite manifestation of Heisenberg's uncertainty principle: given a value of the angle θ, the conjugate variable, l, could have any physically permitted value in the range $0 < l < \infty$. However, for practical purposes, it is necessary to truncate the partial-wave expansion at a small value of the angular momentum, $l_{\max}$. Otherwise, the fitting procedure becomes unreliable. Even with only a few terms in the partial-wave expansion, it is necessary to have accurate angular distributions, measured at a large set of closely spaced angles. Since the phase shifts are functions of the collision energy, it is necessary to repeat the procedure for several energies. The relevant contributions to the scattering amplitude comes from angular momenta $l \lesssim l_{\max} \sim k\bar{R}$ (see section 2.6). Thus, this procedure can only be used at low scattering energies.

Once the 'experimental' phase shifts are determined, the second step can be taken. It is the comparison of these phase shifts with the theoretical ones, obtained from the solution of the radial equation with the potential that is under investigation. This potential could be parametrized in some

[13]Note that in the perturbative limit the scattering amplitude is real.

way or obtained from some theoretical model. The comparison then leads to the best values of the parameters or is used as a check for the model.

As an example, let us consider the very low energy scattering of neutrons from nuclei, or atom-atom scattering, where the inclusion of the angular momenta $l = 0$ and 1 is sufficient. Assuming that the potential is real, we can write $S_l = e^{2i\delta_l}$, with real δ_l, and the amplitude of Eq. (2.54a) can be written,

$$f(\theta) = \frac{1}{2ik} \left[\left(e^{2i\delta_0} - 1\right) + 3\left(e^{2i\delta_1} - 1\right) \cos\theta\right], \tag{8.151}$$

where we have used the explicit form of the Legendre polynomials, $P_0(\cos\theta) = 1$ and $P_1(\cos\theta) = \cos\theta$. The corresponding differential cross section then can be written as

$$\frac{d\sigma}{d\Omega} \equiv |f(\theta)|^2 = \frac{1}{k^2}\left[\sin\delta_0^2 + 9\sin\delta_1^2\cos^2\theta + 6\sin\delta_0\sin\delta_1\cos(\delta_0 - \delta_1)\cos\theta\right]. \tag{8.152}$$

For a given collision energy, the above expression is compared with the experimental angular distribution at a set of angles, $\theta = \theta_1, \theta_2, \ldots.$, and the phase shifts δ_0 and δ_1 are fitted as to minimize the deviation of the theoretical expression with respect to the data. Once these phase shifts are determined, their values can be confronted with the theoretical results for the interaction potential.

In practice, low energy scattering is best described by the parameters of the effective range theory (see section 2.11), which we summarize below. We first cast the scattering amplitude in the form

$$f(\theta) = \sum_{l=0}^{\infty} f_l(\theta), \tag{8.153}$$

with

$$f_l(\theta) = \frac{2l+1}{2ik}\left[e^{2i\delta_l} - 1\right] P_l(\cos\theta) = (2l+1)\frac{e^{i\delta_l}}{k}\sin\delta_l\, P_l(\cos\theta), \tag{8.154}$$

or[14]

$$f_l(\theta) = \left[\frac{2l+1}{k\cot\delta_l - ik}\right] P_l(\cos\theta). \tag{8.155}$$

As mentioned earlier in this section, only angular momenta $l \lesssim l_{\max}$ contributes to the partial-wave expansion of the scattering amplitude. At

[14]We used the identity: $e^{i\delta_l}\sin\delta_l = 1/\left[\cot\delta_l - i\right]$, which can easily be checked by direct substitution.

very low energies, we have: $l_{\max} \sim k\bar{R} \ll 1$. Thus, the expansion can be truncated at $l = 0$. The scattering amplitude, which in this limit is isotropic, reduces to

$$f(\theta) \simeq f_0 = \frac{1}{g(k) - ik}. \tag{8.156}$$

with

$$g(k) = k \cot \delta_0. \tag{8.157}$$

In the low energy limit, the term $k \cot \delta_0$ can be expanded around $k = 0$. The scattering length, a, is given by its zero-th oder expansion through the expression

$$-\frac{1}{a} = \lim_{k \to 0} \left[k \cot \delta_0\right]. \tag{8.158}$$

The effective range formula corresponds to the expansion of $k \cot \delta_0$ to the next order in k^2,

$$k \cot \delta_0 \simeq g^{(2)}(k) = \left[-\frac{1}{a} + \frac{1}{2} r_{\text{eff}}\, k^2\right]. \tag{8.159}$$

The parameter r_{eff} is called the *effective range* and its relation to the s-wave function is given in section 2.11. Within this approximation, the scattering amplitude is

$$f(\theta) \simeq f^{(2)} = \left[-\frac{1}{a} + \frac{1}{2} r_{\text{eff}}\, k^2 - ik\right]^{-1}. \tag{8.160}$$

The importance of the effective range correction to the scattering amplitude (term in k^2) is illustrated in figure 2.6, of section 2.11. It becomes progressively more important as the collision energy increases.

The analysis of low energy scattering data based on the effective range formula uses a generalization of the above procedure, so that it can be applied to energies slightly higher. It contains two improvements. The first is that it takes into account a few of the lowest angular momenta, besides $l = 0$. The second is that the expansion includes also terms of the order k^4. In the case of $l > 0$, one re-write Eq. (8.155) as

$$f_l(\theta) = \left[\frac{2l+1}{k^{-2l}\, g_l(k) - ik}\right] P_l(\cos\theta). \tag{8.161}$$

with

$$g_l(k) = k^{2l+1} \cot \delta_l. \tag{8.162}$$

The factor $g_l(k)$ is then expanded to fourth order around $k = 0$, as [Teichmann (1951); Baye (2004); Hussein *et al.* (2008)]

$$g_l^{(4)}(k) = -\frac{1}{a_l} + \frac{1}{2} r_l \, k^2 - \mathcal{P}_l \, r_l^3 \, k^4. \tag{8.163}$$

Above, a_l and r_l are respectively the generalizations of the scattering length and the effective range to higher partial-waves. The expansion now includes a fourth order contribution, which is given in terms of the *shape parameter*, $\mathcal{P}_l$. Note that, for $l \neq 0$, a_l, r_l and $\mathcal{P}_l$ do not have the dimension of length (L). Their dimensions are respectively, L^{2l+1}, L^{-2l+1} and L^{-4l}.

Using the approximation of Eq. (8.163) in Eqs. (8.161) and (8.153), one obtains a better low-energy approximation for the scattering amplitude and, consequently, for the cross section. As far as low energy scattering is concerned, knowing the parameters a_l, r_l and $\mathcal{P}_l$ is as good as knowing the potential itself. Within this view, the inverse scattering problem in this limit is solved fitting these parameters to the experimental cross section. The extension of the above discussion to collisions of charged particles will be discussed in section 8.4.1.

It is also possible to derive differential equations for the low energy scattering parameters. This method, which offers some advantages, is discussed in details in [Calogero (1967); Newton (2002); Hussein *et al.* (2008)]. For completeness, we turn to the derivation of the Calogero equation.

8.4 The Calogero equation for neutral particle scattering

The determination of the effective range parameters is made through the defining relation,

$$k^{2l+1} \cot \delta_l(k) = -\frac{1}{a_l} + \frac{1}{2} r_l \, k^2 - \mathcal{P}_l \, r_l^3 \, k^4.$$

However, a more direct means is also available which in a way avoids the phase shift and goes directly to what we may call the scattering length function, $a_l(k, r)$, which satisfies a first order non-linear differential equation. This equation was originally derived by Calogero just for the scattering length. Recently, it has been extended to the effective range and the shape parameter [Hussein *et al.* (2008)]. This equation can be derived directly from the general solution of the radial equation, which can be written as,

$$u_l(k, r) = q_l(k, r) \, \hat{j}_l(kr) + p_l(k, r) \, \hat{n}_l(kr), \tag{8.164}$$

where $\hat{\jmath}_l(kr)$ and $\hat{n}_l(kr)$ are the Riccati-Bessel and Riccati-Neumann functions, introduced in chapter 2, and the amplitudes $q_l(k,r)$ and $p_l(k,r)$ are to be determined using the appropriate boundary conditions, as done in chapter 4. In fact these amplitudes can be read off using the Green's function in the l-projected Lippmann-Schwinger equation, Eq. (4.183). The radial wave function can therefore be written as, with the help of Eqs. (2.70) and (2.42),

$$q_l(k,r) = e^{i\delta_l}\cos\delta_l - \frac{1}{k}\int_r^\infty dr'\, \hat{n}_l(kr')\, U(r')\, u_l(k,r') \tag{8.165}$$

$$p_l(k,r) = -\frac{1}{k}\int_0^r dr'\, \hat{\jmath}_l(kr')\, U(r')\, u_l(k,r'). \tag{8.166}$$

It is clear that asymptotically the ratio, $p_l(k,\infty)/q_l(k,\infty)$ is just $\tan\delta_l$, while $q_l(k,r) \to \alpha = e^{i\delta_l}\cos\delta_l$ (see Eqs. (2.32), (2.34) and (2.38)).

Following Flügge [Flugge (1999)], we write

$$u_l(k,r) = q_l(k,r)\Big[\hat{\jmath}_l(kr) + \Delta_l(k,r)\,\hat{n}_l(kr)\Big], \tag{8.167}$$

with the notation

$$\Delta_l(k,r) = \frac{p_l(k,r)}{q_l(k,r)}. \tag{8.168}$$

The Calogero equation is a differential equation obeyed by $\Delta_l(k,r)$. It is straightforward to obtain this equation by differentiation, and using Eqs. (8.165) and (8.166),

$$\frac{d\,\Delta_l(k,r)}{dr} = -\frac{U(r)}{k}\,\frac{u_l^2(k,r)}{q_l^2(k,r)} = -\frac{U(r)}{k}\left[\hat{\jmath}_l(kr) + \Delta_l(k,r)\,\hat{n}(kr)\right]^2, \tag{8.169}$$

were $U(r) = 2\mu V(r)/\hbar^2$. For our purposes, it is convenient to introduce the scattering length function,

$$a_l(k,r) = -\frac{1}{k^{2l+1}}\,\Delta_l(k,r), \tag{8.170}$$

and insert the above expression in Eq. (8.169). We then obtain the Calogero-type equation for the scattering length function,

$$\frac{da_l(k,r)}{dr} - \frac{U(r)}{k^{2l+2}}\left[\hat{\jmath}_l(kr) - k^{2l+1}\,\hat{n}_l(kr)\,a_l(k,r)\right]^2 = 0. \tag{8.171}$$

The above equation is solved with the boundary condition $a_l(k, r=0) = 0$. Its asymptotic limit gives the phase shift, according to the relation

$$\frac{1}{a_l(k, r\to\infty)} = -\,k^{2l+1}\cot\delta_l(k). \tag{8.172}$$

Thus, we can obtain the low-energy scattering parameters for all l by solving one, non-linear, first-order, differential equation. At very low energies we can use the small kr limit of the Bessel functions, $\hat{\jmath}_l(kr) = (kr)^{l+1} / (2l+1)!!$ and $\hat{n}_l(kr) = (2l-1)!!/(kr)^l$ (see Eqs. (2.22a) and (2.22b)), to write the desired low-energy equation,

$$\frac{da_l(0,r)}{dr} - U(r)\left[\frac{r^{2l+1}}{(2l+1)!!} - \frac{(2l-1)!!}{r^l}\, a_l(0,r)\right]^2 = 0. \tag{8.173}$$

The non-linear (8.171) equation is solved numerically, using the boundary conditions cited above. We refer the reader to recent work on its solution [Hussein *et al.* (2008)]. Its $l = 0$ version is,

$$\frac{a_0(k,r)}{dr} - \frac{U(r)}{k^2}\, [\sin(kr) - k\cos(kr)\, a_0(k,r)]^2 = 0, \tag{8.174}$$

giving at $k = 0$ the simple looking equation,

$$\frac{da_0(0,r)}{dr} - U(r)\, [r - a_0(0,r)]^2 = 0. \tag{8.175}$$

An analytical solution of Eq. (8.175) is not possible. One can, however, resort to approximate treatments, such as linearization,

$$\frac{da_0(0,r)}{dr} = U(r)\, \left[r^2 - 2r\, a_0(0,r)\right]. \tag{8.176}$$

The general solution can be written as

$$a_0(0,r) = F_0(r) \int_0^r dr' r'^2\, U(r')\, F_0(r'), \tag{8.177}$$

where F_0 is the function

$$F_0(r) = \exp\left[-\int^r dr''\; 2r''\, U(r'')\right]. \tag{8.178}$$

For the general case of non-zero k and any l, the linearization of Eq. (8.171) leads to

$$\frac{a_l(k,r)}{dr} + 2\,\frac{U(r)}{k}\, \hat{\jmath}_l(kr)\, \hat{n}_l(kr)\, a_l(k,r) - \frac{U(r)}{k^{2l+2}}\, \hat{\jmath}_l^2(kr) = 0. \tag{8.179}$$

The above equation can also be solved. First we cast it in the form

$$\frac{da_l(k,r)}{dr} + B_l(r)\, a_l(k,r) + C_l(r) = 0, \tag{8.180}$$

where

$$B_l(r) = 2\,\frac{U(r)}{k}\, \hat{\jmath}_l(kr)\, \hat{n}_l(kr) \quad \text{and} \quad C_l(r) = -\frac{U(r)}{k^{2l+2}}\, \hat{\jmath}_l^2(kr). \tag{8.181}$$

The general solution for the scattering length for any l can be written as

$$a_l(k,r) = -F_l(r) \int_0^r dr' \; C_l(r')\, F_l(r'), \tag{8.182}$$

where F_l is the function

$$F_l(r) = \exp\left[- \int^r dr'' \; B_l(r'') \right]. \tag{8.183}$$

In the $k \to 0$ limit the functions $B_l(r)$ and $C_l(r)$ take the simpler forms

$$B_l(r) \to 2\, \frac{(2l-1)!!}{(2l+1)!!}\; U(r)\, r^{l+1} \quad \text{and} \quad C_l(r) \to -\frac{1}{(2l+1)!!}\; U(r)\, r^{2l+2}. \tag{8.184}$$

Calogero-type equations for the s-wave effective range, r_0, and shape parameter, $\mathcal{P}_0$, can be derived by expanding the $l = 0$ version of Eq. (8.174), and using the definition,

$$-\frac{1}{a_0(k,r)} = -\frac{1}{a_0(0,r)} + \frac{1}{2}\, r_0(r)\, k^2 - \mathcal{P}_0(r)\, r_0^3\, k^4. \tag{8.185}$$

For details of the derivation and the solutions, we refer the reader to the original reference, [Hussein *et al.* (2008)].

8.4.1 *The Calogero equation for charged particle scattering*

When the scattering particles are charged, the $1/r$ Coulomb interaction has to be added to the potential. The free solutions are now Coulomb wave functions and special care must be taken when performing the low-energy expansion to ensure convergent results for the scattering length, effective range and shape parameters. The appropriate low-energy expansion for nuclear scattering was developed by Teichmann in [Teichmann (1951)]. In this case the l-dependent counterpart of Eq. (8.185), should be modified as,

$$2\,\frac{\omega_l(E)}{(l!)^2\, a_N^{2l+1}} \left[\frac{2}{D_l^{(\mathrm{C})}(E)} + h(\eta) \right] = -\frac{1}{a_l(0,r)} + \frac{1}{2} r_l\, k^2 - \mathcal{P}_l\, r_l^3\, k^4 + O(k^6). \tag{8.186}$$

Above, we have introduced the functions,

$$D_l^{(\mathrm{C})}(E) = \frac{\pi}{2}\, \tan\delta_l(E)\, e^{2\pi\eta}, \tag{8.187}$$

$$\omega_l(E) = \prod_{n=1}^{l} \left(1 + \frac{n^2}{\eta^2} \right), \tag{8.188}$$

$$h(\eta) = \frac{1}{12\eta^2} + \frac{1}{120\eta^4} + O\left(\frac{1}{\eta^6}\right), \tag{8.189}$$

where $\eta = Z_p\, Z_t e^2/4\pi\epsilon_0\hbar v$ is the usual Sommerfeld parameter (see chapter 3), v is the asymptotic velocity, a_N is the so-called nuclear Bohr radius, $a_N = 4\pi\epsilon_0\hbar c/\mu Z_p Z_t e^2$, and μ is the reduced mass. The above equation does indeed give convergent results as η goes to ∞ in the zero-E limit. Further, the equation reduces to that for neutral particles upon setting $\eta = 0$.

The corresponding Calogero equation then follows by use of the Wronskian of the scaled Coulomb wave functions

$$W[\mathcal{G}_l(kr), \mathcal{F}_l(kr)] = \frac{\pi}{2}, \tag{8.190}$$

where $\mathcal{F}_l(kr)$ and $\mathcal{G}_l(kr)$ are related to the regular, $F_l(kr)$, and irregular, $G_l(kr)$, Coulomb wave functions [Abramowitz and Stegun (1972)]

$$\mathcal{F}_l(kr) = \frac{e^{\pi\eta}}{\sqrt{k}}\, F_l(kr), \tag{8.191}$$

and

$$\mathcal{G}_l(kr) = \frac{\pi\, e^{-\pi\eta}}{2\sqrt{k}}\, G_l(kr). \tag{8.192}$$

The Calogero equation for the tangent function can be derived following the same procedure as the one used for the neutral particle scattering and one gets

$$\frac{da_l^c(k,r)}{dr} - \frac{\pi}{2}\, e^{-2\pi\eta}\, U(r)\, \left[\mathcal{F}_l(kr) - \frac{\pi}{2} e^{-2\pi\eta}\, \mathcal{G}_l(kr)\; a_l^c(k,r)\right]^2 = 0. \tag{8.193}$$

Solving this equation, one can take the asymptotic limit of the solution, which is just the modified tangent function of Eq. (8.187). The low energy expansion parameters $a_l + r_l$ and P_l can then be directly determined from Eq. (8.186).

Exercises

(1) a) Show that the radial wave function of Eq. (8.5),

$$\tilde{u}_l(k,r) = \hat{\jmath}_l(kr) + \int_0^r dr'\, \tilde{g}_l(r,r')\, V(r')\, \tilde{u}_l(k,r'),$$

is a solution of the radial equation[15],

$$\frac{\hbar^2}{2\mu}\left[\frac{d^2}{dr^2} + k^2 - \frac{l\,(l+1)}{r^2}\right]\, \tilde{u}_l\,(k,r) = V(r)\; \tilde{u}_l\,(k,r)\,.$$

[15] *Hint*: insert Eq. (8.5) in the LHS of the radial equation and, using the Wronskian relation for $\hat{\jmath}_l(kr)$ and $\hat{n}_l(kr)$ (Eq. (2.65), show that its RHS is reproduced).

Above, $k = \sqrt{2\mu E}/\hbar$, $V(r)$ is a short range potential and $\tilde{g}(r, r')$ is the Green's function of Eq. (8.6).
b) Show that $\tilde{u}_l(k, r)$, as defined in Eq. (8.5), satisfies the boundary conditions:

$$\tilde{u}_l(k, 0) = 0 \quad \text{and} \quad \tilde{u}_l'(k, 0) = \jmath_l'(0).$$

(2) Prove that the angular momentum expansion of the scattering amplitude in 2D scattering is given by Eq. (8.150).

(3) Consider the scattering amplitude

$$f(q) = \frac{2\mu V_0}{\hbar^2 \alpha} \frac{1}{a^2 + \alpha^2},$$

where q is the momentum transfer in $\hbar$ units. Show that the corresponding potential is of Yukawa type, given by the expession[16],

$$V(r) = V_0 \frac{e^{-\alpha r}}{\alpha r}.$$

(4) Derive the Calogero equation for the effective range, using Eq. (8.174). For this purpose, make an expansion in the wave number, k, and collect the terms proportional to k^2.

[16] *Hint*: Use Eq. (8.150) for this purpose.

PART 2

Many-body Scattering

Chapter 9

The Coupled Channel Formalism

The potential scattering theory developed in part I of this book is used to describe the scattering of structureless particles. In chapters 1 to 5, the only degrees of freedom involved in the collision were those associated with the projectile-target separation vector, $\mathbf{r}$. The spin degree of freedom was incorporated in the formalism in chapter 6. This allows the description of the scattering of structureless particles with spin, in the presence of spin-orbit interaction. One could use the formalism of chapter 6 to treat e.g. the scattering of nucleons off nucleons at low energies or electron-atom scattering at extremely low energies.

In general, composite particles, such as atoms, molecules and nuclei, can be excited when they are involved in a collision. In such cases, the scattering process involves the active participation of several degrees of freedom. It is then necessary to generalize the formalism described in part I of this book. When one is only interested in a description of elastic scattering, the influence of the intrinsic degrees of freedom may be simulated by the use of an effective complex potential in a problem of potential scattering. Effective complex potentials were introduced in chapter 7 and will be considered in detail in chapter 10. However, this procedure can only be applied to elastic scattering. It does not lead to theoretical predictions of cross sections for other processes, such as inelastic scattering.

In order to evaluate the cross section for each channel excited in a collision, it is necessary to develop a formalism in which the couplings among the relevant degrees of freedom are explicitly taken into account. This is accomplished by the *coupled channels* (CC) formalism. In this formalism, the wave function of the projectile-target system is expanded in a basis of eigenstates (channels) of the intrinsic Hamiltonian and this expansion is inserted in the Schrödinger equation. Taking scalar product with each intrinsic state, one obtains an infinite set of coupled equations, which cannot be solved exactly. However, it frequently occurs that a finite number of these states (say N) are strongly coupled among themselves, while their couplings with the remaining channels are weaker. One can then resort to the *close coupling* approximation, which consists of solving the problem in the truncated space spanned by the N strongly coupled intrinsic states. Couplings with the remaining channels are handled differently, either through the use of effective channel potentials (see chapter 10) or by perturbative procedures. A very simple case of coupled equations in a scattering problem has already been presented in the first section of chapter 6, where we discussed collisions of particles with spin. For a projectile with spin s colliding with a spinless target the wave function with well defined total angular momentum J contains components with orbital angular momentum $l = J + s, l = j + s - 1, ..., l = |J - s|$. If the orbital angular momentum is not a constant of motion, the radial wave functions associated to partial-waves of the same parity will be coupled. In this case, the coupled equations are similar to the ones appearing in multi-channel scattering problems. In the present chapter, we go beyond the spin degree of freedom and develop a full CC theory in a truncated space, appropriate for the scattering of particles with arbitrary intrinsic structure.

Let us consider the collision described by the total wave function $\Psi^{(+)}(\mathbf{r}, \xi)$ where $\mathbf{r}$ stands for the projectile-target separation vector and ξ for the set of intrinsic coordinates of the projectile-target system. We neglect *re-arrangement channels*, where the collision partners exchange mass. Such channels lead to additional complications, which will be avoided in this text[1]. The collision dynamics in the truncated space is generated by the Hamiltonian

$$\mathbb{H} = K + h + \mathbb{V}, \tag{9.1}$$

where $h \equiv h(\xi, p_\xi)$ is the intrinsic Hamiltonian, $K \equiv -\hbar^2 \nabla^2 / 2\mu$ is the operator corresponding to the kinetic energy of the relative motion and $\mathbb{V} \equiv \mathbb{V}(\mathbf{r}, \xi)$ is the interaction[2]. This equation corresponds to Eq. (1.2) in operator form, using the center of mass frame.

[1] For a detailed study multi-channel problems with re-arrangement channels we refer to [Satchler (1983)] or [Thompson (1988)].

[2] Throughout this chapter, 'blackboard bold ' fonts, e.g., $\mathbb{V}$, denote operators acting in the full $\{\xi, \mathbf{r}\}$ space.

The eigenstates of the intrinsic Hamiltonian satisfy Eq. (1.1), which can be written as

$$h\ |\alpha) = \varepsilon_\alpha\ |\alpha)\,. \tag{9.2}$$

They obey the orthonormality,

$$(\alpha'|\ \alpha) = \delta_{\alpha\alpha'}, \tag{9.3}$$

and the completeness[3],

$$\sum_\alpha |\alpha)\ (\alpha| = \mathbf{1}, \tag{9.4}$$

relations. The label α represents the set of quantum numbers which are needed to fully specify the intrinsic state of the projectile-target system. Above, round brackets indicate scalar products in the intrinsic space.

For practical purposes, the interaction is split as

$$\mathbb{V} = \mathbb{V}^{(1)} + \mathbb{V}^{(2)}, \tag{9.5}$$

where $\mathbb{V}^{(1)}$ is diagonal in channel space. One can then write the expansions

$$\mathbb{V}^{(1)} = \sum_\alpha |\alpha)\ V^{(1)}_\alpha\ (\alpha| \tag{9.6a}$$

$$\mathbb{V}^{(2)} = \sum_{\alpha\alpha'} |\alpha)\ V^{(2)}_{\alpha,\alpha'}\ (\alpha'|\,, \tag{9.6b}$$

with

$$V^{(1)}_\alpha(\mathbf{r}) = \int d\xi\ |\varphi_\alpha(\xi)|^2\ \mathbb{V}^{(1)}(\mathbf{r},\xi) \tag{9.7}$$

and

$$V^{(2)}_{\alpha,\alpha'}(\mathbf{r}) = \int d\xi\, \varphi^*_\alpha(\xi)\ \mathbb{V}^{(2)}(\mathbf{r},\xi)\ \varphi_{\alpha'}(\xi). \tag{9.8}$$

In Eqs. (9.7) and (9.8), $\varphi_\alpha(\xi)$ and $\varphi_{\alpha'}(\xi)$ are, respectively, the wave functions corresponding to the states $|\alpha)$ and $|\alpha')$, in the ξ-representation. Note that the potential $\mathbb{V}^{(1)}$ is arbitrary, except for the condition of being diagonal in channel space. However, once it is chosen, $\mathbb{V}^{(2)}$ is given by the relation $\mathbb{V}^{(2)} = \mathbb{V} - \mathbb{V}^{(1)}$. Frequently, it is convenient to choose $\mathbb{V}^{(1)}$ such that $\mathbb{V}^{(2)}$ is purely off-diagonal. In such cases the components of $\mathbb{V}^{(2)}$ can be written

$$V^{(2)}_{\alpha,\alpha'}(\mathbf{r}) = \int d\xi\ \varphi^*_\alpha(\xi)\ \mathbb{V}(\mathbf{r},\xi)\ \varphi_{\alpha'}(\xi)\ -\ \delta_{\alpha\alpha'}\ V^{(1)}_\alpha(\mathbf{r}). \tag{9.9}$$

[3]At this stage, we are not considering here intrinsic states in the continuum, which correspond to three or more particles in the final state.

To derive the coupled-channel equations, we use the Schrödinger equation,

$$\left[E - \mathbb{H}\right] |\Psi^{(+)}(0\mathbf{k}_0)\rangle = 0, \tag{9.10}$$

and the channel-expansion

$$|\Psi^{(+)}(0\mathbf{k}_0)\rangle = \sum_\alpha |\psi_\alpha^{(+)}(0\mathbf{k}_0)\rangle \; |\alpha) . \tag{9.11}$$

The notation $|\Psi^{(+)}(0\mathbf{k}_0)\rangle$ indicates the scattering state in a collision initiated in the entrance channel ($\alpha = 0$) with wave vector $\mathbf{k}_0$. If the intrinsic state in the incident channel has spin different from zero, one should include the value of its z-component, ν_0, among the labels of the wave functions. That is

$$\left[E - \mathbb{H}\right] |\Psi^{(+)}(0\nu_0\mathbf{k}_0)\rangle = 0, \tag{9.12}$$

and

$$|\Psi^{(+)}(0\nu_0\mathbf{k}_0)\rangle = \sum_\alpha |\psi_\alpha^{(+)}(0\nu_0\mathbf{k}_0)\rangle \; |\alpha) . \tag{9.13}$$

This situation will be discussed when we carry out angular momentum projections. For the moment, we will assume that the intrinsic states have spin zero.

Usually, the energy scale is chosen such that $\varepsilon_0 = 0$. Owing to the off-diagonal part of the interaction, the solution of the Schrödinger equation has components $|\psi_\alpha^{(+)}(0\mathbf{k}_0)\rangle$ for both $\alpha = 0$ and $\alpha \neq 0$.

9.1 The close coupling approximation

For practical purposes, the infinite expansion of Eq. (9.11) is truncated after a finite number of channels (N say). This procedure is known as the *close coupling* approximation. For an accurate calculation, the most relevant channels should be included in the truncated space. To account for the loss of flux through neglected channels, one may include an imaginary part in the channel potentials, $V_\alpha^{(1)}(\mathbf{r})$.

To find the wave function, we write the Hamiltonian as

$$\mathbb{H} = h + \mathbb{H}^{(1)} + \mathbb{V}^{(2)}, \tag{9.14}$$

where

$$\mathbb{H}^{(1)} = K + \mathbb{V}^{(1)}, \tag{9.15}$$

insert Eqs. (9.11) and (9.14) into Eq. (9.10), and take the scalar product with each intrinsic state $(\alpha|$. In this way, we get the set of N coupled differential equations

$$\left[E_\alpha - H_\alpha^{(1)}\right] |\psi_\alpha^{(+)}(0\mathbf{k}_0)\rangle = \sum_{\alpha'=0}^{N-1} V_{\alpha\alpha'}^{(2)} \left|\psi_{\alpha'}^{(+)}(0\mathbf{k}_0)\right\rangle, \quad \alpha = 0, N-1. \quad (9.16)$$

In the coordinate representation, these equations are written[4]

$$\left[E_\alpha + \frac{\hbar^2}{2\mu}\nabla^2 - V_\alpha^{(1)}(\mathbf{r})\right] \psi_\alpha^{(+)}(0\mathbf{k}_0;\mathbf{r}) = \sum_{\alpha'=0}^{N-1} V_{\alpha\alpha'}^{(2)}(\mathbf{r})\, \psi_{\alpha'}^{(+)}(0\mathbf{k}_0;\mathbf{r}). \quad (9.17)$$

Above,

$$E_\alpha = E - \varepsilon_\alpha \quad (9.18)$$

is the energy of the relative motion in channel α and

$$H_\alpha^{(1)} = K + V_\alpha^{(1)}, \quad \alpha = 1, ..., N-1, \quad (9.19)$$

are the diagonal matrix-element of $H^{(1)}$. The diagonal matrix-elements of the potential $\mathbb{V}^{(1)}$,

$$V_\alpha^{(1)} \equiv V_{\alpha,\alpha}^{(1)} = \langle\alpha\,|\mathbb{V}^{(1)}|\,\alpha\rangle\,,$$

are usually called the *channel potentials.*

It is convenient to represent the wave function for neutral particle scattering as

$$\Psi^{(+)}(0\mathbf{k}_0;\mathbf{r}) = \Phi(0\mathbf{k}_0;\mathbf{r}) + \Psi^{\rm sc}(0\mathbf{k}_0;\mathbf{r}), \quad (9.20)$$

where

$$\Phi(0\mathbf{k}_0;\mathbf{r}) = \frac{1}{(2\pi)^{3/2}}\, e^{i\mathbf{k}_0\cdot\mathbf{r}}\; |0) \quad (9.21)$$

describes the incident free particle state and $\Psi^{\rm sc}$ gives the modification in the wave function arising from the potential $\mathbb{V}$. It satisfies the boundary condition[5]

$$\Psi^{\rm sc}(0\mathbf{k}_0;\mathbf{r}\to\infty) = \frac{1}{(2\pi)^{3/2}} \sum_\alpha \left[f_{\alpha,0}(\mathbf{k}_\alpha,\mathbf{k}_0)\,\frac{e^{ik_\alpha r}}{r}\right] |\alpha)\,. \quad (9.22)$$

In this way, the total wave function has the asymptotic form

$$\Psi^{(+)}(0\mathbf{k}_0;\mathbf{r}) \to \frac{1}{(2\pi)^{3/2}} \sum_\alpha \left[e^{i\mathbf{k}_0\cdot\mathbf{r}}\,\delta_{\alpha 0} + f_{\alpha,0}(\mathbf{k}_\alpha,\mathbf{k}_0)\,\frac{e^{ik_\alpha r}}{r}\right] |\alpha)\,. \quad (9.23)$$

[4]We are assuming that the potentials are local in the coordinate space.

[5]In this chapter we use $(\mathbf{k}_\alpha, \mathbf{k}_0)$ as the argument of the scattering amplitude, instead of the scattering angle.

This equation is a straightforward generalization of Eq. (2.3). In each channel there is an outgoing spherical wave with the wave number, $k_\alpha = \sqrt{2\mu E_\alpha/\hbar^2}$. The vector $\mathbf{k}_\alpha$, appearing in the argument of the scattering amplitude[6], is parallel to $\mathbf{r}$ and has modulus k_α. Note that, the elastic channel contains also the incident plane wave $\exp(i\mathbf{k}_0 \cdot \mathbf{r})$.

If the channel potentials include Coulomb terms, the boundary conditions should be modified in order to include Coulomb effects (see chapter 3). Eq. (9.20) takes the form

$$\Psi^{(+)}(0\mathbf{k}_0;\mathbf{r}) = \Phi_{\mathrm{C}}(0\mathbf{k}_0;\mathbf{r}) + \Psi^{\mathrm{sc}}(0\mathbf{k}_0;\mathbf{r}), \tag{9.24}$$

where both Φ_{C} and Ψ^{sc} contain long range distortions arising from the Coulomb field. In Eq. (9.24), Φ_{C} is given by

$$\Phi_{\mathrm{C}}(0\mathbf{k}_0;\mathbf{r}) = \phi_{\mathrm{C}}(\mathbf{k}_0;\mathbf{r})\ |0), \tag{9.25}$$

where $\phi_{\mathrm{C}}(\mathbf{k}_0;\mathbf{r})$ is the wave function for Coulomb scattering in the entrance channel. In the presence of the Coulomb field, the asymptotic form of $\Psi^{(+)}(0\mathbf{k}_0;\mathbf{r})$ becomes

$$\Psi^{(+)}(0\mathbf{k}_0;\mathbf{r}) \to \frac{1}{(2\pi)^{3/2}} \sum_\alpha \left[\phi_{\mathrm{C}}(\mathbf{k}_0;\mathbf{r})\ \delta_{\alpha 0} + \bar{f}_{\alpha,0}(\mathbf{k}_\alpha,\mathbf{k}_0)\ \frac{e^{i\Theta_\alpha(r)}}{r}\right]\ |\alpha). \tag{9.26}$$

Above, $\bar{f}_{\alpha,0}(\mathbf{k}_\alpha,\mathbf{k}_0)$ is the partial scattering amplitude arising from the short range potential and $\Theta_\alpha(r)$ is the asymptotic phase of the outgoing wave

$$\Theta_\alpha(r) = k_\alpha r - \eta_\alpha \ln(2k_\alpha r). \tag{9.27}$$

In the asymptotic limit of Eq. (9.26), ϕ_{C} can be approximated by Eq. (3.28).

The cross sections are evaluated through a trivial generalization of the procedures of chapter 1. The starting point is the definition (see Eq. (1.6)),

$$\frac{d\sigma_\alpha(\mathbf{k}_\alpha,\mathbf{k}_0)}{d\Omega} = \frac{N_\alpha(\mathbf{k}_\alpha,\Delta\Omega)}{J_0 \cdot \Delta\Omega}, \tag{9.28}$$

where J_0 is the incident current of Eq. (1.68) and $N_\alpha(\mathbf{k}_\alpha,\Delta\Omega)$ is the count rate in channel-α on a detector of aperture $\Delta\Omega$ centered at the orientation $\mathbf{k}_\alpha$. Evaluating J_0 and $N_\alpha(\mathbf{k}_\alpha,\Delta\Omega)$ and inserting the results into Eq. (9.28), we get (see problem 2)

$$\frac{d\sigma_\alpha(\mathbf{k}_\alpha,\mathbf{k}_0)}{d\Omega} = \frac{k_\alpha}{k_0}\left|f_{\alpha,0}(\mathbf{k}_\alpha,\mathbf{k}_0)\right|^2. \tag{9.29}$$

[6]In fact, the asymptotic wave function in a closed channel vanishes asymptotically. We postpone this discussion for a forthcoming chapter.

The factor k_α/k_0 accounts for the different speeds of the projectile in the entrance and exit channels, when $E_\alpha \neq E_0 \equiv E$.

If the interaction has a Coulomb term, the scattering amplitude is split as

$$f_{\alpha,0}(\mathbf{k}_\alpha, \mathbf{k}_0) = f_{\rm C}(\mathbf{k}_\alpha, \mathbf{k}_0)\ \delta_{\alpha 0} + \bar{f}_{\alpha,0}(\mathbf{k}_\alpha, \mathbf{k}_0), \tag{9.30}$$

where $f_{\rm C}(\mathbf{k}_\alpha, \mathbf{k}_0)$ is the Coulomb amplitude of Eq. (3.29).

9.1.1 *Absorption in multi-channel scattering*

In general, the potential in multi-channel scattering is complex. Its imaginary part simulates the average attenuation of the incident flux arising from couplings with channels that are not included in the space of the coupled equations[7]. Therefore, the stationary continuity equation breaks down. That is, the divergence of the total current does not vanish. Instead, it is related to the imaginary part of the potential by a simple equation, which we derive below. It is a generalization of the discussion of section 7.1 for the case of multi-channel scattering.

For simplicity, we use in the following the shorthand notation

$$\big|\psi_\alpha^{(+)}(0\mathbf{k}_0)\big\rangle \equiv \big(\alpha\big|\Psi^{(+)}(0\mathbf{k}_0)\big\rangle \to |\psi_\alpha\rangle\,.$$

Evaluating the quantity

$$[\psi_\alpha(\mathbf{r})]^* \times [\text{eq. (9.17)}] - [\psi_\alpha(\mathbf{r})] \times [\text{eq. (9.17)}]^*\,,$$

we get

$$\frac{\hbar}{2\mu i}\left[\psi_\alpha^*(\mathbf{r})\,\boldsymbol{\nabla}^2\,\psi_\alpha(\mathbf{r}) - \psi_\alpha(\mathbf{r})\,\boldsymbol{\nabla}^2\,\psi_\alpha^*(\mathbf{r})\right] = \frac{2}{\hbar}\,W_\alpha(\mathbf{r})\ |\psi_\alpha(\mathbf{r})|^2$$
$$+\frac{1}{i\hbar}\sum_{\alpha'}\Big[\psi_\alpha^*(\mathbf{r})\,V^{(2)}_{\alpha\alpha'}(\mathbf{r})\,\psi_{\alpha'}(\mathbf{r}) - \psi_{\alpha'}^*(\mathbf{r})\ \big(V^{(2)}_{\alpha\alpha'}(\mathbf{r})\big)^*\ \psi_\alpha(\mathbf{r})\Big]\,,$$

where $W_\alpha = \mathrm{Im}\,\{V_\alpha^{(1)}\}$. Writing the LHS of the above equation as

$$\boldsymbol{\nabla}\cdot\left\{\frac{\hbar}{2\mu i}\Big[\psi_\alpha^*(\mathbf{r})\,\boldsymbol{\nabla}\,\psi_\alpha(\mathbf{r}) - \psi_\alpha(\mathbf{r})\,\boldsymbol{\nabla}\,\psi_\alpha^*(\mathbf{r})\Big]\right\} \equiv \boldsymbol{\nabla}\cdot\boldsymbol{j}_\alpha(\mathbf{r}),$$

and summing over α, we obtain

$$\boldsymbol{\nabla}\cdot\sum_\alpha \mathbf{j}_\alpha(\mathbf{r}) = \frac{2}{\hbar}\sum_\alpha W_\alpha(\mathbf{r})\ |\psi_\alpha(\mathbf{r})\,|^2$$
$$+\frac{1}{i\hbar}\sum_{\alpha'\alpha}\psi_\alpha^*(\mathbf{r})\ \Big(\mathbb{V}^{(2)} - \mathbb{V}^{(2)\dagger}\Big)_{\alpha\alpha'}\ \psi_{\alpha'}(\mathbf{r}). \tag{9.31}$$

[7]The concept of complex optical potentials is introduced in the next chapter.

Frequently, the channel-coupling interaction $\mathbb{V}^{(2)}$ is Hermitian. In such cases the second term on the RHS of Eq. (9.31) vanish and one gets the simpler expression

$$\boldsymbol{\nabla}\cdot\sum_{\alpha}\mathbf{j}_{\alpha}(\mathbf{r})=\frac{2}{\hbar}\sum_{\alpha}W_{\alpha}(\mathbf{r})\,|\psi_{\alpha}(\mathbf{r})|^{2}\,. \tag{9.32}$$

Eq. (9.32) leads to a generalized expression for the absorption cross section. By *absorption*, we mean the flux lost to the channels neglected in the truncated expansion of Eq. (9.11). The effects of these channels are usually simulated by the inclusion of an imaginary part, W_α, in the channel potentials. We first calculate the flux of the total current over a spherical surface $\mathcal{S}$ with radius $R\to\infty$. That is,

$$N_a^{\rm total}=-\sum_{\alpha}\int_{\mathcal{S}}d\mathbf{s}\cdot\mathbf{j}_{\alpha}(\mathbf{r})\;=-\sum_{\alpha}\int_{\mathcal{V}}d^3\mathbf{r}\;\boldsymbol{\nabla}\cdot\mathbf{j}_{\alpha}(\mathbf{r}). \tag{9.33}$$

Above, we used the divergence theorem to express the integral over the closed surface $\mathcal{S}$ by an integral over the volume inside this surface, $\mathcal{V}$. Using Eq. (9.32), we get

$$N_a^{\rm total}=\frac{2}{\hbar}\sum_{\alpha}\int_{\mathcal{V}}\mathbf{d^3r}\;(-W_{\alpha}(\mathbf{r}))\;|\psi_{\alpha}(\mathbf{r})|^{2}\equiv\frac{2}{\hbar}\sum_{\alpha}\langle\psi_{\alpha}|\;-W_{\alpha}\;|\psi_{\alpha}\rangle\,.$$

The total absorption cross section is given by the ratio of $N_a^{\rm total}$ to the incident flux of Eq. (1.68). One gets

$$\sigma_a=\frac{1}{|A|^2}\,\frac{2}{\hbar v_0}\sum_{\alpha}\langle\psi_{\alpha}|\;-W_{\alpha}\;|\psi_{\alpha}\rangle\,.$$

Changing $2/\hbar v_0=k_0/E$ and turning back to the full notation, the above equation becomes

$$\sigma_a=\frac{1}{|A|^2}\;\frac{k_0}{E}\sum_{\alpha}\Big\langle\psi_{\alpha}^{(+)}(0\mathbf{k}_0)\Big|\;-W_{\alpha}\;\Big|\psi_{\alpha}^{(+)}(0\mathbf{k}_0)\Big\rangle. \tag{9.34}$$

The factor $1/|A|^2$ guarantees that the absorption cross section is independent of the normalization of $\psi_{\alpha}^{(+)}$. Some authors adopt the normalization with $A=1$ and ommit the factor $|A|^{-2}$ in the expression for σ_a.

We can also define the *total reaction cross section*, $\sigma_{\rm R}$, which is associated with the probability that the system does not remain in the entrance channel after the collision. It is given by

$$\sigma_{\rm R}=\sigma_a+\sum_{\alpha\neq0}\sigma_{\alpha}, \tag{9.35}$$

where σ_α is the integrated cross section

$$\sigma_{\alpha}=\int d\Omega_{\mathbf{k}_{\alpha}}\left[\frac{d\sigma_{\alpha}(\mathbf{k}_{\alpha},\mathbf{k}_0)}{d\Omega}\right]. \tag{9.36}$$

9.2 The Lippmann-Schwinger equations

To derive the Lippmann-Schwinger equations, we first introduce the free particle states, $|\Phi(\alpha\mathbf{k}_\alpha)\rangle$, and the distorted states, $|X^{(\pm)}(\alpha\mathbf{k}_\alpha)\rangle$. The free states are

$$|\Phi(\alpha\mathbf{k}_\alpha)\rangle \equiv |\phi(\mathbf{k}_\alpha)\rangle\ |\alpha)\,, \tag{9.37}$$

where $|\alpha)$ and $|\phi(\mathbf{k}_\alpha)\rangle$ are respectively the intrinsic (see Eq. (9.2)) and the plane wave states. In the coordinate representation we have (using the normalization with $A = (2\pi)^{-3/2}$)

$$\Phi(\alpha\mathbf{k}_\alpha;\xi,\mathbf{r}) = \frac{1}{(2\pi)^{3/2}}\ e^{i\mathbf{k}_\alpha\cdot\mathbf{r}}\ \varphi_\alpha(\xi). \tag{9.38}$$

The distorted states in channel-α can be written as

$$|X^{(\pm)}(\alpha\mathbf{k}_\alpha)\rangle = |\chi^{(\pm)}(\alpha\mathbf{k}_\alpha)\rangle\ |\alpha)\,. \tag{9.39}$$

The distorted waves $|\chi^{(\pm)}(\alpha\mathbf{k}_\alpha)\rangle$ satisfy the equation

$$(E_\alpha - H_\alpha^{(1)})\ |\chi^{(\pm)}(\alpha\mathbf{k}_\alpha)\rangle = 0 \tag{9.40}$$

and have the asymptotic forms[8]

$$\chi^{(+)}(\alpha\mathbf{k}_\alpha;\mathbf{r}) \longrightarrow \frac{1}{(2\pi)^{3/2}}\left[e^{i\mathbf{k}_\alpha\cdot\mathbf{r}} + f_\alpha^{(1)}(\mathbf{k}'_\alpha,\mathbf{k}_\alpha)\,\frac{e^{ik_\alpha r}}{r}\right] \tag{9.41a}$$

$$\chi^{(-)}(\alpha\mathbf{k}_\alpha;\mathbf{r}) \longrightarrow \frac{1}{(2\pi)^{3/2}}\left[e^{i\mathbf{k}_\alpha\cdot\mathbf{r}} + \left(f_\alpha^{(1)}(-\mathbf{k}'_\alpha,\mathbf{k}_\alpha)\right)^*\ \frac{e^{-ik_\alpha r}}{r}\right]. \tag{9.41b}$$

Above, $\mathbf{k}'_\alpha$ is a vector with modulus k_α and the orientation of $\hat{\mathbf{r}}$ and $f_\alpha^{(1)}$ is the scattering amplitude for a collision in channel-α, in the absence of the channel-coupling interaction.

If $H_\alpha^{(1)}$ contains a Coulomb term, the modifications discussed in chapter 3 should be made. In this case, the distorted wave functions have the asymptotic forms

$$\chi^{(+)}(\alpha\mathbf{k}_\alpha;\mathbf{r}) \longrightarrow \phi_{\rm C}(\mathbf{k}_\alpha;\mathbf{r}) + \frac{1}{(2\pi)^{3/2}}\left[\bar{f}_\alpha^{(1)}(\mathbf{k}'_\alpha,\mathbf{k}_\alpha)\ \frac{e^{i\Theta_\alpha(r)}}{r}\right] \tag{9.42a}$$

and

$$\chi^{(-)}(\alpha\mathbf{k}_\alpha;\mathbf{r}) \longrightarrow \phi_{\rm C}^*(-\mathbf{k}_\alpha;\mathbf{r}) + \frac{1}{(2\pi)^{3/2}}\left[\left(\bar{f}_\alpha^{(1)}(-\mathbf{k}'_\alpha,\mathbf{k}_\alpha)\right)^*\ \frac{e^{-i\Theta_\alpha(r)}}{r}\right], \tag{9.43}$$

[8]The asymptotic form of $\chi^{(-)}$ is obtained from the asymptotic form of $\chi^{(+)}$ using its time-reversal properties (see chapter 4).

where Θ_α is given by Eq. (9.27) and $\bar{f}^{(1)}_\alpha$ is the partial scattering amplitude arising from the short range part of $V^{(1)}_\alpha$.

We now define the free Green's functions, $\mathbb{G}^{(\pm)}_0$, the distorted Green's functions, $\mathbb{G}^{(\pm)}_1$, and the full Green's functions, $\mathbb{G}^{(\pm)}$,

$$\mathbb{G}^{(\pm)}_0(E) = \frac{1}{E - K - h \pm i\epsilon}, \tag{9.44a}$$

$$\mathbb{G}^{(\pm)}_1(E) = \frac{1}{E - \mathbb{H}^{(1)} - h \pm i\epsilon}, \tag{9.44b}$$

$$\mathbb{G}^{(\pm)}(E) = \frac{1}{E - \mathbb{H} \pm i\epsilon}. \tag{9.44c}$$

Note that $\mathbb{G}^{(\pm)}_0$ and $\mathbb{G}^{(\pm)}_1$ are diagonal in channel-space while $\mathbb{G}^{(\pm)}$ is not.

Using the same procedures of chapter 4, we obtain the Lippmann-Schwinger equations. For a collision initiated[9] in channel-α with wave vector $\mathbf{k}_\alpha$, the Lippmann-Schwinger equation for the distorted states are

$$|X^{(\pm)}(\alpha\mathbf{k}_\alpha)\rangle = |\Phi(\alpha\mathbf{k}_\alpha)\rangle + \mathbb{G}^{(\pm)}_0(E)\,\mathbb{V}^{(1)}\;|X^{(\pm)}(\alpha\mathbf{k}_\alpha)\rangle \tag{9.45a}$$

$$= \left[1 + \mathbb{G}^{(\pm)}_1(E)\,\mathbb{V}^{(1)}\right]\,|\Phi(\alpha\mathbf{k}_\alpha)\rangle\,. \tag{9.45b}$$

The Lippmann-Schwinger equation for the exact states are

$$|\Psi^{(\pm)}(0\mathbf{k}_0)\rangle = |\Phi(0\mathbf{k}_0)\rangle \;+\; \mathbb{G}^{(\pm)}_0(E)\,\mathbb{V}\;|\Psi^{(\pm)}(0\mathbf{k}_0)\rangle \tag{9.46a}$$

$$= [1 + \mathbb{G}^{(\pm)}(E)\,\mathbb{V}]\;|\Phi(0\mathbf{k}_0)\rangle \tag{9.46b}$$

$$= |X^{(\pm)}(0\mathbf{k}_0)\rangle + \mathbb{G}^{(\pm)}_1(E)\,\mathbb{V}^{(2)}\;|\Psi^{(\pm)}(0\mathbf{k}_0)\rangle\,. \tag{9.46c}$$

Lippmann-Schwinger equations in the **r**-space can be obtained using the channel expansion of Eq. (9.11) and taking scalar product with intrinsic states. Multiplying Eq. (9.46a) by $(\alpha|$, we get

$$\begin{aligned} |\psi^{(\pm)}_\alpha(0\mathbf{k}_0)\rangle &= |\phi(\mathbf{k}_0)\rangle\;\delta_{\alpha 0} + G^{(\pm)}_0(\alpha E)\;\;(\alpha|\,\mathbb{V}|\Psi^{(\pm)}(0\mathbf{k}_0)\rangle \\ &= |\phi(\mathbf{k}_0)\rangle\,\delta_{\alpha 0} + G^{(\pm)}_0(\alpha E)\sum_{\alpha'} V_{\alpha\alpha'}\,|\psi^{(\pm)}_{\alpha'}(0\mathbf{k}_0)\rangle\,. \end{aligned} \tag{9.47}$$

Following the same steps for Eq. (9.46c), we obtain

$$|\psi^{(\pm)}_\alpha(0\mathbf{k}_0)\rangle = |\chi^{(\pm)}(0\mathbf{k}_0)\rangle\;\delta_{\alpha 0} + G^{(\pm)}_1(\alpha E)\;(\alpha|\;\mathbb{V}^{(2)}\,|\Psi^{(\pm)}(\alpha_0\mathbf{k}_0)\rangle,$$

and finally,

$$|\psi^{(\pm)}_\alpha(0\mathbf{k}_0)\rangle = |\chi^{(\pm)}(0\mathbf{k}_0)\rangle\;\delta_{\alpha 0} + G^{(\pm)}_1(\alpha E)\sum_{\alpha'} V^{(2)}_{\alpha\alpha'}\;\left|\psi^{(\pm)}_{\alpha'}(0\mathbf{k}_0)\right\rangle. \tag{9.48}$$

---

[9]Here, we only need plane and distorted waves initiated in the elastic channel. However, waves initiated in excited channels are needed in approximations like the DWBA.

In the coordinate representation, Eqs. (9.47) and (9.48) respectively become

$$\psi_\alpha^{(\pm)}(0\mathbf{k}_0;\mathbf{r}) = \phi(\mathbf{k}_0;\mathbf{r})\,\delta_{\alpha 0} + \int d^3\mathbf{r}'\,G_0^{(\pm)}(\alpha E;\mathbf{r},\mathbf{r}') \sum_{\alpha'} \langle \mathbf{r}'|\,V_{\alpha\alpha'}|\,\psi_{\alpha'}^{(\pm)}(0\mathbf{k}_0)\rangle. \quad (9.49)$$

and

$$\psi_\alpha^{(\pm)}(0\mathbf{k}_0;\mathbf{r}) = \chi^{(\pm)}(0\mathbf{k}_0;\mathbf{r})\,\delta_{\alpha 0} + \int d^3\mathbf{r}'\,G_1^{(\pm)}(\alpha E;\mathbf{r},\mathbf{r}') \sum_{\alpha'} \langle \mathbf{r}'|\,V_{\alpha\alpha'}^{(2)}|\,\psi_{\alpha'}^{(\pm)}(0\mathbf{k}_0)\rangle. \quad (9.50)$$

Above,

$$G_0^{(\pm)}(\alpha E) = (\alpha|\,\frac{1}{E-K-h\pm i\epsilon}\,|\alpha) = \frac{1}{E_\alpha - K \pm i\epsilon}, \quad (9.51)$$

$$G_1^{(\pm)}(\alpha E) = (\alpha|\,\frac{1}{E-\mathbb{H}^{(1)}-h\pm i\epsilon}\,|\alpha) = \frac{1}{E_\alpha - H_\alpha^{(1)} \pm i\epsilon} \quad (9.52)$$

and

$$G_0^{(\pm)}(\alpha E;\mathbf{r},\mathbf{r}') = \langle \mathbf{r}|\,G_0^{(\pm)}(\alpha E)\,|\mathbf{r}'\rangle \quad (9.53)$$

$$G_1^{(\pm)}(\alpha E;\mathbf{r},\mathbf{r}') = \langle \mathbf{r}|\,G_1^{(\pm)}(\alpha E)\,|\mathbf{r}'\rangle\,. \quad (9.54)$$

If the interactions are local in the coordinate space, that is $\langle \mathbf{r}'|\,V_{\alpha\alpha'}|\,\mathbf{r}''\rangle = V_{\alpha\alpha'}(\mathbf{r}')\,\delta(\mathbf{r}'-\mathbf{r}'')$, Eq. (9.49) can be written

$$\psi_\alpha^{(\pm)}(0\mathbf{k}_0;\mathbf{r}) = \phi(\mathbf{k}_0;\mathbf{r})\,\delta_{\alpha 0} + \sum_{\alpha'} \int d^3\mathbf{r}'\,G_0^{(\pm)}(\alpha E;\mathbf{r},\mathbf{r}') \times V_{\alpha\alpha'}(\mathbf{r}')\,\psi_{\alpha'}^{(\pm)}(0\mathbf{k}_0;\mathbf{r}'). \quad (9.55)$$

A similar simplification applies to Eq. (9.50). That is

$$\psi_\alpha^{(\pm)}(0\mathbf{k}_0;\mathbf{r}) = \chi_\alpha^{(\pm)}(\mathbf{k}_0;\mathbf{r})\,\delta_{\alpha 0} + \sum_{\alpha'} \int d^3\mathbf{r}'\,G_1^{(\pm)}(\alpha E;\mathbf{r},\mathbf{r}') \times V_{\alpha\alpha'}^{(2)}(\mathbf{r}')\,\psi_{\alpha'}^{(\pm)}(0\mathbf{k}_0;\mathbf{r}'). \quad (9.56)$$

If the interaction contains Coulomb terms, Eq. (9.55) is not useful. In this case, it is necessary to use Eq. (9.56) and include the long range terms in the potential $\mathbb{V}^{(1)}$.

Using in Eq. (9.49) the explicit form of free particle Green's function (Eq. (4.37)), taking the $r \to \infty$ limit and comparing with Eq. (9.41a), we obtain the relation

$$f_{\alpha,0}(\mathbf{k}_\alpha,\mathbf{k}_0) = -2\pi^2\Big(\frac{2\mu}{\hbar^2}\Big)\Big\langle \Phi(\alpha\mathbf{k}_\alpha)\Big|\mathbb{V}\Big|\Psi^{(+)}(0\mathbf{k}_0)\Big\rangle \quad (9.57)$$

$$= -2\pi^2\Big(\frac{2\mu}{\hbar^2}\Big)\sum_{\alpha'}\Big\langle \phi(\mathbf{k}_\alpha)\Big|V_{\alpha\alpha'}\Big|\psi_{\alpha'}^{(+)}(0\mathbf{k}_0)\Big\rangle. \quad (9.58)$$

From the definition of $|\Phi(\alpha\mathbf{k}_\alpha)\rangle$ (Eq. (9.37)) and the orthonormality relations for intrisic states and for plane waves, we get

$$\big\langle\Phi(\alpha\mathbf{k}_\alpha)\big|\Phi(\alpha'\mathbf{k}'_{\alpha'})\big\rangle = \delta_{\alpha\alpha'}\,\delta\left(\mathbf{k}_\alpha - \mathbf{k}'_{\alpha'}\right). \tag{9.59}$$

If the interactions are real, similar results can be obtained for $X^{(\pm)}(\alpha\mathbf{k}_\alpha)$ and for $\Psi^{(\pm)}(\alpha\mathbf{k}_\alpha)$. That is[10]

$$\left\langle X^{(\pm)}(\alpha\mathbf{k}_\alpha)\middle|X^{(\pm)}(\alpha'\mathbf{k}'_{\alpha'})\right\rangle = \delta_{\alpha\alpha'}\;\delta(\mathbf{k}_\alpha - \mathbf{k}'_{\alpha'}) \tag{9.60}$$

and

$$\left\langle \Psi^{(\pm)}(\alpha\mathbf{k}_\alpha)\middle|\Psi^{(\pm)}(\alpha'\mathbf{k}'_{\alpha'})\right\rangle = \delta_{\alpha\alpha'}\;\delta(\mathbf{k}_\alpha - \mathbf{k}'_{\alpha'}). \tag{9.61}$$

The proofs of Eqs. (9.60) and (9.61) follow exactly the same steps of section 4.1.3. The former is based on Eq. (9.45a) and the hermitian conjugate of Eq. (9.45b), and the latter is based on Eq. (9.46a) and the hermitian conjugate of Eq. (9.46b).

Using the channel expansion of Eq. (9.11), Eq. (9.61) becomes

$$\sum_{\alpha''}\langle\psi^{(+)}_{\alpha''}(\alpha\mathbf{k}_\alpha)|\psi^{(+)}_{\alpha''}(\alpha'\mathbf{k}'_{\alpha'})\rangle = \delta_{\alpha\alpha'}\;\delta(\mathbf{k}_\alpha - \mathbf{k}'_{\alpha'}).$$

The above equation yields (for real potentials)

$$\left\langle \psi^{(+)}_\beta(\alpha\mathbf{k}_\alpha)\middle|\psi^{(+)}_\beta(\alpha'\mathbf{k}_{\alpha'})\right\rangle = \Big|A_\beta\Big|^2\;\delta_{\alpha\alpha'}\,\delta(\mathbf{k}_\alpha - \mathbf{k}'_{\alpha'}), \tag{9.62}$$

with

$$\sum_\beta |A_\beta|^2 = 1. \tag{9.63}$$

The norm $|A_\beta|^2$ expresses the populations of channel β in the end of the scattering process.

---

[10]For complex potentials, these relations become

$$\left\langle \tilde{X}^{(\pm)}(\alpha\mathbf{k}_\alpha)\middle|X^{(\pm)}(\alpha'\mathbf{k}'_{\alpha'})\right\rangle = \delta_{\alpha,\alpha'}\;\delta(\mathbf{k}_\alpha - \mathbf{k}'_{\alpha'}\Big)$$

and

$$\left\langle \tilde{\Psi}^{(\pm)}(\alpha\mathbf{k}_\alpha)\middle|\Psi^{(\pm)}(\alpha'\mathbf{k}'_{\alpha'})\right\rangle = \delta_{\alpha,\alpha'}\;\delta(\mathbf{k}_\alpha - \mathbf{k}'_{\alpha'}),$$

where $\tilde{X}^{(\pm)}$ and $\tilde{\Psi}^{(\pm)}$ are solutions of the Schrödinger equations with the Hamiltonians $\mathbb{H}^{(1)\dagger}$ and $\mathbb{H}^\dagger$, respectively. For details, see chapter 7.

9.2.1 *Møller operators, the T- and the S-matrices*

We can define Møller wave operators similarly to the single channel case,

$$\boldsymbol{\Omega}^{(\pm)} \left|\Phi(\alpha \mathbf{k}_0)\right\rangle = \left|\Psi^{(\pm)}(\alpha \mathbf{k}_0)\right\rangle . \tag{9.64}$$

That is, $\boldsymbol{\Omega}^{(\pm)}$ is the operator which acting on a free state in any channel α yields the scattering state initiated in that same channel. It can easily be checked that it satisfies the Lippmann-Schwinger equations[11]

$$\boldsymbol{\Omega}^{(\pm)}(E) = 1 + \mathbb{G}_0^{(\pm)}(E)\,\mathbb{V}\,\boldsymbol{\Omega}^{(\pm)}(E) \tag{9.65}$$

$$= 1 + \mathbb{G}^{(\pm)}(E)\,\mathbb{V}. \tag{9.66}$$

The transition operator is defined similarly to the single channel case[12]

$$\mathbb{T} = \mathbb{V}\,\boldsymbol{\Omega}^{(+)} . \tag{9.67}$$

Eqs. (9.67), (9.65) and (9.66) lead to the Lippmann-Schwinger equations

$$\mathbb{T} = \mathbb{V} + \mathbb{V}\,\mathbb{G}_0^{(+)}(E)\,\mathbb{T} \tag{9.68}$$

$$= \mathbb{V} + \mathbb{V}\,\mathbb{G}^{(+)}(E)\,\mathbb{V} . \tag{9.69}$$

Taking matrix elements of the transition operator between free states, we have the multi-channel T-matrix

$$T_{\alpha\mathbf{k}_\alpha,0\mathbf{k}_0} = \left\langle\Phi(\alpha\mathbf{k}_\alpha)\right| \mathbb{T} \left|\Phi(0\mathbf{k}_0)\right\rangle = \left\langle\Phi(\alpha\mathbf{k}_\alpha)\right| \mathbb{V} \left|\Psi^{(+)}(0\mathbf{k}_0)\right\rangle , \tag{9.70}$$

or

$$T_{\alpha\mathbf{k}_\alpha,0\mathbf{k}_0} = \sum_{\alpha'} \left\langle\phi(\mathbf{k}_\alpha)\right| V_{\alpha\alpha'} \left|\psi_{\alpha'}^{(+)}(0\mathbf{k}_0)\right\rangle . \tag{9.71}$$

Following the same procedures of section 4.5, one can derive the Gell-Mann & Goldberger relation for multi-channel scattering. One obtains

$$\begin{aligned} T_{\alpha\mathbf{k}_\alpha,0\mathbf{k}_0} &= \left\langle\Phi(\alpha\mathbf{k}_\alpha)\right| \mathbb{V}^{(1)} \left|X^{(+)}(0\mathbf{k}_0)\right\rangle \\ &\quad + \left\langle X^{(-)}(\alpha\mathbf{k}_\alpha)\right| \mathbb{V}^{(2)} \left|\Psi^{(+)}(0\mathbf{k}_0)\right\rangle \end{aligned} \tag{9.72}$$

or

$$\begin{aligned} T_{\alpha\mathbf{k}_\alpha,0\mathbf{k}_0} &= \left\langle\phi\left(\mathbf{k}_\alpha\right)\right| V_\alpha^{(1)} \left|\chi^{(+)}(0\mathbf{k}_0)\right\rangle \, \delta_{\alpha 0} \\ &\quad + \sum_{\alpha'} \left\langle\chi^{(-)}(\alpha\mathbf{k}_\alpha)\right| V_{\alpha\alpha'}^{(2)} \left|\psi_{\alpha'}^{(+)}(0\mathbf{k}_0)\right\rangle . \end{aligned} \tag{9.73}$$

---

[11]As remarked in chapter 4 on the paragraph after Eq. (4.53), the operators $\boldsymbol{\Omega}^{(\pm)}$ are energy-independent. However, their Lippmann-Schwinger equations are only valid for states with a fixed value of the energy, E. To emphasize this point we use the notation $\boldsymbol{\Omega}^{(\pm)}(E)$ in the Lippmann-Schwinger equation.

[12]We denote the transition operator in the full $\{\xi, \mathbf{r}\}$ space by $\mathbb{T}$. Although in chapter 4 this notation was used for the time-reversal operator, there is no room for confusion here.

This expression can easily be evaluated within the DWBA approximation. In this case, one replaces $|\Psi^{(+)}(0\mathbf{k}_0)\rangle \to |X^{(+)}(0\mathbf{k}_0)\rangle$ on the RHS of Eq. (9.72) and get

$$T_{\alpha\mathbf{k}_\alpha,0\mathbf{k}_0} \simeq \langle\phi(\mathbf{k}_\alpha)|\, V_\alpha^{(1)}\, |\chi^{(+)}(0\mathbf{k}_0)\rangle\, \delta_{\alpha 0} + \langle\chi^{(-)}(\alpha\mathbf{k}_\alpha)|\, V_{\alpha 0}^{(2)}\, |\chi^{(+)}(0\mathbf{k}_0)\rangle\,. \tag{9.74}$$

The definition of the scattering operator given in chapter 4 is still valid in multi-channel scattering. That is

$$\mathbb{S} = \mathbf{\Omega}^{(-)\dagger}\, \mathbf{\Omega}^{(+)}. \tag{9.75}$$

Now, however, $\mathbb{S}$ and $\mathbf{\Omega}^{(\pm)}$ are operators acting on a larger space, with intrinsic and relative degrees of freedom. The set of matrix-elements of the scattering operator on-the-energy-shell (S-matrix) can be written

$$S_{\alpha\mathbf{k}_\alpha,0\mathbf{k}_0} = \langle\Phi(\alpha\mathbf{k}_\alpha)|\, \mathbf{\Omega}^{(-)\dagger}\, \mathbf{\Omega}^{(+)}\, |\Phi(0\mathbf{k}_0)\rangle \tag{9.76a}$$

$$= \langle\Psi^{(-)}(\alpha\mathbf{k}_\alpha)\,|\,\Psi^{(+)}(0\mathbf{k}_0)\rangle. \tag{9.76b}$$

For Hermitian Hamiltonians, the S-matrix is unitary and we have

$$\left(\mathbb{S}^\dagger\mathbb{S}\right)_{\alpha\mathbf{k}_\alpha,0\mathbf{k}_0} = \left(\mathbb{S}\mathbb{S}^\dagger\right)_{\alpha\mathbf{k}_\alpha,0\mathbf{k}_0} = \delta\left(\mathbf{k}_\alpha - \mathbf{k}_0\right)\, \delta_{\alpha 0}. \tag{9.77}$$

The relation between the scattering operator and the transition operator on the energy-shell is the same as that for potential scattering (see Eq. (4.83)),

$$\mathbb{S}(E) = 1 - 2\pi i\, \delta\left(E - \mathbb{H}_0\right)\, \mathbb{T}(E), \tag{9.78}$$

except that these operators now act also on the intrinsic space. Eq. (9.78) can be expressed in terms of matrix-elements as

$$S_{\alpha\mathbf{k}_\alpha,0\mathbf{k}_0} = \delta\left(\mathbf{k}_\alpha - \mathbf{k}_0\right)\, \delta_{\alpha 0} - 2\pi i\;\; \delta\left(E_{k_0} + \varepsilon_0 - E_\alpha - \varepsilon_\alpha\right)\, T_{\alpha\mathbf{k}_\alpha,0\mathbf{k}_0}. \tag{9.79}$$

As in potential scattering, the scattering amplitude is proportional to the T-matrix elements on-the-energy-shell. Using Eq. (9.64) in Eq. (9.57), we get

$$f_{\alpha,0}(\mathbf{k}_\alpha, \mathbf{k}_0) = -2\pi^2 \left(\frac{2\mu}{\hbar^2}\right) \langle\Phi(\alpha\mathbf{k}_\alpha)|\, \mathbb{V}\mathbf{\Omega}^{(+)}\, |\Phi(0\mathbf{k}_0)\rangle\,. \tag{9.80}$$

or, according to Eqs. (9.67) and (9.70),

$$f_{\alpha,0}(\mathbf{k}_\alpha, \mathbf{k}_0) = -2\pi^2 \left(\frac{2\mu}{\hbar^2}\right) T_{\alpha\mathbf{k}_\alpha,0\mathbf{k}_0}. \tag{9.81}$$

9.3 Partial-wave expansions

9.3.1 *Collisions of spinless particles*

We now apply the partial-wave expansion method to multi-channel scattering. In this section, we restrict our discussion to spinless[13] collision partners. As in chapter 6, we single out the radial coordinate and look for a basis for the vector space associated with the remaining degrees of freedom. In the present case these degrees of freedom are the angular variables of the projectile-target separation vector, $\hat{\mathbf{r}}$, and the intrinsic coordinates, ξ. The state vectors of this basis are denoted by $\mathcal{Y}_n(\zeta)$, where $\zeta \equiv \{\hat{\mathbf{r}};\xi\}$ and the label n stands for the quantum numbers required to specify the state. For spinless particles, the intrinsic states are scalar so that the rotational invariance of the system yields orbital angular momentum conservation. In this way, the quantum numbers l and m are conserved and the Hamiltonian is diagonal in the spherical harmonics representation. We adopt the basis of the *channel-angle wave functions,* which in this case is the set of states[14]

$$\mathcal{Y}_n(\zeta) \equiv \mathcal{Y}_\alpha^{lm}(\zeta) = i^l \varphi_\alpha(\xi)\, Y_{lm}(\hat{\mathbf{r}}). \tag{9.82}$$

In the multi-channel case, one cannot usually find a basis which diagonalizes the Hamiltonian, as was done in potential scattering. In the $\mathcal{Y}_\alpha^{lm}$ representation, the Hamiltonian is diagonal with respect to the quantum numbers l and m but not with respect to the channel label, α. To stress this difference, we write the former as superscripts and the latter as a subscript. Using the orthogonality relations of the intrinsic states and of the spherical harmonics, one can easily show the orthonormality and the completeness relations, respectively given by

$$(\mathcal{Y}_\alpha^{lm}|\mathcal{Y}_{\alpha'}^{l'm'}) = \delta_{\alpha\alpha'}\, \delta_{ll'}\, \delta_{mm'}, \tag{9.83}$$

and

$$\sum_{\alpha lm} |\mathcal{Y}_\alpha^{lm})\, (\mathcal{Y}_\alpha^{lm}| = \mathbf{1}. \tag{9.84}$$

The free state in channel-α with wave vector $\mathbf{k}_\alpha$ can be expanded in the $\mathcal{Y}_\alpha^{lm}$ basis as

$$\Phi(\alpha\mathbf{k}_\alpha;\xi,\mathbf{r}) = \frac{1}{(2\pi)^{3/2}} \sum_{lm} C_{lm}\,(\alpha\mathbf{k}_\alpha)\ \mathcal{Y}_\alpha^{lm}(\zeta)\ \frac{\hat{\jmath}_l(k_\alpha r)}{k_\alpha r}. \tag{9.85}$$

[13] We use here the term *spin* with the meaning of the total angular momentum associated with the intrinsic degrees of freedom of the collision partners.

[14] As in [Sa 83], we adopt the arbitrary phase i^l in the definition of $\mathcal{Y}_n(\zeta)$. With this phase, the wave function has convenient time-reversal properties.

The coefficient $C_{lm}(\alpha\mathbf{k}_\alpha)$ is the probability amplitude for the state $|\Phi(\alpha\mathbf{k}_\alpha)\rangle$ to have orbital angular momentum l and z-component m. Using in Eq. (9.21) the partial-wave expansion of the plane wave in the incident channel (Eq. (2.44a)) and comparing the result with Eq. (9.85), we find

$$C_{lm}(\alpha\mathbf{k}_0) = 4\pi\, Y^*_{lm}(\hat{\mathbf{k}}_0). \tag{9.86}$$

Note that these coefficients are independent of the incident channel. For this reason, we henceforth suppress the channel index and use the notation $C_{lm}(\mathbf{k}_0)$. Taking the $r \to \infty$ limit in Eq. (9.85), the Ricatti-Bessel functions can be approximated by (see Eq. (6.29)),

$$\left[\frac{\hat{\jmath}_l(kr)}{kr}\right]_{kr\to\infty} = \frac{e^{-ikr}}{r}\left[-\frac{i^l}{2ik}\right] + \frac{e^{ikr}}{r}\left[\frac{i^{-l}}{2ik}\right], \tag{9.87}$$

and we get the asymptotic form of the free state,

$$\Phi(0\mathbf{k}_0;\xi,\mathbf{r}) \to \frac{1}{(2\pi)^{3/2}}\sum_\alpha \left[\frac{e^{-ik_\alpha r}}{r}\,Z^{(\alpha)}_- + \frac{e^{ik_\alpha r}}{r}\,Z^{(\alpha)}_+\right] |\alpha), \tag{9.88}$$

with[15]

$$Z^{(\alpha)}_- = -\,\delta_{\alpha 0}\,\frac{1}{2i\sqrt{k_\alpha k_0}}\sum_{lm} 4\pi\, Y^*_{lm}(\hat{\mathbf{k}}_0)\,Y_{lm}(\hat{\mathbf{r}})\,i^{2l} \tag{9.89a}$$

$$Z^{(\alpha)}_+ = \delta_{\alpha 0}\,\frac{1}{2i\sqrt{k_\alpha k_0}}\sum_{lm} 4\pi\, Y^*_{lm}(\hat{\mathbf{k}}_0)\,Y_{lm}(\hat{\mathbf{r}}). \tag{9.89b}$$

If the channel-potentials include Coulomb terms, we start from Φ_C, instead of Φ. Using in Eq. (9.25) the partial-wave expansion of ϕ_C (Eq. (3.102a)), we get

$$\Phi_C(0\mathbf{k}_0;\xi,\mathbf{r}) = \frac{1}{(2\pi)^{3/2}}\sum_{lm} C_{lm}(\mathbf{k}_0)\,\mathcal{Y}^{lm}_\alpha(\zeta)\,\frac{F_l(\eta_0,r)}{k_0 r}, \tag{9.90}$$

with the coefficients

$$C_{lm}(\mathbf{k}_0) = 4\pi\, Y^*_{lm}(\hat{\mathbf{k}}_0)\,e^{i\sigma_{0l}}. \tag{9.91}$$

Above, F_l and σ_{0l} are respectively the regular wave function and the phase shift for Coulomb scattering in the entrance channel, at the lth partial-wave. In the $r \to \infty$ limit, the Coulomb functions can be replaced by their asymptotic forms. Taking this limit in Eq. (9.90) and using the explicit form of C_{lm}, one obtains

$$\Phi_C(0\mathbf{k}_0;\xi,\mathbf{r}) \to \frac{1}{(2\pi)^{3/2}}\sum_\alpha \left[\frac{e^{-i\,\Theta_\alpha(r)}}{r}\,Z^{(\alpha)}_- + \frac{e^{i\,\Theta_\alpha(r)}}{r}\,Z^{(\alpha)}_+\right] |\alpha), \tag{9.92}$$

[15]Although it would be simpler to write $\delta_{\alpha 0}/k_0$ instead of $\delta_{\alpha 0}/\sqrt{k_0 k_\alpha}$, the latter is more convenient for the extension to the case of the scattering state.

where $Z_{-}^{(\alpha)}$ is given by Eq. (9.89a), and the outgoing amplitude is

$$Z_{+}^{(\alpha)} = \delta_{\alpha 0} \; \frac{1}{2i\sqrt{k_\alpha k_0}} \sum_{lm} 4\pi \, Y_{lm}^{*}(\hat{\mathbf{k}}_\alpha) \, Y_{lm}(\hat{\mathbf{r}}) \; e^{2i\sigma_{\alpha l}} \,. \tag{9.93}$$

In Eq. (9.92), $\Theta_\alpha(r)$ is the asymptotic argument given in Eq. (9.27).

We now consider the partial-wave expansions of the distorted and the full scattering wave functions. The distorted wave in channel-α can be expanded as

$$X^{(+)}(\alpha\mathbf{k}_\alpha;\xi,\mathbf{r}) = \frac{1}{(2\pi)^{3/2}} \sum_{lm} C_{lm}(\mathbf{k}_\alpha) \; \mathcal{Y}_\alpha^{lm}(\zeta) \; \frac{w_{\alpha l}(k_\alpha, r)}{k_\alpha r}. \tag{9.94}$$

Inserting this expansion into Eq. (9.40), taking scalar product with $(\mathcal{Y}_\alpha^{lm}|$ and using Eq. (9.83), we obtain the radial equations

$$\left[E_\alpha + \frac{\hbar^2}{2\mu} \left(\frac{d^2}{dr^2} - \frac{l\,(l+1)}{r^2} \right) - V_{\alpha l}^{(1)}(r) \right] \, w_{\alpha l}(k_\alpha, r) = 0, \tag{9.95}$$

where $V_{\alpha l}^{(1)}(r)$ is the matrix-element

$$V_{\alpha l}^{(1)}(r) = \int d\hat{\mathbf{r}} \; Y_{lm}^{*}(\hat{\mathbf{r}}) \; (\alpha|\, \mathbb{V}^{(1)} \, |\alpha) \; Y_{lm}(\hat{\mathbf{r}}). \tag{9.96}$$

At large distances ($r > \bar{R}$), the radial wave functions can be written

$$w_{\alpha l}(k_\alpha, r > \bar{R}) = \frac{i}{2} \left[\hat{h}_l^{(-)}(k_\alpha r) - S_{\alpha l}^{(1)} \; \hat{h}_l^{(+)}(k_\alpha r) \right], \tag{9.97}$$

where $S_{\alpha l}^{(1)}$ is the lth partial-wave component of the S-matrix associated with the Hamiltonian $H_\alpha^{(1)}$, and $\hat{h}_l^{(\pm)}$ are the Ricatti-Haenkel functions, discussed in chapter 2.

In the partial-wave expansion of $\Psi^{(+)}$ one has to consider that the incident wave in elastic channel will flow to other channels, through the action of the coupling interaction $\mathbb{V}^{(2)}$. In this way, one expands

$$\Psi^{(+)}(0\mathbf{k}_0;\xi,\mathbf{r}) = \frac{1}{(2\pi)^{3/2}} \sum_{lm} \; C_{lm}\left(\mathbf{k}_0\right) \sum_{\alpha} \; \mathcal{Y}_\alpha^{lm}(\zeta) \; \frac{u_{\alpha l,0l}(k_\alpha, r)}{k_0 r}. \tag{9.98}$$

Above, $u_{\alpha l,0l}$ is the radial wave function in channel-α and angular momentum $\hbar l$, populated in a collision initiated in channel $\alpha = 0$ with the same orbital angular momentum. We remark that the observable quantities depend only on the product of $C_{lm}\left(\mathbf{k}_0\right)$ with the radial wave functions. These coefficients then depend on the adopted normalization for the radial wave functions. It is convenient to normalize them so that the incoming part of the radial wave function in the elastic channel coincides with that of the free wave. With this choice, the coefficients $C_{lm}(\mathbf{k}_0)$ of Eq. (9.98) are still given by Eq. (9.85). Adopting this normalization, we obtain the relation

$$\sum_{\beta} \int dr\, u^*_{\beta l,0l}(k_\beta, r)\, u_{\beta l,0l}(k'_\beta, r) = \frac{\pi}{2}\, \delta\left(k_0 - k'_0\right). \tag{9.99}$$

Eq. (9.99) is the multi-channel generalization of the result of Eq. (4.173), obtained for potential scattering. It can be derived through the evaluation of $\langle \Psi^{(+)}(0\mathbf{k}_0) |\, \Psi^{(+)}(0\mathbf{k}'_0)\rangle$, with the help of Eqs. (9.86) and (9.98), and comparison of the result with Eq. (9.61).

Inserting into Eq. (9.98) the coefficients of Eq. (9.86), we get

$$\Psi^{(+)}(0\mathbf{k}_0; \xi, \mathbf{r}) = \frac{1}{(2\pi)^{3/2}} \sum_{lm} 4\pi\, Y^*_{lm}(\hat{\mathbf{k}}_0) \sum_{\alpha} \mathcal{Y}^{lm}_{\alpha}(\zeta)\, \frac{u_{\alpha l,0l}(k_\alpha, r)}{k_0 r}. \tag{9.100}$$

Substituting the values of $\mathcal{Y}^{lm}_{\alpha}(\zeta)$ given in Eq. (9.82) and using Dirac notation for the intrinsic states (removing the representation in ξ), we get

$$\Psi^{(+)}(0\mathbf{k}_0; \mathbf{r}) = \frac{1}{(2\pi)^{3/2}} \sum_{lm} 4\pi\, Y^*_{lm}(\hat{\mathbf{k}}_0)\, Y_{lm}(\hat{\mathbf{r}})\, i^l \sum_{\alpha} \frac{u_{\alpha l,0l}(k_\alpha, r)}{k_0 r}\, |\alpha). \tag{9.101}$$

If one chooses the z-axis along the incident beam, one can use the addition theorem to replace $\sum_m 4\pi\, Y^*_{lm}(\hat{\mathbf{k}}_0)\, Y_{lm}(\hat{\mathbf{r}}) = (2l+1)\, P_l(\cos\theta)$ and the wave function becomes

$$\Psi^{(+)}(0\mathbf{k}_0; \mathbf{r}) = \frac{1}{(2\pi)^{3/2}} \sum_{l} (2l+1)\, P_l(\cos\theta)\, i^l \sum_{\alpha} \frac{u_{\alpha l,0l}(k_\alpha, r)}{k_0 r}\, |\alpha). \tag{9.102}$$

If the channel-potentials contain Coulomb terms, one proceeds similarly but using the coefficients $C_{lm}(\mathbf{k}_0)$ of Eq. (9.91), instead of Eq. (9.86). Eqs. (9.100) and (9.101) then become

$$\Psi^{(+)}(0\mathbf{k}_0; \mathbf{r}) = \frac{1}{(2\pi)^{3/2}} \sum_{lm} 4\pi\, Y^*_{lm}(\hat{\mathbf{k}}_0)\, e^{i\sigma_{0l}} \sum_{\alpha} \mathcal{Y}^{lm}_{\alpha}(\zeta)\, \frac{u_{\alpha l,0l}(k_\alpha, r)}{k_0 r} \tag{9.103}$$

and

$$\Psi^{(+)}(0\mathbf{k}_0; \mathbf{r}) = \frac{1}{(2\pi)^{3/2}} \sum_{lm} 4\pi\, Y^*_{lm}(\hat{\mathbf{k}}_0)\, Y_{lm}(\hat{\mathbf{r}})\, i^l\, e^{i\sigma_{0l}} \sum_{\alpha} \frac{u_{\alpha l,0l}(k_\alpha, r)}{k_0 r}\, |\alpha). \tag{9.104}$$

Inserting Eq. (9.103) into the Schrödinger equation for $\Psi^{(+)}$, taking scalar product with $(\mathcal{Y}_\alpha^{lm}|$ and using the orthogonality relation of Eq. (9.83), we obtain the coupled radial equations

$$\left[E_\alpha + \frac{\hbar^2}{2\mu}\left(\frac{d^2}{dr^2} - \frac{l(l+1)}{r^2}\right) - V_{\alpha l}(r)\right] u_{\alpha l,0l}(k_\alpha, r) = \sum_{\alpha'} \mathcal{V}_{\alpha l,\alpha' l}(r)\, u_{\alpha' l,0l}(k_{\alpha'}, r), \tag{9.105}$$

where[16]

$$\mathcal{V}_{\alpha l,\alpha' l}(r) = \int d\hat{\mathbf{r}}\; Y_{lm}^*(\hat{\mathbf{r}})\; (\alpha|\, \mathbb{V}^{(2)}\, |\alpha')\; Y_{lm}(\hat{\mathbf{r}}). \tag{9.106}$$

Since we are dealing with a spinless system, the rotational invariance of the interaction leads to vanishing off-diagonal matrix-elements in the orbital angular momentum space. At large distances ($r > \bar{R}$) the interaction vanishes and Eq. (9.105) becomes the equation for a free particle in channel-α. Therefore, the radial wave function takes the asymptotic form

$$u_{\alpha l,0l}(k_\alpha, r > \bar{R}) = \frac{i}{2}\left[\delta_{\alpha 0}\, \hat{h}_l^{(-)}(k_\alpha r) - \sqrt{\frac{k_0}{k_\alpha}}\; S_{\alpha l,0l}\; \hat{h}_l^{(+)}(k_\alpha r)\right] \tag{9.107}$$

and in the $r \to \infty$ limit it becomes

$$u_{\alpha l,0l}(k_\alpha, r) \to e^{-ik_0 r}\left[-\,\delta_{\alpha 0}\,\frac{i^l}{2i}\right] + e^{ik_\alpha r}\left[\sqrt{\frac{k_0}{k_\alpha}}\; S_{\alpha l,0l}\,\frac{i^{-l}}{2i}\right]. \tag{9.108}$$

In Eqs. (9.107) and (9.108), $S_{\alpha l,0l}$ is the matrix-element of the S-operator representing the transition from the elastic channel to channel-α at the lth partial-wave, in the propagation with the full Hamiltonian. Note that a different normalization has been adopted for the outgoing spherical wave in excited channels. Instead of $\hat{h}_l^{(+)}(k_\alpha r)$, we write $\sqrt{k_0/k_\alpha}\; \hat{h}_l^{(+)}(k_\alpha r)$. Since we are not considering channels with mass rearrangement, it is equivalent to $\sqrt{v_0/v_\alpha}\; \hat{h}_l^{(+)}(k_\alpha r)$. It is clear that this normalization factor is arbitrary since only its product with $S_{\alpha l,0l}$ is determined by the scattering boundary conditions (Eq. (9.23)). This factor arises naturally if we recall that the coefficient $S_{\alpha l,0l}$ corresponds to a matrix-element of the scattering operator, evaluated between two properly normalized free states of the angular momentum basis. However, it can be justified by the requirement that at asymptotic distances the incoming wave ($\hat{h}_l^{(-)}(k_0 r)$) and the outgoing wave in any channel ($\sqrt{v_0/v_\alpha}\, \hat{h}_l^{(+)}(k_\alpha r)$) have the same current. An immediate

[16]We assume that the interaction potentials are local in the coordinate space.

consequence of this choice is that when the potentials are real ($\mathbb{H}^\dagger = \mathbb{H}$) the unitarity of the S-matrix is expressed as

$$\sum_\alpha |S_{\alpha l,0l}|^2 = 1. \tag{9.109}$$

To obtain this result, one inserts Eq. (9.108) in Eq. (9.32), integrates within a sphere with radius $R \to \infty$ and uses the divergence theorem.

To derive the partial-wave expansion of the scattering amplitude, we proceed as in section 2.3. First, we take the $r \to \infty$ limit in Eq. (9.101) and replace $u_{\alpha l,0l}$ by Eq. (9.108). We get

$$\Psi^{(+)}(0\mathbf{k}_0;\mathbf{r}) \to \frac{1}{(2\pi)^{3/2}} \sum_\alpha \left[\frac{e^{-ik_\alpha r}}{r} Z_-^{(\alpha)} + \frac{e^{ik_\alpha r}}{r} \bar{Z}_+^{(\alpha)} \right] |\alpha) , \tag{9.110}$$

where $Z_-^{(\alpha)}$ is the incoming amplitude given by Eq. (9.89a) and $\bar{Z}_+^{(\alpha)}$ is

$$\bar{Z}_+^{(\alpha)} = \sum_{lm} \frac{4\pi}{2i\sqrt{k_\alpha k_0}} Y_{lm}^*(\hat{\mathbf{k}}_0)\, Y_{lm}(\hat{\mathbf{r}})\, S_{\alpha l,0l}. \tag{9.111}$$

Eq. (9.110) should be compared with Eq. (9.23). Using Bauer's expansion for the plane wave, Eq. (9.23) can be cast in the form

$$\Psi^{(+)}(0\mathbf{k}_0;\mathbf{r}) \to \frac{1}{(2\pi)^{3/2}} \sum_\alpha \left[\frac{e^{-ik_\alpha r}}{r} Z_-^{(\alpha)} + \frac{e^{ik_\alpha r}}{r} \tilde{Z}_+^{(\alpha)} \right] |\alpha) , \tag{9.112}$$

with $Z_-^{(\alpha)}$ given by Eq. (9.89a) and

$$\begin{aligned} \tilde{Z}_+^{(\alpha)} &= f_{\alpha,0}(\mathbf{k}_\alpha, \mathbf{k}_0) + Z_+^{(\alpha)} \\ &= f_{\alpha,0}(\mathbf{k}_\alpha, \mathbf{k}_0) + \delta_{\alpha 0} \sum_{lm} \frac{4\pi}{2i\sqrt{k_\alpha k_0}} Y_{lm}^*(\hat{\mathbf{k}}_0)\, Y_{lm}(\hat{\mathbf{r}}). \end{aligned} \tag{9.113}$$

Comparing $\tilde{Z}_+^{(\alpha)}$ and $\bar{Z}_+^{(\alpha)}$, which should be identical, we obtain the scattering amplitude

$$f_{\alpha,0}(\mathbf{k}_\alpha, \mathbf{k}_0) = \frac{1}{2i\sqrt{k_\alpha k_0}} \sum_{lm} 4\pi\, Y_{lm}(\hat{\mathbf{k}}_\alpha)\, Y_{lm}^*(\hat{\mathbf{k}}_0) \left[S_{\alpha l,0l} - \delta_{\alpha 0} \right], \tag{9.114}$$

where $\hat{\mathbf{k}}_\alpha$ is a unit vector parallel to $\hat{\mathbf{r}}$. Choosing the z-axis parallel to the incident beam and using the addition theorem of spherical harmonics, Eq. (9.114) takes the form

$$f_{\alpha,0}(\mathbf{k}_\alpha, \mathbf{k}_0) = \frac{1}{2i\sqrt{k_\alpha k_0}} \sum_l (2l+1)\; P_l(\cos\theta) \left[S_{\alpha l,0l} - \delta_{\alpha 0} \right], \tag{9.115}$$

where θ is the angle subtended by the vectors $\mathbf{k}_0$ and $\mathbf{k}_\alpha$, and the cross section is given by Eq. (9.29),

$$\frac{d\sigma_\alpha(\mathbf{k}_\alpha, \mathbf{k}_0)}{d\Omega} = \frac{k_\alpha}{k_0} \left| f_{\alpha,0}(\mathbf{k}_\alpha, \mathbf{k}_0) \right|^2.$$

If the channel potentials contain Coulomb terms, one proceeds similarly. The asymptotic form of Eq. (9.26) can be written

$$\Psi^{(+)}(0\mathbf{k}_0; \mathbf{r}) = \frac{1}{(2\pi)^{3/2}} \sum_\alpha \left[\frac{e^{-i\Theta_\alpha(r)}}{r} Z_-^{(\alpha)} + \frac{e^{i\Theta_\alpha(r)}}{r} \bar{Z}_+^{(\alpha)} \right] |\alpha), \quad (9.116)$$

with $\Theta_\alpha(r)$ and $Z_-^{(\alpha)}$ given respectively by Eqs. (9.27) and Eq. (9.89a), and with the outgoing amplitude

$$\bar{Z}_+^{(\alpha)} = \sum_{lm} \frac{4\pi}{2i\sqrt{k_\alpha k_0}} Y_{lm}^*(\hat{\mathbf{k}}_0)\, Y_{lm}(\hat{\mathbf{r}})\, e^{i(\sigma_{0l}+\sigma_{\alpha l})}\, \bar{S}_{\alpha l,0l}. \quad (9.117)$$

The next step is to take the $r \to \infty$ limit of Eq. (9.24), instead of Eq. (9.20). Using the asymptotic partial-wave expansion of $\phi_C(\mathbf{k}_\alpha; \mathbf{r})$(Eq. (3.105)), we get

$$\Psi^{(+)}(0\mathbf{k}_0; \mathbf{r}) = \frac{1}{(2\pi)^{3/2}} \sum_\alpha \left[\frac{e^{-i\Theta_\alpha(r)}}{r} Z_-^{(\alpha)} + \frac{e^{i\Theta_\alpha(r)}}{r} \tilde{Z}_+^{(\alpha)} \right] |\alpha), \quad (9.118)$$

with the same $Z_-^{(\alpha)}$ of Eq. (9.89a) and

$$\tilde{Z}_+^{(\alpha)} = \bar{f}_{\alpha,0}(\mathbf{k}_\alpha, \mathbf{k}_0) + \delta_{\alpha 0} \sum_{lm} \frac{4\pi}{2i\sqrt{k_\alpha k_0}} Y_{lm}^*(\hat{\mathbf{k}}_0)\, Y_{lm}(\hat{\mathbf{r}})\, e^{i(\sigma_{0l}+\sigma_{\alpha l})}. \quad (9.119)$$

The condition $\tilde{Z}_+^{(\alpha)} = \bar{Z}_+^{(\alpha)}$ leads to the partial scattering amplitude

$$\bar{f}_{\alpha,0}(\mathbf{k}_\alpha, \mathbf{k}_0) = \frac{1}{2i\sqrt{k_\alpha k_0}} \sum_{lm} 4\pi\, Y_{lm}^*(\hat{\mathbf{k}}_0)\, Y_{lm}(\hat{\mathbf{k}}_\alpha)\, e^{i(\sigma_{0l}+\sigma_{\alpha l})} \left[\bar{S}_{\alpha l,0l} - \delta_{\alpha 0} \right], \quad (9.120)$$

or

$$\bar{f}_{\alpha,0}(\mathbf{k}_\alpha, \mathbf{k}_0) = \frac{1}{2i\sqrt{k_\alpha k_0}} \sum_l (2l+1)\, P_l(\cos\theta)\, e^{i(\sigma_{0l}+\sigma_{\alpha l})} \left[\bar{S}_{\alpha l,0l} - \delta_{\alpha 0} \right]. \quad (9.121)$$

The full scattering amplitude is then given by

$$f_{\alpha,0}(\mathbf{k}_\alpha, \mathbf{k}_0) = f_{\rm C}(\mathbf{k}_\alpha, \mathbf{k}_0)\, \delta_{\alpha 0} + \bar{f}_{\alpha,0}(\mathbf{k}_\alpha, \mathbf{k}_0), \tag{9.122}$$

where $f_{\rm C}$ is the Coulomb scattering amplitude of Eq. (3.29).

The partial-wave expansion of the total absorption cross section is easily obtained from Eqs. (9.34), (9.11) and[17] (9.102). It is given[18] by

$$\sigma_a = \sum_\alpha \sigma_a^{(\alpha)}, \tag{9.123}$$

where

$$\sigma_a^{(\alpha)} = \frac{\pi}{k_0^2} \sum_l (2l+1)\, \mathcal{T}_l^{(\alpha)} \tag{9.124}$$

corresponds to absorption through channel-α. The absorption factor $\mathcal{T}_l^{(\alpha)}$ is

$$\mathcal{T}_l^{(\alpha)} = \frac{4k_0}{E_0} \int dr\, (-W_\alpha(r))\, \left|u_{\alpha l,0l}(k_\alpha, r)\right|^2 . \tag{9.125}$$

The absorption cross section in each channel is analogous to that in chapter 7, with

$$\mathcal{T}_l^{(\alpha)} = \delta_{\alpha 0} - \left|S_{\alpha l,0l}\right|^2 . \tag{9.126}$$

9.3.1.1 *Partial-wave projected T-matrix and Lippmann-Schwinger equations*

Using the partial-wave expansions of $\Phi(\alpha\mathbf{k}_\alpha;\xi,\mathbf{r})$, $X^{(+)}(\alpha\mathbf{k}_\alpha;\xi,\mathbf{r})$ and $\Psi^{(+)}(0\mathbf{k}_0;\xi,\mathbf{r})$ (Eqs. (9.85), (9.94) and (9.98), respectively) with the coefficients $C_{lm}(\alpha\mathbf{k}_\alpha)$ of Eq. (9.86), we can derive l-projected LS equations. For this purpose, we need also the angular momentum expansions of the Greens' functions,

$$G_0^{(+)}(\alpha E;\mathbf{r},\mathbf{r}') = \frac{1}{rr'} \sum_{l,m} Y_{lm}^*(\hat{\mathbf{r}})\, g_{0,l}^{(+)}(\alpha E; r, r')\, Y_{lm}(\hat{\mathbf{r}}'), \tag{9.127}$$

$$G_1^{(+)}(\alpha E;\mathbf{r},\mathbf{r}') = \frac{1}{rr'} \sum_{l,m} Y_{lm}^*(\hat{\mathbf{r}})\, g_{1,l}^{(+)}(\alpha E; r, r')\, Y_{lm}(\hat{\mathbf{r}}'). \tag{9.128}$$

[17] To be consistent with Eq. (9.11), the wave function should be normalized with $A = 1$. In this way, we actually use Eq. (9.102) multiplied by the factor $(2\pi)^{3/2}$.
[18] We are assuming that $\mathbb{V}^{(2)\dagger} = \mathbb{V}^{(2)}$. Otherwise, off-diagonal contributions in channel-space should be included.

Inserting these expansions into Eqs. (9.46a) and (9.46c), we can easily prove that the radial wave functions satisfy the LS equations

$$u_{\alpha l,0l}(k_\alpha, r) = \hat{\jmath}_l(k_\alpha r)\ \delta_{\alpha 0} + \sum_{\alpha'} \int dr'\ g_{0,l}^{(+)}(\alpha E; r, r')\ V_{\alpha l,\alpha' l}(r)\ u_{\alpha' l,0l}(k_{\alpha'}, r) \quad (9.129)$$

and

$$u_{\alpha l,0l}(k_\alpha, r) = w_{\alpha l}(k_\alpha r)\ \delta_{\alpha 0} + \sum_{\alpha'} \int dr'\ g_{1,l}^{(+)}(\alpha E; r, r')\ \mathcal{V}_{\alpha l,\alpha' l}(r)\ u_{\alpha' l,0l}(k_{\alpha'}, r). \quad (9.130)$$

Above we used the notations,

$$V_{\alpha l,\alpha' l} = \left(\mathcal{Y}_{\alpha}^{lm}\, |\mathbb{V}|\, \mathcal{Y}_{\alpha'}^{lm}\right)$$

and (see Eq. (9.106))

$$\mathcal{V}_{\alpha l,\alpha' l} = \left(\mathcal{Y}_{\alpha}^{lm}\, |\mathbb{V}^{(2)}|\, \mathcal{Y}_{\alpha'}^{lm}\right).$$

One can also derive l-projected versions of the LS equation for the T-matrix. For this purpose we generalize the procedure of section 4.6. First we introduce the normalized eigenstates of the sets of operators $\{\mathbb{H}_0, L^2, L_z\}$, $\{\mathbb{H}_1, L^2, L_z\}$ and $\{\mathbb{H}, L^2, L_z\}$, which we denote respectively by $|\phi(\alpha Elm)\rangle$, $|\chi(\alpha Elm)\rangle$ and $|\psi_\alpha(Elm)\rangle$. In the coordinate representation, they are given by (similarly to section 4.6)

$$\phi\,(\alpha Elm; \zeta) = \sqrt{\frac{2\mu k_\alpha}{\pi \hbar^2}}\ \mathcal{Y}_{\alpha}^{lm}(\zeta)\ \frac{\hat{\jmath}_l(k_\alpha r)}{k_\alpha r} \quad (9.131)$$

$$\chi\,(\alpha Elm; \zeta) = \sqrt{\frac{2\mu k_\alpha}{\pi \hbar^2}}\ \mathcal{Y}_{\alpha}^{lm}(\zeta)\ \frac{w_{\alpha l}(k_\alpha, r)}{k_\alpha r} \quad (9.132)$$

and[19]

$$\psi_\alpha(Elm; \zeta) = \sqrt{\frac{2\mu k_\alpha}{\pi \hbar^2}}\ \sum_{\alpha'} \mathcal{Y}_{\alpha'}^{lm}(\zeta)\ \frac{u_{\alpha' l;\alpha l}(k_{\alpha'}, r)}{k_0 r}. \quad (9.133)$$

These wave functions satisfy the orthonormality relations (see problem 3)

$$\left\langle \phi\,(\alpha Elm)\,\middle|\,\phi\,(\alpha' E' l' m')\,\right\rangle = \delta_{\alpha\alpha'}\ \delta_{ll'}\ \delta_{mm'}\ \delta(E - E') \quad (9.134)$$

$$\left\langle \chi\,(\alpha Elm)\,\middle|\,\chi\,(\alpha' E' l' m')\,\right\rangle = \delta_{\alpha\alpha'}\ \delta_{ll'}\ \delta_{mm'}\ \delta(E - E') \quad (9.135)$$

$$\left\langle \psi_\alpha\,(Elm)\,\middle|\,\psi_\alpha\,(E' l' m')\,\right\rangle = \delta_{ll'}\ \delta_{mm'}\ \delta(E - E'). \quad (9.136)$$

[19]The exact wave functions below are labeled by the eigenvalues of the operators $\mathbb{H}$, L^2 and L_z. In addition, it is necessary to specify the incident channel, α. In contrast with the labels E, l and m, which are constants of motion, the channel label is not. To emphasize this difference, we write the channel label as a subscript.

Now we consider the partial-wave expansion of the T-matrices. Using the expansions of the wave functions and Eqs. (9.131) to (9.136), we can write the on-shell T-matrix elements as

$$T_{\alpha\mathbf{k}_\alpha,0\mathbf{k}_0} = \frac{\hbar^2}{\mu\sqrt{k_0 k_\alpha}} \sum_{lm} Y_{lm}(\hat{\mathbf{k}}_\alpha)\, Y^*_{lm}(\hat{\mathbf{k}}_0)\; T_{\alpha l,0l}(E), \tag{9.137}$$

with the l-projected T-matrix given by

$$T_{\alpha l,0l}(E) = \frac{k_0}{\pi E_0}\sqrt{\frac{k_0}{k_\alpha}} \sum_{\alpha'} \int dr\; \hat{\jmath}_l(k_\alpha r)\, V_{\alpha l,\alpha' l}(r)\; u_{\alpha' l,0l}(k_{\alpha'}, r). \tag{9.138}$$

It can be easily checked that the l-components of the S-matrix are related to $T_{\alpha l,0l}(E)$ by a trivial generalization of the relation for potential scattering, in the form

$$S_{\alpha l,0l}(E) = \delta_{\alpha 0} - 2\pi i\, T_{\alpha l,0l}(E). \tag{9.139}$$

To finish this section, we discuss the partial-wave projection of the Gell-Mann Goldberger relation. We consider the situation where the dominant potential, $\mathbb{V}^{(1)}$, is diagonal in channel space while the complementary potential, $\mathbb{V}^{(2)}$, is not. Carrying out angular momentum projections of the wave functions involved, the l-projected T-matrix can be written as

$$T_{\alpha l,0l}(E) = T^{(1)}_{\alpha l,0l}(E) + T^{(2)}_{\alpha l,0l}(E), \tag{9.140}$$

with

$$T^{(1)}_{\alpha l,0l}(E) = \frac{k_0}{\pi E_0}\left[\int dr\; \hat{\jmath}_l(k_0 r)\, V_{0l}(r)\; w_{0l}(k_0, r)\right] \delta_{\alpha 0} \tag{9.141}$$

and

$$T^{(2)}_{\alpha l,0l}(E) = \frac{k_0}{\pi E_0}\sqrt{\frac{k_0}{k_\alpha}} \sum_{\alpha'} \left[\int dr\; w_{0l}(k_\alpha, r)\, \mathcal{V}_{\alpha l,\alpha' l}(r)\; u_{\alpha' l,0l}(k_{\alpha'}, r)\right]. \tag{9.142}$$

Above, we used the notation

$$V_{\alpha l} = (\mathcal{Y}^{lm}_\alpha|\, \mathbb{V}^{(1)}\, |\mathcal{Y}^{lm}_\alpha) \tag{9.143}$$

for the matrix elements of the potential $\mathbb{V}^{(1)}$.

9.3.2 *Partial-wave expansions with spin*

When the coupled-channel problem involves states with spin, the channel-angle basis should be modified. It should take into account the rotational properties of the Hamiltonian, which would then be different. In this section, we discuss this problem in situations where one of the collision partners has spin zero and the other has not. For more complicated cases where both the projectile and the target have spin different from zero, we refer to text books like [Austern (1970); Newton (2002); Satchler (1983)].

Assuming that the intrinsic Hamiltonian, h, is rotationally invariant, it must commute with the components of the intrinsic angular momentum operators. In this way,

$$\left[h, I^2\right]_- = \left[h, I_z\right]_- = 0,$$

and one can build a basis for the intrisic space with the eigenstates of the operator set $\left\{h, I^2, I_z\right\}$. These states are grouped into multiplets of $(2I+1)$ states, $|\alpha I\,\nu\rangle$, with $\nu = -I, -I+1, \cdots, I$. Since the multiplet is in the same channel, it is necessary to give the quantum numbers α and ν to specify the intrinsic state. The quantum number I is superfluous, since it is contained in α (that is: $I \equiv I_\alpha$). However, we keep it in the notation $|\alpha I \nu)$, as a reminder that ν is the z-projection associated with I.

When there are intrinsic states with spin $I \neq 0$, Eqs. (9.37) and (9.39) should be re-written as

$$|\Phi(\alpha\nu\mathbf{k})\rangle \equiv |\phi(\mathbf{k}_\alpha)\rangle\; |\alpha I \nu)\,, \tag{9.144}$$

$$|X^{(+)}(\alpha\nu\mathbf{k})\rangle \equiv |\chi^{(+)}(\alpha\mathbf{k}_\alpha)\rangle\; |\alpha I \nu)\,. \tag{9.145}$$

The scattering wave function for a collision initiated in the elastic channel with spin I_0, z-component ν_0 and wave vector $\mathbf{k}_0$ is given by

$$|\Psi^{(+)}(0\nu_0\mathbf{k}_0)\rangle = \sum_{\alpha\nu} |\psi^{(+)}_{\alpha\nu}(0\nu_0\mathbf{k}_0)\rangle\; |\alpha I \nu)\,,$$

which is a straightforward generalization of Eq. (9.11). It satisfies the scattering boundary condition

$$\Psi^{(+)}(0\nu_0\mathbf{k}_0;\mathbf{r}) \to \frac{1}{(2\pi)^{3/2}} \sum_{\alpha\nu} |\alpha I \nu) \times \left[\delta_{\alpha 0}\,\delta_{\nu\nu_0}\, e^{i\mathbf{k}_0\cdot\mathbf{r}} + f_{\alpha\nu,0\nu_0}(\mathbf{k}_\alpha,\mathbf{k}_0)\frac{e^{ik_\alpha\, r}}{r}\right]. \tag{9.146}$$

Since $I \equiv I_\alpha$, a sum over I is performed through the index α.

If the channel-potentials contain Coulomb terms, the scattering boundary condition should be modified as in the previous section. That is,

$$\Psi^{(+)}(0\nu_0\mathbf{k}_0;\mathbf{r}) \to \frac{1}{(2\pi)^{3/2}} \sum_{\alpha\nu} \; |\alpha I\nu) \times \left[\delta_{\alpha 0}\,\delta_{\nu\nu_0}\,\phi_{\rm C}(\mathbf{k}_0,\mathbf{r}) + \bar{f}_{\alpha\nu,0\nu_0}(\mathbf{k}_\alpha,\mathbf{k}_0)\,\frac{e^{i\Theta_\alpha(r)}}{r}\right] \quad (9.147)$$

and $\phi_{\rm C}(\mathbf{k}_0,\mathbf{r})$ can be replaced by its $r \to \infty$ limit (Eq. (3.105)).

In the scattering of systems with spin, rotational invariance no longer guarantees conservation of the orbital angular momentum and the set of states of Eq. (9.82) is not convenient for the partial-wave expansions. However, the total angular momentum $\mathbf{J} = \mathbf{L} + \mathbf{I}$ is conserved, similarly to the case of potential scattering with spin. The partial-wave expansions are then made in the basis of the *channel-angle* (or *channel-spin*) wave functions,

$$\mathcal{Y}_{\alpha l}^{\pi JM}(\hat{\mathbf{r}}) = i^l \sum_{m\nu} \langle lmI\nu\,|JM\rangle \; Y_{lm}(\hat{\mathbf{r}}) \; |\alpha I\,\nu)\,, \\ = i^l \sum_{\nu} \langle l(M-\nu)I\nu\,|JM\rangle\, Y_{l(M-\nu)}(\hat{\mathbf{r}}) \; |\alpha I\,\nu)\,, \quad (9.148)$$

where $\langle lmI\nu\,|JM\rangle$ are the usual Clebsch-Gordan coefficients. As before, the quantum numbers πJM, which are constants of motion, are represented as superscripts in $\mathcal{Y}_{\alpha l}^{\pi JM}$. The channel label and the orbital angular momentum, which in general are not conserved, are represented as subscripts. Although the parity quantum number π can be inferred from the channel quantum numbers and the orbital angular momentum, it is convenient to include it in the set of quantum numbers labeling the states of the expansion basis. As will be clear in the next section, it plays a very important role in the selection of the radial wave functions appearing in the coupled channel equations. The states of Eq. (9.148) are very similar to the spin-angle wave functions of section 6.1.1. Using the orthonormality of the intrinsic states and properties of the Clebsch-Gordan coefficients, one gets, respectively, the orthonormality and the completeness relations

$$\left(\mathcal{Y}_{\alpha l}^{\pi JM}\middle|\mathcal{Y}_{\alpha' l'}^{\pi J'M'}\right) = \delta_{\alpha\alpha'}\,\delta_{ll'}\,\delta_{JJ'}\,\delta_{MM'}. \quad (9.149)$$

and

$$\sum_{\alpha l JM} \left|\mathcal{Y}_{\alpha l}^{\pi JM}\right)\left(\mathcal{Y}_{\alpha l}^{\pi JM}\right| = \mathbf{1}. \quad (9.150)$$

The single-channel wave functions $\Phi(\alpha\nu\mathbf{k};r\zeta)$ and $X^{(+)}(\alpha\nu\mathbf{k};r\zeta)$ are expanded as[20]

$$\Phi(\alpha\nu\mathbf{k}_\alpha;r\zeta) = \frac{1}{(2\pi)^{3/2}} \sum_{lJM} C_{JM}\,(\alpha\nu l\mathbf{k}_0)\,\mathcal{Y}_{\alpha l}^{\pi JM}(\zeta)\,\frac{\hat{j}_l(k_\alpha r)}{k_\alpha r} \tag{9.151}$$

and

$$X^{(+)}(\alpha\nu\mathbf{k}_\alpha;r\zeta) = \frac{1}{(2\pi)^{3/2}} \sum_{lJM} C_{JM}\,(\alpha\nu l\mathbf{k}_0)\,\mathcal{Y}_{\alpha l}^{\pi JM}(\zeta)\,\frac{w_{\alpha l}^{J}(k_\alpha,r)}{k_\alpha r}. \tag{9.152}$$

The coefficients $C_{JM}\,(\alpha\nu l\mathbf{k}_0)$ give the amplitudes for the incident wave in channel-α, with spin I and z-projection ν, to have orbital angular momentum l. We want to evaluate these coefficients for the free state in the entrance channel. That is, $\alpha\nu l\,\mathbf{k}_\alpha \to 0\nu_0 l_0\,\mathbf{k}_0$. For this purpose, we write Eq. (9.144) in the coordinate representation and use Bauer's expansion for $\phi(\mathbf{k}_0;\mathbf{r})$ (Eq. (2.44a)) and the relation[21]

$$i^l\,Y_{lm}(\hat{\mathbf{r}})\,|\alpha I\,\nu) = \sum_J \langle J\,(m+\nu)\,|lmI\nu\rangle\,\mathcal{Y}_{\alpha l}^{\pi J(m+\nu)}(\hat{\mathbf{r}}). \tag{9.153}$$

Following the same steps as in the derivation of Eq. (6.35), one obtains

$$C_{JM}\,(0\nu_0 l_0\mathbf{k}_0) = 4\pi\,\langle JM\,|l_0(M-\nu_0)\,I_0\nu_0\rangle\,Y^*_{l_0(M-\nu_0)}(\hat{\mathbf{k}}_0). \tag{9.154}$$

Thus, the angular momentum expansions of Eqs. (9.151) and (9.152) have the explicit forms[22]

$$\Phi(\alpha\nu\mathbf{k}_\alpha;r\zeta) = \frac{1}{(2\pi)^{3/2}} \sum_{JMl_0m_0} 4\pi\,\langle JM\,|l_0m_0\,I_0\nu_0\rangle \times Y^*_{l_0m_0}(\hat{\mathbf{k}}_0)\,\mathcal{Y}_{\alpha l}^{\pi JM}(\zeta)\,\frac{\hat{j}_l(k_\alpha r)}{k_\alpha r}. \tag{9.155}$$

and

$$X^{(+)}(\alpha\nu\mathbf{k}_\alpha;r\zeta) = \frac{1}{(2\pi)^{3/2}} \sum_{JMl_0m_0} 4\pi\,\langle JM\,|l_0m_0\,I_0\nu_0\rangle \times Y^*_{l_0m_0}(\hat{\mathbf{k}}_0)\,\mathcal{Y}_{\alpha l}^{\pi JM}(\zeta)\,\frac{w_{\alpha l}^{J}(k_\alpha,r)}{k_\alpha r}. \tag{9.156}$$

---

[20]The degrees of freedom of the system can be represented by the set of coordinates $\{\hat{\mathbf{r}},\xi\}$ or by the equivalent set $\{r,\zeta\}$.

[21]This relation is the inverse of Eq. (9.148). It can easily be derived with the help of the properties of the Clebsch-Gordan coefficients. Since the orbital angular momentum remains the same as that of the incident state (this will not happen in the partial-wave expansion of the scattering state), we represent it by l_0.

[22]Although the Clebsh-Gordan coefficients impose that $m_0 = M - \nu_0$, it is convenient to keep this constraint implicitly and sum over m_0.

Taking the $r \to \infty$ limit in Eq. (9.151), we replace $\hat{\jmath}_{l_0}(k_0 r)$ by its asymptotic form and get

$$\Phi(0\nu_0\mathbf{k}_0, \mathbf{r}) \to \frac{1}{(2\pi)^{3/2}} \sum_{\alpha\nu} \left[\frac{e^{-ik_\alpha r}}{r}\, Z_-^{(\alpha\nu)} + \frac{e^{ik_\alpha r}}{r}\, Z_+^{(\alpha\nu)} \right] |\alpha I \nu), \quad (9.157)$$

with

$$Z_-^{(\alpha\nu)} = -\,\delta_{\alpha 0}\,\delta_{l l_0}\,\delta_{\nu\nu_0}\,\frac{4\pi}{2i\sqrt{k_\alpha k_0}} \sum_l Y^*_{l(M-\nu_0)}(\hat{\mathbf{k}}_0)\, Y_{l(M-\nu_0)}(\hat{\mathbf{r}})\; i^{2l_0} \quad (9.158)$$

and

$$Z_+^{(\alpha\nu)} = \delta_{\alpha 0}\,\delta_{l l_0}\,\delta_{\nu\nu_0}\,\frac{4\pi}{2i\sqrt{k_\alpha k_0}} \sum_l Y^*_{l(M-\nu_0)}(\hat{\mathbf{k}}_0)\, Y_{l(M-\nu_0)}(\hat{\mathbf{r}})\,. \quad (9.159)$$

For potentials with a Coulomb part, the incident wave function $\Phi_{\rm C}$ has the asymptotic form

$$\Phi_{\rm C}(0\nu_0\mathbf{k}_0, \mathbf{r}) \to \frac{1}{(2\pi)^{3/2}} \sum_{\alpha\nu} \left[\frac{e^{-i\,\Theta_\alpha(r)}}{r}\, Z_-^{(\alpha\nu)} + \frac{e^{i\,\Theta_\alpha(r)}}{r}\, Z_+^{(\alpha\nu)} \right] |\alpha I \nu)\,,$$

with $\Theta_\alpha(r)$ given by Eq. (9.27), $Z_-^{(\alpha\nu)}$ still given by Eq. (9.158) and

$$Z_+^{(\alpha\nu)} = \delta_{\alpha 0}\,\delta_{l l_0}\,\delta_{\nu\nu_0}\,\frac{4\pi}{2i\sqrt{k_\alpha k_0}} \sum_l Y^*_{l(M-\nu_0)}(\hat{\mathbf{k}}_0)\, Y_{l(M-\nu_0)}(\hat{\mathbf{r}})\, e^{2i\sigma_{0l_0}}\,. \quad (9.160)$$

The expansion coefficients in Eq. (9.152) are the same as those in Eq. (9.151), since $X^{(+)}$ and Φ have the same incoming terms. Using Eq. (9.152) in the Schrödinger equation for $X^{(+)}$ and taking scalar product with $(\mathcal{Y}_{\alpha l}^{\pi JM}|$, we get the radial equation

$$\left[E_\alpha + \frac{\hbar^2}{2\mu}\left(\frac{d^2}{dr^2} - \frac{l(l+1)}{r^2} \right) - V_{\alpha l}^J(r) \right] w_{\alpha l}^J(k_\alpha, r) = 0, \quad (9.161)$$

where

$$V_{\alpha l}^J = \left(\mathcal{Y}_{\alpha l}^{\pi JM}\right| \mathbb{V}^{(1)} \left|\mathcal{Y}_{\alpha l}^{\pi JM}\right)\,. \quad (9.162)$$

In Eqs. (9.161) and (9.162), we are assuming that the Hamiltonian $\mathbb{H}^{(1)}$ conserves the orbital angular momentum. In this way, any off-diagonal term in l-space is included in the interaction $\mathbb{V}^{(2)}$.

The wave function $w^J_{\alpha l}(k_\alpha, r)$ has the asymptotic form of Eq. (9.108), except for the J-dependence of the S-matrix. Using the same notation of the previous section, we replace $S^{(1)}_{\alpha l,0l} \to S^{(1)J}_{\alpha l,0l}$. If the channel Hamiltonians contain Coulomb terms, the modifications discussed in chapter 3 are required.

The partial-wave expansion of $\Psi^{(+)}(0\nu_0\mathbf{k}_0;\mathbf{r})$ is more complicated. Owing to the fact that the interaction has off-diagonal components both in the α- and in the l-spaces, an incident state $\mathcal{Y}^{\pi JM}_{0l_0}(\zeta)$ may emerge as $\mathcal{Y}^{\pi JM}_{\alpha l}(\zeta)$. That is, in a different channel and with different orbital angular momentum. The reasons presented in the previous section lead us to keep the same coefficients in the expansion of $\Psi^{(+)}$ and write

$$\Psi^{(+)}(0\nu_0\mathbf{k}_0;r\zeta) = \frac{1}{(2\pi)^{3/2}} \sum_{l_0\alpha lJM} C_{JM}\left(0\nu_0 l_0\mathbf{k}_0\right) \mathcal{Y}^{\pi JM}_{\alpha l}(\zeta)\; \frac{u^J_{\alpha l,0l_0}(k_\alpha, r)}{k_0 r}, \tag{9.163}$$

or, inserting the explicit form of the coefficients $C_{JM}\left(0\nu_0 l_0\mathbf{k}_0\right)$,

$$\Psi^{(+)}(0\nu_0\mathbf{k}_0;r\zeta) = \frac{1}{(2\pi)^{3/2}} \sum_{l_0\alpha lJM} 4\pi\, \langle JM\,|l_0(M-\nu_0)\,I_0\nu_0\rangle\; Y^*_{l_0\,(M-\nu_0)}(\hat{\mathbf{k}}_0) \times \mathcal{Y}^{\pi JM}_{\alpha l}(\zeta)\; \frac{u^J_{\alpha l,0l_0}(k_\alpha, r)}{k_0 r}. \tag{9.164}$$

In some situations it is convenient to sum over the z-component of l_0, denoted by m_0, leaving the restriction $m_0 = M - \nu_0$ implicit in the Clebsh-Gordan coefficients. That is,

$$\Psi^{(+)}(0\nu_0\mathbf{k}_0;r\zeta) = \frac{1}{(2\pi)^{3/2}} \sum_{l_0 m_0 JM} 4\pi\, \langle JM\,|l_0 m_0 I_0\nu_0\rangle\; Y^*_{l_0 m_0}(\hat{\mathbf{k}}_0) \times \sum_{\alpha l} \mathcal{Y}^{\pi JM}_{\alpha l}(\zeta)\; \frac{u^J_{\alpha l,0l_0}(k_\alpha, r)}{k_0 r}. \tag{9.165}$$

Inserting the above expansion into the full Schrödinger equation (Eq. (9.10)) and taking scalar product with $\left(\mathcal{Y}^{\pi JM}_{\alpha l}\right|$, one obtains the set of coupled equations

$$\left[E_\alpha + \frac{\hbar^2}{2\mu}\left(\frac{d^2}{dr^2} - \frac{l(l+1)}{r^2}\right) - V^J_{\alpha l}(r)\right] u^J_{\alpha l,0l_0}(k_\alpha, r) = \sum_{\alpha' l'} \mathcal{V}^J_{\alpha l,\alpha' l'}(r)\; u^J_{\alpha' l',0l_0}(k_{\alpha'}, r), \tag{9.166}$$

where α and α' run over all channels included in the calculation, and

$$\mathcal{V}^{J}_{\alpha l,\alpha' l'} = \left(\mathcal{Y}^{\pi JM}_{\alpha l}\right| \mathbb{V}^{(2)} \left|\mathcal{Y}^{\pi JM}_{\alpha' l'}\right) . \tag{9.167}$$

In Eq. (9.166), $u^{J}_{\alpha l,0l_0}(k_\alpha, r)$ stands for the radial wave function with total angular momentum J, incident in the elastic channel with orbital angular momentum l_0 and emerging in channel α, with orbital angular momentum l. Since the total angular momentum is a conserved quantity, the coupled equations do not mix different J-values. The same thing happens with respect to the total parity, π. The important issue of determining which l_0, α and l are coupled in the equations for a given pair $\{J, \pi\}$ will be considered in the next section.

The requirement that the scattering states $|\Psi(0\nu_0\mathbf{k}_0)\rangle$ form an orthonormal set (see Eq. (9.61)) imposes two conditions on the normalization of the radial wave functions. The first is that the sum or radial integrals

$$\sum_{\beta l} \int dr\, u^{J*}_{\beta l,0l_0}(k_\beta, r)\, u^{J}_{\beta l,0l'_0}(k'_\beta, r) \tag{9.168}$$

be independent of J, for any l_0 and l'_0. The second is that for $l_0 = l'_0$ they are given by the the orthonormality relation

$$\sum_{\beta l} \int dr\, u^{J*}_{\beta l,0l_0}(k_\beta, r)\, u^{J}_{\beta l,0l_0}(k'_\beta, r) = \frac{\pi}{2}\, \delta\left(k_0 - k'_0\right) . \tag{9.169}$$

These properties guarantee that the expansion of Eq. (9.165) is consistent with the orthonormality relation of Eq. (9.61) (see problem 4).

We now derive the partial-wave expansion of the scattering amplitude. First, we express Eq. (9.164) in terms of products like $Y_{lm}(\hat{\mathbf{r}})\, |\alpha I\nu)$, with the help of Eq. (9.148). We then take the $r \to \infty$ limit, and write the radial wave as

$$u^{J}_{\alpha l,0l_0}(k_\alpha, r \to \infty) = e^{-ik_\alpha r}\left[-\,\delta_{\alpha 0}\,\delta_{ll_0}\,\frac{i^l}{2i}\right] + e^{ik_\alpha r}\left[\sqrt{\frac{k_0}{k_\alpha}}\, S^{J}_{\alpha l,0l_0}\,\frac{i^{-l}}{2i}\right] . \tag{9.170}$$

Or, for potentials with a Coulomb part,

$$u^{J}_{\alpha l,0l_0}(k_\alpha, r \to \infty) = e^{-i\Theta_\alpha(r)}\left[-\,\delta_{\alpha 0}\,\delta_{ll_0}\,\frac{i^l}{2i}e^{-i\sigma_{\alpha l}}\right] + e^{i\Theta_\alpha(r)}\left[\sqrt{\frac{k_0}{k_\alpha}}\, \bar{S}^{J}_{\alpha l,0l_0}\,\frac{i^{-l}}{2i}\, e^{i\sigma_{\alpha l}}\right] . \tag{9.171}$$

Using Eq. (9.170) in Eq. (9.164), we get

$$\Psi^{(+)}(0\nu_0\mathbf{k}_0,\mathbf{r}) \to \frac{1}{(2\pi)^{3/2}}\sum_{\alpha\nu}\left[\frac{e^{-ik_\alpha r}}{r}\,Z_-^{(\alpha\nu)} + \frac{e^{ik_\alpha r}}{r}\,\bar{Z}_+^{(\alpha\nu)}\right]\,|\alpha I\nu\rangle\,, \tag{9.172}$$

with $Z_-^{(\alpha\nu)}$ given by Eq. (9.158) and

$$\bar{Z}_+^{(\alpha\nu)} = \frac{4\pi}{2i\sqrt{k_\alpha k_0}}\sum_{Jl_0m_0l} Y^*_{l_0m_0}(\hat{\mathbf{k}}_0)\,Y_{l(m_0+\nu_0-\nu)}(\hat{\mathbf{r}})\,\langle l_0m_0I_0\nu_0\,|J\,(m_0+\nu_0)\rangle$$
$$\times\, S^J_{\alpha l,0l_0}\,\langle J\,(m_0+\nu_0)\,|l\,(m_0+\nu_0-\nu)\,I\nu\rangle\,. \tag{9.173}$$

Proceeding as in the previous section, we write the partial-wave expansion of the scattering boundary condition given in Eq. (9.146). Using Eq. (2.43b) it can be expressed as in Eq. (9.110), with the same $Z_-^{(\alpha\nu)}$ and the new outgoing amplitude

$$\tilde{Z}_+^{(\alpha\nu)} = f_{\alpha\nu,0\nu_0}\left(\hat{\mathbf{k}}_\alpha,\hat{\mathbf{k}}_0\right) + \delta_{\alpha 0}\,\delta_{\nu\nu_0}\frac{4\pi}{2i\sqrt{k_\alpha k_0}}\sum_{lm} Y^*_{lm}(\hat{\mathbf{k}}_\alpha)\,Y_{lm}(\hat{\mathbf{r}})\,.$$

Setting $\tilde{Z}_+^{(\alpha\nu)} = \bar{Z}_+^{(\alpha\nu)}$, we obtain the partial-wave expansion of the scattering amplitude

$$f_{\alpha\nu,0\nu_0}\left(\hat{\mathbf{k}}_\alpha,\hat{\mathbf{k}}_0\right) = \frac{1}{2i\sqrt{k_\alpha k_0}}\sum_{Jl_0m_0l} 4\pi\; Y_{l(m_0+\nu_0-\nu)}(\hat{\mathbf{k}}_\alpha)\,Y^*_{l_0m_0}(\hat{\mathbf{k}}_0)$$
$$\times\big\langle J\,(m_0+\nu_0)\big|l_0m_0I_0\nu_0\big\rangle\big\langle l\,(m_0+\nu_0-\nu)I\nu\big|J\,(m_0+\nu_0)\big\rangle$$
$$\times\left[S^J_{\alpha l,0l_0} - \delta_{\alpha 0}\,\delta_{\nu\nu_0}\delta_{ll_0}\right]. \tag{9.174}$$

Choosing the z-axis along the direction of the incident beam, we can use the relation $Y_{l_0m_0}(\hat{\mathbf{k}}_0) \equiv Y_{l_0m_0}(\mathbf{0}) = \delta_{0m_0}\sqrt{(2l_0+1)/4\pi}$ and the above equation takes the simpler form

$$f_{\alpha\nu,0\nu_0}\left(\hat{\mathbf{k}}_\alpha,\hat{\mathbf{k}}_0\right) = \frac{1}{2i\sqrt{k_\alpha k_0}}\sum_{Jl_0l}\sqrt{4\pi\,(2l_0+1)}\;\,Y_{l(\nu_0-\nu)}(\hat{\mathbf{k}}_\alpha)$$
$$\times\,\big\langle J\nu_0\big|l_0 0I_0\nu_0\big\rangle\,\big\langle l\,(\nu_0-\nu)\,I\nu\big|J\nu_0\big\rangle$$
$$\times\,\left[S^J_{\alpha l,0l_0} - \delta_{\alpha 0}\,\delta_{\nu\nu_0}\delta_{ll_0}\right]. \tag{9.175}$$

If the channel potentials contain Coulomb terms, the same changes of the previous section must be performed. Choosing the z-axis along the incident beam and following the same procedures one gets the partial scattering amplitude

$$\bar{f}_{\alpha\nu,0\nu_0}(\hat{\mathbf{k}}_\alpha) = \frac{1}{2i\sqrt{k_\alpha k_0}} \sum_{l_0 l J} \sqrt{4\pi(2l_0+1)}\; Y_{l(\nu_0-\nu)}(\hat{\mathbf{k}}_\alpha)\; e^{i\left(\sigma_{\alpha l}+\sigma_{0 l_0}\right)}$$
$$\times \left\langle J\nu_0 \middle| l_0 0 I_0 \nu_0 \right\rangle \left\langle l\,(\nu_0-\nu)\, I\nu \middle| J\nu_0 \right\rangle$$
$$\times \left[\bar{S}^J_{\alpha l,0l_0} - \delta_{\alpha 0}\, \delta_{\nu\nu_0} \delta_{ll_0}\right], \tag{9.176}$$

and the full scattering amplitude is given by Eq. (9.122).

The spin-dependent cross sections are given by Eq. (9.29). Inclusive cross sections for unpolarized beams are obtained similarly to Eq. (6.14). That is

$$\frac{d\sigma_\alpha\left(\mathbf{k}_\alpha, \mathbf{k}_0\right)}{d\Omega} = \frac{k_\alpha}{k_0}\left(\frac{1}{2I_0+1}\right) \sum_{\nu\nu_0} \left|f_{\alpha\nu,0\nu_0}\left(\mathbf{k}_\alpha, \mathbf{k}_0\right)\right|^2. \tag{9.177}$$

The derivation of the total absorption cross section is analogous to the case of spinless intrinsic states. In the usual case of an absorptive potential that conserves the orbital angular momentum and is diagonal in channel space, the cross section is given by the expression

$$\sigma_a = |A|^{-2}\, \frac{k_0}{E}\, \left\langle \Psi^{(+)}(0\nu_0\mathbf{k}_0) \right| - \mathbb{W},\ \left| \Psi^{(+)}(0\nu_0\mathbf{k}_0) \right\rangle,$$

where $\mathbb{W}$ is the imaginary part of the total potential. Replacing the scattering wave function by its angular momentum expansion (Eq. (9.165)) and using the orthonormality of the spin-channel wave functions, we get

$$\sigma_a = \sum_\alpha \sigma_a^{(\alpha)} \quad \text{with} \quad \sigma_a^{(\alpha)} = \frac{\pi}{k_0^2} \sum_J\, (2J+1)\; \mathcal{T}^J_\alpha . \tag{9.178}$$

However, here the absorption factor is more complicated. After some algebra, one obtains

$$\mathcal{T}^J_\alpha = \frac{4k_0}{E_0\,(2J+1)\,(2I_0+1)} \sum_{ll_0l_0'\nu_0} \sqrt{(2l_0+1)\,(2l_0'+1)}\, \left\langle l_0 0 I_0 \nu_0 \middle| J\nu_0 \right\rangle$$
$$\times \left\langle J\nu_0 \middle| l_0' 0 I_0 \nu_0 \right\rangle \int dr\, u^{J\,*}_{\alpha l,0l_0}(k_\alpha, r)\, W_{\alpha l}(r)\, u^J_{\alpha l,0l_0'}(k_\alpha, r), \tag{9.179}$$

where

$$W_{\alpha l} = \left(\mathcal{Y}^{\pi JM}_{\alpha l} \,|\, \mathbb{W} \,|\, \mathcal{Y}^{\pi JM}_{\alpha l}\right). \tag{9.180}$$

9.3.2.1 *Angular momentum projected Lippmann-Schwinger equations with spin*

It is convenient to derive Lippmann-Schwinger equations for the radial wave functions, as we did in the case of spinless collision partners. The LS equation involving the exact and the free states can be derived with the help of Eqs. (9.151), (9.165) and (9.127). Inserting these expansions into Eq. (9.46a) and carrying out some algebra, one gets (exercise 5a)

$$u^J_{\alpha l,0l_0}(k_\alpha, r) = \hat{\jmath}_l(k_0 r)\;\delta_{\alpha 0}\,\delta_{ll_0} + \sum_{\alpha' l'} \int dr'\; g^{(+)}_{0,l}(\alpha E; r, r')\, V^J_{\alpha l,\alpha' l'}(r')\; u^J_{\alpha' l',0l_0}(k_{\alpha'}, r'), \tag{9.181}$$

where $V^J_{\alpha l,\alpha' l'} = \left(\mathcal{Y}^{\pi JM}_{\alpha l}\right| \mathbb{V} \left|\mathcal{Y}^{\pi JM}_{\alpha' l'}\right)$ are the matrix-elements of the full interaction.

The derivation of the LS equation involving the exact solution and the distorted wave is similar. In this case we use the expansions of Eqs. (9.152), (9.165) and (9.128) in Eq. (9.46c). The result is (exercise 5b)

$$u^J_{\alpha l,0l_0}(k_\alpha, r) = w_{0l_0}(k_0, r)\;\delta_{\alpha 0}\delta_{ll_0} + \sum_{\alpha' l'} \int dr'\; g^{(+)}_{1,l}(\alpha E; r, r')\, \mathcal{V}^J_{\alpha l,\alpha' l'}(r')\; u^J_{\alpha' l',0l_0}(k_{\alpha'}, r'), \tag{9.182}$$

with $\mathcal{V}^J_{\alpha l,\alpha' l'}(r')$ given by Eq. (9.167)

Now we carry out the angular momentum projection of the T-matrix for systems with spin, which, in this case is related with the scattering amplitude by the equation

$$T_{\alpha\nu\mathbf{k}_\alpha,0\nu_0\mathbf{k}_0} = -\frac{\hbar^2}{4\pi^2\mu}\; f_{\alpha\nu_\alpha,0\nu_0}(\mathbf{k}_\alpha, \mathbf{k}_0). \tag{9.183}$$

For this purpose, we define normalized eigenstates of energy and total angular momentum. They are straightforward generalizations of Eqs. (9.131), (9.132) and (9.133):

$$\phi\,(\alpha ElJM;\zeta) = \sqrt{\frac{2\mu k_\alpha}{\pi\hbar^2}}\;\; \mathcal{Y}^{\pi JM}_{\alpha l}(\zeta)\;\; \frac{\hat{\jmath}_l(k_\alpha r)}{k_\alpha r} \tag{9.184}$$

$$\chi\,(\alpha ElJM;\zeta) = \sqrt{\frac{2\mu k_\alpha}{\pi\hbar^2}}\;\; \mathcal{Y}^{\pi JM}_{\alpha l}(\zeta)\;\; \frac{w_{\alpha l}(k_\alpha, r)}{k_\alpha r} \tag{9.185}$$

and

$$\psi_{\alpha l}(EJM;\zeta) = \sqrt{\frac{2\mu k_\alpha}{\pi\hbar^2}} \sum_{\alpha' l'} \mathcal{Y}^{\pi JM}_{\alpha' l'}(\zeta) \; \frac{u_{\alpha' l';\alpha l}(k_{\alpha'}, r)}{k_\alpha r}. \tag{9.186}$$

Similarly to the case of spinless systems, these wave functions satisfy the orthonormality relations (see exercise 6),

$$\langle \phi\,(\alpha ElJM)\,|\phi\,(\alpha' E' l' J' M')\,\rangle = \delta_{\alpha\alpha'}\,\delta_{ll'}\,\delta_{JJ'}\,\delta_{MM'}\,\delta(E-E'), \tag{9.187}$$

$$\langle \chi\,(\alpha ElJM)\,|\chi\,(\alpha' E' l' J' M')\,\rangle = \delta_{\alpha\alpha'}\,\delta_{ll'}\,\delta_{JJ'}\,\delta_{MM'}\,\delta(E-E') \tag{9.188}$$

and

$$\langle \psi_{\alpha l}\,(EJM)\,|\psi_{\alpha l}\,(E' J' M')\,\rangle = \delta_{JJ'}\,\delta_{MM'}\,\delta(E-E'). \tag{9.189}$$

The angular momentum projected T-matrix elements are defined as

$$\begin{aligned} T^J_{\alpha l,0l_0}(E) &\equiv \langle \phi\,(\alpha ElJM)\,|\,\mathbb{T}\,|\phi\,(0El_0JM)\rangle \\ &= \langle \phi\,(\alpha ElJM)\,|\,\mathbb{V}\,|\psi_{0l_0}\,(EJM)\rangle\,. \end{aligned} \tag{9.190}$$

They can be calculated using the explicit forms of $\phi\,(\alpha ElJM;\zeta)$ and $\psi_{0l_0}\,(EJM;\zeta)$, given respectively in Eqs. (9.184) and (9.186). One gets

$$T^J_{\alpha l,0l_0}(E) = \frac{k_0}{\pi E_0}\sqrt{\frac{k_0}{k_\alpha}} \sum_{\alpha\alpha'} \int dr\,\hat{\jmath}_l(k_\alpha r)\,V^J_{\alpha l,\alpha' l'}(r)\;u^J_{\alpha' l',0l_0}(k_{\alpha'}, r), \tag{9.191}$$

where $V^J_{\alpha l,\alpha' l'} = \big(\mathcal{Y}^{\pi JM}_{\alpha l}\,|\,\mathbb{V}\,|\,\mathcal{Y}^{\pi JM}_{\alpha' l'}\big)$.

Now, we have to relate the T-matrix,

$$T_{\alpha\nu_\alpha\mathbf{k}_\alpha,0\nu_0\mathbf{k}_0} \equiv \langle \Phi(\alpha\nu_\alpha\mathbf{k}_\alpha)|\,\mathbb{V}\,|\Psi^{(+)}(0\nu_0\mathbf{k}_0)\rangle\,,$$

with its angular momentum components. For this purpose, we re-write the angular momentum expansion of $\Phi(\alpha\nu_\alpha\mathbf{k}_\alpha)$ in terms of $\phi(\alpha ElJM)$ and that of $\Psi^{(+)}(0\nu_0\mathbf{k}_0)$ in terms of $\psi_{0l_0}(EJM)$. This can be easily done using Eqs. (9.155) and (9.184), for the former, and (9.164) and (9.186) for the latter. Carrying out these procedures, we get,

$$\begin{aligned} T_{\alpha\nu\mathbf{k}_\alpha,0\nu_0\mathbf{k}_0} = \frac{\hbar^2}{\mu\sqrt{k_\alpha k_0}} &\sum_{Jl_0m_0l} Y_{l(m_0+\nu_0-\nu)}(\hat{\mathbf{k}}_0)\;Y^*_{l_0m_0}(\hat{\mathbf{k}}_\alpha) \\ &\times \langle J(m_0+\nu_0)|l(m_0+\nu_0-\nu)I\nu\rangle \\ &\times \langle l_0m_0I_0\nu_0|J(m_0+\nu_0)\rangle\;T^J_{\alpha l,0l_0}(E). \end{aligned} \tag{9.192}$$

One can also derive the angular momentum projected two-potential formula for systems with spin. Considering the simplest situation where $\mathbb{V}^{(1)}$ is diagonal in channel space and conserves orbital angular momentum, one gets

$$T^J_{\alpha l,0l_0}(E) = \delta_{\alpha 0}\,\delta_{ll_0}\; T^{(1)J}_{0l_0,0l_0}(E) \;+\; T^{(2)J}_{\alpha l,0l_0}(E). \tag{9.193}$$

Above the partial T-matrices are,

$$T^{(1)J}_{0l_0,0l_0}(E) = \frac{k_0}{\pi E_0}\int dr\,\hat{\jmath}_{l_0}(k_0 r)\,V^J_{0l_0}(r)\;w^J_{0l_0}(k_0,r) \tag{9.194}$$

and

$$T^J_{\alpha l,0l_0}(E) = \frac{k_0}{\pi E_0}\sqrt{\frac{k_0}{k_\alpha}}\sum_{\alpha' l'}\int dr\,w_{\alpha l}(k_\alpha,r)\,\mathcal{V}^J_{\alpha l,\alpha' l'}(r)\;u^J_{\alpha' l',0l_0}(k_{\alpha'},r), \tag{9.195}$$

with the usual notation,

$$V^J_{\alpha l} = \left(\mathcal{Y}^{\pi JM}_{\alpha l}\,|\,\mathbb{V}^{(1)}\,|\,\mathcal{Y}^{\pi JM}_{\alpha l}\right), \quad \mathcal{V}^J_{\alpha l,\alpha' l'} = \left(\mathcal{Y}^{\pi JM}_{\alpha l}\,|\,\mathbb{V}^{(2)}\,|\,\mathcal{Y}^{\pi JM}_{\alpha' l'}\right).$$

The final point to address in this sub-section is the relation between the angular momentum projected S- and T-matrices. Proceeding as in chapter 4, one finds the straightforward generalization of the case of spinless systems,

$$S^J_{\alpha l,0l_0}(E) = \delta_{\alpha 0}\,\delta_{ll_0} \;-\; 2\pi i\; T^J_{\alpha l,0l_0}(E). \tag{9.196}$$

9.3.3 *Evaluation of matrix-elements*

In this section we consider the calculation of the matrix-elements of the channel-coupling interaction. For this purpose, we follow Satchler in [Satchler (1983)]. We restrict our discussion to potentials of the type $\mathbb{V}^{(2)}(r,\hat{\mathbf{r}},\xi)$, which do not depend on momenta. In this case, the angular dependence of the interaction can be expanded in the spherical harmonics basis, or, equivalently, in its hermitian conjugate, as[23]

$$\mathbb{V}^{(2)}(\xi,r,\hat{\mathbf{r}}) = \sum_{\lambda\mu}\left[i^\lambda\,Y_{\lambda\mu}(\hat{\mathbf{r}})\right]^*\,V_{\lambda\mu}(\xi,r). \tag{9.197}$$

[23]We recall the notation of Eq. (9.5), according to which $\mathbb{V}^{(2)}(r,\hat{\mathbf{r}},\xi)$ stands for the part of the projectile-target interaction, $\mathbb{V}(r,\hat{\mathbf{r}},\xi)$, that has off-diagonal matrix-elements in channel space.

The phase[24] i^λ is used to give convenient time-reversal properties, since the channel-angle basis is defined with a similar factor. This expansion is very convenient since the coupling is frequently dominated by a single term or by a few terms. An important feature of the functions $V_{\lambda\mu}(\xi, r)$ is that they transform under rotation as the components of a spherical tensor of rank λ. Since the potential must be a scalar under rotations, the RHS of Eq. (9.197) must have the form of a scalar product of two spherical tensors. In this way, $V_{\lambda\mu}$ should transform contragrediently to $Y^*_{\lambda\mu}(\hat{\mathbf{r}})$. That is, it should transform as $Y_{\lambda\mu}(\hat{\mathbf{r}})$. The components $V_{\lambda\mu}$ are given by

$$V_{\lambda\mu}(\xi, r) = i^\lambda \int d\hat{\mathbf{r}}\; V(\xi, r, \hat{\mathbf{r}})\, Y_{\lambda\mu}(\hat{\mathbf{r}}). \tag{9.198}$$

The potential of Eq. (9.197) corresponds to a sum of scalar products of tensor operators acting on commuting spaces, so that we can write

$$\mathbb{V}^{(2)}(\xi, r, \hat{\mathbf{r}}) = \sum_\lambda i^{-\lambda}\, (\mathbf{Y}_\lambda \cdot \mathbf{V}_\lambda)\,, \tag{9.199}$$

where $\mathbf{Y}_{(\lambda)}$ ($\mathbf{V}_{(\lambda)}$) represents the tensor operator of rank λ with components $Y_{\lambda\mu}$ ($V_{\lambda\mu}$).

The matrix elements of $\mathbb{V}^{(2)}$ in the $\left|\mathcal{Y}^{\pi JM}_{\alpha l}\right)$ basis can be written as

$$\begin{aligned}
\mathcal{V}^J_{\alpha' l', \alpha l}(r) &= \sum_\lambda i^{-\lambda}\, \left(\mathcal{Y}^{\pi JM}_{\alpha' l'}\right| (\mathbf{Y}_\lambda \cdot \mathbf{V}_\lambda) \left|\mathcal{Y}^{\pi JM}_{\alpha l}\right) \\
&= \sum_\lambda i^{l - l' - \lambda}\, (\alpha' l' I_{\alpha'} JM|\, (\mathbf{Y}_\lambda \cdot \mathbf{V}_\lambda)\, |\alpha l I_\alpha JM)\,,
\end{aligned} \tag{9.200}$$

where

$$|\alpha l I_\alpha JM) = i^{-l} \left|\mathcal{Y}^{\pi JM}_{\alpha l}\right)\,. \tag{9.201}$$

The matrix-elements $(\alpha' l' I_{\alpha'} JM|\, (\mathbf{Y}_\lambda \cdot \mathbf{V}_\lambda)\, |\alpha l I_\alpha JM)$, which are independent of M, can be evaluated with the help of the Wigner-Eckart's theorem and properties of Clebsch-Gordan coefficients, using the fact that $\mathbf{Y}_\lambda$ and $\mathbf{V}_\lambda$ are operators acting on different subspaces. They are given as (see Eq. (7.1.6) of [Edmonds (1974)], p. 111)

$$\begin{aligned}
&(\alpha' l' I_{\alpha'} JM|\, (\mathbf{Y}_\lambda \cdot \mathbf{V}_\lambda)\, |\alpha l I_\alpha JM) \\
&\quad = (-)^{l + I_{\alpha'} + J} \begin{Bmatrix} J & I_{\alpha'} & l' \\ \lambda & l & I_\alpha \end{Bmatrix} (l' \|\, \mathbf{Y}_{(\lambda)}\, \| l)\; (\alpha' I_{\alpha'} \|\, \mathbf{V}_{(\lambda)}\, \| \alpha I_\alpha)\,,
\end{aligned} \tag{9.202}$$

[24] This phase, or any multiplicative factor, can be introduced arbitrarily. It corresponds to redefining the coefficients $V_{\lambda\mu}$.

where $(l' \| \mathbf{Y}_{(\lambda)} \| l)$ and $(\alpha' I_{\alpha'} \| \mathbf{V}_{(\lambda)} \| \alpha I_\alpha)$ are the reduced matrix-elements[25] of the tensor operators $\mathbf{Y}_{(\lambda)}$ and $\mathbf{V}_{(\lambda)}$, appearing in Wigner-Eckart's theorem, and $\begin{Bmatrix} J & I_{\alpha'} & l' \\ \lambda & l & I_\alpha \end{Bmatrix}$ are the 6-J coefficients. The reduced matrix-elements of $\mathbf{Y}_{(\lambda)}$ can easily be evaluated and one gets (see [Edmonds (1974)], Eq. (5.4.5))

$$(l' \| \mathbf{Y}_{(\lambda)} \| l) = (-)^{l'} \frac{\hat{l'}\,\hat{l}\,\hat{\lambda}}{\sqrt{4\pi}} \begin{pmatrix} l' & \lambda & l \\ 0 & 0 & 0 \end{pmatrix}, \tag{9.203}$$

where $\hat{k} = \sqrt{2k+1}$ and $\begin{pmatrix} l' & \lambda & l \\ 0 & 0 & 0 \end{pmatrix}$ are the usual 3-J coefficients. Using Eqs. (9.202) and (9.203) in Eq. (9.200), we obtain

$$\mathcal{V}^J_{\alpha' l', \alpha l}(r) = \sum_\lambda (-)^{J + I_{\alpha'} + \frac{1}{2}(l' - l - \lambda)} \frac{\hat{l'}\,\hat{l}\,\hat{\lambda}}{\sqrt{4\pi}} \times \begin{pmatrix} l' & \lambda & l \\ 0 & 0 & 0 \end{pmatrix} \begin{Bmatrix} J & I_{\alpha'} & l' \\ \lambda & l & I_\alpha \end{Bmatrix} (\alpha' I_{\alpha'} \| \mathbf{V}_{(\lambda)} \| \alpha I_\alpha) . \tag{9.204}$$

Note that the reduced matrix-elements $(\alpha' I_{\alpha'} \| \mathbf{V}_{(\lambda)} \| \alpha I_\alpha)$ are functions of r, as will be clear in the example considered below.

Let us now discuss, as an example, the Coulomb excitation of a composite target particle with total charge Q_T, induced by the dipole interaction with a point-like projectile with charge Q_P. The target contains N elementary particles with charges $q_1, q_2, ..., q_N$ and position vectors (with respect to the target's center of mass) $\mathbf{r}'_1, \mathbf{r}'_2, ..., \mathbf{r}'_N$, as illustrated in figure 9.1. In

[25] Note that different definitions of the reduced elements are used in the literature. It depends on the particular formulation of the Wigner-Eckart's theorem. Brink and Satchler [Brink and Satchler (1971)] adopts the form

$$\langle \alpha' j' m' | T_{\lambda\mu} | \alpha j m \rangle = (-)^{2\lambda} (\alpha' j' \| \mathbf{T}_{(\lambda)} \| \alpha j)_{BS} \langle j m \lambda \mu | j' m' \rangle,$$

while Edmonds uses

$$\langle \alpha' j' m' | T_{\lambda\mu} | \alpha j m \rangle = (-)^{j' - m'} (\alpha' j' \| \mathbf{T}_{(\lambda)} \| \alpha j)_E \begin{pmatrix} j' & \lambda & j \\ -m' & \mu & m \end{pmatrix}$$
$$= \frac{(-)^{\lambda - j + j'}}{\sqrt{2j'+1}} (\alpha' j' \| \mathbf{T}_{(\lambda)} \| \alpha j)_E \langle \lambda \mu j m | j' m' \rangle .$$

It can be easily shown that

$$(\alpha' j' \| \mathbf{T}_{(\lambda)} \| \alpha j)_{BS} = \frac{(\alpha' j' \| \mathbf{T}_{(\lambda)} \| \alpha j)_E}{\sqrt{2j'+1}} .$$

In the present text, we use the definition of [Edmonds (1974)].

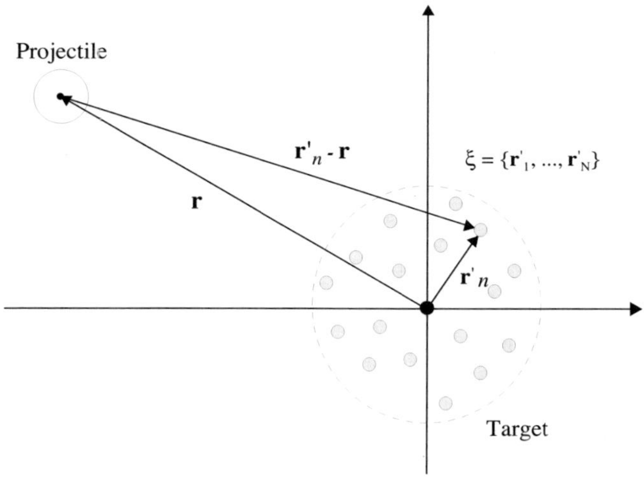

Fig. 9.1 Schematic representation of the collision of a point-like projectile with a composite target.

this case, the channel coupling potential is

$$\mathbb{V}^{(2)}(\mathbf{r},\xi) = \frac{1}{4\pi\epsilon_0}\sum_{n=1}^{N}\frac{\mathrm{Q_P}\,q_n}{|\mathbf{r}'_n - \mathbf{r}|} - \frac{1}{4\pi\epsilon_0}\frac{\mathrm{Q_P}\,\mathrm{Q_T}}{r}. \tag{9.205}$$

Above, $\xi \equiv \{\mathbf{r}'_1, \mathbf{r}'_2, ..., \mathbf{r}'_N\}$ stands for the intrinsic coordinates, $\mathrm{Q_T} = \sum_{n=1}^{N} q_n$ and $|\mathbf{r}'_n - \mathbf{r}|$ is the distance between the nth particle in the target and the projectile. Since the monopole-monopole Coulomb interaction, $\mathrm{Q_P}\,\mathrm{Q_T}\,/\,4\pi\epsilon_0\,r$, is included in the diagonal part of the interaction, $\mathbb{V}^{(1)}$, it is subtracted from $\mathbb{V}^{(2)}$.

Performing the multipole expansion of the Coulomb interaction[26], we get

$$\mathbb{V}^{(2)}(\mathbf{r},\xi) = \frac{\mathrm{Q_P}}{4\pi\epsilon_0}\sum_{\lambda=1}^{\infty}\frac{1}{r^{\lambda+1}}\left[\frac{4\pi}{2\lambda+1}\right]\sum_{\mu=-\lambda}^{\lambda} Y^*_{\lambda\mu}(\hat{\mathbf{r}})\,Q_{\lambda\mu}, \tag{9.206}$$

where $Q_{\lambda\mu}$ are the intrinsic multipole operators

$$Q_{\lambda\mu} = \sum_{n=1}^{N} q_n\,(r'_n)^{\lambda}\,Y_{\lambda\mu}(\hat{\mathbf{r}}'_n). \tag{9.207}$$

[26]See, e.g., [Ring and Schuck (1980)], appendix B.

Eq. (9.206) can be cast in the general form of Eq. (9.197), with

$$V_{\lambda\mu} = \frac{1}{4\pi\epsilon_0}\left[4\pi\, i^{\lambda}\, \frac{\mathsf{Q}_{\mathrm{P}}}{2\lambda+1}\right]\frac{1}{r^{\lambda+1}}\, Q_{\lambda\mu}. \tag{9.208}$$

In the present section, we consider the simple case where the interaction is dominated by the dipole term, so that one can neglect terms[27] with $\lambda > 1$ in Eq. (9.206). As an illustration, we evaluate matrix-elements between two states: one with quantum numbers $\{\alpha = 0, I_0 = 0, \pi_0 = +\}$ and the other with quantum numbers $\{\alpha = 1, I = 1, \pi = -\}$. It could correspond, for example, to the ground state and the Giant Dipole Resonance in an even-even nucleus (even numbers of protons and neutrons). Using Eqs. (9.208) and (9.204), it can be readily checked that diagonal matrix-elements of the dipole interaction vanish[28] and the matrix-elements between the above mentioned two states are

$$\mathcal{V}^{J}_{1l,0l_0}(r) = \frac{1}{4\pi\epsilon_0}\,\mathsf{Q}_{\mathrm{P}}\,\frac{(1\|\,Q_1\,\|0)}{r^2}\,(-)^{J+1}\, i^{l-l_0}\,\hat{l}\,\hat{l}_0\,\sqrt{\frac{4\pi}{3}}$$
$$\times\begin{pmatrix} l & 1 & l_0 \\ 0 & 0 & 0 \end{pmatrix}\begin{Bmatrix} J & 1 & l \\ 1 & l_0 & 0 \end{Bmatrix}. \tag{9.209}$$

Since the ground state of the target has spin zero, the total angular momentum in the entrance channel is equal to the orbital angular momentum, l_0. Thus, owing to the conservation of the total angular momentum, one can replace in the above equation $J = l_0$.

For hermitian coupling interactions, as that of the present example, the matrix-elements for the inverse transition can be obtained from the above equation through the relation,

$$\mathcal{V}^{J}_{0l_0,1l}(r) = \left[\mathcal{V}^{J}_{1l,0l_0}(r)\right]^{*}.$$

Usually, coupling matrix-elements are expressed in terms of reduced transition probabilities, rather than reduced matrix-element. For this purpose, one uses the relation (see e.g. [Ring and Schuck (1980)]),

$$|(I_f\|\,Q_\lambda\,\|I_i)|^2 = (2I_i+1)\; B(E\lambda, I_i \to I_f). \tag{9.210}$$

In the present case, we can replace in Eq. (9.209)

$$(1\|\,Q_1\,\|0) = \sqrt{B(E1, 0\to 1)}.$$

---

[27] For a comprehensive discussion of multipoles of all orders and realistic intrinsic states, we refer to [Tamura (1965)].

[28] This is true for the diagonal matrix-elements of any odd electric multipole operator. It arises from the 3J coefficient in Eq. (9.204).

Note that Eq. (9.210) does not specify the phase of the reduced matrix-elements. This is not a serious problem. The phase of $(I_f \| Q_\lambda \| I_i)$ is arbitrary because it depends on the particular choice of phase for the intrinsic states. However, once these phases are chosen, the reduced matrix-elements $(I_f \| Q_\lambda \| I_i)$ are fully determined. It can be easily shown that (see e.g. [Edmonds (1974)])

$$(I_i \| Q_\lambda \| I_f) = (-)^{I_i - I_f} \, (I_f \| Q_\lambda \| I_i)^* \,. \tag{9.211}$$

To close this section we briefly discuss the set of states involved in a simple two-channel Coupled-Channel problem. We consider the close coupling approximation in the excitation due to the electric dipole interaction. In this case, the calculation only involves the ground state channel and the dipole resonance. The coupled radial equations for a given $\{J, \pi\}$ pair involve no more than three states. Since the entrance channel has spin zero, only $l_0 = J$ is allowed. On the other hand, total angular momentum and parity conservation alows in the excited channel $l = J + 1$ and $l = J - 1$. Owing to the negative parity of the excited state, the partial-wave $l = J$ violates parity conservation. In this way, the coupled-channel equations relate the radial wave funtions $u^J_{0J,0J}(r)$, $u^J_{1\,J+1,0J}(r)$ and $u^J_{1\,J-1,0J}(r)$. The only exceptions are the equations for $J = 0$. In the case of $\{J = 0, \pi = +\}$, only the wave functions $u^0_{00,00}(r)$ and $u^0_{11,00}(r)$ are involved. For $\{J = 0, \pi = -\}$ the coupled-channel equations are meaningless, since there is no incident wave.

9.4 Coupling with continuum states

In some situations intrinsic states in the continuum play an important role in the collision. These states correspond to the dissociation of a nucleus or a molecule or to the ionization of an atom. Examples of such processes can be found in ion-atom, ion-molecule and nucleus-nucleus collisions. Channels with one unbound electron are very important in the calculation of electron loss in the projectile and stopping power for intermediate-energy light ions penetrating H and He targets [Schiwietz (1990); Grande *et al.* (1996)]. In nuclear physics, these states corresponds to the breakup channel. This channel is very important in collisions of weakly bound projectiles and they have a strong influence on the cross sections for several processes, e.g. elastic scattering, breakup and fusion (for a review see [Canto *et al.* (2006)]).

When the coupling to the unbound channels is very strong, perturbative approximations cannot be used and one has to resort to the coupled channel method. However, in such cases the application of this method is very complicated because the channel label becomes a continuous variable. The number of coupled equations is infinite, even if the channel space is truncated at some maximal excitation energy. It is then necessary to introduce other approximations. In the present section we discuss an extension of the coupled-channel method to include channels in the continuum. This approximation is known as the *Continuum Discretized Coupled Channel* method (CCDC). The CDCC method was introduced [Sakuragi *et al.* (1983, 1986); Austern *et al.* (1987)] to describe stripping reactions in collisions of ^{2}H projectiles. More recently, it was extended to the description of other reaction channels and other nuclear reactions involving weakly bound projectiles (see, e.g., [Diaz-Torres and Thompson (2002); Nunes and Thompson (1999); Tostevin *et al.* (2001)]). We give below a description of the CDCC approximation. For simplicity, we consider in this section the simpler case of two spinless collision partners. For a detailed description including spin and the detailed evaluation of breakup cross sections, we refer to [Tostevin *et al.* (2001)].

Let us consider the situation where the projectile-target interaction leads to the dissociation of one of the collision partners, say the target[29]. We assume further that the target dissociates into only two fragments. Applications of the CDCC method when the dissociation produces three or more fragments are extremely complicated from the computational point of view and there are only a few applications in the literature (some examples are [Matsumoto *et al.* (2004); Rodrígues-Gallardo *et al.* (2008, 2009); Pires *et al.* (2011)]). The coordinates of the system are represented in figure 9.2. In the present example we neglect other intrinsic degrees of freedom. However, their inclusion leads only to a larger number of coupled channel. In this way, there would be no significant complication except for the computational effort required for the solution of the coupled equations.

We expand the full scattering wave function in the set of bound and unbound eigenstates of the intrinsic Hamiltonian of the target, h. In principle, one should use the rotational invariance of h, to take simultaneous eigenstates of operators $\{h, I^2, I_z\}$, where I^2 and I_z are angular momentum operators associated with the relative motion of the two fragments

[29]Projectile dissociation, which is the usual situation in nuclear physics, can be handled similary.

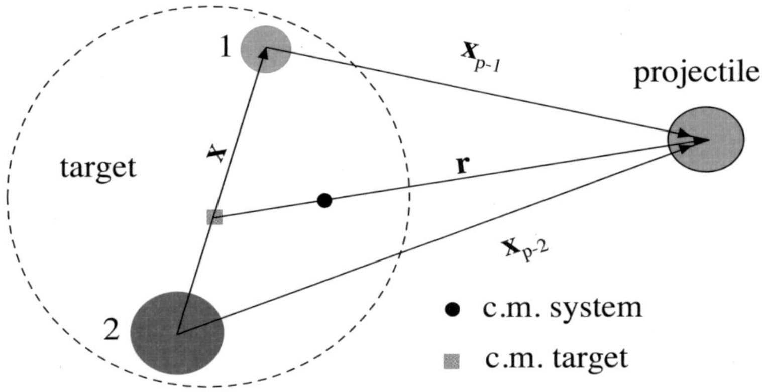

Fig. 9.2 Coordinates describing the collision of a structureless projectile with a two-particle target.

inside the target. However, to keep our example as simple as possible[30], we consider only eigenstates of the target with $I = 0$. The spectrum of the operator h has bound states and states in the continuum. These states satisfy respectively the equations,

$$h\ \varphi_n(x) = \varepsilon_n\ \varphi_n(x), \tag{9.212}$$

$$h\ \varphi_\varepsilon(x) = \varepsilon\ \varphi_\varepsilon(x), \tag{9.213}$$

where x is the distance between fragments 1 and 2. The scattering wave function can then be written as,

$$\Psi^{(+)}(\mathbf{k};\mathbf{r},x) = \sum_{nlm} Y_{lm}(\hat{\mathbf{r}})\,\frac{u_{nl,0l}(k_n,r)}{k_0 r}\,\varphi_n(x) + \sum_{lm} Y_{lm}(\hat{\mathbf{r}}) \int d\varepsilon\,\frac{u_{\varepsilon l,0l}(k_\varepsilon,r)}{k_0 r}\,\varphi_\varepsilon(x). \tag{9.214}$$

The above wave function should satisfy the Schrödinger equation

$$\mathbb{H}\,\Psi^{(+)}(\mathbf{k};\mathbf{r},x) = E\,\Psi^{(+)}(\mathbf{k};\mathbf{r},x), \tag{9.215}$$

where $\mathbb{H}$ is the full Hamiltonian of the system,

$$\mathbb{H}(\mathbf{r},\mathbf{x}) = K(\mathbf{p_r}) + h(\mathbf{x},\mathbf{p_x}) + \mathbb{V}(\mathbf{x},\mathbf{r}). \tag{9.216}$$

[30]This simplification could be easily removed and the only price to pay would be the inclusion of some Clebsch-Gordan coefficients.

Above, K is the kinetic energy operator associated with the projectile-target relative motion and $\mathbb{V}$ is the interaction between the projectile and the two fragments of the target. It is given by

$$\mathbb{V} = V_{\rm p-1}(\mathbf{x}_{\rm p-1}) + V_{\rm p-2}(\mathbf{x}_{\rm p-2}), \tag{9.217}$$

where the vectors $\mathbf{x}_{\rm p-1}$ and $\mathbf{x}_{\rm p-1}$ are represented in figure 9.2. These vectors can be expressed in terms of $\mathbf{r}$ and $\mathbf{x}$ through the transformations

$$\mathbf{x}_{\rm p-1} = \mathbf{r} - \frac{m_2}{m_{\rm T}}\,\mathbf{x} \quad \text{and} \quad \mathbf{x}_{\rm p-2} = \mathbf{r} + \frac{m_1}{m_{\rm T}}\,\mathbf{x}. \tag{9.218}$$

Above, $m_{\rm T}$ is the target mass and m_1 and m_2 are the masses of the two fragments. The intrinsic Hamiltonian, h, is the sum of the kinetic energy of the relative motion of the two fragments and their interaction potential, $V_{1-2}(x)$. We now proceed similarly to section 9.3.2. We insert the expansion of Eq. (9.214) into the Schrödinger equation and take scalar product from the left with each of the eigenstates of h. We then get the set of coupled equations

$$\left[E_n \;+\; \frac{\hbar^2}{2\mu}\left(\frac{d^2}{dr^2} - \frac{l\,(l+1)}{r^2}\right) - V_0(r)\right] u_{nl,0l}(k_n, r)$$
$$= \sum_{n'} \mathcal{V}_{nl,n'l}(r)\, u_{n'l,0l}(k_{n'}, r) \;+\; \int d\varepsilon\; \mathcal{V}_{nl,\varepsilon l}(r)\, u_{\varepsilon l,0l}(k_\varepsilon, r) \tag{9.219}$$

and

$$\left[E_\varepsilon \;+\; \frac{\hbar^2}{2\mu}\left(\frac{d^2}{dr^2} - \frac{l\,(l+1)}{r^2}\right) - V_0(r)\right] u_{\varepsilon l,0l}(k_\varepsilon, r)$$
$$= \sum_{n} \mathcal{V}_{\varepsilon l,nl}(r)\, u_{nl,0l}(k_n, r) \;+\; \int d\varepsilon'\; \mathcal{V}_{\varepsilon l,\varepsilon' l}(r)\, u_{\varepsilon' l,0l}(k_{\varepsilon'}, r). \tag{9.220}$$

We have used the following notation:

$$E_n = E - \varepsilon_n, \tag{9.221}$$
$$E_\varepsilon = E - \varepsilon, \tag{9.222}$$
$$V_0(r) = \langle Y_{lm} \mid \mathbb{V}(\mathbf{r}, 0) \mid Y_{lm}\rangle\,, \tag{9.223}$$
$$\mathcal{V}_{nl,\varepsilon l}(r) = \langle Y_{lm}\,\varphi_n \mid \mathbb{V}(\mathbf{r}, \mathbf{x}) \mid Y_{lm}\,\varphi_\varepsilon\rangle \;-\; V_0(r) \tag{9.224}$$

and an analogous notation for the remaining matrix-elements.

Eqs. (9.219) and (9.220) represent an infinite set of coupled differential equations, which cannot be solved. Truncating the set of bound channels, as in the previous sections, we get a finite set of equations in (9.219)

($n = 0, ..., N_B - 1$). However, the equations for the unbound channels, represented by (9.220), remains an infinite set. Since the index ε is continuous, we remain with infinite equations if we truncate the excitation energy at some maximal value, ε_{max}. It is then necessary to resort to additional approximations. The CDCC method consists of truncating the continuum at some maximal energy ε_{max} and then replace the continuum states in the $\varepsilon_{max} > \varepsilon > 0$ range by the set of wave packets

$$\phi_\beta(x) = \int d\varepsilon \; \Gamma(\varepsilon - \varepsilon_\beta) \, \varphi_\varepsilon(x), \quad \beta = 1, ..., N_C. \tag{9.225}$$

The packets in the energy space, $\Gamma(\varepsilon - \varepsilon_\beta)$, are centered at ε_β and have a width Δ_β. It is convenient to normalize them so that they satisfy the orthonormality relations

$$\int d\varepsilon \; \Gamma(\varepsilon - \varepsilon_\beta) \, \Gamma(\varepsilon - \varepsilon_{\beta'}) = \delta_{\beta\beta'}. \tag{9.226}$$

As in the CDCC the continuum of the dissociated collision partner is represented by a finite set of N_C channels, unbound channels are treated in the same way as bound channels. In this way, one obtains a set of $N = N_B + N_C$ coupled equations, analogous to the ones discussed in previous sections.

We point out that the wave functions $\phi_\beta(x)$ are not exact eigenstates of the intrinsic Hamiltonian, since they are a superpositions of eigenstates with different eigenvalues. On the other hand the wave functions $\phi_\beta(x)$ are localized, whereas the eigenstates $\varphi_\varepsilon(x)$ are not. However, these wave functions remain with appreciable values, even at very large distance from the center of the target. Owing to the Uncertainty Principle, the range of $\phi_\beta(x)$ decreases as the width Δ in the energy (momentum) space increases. This is illustrated in the example of figure 9.3, which shows the continuum discretized wave function of [Bertulani and Canto (1992)]. It is based on a simplified model for the breakup of ^{11}Li, which describes this nucleus as constituted by a ^{9}Li core plus a *di-neutron.* The wave functions then represent the motion of the *di-neutron* with respect to ^{9}Li. The discretization procedure used an equally spaced mesh in the energy space, with width Δ. The packets are centered at the energies $\varepsilon_\beta = (\beta + 1/2)\, \Delta$. The figure show the wave functions $\phi_\beta(x)$ for $\varepsilon_\beta = 200$ keV and three different values of the width. The results were obtained with the simplest possible generator function, which is like an histogram, as discussed in [Bertulani and Canto (1992)]. That is,

$$\Gamma(\varepsilon - \varepsilon_\beta) = \frac{1}{\sqrt{\Delta}}, \qquad \text{for } \; \varepsilon_\beta + \Delta/2 \geq \varepsilon \geq \varepsilon_\beta - \Delta/2 \tag{9.227}$$

$$= 0, \qquad \text{for } \varepsilon < \varepsilon_\beta - \Delta/2 \;\text{ or }\; \varepsilon > \varepsilon_\beta + \Delta/2. \tag{9.228}$$

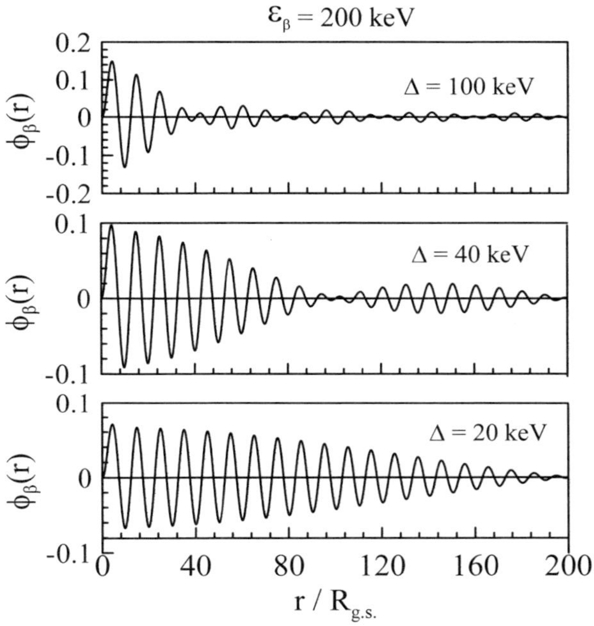

Fig. 9.3 Continuum discretized intrinsic wave function generated by a wave packet $\Gamma(\varepsilon - \varepsilon_\beta)$, centered at $\varepsilon_\beta = 200$ keV. The figure shows results for three different widths. The details are given in the text.

We consider the wave functions for the widths $\Delta = 100, 40$ and 20 keV. The wave functions are shown as functions of the distance between the *di-neutron* and the ^{9}Li core, normalized with respect to the average distance for the ground state of ^{11}Li, which is [Shulgina *et al.* (2009)] $R_{\text{g.s.}} \sim 5.0$ fm. In all cases the range of $\phi_\beta(x)$ is over one order of magnitude larger than that for the ^{11}Li ground state. As expected, the range grows as Δ decreases. We remark that the beats appearing in the case of $\Delta = 20$ keV arises from the sharp edges of the adopted shape of Γ. It can be eliminated through the use of a smooth functions [Bertulani and Canto (1992)].

Several implementations of the CDCC method can be found in the literature. In most cases, they consider the scattering of weakly bound nuclei. In such cases, the CDCC method is used to describe the breakup of the weakly bound collision partner and the influence of the breakup channel on various processes. In collisions of typical tightly bound nuclei at near-barrier energies, the breakup channel does not play a relevant role and the continuum

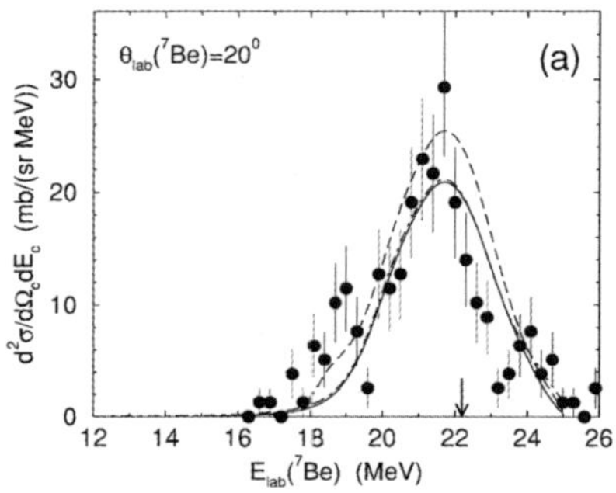

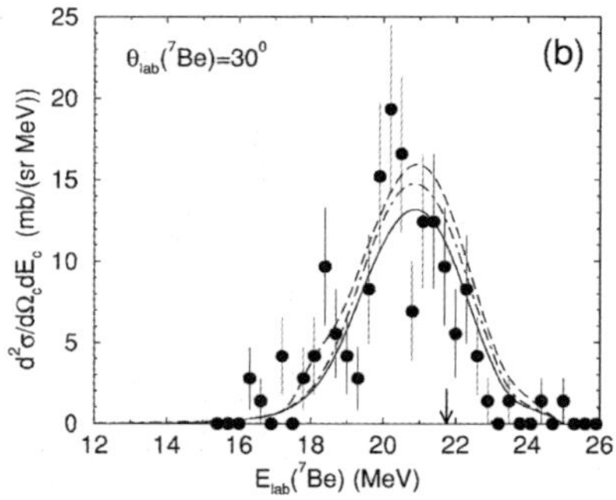

Fig. 9.4 Energy distribution of the ^{7}Be fragment produced in the ^{8}B + ^{58}Ni collision, at two angles in the laboratory frame. The calculations and the data are respectively from [Tostevin *et al.* (2001)] and [Kolata *et al.* (2001)].

can be ignored. Several discretization procedures have been adopted. Frequently, one uses an equally spaced mesh in momentum space or an equally spaced mesh in the energy space (quadratic in momentum space). However, in some situations one uses wave packets of variable width. This is particularly important in the case of resonant breakup, where a finer mesh around the resonance energy is needed.

Some CDCC calculations lead to accurate descriptions of the experimental data. We illustrate this point with the calculations of Tostevin, Nunes and Thompson [Tostevin *et al.* (2001)] of the breakup cross section of ^{8}B in collisions with a ^{58}Ni target. The Coulomb and the nuclear fields of the target lead to the dissociation of the projectile into two fragments: ^{7}Be and a proton. The figure shows experimental energy distributions of the ^{8}B fragment at two laboratory angles, 20 and 30 degrees. The theoretical predictions of CDCC calculations are represented by the solid and the dashed lines. The difference between the two theoretical results arise from different choices of the interaction (see [Tostevin *et al.* (2001)]). It is clear that the agreement between theory and experiment is very good.

A variant of the CDCC method has been used to describe the ionization of hydrogen, in collisions with electrons [Bray and Stelbovics (1992); Bray (2002)]. Other authors used the same discretization procedures in semiclassical approximations. The projectile-target motion is treated by classical mechanics whereas the intrinsic degrees of freedom are handled as a time-dependent quantum mechanics problem. This method will be discussed in detail in chapter 11. The applications are both in the fields of atomic and nuclear physics. In Atomic Physics, they describe the ionization process in ion-atom collisions [Schiwietz (1990)] whereas in Nuclear Physics they are used to study the breakup of stable and unstable weakly bound nuclei [Bertulani and Canto (1992); Marta *et al.* (2002, 2008)].

9.5 Solution of the coupled channel equations

Let us consider a coupled channel problem within the close-coupling approximation, in a truncated space with dimension $N_{\rm ch}$. This space is spanned by the channels labelled by $\alpha = 0, ..., N_{\rm ch} - 1$, which have spins I_α and intrinsic parities π_α. As in the previous sections, the entrance channel is represented by $\alpha = 0$ and the quantum numbers associated with it are characterized by the subscript 0 (I_0, l_0). To simplify the notation, the orbital angular momentum associated with a channel-α will be represented by[31] l. The discussion below applies to traditional CC problems, involving only bound channels, and also to CC problems with the continuum discretization described in the previous section.

As we pointed out before, the Hamiltonian is not a diagonal matrix in the $\mathcal{Y}_{\alpha l}^{\pi;JM}$ representation. However, the non-zero matrix-elements are grouped in blocks around the diagonal. Each one corresponds to an invariant subspace and is characterized by the constants of motion $\{J, \pi\}$. That is, one gets a set of independent coupled-channel equations for each of the combinations: $\{J, \pi\} = \{0, +\}, \{0, -\}, \{1, +\}, \{1, -\}, ..., \{J_{\max}, -\}, \{J_{\max}, +\}$. We consider in this section the most general situation, where the total potential is the sum of a short-range interaction, with range $\bar{R}$, plus a long range Coulomb term. The cut-off angular momentum $J_{\max}$ must be large enough for the dependence of the cross sections on its particular value to be negligible. Typically, for interactions[32] with a range $\bar{R}$, it should be of the order of $l_{\max} + I_{\max}$, with $l_{\max} = k_0\bar{R}$ and $I_{\max}$ the largest spin among all the channels involved.

The first step to solve the coupled equations for $\{J, \pi\}$ is to identify the radial wave functions, $u_{\alpha l, 0l_0}^{J}$, relevant to the calculation[33]. The quantum

[31] One should have in mind that, while the channel spin I_α is determined by the channel label α, the orbital angular momentum in channel-α, l, is a variable which can take any integer value from 0 to ∞.
[32] The long range of the Coulomb field does not matter, since Coulomb wave functions are used for $r > \bar{R}$.
[33] A detailed discussion of this problem is presented in [Satchler (1983)].

numbers l, I_α and π_α, should be consistent with the conserved parity and angular momentum values, $\{J, \pi\}$. In this way, l and I_α must satisfy two conditions. The first, which results from total angular momentum conservation, is

$$l + I_\alpha \geq J \geq |l - I_\alpha| \,. \tag{9.229}$$

The second, which expresses parity conservation, is

$$\pi_\alpha (-)^l = \pi. \tag{9.230}$$

The above condition establishes that only even partial-waves are allowed in channels with $\pi_\alpha = \pi$ and only odd waves are allowed when $\pi_\alpha = -\pi$. The set of coupled equations for $\{J, \pi\}$ involves the radial wave functions $u^J_{\alpha l, 0 l_0}(r)$, with $\alpha = 0, ..., N_{\mathrm{ch}} - 1$ and $l_0, l = 0, .., l_{\max}$. Of course, when the quantum numbers of the exit channel, αl, violate Eq. (9.229) or Eq. (9.230), the coupling matrix elements for this channel vanish and it can be excluded from the calculation. There are many methods to solve the coupled-channel equations [Buck *et al.* (1963); Melkanoff *et al.* (1966); Rhoades-Brown *et al.* (1980); Thompson (1988); Breitschaft *et al.* (1995); Raynal (1996)]. We review below the method usually called *the traditional method* [Buck *et al.* (1963)].

Let us consider the situation where the coupled-equations for $\{J, \pi\}$ involve radial wave functions labeled by $i = 1, 2, ..., N$. For simplicity of notation, we use a single index (i) to represent the quantum numbers α and l. We assume that the lowest i-values, $i = 1, 2, ..., M$, represent the M radial wave functions in the ground state ($\alpha = 0$). They differ only by the value of the orbital angular momentum, l. A generalization of the numerical method used in chapter 2 to our case of coupled equations encounters an important difficulty. We do not know the wave functions and their derivatives at $r = 0$, which would be necessary for numerical integration starting from this point. We recall that in the case of potential scattering we had the initial conditions

$$u_l(0) = 0, \qquad \frac{du_l(0)}{dr} = c,$$

where c is an arbitrary constant which is irrelevant for the phase shifts or the S-matrix elements. Here the situation is quite different. Although all radial wave functions vanish at $r = 0$, we do not know their derivatives there. We know however the behavior of the radial wave functions at asymptotic distances (Eq. (9.171)). This information is sufficient to determine the scattering wave function but now it is more difficult to handle the problem. It is necessary to resort to more complicated procedures. The method of [Buck *et al.* (1963)] consists of solving N times the set of N coupled equations, using in each case a linearly independent set of boundary conditions (radial wave functions and derivatives at $r = 0$). We use the notation $u_i^{(j)}$ to represent the ith radial wave function, obtained from the jth set of boundary conditions. The N linearly independent sets of boundary conditions can be built by a simple rule, like

$$u_i^{(j)}(0) = 0, \qquad \left[\frac{du_i^{(j)}(r)}{dr}\right]_{r=0} = B\ \delta_{i\,j}, \tag{9.231}$$

with $j = 1, ..., N$. Similarly to the case of potential scattering, B is an arbitrary constant which only affects the overall normalization of the wave function.

The general solutions for the radial wave functions $u_i(r)$ can be written as linear combinations of the above functions, in the form

$$u_i(r) = \sum_{j=1}^{N} X_j\, u_i^{(j)}(r). \tag{9.232}$$

One should then look for the coefficients X_j that generate radial wave functions with the appropriate asymptotic behavior, given by Eq. (9.171). In order to evaluate cross sections, we need the S-matrix elements $S^J_{\alpha l,0l_0}$. They are given by the asymptotic form of the radial wave functions, $u^J_{\alpha l,0l_0}$, which satisfy the coupled channel equations.

It is important to observe that each set of coupled equations is characterized by the total angular momentum J and by the quantum numbers of the entrance channel, $\alpha = 0$ and l_0. Whereas the entrance channel is always $\alpha = 0$, l_0 can take any of the M values allowed by the selection rules (Eqs. (9.229) and (9.230)). in this way, each of the M possible values of l_0 leads to an independent set of coupled differential equations. They have as solutions the radial wave functions $u^J_{\alpha l,0l_0}(k_\alpha, r)$, labeled by that particular value of l_0. It is convenient to organize the labels i so that $i = 1$ is assigned to the entrance channel with orbital angular moment l_0. That is, $i = 1 \equiv \{0, l_0\}$. When one changes $l_0 \to l_0'$, to get equations for radial wave functions with a different initial angular momentum $(u^J_{\alpha l,0l_0'}(k_\alpha, r))$, one reorganizes the labels so that $i = 1 \equiv \{0, l_0'\}$.

To solve the differential equations for a particular l_0, one evaluates logarithmic derivatives at $r = \bar{R}$, using Eq. (9.232), and the results are matched with their asymptotic forms. From Eq. (9.232), we have

$$\mathcal{L}_i^{\mathrm{I}} = \frac{\sum_{j=1}^{N} X_j \; b_i^{(j)}}{\sum_{j=1}^{N} X_j \; a_i^{(j)}}. \tag{9.233}$$

We use the short-hand notation

$$a_i^{(j)} = u_i^{(j)}(\bar{R}); \quad b_i^{(j)} = \bar{R} \left[du_i^{(j)}(r)/dr\right]_{r=\bar{R}}. \tag{9.234}$$

In Eq. (9.233), $\mathcal{L}_i^{\mathrm{I}}$ means the logarithmic derivative of the state i in the internal region ($\mathrm{I} \Rightarrow r \leq \bar{R}$), taken at the matching radius. On the other hand, the corresponding logarithmic derivative can be calculated from the asymptotic form of the radial wave functions ($\mathrm{II} \Rightarrow r > \bar{R}$),

$$u_{i=1}^{\mathrm{II}}(r) = \frac{i}{2} \left[H_{l_0}^{(-)}(k_0 r) - \bar{S}_{0l_0,0l_0}^{J} \; H_{l_0}^{(+)}(k_0 r)\right], \tag{9.235}$$

$$u_{i\neq 1}^{\mathrm{II}}(r) = -\frac{i}{2} \sqrt{\frac{k_0}{k_i}} \; \bar{S}_{\alpha_i l_i,0l_0}^{J} \; H_{l_i}^{(+)}(k_i r). \tag{9.236}$$

Above, $H_l^{(\pm)}$ are the Coulomb functions with outgoing $_{(+)}$ and ingoing $_{(-)}$ asymptotic behaviors (see chapter 3) and $\bar{S}_{\alpha_i l_i,0l_0}^{J}$ is the reduced S-matrix, associated with the short-range part of the potential. Note that the incident state ($i = 1$) contains ingoing and outgoing spherical waves whereas the remaining states ($i \neq 1$) contains exclusively outgoing waves.

Evaluating the asymptotic logarithmic derivative, we obtain

$$\mathcal{L}_i^{\mathrm{II}} = k_i \bar{R} \left[\frac{H_{l_i}^{(-)'}(k_i\bar{R}) \; \delta_{i1} - \sqrt{k_0/k_i} \;\; \bar{S}_{\alpha_i l_i,0l_0}^{J} \, H_{l_i}^{(+)'}(k_i\bar{R})}{H_{l_i}^{(-)}(k_i\bar{R}) \; \delta_{i1} - \sqrt{k_0/k_i} \;\; \bar{S}_{\alpha_i l_i,0l_0}^{J} \, H_{l_i}^{(+)}(k_i\bar{R})}\right], \tag{9.237}$$

where primes stand for derivatives with respect to the argument. For $i \neq 1$ the wave function cannot contain an incident wave and the above equation reduces to

$$\mathcal{L}_{i\neq 1}^{\mathrm{II}} = k_i \bar{R} \left[\frac{H_{l_i}^{(+)'}(k_i\bar{R})}{H_{l_i}^{(+)}(k_i\bar{R})}\right]. \tag{9.238}$$

Note that $\mathcal{L}_{i\neq 1}^{\mathrm{II}}$ are well determined numbers that do not depend on l_0.

The matching condition

$$\mathcal{L}_i^{\mathrm{I}} = \mathcal{L}_i^{\mathrm{II}} \tag{9.239}$$

leads to N equations of the form,

$$\sum_{j=1}^{N} \left[b_i^{(j)} - \mathcal{L}_i^{\mathrm{II}} \, a_i^{(j)}\right] X_j = 0, \quad i = 1, .., , N. \tag{9.240}$$

involving the N unknowns $X_1, X_2, ...X_N$. However, the equation for $i = 1$ is useless because $\mathcal{L}_1^{\text{II}}$ depends on the unknown element of the S-matrix, $\bar{S}^J_{0l_0,0l_0}$ (see Eq. (9.237)). Excluding $i = 0$ from the set of Eq. (9.240), we end up with $N - 1$ equations for N unknowns. This is not a serious problem because one can chose one of the N unknowns arbitrarily, The choice affects only the overall normalization of the radial wave function. This normalization can be changed later, multiplying the wave function by a properly chosen constant. Thus, we set $X_1 = 1$. In this way we get the new system of $N - 1$ inhomogeneous linear equations,

$$\sum_{j=2}^{N} \left[b_i^{(j)} - \mathcal{L}_i^{\text{II}}\, a_i^{(j)} \right] X_j = b_i^{(1)} - \mathcal{L}_i^{\text{II}}\, a_i^{(1)}, \quad i = 2, .., , N, \tag{9.241}$$

which can be easily solved by usual numerical procedures to handle systems of linear equations.

Having the coefficients $X_1, X_2, ..., X_N$, we evaluate the components of the S-matrix. With these coefficients we can determine any wave function in the internal region and the corresponding logarithmic derivatives. Proceeding as in chapter 2, we get the elastic S-matrix,

$$\bar{S}^J_{0l_0,0l_0} = \frac{\mathcal{L}_1^{\text{I}}\, H_{l_0}^{(-)}\left(k_0 \bar{R}\right) - k_0 \bar{R}\; H_{l_0}^{(-)'}\left(k_0 \bar{R}\right)}{\mathcal{L}_1^{\text{I}} H_{l_0}^{(+)}\left(k_0 \bar{R}\right) - k_0 \bar{R}\, H_{l_0}^{(+)'}\left(k_0 \bar{R}\right)}. \tag{9.242}$$

Before we calculate the other component of the S-matrix, we should modify the arbitrarily chosen normalization. The elastic radial wave function should be renormalized in order to satisfy the condition

$$u_1(\bar{R}) \equiv u^J_{0l_0,0l_0}(k_0, \bar{R}) = \frac{i}{2} \left[H_{l_0}^{(-)}\left(k_0 \bar{R}\right) - \bar{S}^J_{0l_0,0l_0}\, H_{l_0}^{(+)}\left(k_0 \bar{R}\right) \right].$$

This can be achieved through the multiplication by the constant

$$C = \frac{i}{2} \left[\frac{H_{l_0}^{(-)}\left(k_0 \bar{R}\right) - \bar{S}^J_{0l_0,0l_0}\, H_{l_0}^{(+)}\left(k_0 \bar{R}\right)}{\sum_{j=1}^{N} X_j\, u_1^{(j)}(\bar{R})} \right].$$

The coefficient X_j are then redefined as

$$X_j \rightarrow \bar{X}_j = C\, X_j, \tag{9.243}$$

and the radial wave functions become

$$u_i(r) \equiv u^J_{\alpha_i l_i,0l_0}(k_{\alpha_i}, r) = \sum_{j=1}^{N} \bar{X}_j\, u_i^{(j)}(r). \tag{9.244}$$

The remaining components of the S-matrix are then determined matching the internal (Eq. (9.244)) and the external (Eq. (9.236)) forms of the radial wave functions at the matching radius, $r = \bar{R}$. We get

$$\bar{S}^J_{\alpha_i l_i, 0 l_0} = 2i \sqrt{\frac{k_i}{k_0}} \left[\frac{\sum_{j=1}^N \bar{X}_j \, u_i^{(j)}(\bar{R})}{H^{(+)}_{l_i}\left(k_i \bar{R}\right)} \right], \quad \text{for } i \neq 1. \tag{9.245}$$

We have outlined the procedure to find the radial wave functions and the S-matrix for a particular choice of the initial angular momentum, l_0. Extending the calculations to other values of l_0, which are needed for the evaluations of cross sections, is very straightforward. As mentioned earlier in this section, one reorganizes the labels of the states so that the new l_0 is assigned to $i = 1$ and then uses the same equations. Carrying out this procedure for each of the M possible values of the initial orbital angular momentum, one obtains all the matrix elements required for the calculation of cross sections.

The *traditional method* gives accurate solutions to coupled-channel problems, provided that the truncated space is large enough to include the most relevant couplings. The effects of the remaining channels, open or closed, can be taken into account approximately. Usually, they are included through effective potentials (see chapter 10). However, the numerical calculations involved may be very time-consuming. This imposes restrictions on the number of channels that can be included in the calculations. There may be also difficulties related with the convergence of the calculations, specially in the cases of coupling potentials with long-range (like e.g., Coulomb dipole) and in collisions involving weakly bound nuclei [Nunes and Thompson (1998)]. In such cases the breakup channel plays an important role, leading to final states in the continuum. This situation is discussed in section 9.4. Furthermore, the solution of the coupled radial equations becomes inviable when there are couplings with non-local matrix elements. That is,

$$\mathcal{V}^J_{\alpha l, \alpha' l'}(r, r') \neq \mathcal{V}^J_{\alpha l, \alpha' l'}(r, r) \, \delta(r - r'). \tag{9.246}$$

This may occur in collisions where rearrangement channels are important. In such cases, other methods must be adopted.

As a final remark to this section, we should mention that the coupled channel problem can also be solved using the R-matrix method, which will be discussed in section 12.8 (for details we refer to the recent review of [Descouvemont and Baye (2010)]).

Exercises

(1) Starting from the orthonormality of the free states,

$$\left\langle \Phi(\alpha \mathbf{k}_\alpha) \middle| \Phi(\alpha' \mathbf{k}'_{\alpha'}) \right\rangle = \delta_{\alpha\alpha'}\, \delta\left(\mathbf{k}_\alpha - \mathbf{k}'_{\alpha'}\right),$$

prove that the exact scattering wave functions satisfy the relation[34]

$$\left\langle \Psi(\alpha \mathbf{k}_\alpha) \middle| \Psi(\alpha' \mathbf{k}'_{\alpha'}) \right\rangle = \delta_{\alpha\alpha'}\, \delta\left(\mathbf{k}_\alpha - \mathbf{k}'_{\alpha'}\right).$$

(2) Prove that the cross section for channel-α in multichannel scattering is given by Eq. (9.29).

(3) Consider the eigenfunctions of the sets of operators: $\{\mathbb{H}_0, L^2, L_z\}$, $\{\mathbb{H}_1, L^2, L_z\}$ and $\{\mathbb{H}, L^2, L_z\}$, as defined respectively in Eqs. (9.131), (9.132) and (9.133).

a) Prove the orthogonality relations,

$$\left\langle \phi\left(\alpha E l m\right) \middle| \phi\left(\alpha' E' l' m'\right) \right\rangle = \delta_{\alpha\alpha'}\ \delta_{ll'}\ \delta_{mm'}\ \delta(E - E')$$
$$\left\langle \chi\left(\alpha E l m\right) \middle| \chi\left(\alpha' E' l' m'\right) \right\rangle = \delta_{\alpha\alpha'}\ \delta_{ll'}\ \delta_{mm'}\ \delta(E - E').$$

b) Assuming that the exact radial wave functions are normalized according to the equation

$$\sum_\alpha \int dr\, u^*_{\alpha l, 0l}(k_\alpha, r)\, u_{\alpha l, 0l}(k'_\alpha, r) = \frac{\pi}{2}\, \delta\left(k - k'\right), \tag{9.247}$$

where $k = \sqrt{2\mu E/\hbar^2}$ and $k_\alpha = \sqrt{2\mu(E - \varepsilon_\alpha)/\hbar^2}$, prove the relation:

$$\left\langle \psi_0\left(E l m\right) \middle| \psi_0\left(E' l' m'\right) \right\rangle = \delta_{ll'}\ \delta_{mm'}\ \delta(E - E').$$

c) Show that the normalization of Eq. (9.247) guarantees that the angular momentum expansion of the exact wave function (Eq. (9.100)) satisfies the orthonormality relation of Eq. (9.61),

$$\left\langle \Psi^{(\pm)}(0\mathbf{k}_0) \middle| \Psi^{(\pm)}(0\mathbf{k}'_0) \right\rangle = \delta(\mathbf{k}_0 - \mathbf{k}'_0).$$

(4) Show that the partial wave expansion for the scattering state of a system with spin (Eq. (9.165)) satisfies the orthonormality relation,

$$\left\langle \Psi^{(+)}(0\nu_0\mathbf{k}_0) \middle| \Psi^{(+)}(0\nu_0\mathbf{k}'_0) \right\rangle = \delta_{\nu_0\nu'_0}\ \delta(\mathbf{k}_0 - \mathbf{k}'_0),$$

[34] *Hint*: Follow the procedures of section 4.1.3, adapting it to multichannel scattering.

under two conditions. The first is that the radial wave functions for the same incident orbital angular momentum should satisfy the orthonormality relation

$$\sum_{\beta l} \int dr\, u^{J*}_{\beta l,0l_0}(k_\beta, r)\, u^{J}_{\beta l,0l_0}(k'_\beta, r) = \frac{\pi}{2}\, \delta\,(k_0 - k'_0)\,.$$

The second is that for different incident angular momentum ($l_0 \neq l'_0$) one should have:

$$\sum_{\beta l} \int dr\, u^{J*}_{\beta l,0l_0}(k_\beta, r)\, u^{J}_{\beta l,0l'_0}(k'_\beta, r) = C(l_0, l'_0),$$

where $C(l_0, l'_0)$ is independent of J.

(5) Derive the Lippmann-Schwinger equations:

$$u^J_{\alpha l,0l_0}(k_\alpha, r) = \hat{\jmath}_l(k_0, r)\, \delta_{\alpha 0}\delta_{ll_0} + \sum_{\alpha' l'} \int dr'\; g^{(+)}_{0,l}(\alpha E; r, r')\, \mathcal{V}^J_{\alpha l,\alpha' l'}(r')\; u^J_{\alpha' l',0l_0}(k_\alpha, r').$$

and

$$u^J_{\alpha l,0l_0}(k_\alpha, r) = w_{0l_0}(k_0, r)\, \delta_{\alpha 0}\delta_{ll_0} + \sum_{\alpha' l'} \int dr'\; g^{(+)}_{1,l}(\alpha E; r, r')\, \mathcal{V}^J_{\alpha l,\alpha' l'}(r')\; u^J_{\alpha' l',0l_0}(k_\alpha, r').$$

(6) Assuming that the radial integrals satisfy the conditions of exercise 4, prove that the wave functions of Eqs. (9.184), (9.185) and (9.186) satisfy the orthonormality relations

$$\begin{aligned}
\big\langle \phi\,(\alpha ElJM)\,\big|\phi\,(\alpha' E' l' J' M')\,\big\rangle &= \delta_{\alpha\alpha'}\,\delta_{ll'}\,\delta_{JJ'}\,\delta_{MM'}\,\delta(E - E'),\\
\big\langle \chi\,(\alpha ElJM)\,\big|\chi\,(\alpha' E' l' J' M')\,\big\rangle &= \delta_{\alpha\alpha'}\,\delta_{ll'}\,\delta_{JJ'}\,\delta_{MM'}\,\delta(E - E')\\
\big\langle \psi_{\alpha l}\,(\alpha EJM)\,\big|\psi_{\alpha l}\,(\alpha' E' J' M')\,\big\rangle &= \delta_{JJ'}\,\delta_{MM'}\,\delta(E - E').
\end{aligned}$$

(7) Show that the angular momentum expansions of $T_{\alpha\nu\mathbf{k}_\alpha,0\nu_0\mathbf{k}_0}$ (Eq. (9.192)) and $f_{\alpha\nu,0\nu_0}(\mathbf{k}_\alpha, \mathbf{k}_0)$ (Eq. (9.174)) are consistent with the relation between the T-matrix and the scattering amplitude (Eq. (9.183)).

(8) Consider the S-wave scattering of a spinless system with two channels with the same channel potential, $V(r)$, and collision energy E. Assume that the channels are coupled by the constant potential, $V_{12}(r) = V_{21}(r) = F$ and the excitation energy of the inelastic channel

is negligible ($E_1 = E_2 = E$). Show that within these approximations the normalized scattering states $\psi_1(r)$ and $\psi_2(r)$ can be written as,

$$\psi_1(r) = \frac{1}{\sqrt{2}} \left[\chi_+(r) + \chi_-(r)\right] \quad \text{and} \quad \psi_2(r) = \frac{1}{\sqrt{2}} \left[\chi_+(r) - \chi_-(r)\right],$$

where $\chi_+(r)$ and $\chi_-(r)$ are normalized distorted waves with the collision energies $E - F$ and $E + F$, respectively.

Chapter 10

Few-channel Description of Many-body Scattering

The many-channel description of scattering presented in the previous chapter, can be recast into an effective description in the space of a small number of channels or, more conveniently, of a single channel. The concepts used in this connection are of great practical value in many fields of physics and certainly warrants detailed discussion here. The most practical way of performing this reduction is the Feshbach projection operator calculus. In the following we give a detailed analysis of this calculus and introduce the concept of effective potentials, commonly referred to, especially in nuclear physics, as optical and polarization potentials. These names are used loosely, having different meanings for different authors. This point will be addressed later in this chapter.

The layout of this chapter is the following. In section 10.1 we develop the Feshbach projection operator calculus. In section 10.2 we discuss prompt processes and resonances within Feshbach's formalism. In section 10.3, we introduce the optical model and the energy-averaging procedure leading to optical potentials, which are frequently used in nuclear physics. We then discuss the main methods to calculate optical potentials in nucleon-nucleus and nucleus-nucleus collisions. Optical potentials aim at reproducing average coupling effects, resulting from the influence of many channels. However, when some specific channel, or group of channels, couples very strongly with the elastic channel, it is necessary to include an additional term in the effective potential. This term is called the *dynamic polarization potential* and it is discussed in section 10.5. Usually, the optical and polarization potentials are energy-dependent. Consequently, the corresponding Schrödinger equation in such cases must be handled with special care. This problem is addressed in section 10.6.

10.1 The Feshbach calculus

We start from the multi-channel Schrödinger equation in many-body scattering,

$$(E - \mathbb{H}) \left|\Psi^{(+)}(0\mathbf{k}_0)\right\rangle = 0, \tag{10.1}$$

where $\mathbb{H}$ is the Hamiltonian of Eq. (9.1),

$$\mathbb{H} = h + K + \mathbb{V}.$$

As in chapter 9, h is the intrinsic Hamiltonian of the projectile-target system, with eigenstates $|\alpha)$ and eigenvalues ε_α, K is the kinetic energy operator for the relative motion, $\mathbb{V}$ is the projectile-target interaction, which may have non-vanishing diagonal and off-diagonal matrix elements in channel space, and $|\Psi^{(+)}(0\mathbf{k}_0)\rangle$ is the scattering state in a collision initiated in the elastic channel ($\alpha = 0$), with wave vector $\mathbf{k}_0$. This wave function can be expanded in channel space as

$$\left|\Psi^{(+)}(0\mathbf{k}_0)\right\rangle = \sum_{\alpha=0}^{\infty} \left|\psi_\alpha(0\mathbf{k}_0)\right\rangle \otimes |\alpha) \equiv \sum_{\alpha=0}^{\infty} |\psi_\alpha\rangle \, |\alpha) \,. \tag{10.2}$$

For simplicity of notation, we henceforth omit the symbol '$\otimes$', which indicates the tensor product of the intrinsic space and the space of collision degrees of freedom, and drop the explicit indication that the collision is initiated in the elastic channel. We omit also the wave vector of the entrance channel, except where it becomes essential.

Now we split the channel space into two partitions, a partition containing N channels, $\alpha = 0, 1, ..., N-1$, and a second partition containing the remaining ones, $\beta = N, N+1, ...$. The channels included in each partition depend on the particular problem being studied. In general, the first partition corresponds to the subspace of channels we are interested in. In some cases, this partition includes only the elastic channel. On the other hand, the states in the second partition are of lesser interest. However, their influence on the channels of interest cannot be neglected. Frequently, some (or even all) of the channels in the second partition are closed.

The next step is to introduce operators that projects onto these two subspaces. Since the intrinsic states are orthogonal, the projectors are immediately obtained as,

$$P = \sum_{\alpha=0}^{N-1} |\alpha)\,(\alpha| \,; \quad Q = \sum_{\beta=N}^{\infty} |\beta)\,(\beta| \,. \tag{10.3}$$

The projectors P and Q have the usual properties[1]:

$$P^2 = P, \quad Q^2 = Q, \quad PQ = QP = 0 \quad \text{and} \quad P + Q = 1. \tag{10.4}$$

Acting with these projectors on the expansion of Eq. (10.2) one gets $|\Psi^{(+)}\rangle = |\Psi_{\rm P}^{(+)}\rangle + |\Psi_{\rm Q}\rangle$, with[2]

$$|\Psi_{\rm P}^{(+)}\rangle \equiv P\,|\Psi^{(+)}\rangle = \sum_{\alpha=0}^{N-1} |\psi_\alpha\rangle\,|\alpha) \tag{10.5}$$

and

$$|\Psi_{\rm Q}\rangle \equiv Q\,|\Psi^{(+)}\rangle = \sum_{\beta=N}^{\infty} |\psi_\beta\rangle\,|\beta)\,. \tag{10.6}$$

Applying the projectors P and Q on the many-body Schrödinger equation and using the properties of Eq. (10.4), we get two equations coupling the P the Q-subspaces:

$$[E - \mathbb{H}_{\rm PP}]\;|\Psi_{\rm P}^{(+)}\rangle = \mathbb{H}_{\rm PQ}\;|\Psi_{\rm Q}\rangle \tag{10.7}$$

$$[E - \mathbb{H}_{\rm QQ}]\;|\Psi_{\rm Q}\rangle = \mathbb{H}_{\rm QP}\;|\Psi_{\rm P}^{(+)}\rangle\,, \tag{10.8}$$

where

$$\mathbb{H}_{\rm PP} \equiv P\mathbb{H}P, \quad \mathbb{H}_{\rm QQ} \equiv Q\mathbb{H}Q, \quad \mathbb{H}_{\rm PQ} \equiv P\mathbb{H}Q \quad \text{and} \quad \mathbb{H}_{\rm QP} \equiv Q\mathbb{H}P. \tag{10.9}$$

The equations coupling the two subspaces can be reduced to a single equation within the subspace of main interest with an effective Hamiltonian that includes the effects of couplings with the channels outside the P-subspace. To derive this equation, we first evaluate the state $|\Psi_{\rm Q}\rangle$ that appears in Eq. (10.8). Applying the inverse of the operator within square brackets to both sides of this equation, we get

$$|\Psi_{\rm Q}\rangle = \left[\frac{1}{E - \mathbb{H}_{\rm QQ} + i\epsilon}\right]\mathbb{H}_{\rm QP}\;|\Psi_{\rm P}^{(+)}\rangle\,. \tag{10.10}$$

Above, the term $i\epsilon$ should be dropped when the partition Q contains exclusively closed channels. As discussed in chapter 4, it avoids divergences and guarantees the outgoing behavior of the scattered waves in the open channels of the Q-subspace.

Using a similar procedure, one gets the formal solution of Eq. (10.7),

$$|\Psi_{\rm P}^{(+)}\rangle = |X_{\rm P}^{(+)}\rangle + \left[\frac{1}{E - \mathbb{H}_{\rm PP} + i\epsilon}\right]\mathbb{H}_{\rm PQ}\;|\Psi_{\rm Q}\rangle\,. \tag{10.11}$$

[1]In practical situations, the infinite series in the definition of Q is truncated. In this case, the relation $P + Q = 1$ cannot be satisfied exactly. However, it is approximately satisfied if all relevant channels are taken into account

[2]Since in the most relevant situations all channels in the Q-subspace are closed, we omit the superscript $(+)$ in the wave function projected onto this subspace.

Since the collision is initiated in the P-subspace, it is necessary to add the *homogeneous* term $\left|X_{\rm P}^{(+)}\right\rangle$, which corresponds to the solution of the scattering problem when the two subspaces are decoupled.

Inserting into Eq. (10.7) the explicit expression for $|\Psi_{\rm Q}\rangle$ (Eq. (10.10)), we obtain the equation within the P-subspace,

$$[E - \mathbb{H}_{\rm eff}] \, |\Psi_{\rm P}^{(+)}\rangle = 0. \tag{10.12}$$

Above, $\mathbb{H}_{\rm eff}$ is the effective Hamiltonian

$$\mathbb{H}_{\rm eff} = \mathbb{H}_{\rm PP} + \mathbb{H}_{\rm PQ} \, \mathbb{G}_{\rm Q}^{(+)}(E) \, \mathbb{H}_{\rm QP}, \tag{10.13}$$

with

$$\mathbb{G}_{\rm Q}^{(+)}(E) = \frac{1}{E - \mathbb{H}_{\rm QQ} + i\epsilon}. \tag{10.14}$$

The physical meaning of $\mathbb{H}_{\rm eff}$ is very clear. It is a sum of two terms. The first term is built exclusively with matrix elements between states contained in the P-subspace. In this way, if one neglects the second term, the resulting equation is equivalent to the coupled channel problem within a truncated space (the P-subspace). Usually, the partitions are chosen so that this term accounts for prompt processes. In the simple case where P projects onto the elastic channel, that is $P = |0)\,(0|$, the term $\mathbb{H}_{\rm PP}$ is the sum of the kinetic energy operator with a folding potential in the elastic channel. The second term in Eq. (10.13) expresses the influence of the channels in the Q-subspace on the collision dynamics of the P-subspace. It is usually called (see e.g. [Satchler (1983)]) the *dynamic polarization potential*. It is a product of three factors. The first factor represents a transition from the P to the Q-subspaces. The intermediate factor, $\mathbb{G}_{\rm Q}^{(+)}(E)$, is the propagator within the Q-subspace. Usually, this space contains a large number of channels and $\mathbb{G}_{\rm Q}^{(+)}(E)$ contains all couplings among them. Thus, it gives rise to time delay. Finally, the third factor gives the transition back to the P-subspace.

In general, the operator $\mathbb{H}_{\rm eff}$ is not hermitian, even when the many-body Hamiltonian, $\mathbb{H}$, is. When $\mathbb{V}$ is hermitian, $\mathbb{H}_{\rm PP}$ is real. However, the second term in Eq. (10.13) contains an imaginary part, unless all channels in the Q-subspace are closed. In this case, there can be no loss of flux from the P-subspace. The flux diverted into the Q-subspace returns to P, after some time delay. Thus, the term $i\epsilon$ in $\mathbb{G}_{\rm Q}^{(+)}(E)$ should be dropped. However, this time delayed amplitude is not considered in the description of prompt processes. As we will discuss in section 10.3, one usually takes energy averages that eliminate the delayed contribution in the optical model cross section. This contribution is then estimated by the Hauser-Feshbach formula, which will be addressed in chapter 12. The energy averaging procedure leads to a complex effective potential and its imaginary part represents the neglected delayed contribution to the scattering amplitude.

10.1.1 *The dispersion relation for* $\mathbb{H}_{\text{eff}}$

As discussed in the previous section, the inclusion of open channels in the partition Q leads to a complex effective potential. In such cases, the real and imaginary parts of $\mathbb{H}_{\text{eff}}$ can be evaluated separately, with the help of the relation (see e.g. [Rodberg and Thaler (1967)]),

$$\lim_{\epsilon\to 0}\left[\frac{1}{x\pm i\epsilon}\right]=\mathcal{P}\left(\frac{1}{x}\right)\mp i\pi\,\delta(x), \tag{10.15}$$

with $\mathcal{P}$ standing for Cauchy's principal value. Using this relation for the propagator in the Q-subspace (Eq. (10.14)), one obtains an important dispersion relation connecting the real and imaginary parts of the effective Hamiltonian. This relation is derived below.

The first step is to write the spectral representation of the propagator in the Q-subspace. One uses the eigenstates of $\mathbb{H}_{\text{QQ}}$ in the discrete $(|X_{\text{Q}}(n)\rangle)$ and continuum $(|X_{\text{Q}}(E,\alpha)\rangle)$ parts of the spectrum. They satisfy the Schrödinger equations,

$$\mathbb{H}_{\text{QQ}}\,|X_{\text{Q}}(n)\rangle=\mathcal{E}_n\,|X_{\text{Q}}(n)\rangle \tag{10.16}$$

and

$$\mathbb{H}_{\text{QQ}}\,|X_{\text{Q}}(E,\alpha)\rangle=E\,|X_{\text{Q}}(E,\alpha)\rangle\,. \tag{10.17}$$

Above, n stands for all quantum numbers specifying the bound state $|X_{\text{Q}}(n)\rangle$ and α represent the quantum numbers which, together with E, label the eigenstates in the continuum. The spectral representation of $\mathbb{G}_{\text{Q}}^{(+)}$ then gives

$$\mathbb{G}_{\text{Q}}^{(+)}(E)=\sum_n\frac{|X_{\text{Q}}(n)\rangle\,\langle X_{\text{Q}}(n)|}{E-\mathcal{E}_n}+\sum_\alpha\int dE'\,\frac{|X_{\text{Q}}(E',\alpha)\rangle\,\langle X_{\text{Q}}(E',\alpha)|}{E-E'+i\epsilon}. \tag{10.18}$$

The real and imaginary parts of the above propagator can be calculated separately with the help of Eq. (10.15). One gets

$$\begin{aligned}\text{Re}\,\left\{\mathbb{G}_{\text{Q}}^{(+)}(E)\right\}=&\sum_n\frac{|X_{\text{Q}}(n)\rangle\,\langle X_{\text{Q}}(n)|}{E-\mathcal{E}_n}\\&+\sum_\alpha\mathcal{P}\int dE'\,\frac{|X_{\text{Q}}(E',\alpha)\rangle\,\langle X_{\text{Q}}(E',\alpha)|}{E-E'}\end{aligned} \tag{10.19}$$

and

$$\text{Im}\left\{\mathbb{G}_{\text{Q}}^{(+)}(E)\right\} = -\pi \sum_{\alpha} |X_{\text{Q}}(E,\alpha)\rangle \langle X_{\text{Q}}(E,\alpha)| \,. \tag{10.20}$$

Using the above results in Eq. (10.13), one obtains

$$\begin{aligned}\text{Re}\,\{\mathbb{H}_{\text{eff}}\} = \mathbb{H}_{\text{PP}} &+ \sum_{n} \frac{\mathbb{H}_{\text{PQ}}\,|X_{\text{Q}}(n)\rangle\,\langle X_{\text{Q}}(n)|\,\mathbb{H}_{\text{QP}}}{E - \mathcal{E}_n} \\ &+ \mathcal{P}\int dE' \sum_{\alpha} \frac{\mathbb{H}_{\text{PQ}}\,|X_{\text{Q}}(E',\alpha)\rangle\,\langle X_{\text{Q}}(E',\alpha)|\,\mathbb{H}_{\text{QP}}}{E - E'}\end{aligned} \tag{10.21}$$

and

$$\text{Im}\,\{\mathbb{H}_{\text{eff}}\} = -\pi \sum_{\alpha} \mathbb{H}_{\text{PQ}}\,|X_{\text{Q}}(E,\alpha)\rangle\,\langle X_{\text{Q}}(E,\alpha)|\,\mathbb{H}_{\text{QP}}. \tag{10.22}$$

The dispersion relation can be immediately obtained using the above equation in the integral of Eq. (10.21). One gets

$$\begin{aligned}\text{Re}\,\{\mathbb{H}_{\text{eff}}\} = \mathbb{H}_{\text{PP}} &+ \sum_{n} \frac{\mathbb{H}_{\text{PQ}}\,|X_{\text{Q}}(n)\rangle\,\langle X_{\text{Q}}(n)|\,\mathbb{H}_{\text{QP}}}{E - \mathcal{E}_n} \\ &- \frac{1}{\pi}\mathcal{P}\int dE' \,\frac{\text{Im}\,\{\mathbb{H}_{\text{eff}}\}}{E - E'}.\end{aligned} \tag{10.23}$$

Note that in Eq. (10.23) the energy dependence of the effective potential appears explicitly. The dispersion relation is reminiscent of the Kramers-Kronig relation in optics, where the function is the frequency-dependent index of refraction. An important consequence of the dispersion relation in Nuclear Physics is the threshold anomaly, discussed in section 10.5.4.

The P-projected Shrödinger equation is a convenient starting point for developing approximate method to tackle the scattering of complex many-body systems.

10.2 Prompt processes, time delay and resonances

Feshbach's projector technique lead to an effective Hamiltonian that can be written as

$$\mathbb{H}_{\text{eff}} = \mathbb{H}_{\text{PP}} + \mathbb{V}_{\text{pol}} \equiv h + K + \mathbb{V}_{\text{PP}} + \mathbb{V}_{\text{pol}}. \tag{10.24}$$

Above, we have introduced the so called *polarization potential* in the P-subspace,

$$\mathbb{V}_{\text{pol}} \equiv \mathbb{V}_{\text{PQ}}\,\mathbb{G}_{\text{Q}}^{(+)}(E)\,\mathbb{V}_{\text{QP}}. \tag{10.25}$$

This potential accounts for the influence of the Q-subspace on the dynamics of the P-subspace. The partitions P and Q are arbitrary and the results of the previous sections are valid for any particular choice of the channels included in each of them. In the original formulation of Feshbach, the subspace P was associated with prompt processes. It contained the elastic channel and possibly a small number of non-elastic channels with intrinsic structure similar to that of the ground state. In this way, the excitation of an inelastic channel included in the P-subspace could take place through the action of the coupling interaction $\mathbb{V}_{\rm PP}$ in a single step or, at most, in a few steps. On the other hand, the subspace Q contained a large number of closed channels (and possibly a few open channels), coupled among themselves by $\mathbb{V}_{\rm QQ}$. In this way, once the system reaches the Q-subspace, through the action of the interaction $\mathbb{V}_{\rm PQ}$, the operator $\mathbb{V}_{\rm QQ}$ contained in the propagator $\mathbb{G}_{\rm Q}^{(+)}$ yields multi-step transitions among the channels in the Q-subspace. Eventually, the system may re-emerge in the P-subspace after a long time delay. This gives rise to long-lived resonances.

10.2.1 *Prompt processes: direct reactions*

Neglecting completely the time-delayed processes corresponds to ignoring the couplings between the subspaces P and the Q. Thus, the effective Hamiltonian reduces to $\mathbb{H}_{\rm PP}$. If P contains exclusively the elastic channel, the collision reduces to a trivial problem of potential scattering. Otherwise, the scattering state, $\big|X_{\rm P}^{(+)}(0\mathbf{k}_0)\big\rangle$, is the solution of a set of coupled-channel equations, as the ones considered in the previous chapter. Formally, we can write the Schrödinger equation,

$$\mathbb{H}_{\rm PP}\,|X_{\rm P}^{(+)}(0\mathbf{k}_0)\rangle = E\,|X_{\rm P}^{(+)}(0\mathbf{k}_0)\rangle\,. \tag{10.26}$$

In general, the above is not a single equation but rather a set of coupled equations. This becomes clear if one performs the channel expansion

$$|X_{\rm P}^{(+)}(0\mathbf{k}_0)\rangle = \sum_\alpha \big|X_{{\rm P},\alpha}^{(+)}(0\mathbf{k}_0)\big\rangle = \sum_\alpha |\chi_\alpha^{(+)}(0\mathbf{k}_0)\rangle\,|\alpha)\,, \tag{10.27}$$

where the summation runs over all channels included in the P. Inserting this expansion into Eq. (10.26) and taking scalar product with each of the intrinsic states, one gets the set of coupled channel equations,

$$\Big[E_\alpha - K - V_{\alpha\alpha}\Big]\,|\chi_\alpha^{(+)}(0\mathbf{k}_0)\rangle = \sum_{\alpha'} V_{\alpha\alpha'}\,\big|\chi_{\alpha'}^{(+)}(0\mathbf{k}_0)\big\rangle\,. \tag{10.28}$$

Above, $E_\alpha = E - \varepsilon_\alpha$, with ε_α representing the eigenvalues of the intrinsic Hamiltonian, h, and

$$V_{\alpha\alpha'} = (\alpha|\,\mathbb{V}_{\rm PP}\,|\alpha')\,. \tag{10.29}$$

The T-matrix for a final state in channel-α with wave number $\mathbf{k}_\alpha$, in a collision initiated in the elastic channel with wave number $\mathbf{k}_0$, $T^{\mathrm{P}}_{\alpha\mathbf{k}_\alpha,0\mathbf{k}_0}$, is given by[3] (see section 9.2.1)

$$\begin{aligned} T^{\mathrm{P}}_{\alpha\mathbf{k}_\alpha,0\mathbf{k}_0} &= \big\langle \Phi(\alpha\mathbf{k}_\alpha)\big| \mathbb{V}_{\mathrm{PP}} \big|X^{(+)}_{\mathrm{P}}(0\mathbf{k}_0)\big\rangle \\ &= \sum_{\alpha'} \big\langle \phi(\mathbf{k}_\alpha)\big| \mathbb{V}_{\alpha\alpha'} \big|\chi^{(+)}_{\alpha'}(0\mathbf{k}_0)\big\rangle . \end{aligned} \tag{10.30}$$

Above, we have used the notation of chapter 9, where $|\Phi(\alpha\mathbf{k}_\alpha)\rangle = |\phi(\mathbf{k}_\alpha)\rangle\,|\alpha)$, with $|\phi(\mathbf{k}_\alpha)\rangle$ representing a plane wave[4] with wave vector $\mathbf{k}_\alpha$. Since the T-matrix of Eq. (10.30) is associated with prompt processes, it should vary slowly with the collision energy. Thus, it may only exhibit short-lived resonances, with widths like those found in typical resonances of potential scattering.

A simpler situation occurs when the P-partition contains exclusively the elastic channel. In this case the projector P is given by $P = |0)\,(0|$. The set of coupled-channel equations then reduces to the single Schrödinger equation

$$\Big[E - K - V_0\Big]\; |\chi^{(+)}(\mathbf{k})\rangle = 0. \tag{10.31}$$

In this case, $E_\alpha = E_0 = E$ and $V_0 \equiv (0|\,\mathbb{V}\,|0)$ is just a real folding potential.

10.2.2 *Long-lived resonances*

When the channels in the P-subspace couple with closed channels with complicated intrinsic structure, the incident wave re-emerges in an open channel after a long time delay. This gives rise to long-lived resonances. In this section, we consider the situation where the Q-subspace contains only closed channels, giving rise to sharp isolated resonances[5]. In this case, the term $i\epsilon$ in its propagator is dropped and it takes the form

$$\mathbb{G}^{(+)}_{\mathrm{Q}}(E) \to \mathbb{G}_{\mathrm{Q}}(E) = \frac{1}{E - \mathbb{H}_{\mathrm{QQ}}}. \tag{10.32}$$

The effective Hamiltonian then becomes

$$\mathbb{H}_{\mathrm{eff}} = h + K + \mathbb{V}_{\mathrm{PP}} + \mathbb{V}_{\mathrm{pol}}, \tag{10.33}$$

[3]The superscript 'P' in T^{P} indicates that it is the T-matrix restricted to the P-subspace (when $\mathbb{V}_{\mathrm{pol}}$ is neglected).

[4]If the interaction contains Coulomb terms, it is necessary to make the modifications discussed in chapter 9.

[5]For a more comprehensive discussion involving overlapping resonances, we refer to [Feshbach (1992); Levin and Feshbach (1973)].

with

$$\mathbb{V}_{\rm pol} = \mathbb{V}_{\rm PQ} \, \frac{1}{E - \mathbb{H}_{\rm QQ}} \, \mathbb{V}_{\rm QP}. \tag{10.34}$$

We can use the spectral representation for the propagator in the space of closed channels, $\mathbb{G}_{\rm Q}$,

$$\mathbb{G}_{\rm Q}(E) = \sum_n |X_{\rm Q}(n)\rangle \, \frac{1}{E - \mathcal{E}_n} \, \langle X_{\rm Q}(n)| \, , \tag{10.35}$$

where $X_{\rm Q}(n)$ are the bound eigenstates of $\mathbb{H}_{\rm QQ}$, defined by Eq. (10.16). Using this result in Eq. (10.34), the polarization potential can be expressed as the sum of separable terms

$$\mathbb{V}_{\rm pol} = \sum_n \, \mathbb{V}_{\rm PQ} \, |X_{\rm Q}(n)\rangle \, \frac{1}{E - \mathcal{E}_n} \, \langle X_{\rm Q}(n)| \, \mathbb{V}_{\rm QP}. \tag{10.36}$$

We now investigate the behavior of the T-matrix associated with the Hamiltonian $\mathbb{H}_{\rm eff}$. The Schrödinger equation for the scattering state can be put in the form

$$\Big[E - \mathbb{H}_{\rm PP} - \mathbb{V}_{\rm pol}\Big] \, \big|\Psi_{\rm P}^{(+)}(0\mathbf{k}_0)\big\rangle = 0, \tag{10.37}$$

where

$$\mathbb{H}_{\rm PP} = h + K + \mathbb{V}_{\rm PP} \tag{10.38}$$

is the Hamiltonian in the absence of the polarization potential, considered in our discussion of prompt processes (previous sub-section).

To evaluate the T-matrix, we use the two-potential formula of Gell-Mann & Goldberger (section 4.5), for many-body scattering (section 9.2.1). In the present case, the dominant and the complementary potentials of Gell-Mann & Goldberber theory are respectively $\mathbb{V}^{(1)} = \mathbb{V}_{\rm PP}$ and $\mathbb{V}^{(2)} = \mathbb{V}_{\rm pol}$. Using Eq. (9.72), we get

$$T_{\alpha\mathbf{k}_\alpha,0\mathbf{k}_0} = T^{\rm P}_{\alpha\mathbf{k}_\alpha,0\mathbf{k}_0} + T^{\rm pol}_{\alpha\mathbf{k}_\alpha,0\mathbf{k}_0}, \tag{10.39}$$

where $T^{\rm P}_{\alpha\mathbf{k}_\alpha,0\mathbf{k}_0}$ is the prompt contribution to the T-matrix, given in Eq. (10.30), and $T^{\rm pol}_{\alpha\mathbf{k}_\alpha,0\mathbf{k}_0}$ is its time-delayed component. However, the fact that the potential $\mathbb{V}_{\rm PP}$ alone gives rise to a coupled-channel problem renders the study of the full T-matrix too complicated. To avoid this difficulty, we make the simplifying assumption that the potential $\mathbb{V}_{\rm PP}$ does not couple the channels in the P partition among themselves. In this way, the channel expansion of the distorted wave in a collision initiated in channel-α reduces to the single term

$$|X_{\rm P}^{(+)}(\alpha\mathbf{k}_\alpha)\rangle = \big|X_{{\rm P},\alpha}^{(+)}(\alpha\mathbf{k}_\alpha)\big\rangle \, . \tag{10.40}$$

That is, in an expansion like Eq. (10.27) for a collision initiated in channel-α (instead of the elastic channel), one should set

$$\left|\chi_{\alpha'}^{(+)}(\mathbf{k}_\alpha\right\rangle = \delta_{\alpha'\alpha} \left|\chi_{\alpha}^{(+)}(\mathbf{k}_\alpha)\right\rangle . \tag{10.41}$$

The two-potential formula involves also distorted final states with ingoing wave boundary condition, $\langle\chi_{\alpha}^{(-)}(\mathbf{k}_\alpha)|$.

Owing to the neglect of couplings among the states of the P-partition, the prompt component of the T-matrix of Eq. (10.30) takes the simple form

$$T^{\mathrm{P}}_{\alpha\mathbf{k}_\alpha,0\mathbf{k}_0} = \delta_{\alpha,0}\; T^{\mathrm{P}}_{0\mathbf{k},0\mathbf{k}_0} = \delta_{\alpha,0}\; \left\langle\phi(\mathbf{k})\right| V_{00} \left|\chi_0^{(+)}(\mathbf{k}_0)\right\rangle . \tag{10.42}$$

We now evaluate the delayed component. It is given by Eq. (9.72),

$$T^{\mathrm{pol}}_{\alpha\mathbf{k}_\alpha,0\mathbf{k}_0} = \left\langle X^{(-)}_{\mathrm{P},\alpha}(\alpha\mathbf{k}_\alpha)\right| \mathbb{V}_{\mathrm{pol}} \left|\Psi^{(+)}(0\mathbf{k}_0)\right\rangle . \tag{10.43}$$

Using the explicit form of the polarization potential, given in Eq. (10.36), we get

$$T^{\mathrm{pol}}_{\alpha\mathbf{k}_\alpha,0\mathbf{k}_0} = \sum_n \left\langle X^{(-)}_{\mathrm{P},\alpha}(\alpha\mathbf{k}_\alpha)\right| \mathbb{V}_{\mathrm{PQ}} \left|X_{\mathrm{Q}}(n)\right\rangle \times \frac{1}{E-\mathcal{E}_n} \left\langle X_{\mathrm{Q}}(n)\right| \mathbb{V}_{\mathrm{QP}} \left|\Psi^{(+)}(0\mathbf{k}_0)\right\rangle . \tag{10.44}$$

In principle, the matrix-element in the first line of the above equation is easy to evaluate because it only depends on the states $X^{(-)}_{\mathrm{P},\alpha}(\alpha\mathbf{k}_\alpha)$ and $X_{\mathrm{Q}}(n)$, which do not contain the effects of couplings between the two subspaces. The difficulty then resides in the calculation of the last factor at the RHS of Eq. (10.44), which depends on the exact scattering state, $|\Psi^{(+)}(0\mathbf{k}_0)\rangle$. This factor, which we will denote by Z_n, is

$$Z_n = \left\langle X_{\mathrm{Q}}(n)\middle|\mathbb{V}_{\mathrm{QP}}\middle|\Psi^{(+)}(0\mathbf{k}_0)\right\rangle . \tag{10.45}$$

To evaluate Z_n, we resort to the distorted wave Lippmann-Schwinger equation

$$|\Psi^{(+)}_{\mathrm{P}}(0\mathbf{k})\rangle = |X^{(+)}_{\mathrm{P}}(0\mathbf{k})\rangle + \mathbb{G}^{(+)}_{\mathrm{P}}\, \mathbb{V}_{\mathrm{pol}} \, |\Psi^{(+)}_{\mathrm{P}}(0\mathbf{k})\rangle , \tag{10.46}$$

with the Green's function in the P-subspace given by

$$\mathbb{G}^{(+)}_{\mathrm{P}}(E) = \frac{1}{E - \mathbb{H}_{\mathrm{PP}} + i\epsilon}. \tag{10.47}$$

Using the explicit form of the polarization potential (Eq. (10.36)) in Eq. (10.46) we get

$$|\Psi^{(+)}_{\mathrm{P}}(0\mathbf{k})\rangle = |X^{(+)}_{\mathrm{P}}(0\mathbf{k})\rangle + \sum_n \mathbb{G}^{(+)}_{\mathrm{P}}\, \mathbb{V}_{\mathrm{PQ}} \, |X_{\mathrm{Q}}(n)\rangle \times \frac{1}{E-\mathcal{E}_n} \left\langle X_{\mathrm{Q}}(n)\middle|\mathbb{V}_{\mathrm{QP}}\middle|\Psi^{(+)}_{\mathrm{P}}(0\mathbf{k})\right\rangle . \tag{10.48}$$

To solve this equation, we must calculate the factor Z_n of Eq. (10.45), which is precisely the one required[6] to evaluate the T-matrix of Eq. (10.44). This is an easy task because the potential $\mathbb{V}_{\text{pol}}$ is a sum of separable terms. In such cases we can follow the procedure described in section 4.4. Let us consider the simpler situation where the T-matrix exhibits isolated resonances. For energies close to one of the eigenvalues of $\mathbb{H}_{\text{QQ}}$, say $E \simeq \mathcal{E}_n$, only one term in the sum over n appearing in Eq. (10.48) must be taken into account. Following the traditional procedure to handle separable potentials (see section 4.4), we multiply from the left both sides of Eq. (10.48) with $\langle X_{\text{Q}}(n)|\,\mathbb{V}_{\text{QP}}$. We get the equation

$$Z_n = \langle X_{\text{Q}}(n)|\,\mathbb{V}_{\text{QP}}\,|X_{\text{P}}^{(+)}(0\mathbf{k})\rangle + \langle X_{\text{Q}}(n)|\;\mathbb{V}_{\text{QP}}\;\mathbb{G}_{\text{P}}^{(+)}\;\mathbb{V}_{\text{PQ}}\;|X_{\text{Q}}(n)\rangle\;\frac{Z_n}{E-\mathcal{E}_n}, \tag{10.49}$$

which can be immediately solved for Z_n, and the result is

$$Z_n = \frac{\langle X_{\text{Q}}(n)|\,\mathbb{V}_{\text{QP}}\,\big|X_{\text{P}}^{(+)}(0\mathbf{k})\big\rangle}{E-\mathcal{E}_n - \langle X_{\text{Q}}(n)|\;\mathbb{V}_{\text{QP}}\;\mathbb{G}_{\text{P}}^{(+)}\;\mathbb{V}_{\text{PQ}}\;|X_{\text{Q}}(n)\rangle}\;(E-\mathcal{E}_n)\,. \tag{10.50}$$

It is convenient to write

$$\mathbb{G}_{\text{P}}^{(+)} \equiv \frac{1}{E-\mathbb{H}_{\text{PP}}+i\epsilon} = \mathcal{P}\left(\frac{1}{E-\mathbb{H}_{\text{PP}}}\right) - i\pi\;\delta\left(E-\mathbb{H}_{\text{PP}}\right), \tag{10.51}$$

and evaluate separately the contributions from the on-shell and off-shell parts of the Green's function to $\langle X_{\text{Q}}(n)|\;\mathbb{V}_{\text{QP}}\;\mathbb{G}_{\text{P}}^{(+)}\;\mathbb{V}_{\text{PQ}}\;|X_{\text{Q}}(n)\rangle$. Assuming that $\mathbb{V}_{\text{PQ}}$, $\mathbb{V}_{\text{QP}}$ and $X_{\text{Q}}(n)$ are real, the contributions of the on-shell and off-shell parts give rise to one real and one imaginary term, respectively. Using the notations

$$\Gamma_n = -2\;\text{Im}\Big\{\;\langle X_{\text{Q}}(n)|\;\mathbb{V}_{\text{QP}}\;\mathbb{G}_{\text{P}}^{(+)}\;\mathbb{V}_{\text{PQ}}\;|X_{\text{Q}}(n)\rangle\;\Big\} \tag{10.52}$$

and

$$\Delta_n = \;\text{Re}\Big\{\;\langle X_{\text{Q}}(n)|\;\mathbb{V}_{\text{QP}}\;\mathbb{G}_{\text{P}}^{(+)}\;\mathbb{V}_{\text{PQ}}\;|X_{\text{Q}}(n)\rangle\;\Big\} \tag{10.53}$$

in Eq. (10.50), we get

$$Z_n = \frac{\big\langle X_{\text{Q}}(n)\big|\mathbb{V}_{\text{QP}}\big|X_{\text{P}}^{(+)}(0\mathbf{k}_0)\big\rangle}{(E-E_n)+i\,\Gamma_n/2}\;(E-\mathcal{E}_n)\,, \tag{10.54}$$

where

$$E_n = \mathcal{E}_n + \Delta_n. \tag{10.55}$$

[6]Note that $\mathbb{V}_{\text{QP}}\left|\Psi_{\text{P}}^{(+)}(0\mathbf{k}_0)\right\rangle = \mathbb{V}_{\text{QP}}\left|\Psi^{(+)}(0\mathbf{k}_0)\right\rangle$ because $\mathbb{V}_{\text{QP}} = Q\mathbb{V}P$ and $P^2 = P$.

The expression for the width Γ_n can be simplified if one writes the explicit form of the imaginary part appearing in Eq. (10.52) in terms of the on-shell part of the Green's function (see Eq. (10.51)) and uses the spectral representation of $\delta(E - \mathbb{H}_{PP})$ in the case of a single doorway. One gets

$$\Gamma_n = 2\pi \ \langle X_Q(n)| \ \mathbb{V}_{QP} \ \delta(E - \mathbb{H}_{PP}) \ \mathbb{V}_{PQ} \ |X_Q(n)\rangle \, . \tag{10.56}$$

Inserting the identity,

$$\mathbb{I} = \sum_\alpha \int d^3\mathbf{k}' \ |X^{(+)}_{P,\alpha}(\mathbf{k}')\rangle \ |\alpha)\ (\alpha| \ \langle X^{(+)}_{P,\alpha}(\mathbf{k}')| \, ,$$

between the operators above, we get

$$\Gamma_n = \sum_\alpha \Gamma_{n,\alpha}, \tag{10.57}$$

where $\Gamma_{n,\alpha}$ are the partial widths given by,

$$\Gamma_{n,\alpha} = 2\pi \int d^3\mathbf{k}' \ \left| \left\langle X_Q(n) \right| \mathbb{H}_{Q\alpha} \left| X^{(+)}_{P,\alpha}(\mathbf{k}') \right\rangle \right|^2 \ \delta(E_k - E_{k'}) \, . \tag{10.58}$$

Inserting the above expression into Eq. (10.48) for an isolated resonance, we get

$$T^{\rm pol}_{\alpha\mathbf{k}_\alpha, 0\mathbf{k}_0} = \frac{\langle X^{(-)}_P(\alpha\mathbf{k}_\alpha)| \, \mathbb{V}_{PQ} | X_Q(n)\rangle \ \langle X_Q(n)|\mathbb{V}_{QP}|X^{(+)}_P(0\mathbf{k}_0)\rangle}{(E - E_n) + i\,\Gamma_n/2}. \tag{10.59}$$

If we assume that the Hamiltonians $\mathbb{H}_{PP}$, $\mathbb{H}_{QQ}$ and the couplings $\mathbb{V}_{PQ}$, $\mathbb{V}_{QP}$ are time-reversal invariant, one can use the relation

$$\langle X_Q(n)|\mathbb{V}_{QP}|X^{(+)}_P(\alpha\mathbf{k}_\alpha)\rangle = \langle X^{(-)}(\alpha, -\mathbf{k}_\alpha)| \, \mathbb{V}_{PQ}|X_Q(n)\rangle \, , \tag{10.60}$$

and Eq. (10.59) becomes

$$T^{\rm pol}_{\alpha\mathbf{k}_\alpha, 0\mathbf{k}_0} = \frac{\langle X^{(-)}_P(\alpha\mathbf{k}_\alpha)| \, \mathbb{V}_{PQ} | X_Q(n)\rangle \ \langle X^{(-)}_P(0, -\mathbf{k}_0)| \, \mathbb{V}_{PQ}|X_Q(n)\rangle}{(E - E_n) + i\,\Gamma_n/2}. \tag{10.61}$$

The proof of Eq. (10.60) is given below. First we call $|A\rangle = \mathbb{V}_{PQ}\,|X^{(+)}_P(0\mathbf{k}_0)\rangle$ and denote the time-reversed states corresponding to $|X_Q(n)\rangle$ and $|A\rangle$ by $|\bar{X}_Q(n)\rangle$ and $|\bar{A}\rangle$. That is,

$$|\bar{X}_Q(n)\rangle = \mathbb{T}\,|X_Q(n)\rangle \quad \text{and} \quad |\bar{A}\rangle = \mathbb{T}\,|A\rangle \, , \tag{10.62}$$

with $\mathbb{T}$ standing for the time-reversal operator. Now we recall that time-reversal is an anti-unitary operator and thus the transformed states have the property

$$\Big\langle X_{\rm Q}(n)\Big|A\Big\rangle = \Big\langle \bar{X}_{\rm Q}(n)\Big|\bar{A}\Big\rangle^* . \tag{10.63}$$

For isolated resonances the bound eigen-functions of the Hamiltonian $\mathbb{H}$ are real functions. Therefore, the time-reversal operation leaves them unchanged, so that $\big|\bar{X}_{\rm Q}(n)\big\rangle = |X_{\rm Q}(n)\rangle$. Using this result and writing the explicit form of $\big|\bar{A}\big\rangle$, Eq. (10.63) becomes

$$\Big\langle X_{\rm Q}(n)\Big|A\Big\rangle = \Big\langle X_{\rm Q}(n)\Big|\mathbb{T}\,\mathbb{V}_{\rm QP}\Big|X_{\rm P}^{(+)}(0\mathbf{k}_0)\Big\rangle^* . \tag{10.64}$$

If the interaction is time-reversal invariant, it commutes with the time-reversal operator. Thus we can replace in the above equation: $\mathbb{T}\,\mathbb{V}_{\rm QP} \to \mathbb{V}_{\rm QP}\,\mathbb{T}$, and write

$$\Big\langle X_{\rm Q}(n)\Big|A\Big\rangle = \Big\langle X_{\rm Q}(n)\Big|\mathbb{V}_{\rm QP}\,\mathbb{T}\Big|X_{\rm P}^{(+)}(0\mathbf{k}_0)\Big\rangle^* . \tag{10.65}$$

According to Eq. (4.122), the time-reversal conjugate of the scattering state is $\mathbb{T}\,\big|X_{\rm P}^{(+)}(0\mathbf{k}_0)\big\rangle = \big|X_{\rm P}^{(-)}(0,-\mathbf{k}_0)\big\rangle$. Using this result and evaluating the complex conjugate indicated on the RHS of Eq. (10.65), we get the desired relation,

$$\Big\langle X_{\rm Q}(n)\Big|A\Big\rangle = \Big\langle X_{\rm P}^{(-)}(0,-\mathbf{k}_0)\Big|\mathbb{V}_{\rm PQ}\Big|X_{\rm Q}(n)\Big\rangle^* . \tag{10.66}$$

In the simple case of spinless intrinsic states and interactions that conserve the orbital angular momentum, one expands the T-matrix in partial-waves and the l-components are given in terms of the l-components of the matrix-elements appearing in the numerator of Eq. (10.59). That is [Levin and Feshbach (1973)]

$$\big\langle X_{{\rm P},\alpha}^{(-)}(\alpha\mathbf{k}_\alpha)\big|\,\mathbb{V}_{\rm PQ}\big|X_{\rm Q}(n)\big\rangle \to \big\langle X_{{\rm P},\alpha}^{(-)}(\alpha\mathbf{k}_\alpha)\big|\,\mathbb{V}_{\rm PQ}\big|X_{\rm Q}(n)\big\rangle_l = \frac{e^{-i\delta_\alpha}}{\sqrt{2\pi}}\sqrt{\Gamma_{n,\alpha}}. \tag{10.67}$$

Accordingly, the l-projected resonant part of the T-matrix takes the form

$$T_{\alpha,0}^{\rm pol} = \frac{e^{i(\delta_0+\delta_\alpha)}}{2\pi}\left[\frac{\sqrt{\Gamma_{n,0}\cdot\Gamma_{n,\alpha}}}{(E-E_n)+i\,\Gamma_n/2}\right] . \tag{10.68}$$

Above, the indices 0 and α label the initial and final channels, and also the initial and final angular momenta in the partial wave expansion. Of course, owing to our assumption that the interactions conserve orbital angular momentum, one must have $l_\alpha = l_0$. The above form of the resonant T-matrix is universally known as the Breit-Wigner form.

The full T-matrix then becomes

$$T_{\alpha,0} = T^{\mathrm{P}}_{0,0}\ \delta_{\alpha 0} + \frac{e^{i(\delta_0+\delta_\alpha)}}{2\pi} \left[\frac{\sqrt{\Gamma_{n,0}\cdot\Gamma_{n,\alpha}}}{(E-E_n)+i\,\Gamma_n/2}\right] \tag{10.69}$$

or, in terms of the S-matrices associated with $\mathbb{H}_{\mathrm{PP}}$, $S^{\mathrm{P}}_{j,j} = \exp[2i\delta_j]$,

$$T_{\alpha,0} = T^{\mathrm{P}}_{0,0}\ \delta_{\alpha 0} + \frac{1}{2\pi}\sqrt{S^{\mathrm{P}}_{\alpha,\alpha}} \left[\frac{\sqrt{\Gamma_{n,0}\cdot\Gamma_{n,\alpha}}}{(E-E_n)+i\,\Gamma_n/2}\right] \sqrt{S^{\mathrm{P}}_{0,0}}. \tag{10.70}$$

Since we are assuming that the Hamiltonian $\mathbb{H}_{\mathrm{PP}}$ is diagonal in channel space and conserves the orbital angular momentum, the non-resonant part of the T-matrix, $T^{\mathrm{P}}_{\alpha,0}$, is diagonal.

The above result can immediately be extended to the case of several isolated resonances. It becomes

$$T_{\alpha,0} = T^{\mathrm{P}}_{0,0}\ \delta_{\alpha 0} + \frac{1}{2\pi}\sqrt{S^{\mathrm{P}}_{\alpha,\alpha}} \left[\sum_n \frac{\sqrt{\Gamma_{n,0}\cdot\Gamma_{n,\alpha}}}{(E-E_n)+i\,\Gamma_n/2}\right] \sqrt{S^{\mathrm{P}}_{0,0}}. \tag{10.71}$$

In the more general case of overlapping resonances the resonance wave functions are not real and $|\bar{X}_{\mathrm{Q}}(n)\rangle$ is not equal to $|X_{\mathrm{Q}}(n)\rangle$. However, even in this case, as long as time-reversal-symmetry is obeyed, the matrix elements appearing in the resonant part of the T-matrix are such as to guarantee that the T-matrix is symmetric and detailed balance holds (see below).

From the definition of the S-matrix in chapter 4, we have,

$$S_{\alpha,0} = 1 - 2\pi i\ T_{\alpha,0}. \tag{10.72}$$

For the T-matrix of Eq. (10.71), the S-matrix becomes

$$\begin{aligned} S_{\alpha,0} &= 1 - 2\pi i\ \ T^{\mathrm{P}}_{\alpha,0}\ \delta_{\alpha,0} \\ &\quad - i\ \sqrt{S^{\mathrm{P}}_{\alpha,\alpha}} \left[\sum_n \frac{\sqrt{\Gamma_{n,0}\cdot\Gamma_{n,\alpha}}}{(E-E_n)+i\,\Gamma_n/2}\right] \sqrt{S^{\mathrm{P}}_{0,0}}. \end{aligned} \tag{10.73}$$

An important property of the above S-matrix is its symmetry,

$$S_{\alpha,\beta} = S_{\beta,\alpha}, \tag{10.74}$$

which is a direct consequence of time reversal symmetry. This symmetry of the S-matrix or the T-matrix guarantees the principle of detailed balance in the sense that the cross section for the reaction that involves the transition,

$\alpha \to \beta$, is, aside from simple kinematical factors, directly related to that for the reaction of the inverse transition, $\beta \to \alpha$.

It is convenient to write the S-matrix as a sum of a potential scattering term, $S^{\rm P}$, and a resonance term, $S^{\rm pol}$, in the form

$$S = S^{\rm P} + S^{\rm pol}, \tag{10.75}$$

where

$$S^{\rm P}_{\alpha,0} = 1 - 2\pi i\, \delta_{\alpha,0}\, T^{\rm P}_{\alpha,0} \tag{10.76}$$

and

$$S^{\rm pol}_{\alpha,0} = -2\pi i\; T^{\rm pol}_{\alpha,0}. \tag{10.77}$$

In the situation where the P-subspace contains only the elastic channel and $Q + P$ exhausts the channel space, the above S-matrix is unitary. That is

$$\sum_{\beta} S_{\alpha,\beta}\; S^{*}_{\beta,\alpha'} = \delta_{\alpha\alpha'}. \tag{10.78}$$

or simply

$$SS^{\dagger} = 1. \tag{10.79}$$

Using the decomposition of Eq. (10.75), the above equation becomes

$$S^{\rm P}\, S^{\rm P\,\dagger} + S^{\rm pol}\, S^{\rm pol\,\dagger} + S^{\rm P}\, S^{\rm pol\,\dagger} + S^{\rm pol}\, S^{\rm P\,\dagger} = 1. \tag{10.80}$$

But the potential scattering S-matrix, $S^{\rm P}$, is itself unitary on account of the hermiticity of $\mathbb{H}_{\rm PP}$, which implies $S^{\rm P}\, S^{\rm P\,\dagger} = 1$. Accordingly we arrive at the following important relation concerning the resonant S-matrix,

$$S^{\rm pol}\, S^{\rm pol\,\dagger} + S^{\rm P}\, S^{\rm pol\,\dagger} + S^{\rm pol}\; S^{\rm P\,\dagger} = 0. \tag{10.81}$$

For the general case of non-elastic transition involving several isolated resonances,

$$\sum_{\gamma} S^{\rm pol}_{\alpha,\gamma}\; S^{\rm pol\,*}_{\beta,\gamma} + S^{\rm P}_{\alpha,\alpha}\; S^{\rm pol\,*}_{\alpha,\beta} + S^{\rm pol}_{\alpha,\beta}\, S^{\rm P\,*}_{\beta,\beta} = 0, \tag{10.82}$$

which can easily be verified to be an identity *if* the resonance part of the S-matrix has the general form of the third term of Eq. (10.73), and only one resonance is involved. Otherwise, the relation above can be considered as a constraining condition on the resonant contribution.

Owing to its wide use in the literature on atomic, and nuclear scattering, we give below the elastic scattering S-matrix element. The resonant part of the elastic T-matrix for an isolated resonance is (second term on the RHS of Eq. (10.70), dropping the sum over n)

$$T^{\text{pol}}_{0,0} = \frac{1}{2\pi}\, S^{\text{P}}_{0,0} \left[\frac{\Gamma_{n,0}}{E - E_n + i\,\Gamma_n/2}\right]. \tag{10.83}$$

Thus, according to Eq. (10.77), we may write

$$S^{\text{pol}}_{0,0} = -i\, S^{\text{P}}_{0,0} \left[\frac{\Gamma_{n,0}}{E - E_n + i\,\Gamma_n/2}\right], \tag{10.84}$$

and the full S-matrix becomes

$$S_{0,0} = S^{\text{P}}_{0,0} \left[1 - i\, \frac{\Gamma_{n,0}}{E - E_n + i\,\Gamma_n/2}\right]. \tag{10.85}$$

The resonant component $S^{\text{pol}}_{0,0}$ is a typical Breit-Wigner shaped S-matrix element.

To evaluate the elastic cross section, it is convenient to use a more compact notation for the S-matrix elements. First, we recall that the indices 0 and α label the channel and the orbital angular momentum, that is $0 \equiv \{0, l_0\}$ and $\alpha \equiv \{\alpha, l_\alpha\}$. Thus, we can write explicitly: $S_{\alpha,0} \equiv S_{\alpha l_\alpha, 0 l_0}$. Since we are assuming that the orbital angular momentum is conserved, $S_{\alpha l_\alpha, 0 l_0} = S_{\alpha l_0, 0 l_0}\, \delta_{l_\alpha, l_0}$. Therefore, the elastic amplitude will only involve S-matrix elements of the form $S_{0 l_0, 0 l_0}$. For these matrix elements, we can adopt the simpler notation $S_{0l,0l} = S_l$. In this case, the resonant amplitude of Eq. (10.84) becomes

$$S^{\text{pol}}_{l} = -\,i\, S^{\text{P}}_{l} \left[\frac{\Gamma_{n,0}(l_n)}{E - E_n + i\,\Gamma_n/2}\right] \delta_{l,l_n}. \tag{10.86}$$

Above, we used the notation $\Gamma_{0,n}(l_n)$ to remind that $\Gamma_{0,n}$ depends on the orbital angular momentum of the resonant state $X_{\text{Q}}(n)$. The Kronecker delta, δ_{ll_n}, comes from the definition of the partial width (Eq. (10.58)) and the conservation of orbital angular momentum.

The scattering amplitude can be evaluated through the partial-wave expansion (see chapter 2)

$$f(\theta) = \frac{1}{2ik} \sum_l (2l+1)\, [S_l - 1]\, P_l(\cos\theta). \tag{10.87}$$

Inserting in the above equation the S-matrix of Eqs. (10.75) to (10.86), we get

$$f(\theta) = f^{\text{P}}(\theta) + f^{\text{pol}}(\theta), \tag{10.88}$$

with

$$f^{\mathrm{P}}(\theta) = \frac{1}{2ik} \sum_{l} (2l+1) \left[S_l^{\mathrm{P}} - 1\right] P_l(\cos\theta) \tag{10.89}$$

and

$$f^{\mathrm{pol}}(\theta) = -\frac{1}{2k} (2l_n+1)\, S_{l_n}^{\mathrm{P}} \left[\frac{\Gamma_{n,0}(l_n)}{E - E_n + i\,\Gamma_n/2}\right] P_{l_n}(\cos\theta)\,. \tag{10.90}$$

The resonant scattering amplitude for several isolated resonances, can trivially be obtained from the above equation, performing a summation over n.

At very low energies, only the s-wave contribution has to be taken into account. In the case of a resonance with $l_n = 0$, the scattering amplitude becomes

$$f = \frac{1}{2ik}\left[e^{2i\delta_0^{\mathrm{P}}}\left(1 - i\frac{\Gamma_n}{E - E_n + i\Gamma_n/2}\right) - 1\right], \tag{10.91}$$

or

$$f = \frac{1}{2ik}\left[e^{2i\delta_0^{\mathrm{P}}}\left(\frac{E - E_n - i\Gamma_n/2}{E - E_n + i\Gamma_n/2}\right) - 1\right]. \tag{10.92}$$

The above formula can be expressed in terms of a total (non-resonant plus resonant) s-phase shift as,

$$f = \frac{1}{2ik}\left[e^{2i\delta_0} - 1\right], \tag{10.93}$$

where

$$\delta_0 = \delta_0^{\mathrm{P}} + \delta_0^{\mathrm{res}} \tag{10.94}$$

and

$$\delta_0^{\mathrm{res}} = \tan^{-1}\left[\frac{\Gamma_n/2}{E_n - E}\right]. \tag{10.95}$$

Clearly, the phase shift attains the value of $\pi/2$, whenever the energy becomes equal to the Q-resonance energy, E_n. The background, potential, phase shift would also attain such a value if the energy is close to that of a *potential* resonance such as the ones discussed in section 2.9.

The Q-resonances discussed above are collectively referred to as *Feshbach Resonances* (FR). These, many-body, resonances are of great importance both in nuclear and atomic physics. In nuclear reactions, the FR are called the Compound Nucleus resonances, while the FR in atom-atom collisions are simply called as Feshbach resonances. They are of paramount practical importance in the physics of cold atom gases and in particular in the creation and manipulation of Bose-Einstein condensates in these gases. Further discussion of FR in atom-atom collision will be given in chapter 12.

In very low energy nuclear reactions induced by thermal or epithermal neutrons, the only process which can occur is elastic scattering. The capture of the neutron by a target such as ^{238}U, as usually occurs in reactors, produces the compound nuclear resonances in ^{239}U, which are described by an expression such as the above. These resonances are labeled by their parity and angular momentum, which for thermal neutrons is just $l = 0$. Since the neutron has a spin 1/2, it is this angular momentum which couples with the spin of the low lying states in ^{238}U to give the resonances spins. In figure 10.1, we show the total cross section of the n + ^{238}U system at very low energies. The data are from [Firk *et al.* (1960)]. We see clearly the isolated nature of the resonances. Another example of resonances which occur in more complex nuclear systems is the so-called quasi-molecular resonances in the scattering of carbon on carbon. The intermediate states here are complex quasi-bound states in magnesium. Thus we may think of the elastic scattering cross section at a given angle as a function of the center of mass energy,

$$^{12}\mathrm{C} + ^{12}\mathrm{C} \ \rightarrow \ ^{24}\mathrm{Mg} \ \rightarrow \ ^{12}\mathrm{C} + ^{12}\mathrm{C}. \tag{10.96}$$

In figure 10.2, we show the data points, taken from [Erb and Bromley (1984)]. As customary in low energy fusion research of relevance to astrophysics, the cross section is reduced according to the following,

$$\mathcal{S}(E) = [E\,\sigma_{\mathrm{F}}(E)]\ e^{2\pi\eta}, \tag{10.97}$$

where η is the usual Sommerfeld parameter (see chapter 3). The function $\mathcal{S}(E)$ is usually referred to as the *astrophysical S-factor*, or *the nuclear structure factor*. This function, which will be discussed in detail in chapter 12, has the advantage over $\sigma_F(E)$, in that the very steep drop at low energies is removed. Clearly we see resonances of the type described above. However, a more careful examination of the magnesium resonances will show that their widths are not just the sum of the partial widths, but contain other pieces which indicates decay to closed channels, namely more complicated states of the di-carbon system. This phenomenon, known as intermediate structure, will be discussed further in the context of doorway states. We remind the reader that the resonances we are discussing here are isolated resonances and accordingly they should satisfy the unitarity sum rule

$$\sum_{\alpha} \Gamma_{n,\alpha} = \Gamma_n. \tag{10.98}$$

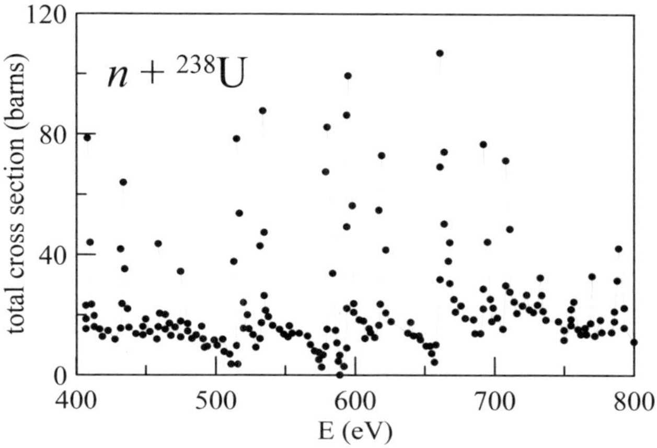

Fig. 10.1 Total cross section for thermal neutrons capture by ^{238}U, as a function of the neutron energy. The data are taken from [Firk *et al.* (1960)].

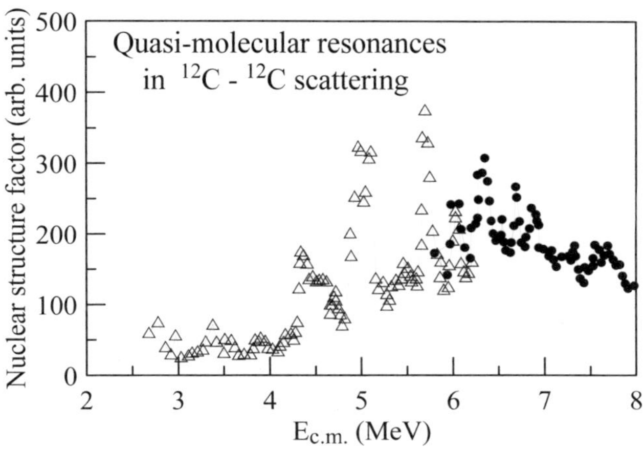

Fig. 10.2 The astrophysical S-factor for the $^{12}\mathrm{C} + ^{12}\mathrm{C}$ system. See the text and [Erb and Bromley (1984)] for further information about the data.

10.3 Energy averages and the optical model

One of the major concepts used in nuclear physics is that of the optical potential. The idea is to perform several approximations on V_{eff} to make it appropriate for use in the description of scattering in the subspace P. If P

projects onto the elastic channel only, then V_{eff} is a one-channel operator. It describes how the elastic channel is depleted *on the average* owing to the coupling with all other channels. Of course an exact calculation of V_{eff} would be as difficult as solving the exact problem. Thus, it is necessary to resort to approximations.

The basic idea of the optical model is to neglect rapid energy-oscillations of the elastic S-matrix, which corresponds to delayed scattering processes. For this purpose, we look for an average scattering state, given by

$$\left|\, \overline{\Psi_{\text{P}}^{(+)}(E)} \right\rangle = \int \rho(E-E')\, |\Psi_{\text{P}}^{(+)}(E')\rangle \; dE'. \qquad (10.99)$$

The weight function $\rho(E-E')$ is centered at $E' = E$ and has a width I which should be large enough to contain several sharp resonances. On the other hand, it should be such that the variation of $\left|\, \overline{\Psi_{\text{P}}^{(+)}(E)} \right\rangle$ over the interval I is small. For simplicity, we make the assumption that the P-subspace contains only the elastic channel[7], that is $P = |0)\,(0|$, and use a compact notation in which the explicit energy dependence and the outgoing nature of the scattering state are suppressed.

With the above assumption, the optical model wave function for the single channel scattering problem is

$$|\psi_{\text{opt}}\rangle = \big(0 \,\big|\, \overline{\Psi_{\text{P}}}\, \big\rangle. \qquad (10.100)$$

The optical potential, V_{opt}, is then defined by the condition that the average wave function satisfies the Schrödinger equation

$$\big[E - K - V_{\text{opt}}\big]\; |\psi_{\text{opt}}\rangle = 0. \qquad (10.101)$$

We now calculate the optical potential. A first attempt would be to take energy-averages on the Schrödinger equation with the effective potential (see Eq. (10.37)),

$$\big[E - \mathbb{H}_{\text{PP}} - \mathbb{V}_{\text{pol}}\big]\; |\Psi_{\text{P}}\rangle = 0. \qquad (10.102)$$

However, this procedure is not appropriate because both $\mathbb{V}_{\text{pol}}$ and $|\Psi_{\text{P}}\rangle$ present sharp energy-oscillations in the neighborhood of the resonances. Thus, one cannot average the wave function separately. It is then necessary to start from the equations coupling $|\Psi_{\text{P}}\rangle$ and $|\Psi_{\text{Q}}\rangle$. The formal solution of Eq. (10.7) is

$$|\Psi_{\text{P}}\rangle = |X_{\text{P}}\rangle \;+\; \mathbb{G}_{\text{P}}^{(+)}\, \mathbb{H}_{\text{PQ}}\, |\Psi_{\text{Q}}\rangle\,, \qquad (10.103)$$

[7]For a more general derivation, we refer to [Levin and Feshbach (1973); Feshbach (1992)].

where $|X_{\rm P}\rangle$ is the corresponding solution of the scattering problem when the couplings with the Q-subspace are neglected. That is, it satisfies the equation

$$[E - \mathbb{H}_{\rm PP}]\,|X_{\rm P}\rangle = 0. \tag{10.104}$$

Taking energy averages of Eq. (10.103), we get

$$\left|\,\overline{\Psi_{\rm P}}\,\right\rangle = |X_{\rm P}\rangle \;+\; \mathbb{G}_{\rm P}^{(+)}\,\mathbb{H}_{\rm PQ}\left|\,\overline{\Psi_{\rm Q}}\,\right\rangle. \tag{10.105}$$

Since $|X_{\rm P}\rangle\,, \mathbb{G}_{\rm P}^{(+)}$ and $\mathbb{H}_{\rm PQ}$ varies very slowly with energy, they are not affected by the averaging procedure. Thus, they can be moved out of integrals like that of Eq. (10.99). Inserting Eq. (10.103) into Eq. (10.8), we obtain

$$[E - \mathbb{H}_{\rm QQ}]\,|\Psi_{\rm Q}\rangle = \mathbb{H}_{\rm QP}\,|X_{\rm P}\rangle + \mathbb{H}_{\rm QP}\,\mathbb{G}_{\rm P}^{(+)}\,\mathbb{H}_{\rm PQ}\,|\Psi_{\rm Q}\rangle\,, \tag{10.106}$$

or

$$[E - \mathbb{H}_{\rm QQ} - \mathbb{W}_{\rm Q}]\,|\Psi_{\rm Q}\rangle = \mathbb{H}_{\rm QP}\,|X_{\rm P}\rangle\,. \tag{10.107}$$

Above, $\mathbb{W}_{\rm Q}$ is the polarization potential in the Q-subspace, given by

$$\mathbb{W}_{\rm Q} = \mathbb{H}_{\rm QP}\,\mathbb{G}_{\rm P}^{(+)}\,\mathbb{H}_{\rm PQ}. \tag{10.108}$$

Multiplying Eq. (10.107) from the left with the Green's operator

$$\mathcal{G}_{\rm Q} = \frac{1}{E - \mathbb{H}_{\rm QQ} - \mathbb{W}_{\rm Q}}, \tag{10.109}$$

we get

$$|\Psi_{\rm Q}\rangle = \mathcal{G}_{\rm Q}\,\mathbb{H}_{\rm QP}\,|X_{\rm P}\rangle \tag{10.110}$$

and the energy-averaged version of the above equation is

$$\left|\,\overline{\Psi_{\rm Q}}\,\right\rangle = \overline{\mathcal{G}_{\rm Q}}\,\mathbb{H}_{\rm QP}\,|X_{\rm P}\rangle\,. \tag{10.111}$$

Inserting this equation into Eq. (10.105), we obtain

$$\left|\,\overline{\Psi_{\rm P}}\,\right\rangle = \left[1 \;+\; \mathbb{G}_{\rm P}^{(+)}\,\mathbb{H}_{\rm PQ}\,\overline{\mathcal{G}_{\rm Q}}\,\mathbb{H}_{\rm QP}\right]|X_{\rm P}\rangle \tag{10.112}$$

and thus

$$|X_{\rm P}\rangle = \frac{1}{1 \;+\; \mathbb{G}_{\rm P}^{(+)}\,\mathbb{H}_{\rm PQ}\,\overline{\mathcal{G}_{\rm Q}}\,\mathbb{H}_{\rm QP}}\left|\,\overline{\Psi_{\rm P}}\,\right\rangle. \tag{10.113}$$

Using this result in Eq. (10.111), we get

$$\left|\,\overline{\Psi_{\rm Q}}\,\right\rangle = \overline{\mathcal{G}_{\rm Q}}\,\mathbb{H}_{\rm QP}\,\frac{1}{1 \;+\; \mathbb{G}_{\rm P}^{(+)}\,\mathbb{H}_{\rm PQ}\,\overline{\mathcal{G}_{\rm Q}}\,\mathbb{H}_{\rm QP}}\left|\,\overline{\Psi_{\rm P}}\,\right\rangle. \tag{10.114}$$

Now we multiply Eq. (10.105) from the left with the operator $E - \mathbb{H}_{\text{PP}}$. The first term on the RHS vanishes because $|X\rangle$ satisfies Eq. (10.104). The remaining terms can be regrouped in the equation

$$\Big[E - K - \mathbb{V}_{\text{opt}}\Big] \,|\,\overline{\Psi_{\text{P}}}\,\rangle = 0. \tag{10.115}$$

We have introduced the optical potential, $\mathbb{V}_{\text{opt}}$, which is defined as $\mathbb{V}_{\text{PP}}$ plus the energy-averaged polarization potential associated with the Q-subspace. It is given by the expression,

$$\mathbb{V}_{\text{opt}} \equiv \mathbb{V}_{\text{PP}} + \overline{\mathbb{V}}_{\text{pol}} = \mathbb{V}_{\text{PP}} + \mathbb{H}_{\text{PQ}}\,\overline{\mathcal{G}_{\text{Q}}}\,\mathbb{H}_{\text{QP}}\,\frac{1}{1\ +\ \mathbb{G}_{\text{P}}^{(+)}\,\mathbb{H}_{\text{PQ}}\,\overline{\mathcal{G}_{\text{Q}}}\,\mathbb{H}_{\text{QP}}}. \tag{10.116}$$

The above equation can be simplified if we use some relations of operators. The first one is [Levin and Feshbach (1973)]

$$AB\,\frac{1}{1+CAB} = A\,\frac{1}{1+BCA}\,B \tag{10.117}$$

for

$$A = \overline{\mathcal{G}_{\text{Q}}}, \qquad B = \mathbb{H}_{\text{QP}} \quad \text{and} \quad C = \mathbb{G}_{\text{P}}^{(+)}\,\mathbb{H}_{\text{PQ}}. \tag{10.118}$$

The optical potential then becomes

$$\mathbb{V}_{\text{opt}} = \mathbb{V}_{\text{PP}} + \mathbb{H}_{\text{PQ}}\left[\overline{\mathcal{G}_{\text{Q}}}\,\frac{1}{1\ +\ \mathbb{H}_{\text{QP}}\,\mathbb{G}_{\text{P}}^{(+)}\,\mathbb{H}_{\text{PQ}}\,\overline{\mathcal{G}_{\text{Q}}}}\right]\mathbb{H}_{\text{QP}} \tag{10.119}$$

or, according to Eq. (10.108)

$$\mathbb{V}_{\text{opt}} = \mathbb{V}_{\text{PP}} + \mathbb{H}_{\text{PQ}}\left[\overline{\mathcal{G}_{\text{Q}}}\ \frac{1}{1\ +\ \mathbb{W}_{\text{Q}}\,\overline{\mathcal{G}_{\text{Q}}}}\right]\ \mathbb{H}_{\text{QP}} \tag{10.120}$$

Now we use the identity

$$A\,B^{-1} = \left[B\,A^{-1}\right]^{-1}, \tag{10.121}$$

within the square brackets. Taking

$$A = \overline{\mathcal{G}_{\text{Q}}} \quad \text{and} \quad B = 1\ +\ \mathbb{G}_{\text{P}}^{(+)}\,\mathbb{H}_{\text{PQ}}\,\overline{\mathcal{G}_{\text{Q}}}\,, \tag{10.122}$$

we get

$$\mathbb{V}_{\text{opt}} = \mathbb{V}_{\text{PP}} + \mathbb{H}_{\text{PQ}}\left[\ \left(1\ +\ \mathbb{W}_{\text{Q}}\,\overline{\mathcal{G}_{\text{Q}}}\right)\ \left(\overline{\mathcal{G}_{\text{Q}}}\right)^{-1}\ \right]^{-1}\mathbb{H}_{\text{QP}}. \tag{10.123}$$

or

$$\mathbb{V}_{\text{opt}} = \mathbb{V}_{\text{PP}} + \mathbb{H}_{\text{PQ}}\left[\frac{1}{\left(\overline{\mathcal{G}_{\text{Q}}}\right)^{-1}\ +\ \mathbb{W}_{\text{Q}}}\right]\mathbb{H}_{\text{QP}}. \tag{10.124}$$

An important feature of the optical potential is that the energy averaging procedure leads to a complex potential, even when the Q-subspace contains only closed channels. To see this, we need to specify the weight function. One possible choice is the Breit-Wigner form,

$$\rho(E - E') = \frac{I}{2\pi} \left[\frac{1}{(E - E')^2 + (I/2)^2} \right]. \tag{10.125}$$

The calculation of the energy averages can now be performed using the method of residues, once we accept that $I \gg |\mathrm{Im}\, \mathbb{W}_{\mathrm{Q}}|$. Since $\mathrm{Im}\, \mathbb{W}_{\mathrm{Q}}$ is negative, the contour of integration is chosen to be the half circle in the upper half of the complex energy plane. The only pole that contributes is at $E' = E + i\, I/2$. Accordingly,

$$\begin{aligned} \overline{\mathcal{G}_{\mathrm{Q}}} &= \frac{I}{2\pi} \int dE' \left[\frac{1}{(E - E')^2 + (I/2)^2} \right] \frac{1}{E' - \mathbb{H}_{\mathrm{QQ}} - \mathbb{W}_{\mathrm{Q}}} \\ &= \frac{1}{E + i\, I/2 - \mathbb{H}_{\mathrm{QQ}} - \mathbb{W}_{\mathrm{Q}}}. \end{aligned} \tag{10.126}$$

Using this result in Eq. (10.124), the term $\mathbb{W}_{\mathrm{Q}}$ is canceled out and the denominator reduces to $E + i\, I/2 - \mathbb{H}_{\mathrm{QQ}}$, which is complex. Thus the optical potential arising from the coupling to the closed channels contained in Q is complex and is given by,

$$\mathbb{V}_{\mathrm{opt}} = \mathbb{V}_{\mathrm{PP}} + \mathbb{H}_{\mathrm{PQ}} \left[\frac{1}{E + i\, I/2 - \mathbb{H}_{\mathrm{QQ}}} \right] \mathbb{H}_{\mathrm{QP}}. \tag{10.127}$$

The energy-averaged scattering wave function thus satisfies the equation,

$$\left[E - K - \mathbb{V}_{\mathrm{opt}} \right] \left| \overline{\Psi_{\mathrm{P}}} \right\rangle = 0. \tag{10.128}$$

With our assumption that the P-subspace contains only the elastic channel, one can multiply Eq. (10.128) from the left with $(0|$ and obtain the simple equation involving only the collision degrees of freedom

$$[E - K - V_{\mathrm{opt}}]\, |\psi_{\mathrm{opt}}\rangle = 0, \tag{10.129}$$

with

$$V_{\mathrm{opt}} = (0|\, \mathbb{V}_{\mathrm{PP}}\, |0) + (0|\, \overline{\mathbb{V}}_{\mathrm{pol}}\, |0)\,. \tag{10.130}$$

It is interesting that after the above lengthy calculation, one reaches a very simple prescription of how to obtain the optical potential operator.

The exact two coupled set of equations, Eqs. (10.7) and (10.8), are merely replaced by,

$$\left[E - \mathbb{H}_{\rm PP}\right] \left|\overline{\Psi_{\rm P}^{(+)}}\right\rangle = \mathbb{H}_{\rm PQ} \left|\overline{\Psi_{\rm Q}}\right\rangle \tag{10.131}$$

and

$$\left[E + i\frac{I}{2} - \mathbb{H}_{\rm QQ}\right] \left|\overline{\Psi_{\rm Q}}\right\rangle = \mathbb{H}_{\rm QP} \left|\overline{\Psi_{\rm P}^{(+)}}\right\rangle. \tag{10.132}$$

Accordingly, the equation for the energy averaged P-subspace wave function,

$$\left[E - K - \mathbb{V}_{\rm PP} - \mathbb{H}_{\rm PQ}\left(\frac{1}{E + i\,I/2 - \mathbb{H}_{\rm QQ}}\right)\mathbb{H}_{\rm QP}\right] \left|\overline{\Psi_{\rm P}^{(+)}}\right\rangle = 0. \tag{10.133}$$

The presence of the imaginary term $i\,I/2$ guarantees that the rapid energy fluctuations present in the un-averaged polarization potential of Eq. (10.34) are removed. What happens if one chooses a different weight function? It would seem physically that regardless to the type of weight function used to perform the averaging, be it a Breit-Wigner or any other function centered at $E' = E$ and width $I \gg |\mathrm{Im}\, \mathbb{W}_Q|$, the rapid energy variation is removed. We may thus conjecture for the average wave function in the P-subspace the following,

$$\left[E - K - \mathbb{V}_{\rm PP} - \mathbb{H}_{\rm PQ}\ \overline{(E - \mathbb{H}_{\rm QQ})^{-1}}\ \mathbb{H}_{\rm QP}\right] \left|\overline{\Psi_{\rm P}^{(+)}}\right\rangle = 0. \tag{10.134}$$

One final note: the difference between the exact wave function $|\Psi_{\rm P}^{(+)}\rangle$, and the average one, $|\overline{\Psi_{\rm P}^{(+)}}\rangle$, which we shall call the fluctuation part of the wave function, $|\Psi_{\rm P}^{\rm fl}\rangle$, must, by construction average to zero,

$$\left|\overline{\Psi_{\rm P}^{\rm fl}}\right\rangle = 0. \tag{10.135}$$

However, we shall see in chapter 12 that the fluctuation wave function has a very important rôle in determining the energy averaged cross section.

10.4 Optical potentials

Measurements of the elastic cross section with high energy resolution show sharp fluctuations with the collision energy. These oscillations depend strongly on the collision partners, even for very similar systems. The abrupt energy oscillations arise from couplings with long-lived resonances, which are determined by particular details of the nuclear structure of the projectile and/or the target. Thus, making theoretical predictions of such cross sections is an extremely hard task. On the other hand, scattering data with lower resolutions exhibit a smooth energy dependence, and the cross sections for similar systems show minor differences. This is because the resolution of the experiment washes out the sharp resonances. Thus, only the average behavior of the cross sections survives. Therefore, data of this kind are suitable for comparisons with energy averaged theoretical predictions. However, there is an important difference: averaging the effective potential is not the same as averaging the cross section. The latter contains an additional contribution which is associated with the decay of the long-lived resonances through the elastic channel. This contribution, which may be relevant in low energy collisions, can be estimated by the statistical model (see chapter 12). In this section we discuss the main features of the optical potential and their cross sections.

Optical potentials are widely used in Nuclear Physics. However, a direct evaluation of $\overline{\mathbb{V}}_{\rm pol}$, as described in the previous section, is extremely complicated. For this reason, it is hardly carried out. Instead, approximate or phenomenological approaches are adopted. In some cases, one looks for a parametrized form of the complex optical potential that changes slowly with the collision energy and with the mass and charge of the collision partners. Potentials with such characteristics are quite useful in the description of the gross properties of elastic scattering and for generating distorted waves for DWBA approximations in inelastic scattering or transfer reactions. In some studies, the whole optical potential is treated phenomenologically. Other authors evaluate $V_{\rm P} = (0|\mathbb{V}_{\rm PP}|0)$ using some physical model and treat the term $(0|\,\overline{\mathbb{V}}_{\rm pol}\,|0)$ phenomenologically. In nuclear physics, the optical potential always contains a short-range part, arising from nuclear forces. If the projectile and the target are both charged, it contains also a long-range Coulomb term. This term is usually approximated by the potential of a point charge interacting with an homogeneous spherically symmetric charge distribution, as discussed in section 3.4.1 (see Eq. (3.130)). However, some authors use better approximations for the Coulomb potential.

In the next two sub-sections we discuss some physical models used to estimate the real part of the nuclear contribution to the optical potential.

10.4.1 *Optical potentials in nucleon-nucleus collisions*

Soon after the introduction of the optical model, Perey and Buck [Perey and Buck (1962)] developed a phenomenological method to evaluate optical potentials in the scattering of low-energy neutrons from nuclei. This potential is in general non-local due to the exchange effet. They adopted the particular non-local model for the optical potential,

$$V_{\rm opt}(\mathbf{r},\mathbf{r}') = V_{\rm NL}(\bar{r}) \times F\,(s)\,, \tag{10.136}$$

where $\bar{r}$ is the average projectile-target separation,

$$\bar{r} = |\bar{\mathbf{r}}| = \frac{|\mathbf{r}+\mathbf{r}'|}{2} \tag{10.137}$$

and s is the modulus of the non-locality,

$$s = |\mathbf{s}| = |\mathbf{r}-\mathbf{r}'|\,. \tag{10.138}$$

The non-locality factor was approximated by the gaussian function

$$F(s) = \frac{1}{\pi^{3/2}\,\beta^3}\; e^{-s^2/\beta^2}. \tag{10.139}$$

It was given in terms of the range parameter, β, and was normalized according to the condition

$$\int F(s)\; d^3\mathbf{s} = 1. \tag{10.140}$$

The potential, $V_{\rm NL}(\bar{r})$ was assumed to be the complex function

$$V_{\rm NL}(\bar{r}) = U_{\rm NL}(\bar{r}) \,+\, i\,W_{\rm NL}(\bar{r}). \tag{10.141}$$

Its real part, $U_{\rm NL}(\bar{r})$, was parametrized as

$$U_{\rm NL}(\bar{r}) = -U_{\rm vol}\;f_{\rm vol}(\bar{r}), \tag{10.142}$$

where $U_{\rm vol}$ is a strength parameter and $f_{\rm vol}(\bar{r})$ gives its $\bar{r}$-dependence. They used the Woods-Saxon shape

$$f_{\rm vol}(\bar{r}) = \frac{1}{1+\exp\left[\left(\bar{r}-R\right)/a_{\rm r}\right]}. \tag{10.143}$$

Thus, the real part of the potential was expressed in terms of three parameters. Besides its strength, $U_{\rm vol}$, there were the radius parameter, R, and the diffusivity, $a_{\rm r}$. The former was related with the mass number of the target as $R = r_0\,A_{\rm T}^{1/3}$. As an illustration, we show in figure 10.3 the function $f_{\rm vol}(\bar{r})$ appropriate for the $n+^{208}$Pb collision. It is represented by the solid line. In this example, we used $r_0 = 1.2$ fm and $a_{\rm r} = 0.6$ fm. Note that

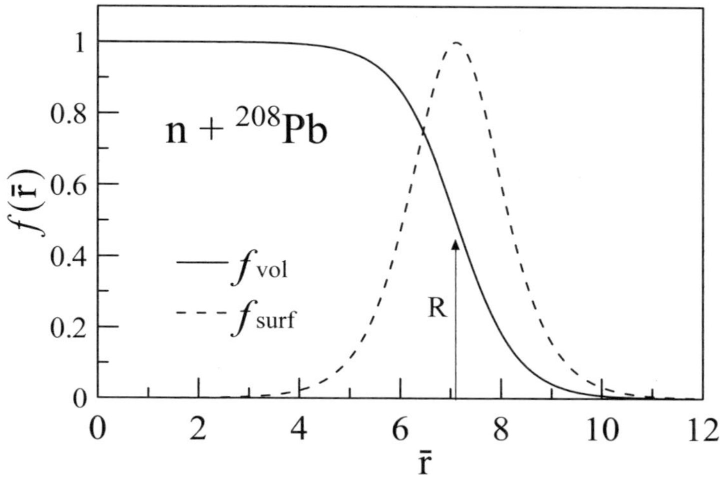

Fig. 10.3 The shape functions of the diagonal term of the optical potential in the non-local parametrization of Perey and Buck [Perey and Buck (1962)]. In this illustration, we use the same parameters for the volume and surface terms. The value of the radius parameter, R, is indicated by an up-arrow. For details, see the text.

this function is roughly constant over the nuclear volume[8]. It decreases progressively as $\bar{r}$ approaches the nuclear surface, reaching the value 0.5 at $\bar{r} = R$. For $\bar{r} \gg R$, it goes exponentially to zero, with the slope given by the diffuseness parameter.

The imaginary part, $W_{\rm NL}(\bar{r})$, was given by the sum of a volume term, proportional to $f_{\rm vol}(\bar{r})$, and a term peaked at the nuclear surface. That is,

$$W_{\rm NL}(\bar{r}) = -\left[W_{\rm vol}\ f_{\rm vol}(\bar{r}) + W_{\rm surf}\ f_{\rm surf}(\bar{r})\right]. \tag{10.144}$$

Above, $W_{\rm vol}$ and $W_{\rm surf}$ are respectively the strengths of volume and surface absorptions, $f_{\rm vol}(\bar{r})$ is the function of Eq. (10.143), and $f_{\rm surf}(\bar{r})$ is a function peaked at the surface, represented by the derivative of a Woods-Saxon form factor[9],

$$f_{\rm surf}(\bar{r}) = 4\ \frac{\exp\left[\left(\bar{r} - R\right)/a_{\rm i}\right]}{\left[1 + \exp\left[\left(\bar{r} - R\right)/a_{\rm i}\right]\right]^2}. \tag{10.145}$$

[8]For this reason we use the subscript 'vol' (volume) to denote the form factor of the real part of the potential.

[9]Note that the normalization of the form factor is arbitrary, since it is multiplied by the strength $W_{\rm surf}$. Thus, any change in $f_{\rm surf}$ can be compensated by $W_{\rm surf}$. With the particular choice of the factor 4, the form factor has the property: $f_{\rm surf}(\bar{r} = R) = 1$.

An illustration of this function for the $n+^{208}$Pb system is the dashed line in figure 10.3. It is clearly peaked at the surface region. Note that in this illustration we have used the same diffuseness parameters for volume and surface absorption functions. Usually, different values are used in each case.

The optical potential of Perey and Buck contained also a local spin-orbit term with a complex strength, given by

$$V_{\rm so}(r) = -\,(U_{\rm so} + i\,W_{\rm so})\;f_{\rm so}(r)\;\mathbf{l}\cdot\boldsymbol{\sigma}. \tag{10.146}$$

Above, $\mathbf{l}$ is the orbital angular momentum of the incident nucleon and $\boldsymbol{\sigma}$ is a vector operator having as components the Pauli matrices. The shape function $f_{\rm so}(r)$ was given by the expression

$$f_{\rm so}(r) = \left(\frac{\hbar}{2m_N c}\right)^2 \frac{1}{a_{\rm so}\,r}\;\frac{\exp\left[\left(\bar{r}-R\right)/a_{\rm so}\right]}{\left[1+\exp\left[\left(\bar{r}-R\right)/a_{\rm so}\right]\right]^2}, \tag{10.147}$$

where m_N and c are respectively the nucleon mass and the speed of light. Note that the above spin-orbit potential is peaked at the nuclear surface. The inclusion of a spin-orbit term was essential for the description of spin polarization phenomena measurements (see section 6.2).

The non-local potential of Perey and Buck [Perey and Buck (1962)] was shown to describe accurately a large volume of low energy neutron scattering data for different target nuclei. The parameters resulting from the fits have the important advantage of depending weakly on the collision energy. The dependence is much weaker than that of parameters obtained in fits with standard local potentials. Using the energy independent parameters: $U_{\rm vol} = -71$ MeV, $W_{\rm vol} = 0$, $W_{\rm surf} = -15$ MeV, $U_{\rm so} = -7.2$ MeV, $W_{\rm so} = 0$, $r_0 = 1.22$ fm, $a_{\rm r} = 0.65$ fm, $a_{\rm i} = 0.47$ fm, $a_{\rm so} = 0.7$ fm and $\beta = 0.85$ fm, Perey and Buck got good overall agreement with low energy ($E = 1 - 25$ MeV) neutron scattering data. They studied the differential, total and reaction cross sections, in collisions with Si, S, Ca, Ti, Ba, Pb and U target nuclei.

A similar analysis was carried out by Shultz and Wiebicke [Schultz and Wiebicke (1966)] for proton scattering. They obtained good agreement with the same kind of data using the set of paramters: $U_{\rm vol} = -78$ MeV, $W_{\rm vol} = 0$, $W_{\rm surf} = -11$ MeV, $U_{\rm so} = -5.3$ MeV, $W_{\rm so} = 0$, $r_0 = 1.22$ fm, $a_{\rm r} = 0.65$ fm, $a_{\rm i} = 0.47$ fm, $a_{\rm so} = 0.65$ fm and $\beta = 0.90$ fm. It has been found that at low energies ($E \lesssim 25$ MeV) the absorption is mainly at the surface while at higher energies volume absorption dominates.

We remark that some energy dependence should be expected even in non-local optical potentials introduced by Perey and Buck. It stems from the Green's function in the definition of the polarization potential. Although its contribution to the optical potential is averaged over energy, some weak energy dependence is likely to remain. The situation is rather different in the case of trivially equivalent local optical potentials. These potentials contain a spurious energy dependence introduced by the procedure to eliminate non-locality[10]. The trivially local optical potential is obtained from the non-local potential through the equation

$$V_{\text{opt}}(r) = \frac{1}{\psi(r)} \int dr' \; V_{\text{eff}}(r,r')\, \psi(r'), \tag{10.148}$$

where $V_{\text{eff}} = (0|\,\mathbb{V}_{\text{eff}}\,|0)$ and $\psi(r)$ is the exact scattering wave function in the elastic channel. Since $\psi(r)$ is energy and angular momentum dependent, these dependences are carried over to V_{opt}. The use of a non-local parametrization reduces these dependences to the ones intrinsically contained in the expression of Feshbach's effective potential.

Non-local optical potentials have one disadvantage: they cannot be used directly in standard optical model codes for data analysis. To avoid this problem, Perey and Buck [Perey and Buck (1962)] developed a method to construct a local potential containing the effects of non-locality. This equivalent local potential, $V_{\text{L}}(r)$, is related to their non-local potential, $V_{\text{NL}}(r)$, by the equation

$$V_{\text{L}}(r) \times \exp\left[\frac{m\beta^2}{2\hbar^2}\,(E - V_{\text{L}}(r))\right] = V_{\text{NL}}(r). \tag{10.149}$$

Note that the above equation introduces a dependence of the optical potential on the local wave number. This dependence can be seen more clearly if it is cast in the form

$$V_{\text{L}}(r) \times \exp\left[\frac{\beta^2 k^2(r)}{4}\right] = V_{\text{NL}}(r), \tag{10.150}$$

with

$$k(r) = \frac{\sqrt{2m\,[E - V_{\text{NL}}(r)]}}{\hbar}. \tag{10.151}$$

The local potential of Eq. (10.149) contains a strong energy dependence and it depends also on the target mass. A similar situation occurs in phenomenological studies with local potentials. However, here the energy dependence is controlled by Eq. (10.149). It is then possible to include the solution of the above equation in the optical model analysis code, and carry

[10]This procedure will be further discussed in section 10.5, in connection with polarization potentials

out the parameter search for the original non-local potential. If the local potential is parametrized directly, the fit leads to much stronger energy dependences. As an illustration, we give below the energy-dependent parameters of the local optical potential, obtained in the phenomenological analysis of Wilmore and Hodgson [Wilmore and Hodgson (1964)] for the same data of [Perey and Buck (1962)]:

$$\begin{aligned} U_{\rm vol} &= -\left(47.1 - 0.267\,E - 0.00118\,E^2\right) \\ W_{\rm surf} &= -\left(9.52 - 0.53\,E\right) \\ r_{0\rm r} &= 1.322 - 0.00076/A + 4\times10^{-6}\,A^2 - 8\times10^{-9}\,A^3 \\ r_{0\rm i} &= 1.266 - 0.00037/A + 2\times10^{-6}\,A^2 - 4\times10^{-9}\,A^3 \\ a_{\rm r} &= 0.66 \quad \text{and} \quad a_{\rm i} = 0.48. \end{aligned}$$

Above, the energies are given in MeV and the lengths in fm. Note that this analysis uses different radius parameters for the real and imaginary potentials ($r_{0\rm r}$ $r_{0\rm i}$ instead of r_0) and neglects the spin-orbit interaction.

We remark that there is an important difference between phenomenological local potentials and the local potentials derived from non-local ones, through the use of Eq. (10.149). Since this equation is solved locally, the local potentials of Perey and Buck deviate from the Woods-Saxon shape. Consequently, the two approaches lead to different radial wave functions inside the range of the nuclear potential. However, these wave functions should have similar behaviors in the asymptotic region, so that their phase-shifts would be close. Otherwise, they would not give good fits to the experimental cross sections. This means that the two potentials are phase-shift equivalent but **not** wave function equivalent. Therefore, the local potentials derived directly by optical model analysis may lead to inaccurate distorted waves.

Several other phenomenological studies of nucleon-nucleus optical potentials have been performed. For a comprehensive review of the early optical model analysis we refer to [Hodgson (1967)]. Becchetti and Greenlees [Becchetti and Greenless (1969)] performed a very thorough phenomenological study of the optical potentials for low-energy ($E < 50$ MeV) proton and neutron scattering in collisions with heavy target nuclei ($A > 40$). Although their potentials have been derived in the sixties, they are used even in very recent nuclear reaction calculations (e.g. [Tostevin *et al.* (2001); Lubian and Nunes (2007); Lubian *et al.* (2009)]). More recent phenomenological optical potentials for nucleon scattering can be found in [Mayer *et al.* (1981a,b, 1983); Glover (1985)].

There are optical model analysis that involve simultaneously the elastic and one or more inelastic channels, in a set of coupled equations. In these cases, the effects of the strong couplings with open inelastic channels are included in the coupled equations. Thus, they do not have to be included in the phenomenological optical potential. Consequently, fitting data using this procedure reduces the energy dependence of the optical model parameters. This has been done, for example, in [Cole *et al.* (1966)].

Some studies of the nucleon-nucleus optical potential performed in more recent years adopt a more fundamental approach. They approximate the real part of the optical potential by the projection of the effective interaction onto the P-subspace. Writing the optical potential as

$$V_{\rm opt} = U_{\rm opt} + i\, W_{\rm opt}, \tag{10.152}$$

they set

$$U_{\rm opt} \simeq (0\,|\,V_{\rm PP}\,|\,0) \equiv (\Phi_{\rm T}\,|\,V_{\rm PP}\,|\,\Phi_{\rm T})\,, \tag{10.153}$$

and fit $W_{\rm opt}$ to experimental data. This amounts to neglecting the contribution from the second term at the RHS of Eq. (10.130) to the real part of the optical potential. In Eq. (10.153), $\Phi_{\rm T}(\mathbf{r}_1, ..., \mathbf{r}_{A_{\rm T}})$ is the Slater determinant representing the ground state of the target nucleus and the full scattering wave function is

$$\Psi^{(+)}(\mathbf{r}; \mathbf{r}_1, ..., \mathbf{r}_{A_{\rm T}}) = \mathcal{A}\left[\psi^{(+)}(\mathbf{r})\ \Phi_{\rm T}(\mathbf{r}_1, ..., \mathbf{r}_{A_{\rm T}})\right], \tag{10.154}$$

where $\mathcal{A}$ is the anti-symmetrizer operator that exchanges the incident nucleon with the identical nucleons of the target. If one neglects exchange, the above potential is given by the folding model [Satchler and Love (1979)] as

$$U_{\rm opt}(\mathbf{r}) = \int d\mathbf{r}'\, \rho_{\rm T}(\mathbf{r}')\, v(\mathbf{r} - \mathbf{r}'). \tag{10.155}$$

Above, $\rho_{\rm T}(\mathbf{r}')$ is the one-body density matrix of the target and v is an effective nucleon-nucleon interaction. If exchange is taken into account, the optical potential gets a non-local term. Then, the real part of the optical potential becomes [Amos *et al.* (2000); Kim *et al.* (2008)]

$$U_{\rm opt}(\mathbf{r}, \mathbf{r}') = \delta\,(\mathbf{r} - \mathbf{r}')\ U_{\rm d}(\mathbf{r}) + U_{\rm ex}(\mathbf{r}, \mathbf{r}') \tag{10.156}$$

with

$$U_{\rm d}(\mathbf{r}) = \int d\mathbf{r}'\, \rho_{\rm T}(\mathbf{r}')\, v_d(\mathbf{r} - \mathbf{r}') \tag{10.157}$$

and

$$U_{\rm ex}(\mathbf{r},\mathbf{r}') = \sum_n \zeta_n\, \varphi_n(\mathbf{r})\, v_{ex}(\mathbf{r}-\mathbf{r}')\, \varphi_n(\mathbf{r}'). \tag{10.158}$$

Above, φ_n are the single particle orbitals in the Slater determinants and ζ_n is the occupancy of this state. The potentials v_d and v_{ex} correspond respectively to the direct and exchange parts of the nucleon-nucleon interaction. For practical purposes, it is convenient to avoid non-localities. This can be done though the replacement of the non-local potential by a density and energy dependent potential. This potential is determined in terms of the local wave number, similarly to the procedure introduced by Perey and Buck [Perey and Buck (1962)], discussed above.

The calculation of the potential by the method outlined above requires reasonable assumptions for the nucleon-nucleon interaction and for the target densities. In general, the effective nucleon-nucleon interaction has a central part, and also spin-orbit and tensor parts. The central part is usually given by some kind of G-matrix [Kuo and Brown (1966)] approach. At low scattering energies (a few tens of MeV), it is assumed to be similar to a G-matrix of two nucleons bound near the Fermi surface. A very popular version is the Michigan's M3Y interaction [Bertsch *et al.* (1977); Anantaraman *et al.* (1983)]. It is a sum of three Yukawa functions of different ranges, with parameters fitted to reproduce G-matrix elements of realistic nucleon-nucleon interactions in an oscillator basis.

The second important ingredient in the folding model is the target density. It can be obtained through Hartree-Fock calculations or extracted from electron scattering data. For collision energies of a few tens of MeV, the adopted effective interaction is real. In this case, the absorption cannot be properly included through a complex effective interaction. It is then necessary to use a phenomenological approach to the imaginary part of the potential. Studies of the optical potential along this line have been carried out by several authors. Usually, the optical potential is used in the context of a coupled channel calculation including the main excited channel. An example of a successful optical model analysis of this kind can be found in [Khoa *et al.* (2002)].

At higher collision energies (a few hundreds of MeV), the imaginary potential can be obtained directly from a complex effective nucleon-nucleon interaction [Sinha (1975); Brieva and Rook (1977a,b, 1978); Amos *et al.* (2000); Khoa *et al.* (2007)]. Several optical model calculations have been performed for proton scattering in this energy range, for several targets [Amos *et al.* (2000); Canton *et al.* (2005); Amos *et al.* (2005)]. In these calculations, the non-locality of the optical potential is kept and its origin is discussed in detail in [Fraser *et al.* (2008)]. The agreement with the data is very good.

10.4.2 *Optical potentials in nucleus-nucleus collisions*

The nucleus-nucleus optical potentials are more complicated. The two terms that define the optical potential in Eq. (10.130), $(0\,|\mathbb{V}_{\rm PP}|\,0)$ and specially $\left(0\,\middle|\overline{\mathbb{V}}_{\rm pol}\middle|\,0\right)$, are harder to calculate. As a consequence, the latter term is treated phenomenologically. Usually, its imaginary part is determined by fitting procedures and its real part is neglected. One hopes to take its effect into account through small changes in the strength of the dominant term, $\mathrm{Re}\,\{(0\,|\mathbb{V}_{\rm PP}|\,0)\}$. In general, the nucleon-nucleon interaction is real and consequently one has,

$$U_{\rm opt} \simeq (0\,|\,\mathbb{V}_{\rm PP}\,|\,0)\,.$$

Frequently, the real part of the optical potential is approximated by the folding of a nucleon-nucleon interaction with the densities of the projectile and target[11], as

$$U_{\rm opt}(\mathbf{r}) = \int d\mathbf{r}'\, d\mathbf{r}''\;\rho_{\rm P}(\mathbf{r}')\, v(\mathbf{r}-\mathbf{r}'+\mathbf{r}'')\,\rho_{\rm T}(\mathbf{r}''). \tag{10.159}$$

Above, $\rho_{\rm P}$ and $\rho_{\rm T}$ are respectively the matter densities in the projectile and the target and v is a realistic nucleon-nucleon effective interaction. Usually, the non-localities arising from exchange are mocked up by energy and density dependences in a local potential. We give below a discussion about some derivations of the nucleus-nucleus optical potential along these lines.

Before we start a detailed discussion of the double folding potential of Eq. (10.159), we briefly describe the intuitive attempt to simplify Eq. (10.159) by replacing the integral over $d\mathbf{r}''$ by a nucleon-target optical potential. One could then proceed phenomenologically, using a nucleon-target potential that fits nucleon scattering data. In this way, one would

[11] In fact the anti-symmetrization operator acting on the full scattering wave function leads to complicated non-local term, as pointed out in section 6.3.3. Here they are not fully taken into account. The non-locality is approximated by an equivalent local potential. For this reason, the optical potential is taken to be local.

end up with the single folding expression

$$U_{\rm opt}(\mathbf{r}) = \int d\mathbf{r}' \; \rho_{\rm P}(\mathbf{r}')\, U_{\rm n-T}(\mathbf{r}-\mathbf{r}'). \qquad (10.160)$$

However, nucleus-nucleus optical potentials determined in this way turned out to be too attractive. They overestimate the strength of the real potential required to fit nucleus-nucleus scattering data by a factor of ~ 2 [Satchler and Love (1979)].

We should also mention that some authors adopt phenomenological procedures for both the real and the imaginary parts of the optical potential. However, it is more difficult than in the case of nucleon-nucleus potentials. Here the dependence on the collision partners is much more pronounced. Nevertheless, some phenomenological optical potentials are very successful in a limited mass range. An example is the E18 strong absorption potential [Cramer *et al.* (1976)], which has been widely used in the eighties for ^{16}O + ^{28}Si scattering. Other phenomenological potentials can be applied to a broader set of projectile-target systems, as the one proposed by Christensen and Winther [Christensen and Winther (1976)]. Using a semiclassical approximation for elastic scattering, these authors have shown that the cross section depends basically on the value of the real potential at a particular projectile-target separation, which is slightly smaller than the distance of closest approach in a classical trajectory leading to the rainbow angle. Then, comparing theoretical predictions with data, they checked a few models for the optical potential and proposed an exponential parametrization for the nuclear potential. Their potential is

$$U_{\rm opt}(r) = U_0 \; e^{-(r-R_0)/a}, \qquad (10.161)$$

where

$$U_0 = -50 \left[\frac{R_{\rm P}\, R_{\rm T}}{R_0}\right] \frac{\rm MeV}{\rm fm}, \quad a = 0.63\,{\rm fm}, \quad R_0 = R_{\rm P} + R_{\rm T}, \qquad (10.162)$$

with

$$R_i = 1.233\, A_i^{1/3} - 0.978\, A_i^{-1/3}, \quad i = {\rm P, T}. \qquad (10.163)$$

However, the most successful treatments of the nucleus-nucleus optical potential are implementations of the double-folding model, or can be derived from it by introducing approximations. Since the double-folding integral of Eq. (10.159) can only be evaluated with the help of computer codes, it is convenient to develop approximations leading to simpler expressions. The potential of Akyüz and Winther [Akyüz and Winther (1981)] and the proximity potential [Blocki *et al.* (1977)], which are discussed below, are among the most successful potentials of this kind.

Krappe, Nix and Sierk [Krappe *et al.* (1979a,b)] evaluated the folding potential of Eq. (10.159) using sharp densities for the collision partners (*liquid drops*) and a superposition of two Yukawa functions for the nucleon-nucleon interaction. With two Yukawas with different ranges and strengths, they were able to obtain a potential with the appropriate tail and satisfying the saturation condition at small projectile-target separations. Akyüz and Winther [Akyüz and Winther (1981)] went one step further, performing a similar calculation with diffuse densities. They obtained simple expressions for the nucleus-nucleus optical potential taking advantage of the fact that convolution integrals of exponential and Yukawa functions can be evaluated analytically. The nuclear densities were approximated by the expression,

$$\rho_i(r) = \frac{\rho_{0i}\,\kappa_i^2}{4\pi} \int d\mathbf{r}'\,\Theta(R_i - r)\,\frac{e^{\kappa_i r'}}{r'}, \quad \text{with } i = \mathrm{P}, \mathrm{T}. \tag{10.164}$$

The parameters ρ_{0i}, R_i and κ_i of the above densities are respectively associated with the central density, radius and diffusivity of the collision partners, and $\Theta(R_i - r)$ is the usual step function. The integral of Eq. (10.164) can be evaluated analytically and the result is

$$\begin{aligned} \rho_i(r) &= \rho_{0i}\left[1 - e^{-\kappa_i R_i}\,[1 + \kappa_i R_i]\,\frac{\sinh(\kappa_i r)}{\kappa_i r}\right], \quad &\text{for } r \le R_i, \\ &= \rho_{0i}\left[\frac{e^{-\kappa_i r}}{\kappa_i r}\,[\kappa_i r\cosh(\kappa_i r) - \sinh(\kappa_i r)]\right], \quad &\text{for } r > R_i. \end{aligned} \tag{10.165}$$

Akyüz and Winther [Akyüz and Winther (1981)] have shown that the above approximation is very close to the usual fermi density,

$$\rho_i(r) = \frac{\rho_{0i}}{1 + \exp\left[-\left(r - \mathcal{R}_i\right)/a_i\right]}, \tag{10.166}$$

if one uses in Eqs. (10.164) and (10.165) the parameters,

$$\kappa_i = 0.67\left(\frac{1.2}{\mathcal{R}_i} + \frac{1}{a_i}\right) \tag{10.167}$$

$$R_i = \mathcal{R}_i + 0.5\,a_i - 0.12\ \mathrm{fm}. \tag{10.168}$$

The Akyüz-Winther potential is calculated by the double folding procedure of Eq. (10.159), with the projectile and target densities parametrized as in Eq. (10.165) and using the M3Y interaction [Bertsch *et al.* (1977)].

Since this interaction is a combination of Yukawa functions plus a zero-range term that mocks up two-nucleon exchange (*knock-on exchange*), the double folding integrals can be evaluated analytically. The potential becomes particularly simple if one neglects differences between protons and neutrons, regarding the nucleon-nucleon forces and matter densities, and considers the interaction only in the surface region ($r \gtrsim R_{\rm P} + R_{\rm T}$), which has greatest influence on elastic scattering. One gets

$$U_{\rm opt}(r) = -\,65.4 \left[\frac{R_{\rm P}\,R_{\rm T}}{R}\right]\, e^{-\kappa\,(r-R)}\ \ {\rm MeV/fm}, \tag{10.169}$$

where,

$$R = R_{\rm P} + R_{\rm T}. \tag{10.170}$$

The best results are obtained with the parameters,

$$R_i = \left[1.2\,A_i^{1/3} - 0.35\right]\ {\rm fm}, \quad i = {\rm P, T}, \tag{10.171}$$

$$\kappa = \left[1.16 + 0.56\left(\frac{1}{A_{\rm P}^{1/3}} + \frac{1}{A_{\rm T}^{1/3}}\right)\right]\ {\rm fm}^{-1}. \tag{10.172}$$

In order to have a more reasonable shape inside the Coulomb barrier, the exponential potential is usually replaced by a Woods-Saxon function and Eq. (10.169) becomes,

$$U_{\rm opt}(r) = -\,65.4 \left[\frac{R_{\rm P}\,R_{\rm T}}{R}\right]\ \frac{1}{1+\exp\left[\kappa\,(r-R)\right]}\ \ {\rm MeV/fm}. \tag{10.173}$$

This potential became very popular in the nuclear physics community, owing to its simplicity and its applicability to nuclear reactions involving collision partners in different mass ranges.

An equally simple potential with broad applicability in nuclear reactions is the one given by the proximity model [Blocki *et al.* (1977)]. In this model, the optical potential between two nuclei is given in terms of the interaction energy per unit area of two parallel semi-infinite slabs of nuclear matter placed at a distance D, $\mathcal{E}(D)$. Since the surfaces are diffuse, the distance D is measured with respect to the point where the density is one half of its maximal value. Let us consider two spherical nuclei with radii $R_{\rm P}$ and $R_{\rm T}$, in the situation represented in figure 10.4. The nuclear surfaces are separated by the distance s and the distance between their centers is $r = R_{\rm P} + R_{\rm T} + s$. In our cylindrical coordinate system, the z-axis is along the line joining the centers. The remaining coordinates are the modulus of

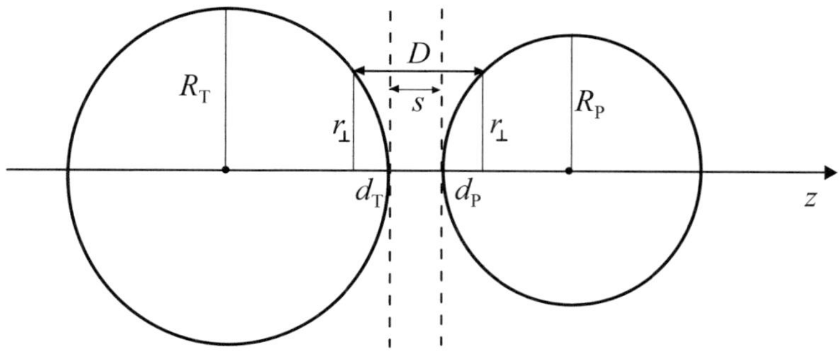

Fig. 10.4 The geometry of two nuclei separated by the distance $r = R_P + R_T + s$. The circles represent the region where the nuclear densities reach one-half of their central values.

the vector perpendicular to this axis and the azimuthal angle, φ. In the proximity model, the optical potential is approximated by the integral

$$U_{\text{opt}}(r) = 2\pi \int_0^\infty r_\perp \, dr_\perp \; E(r_\perp), \tag{10.174}$$

where $E(r_\perp)$ is the projectile-target interaction energy per unity area associated with the nuclear matter within a cylinder parallel to the z-axis with transverse coordinate $r_\perp$. The factor 2π in the above equation comes from the integration over φ and the cylindrical symmetry. Assuming that the nuclear radii are much larger than the diffusivities[12], $E(r_\perp)$ can be approximated by the interaction energy per unit area of two semi-infinite slabs of nuclear matter perpendicular to the z-axis and separated by a distance D, $\mathcal{E}(D)$. The variables D and $r_\perp$ are represented in figure 10.4. Writing

$$D = s + d_P + d_T, \tag{10.175}$$

the distances d_P and d_T can be trivially obtained as

$$d_P = \sqrt{R_P^2 + r_\perp^2} - R_P \quad \text{and} \quad d_T = \sqrt{R_T^2 + r_\perp^2} - R_T. \tag{10.176}$$

For leptodermous nuclei interacting through short range forces, the potential vanishes unless the distance between the surfaces is much smaller than nuclear radii. This condition implies that in the relevant region for the interaction the conditions $r_\perp/R_P \ll 1$ and $r_\perp/R_T \ll 1$ are satisfied. The above equations can then be expanded to lowest order in $r_\perp/R_P$ and $r_\perp/R_T$. Using this approximation and inserting the result into Eq. (10.174), we obtain the relation

$$D(r_\perp) = s + \frac{1}{2\,\overline{R}}\, r_\perp^2, \tag{10.177}$$

[12]This is the definition of a *leptodermous system* [Blocki *et al.* (1977)].

where $\overline{R}$ is the geometric radius,

$$\overline{R} = \frac{R_{\rm P}\, R_{\rm T}}{R_{\rm P} + R_{\rm T}}. \tag{10.178}$$

Now we can change variable in the integral giving the proximity potential. Differentiating Eq. (10.177) (for changes of $r_\perp$), we get

$$dD = \frac{1}{\overline{R}}\, r_\perp\, dr_\perp \tag{10.179}$$

and using this result in Eq. (10.174) we obtain the potential

$$U_{\rm opt}(r) = 2\pi\, \overline{R} \int_s^\infty dD\; \mathcal{E}\,(D)\,. \tag{10.180}$$

If the saturating properties of nuclear matter are properly accounted for, one expects that the energy per unit area has its maximal strength at the separation $s = 0$. Is this case, it should reduce to

$$\mathcal{E}\,(0) = 2\,\gamma, \tag{10.181}$$

where γ is the surface tension.

It is convenient to express the proximity potential in terms of dimensionless variables. Changing $D \to \zeta = D/b$, where b is a diffusivity parameter, and $\mathcal{E}\,(D) \to \mathcal{E}\,(\zeta)\,/2\,\gamma$, Eq. (10.180) becomes

$$U_{\rm opt}(r) = 4\pi\gamma\, b\, \overline{R}\;\; \Phi(\zeta), \tag{10.182}$$

where $\Phi(\zeta)$ is the universal function

$$\Phi(\zeta) = \int_{s/b}^\infty d\zeta\; \frac{\mathcal{E}\,(\zeta b)}{2\gamma}. \tag{10.183}$$

Blocki *et al.* [Blocki *et al.* (1977)] carried out a systematic study of the function $\Phi(\zeta)$ for several projectile-target combinations. They used the phenomenological momentum dependent nucleon-nucleon interaction of Seyler and Blanchard [Seyler and Blanchard (1963)], and handled the nuclear many-body problem by the Thomas-Fermi approximation[13]. They concluded that it can be accurately parametrized as

$$\begin{aligned} \Phi(\zeta) &= -0.5\;(\zeta - 2.54)^2 - 0.0853\,(\zeta - 2.54)^3\,, && \text{for } \zeta \le 1.2511 \\ &= -3.437\, e^{-\zeta/0.75} && \text{for } \zeta > 1.2511. \end{aligned} \tag{10.184}$$

[13]The details of the calculation can be found in [Randrup (1975, 1976)].

For the surface tension they adopted the expression

$$\gamma = 0.95\left[1 - 1.8\left(\frac{N_{\rm P} - Z_{\rm P}}{A_{\rm P}}\right)\left(\frac{N_{\rm T} - Z_{\rm T}}{A_{\rm T}}\right)\right]\ \frac{\rm MeV}{\rm fm^2}, \qquad (10.185)$$

the nuclear radii used in Eq. (10.178) were given by

$$R_i = \left[1.28\,A_i^{1/3} - 0.76 + 0.8\,A_i^{-1/3}\right]\ \text{fm}, \quad \text{with} \quad i = \text{P}, \text{T}, \qquad (10.186)$$

and the diffuseness parameter was taken as

$$b = 1\ \text{fm}. \qquad (10.187)$$

One last remark about the proximity potential is that it can be derived from the folding model using some approximations. Although this point is not clear in its original formulation, it has been proved in [Brink and Stancu (1978); Baltz and Bayman (1982)].

Bass [Bass (1977)] proposed a phenomenological expression for the nucleus-nucleus potential based on the proximity formula of Eq. (10.182). His potential was devised to describe fusion barriers, used in calculations of heavy ion fusion cross sections. He wrote the nuclear contribution to the real potential as

$$U_{\rm opt}(r) = \overline{R}\ g(s) \quad \text{with} \quad s = r - R_{\rm P} - R_{\rm T}, \qquad (10.188)$$

and treated $g(s)$ as a phenomenological system-independent function, to be determined from experimental fusion excitation functions at above-barrier energies. He assumed that the fusion cross section could be approximated by the classical equation

$$\sigma_{\rm F}(E) = \pi\, R_{\rm B}^2\left[1 - \frac{V_{\rm B}}{E}\right], \qquad (10.189)$$

and determined the barrier parameters for several systems comparing the above expression with experimental data. In this way, he determined several pairs of barrier parameters $\{R_{\rm B}; V_{\rm B} \equiv V(R_{\rm B})\}$, each one corresponding to one pair $\{s; g(s)\}$, according to Eq. (10.188). He then fitted these points using the function

$$g(s) = \frac{1}{A\exp(s/d_1) + B\exp(s/d_2)}, \qquad (10.190)$$

with the parameters

$$A = 0.0300\,\text{fm/MeV}, \ \ B = 0.0061\,\text{fm/MeV}, \ \ d_1 = 3.3\,\text{fm} \ \text{and} \ d_2 = 0.65\,\text{fm}. \qquad (10.191)$$

The nuclear radii appearing in Eq. (10.188) were given by the expression

$$R_i = \left[1.16\ A_i^{1/3} - 1.39\ A_i^{-1/3}\right]\ \text{fm}, \quad \text{with} \quad i = \text{P}, \text{T}. \tag{10.192}$$

We now consider more realistic folding calculations of the optical potential, resorting to numerical methods. These calculations, use better estimates of matter densities of the projectile and the target and adopt realistic nucleon-nucleon interactions[14]. The most reliable distributions are those obtained from electron scattering data. However, these experiments give only charge distributions. That is, they yield only proton densities. Thus, the neutron densities cannot be determined in this way. In the case of light nuclei with $N \simeq Z$, it is a reasonable assumption to take the same densities for protons and neutrons. For heavier nuclei, with N considerably larger then Z, one has to make some assumption about the distribution of the neutron excess. Another possibility is to rely on theoretical determinations of the nuclear density, based on the independent particle model, Hartree-Fock or Hartree-Fock-Bogoliubov calculations. In the cases of stable weakly bound nuclei, like ^{6}Li or ^{9}Be, and of radioactive halo nuclei, like ^{11}Li or ^{11}Be, it is essential that the neutron density takes into account the low binding energy of the valence particles.

The other fundamental quantity in the calculation of the double-folding potential is the nucleon-nucleon interaction. Satchler and Love [Satchler and Love (1979)] showed that good descriptions of the elastic scattering data at near-barrier energies can be obtained with the M3Y nucleon-nucleon interaction, discussed in the previous section. In fact, the folding potentials obtained in this way turned out to have nearly the same tail as the other ones previously discussed in this section. This is illustrated in figure 10.5, in the case of $^{16}\text{O} + ^{208}\text{Pb}$ scattering. We compare the Akÿuz-Winther, Christensen-Winther[15], proximity, Bass and double-folding potentials (using the M3Y interaction). Although these potentials are rather different at small projectile-target separations, their tails are quite close. Since elastic scattering at near-barrier energies can only probe the real potential in the surface region [Brandan and Satchler (1997)], these five potentials can reproduce angular distributions in this energy range equally well. In figure 10.6 we show the potential barriers for the same potentials of figure

[14]For a review of the early calculations along these lines we refer to [Satchler and Love (1979)].

[15]We use the parametrization suggested in Broglia and Winther's book [Broglia and Winther (1991)], instead of the ones of the original Christensen and Winther's paper [Christensen and Winther (1976)]

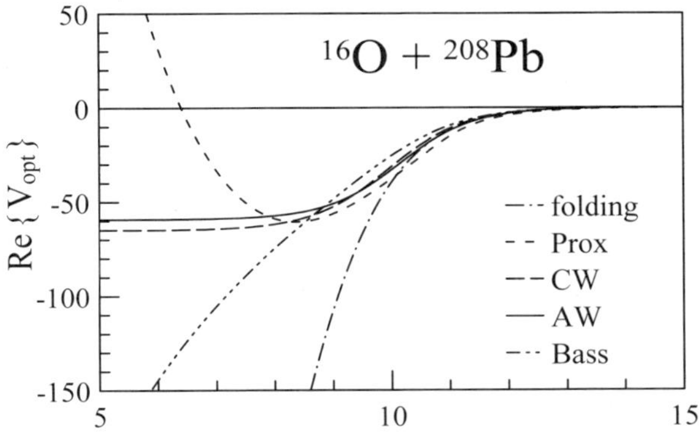

Fig. 10.5 Comparison of the double-folding potential using the M3Y interaction with the proximity, Bass, Christensen-Winther and Akyüz-Winther potentials. The meaning of each curve in indicated in the legend. For details, see the text.

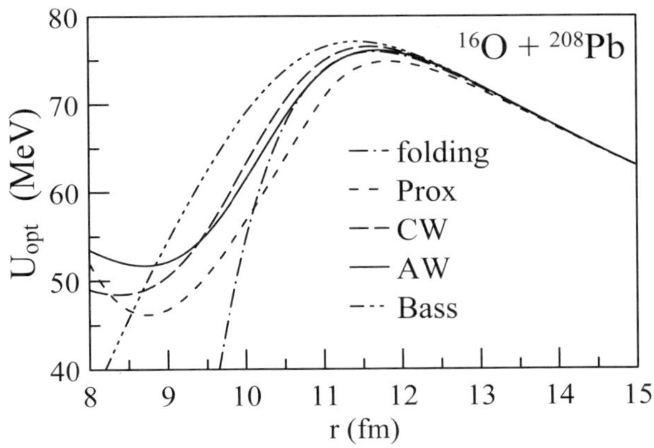

Fig. 10.6 Comparison of the potential barriers for the same potentials of the previous figure.

10.5. The surface region corresponds to the external side of the barrier, where these potentials are very similar.

As mentioned above, elastic scattering at near-barrier energies cannot probe the nucleus-nucleus potential inside the Coulomb barrier. However, other kind of data can shed light on the nuclear potential in this inner region. One example is the scattering of some tightly bound light systems at collision energies well above the barrier. In this case absorption is not strong enough to dampen the far side branch of the scattering amplitude (see section 7.6). Thus, one may observe the nuclear rainbow, which is dominated by the attraction of the optical potential (this is the case of figure 7.7, of chapter 7). Investigations of this kind have been performed by several authors and neat nuclear rainbow patterns have been identified in the scattering of α-particles from different targets (see e.g. [Put and Paans (1977)]) and in the scattering of some tightly bound light systems, like ^{12}C + ^{12}C [McVoy and Brandan (1992)] (and references therein) and ^{16}O + ^{16}O [Stiliaris *et al.* (1989); Bohlen *et al.* (1993); Bartnitzky *et al.* (1996); Kondo *et al.* (1996)], at conveniently chosen collision energies. Details can be found in the recent review article of Khoa *et al.* [Khoa *et al.* (2007)], where it is shown that the rainbow patterns observed in the elastic cross section for these collisions can be reproduced by calculations with the double-folding model with a modified M3Y interaction. The modifications are the following: (*i*) the zero range approximation for the single nucleon knock-on exchange is replaced by a local interaction dependent on the local momentum, (*ii*) the M3Y interaction is multiplied by a density dependent factor. This factor was written in terms of a few parameters, which were determined by the requirement that the interaction reproduces correct binding energy and the central density of nuclear matter in a Hartree-Fock calculation. They performed calculations with the Reid and the Paris versions [Bertsch *et al.* (1977)] of the M3Y interactions, adopting different parametrizations for the density dependent factor. These calculations, lead to equations of state with different nuclear compressibilities. The best results were obtained with compressibilities in the range $K \approx 230 - 260$ MeV. Khoa *et al.* [Khoa *et al.* (2007)] pointed out that it is necessary to use a deep potential to reproduce the data. It should be much deeper than the Akÿuz-Winther but not as deep as the potential obtained by the double folding model with the standard M3Y interaction. The introduction of density dependence leads to some repulsion, which gives the appropriate strength to the nucleus-nucleus potential.

Another approximate treatment of the effects of the Pauli Principle (exchange effects) that reduces the attraction of the folding potential was developed by the São Paulo group [Cândido-Ribeiro *et al.* (1997); Chamon *et al.* (1997, 2002)], and it is known as the *São Paulo Potential*. These authors base their treatment on a generalization of the Perey-Buck method [Perey and Buck (1962)], as suggested for α-nucleus scattering by Jackson and

Jonhson [Jackson and Johnson (1974)]. The nucleus-nucleus potential is approximated by a separable non-local function (see section 4.4), parametrized as

$$U_{\text{opt}}(\mathbf{r}, \mathbf{r}') = U_{\text{f}}(\bar{r}) \times F(s), \tag{10.193}$$

where U_{f} is the direct part of the double-folding potential, evaluated with realistic nucleon-nucleon interaction and nuclear densities, and the non-locality factor is given by the gaussian function

$$F(s) = e^{-s^2/\beta^2}. \tag{10.194}$$

Above, $\bar{r}$ and s are the variables,

$$\bar{r} = \left| \frac{\mathbf{r} + \mathbf{r}'}{2} \right| \quad \text{and} \quad s = |\mathbf{r} - \mathbf{r}'| . \tag{10.195}$$

The range of non-locality, β, is given by an adjustable parameter b and the scaling rule

$$\beta = \frac{m_0}{\mu}\, b, \tag{10.196}$$

with m_0 and μ representing respectively the nucleon mass and the reduced mass of the projectile-target system. Note that this scaling rule is consistent with the range of the non-locality of the Generator Coordinate method's kernels [Brink (1966)], from which the non-local potential of the resonating group method can be derived.

In order to determine scattering wave functions and elastic cross sections with the São Paulo potential, it is necessary to solve an integro-differential equation with the general form of Eq. (4.129). As we pointed out in our discussion of the Perey and Buck nucleon-nucleus potential [Perey and Buck (1962)], this kind of equation cannot be solved using typical optical model codes. Thus, it is convenient to convert this potential into an energy-dependent local one. This procedure was carried out by Perey and Buck, in the case of nucleon-nucleus potentials. A similar procedure was adopted for the São Paulo potential. The equivalent local potential was derived through a simple equation which depends on the local velocity of the projectile-target relative motion [Chamon *et al.* (2002)]. Cross sections obtained in this way were shown to be very close to the ones obtained with the original non-local potential [Cândido-Ribeiro *et al.* (1997); Chamon *et al.* (1997)]. The SPP has been applied to the scattering of many systems, over a broad mass range. In their calculations, Chamon *et al.* [Chamon *et al.* (2002)] used the M3Y interaction with one modification. Since the São Paulo potential

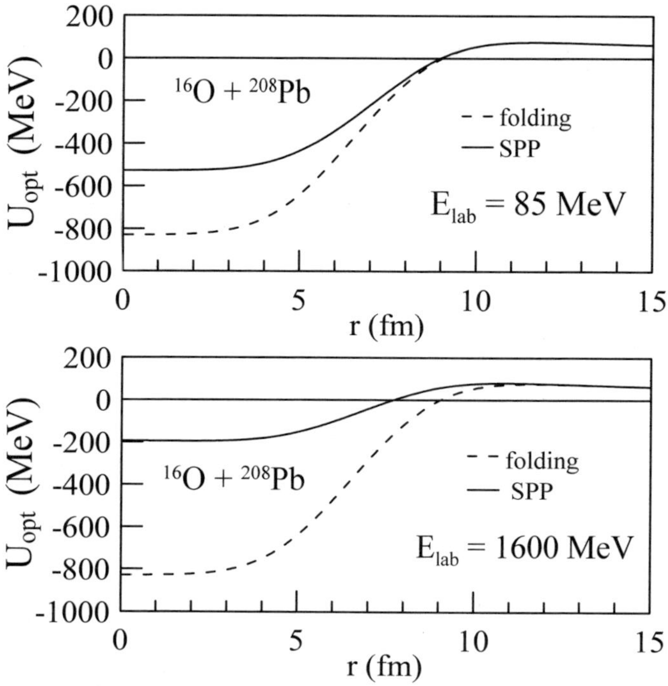

Fig. 10.7 Influence of non-locality in the São Paulo potential (SSP) [Chamon (2010)]. The solid lines correspond to the São Paulo potentials with effects of anti-symmetrization and the dashed lines are the usual folding potential (SSP with the exchange term switched off). Results for $E_{\rm lab} = 85$ MeV and $E_{\rm lab} = 1600$ MeV are shown in the top and in the bottom panels, respectively. For details, see the text.

already contained the energy dependence associated with non-locality, it would be inconsistent to keep the energy dependence of the single-nucleon exchange term of the M3Y interaction. To avoid this problem, they kept its strength frozen at its value for a collision energy of 10 MeV per nucleon. The authors adopted a parametrization of nuclear densities based on a systematic study of scattering data and realistic microscopic calculations. In this way, their potential becomes parameter free.

The energy-dependence associated with non-locality in the São Paulo potential is illustrated in figure 10.7, for $^{16}\text{O} + {}^{208}\text{Pb}$ scattering. We consider two different collision energies: one near the Coulomb barrier, $E_{\rm lab} = 85$ MeV, and one well above it, $E_{\rm lab} = 1600$ MeV. The folding potentials, $U_{\rm f}$, which correspond to setting in Eq. (10.193) $F(s) = 1$, are represented by dashed lines. The equivalent local potentials, which contain the effects of anti-symmetrization, are represented by solid lines. For both energies the depth of the potential is strongly reduced by non-local effects. At $E_{\rm lab} = 85$ MeV (top panel), the neglect of anti-symmetrization only affects the potential depth. The Coulomb barriers for the two potentials are very closely the same. Thus, elastic scattering and fusion reactions at near-barrier energies are not expected to be affected by non-locality. The influence of non-locality at $E_{\rm lab} = 1600$ MeV, illustrated in the bottom panel, is much more pronounced. The reduction of the potential depth is considerably stronger and even the potential barrier is modified. The barrier is higher and thicker when non-locality is taken into account.

Experimental information about the nucleus-nucleus potentials can also be extracted from fusion data at deep sub-barrier energies. Recently, fusion cross sections for several systems were measured with great accuracy at very low energies. The experiments were made at energies reaching several MeV below the Coulomb barrier, where the cross section is as low as few nanobarns [Jiang *et al.* (2002, 2004); Dasgupta *et al.* (2007); Stefanini *et al.* (2009)]. These data showed an unexpected behavior, deviating from the characteristic exponential decrease usually observed below the Coulomb barrier. At deep sub-barrier energies, the cross section was shown to decrease much faster than this exponential trend. This effect can be seen more clearly if one eliminates the exponential decrease, showing the energy dependence of the astrophysical S-factor, instead of the fusion cross section. This factor, introduced in Eq. (10.97), is

$$\mathcal{S}(E) = [E\,\sigma_{\rm F}(E)]\;e^{2\pi\eta},$$

where η is the usual Sommerfeld parameter (see chapter 3). This phenomenon is illustrated in figure 10.8, which shows the S-factor as a function of the collision energy. The data, taken from [Jiang *et al.* (2004)], are normalized by the factor $\exp\,[2\pi\eta_0]$, where η_0 is a parameter introduced by the authors with the purpose of bringing to the same scale their results for different systems. The dashed and the solid lines are the coupled channel calculations of Mişicu and Esbensen [Mişicu and Esbensen (2006, 2007)], including all the relevant non-elastic channels. The dashed line was obtained using the Akÿuz-Winther potential. It exhibits the expected trend of growing as the energy decreases. This trend is not consistent with the data, which exhibit a maximum at $E_{\rm c.m.} \simeq 88$ MeV, and decrease at lower energies. Dasso and Pollarolo [Dasso and Pollarolo (2003)] showed that the deviation from the exponential trend can be traced back to a shallow

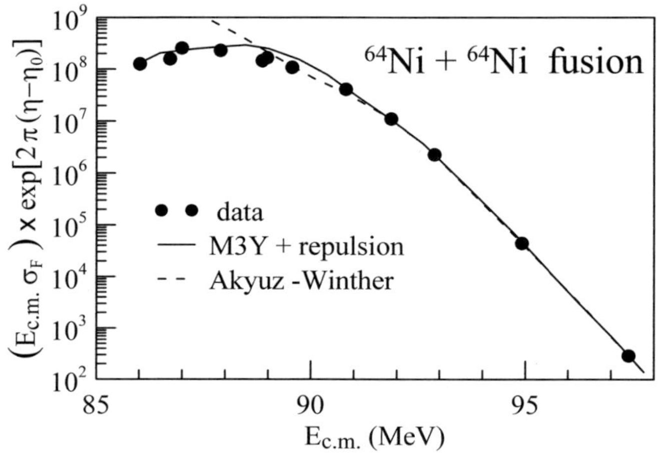

Fig. 10.8 The astrophysical S-factor for the ^{64}Ni + ^{64}Ni fusion. The data are from [Jiang *et al.* (2004)] and the calculations are from [Mişicu and Esbensen (2006)]. For details see the text.

potential. The Akÿuz-Winther and folding potentials obtained with the usual nucleon-nucleon interactions [Satchler and Love (1979); Khoa *et al.* (2007); Chamon *et al.* (2002)] lead to optical potentials with a deep pocket. The shallower potentials obtained with density dependent nucleon-nucleon interactions, which were successful in the description of rainbow scattering of light systems [Khoa *et al.* (2007)], were also unable to reproduce the sharp decrease of the cross section at deep sub-barrier energies [Gontchar *et al.* (2004)]. However, this anomalous behavior at deep sub-barrier energies was well reproduced with the modified folding potential of [Mişicu and Esbensen (2006); Esbensen and Mişicu (2007); Mişicu and Esbensen (2007)]. This is illustrated in figure 10.8, in the case of ^{64}Ni + ^{64}Ni fusion, where the results of the calculation with this interaction is represented by the solid line [Mişicu and Esbensen (2006)]. In their calculations, Mişiku and Esbensen used a folding potential based on a M3Y interaction but adding to it a repulsive core. The strength of the core was calibrated to reproduce the nuclear incompressibility at full overlap.

The above folding potentials are based on the sudden limit, in which the nuclear densities are kept frozen, even when the nuclei overlap strongly. This is a very strong assumption. Ichikawa, Hagino and Iwamoto [Ichikawa *et al.* (2007a,b, 2009)] proposed a model in which the sudden limit is adopted only while the nuclei are far apart, and the nuclear densities overlap weakly. At the point where the nuclear surfaces touch each other, they assume that the collision follows the adiabatic limit. Using this hybrid model, they were able to describe the sharp decrease of the fusion cross sections at deep sub-barrier energies, for several systems.

10.5 Dynamic polarization potentials

The optical potential reproduces, on the average, the effects of channel couplings on the elastic wave function, or on the wave functions of a selected group of channels (the P-subspace). Of course, the optical potential cannot account for strong couplings resulting from specific nuclear properties of the collision partners. This happens, for example, in heavy collisions involving a highly deformed nucleus. In such cases there are strong couplings between the elastic channel and the channels associated with rotational excitations of the deformed nucleus (see section 10.5.2). It is then necessary to include these channels in the P-subspace and perform a coupled channel calculation. However, when one is interested mainly in the elastic channel, one can obtain the same wave function, ψ, from a single-channel Schrödinger equation, with the modified potential

$$V_{\text{opt}} \rightarrow V_{\text{opt}} + V_{\text{pol}}. \tag{10.197}$$

That is, the energy averaged elastic wave function satisfies the equation

$$\left[E - K - V_{\text{opt}} - V_{\text{pol}}\right] \left|\psi^{(+)}\right\rangle = 0. \tag{10.198}$$

This equation can be taken as the definition of the polarization potential, V_{pol}. To evaluate V_{pol}, we split the P-subspace as

$$P = P_0 + P_1, \tag{10.199}$$

where

$$P_0 = |0)\ (0| \tag{10.200}$$

contains only the elastic channel and

$$P_1 = \sum_{\alpha=1}^{N-1} |\alpha)\ (\alpha| \tag{10.201}$$

includes the remaining $N-1$ channels spanning the P-subspace. The effective potential in the P_0 space can be written

$$\mathbb{H}_{\text{eff}} = \mathbb{H}_{\text{P}_0\text{P}_0} + \mathbb{V}_{\text{opt,P}_0} + \mathbb{H}_{\text{P}_0\text{P}_1}\ \mathbb{G}^{(+)}_{\text{P}_1\text{P}_1}\ \mathbb{H}_{\text{P}_1\text{P}_0}, \tag{10.202}$$

where the Green's function in the P_1 subspace is given by,

$$\mathbb{G}^{(+)}_{\mathrm{P_1P_1}} = \frac{1}{E - \mathbb{H}_{\mathrm{P_1P_1}} - \mathbb{V}_{\mathrm{opt,P_1}} + i\epsilon}. \tag{10.203}$$

Above, we have distinguished the optical potentials in the elastic channel from the one acting on the remaining channels (projected by P_1). They are given by,

$$\mathbb{V}_{\mathrm{opt,P_0}} = |0)\ (0|\,\mathbb{V}_{\mathrm{opt}}\,|0)\ (0| \tag{10.204}$$

and

$$\mathbb{V}_{\mathrm{opt,P_1}} = \sum_{\alpha=1}^{N-1} |\alpha)\ (\alpha|\,\mathbb{V}_{\mathrm{opt}}\,|\alpha)\ (\alpha|\,. \tag{10.205}$$

Thus we can write for the energy-averaged effective Hamiltonian in the P_0-space[16],

$$\overline{\mathbb{H}}_{\mathrm{eff}} = K + \mathbb{V}_{\mathrm{opt,P_0}} + \mathbb{V}_{\mathrm{pol}} = K + \mathbb{V}_{\mathrm{eff}}. \tag{10.206}$$

Above,

$$\mathbb{V}_{\mathrm{eff}} = \mathbb{V}_{\mathrm{opt,P_0}} + \mathbb{V}_{\mathrm{pol}} \tag{10.207}$$

and

$$\mathbb{V}_{\mathrm{pol}} = \mathbb{V}_{\mathrm{P_0P_1}}\ \mathbb{G}^{(+)}_{\mathrm{P_1P_1}}\ \mathbb{V}_{\mathrm{P_1P_0}} \tag{10.208}$$

is the *dynamic polarization potential* arising from the coupling of the elastic channel with the channels in the subspace P_1.

The optical and the polarization potentials appearing in Eq. (10.198) are the projections of the above operators onto the elastic channel. That is,

$$V_{\mathrm{opt}} = (0|\ \mathbb{V}_{\mathrm{opt}}\ |0) \tag{10.209}$$

and

$$V_{\mathrm{pol}} = (0|\ \mathbb{V}_{\mathrm{P_0P_1}}\ \mathbb{G}^{(+)}_{\mathrm{P_1P_1}}\ \mathbb{V}_{\mathrm{P_1P_0}}\ |0)\,. \tag{10.210}$$

In contrast to the optical potential, for which one can obtain a systematic description changing slowly with the collision energy and the mass of the system, the polarization potential depends critically on the particular

[16]Here, the energy average is taken only on part of the polarization potential representing transitions to the Q-subspace. No energy average is performed for the polarizations associated with couplings with P_1.

collision under investigation. Usually, it is calculated using some approximation for Eq. (10.210). Later in this section, we give a few examples of approximate calculations of this potential.

With the inclusion of a complex polarization potential in the Schrödinger equation for the elastic channel, the absorption cross section must be re-interpreted. So far, it corresponded to the energy-averaged cross section for the excitation of long lived resonances, associated with channels in the Q-subspace. Now, the situation is more complicated because the polarization potential may also have an imaginary part. The total flux removed from the incident channel along the collision is then measured by the total *reaction* cross section, given by (see Eq. (7.9))

$$\sigma_{\rm R} = \left|\frac{1}{A}\right|^2 \frac{k}{E} \left\langle \psi^{(+)} \right| - \mathrm{Im}\big\{V_{\rm opt} + V_{\rm pol}\big\} \left| \psi^{(+)} \right\rangle . \tag{10.211}$$

Thus, the reaction cross section is the sum of two terms. The first,

$$\sigma_{\rm a} = \left|\frac{1}{A}\right|^2 \frac{k}{E} \left\langle \psi^{(+)} \right| - \mathrm{Im}\big\{V_{\rm opt}\big\} \left| \psi^{(+)} \right\rangle , \tag{10.212}$$

corresponds to the flux absorbed from the incident channel through the direct action of the optical potential. The second, is associated with prompt processes. It is given by

$$\sigma_{\rm p} = \left|\frac{1}{A}\right|^2 \frac{k}{E} \left\langle \psi^{(+)} \right| - \mathrm{Im}\big\{V_{\rm pol}\big\} \left| \psi^{(+)} \right\rangle \tag{10.213}$$

and represents the fraction of the incident flux lost to the P_1-space. It leads either to inelastic scattering (or transfer reactions) or to absorption by the optical potential in non-elastic channels. Although these contributions can be formally separated (see excercise 4), in practical implementations of the polarization potential approach they are mixed together. Separate contributions from these processes can be easily evaluated when the coupled channel equations involving the same channels are solved directly (see chapter 9, Eqs. (9.34) to (9.36)).

In the case of optical plus polarization potentials, discussed above, the optical theorem (see section 7.1) becomes

$$\sigma_{\rm tot} \equiv \sigma_{\rm el} + \sigma_{\rm R} = \frac{4\pi}{k}\, \mathrm{Im}\left\{f(0)\right\} , \tag{10.214}$$

with $f(0)$ representing the forward scattering amplitude associated with the scattering solution of Eq. (10.198).

10.5.1 *Adiabatic polarization potentials*

The adiabatic limit corresponds to the situation where the coupling interaction connects the elastic channel only with bound channels with very high excitation energies, ε_n. In such cases, the characteristic period of the excitation, $\hbar/\varepsilon_n$ is much smaller than the collision time. The P_1 Green's function is then approximated by,

$$\mathbb{G}^{(+)}_{\mathrm{P_1P_1}} = -\frac{1}{h_{\mathrm{P_1P_1}}}. \qquad (10.215)$$

Above, $h_{\mathrm{P_1P_1}}$ is the P_1-projected intrinsic Hamiltonian of the non-interacting projectile + target system[17] $h \equiv h^{(\mathrm{p})} + h^{(\mathrm{t})}$, which has eigenstates $|\varphi_n^{(\mathrm{p})}\varphi_m^{(\mathrm{t})}\rangle$, where

$$h^{(\mathrm{p})}\,|\varphi_n^{(\mathrm{p})}\rangle = \varepsilon_n^{(\mathrm{p})}\,|\varphi_n^{(\mathrm{p})}\rangle \quad \text{and} \quad h^{(\mathrm{t})}\,|\varphi_n^{(\mathrm{t})}\rangle = \varepsilon_m^{(\mathrm{t})}\,|\varphi_m^{(\mathrm{t})}\rangle. \qquad (10.216)$$

Using the spectral representation and taking matrix elements in the coordinate space, the Green's function of Eq. (10.215) becomes,

$$\langle \mathbf{r}|\,\mathbb{G}^{(+)}_{\mathrm{P_1P_1}}\,|\mathbf{r}'\rangle = -\sum\!\!\!\!\!\!\int_n \sum\!\!\!\!\!\!\int_m \frac{|\varphi_n^{(\mathrm{p})}\,\varphi_m^{(\mathrm{t})}\rangle\;\langle\varphi_n^{(\mathrm{p})}\,\varphi_m^{(\mathrm{t})}|}{\varepsilon_n^{(\mathrm{p})} + \varepsilon_m^{(\mathrm{t})}}\,\delta(\mathbf{r}-\mathbf{r}'). \qquad (10.217)$$

Inserting the above Green's function into Eq. (10.210) and taking the coordinate representation, we get the polarization potential,

$$\mathbb{V}_{\mathrm{pol}}(\mathbf{r}) = -\sum\!\!\!\!\!\!\int_n \sum\!\!\!\!\!\!\int_m \frac{\langle\varphi_0^{(\mathrm{p})}\,\varphi_0^{(\mathrm{t})}|\,\mathbb{V}_{\mathrm{PP_1}}\,|\varphi_n^{(\mathrm{p})}\,\varphi_m^{(\mathrm{t})}\rangle(\mathbf{r})\;\langle\varphi_n^{(\mathrm{p})}\,\varphi_m^{(\mathrm{t})}|\,\mathbb{V}_{\mathrm{P_1P}}\,|\varphi_0^{(\mathrm{p})}\,\varphi_0^{(\mathrm{t})}\rangle(\mathbf{r})}{\varepsilon_n^{(\mathrm{p})} + \varepsilon_m^{(\mathrm{t})}}, \qquad (10.218)$$

where the superposition of the sum and integral symbols means summing over the discrete part of the spectrum and integrating over its continuous part. Using the compact notation,

$$\left(\varphi_0^{(\mathrm{p})}\,\varphi_0^{(\mathrm{t})}\right|\;\mathbb{V}_{\mathrm{PP_1}}\;\left|\varphi_n^{(\mathrm{p})}\,\varphi_m^{(\mathrm{t})}\right)(\mathbf{r}) \equiv V_{00;nm}(\mathbf{r}), \qquad (10.219)$$

the polarization potential takes the form

$$\mathbb{V}_{\mathrm{pol}}(\mathbf{r}) = -\sum\!\!\!\!\!\!\int_n \sum\!\!\!\!\!\!\int_m \frac{V_{00;nm}(\mathbf{r})\;V_{nm;00}(\mathbf{r})}{\varepsilon_n^{(\mathrm{p})} + \varepsilon_m^{(\mathrm{t})}}. \qquad (10.220)$$

In most applications in nuclear, atomic and molecular physics, $V_{01}(\mathbf{r}) = V_{10}(\mathbf{r})$. Thus, one gets the dispersive, real and attractive polarization potential

$$V_{\mathrm{pol}}(\mathbf{r}) = -\sum\!\!\!\!\!\!\int_n \sum\!\!\!\!\!\!\int_m \frac{|\,V_{00;nm}(\mathbf{r})\,|^2}{\varepsilon_n^{(\mathrm{p})} + \varepsilon_m^{(\mathrm{t})}}. \qquad (10.221)$$

[17]To avoid confusion with the notation for the projectors (P), we use small letters to represent the projectile and the target.

We point out that the energies $\varepsilon_n^{(\mathrm{p})}$ and $\varepsilon_m^{(\mathrm{t})}$ are respectively measured with respect to the ground state energies of the projectile and the target, which can be taken to be zero. Clearly, the polarization potential of Eq. (10.221) can also be derived from second order perturbation theory. We opted for the reaction theory approach in order to have a unified picture of the effects of open and closed channels on the elastic channel.

Frequently, the projectile-target interaction is of electric nature[18], with the general form

$$V_{\mathrm{EM}}(\mathbf{r}) = \int\int d\mathbf{r}'_{\mathrm{p}}\, d\mathbf{r}'_{\mathrm{t}}\, \frac{\rho_{\mathrm{p}}(\mathbf{r}'_{\mathrm{p}})\,\rho_{\mathrm{t}}(\mathbf{r}'_{\mathrm{t}}) - (1/c^2)\, j_{\mathrm{p}}(\mathbf{r}'_{\mathrm{p}})\, j_{\mathrm{t}}(\mathbf{r}'_{\mathrm{t}})}{\left|\mathbf{r} + \mathbf{r}'_{\mathrm{p}} - \mathbf{r}'_{\mathrm{t}}\right|}. \tag{10.222}$$

For practical purposes, the electromagnetic potential is expanded in multipoles, and the scattering process is dominated by the lowest ones. The above interaction is then expressed as a sum of multipoles through the expansion of the denominator inside the integral [Jackson (1975)]. One finds for the electric terms (the first term in the numerator) (see [Alder and Winther (1975)]),

$$\begin{aligned} V_{\mathrm{EM}}(\mathbf{r}) = & \sum_{\lambda_{\mathrm{p}}\lambda_{\mathrm{t}}\mu_{\mathrm{p}}\mu_{\mathrm{t}}} c(\lambda_{\mathrm{p}}, \lambda_{\mathrm{t}}) \begin{pmatrix} \lambda_{\mathrm{p}} & \lambda_{\mathrm{t}} & \lambda_{\mathrm{p}} + \lambda_{\mathrm{t}} \\ \mu_{\mathrm{p}} & \mu_{\mathrm{t}} & -(\mu_{\mathrm{p}} + \mu_{\mathrm{t}}) \end{pmatrix} \mathfrak{M}_{\mathrm{p}}(E\lambda_{\mathrm{p}}, \mu_{\mathrm{p}}) \\ & \times \mathfrak{M}_{\mathrm{t}}(E\lambda_{\mathrm{t}}, \mu_{\mathrm{t}})\, \frac{1}{r^{\lambda_{\mathrm{p}}+\lambda_{\mathrm{t}}+1}}\, Y_{\lambda_{\mathrm{p}}+\lambda_{\mathrm{t}}, -(\mu_{\mathrm{p}}+\mu_{\mathrm{t}})}(\hat{\mathbf{r}}), \end{aligned} \tag{10.223}$$

where we have introduced the following,

$$c(\lambda_{\mathrm{p}}, \lambda_{\mathrm{t}}) = (4\pi)^{3/2}(-1)^{\lambda_{\mathrm{t}}} \sqrt{\frac{(2\lambda_{\mathrm{p}} + 2\lambda_{\mathrm{t}})!}{(2\lambda_{\mathrm{p}} + 1)!(2\lambda_{\mathrm{t}} + 1)!}} \tag{10.224}$$

and the electric multipole moments,

$$\mathfrak{M}(E\lambda, \mu) = \int d\mathbf{r}\ \rho(\mathbf{r})\, r^{\lambda}\, Y_{\lambda\mu}(\hat{\mathbf{r}}), \tag{10.225}$$

which can be rewritten as a finite sum over the charges. Using the relation

$$\rho(\mathbf{r}) = \sum_{i=1}^{Z} e\ \delta(\mathbf{r} - \mathbf{r}_i), \tag{10.226}$$

we get

$$\mathfrak{M}(E\lambda, \mu) = \sum_{i=1}^{Z} e\ r_i^{\lambda}\, Y_{\lambda\mu}(\hat{\mathbf{r}}_i). \tag{10.227}$$

[18]This is the case of collisions of atoms, ions and molecules. It is also the case of nuclear collisions at energies well below the Coulomb barrier, or collisions with large impact parameters. In such cases the projectile remains outside the range of the short-range nuclear forces.

In the case of pure electrostatic interaction, the matrix element $V_{00;nm}(\mathbf{r})$, is given by

$$V_{00;nm}(\mathbf{r}) = \left(\varphi_0^{(\mathrm{p})}\,\varphi_0^{(\mathrm{t})}\right| V_{\mathrm{EM}} \left|\varphi_n^{(\mathrm{p})}\,\varphi_m^{(\mathrm{t})}\right)(\mathbf{r}), \tag{10.228}$$

with V_{EM} given by Eq. (10.223). The calculation of the above matrix elements requires knowledge of angular momentum coupling algebra but can be easily done. The final result for the adiabatic polarization potential is,

$$V_{\mathrm{pol}}(r) = -16\pi^2 \sum_{\lambda_\mathrm{p}=1}^{\infty} \sum_{\lambda_\mathrm{t}=1}^{\infty} \frac{(2\lambda_\mathrm{p}+2\lambda_\mathrm{t})!}{(2\lambda_\mathrm{p}+1)!\,(2\lambda_\mathrm{t}+1)!} \frac{1}{r^{2(\lambda_\mathrm{p}+\lambda_\mathrm{t}+1)}}$$
$$\times \sum\!\!\!\!\!\!\int_n \sum\!\!\!\!\!\!\int_m \frac{\left|\left\langle\varphi_0^{(\mathrm{p})}\left|\mathfrak{M}_\mathrm{p}(E\lambda_\mathrm{p},0)\right|\varphi_n^{(\mathrm{p})}\right\rangle\right|^2 \left|\left\langle\varphi_0^{(\mathrm{t})}\left|\mathfrak{M}_\mathrm{t}(E\lambda_\mathrm{t},0)\right|\varphi_m^{(\mathrm{t})}\right\rangle\right|^2}{\varepsilon_n^{(\mathrm{p})}+\varepsilon_m^{(\mathrm{t})}}. \tag{10.229}$$

For completeness we present in the following an outline of the procedures used to evaluate double sum/integral over the virtually excited multipole states as given in [Dalgarno and Davidson (1966); Dalgarno (1967)]. The presence of the two excitation energies (of the two atoms) in the denominator of Eq. (10.229) requires one to use the so-called two-center model. An alternative method is the single-center method which relies on the following integral identity,

$$\frac{1}{a+b} = \frac{2}{\pi}\int_0^\infty dx\,\frac{ab}{(a^2+x^2)(b^2+x^2)}. \tag{10.230}$$

Defining the dynamic multipole frequency-dependent polarizabilities,

$$\alpha^{(\lambda)}(\omega) = \frac{8\pi}{2\lambda+1} \sum\!\!\!\!\!\!\int_n \frac{\varepsilon_n \left|\left\langle\varphi_0\left|\mathfrak{M}(E\lambda,0)\right|\varphi_n\right\rangle\right|^2}{\varepsilon_n^2-\omega^2}, \tag{10.231}$$

the final form of the adiabatic polarization potential is,

$$V_{\mathrm{pol}}(r) = -\frac{2}{\pi} \sum_{\lambda_\mathrm{p}=1}^{\infty} \sum_{\lambda_\mathrm{t}=1}^{\infty} \frac{(2\lambda_\mathrm{p}+2\lambda_\mathrm{t})!}{(2\lambda_\mathrm{p}+1)!\,(2\lambda_\mathrm{t}+1)!} \frac{1}{r^{2(\lambda_\mathrm{p}+\lambda_\mathrm{t}+1)}}$$
$$\times \int_0^\infty d\omega\;\alpha_\mathrm{p}^{(\lambda_\mathrm{p})}(i\omega)\;\alpha_\mathrm{t}^{(\lambda_\mathrm{t})}(i\omega). \tag{10.232}$$

Equation (10.229) can also be used to calculate the induced interaction between an ion with net charge, say, q_p, and a neutral atom. In such a case the coupling is of the monopole-dipole type and the moment $\mathfrak{M}_\mathrm{p}(E\lambda_\mathrm{p} = 0, 0)$ is just $q_\mathrm{p}/\sqrt{4\pi}$. Then we have,

$$V_{\mathrm{pol}}(r) = -\frac{1}{2}\,q_\mathrm{p}^2 \sum_{\lambda_\mathrm{t}=1}^{\infty} \frac{1}{r^{2(\lambda_\mathrm{t}+1)}}\,\alpha_\mathrm{t}^{\lambda_\mathrm{t}}(0), \tag{10.233}$$

where we have introduced the dynamic multipole polarizabilities calculated at zero frequency, $\alpha^{(\lambda)}(0)$,

$$\alpha^{(\lambda)}(0) = \frac{8\pi}{2\lambda+1} \sum\!\!\!\!\!\!\int_n \frac{\left| \langle \varphi_0 \left| \mathfrak{M}(E\lambda, 0) \right| \varphi_n \rangle \right|^2}{\varepsilon_n}. \tag{10.234}$$

The leading term of the ion-atom potential ($\lambda_{\rm t} = 1$) is given by the dipole moment of the target,

$$\mathfrak{M}_{\rm t}(E1, 0) = \sqrt{\frac{1}{4\pi}} \sum_{i=1}^{Z} e\, \mathbf{r}_i = \sqrt{\frac{1}{4\pi}}\, \mathbf{D}, \tag{10.235}$$

with $\mathbf{D}$ being the dipole operator for the target. Similar expressions of the higher multipolarity moments can be expressed in terms of the quadrupole, octupole, etc, moments operators.

Thus, the leading terms of the long-range interaction in the ion-atom or atom-atom systems can be summarized as follows,

$$V_{\rm pol}(r) = -\left[\frac{1}{2}\, q_{\rm p}^2\, \alpha_{\rm t}^{(1)}(0)\right] \frac{1}{r^4} - \left[C_{\rm VW} + \frac{1}{2}\, q_{\rm p}^2\, \alpha_{\rm t}^{(2)}(0)\right] \frac{1}{r^6} + \cdots. \tag{10.236}$$

Above, $C_{\rm VW}$ is the van der Waals coefficient, given by,

$$C_{\rm VW} = -\frac{3}{\pi} \int_0^{\infty} d\omega\, \alpha_{\rm p}^{(1)}(i\omega)\, \alpha_{\rm t}^{(1)}(i\omega). \tag{10.237}$$

For examples of numerical values of the van der Waals coefficient and the polarizabilities at zero frequency for several systems, we refer to [McDowell and Coleman (1970)].

The calculation of the dynamic dipole polarizability relies on sum rules and the solution of an appropriate differential equation. The coefficients $C_{\rm VW}$ are calculated using sum rules and dipole polarizabilities (for details, see the works of Dalgarno [Dalgarno (1967)] and Dalgarno and Davidson [Dalgarno and Davidson (1966)]). To start with, we recall the convention used in atomic physics, and introduce the electric dipole oscillator strength,

$$f_n = \frac{2}{3}\, \varepsilon_n\, \left| \langle \varphi_0 \left| \mathbf{D} \right| \varphi_n \rangle \right|^2 \equiv \frac{2}{3}\, \varepsilon_n \left| \sum_{i=1}^{Z} \langle \varphi_0 \left| e\, \mathbf{r}_i \right| \varphi_n \rangle \right|^2, \tag{10.238}$$

with the energy-weighted sum rules,

$$\mathbb{S}(0) = \sum\!\!\!\!\!\!\int_n f_n = \frac{2}{3}\, \langle \varphi_0 \left| [\mathbf{D}, [H, \mathbf{D}]] \right| \varphi_0 \rangle = Z \tag{10.239}$$

where Z is the total number of electrons of the neutral atom. Further, we have

$$\mathbb{S}(-1) = \sum\!\!\!\!\!\!\int_n \frac{f_n}{\varepsilon_n} = \frac{2}{3} \langle \varphi_0^{\mathrm{T}} |\mathbf{D}^2| \varphi_0^{\mathrm{T}} \rangle = \frac{2}{\pi} \int_0^{\infty} d\omega \, \alpha^{(1)}(i\omega) \tag{10.240}$$

and

$$\mathbb{S}(-2) = \sum\!\!\!\!\!\!\int_n \frac{f_n}{\varepsilon_n^2} = \alpha^{(1)}(0). \tag{10.241}$$

Dalgarno and collaborators have employed the above sum rules and the available data of the dipole polarizability, $\mathbb{S}(-2)$, and taken into account other multi-polarities (dipole-quadrupole, quadrupole-quadrupole, etc). They have obtained for the long range interaction between neutral atoms the following,

$$V_{\mathrm{pol}}(r) = -\frac{C_6}{r^6} - \frac{C_8}{r^8} - \frac{C_{10}}{r^{10}} \cdots . \tag{10.242}$$

Retardation effects arising from a proper Quantum Electrodynamic treatment, introduce corrections to the above potential, Eq. (10.242), that can be represented by a multiplicative function of r. These corrections were first calculated by Casimir and Polder [Casimir and Polder (1948)]. Taking into account the retardation effects, the C_i potential becomes

$$V_{\mathrm{pol}}(r) = \frac{C_i}{r^i} F_i(r). \tag{10.243}$$

The final effect of, say, $F_6(r)$ is to bring in a $1/r$ dependence, making the van der Waals force go as $1/r^7$.

In the actual analysis of ion-atom collisions, the short range behavior of the interaction is also required. The conventional calculation relies on the use of the Born-Oppenhiemer method employed for the combined two-atom system and the numerical results are represented by a potential that behaves exponentially as a function of the separation distance. One may then write a theoretically motivated ion-atom interaction having the general form,

$$V(r) = f(r)\, e^{-\beta r} + V_{\mathrm{pol}}(r), \tag{10.244}$$

where β is a constant, and the function $f(r)$ is such that

$$V\,(r \to 0) = E_{\mathrm{p+t}}(r = 0). \tag{10.245}$$

Above, $E_{\mathrm{p+t}}(r = 0)$ is the energy of the joint two-atom system. In practice, a simpler semi-empirical interaction is employed, with the correct appropriate long range behavior dictated by the dipole polarizabilitites of the atoms involved. They are given by the Lennard-Jones parametrizations. For neutral atoms and for ion-atom they are given respectively by [McDowell and Coleman (1970)]

$$V_{\rm LJ}(r) = V_{\rm min}\left[\left(\frac{r_0}{r}\right)^{12} - 2\left(\frac{r_0}{r}\right)^6\right] \tag{10.246}$$

and

$$V_{\rm LJ}(r) = \frac{V_{\rm min}}{2}\left[\left(\frac{r_0}{r}\right)^{12} - 3\left(\frac{r_0}{r}\right)^4\right]. \tag{10.247}$$

In the above, r_0 and $V_{\rm min}$ are, respectively, the position and depth of the minimum of $V(r)$.

Dipole polarizability is a general property of many-body systems. In atoms, the mutual polarizabilities of the two neutral systems is responsible for the dominant long-range interaction. And as we have seen, the polarizabilities and the van de Waals coefficients involve an infinite integral over the frequency. This is a consequence of the lack of any collective dipole excitation that would exhaust the energy-weighted sum rule. In contrast, in nuclei there is no need for the frequency integral as there is the well known Giant Dipole Resonance (GDR), which dominates the dipole strength. This state sits at quite high excitation energy, well above the particle emission threshold, and thus the name "resonance". In the calculation of the adiabatic polarization potential, the GDR is only virtually excited and since there are no low lying dipole excitations in the spectra of most stable nuclei, one can replace the sum/integral in Eq. (10.233) by just one term, the GDR, and employ the sum rule of Eq. (10.239). The customary procedure is the following. First rewrite an equivalent version of Eq. (10.233) for the dipole case, $\lambda = 1$, as

$$\alpha^{(\lambda)}(0) = \frac{8\pi}{3}\sum\!\!\!\!\!\!\int_n \frac{\varepsilon_n \left|\langle\psi_0\left|\mathfrak{M}(E1,0)\right|\psi_n\rangle\right|^2}{\varepsilon_n^2}. \tag{10.248}$$

Next, we apply the nuclear concept of the dominance of the GDR, and extract the denominator, $\varepsilon_n^2 = E^2_{\rm DGR}$, from the n sum-integral,

$$\alpha^{(\lambda)}(0) = \frac{1}{E^2_{\rm GDR}}\sum\!\!\!\!\!\!\int_n f_n. \tag{10.249}$$

The oscillator strength function, f_n, is given as in the atomic case by,

$$f_n = \frac{2}{3}\varepsilon_n\left|\sum_{i=1}^{Z}\left\langle\varphi_0^{(t)}\left|e\,\mathbf{r}_i\right|\varphi_n^{(t)}\right\rangle\right|^2. \tag{10.250}$$

Since the position vectors $\mathbf{r}_i$ refer to protons, one has to be careful in specifying the coordinate system that is best suited for the calculation. The nuclear GDR refers to a collective oscillation of the protons against the neutrons. As such the dipole operator must contain the position vectors of the protons with respect to the center of mass of the oscillating nucleus (in the present example it is assumed to be the target). This implies either changing the definition of the vector $\mathbf{r}_i$ or assigning effective charges to both protons and neutrons and let the sum over i run over all the nucleons. In either case the final result is quite simple, owing to the well known sum rule. Accordingly, the nuclear adiabatic dipole polarization potential is given by [Alder and Winther (1975)],

$$V_{\mathrm{pol}}(r) = -\left[\frac{4\pi e^2}{9}\left(\frac{Z_{\mathrm{p}}^2\,\mathbb{S}_{\mathrm{t}}(0)}{E_{\mathrm{GDR}}^2}\right)\right]\frac{1}{r^4}, \tag{10.251}$$

where $\mathbb{S}(0)$ is the classical nuclear dipole sum rule,

$$\mathbb{S}(0) = 14.8\,\frac{NZ}{A}\,\mathrm{e}^2\,\mathrm{fm}^2\,\mathrm{MeV}. \tag{10.252}$$

The position of the GDR, E_{GDR}, is obtained from systematics of nuclear photo-absorption cross section and is usually given by

$$E_{\mathrm{GDR}} \simeq \frac{80}{A^{1/3}}\ \mathrm{MeV}. \tag{10.253}$$

The situation changes when dealing with neutron-rich or proton-rich unstable nuclei. These systems have been the subject of intense experimental and theoretical investigation over the last two decades and it has now been well established that owing to the excess neutrons (or protons), the dipole strength is dramatically changed as low lying collective excitations is now present. This is a result of the possibility of setting the excess nucleons to oscillate against the stable core, giving rise to what has been coined the Pygmy Giant Dipole or Steinwedel-Jensen Resonance (PGDR), which would greatly influence the scattering observables and requires a more careful treatment than the one bestowed to the GDR. In light unstable neutron rich nuclei, such as ^{11}Be, ^{11}Li, or proton rich nuclei like ^{8}B, the PGDR is not really a resonance, as it sits very near to the breakup threshold close to zero energy. Nevertheless these PDGR in light nuclei are critical for the understanding of the collision observables when scattered off light and/or heavy stable target nuclei. The adiabatic treatment is not applicable here and one must resort to a detailed analysis of the polarization potential within, e.g., the sudden limit, as given in the next section.

The nuclear adiabatic potential is but a small correction in the large distance interaction between nuclei, when compared to the dominant monopole-monople Coulomb interaction. On the other hand, in the atom-atom interaction, the adiabatic potential *is* the dominant one, and its detailed knowledge is important in delineating the properties of gases and the understanding of atomic collisions as well as in molecular structure.

10.5.1.1 *Polarization of the neutron*

Let us now consider the case of a neutron beam colliding with target nuclei with atomic number Z_{T}. This hadronic projectile has a net charge of zero, but is constituted by three charged quarks; one up and two downs. The quarks have the charges: $q_{\mathrm{up}} = 2e/3$ and $q_{\mathrm{down}} = -e/3$, with e standing for the absolute value of the electronic charge. Accordingly, in addition to the nuclear interaction, it can interact electrically with a charged target, as in the case of ion-atom collisions, through its dipole polarizability,

$$\begin{aligned} V_{\text{electric}}(r) &= \frac{A_4}{r^4}, \qquad r > R, \\ &= 0, \qquad r < R. \end{aligned} \tag{10.254}$$

Above, A_4 is the constant,

$$A_4 = -\frac{1}{2}\, Z_{\mathrm{T}}^2\, e^2\, \alpha_n, \tag{10.255}$$

where α_n is the electric dipole polarizability of the neutron and R is the radius of the charge distribution of the target nucleus. We have approximated V_{electric} at $r < R$ to be zero, as in this region the nuclear interaction dominates.

The means to determine α_n is through a careful measurement of the elastic scattering angular distribution from a heavy target nucleus at low energies. The nuclear contribution to the scattering at these energies is accounted for by the effective range theory of section 2.11.2. Taking only the scattering length for the nuclear scattering, the amplitude becomes

$$f_{\text{nucl}}(\theta) \to f_{\text{nucl}} = -a. \tag{10.256}$$

We now calculate the effect of the dipole polarizability using the first order Born approximation,

$$f_\alpha(\theta) = -2\pi^2\, \frac{2\mu}{\hbar^2}\, A_4 \int d^3\mathbf{r}\; \phi^*(\mathbf{k}';\mathbf{r})\; \frac{1}{r^4}\; \phi(\mathbf{k};\mathbf{r}), \tag{10.257}$$

with θ standing for the angle between $\mathbf{k}$ and $\mathbf{k}'$. The matrix-element above can be evaluated using the plane waves like

$$\phi(\mathbf{k};\mathbf{r}) = \frac{1}{(2\pi)^{2/3}}\, e^{i\mathbf{k}\cdot\mathbf{r}}. \tag{10.258}$$

We get,

$$f_\alpha(q) = \frac{m}{2\hbar^2}\,\frac{Z_{\rm T}^2 e^2}{R}\,\alpha_n\,\left[-\frac{1}{2}\pi qR + \frac{\sin qR}{qR} + \cos qR + qR\int_0^{qR}\frac{\sin t}{t}dt\right], \tag{10.259}$$

where $q = 2k\sin(\theta/2)$ is the modulus of the momentum transfer divided by $\hbar$. Note that we have approximated the reduced mass by the neutron mass, m. The linear term in q is a characteristic feature of the first Born term, representing the effect of a r^{-4} potential. Further, this term is independent of the radius R. It is this aspect which is explored in the determination of the neutron polarizability. Neglecting the $|f_\alpha(q)|^2$ term, the cross section is given by,

$$\frac{d\sigma}{d\Omega} = |f_{\rm nuc} + f_\alpha(q)|^2 \simeq a^2 - 2a\,\frac{\mu}{2\hbar^2}\,\alpha_n\,\frac{Z_{\rm t}^2 e^2}{R}$$
$$\times\left[-\frac{1}{2}\pi qR + \frac{\sin qR}{qR} + \cos qR + qR\int_0^{qR}\frac{\sin t}{t}dt\right]. \tag{10.260}$$

Thaler [Thaler (1959)], who was the first to suggest this method of determining the neutron polarizability, made an expansion in terms of Legendre polynomials and rewrote the above cross section in the following equivalent form,

$$\frac{d\sigma}{d\Omega} = \frac{\sigma_{\rm tot}}{4\pi}\,\left[1 + \omega_1 P_1(\cos\theta) + \omega_2 P_2(\cos\theta) + \cdots\right], \tag{10.261}$$

where

$$\omega_l = -\frac{2l+1}{(2l+1)!!}\,\frac{2\pi m}{\hbar^2\, a}\,Z_{\rm T}^2\, e^2\, k\,\alpha_n \tag{10.262}$$

and $\sigma_{\rm tot} = 4\pi a^2$ is the total neutron cross section for $\alpha_n = 0$. Notice the linear dependence of ω_l on k. This feature was explored by Thaler [Thaler (1959)] to obtain the the electric dipole polarizability of the neutron from the experimental neutron scattering work of Langsdorf, Lane and Monahan [Langsdorj Jr. *et al.* (1957)] . This latter work involves the scattering of the neutron from a variety of medium and heavy target nuclei. The value estimated by Thaler from these data was in the range $0 < \alpha_n < 20\times10^{-4}$ fm^3. The currently accepted value of the polarizability is $(11.6\pm1.5)\times10^{-4}$ fm^3.

10.5.2 *Polarization potentials in the sudden limit*

There are many situations, especially in nuclear physics, where the elastic channel couples strongly with inelastic channels with very low excitation energies. Typical examples are collisions involving a highly deformed nucleus, like a rare earth or an actinide. In such cases the elastic channel is strongly coupled with inelastic channels associated with the states in the rotational band based on the ground state. In these situations, the collision times are much shorter than the characteristic period of the excitations. One can then use the sudden approximation, which consists of neglecting the excitation energies in the propagator in the P_1-space. In some calculations, the approximation is improved through inclusion of the adiabaticity factor [Alder and Winther (1975)].

The earliest calculations of polarization potentials in Nuclear Physics [Love *et al.* (1977); Baltz *et al.* (1978, 1979)] (see also the review article [Baltz *et al.* (1984)]) were based on simple approximations for the Green's function of Eq. (10.210). They investigated the influence of Coulomb excitation on the elastic scattering of light-heavy ions from deformed nuclei, considering collisions of spherical ^{16}O projectiles with the deformed ^{184}W target nucleus. In this case, there is strong coupling with the $I^\pi = 2^+$ channel, corresponding to the member of the rotational band of ^{184}W with lowest excitation energy. The excitation of this channel corresponds mainly to first order transitions, induced by the quadrupole component of the Coulomb interaction. Since the charge of the ^{16}O projectile is not very high, the excitation of states with $I > 2$ is unlikely for this system. They can only take place through multistep Coulomb excitation, or through the action of high ($\lambda > 2$) multipole components of the Coulomb field. Thus, in the works of [Love *et al.* (1977); Baltz *et al.* (1978, 1979)] only the elastic and the 2^+ channels were included in the calculations. Owing to the long reach of the Coulomb quadrupole coupling, absorption into the inelastic 2^+ channel can take place even for distant collisions, associated with small scattering angles. As an example of a calculation of the polarization potential and its application to predict experimental data, we discuss in detail the work of Baltz *et al.* [Baltz *et al.* (1978, 1979)].

According to the rotational model for the deformed target (see e.g. [Bohr and Mottelson (1998)]), the intrinsic states are the symmetric top eigenfunctions [Brink and Satchler (1971)], $|I\nu K)$, where I is the intrinsic angular momentum quantum number and ν and K are respectively its z-projection in the laboratory frame and its projection onto the symmetry axis in the intrinsic frame. In the case of the ground state rotational band of an even-even (even numbers of protons and neutrons) target, $K = 0$, $I = 0, 2, 4, 6, ...$, $\nu = -I, ..., I$ and $\pi = +$. In the notation of chapter

9, α labels the intrinsic states in the $K = 0$ band. In the calculations of [Love *et al.* (1977); Baltz *et al.* (1978, 1979)], only the ground state, $|0) \equiv |K = 0, I = 0, \nu = 0)$, and the first excited states, $|2\nu) \equiv |K = 0, I = 2, \nu)$, were taken into account. In the present example, the channel spin, I_α, is sufficient to specify the channel and the projectors P and P_1 can be written

$$P = \left|\mathcal{Y}_{0l}^{l0}\right) \left(\mathcal{Y}_{0l}^{l0}\right| \quad \text{and} \quad P_1 = \sum_{JMl} \left|\mathcal{Y}_{2l}^{JM}\right) \left(\mathcal{Y}_{2l}^{JM}\right|. \tag{10.263}$$

Above, $\left|\mathcal{Y}_{I_\alpha l}^{JM}\right\rangle$ are the channel-spin wave functions (see chapter 9) and the subscripts 0 and 2 stand respectively for the spins of the ground and excited states. Since the ground state has spin zero, we set $J = l$ in the expression for the projector P. The Green's function $\mathbb{G}_{\mathrm{P}_1}^{(+)}$ has the spectral representation

$$\mathbb{G}_{\mathrm{P}_1}^{(+)} = \sum_{JMl} \left|\mathcal{Y}_{2l}^{JM}\right) g_{\mathrm{opt},l}^{(+)} \left(\mathcal{Y}_{2l}^{JM}\right|. \tag{10.264}$$

Above, $g_{\mathrm{opt},l}^{(+)}$ is the l-projected optical Green's function, which is associated with the Hamiltonian $K_l + V_{\mathrm{opt}}$. In collisions at low energies or at small scattering angles, the classical turning point is beyond the reach of the nuclear optical potential. One can then approximate $g_{\mathrm{opt},l}^{(+)}$ by the corresponding Coulomb Green's function [Baltz *et al.* (1978, 1979)]. In the coordinate representation, we write (see Eq. (4.256))

$$g_{\mathrm{opt},l}^{(+)}(E_2; r, r') \simeq g_{\mathrm{C},l}^{(+)}(E_2; r, r') = -\frac{2\mu}{\hbar^2 k} \left[F_l(\eta_2, k_2 r_<)\, H_l^{(+)}(\eta_2, k_2 r_>)\right], \tag{10.265}$$

where $E_2 = E - \varepsilon_{2^+}$ is the energy of the relative motion in the 2^+ channel, η_2 is the corresponding Sommerfeld parameter and F_l and $H_l^{(+)}$ are respectively the regular and the outgoing Coulomb functions, introduced in chapter 3. Dropping the subscripts P_0 and P_1, which are unnecessary in the present context, and using the above results in the angular momentum projected version of Eq. (10.210), we get the polarization potential

$$V_{\mathrm{pol},l} = \sum_{l'} \left(\mathcal{Y}_{0l}^{l0}\right| \mathbb{V} \left|\mathcal{Y}_{2l'}^{l0}\right) g_{\mathrm{C},l'}^{(+)}(E_2) \left(\mathcal{Y}_{2l'}^{l0}\right| \mathbb{V} \left|\mathcal{Y}_{0l}^{l0}\right). \tag{10.266}$$

Above, we used the conservation of the total angular momentum to eliminate the summation over J and M. Adopting the notation of section 9.3.3, $\left(\mathcal{Y}_{2l'}^{l0}\right| \mathbb{V} \left|\mathcal{Y}_{0l}^{l0}\right) = \mathcal{V}_{2l',0l}^{l}$ and using Eqs. (9.204) and (9.208), with $\mathrm{Q}_{\mathrm{P}} = Z_{\mathrm{P}}\, e$, we get

$$\mathcal{V}_{2l',0l}^{l}(r) = -\frac{1}{4\pi\epsilon_0} \frac{4\pi}{5} \frac{Z_{\mathrm{P}}\, e}{r'^3} e^{2i\phi} \frac{\sqrt{(2l'+1)\,(2l+1)}\,\sqrt{5}}{\sqrt{4\pi}} \times \begin{pmatrix} l' & 2 & l \\ 0 & 0 & 0 \end{pmatrix} \begin{Bmatrix} l & 2 & l' \\ 2 & l & 0 \end{Bmatrix} (2\|\, Q_2 \,\|0), \tag{10.267}$$

with the phase ϕ given by

$$\phi = l + 2 + \frac{1}{2}(l' - l - 2). \tag{10.268}$$

To evaluate the polarization potential, we use the explicit form of the 6J coefficient (see [Edmonds (1974)])

$$\begin{Bmatrix} l & 2 & l' \\ 2 & l & 0 \end{Bmatrix} = \frac{1}{\sqrt{5\,(2l'+1)}},$$

express the 3J coefficient in terms of Clebsch-Gordan coefficients,

$$\begin{pmatrix} l' & 2 & l \\ 0 & 0 & 0 \end{pmatrix} = \frac{1}{\sqrt{2l'+1}}\ \langle l020|l'0\rangle\,,$$

replace the reduced matrix-element by the corresponding reduced transition probability (see Eq. (9.210))

$$(2\|\ Q_2\ \|0) = \sqrt{B(E2, 0 \to 2)}, \tag{10.269}$$

and use the relation

$$\mathcal{V}^{l}_{0l,2l'} = \left[\mathcal{V}^{l}_{2l',0l}\right]^*. \tag{10.270}$$

Writing Eq. (10.266) in the coordinate representation and assuming that the coupling interaction is local, we obtain

$$\begin{aligned} V_{\mathrm{pol},l}(r,r') = &-\frac{1}{4\pi\epsilon_0}\left[\frac{8\pi\mu}{25\,k\hbar^2}\ Z_{\mathrm{P}}^2\,e^2\ B(E2,0\to 2)\right] \\ &\times\frac{1}{r^3\,r'^3}\sum_{l'=l,l\pm 2}\left|\langle l020|l'0\rangle\right|^2\ F_l(\eta_2,k_2r_<)\,H_l^{(+)}(\eta_2,k_2r_>). \end{aligned} \tag{10.271}$$

In spite of the local nature of the coupling interaction, the potential of Eq. (10.271) is non-local owing to the non-locality of the Green's function. To avoid non-localities, one usually defines trivially equivalent local potentials. If $\psi_l(r)$ is the exact scattering solution of the Schrödinger equation for the angular momentum projected Hamiltonian $H_l = K_l + V_{\mathrm{opt}}(r) + V_{\mathrm{pol},l}(r,r')$, the equivalent local potential, $\overline{V}_{\mathrm{pol},l}(r)$, is defined by the relation[19]

$$\overline{V}_{\mathrm{pol},l}(r) = \frac{1}{\psi_l(r)}\int dr'\,V_{\mathrm{pol},l}(r,r')\ \psi_l(r'). \tag{10.272}$$

[19]Note that in this section we use overline ($\overline{V}_{\mathrm{pol},l}$) to indicate the local version of the polarization potential. It should not be confused with the energy averages performed in the derivation of optical potentials, for which we used this notation in a previous section.

In works of ([Baltz *et al.* (1978, 1979)]), the above local potentials were evaluated with the perturbative approximation $\psi_l \to F_l$. That is,

$$\overline{V}_{\text{pol},l}(r) = \frac{1}{F_l(\eta, kr)} \int dr' \, V_{\text{pol},l}(r, r') \; F_l(\eta, kr'). \tag{10.273}$$

The trivially equivalent local potential was then obtained using several approximations. First, the Coulomb Green's function was calculated within the sudden limit. It consists of neglecting the relative energy loss arising from the excitation of the 2^+ state. In this way, one evaluates Coulomb functions with the Sommerfeld parameter and the wave number corresponding to the elastic channel, instead of η_2 and k_2. To account for adiabatic effects, this approximate polarization potential was multiplied by the adiabaticity factor of Alder and Winther [Alder and Winther (1975)], $g(\xi)$. The second approximation was to neglect the off-shell contribution of the Green's function. It corresponds to replacing in Eq. (10.271)

$$H_l^{(+)}(\eta, kr_>) \equiv G_l\,(\eta, kr_>) + iF_l\,(\eta, kr_>) \to iF_l\,(\eta, kr_>)\,. \tag{10.274}$$

It was argued that the off-shell part of the Green's function gives rise to a rapidly oscillating real potential, which is a negligible correction to the optical potential. With this approximation, the polarization potential becomes the sum of separable imaginary potentials. Using the explicit values of the Clebsch-Gordan coefficients: $|\langle l020|l0\rangle|^2 = 1/4$ and $|\langle l020|(l \pm 2)0\rangle|^2 = 3/8$, one gets

$$\overline{V}_{\text{pol},l}(r) = -\,i\;\frac{A}{r^3}\;\left[I_0(r) + I_+(r) + I_-(r)\right], \tag{10.275}$$

with

$$A = \frac{1}{4\pi\epsilon_0}\left[\frac{2\pi\mu}{25\,k\hbar^2}\;Z_{\rm P}^2\,e^2\;B(E2, 0 \to 2)\right]\;g(\xi), \tag{10.276}$$

$$I_0(r) = \int_0^\infty dr'\;F_l(\eta, kr')\;\frac{1}{r'^3}\;F_l(\eta, kr') \tag{10.277}$$

and

$$I_\pm(r) = \frac{3}{2}\;\frac{F_{l\pm 2}(\eta, kr)}{F_l(\eta, kr)}\;\int_0^\infty dr'\;F_{l\pm 2}(\eta, kr')\;\frac{1}{r'^3}\;F_l(\eta, kr'). \tag{10.278}$$

Using standard integrals of Coulomb functions [Alder and Winther (1975)] and convenient semiclassical approximations, the polarization potential takes the simpler form

$$\overline{V}_{\text{pol},l}(r) = -\,i\;\frac{A}{r^3}\;\left[B_0(l, E) + \frac{B_1(l, E)}{r} + \frac{B_2(l, E)}{r^2}\right], \tag{10.279}$$

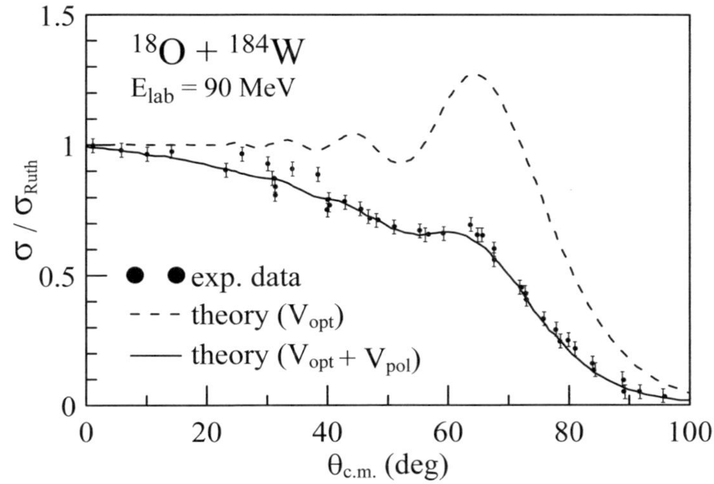

Fig. 10.9 Elastic cross section in collisions of $E_{\rm lab} = 90$ MeV ^{18}O projectiles with a ^{184}W target. The calculations are from [Baltz *et al.* (1978, 1979)] and the data are from [Thorn *et al.* (1977)]. For more details, see the text and [Baltz *et al.* (1978, 1979)].

with the l- and E-dependent strengths

$$B_0(l,E) = \frac{\eta^2 k^2 \left(3\hat{l}^2 + \eta^2\right)}{\hat{l}^2 \left(\hat{l}^2 + \eta^2\right)^2} - \frac{\eta k^2}{\hat{l}^3} \tan^{-1}\left(\hat{l}/\eta\right) \tag{10.280}$$

$$B_1(l,E) = \frac{6\eta k \hat{l}^2}{\left(\eta^2 + \hat{l}^2\right)^2} \tag{10.281}$$

$$B_2(l,E) = \frac{2\hat{l}^4}{\left(\eta^2 + \hat{l}^2\right)^2}, \tag{10.282}$$

where $\hat{l} = l + {}^1\!/_2$.

Despite the approximations involved in the derivation of Eqs. (10.279), the elastic cross section predicted with this polarization potential is quite close to the data. This is illustrated in figure 10.9. The dashed line corresponds to the results of a calculation including only the optical potential whereas the solid line includes also the polarization potential. One sees that the calculation without the polarization potential is very different from the data of [Thorn *et al.* (1977)] (solid circles). It exhibits a typical Fresnel pattern (see figure 7.4), which is not present in the experimental angular distribution. The data is much smaller than the predictions of this calculation, even at small angles. This difference is due to the absorption associated with Coulomb excitation of rotational states. Owing to the long range of the coupling, this effect can be observed even at small angles, corresponding to distant collisions. When the polarization potential of Eq. (10.279) is included in the calculation (solid line), the agreement with the data becomes very good. We conclude that, despite the approximations involved, the polarization potential of [Baltz *et al.* (1978, 1979)] accounts very well for the absorption due to the Coulomb excitation of the 2^+ state in ^{184}W.

In collisions of heavy projectiles (^{40}Ca or heavier) with deformed targets, the situation is more difficult. In such cases, multi-step Coulomb excitation becomes important and higher members of the rotational band have to be included in the P_1-space. The calculations become much more involved. It is then more convenient to resort to non-perturbative approximations. One possibility is to use the classical trajectory approximation in many-body scattering, which will be discussed in detail in chapter 11. In this method the projectile-target relative motion is treated classically and the intrinsic state is approximated as

$$|\psi(t)\rangle = \sum_\alpha a_\alpha(t)\; e^{-i\varepsilon_\alpha t/\hbar}\; |\alpha\rangle\,, \tag{10.283}$$

where ε_α and $|\alpha\rangle$ are eigenvalues and eigenstates of the intrinsic Hamiltonian. The above wave function is then inserted into the time-dependent Schrödinger equation with the potential $V(t) \equiv V(r(t)) = V\left(|\mathbf{r}(t)|\right)$, with $\mathbf{r}(t)$ representing the classical trajectory. This procedure leads to the set of coupled 1st-order differential equations for the amplitudes,

$$i\hbar\,\frac{da_\alpha(t)}{dt} = e^{i(\varepsilon_\alpha-\varepsilon_\beta)t/\hbar}\; V_{\alpha\beta}\,(t)\; a_\beta(t), \quad \alpha,\beta = 0,1,....\,, \tag{10.284}$$

which should be solved with the initial conditions: $a_\alpha(t \to \; -\infty) = \delta_{\alpha 0}$. This set of equations is a semiclassical time-dependent version of the coupled channel equations.

Introducing the polarization potential which accounts for channel coupling effects on the elastic amplitude (yet to be determined), and taking the sudden limit ($\varepsilon_\alpha \simeq \varepsilon_\beta \simeq 0$) one gets the single equation[20],

$$i\hbar\,\frac{da(t)}{dt} = V_{\rm pol}(t)\; a(t). \tag{10.285}$$

[20]Since we are only dealing with the elastic amplitude, we drop the subscript in $a_0(t)$.

Eq. (10.285) can be re-written as

$$V_{\rm pol}(t) = i\hbar \, \frac{d}{dt} \ln \left\{ a(t) \right\} . \tag{10.286}$$

This equation can be used to derive the polarization potential from the elastic amplitude. For this purpose, it is necessary to express the time as a function of r. This amounts to finding the inverse of the function $r(t)$: $t(r)$. However, $t(r)$ is double-valued. A given distance r is reached twice: once on the incoming branch of the trajectory and once on the outgoing one. Thus, the inversion procedure leads to two different functions, $t^-(r)$ and $t^+(r)$. We use the superscripts '-' and '+' to denote the ingoing and the outgoing branches, respectively. The velocity at the separation r, $\dot{r}(r)$, is different in each branch. For a Coulomb trajectory, which is symmetric, these velocities satisfy the relation: $\dot{r}^+(r) = -\dot{r}^-(r)$. Performing the change of variable for the branches $t^-(r)$ and $t^+(r)$, one gets two different polarization potentials, $V^-_{\rm pol}(r)$ and $V^+_{\rm pol}(r)$, given by

$$V^{\pm}_{\rm pol}(r) = i\hbar \, \dot{r}^{\pm}(r) \, \frac{d}{dr} \ln \left\{ a^{\pm}(r) \right\} , \tag{10.287}$$

where $a^{\pm}(r) \equiv a\left(t^{\pm}(r)\right)$.

In [Donangelo *et al.* (1979)] it was shown that the polarization potential leading to the regular scattering solution is the average,

$$V_{\rm pol} = \frac{V^-_{\rm pol} + V^+_{\rm pol}}{2} . \tag{10.288}$$

Thus, for a trajectory with angular momentum $\hbar l$, the polarization potential is

$$\overline{V}_{{\rm pol},l}(r) = i \, \frac{\hbar}{2} \, \dot{r}^+_l \left[\frac{d}{dr} \left(\ln a^+_l(r) \right) - \frac{d}{dr} \left(\ln a^-_l(r) \right) \right] . \tag{10.289}$$

The l-dependent polarization potentials were then evaluated using in the above equation a simple approximation for the elastic amplitude, $a^{\pm}(r)$.

In the early 90's, the role of continuum states of the projectile in nuclear collisions started to attract considerable interest. This was due to the fact that radioactive beams, like ^{11}Li, became available. Since the breakup threshold for such nuclei is extremely low (0.33 MeV for ^{11}Li) the influence of breakup on elastic scattering and on other reaction channels is very important. Of course, this situation could be handled by the continuum discretized coupled channel method, discussed in section 9.4. However, the computer resources required for this kind of calculations were usually unavailable. The situation was critical in collisions of weakly bound projectiles with very heavy targets, where Coulomb breakup is very important. In such cases, the breakup process is dominated by the long ranged dipole coupling and the calculations converge very slowly.

The first estimates of breakup polarization potentials were made for collisions of light weakly bound systems, where the breakup process was mainly due to nuclear couplings [Canto *et al.* (1991, 1992); Hussein *et al.* (1993)] (see also [Takigawa *et al.* (1993)]). For heavier weakly bound systems, the situation is quite different. In this case Coulomb coupling dominates. Polarization potentials associated with Coulomb breakup were first derived in [Andrés *et al.* (1994); Gómez-Camacho *et al.* (1994)]. These works were based on a semiclassical description of the collision using the sudden limit. Deviations from this limit were then taken into account through a commonly used procedure in Coulomb excitation. The potential was corrected by a multiplicative function of the adiabaticity parameter [Alder and Winther (1975)], $\xi = a_0\,\varepsilon^*/\hbar v$, where ε^* is the excitation energy, a_0 is half the distance of closest approach in a head-on collision and v is the asymptotic velocity. Since this approach leads to a purely imaginary potential, the authors evaluated its real part using the dispersion relation, discussed in section 10.1.1. This theory was used to calculate the breakup polarization potential in the collision of ^{11}Li projectiles with a ^{208}Pb target. The resulting polarization potential was written as

$$V_{\rm pol}(r) = V_0(r)\ [\ g\,(z,\xi) + i\,f\,(z,\xi)\ ]\,, \qquad (10.290)$$

where

$$V_0(r) = -\left[\frac{4\pi\,Z_{\rm T}^2 e^2\,B(E1, 0\to 1)}{9\,\hbar v}\right]\ \frac{1}{(r-a_0)^2\,r}, \qquad (10.291)$$

$$f\,(z,\xi) = 4\,\xi^2 z^2\,e^{-\pi\xi}\,K''_{2i\xi}\,(2\xi z)\,, \qquad (10.292)$$

$$g\,(z,\xi) = \frac{1}{\pi}\,\mathcal{P}\int_{-\infty}^{\infty}\frac{f\,(z,\xi')}{\xi-\xi'}\,d\xi'. \qquad (10.293)$$

Above, K'' is the second derivative of the modified Bessel function [Abramowitz and Stegun (1972)] and the argument z is given by

$$z = \frac{r}{a_0} - 1. \qquad (10.294)$$

Other derivations of breakup polarization potentials for weakly bound systems were presented in [Canto *et al.* (1995)] and [Ibraheem and Bonaccorso (2005)]. The former is based in quantum mechanics in the sudden limit and calculates the polarization potential using perturbation theory to lowest order. The latter extracts the potential from a semiclassical calculation of the phase-shifts.

10.5.3 *Realistic polarization potentials*

The polarization potentials discussed in the previous subsections are based on qualitative models and/or account for the couplings with a small number of channels. In this section we discuss the polarization potential proposed by Thompson *et al.* [Thompson *et al.* (1989)] which is based on full coupled-channel calculations and can accurately handle the influence of large numbers of channels, including channels in the continuum.

For the derivation of this potential, it is necessary to solve the coupled-channel problem in advance. Thus, this approach to the scattering problem does not have the advantage of avoiding the complications of a full coupled-channel calculation, as the ones described in the previous section. However, this potential leads to simple interpretations of coupled channel effects in the context of a single channel space. Besides, using this polarization potential one can eliminate the influence of a large number of channels and perform coupled channel calculations in a reduced space, containing only a few channels of major interest.

We start with the definition of the exact l- and E-dependent polarization potentials for a given coupled channel problem. For simplicity, we consider the case of the collision of spinless particles in the space of N channels $\alpha = 0, 1, ..., N-1$. We assume also that all intrinsic states involved in the calculation have spin zero. The elastic radial wave function at the lth-partial wave, satisfies the equation (see Eq. (9.105))

$$\Big[E - H_l\Big]\, u_{0l,0l}(k_0, r) - \sum_{\alpha \neq 0} \mathcal{V}_{0l,\alpha l}(r)\, u_{\alpha l,0l}(k_\alpha, r) = 0, \tag{10.295}$$

where $\mathcal{V}_{0l,\alpha l}(r)$ are the off-diagonal matrix-elements of the interaction and

$$H_l = -\frac{\hbar^2}{2\mu}\left(\frac{d^2}{dr^2} - \frac{l\,(l+1)}{r^2}\right) + V_l(r), \tag{10.296}$$

with $V_l(r) = \mathcal{V}_{0l,0l}(r)$. If other channels where eliminated through the introduction of an optical potential one should take: $V_l(r) = V_{\text{opt},l}(r)$. The influence of channel coupling on the elastic wave function is represented by the last term on the LHS of Eq. (10.295). The *exact* polarization potential is defined as the potential which produces the same effect in a single channel equation for the elastic wave function. Using the short-hand notation $\varphi_l(r) = u_{0l,0l}(k_{0,r})$, the l-dependent polarization potential is defined by the equation

$$\left[E_0 - H_l - V_{\mathrm{pol},l}(r)\right] \varphi_l(r) = 0. \tag{10.297}$$

Comparing Eqs. (10.295) and (10.297), we immediately get

$$V_{\mathrm{pol},l}(r) = \frac{1}{\varphi_l(r)} \sum_{\alpha \neq 0} \mathcal{V}_{0l,\alpha l}(r)\, u_{\alpha l,0l}(k_\alpha, r). \tag{10.298}$$

The above potential has one serious disadvantage: it has poles at the zeroes of the elastic wave function, which is in the denominator. Furthermore, it has the inconvenient feature of being angular momentum dependent. To avoid these shortcomings, Thompson *et al.* [Thompson *et al.* (1989)] proposed the approximate potential

$$V_{\mathrm{pol}}(r) = \frac{\sum_l \mathrm{w}_l(r)\, V_{\mathrm{pol},l}(r)}{\sum_l \mathrm{w}_l(r)}, \tag{10.299}$$

where

$$\mathrm{w}_l(r) = (2l+1) \left[1 - \left|\bar{S}_l\right|^2\right] \left|\varphi_l(r)\right|^2. \tag{10.300}$$

Above, $\bar{S}_l$ is the part of the S-matrix associated with the short-range interaction (see section 3.4).

The method of Thompson *et al.* [Thompson *et al.* (1989)] is very powerful, leading to polarization potentials that account very well for the effects of couplings of any kind. As an illustration, we discuss the influence of the breakup channel in the elastic scattering of ^{8}B in collisions with a ^{58}Ni target. Owing to the very low threshold energy for the breakup reaction $^8\mathrm{B} + {}^{58}\mathrm{Ni} \rightarrow {}^7\mathrm{Be} + p + {}^{58}\mathrm{Ni}$ ($E_{\mathrm{th}} = 0.138$ MeV), the elastic channel is strongly coupled with the breakup one. Thus, this effect should be accounted for in the calculations. This can be done by the CDCC method, discussed in section 9.4. Alternatively, this influence can be expressed as a polarization potential in the elastic channel. A CDCC calculation for this system has been performed by Lubian and Nunes [Lubian and Nunes (2007)], and the polarization potential according to the prescription of Thompson *et al.* has been derived. Figure 10.10 shows the elastic cross section in $^8\mathrm{B}+^{58}\mathrm{Ni}$ scattering obtained by a CDCC calculation (solid

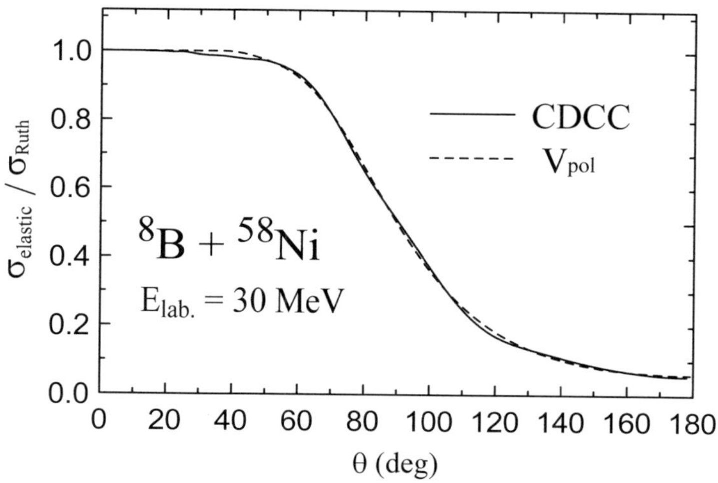

Fig. 10.10 Comparison of the elastic scattering cross sections obtained with a CDCC calculation (solid line) and with a single channel calculation with the polarization potential that includes the effects of couplings with inelastic and breakup channels. See text for details.

line) and by a single channel calculation including the polarization potential (dashed line) [Lubian (2010)]. The agreement between the two calculations is quite impressive. Thus, the coupled channel effects on elastic scattering are well accounted for. Although the evaluation of the polarization potential is based on the CDCC calculation itself, this potential may be very useful. It allows investigations of the influence of other channels, while keeping the channel space within a reasonable size.

10.5.4 *Threshold anomaly*

The dispersion relation connecting the real and the imaginary parts of the polarization potential (Eq. (10.23)) has an interesting consequence on the potentials that describe heavy ion scattering. This property is discussed in this section.

According to Feshbach's formalism, the channel space in a scattering problem can be split into two subspaces: P and Q. The influence of the Q-subspace on the channels in the P-subspace is then given by an effective potential. Taking energy averages of the effective potential, one obtains the optical potential, which greatly simplifies the description of the collision. The coupled channel problem is then reduced to the P space. Further simplifications can be achieved if one splits the P-subspace into two parts: (*i*) P_0, containing only the elastic channel, and (*ii*) P_1, containing the remaining ones. The coupled channel equations are then equivalent to a Schrödinger equation in a single channel, with the local potential[21]

$$V(r) = V_{\rm opt}(r) + V_{\rm pol}(r). \tag{10.301}$$

Above, $V_{\rm opt}(r)$ and $V_{\rm pol}(r)$ are complex potentials, that can be written,

$$V_{\rm opt}(r) = U_{\rm opt}(r) + i\,W_{\rm opt}(r) \tag{10.302}$$

$$V_{\rm pol}(r) = U_{\rm pol}(r) + i\,W_{\rm pol}(r). \tag{10.303}$$

Expressing the Green's function in the P_1-space in terms of its on-shell and off-shell parts, one gets a dispersion relation for the polarization potential, similar to Eq. (10.23). In this case, one drops the contribution from bound eigenstates and obtains

$$U_{\rm pol}(E;r) = U(E;r) - U_{\rm opt}(E;r) = \frac{1}{\pi}\mathcal{P}\int dE'\,\frac{W_{\rm pol}(E';r)}{E'-E}, \tag{10.304}$$

where $U(E;r)$ is the real part of the total potential. This relation holds for any value of r. The energy-dependence of the polarization potential, which comes from the Green's function, was explicitly represented in Eq. (10.304)[22].

The imaginary part of the full nuclear interaction, $W(r)$, has contributions from the optical and the polarization potentials. The former, $W_{\rm opt}$, accounts for the strong absorption arising from the fusion process. It is volumetric and falls rapidly to zero beyond the strong absorption radius, $R_{\rm sa}$, which is smaller than the radius of the Coulomb barrier. The latter, $W_{\rm pol}$, is associated with absorption into inelastic and transfer channels. It acts only on the surface region. As the collision energy decreased below the Coulomb barrier, the excitation of these channels becomes very unlikely. Accordingly, $W_{\rm pol}$ goes to zero. The sharp decrease of $W_{\rm pol}$ leads to an interesting behavior on the real part of the polarization potential, which has been observed for several nuclear systems. This behavior is known as the *threshold anomaly* [Nagarajan *et al.* (1985)]. We may venture to add here that it is unfortunate that the dispersion relation effect has come to be known by this name, since it should have been expected from the general physics principle of causality.

[21]We assume that the optical and the polarization potentials depend on E but are independent of l.

[22]If there is also an artificial energy dependence arising from the use of a local-equivalent polarization potential, it is ignored in this discussion.

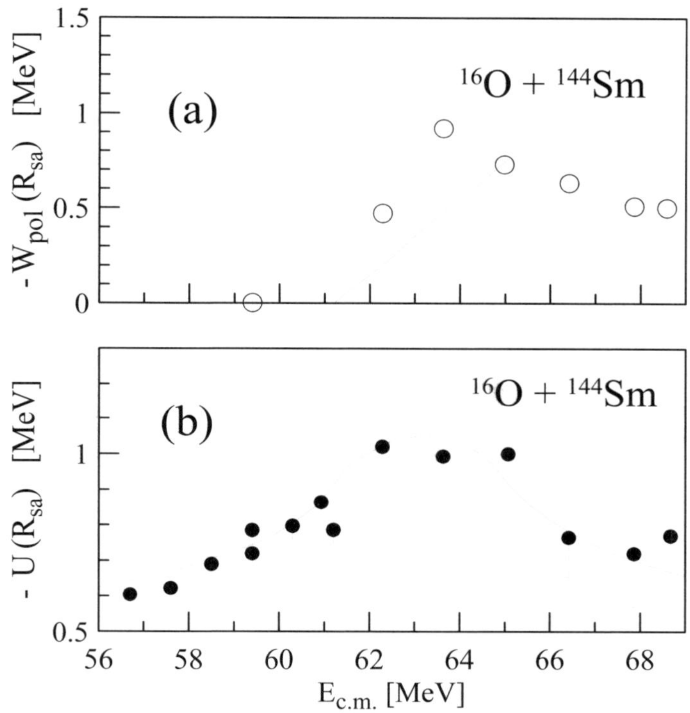

Fig. 10.11 Threshold anomaly in ^{16}O + ^{144}Sm scattering. The figure shows the imaginary part of the polarization potential (panel (a)) and the real part of the total nuclear potential (panel (b)), evaluated at the strong absorption radius of the optical potential, $R_{\mathrm{sa}} = 11.8$ fm. The circles were obtained fitting the potentials to experimental data and the solid lines are described in the text. Further details of this calculation are given in [Abriola *et al.* (1989)].

One example of the threshold anomaly is given in the study of Abriola *et al.* [Abriola *et al.* (1989)] of $^{16}\mathrm{O} +^{144}$ Sm scattering. These authors determined the parameters of the optical and polarization potentials, fitting elastic scattering and fusion data at near-barrier energies. The resulting potentials evaluated at the strong absorption radius are represented in figure 10.11 as functions of the collision energy by solid (real part) and open (imaginary part) circles. Panel (a) shows the imaginary part of the polarization potential whereas panel (b) shows the real part of the total nuclear potential. One observes in panel (a) a sharp decrease of $W_{\mathrm{pol}}(E; R_{\mathrm{sa}})$. It is accompanied by a bell-shaped maximum in the real part of the potential, shown in panel (b). Since U_{opt} is taken to be energy-independent, this maximum should arise from $U_{\rm pol}(E; R_{\rm sa})$. This behavior results from the dispersion relation. To justify this statement, the energy-dependence of $W_{\rm pol}(E; R_{\rm sa})$ was approximated by the straight lines shown in panel (a)). The real part of the polarization potential was then evaluated through the integral of Eq. (10.304), and $U(E; R_{\rm sa})$ was determined summing $U_{\rm opt}(R_{\rm sa})$. The result is the solid line in panel (b), which approximately reproduces the solid circles, obtained by the fitting procedure.

10.6 Scattering from energy-dependent potentials

One subtle issue in the one-channel description of scattering is the appearance of the energy in the effective interaction itself,

$$V_{\rm eff}(E) = V_{\rm opt} + V_{\rm pol}(E). \tag{10.305}$$

The energy dependence arises from the Green's operator, $\mathbb{G}_{\rm P_1}$, with which the polarization potential is built (see Eq. (10.210)), and in some cases also from the optical potential. This amounts to dealing with an effective one-channel Schrödinger equation of the form

$$\Big[E - H(E_k)\Big] \, |\psi^{(+)}(\mathbf{k})\rangle = 0, \tag{10.306}$$

where

$$H(E_k) = K + V_{\rm eff}(E_k) \tag{10.307}$$

and $E_k = \hbar^2 k^2/2\mu$. The solution of Eq. (10.306) and of the corresponding Lippmann-Schwinger equation can be found by treating E_k as a parameter and not as an eigenvalue. Similarly, the ingoing wave solution satisfies the equation

$$\Big[E - H^\dagger(E_k)\Big] \, |\psi^{(-)}(\mathbf{k})\rangle = 0, \tag{10.308}$$

which can also be solved as before. Thus, the cross section can be evaluated just like in the case of energy-independent potentials.

However, the formal scattering discussion of section 7.2 must be modified. The dual solutions, $\big|\tilde{\psi}^{(+)}(\mathbf{k})\big\rangle$ and $\big|\tilde{\psi}^{(-)}(\mathbf{k})\big\rangle$, which are needed in the completeness and orthogonality relations, are no longer given by the Lippmann-Schwinger equations of section 7.2. Rather, they have to be found from complicated integral equations. To be specific, we have,

$$\big\langle \tilde{\psi}^{(+)}(\mathbf{k}') \big| \psi^{(+)}(\mathbf{k}) \big\rangle = \delta(\mathbf{k}' - \mathbf{k}), \tag{10.309}$$

which is the starting point for the obtention of the dual wave function. As in chapter 4, the physical scattering wave function itself satisfies the Lippmann-Schwinger equation,

$$|\psi^{(+)}(\mathbf{k})\rangle = \big|\phi(\mathbf{k})\big\rangle + G_0^{(+)}(E_k)\, V_{\text{eff}}(E_k)\; |\psi^{(+)}(\mathbf{k})\rangle\,, \tag{10.310}$$

where $G_0^{(+)}(E_k)$ is the free particle Green's operator[23], evaluated in section 4.1. Using the notation of Eq. (4.90), we can write the scattering state as the sum of the incident wave plus a scattered wave, that is,

$$|\psi^{(+)}(\mathbf{k})\rangle = \big|\phi(\mathbf{k})\big\rangle + |\psi^{\text{sc}}(\mathbf{k})\rangle\,, \tag{10.311}$$

with

$$|\psi^{\text{sc}}(\mathbf{k})\rangle = G_0^{(+)}(E_k)\, V_{\text{eff}}(E_k)\; |\psi^{(+)}(\mathbf{k})\rangle\,. \tag{10.312}$$

In the momentum representation, Eq. (10.311) becomes

$$\Big\langle \phi(\mathbf{q}) \Big| \psi^{(+)}(\mathbf{k}) \Big\rangle = \delta(\mathbf{q}-\mathbf{k}) + \varphi_{\mathbf{k}}^{(+)}(\mathbf{q}), \tag{10.313}$$

where

$$\varphi_{\mathbf{k}}^{(+)}(\mathbf{q}) \equiv \Big\langle \phi(\mathbf{q}) \Big| \psi^{\text{sc}}(\mathbf{k}) \Big\rangle \tag{10.314}$$

is the scattered wave in momentum representation. Formally, one can write a similar relation for the dual bra states[24],

$$\Big\langle \tilde{\psi}^{(+)}(\mathbf{k}') \Big| \phi(\mathbf{q}) \Big\rangle = \delta(\mathbf{q}-\mathbf{k}') - \tilde{\varphi}_{\mathbf{k}'}^{(+)*}(\mathbf{q}). \tag{10.315}$$

The first term in the RHS of the above equation corresponds to the situation where the potential V_{eff} is set equal to zero. The second term is the dual state of the wave scattered by the potential V_{eff}, in the momentum representation. It should be determined from the orthogonality relation of Eq. (10.309). For this purpose, we carry out the scalar product in the momentum representation,

$$\Big\langle \tilde{\psi}^{(+)}(\mathbf{k}') \Big| \psi^{(+)}(\mathbf{k}) \Big\rangle = \int d\mathbf{q}\, \Big\langle \tilde{\psi}^{(+)}(\mathbf{k}') \Big| \phi(\mathbf{q}) \Big\rangle\, \Big\langle \phi(\mathbf{q}) \Big| \psi^{(+)}(\mathbf{k}) \Big\rangle = \delta(\mathbf{k}'-\mathbf{k}), \tag{10.316}$$

[23]In many applications, mainly in Nuclear Physics, the effective potential contains a Coulomb term. In such cases, one should use the two-potential Lippmann-Schwinger equation, with the Coulomb term playing the role of the dominant potential (V_1 - in the notation of section 4.5). The free particle's Green's operator should then be replaced by the Coulomb one.

[24]We follow [Hussein and Moniz (1984)] and define the dual scattering state with a minus sign. This choice is completely arbitrarily and has no relevant consequences.

and use Eqs. (10.315) and (10.313). We get the integral equation

$$\tilde{\varphi}^{(+)}_{\mathbf{k}'}(\mathbf{k}) = \varphi^{(+)}_{\mathbf{k}}(\mathbf{k}') - \int d\mathbf{q}\, \tilde{\varphi}^{(+)*}_{\mathbf{k}'}(\mathbf{q})\, \varphi^{(+)}_{\mathbf{k}}(\mathbf{q}). \tag{10.317}$$

Note the important difference between $\tilde{\varphi}^{(+)}_{\mathbf{k}'}(\mathbf{k})$ and $\varphi^{(+)}_{\mathbf{k}}(\mathbf{k}')$: the momentum vector labeling the state is interchanged with that of the representation when one goes from $\tilde{\varphi}^*$ to φ. The above equation can be solved by iteration or other standard technique to handle integral equations. An analogous procedure can be used to obtain the dual states of the ingoing solution, $\tilde{\varphi}^{(-)}_{\mathbf{k}'}(\mathbf{k})$.

The dual states in the scattering from energy dependent potential in the momentum representation play a central role in the derivation of energy independent effective potentials. For detail, we refer to [Hussein and Moniz (1984)].

Exercises

(1) Show that the P-projected full Green's function, $\mathcal{G}^{(+)}_{\mathrm{P}} = P\,[E - \mathbb{H} + i\epsilon]^{-1}\,P$, is related to the P- and Q-projected decoupled Green's functions, $\mathbb{G}^{(+)}_{\mathrm{P}} = P\,[E - \mathbb{H}_{\mathrm{PP}} + i\epsilon]^{-1}\,P$ and $\mathbb{G}^{(+)}_{\mathrm{Q}} = Q\,[E - \mathbb{H}_{\mathrm{QQ}} + i\epsilon]^{-1}\,Q$, through the equation,

$$\mathcal{G}^{(+)}_{\mathrm{P}} = \frac{1}{\left(\mathbb{G}^{(+)}_{\mathrm{P}}\right)^{-1} - \mathbb{V}_{\mathrm{PQ}}\,\mathbb{G}^{(+)}_{\mathrm{Q}}\,\mathbb{V}_{\mathrm{QP}}}.$$

(2) With the aid of operator identities, calculate the imaginary part of $\mathcal{G}^{(+)}_{\mathrm{P}} \equiv P\,\mathbb{G}^{(+)}P$ and show that it can be written as,

$$\begin{aligned}\mathrm{Im}\,\{\mathcal{G}^{(+)}_{\mathrm{P}}\} = &-\pi\,\left[1 + \mathbb{G}^{(-)}_{\mathrm{P}}\,T^{\dagger}_{\mathrm{PP}}\right]\,\delta\,(E - \mathbb{H}_{\mathrm{PP}})\,\left[1 + T_{\mathrm{PP}}\,\mathbb{G}^{(+)}_{\mathrm{P}}\right] \\ &-\pi\,(\mathcal{G}^{(+)}_{\mathrm{P}})^{\dagger}\,\mathbb{V}_{\mathrm{PQ}}\,\delta\,(E - \mathbb{H}_{\mathrm{QQ}})\,\mathbb{V}_{\mathrm{QP}}\,\mathcal{G}^{(+)}_{\mathrm{P}}.\end{aligned}$$

where T_{PP} is the exact T-matrix in the P-space.

(3) Calculate the total cross section for resonant reactions. Use Eqs. (10.88), (10.89), (10.90), and employ the optical theorem,

$$\sigma_{\mathrm{tot}} = \frac{4\pi}{k}\,\mathrm{Im}f(\theta = 0).$$

Show that σ_{tot} exhibits the type of resonant structure shown in figures 10.1 and 10.2.

(4) Consider the P-space Green's function, $\mathcal{G}_{\rm el}^{(+)} \equiv P_0\, G^{(+)} P_0$.

a) Project the elastic channel part of it and show that it can be written as

$$\mathcal{G}_{\rm el}^{(+)} = \frac{1}{E - H_0 - V_{\rm opt} - V_{\rm pol} + i\varepsilon}.$$

Above, P_0 is the projector onto the elastic channel, H_0 is the kinetic energy operator, $V_{\rm opt}$ is the optical potential arising from eliminating the Q-space, while $V_{\rm pol}$ is the one which arises from eliminating the open channels contained in the P-space.

b) Show that the absorption cross section is composed of three distinct terms. The one coming from the Q-space, the one arising from the Q-space reached through the coupled open channels and he last one is the direct, inelastic contribution.

(5) Consider the scattering of a Hydrogen atom from an anti-Hydrogen atom (bound state of a positron and an anti-proton). What is the Van der Waals interaction between the atom and the anti-atom? Argue that it is the same (with equal sign) as that in the atom-atom case. Argue further that there is no repulsion at short distances, but rather very strong absorption owing to the annihilation mechanism.

(6) a) Show that the first Born amplitude for a r^{-4} potential, when expanded in powers of the momentum transfer, contains only ONE odd power of q.

b) Show that the Born amplitude for a short-range potential contains only even powers of q.

Chapter 11

Approximate Methods in Many-body Scattering

In chapter 9 we introduced the coupled channel formalism and the close coupling approximation. We discussed the solution of the coupled channel equations, which may be very complicated, specially when it involves channels in the continuum. Here we discuss several methods that can be used to find approximate solutions of the coupled channel problem.

In this chapter, we restrict our discussion to the situations considered in chapter 9, where the channel expansion is valid. In some collisions the intrinsic states of the projectile and the target are not meaningful separately, and the channel expansion breaks down. This is the case of atom-atom collisions at very low velocities, where the states of the electronic cloud evolve adiabatically to the orbitals of a two-center (projectile and target) potential, as in a molecule. Such cases require the use of more complicated approaches, as *ab initio* calculations (see e.g. [Macias and Riera (1982)]). A similar situation takes place in nuclear collisions when the projectile and target densities overlap strongly. If the outcome can only be fusion, the details of the reaction dynamics are not relevant, and, in that case, the inclusion of a short-range imaginary potential handles the situation. However, in collisions of very heavy systems, where the outcome may be the exchange of a few or many nucleons (deep inelastic processes or fusion-fission), it is necessary to resort to time-dependent Hartree-Fock calculations [Strayer *et al.* (1983)] or to phenomenological macroscopic models [Swiatecki (1980, 1981); Randrup and Swiatecki (1984); Zagrevaev and Greiner (2005)], based on classical physics. Collisions of this kind will not be considered in this book.

11.1 Perturbative approximations

As we have shown in chapter 9, the collision dynamics in many-body scattering is governed by a Hamiltonian with the general form

$$\mathbb{H} = h + K + \mathbb{V}, \tag{11.1}$$

where h is the intrinsic Hamiltonian, having eigenvalues ε_α and eigenstates φ_α ($\alpha = 0$ represents the elastic channel), K stands for the kinetic energy of the projectile-target relative motion and $\mathbb{V}$ is the projectile-target interaction potential, which acts on the space of the collision ($\mathbf{r}$) and the intrinsic (ξ) degrees of freedom. The many-body scattering state, $|\Psi^{(+)}(0, \mathbf{k}_0)\rangle$, for a collision initiated with the system in its intrinsic ground state and with wave vector $\mathbf{k}_0$ is written as the channel-expansion

$$|\Psi^{(+)}(0, \mathbf{k}_0)\rangle = \sum_{\alpha=0}^{\infty} |\psi_\alpha^{(+)}(0, \mathbf{k}_0)\rangle \;\; |\alpha) \, . \tag{11.2}$$

For practical purposes, the infinite series of the above expansion is approximated by a finite set of N channels (close coupling approximation) and the relative wave function in each channel, $|\psi_\alpha^{(+)}(0, \mathbf{k}_0)\rangle$, is determined by the coupled channel equations (Eq. (9.16)). From these wave functions one determines the transition operator.

The scattering amplitude for a particular channel α with final wave vector $\mathbf{k}_\alpha$ is then given by

$$f_{\alpha,0}(\mathbf{k}_\alpha, \mathbf{k}_0) = -\, 2\pi^2 \left(\frac{2\mu}{\hbar^2}\right) T_{\alpha\mathbf{k}_\alpha, 0\mathbf{k}_0} \tag{11.3}$$

and the corresponding cross section is given by

$$\frac{d\sigma_\alpha(\mathbf{k}_\alpha, \mathbf{k}_0)}{d\Omega} = \frac{\mathbf{k}_\alpha}{\mathbf{k}_0} \; |f_{\alpha,0}(\mathbf{k}_\alpha, \mathbf{k}_0)|^2 \, . \tag{11.4}$$

In Eq. (11.3),

$$T_{\alpha\mathbf{k}_\alpha, 0\mathbf{k}_0} = \langle\Phi(\alpha\mathbf{k}_\alpha)|\, \mathbb{T}\, |\Phi(0\mathbf{k}_0)\rangle = \langle\Phi(\alpha\mathbf{k}_\alpha)|\, \mathbb{V}\, |\Psi^{(+)}(0\mathbf{k}_0)\rangle \tag{11.5}$$

stands for the matrix-elements of the transition operator between states of the form $|\Phi(\alpha\mathbf{k}_\alpha\rangle \equiv |\phi(\mathbf{k}_\alpha\rangle\; |\alpha)$, where $|\phi(\mathbf{k}_\alpha\rangle$ is a plane wave with wave number $\mathbf{k}_\alpha$. That is, a vector with modulus $k_\alpha = \sqrt{2\mu(E - \varepsilon_\alpha)}/\hbar$ and forming an angle θ with the initial momentum $\mathbf{k}_0$.

In the next two sub-sections we will discuss the perturbative approximations based on the above formulation of the scattering problem.

11.1.1 *The Plane Waves Born Approximation*

The transition operator of Eq. (11.5) is given by the Lippmann-Schwinger equation (see chapter 9, Eq. (9.68))

$$\begin{aligned} \mathbb{T} &= \mathbb{V} + \mathbb{V}\,\mathbb{G}_0^{(+)}(E)\,\mathbb{T} \\ &= \mathbb{V} + \mathbb{V}\,\mathbb{G}_0^{(+)}(E)\,\mathbb{V} + \mathbb{V}\,\mathbb{G}_0^{(+)}(E)\,\mathbb{V}\,\mathbb{G}_0^{(+)}(E)\,\mathbb{V} + \cdots . \end{aligned} \tag{11.6}$$

The first plane wave Born approximation (PWBA), also called of first Born approximation or simply Born approximation, consists of truncating the Born series after the first term. That is, setting

$$\mathbb{T} \simeq \mathbb{V}. \tag{11.7}$$

In this way, the matrix-elements of Eq. (11.5) become

$$T_{\alpha\mathbf{k}_\alpha,0\mathbf{k}_0} = \langle\phi(\mathbf{k}_\alpha)|\,V_{\alpha 0}\,|\phi(\mathbf{k}_0)\rangle \tag{11.8}$$

$$= \frac{1}{(2\pi)^3}\int d^3\mathbf{r}\,e^{-i(\mathbf{k}_\alpha-\mathbf{k}_0)\cdot\mathbf{r}}\;V_{\alpha 0}(\mathbf{r}), \tag{11.9}$$

where

$$V_{\alpha 0}(\mathbf{r}) = \int d\xi\;\varphi_\alpha^*(\xi)\,\mathbb{V}(\mathbf{r},\xi)\,\varphi_0(\xi). \tag{11.10}$$

The first Born approximation can be used when the strength of the scattered wave in any excited channel is much smaller than that of the incident wave. This happens when the projectile-target interaction potential is weak compared with the collision energy. In nuclear physics this approximation cannot be used, with the exception of collisions involving neutrons. For charged collision partners, this approximation breaks down owing to the long range of the Coulomb potential. On the other hand, it may be used in some atomic and molecular collisions. We illustrate the use of the first Born approximation in the case of the $H^+ - H^0$ inelastic scattering at high velocities (compared with the orbital velocity of the electron). Figure 11.1 shows the summed inelastic cross sections for excitation of the 2s and the 2p orbitals in the hydrogen atom, as a function of the collision energy in the center of mass frame. The calculations used in the figure are from [Bransden and Joachain (1983)] and the data are from [Park *et al.* (1975)]. The dashed line was obtained with a first order Born approximation and the solid line corresponds to a coupled channel calculation which does not

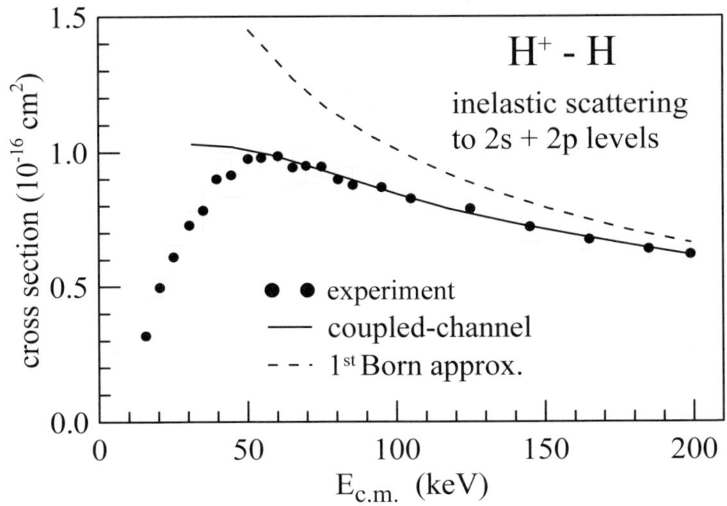

Fig. 11.1 Inelastic cross sections for excitation of the 2s and 2p orbits in the $H^+ - H$ collision. For details see the text.

include rearrangement channels. The data are renormalized to reproduce the result of the coupled channel calculation at 200 keV [Park *et al.* (1976)]. The cross section obtained with the Born approximation is in reasonable agreement with the experiment at the higher energies but it is quite different from the data in the energy regions around and below the maximum at 50 keV. On the other hand, the coupled channel calculation reproduces the data very well above the maximum. At $E_{\mathrm{c.m.}} < 50$ keV, the charge exchange channels, neglected in this calculation, become very important and the theoretical curve fails to reproduce the experimental cross section. We should mention that in this example the Born approximation was made in a semiclassical version of the coupled channel equation, which will be discussed later in this chapter. However, the results of this semiclassical approximation are very close to the ones of a full quantum mechanical coupled channel calculation [Park *et al.* (1976)].

To close this section, we mention the situation where the interaction is weak but the cross section for the channel of interest cannot be treated by the first Born approximation. One example is the inelastic scattering to an intrinsic state with $J^\pi = 4^+$ when the incident channel has $J^\pi = 0^+$, and the channel coupling interaction is dominated by the electric quadrupole term. Since the quadrupole coupling can only exchange two angular momentum units between the relative and the intrinsic spaces, the $J^\pi = 4^+$ channel cannot be reached in a single step. However, this channel could be excited by a two step process, going through an intermediate $J^\pi = 2^+$ channel, if it is available. In such cases, it is necessary to consider the next term in the Born series, performing a second order Born approximation.

11.1.2 *The Distorted Waves Born Approximation*

The distorted waves Born approximation is widely used in atomic and in nuclear collisions. It is based on the Gell-Mann Goldeberger relation in many-body scattering (see section 9.2.1). If the total potential, $\mathbb{V}$, is written as

$$\mathbb{V} = \mathbb{V}^{(1)} + \mathbb{V}^{(2)}, \tag{11.11}$$

where, for simplicity, $\mathbb{V}^{(1)}$ is assumed to be diagonal in channel space, the matrix-elements of the transition operator is given by Eq. (9.73):

$$\begin{aligned} T_{\beta \mathbf{k}_\beta, 0\mathbf{k}_0} &= \langle \phi\left(\mathbf{k}_\beta\right) | \, V_\beta^{(1)} \, | \chi^{(+)}(0\mathbf{k}_0) \rangle \; \delta_{\beta 0} \\ &\quad + \sum_\alpha \langle \chi^{(-)}(\beta \mathbf{k}_\beta) | \, V_{\beta\alpha}^{(2)} \, | \, \psi_\alpha^{(+)}(0\mathbf{k}_0) \rangle \, . \end{aligned} \tag{11.12}$$

Above, $\phi\left(\mathbf{k}_\beta\right)$ represents a plane wave with wave vector $\mathbf{k}_\beta$ and $\chi^{(-)}(\beta\mathbf{k}_\beta)$ and $\chi^{(+)}(0\mathbf{k}_0)$ are waves distorted by the potential $\mathbb{V}^{(1)}$ (that is, with $\mathbb{V}^{(2)}$ switched off). They satisfy Schrödinger equations of general the form

$$\left[E_\alpha - K - V_\alpha^{(1)}\right] \left|\chi^{(\pm)}(\alpha \mathbf{k}_\alpha)\right\rangle = 0 \tag{11.13}$$

($\alpha = 0$ for the entrance channel and $\alpha = \beta$ for the exit channel), and have the asymptotic behaviors[1]

$$\chi^{(+)}(\alpha \mathbf{k}_\alpha; \mathbf{r}) \longrightarrow \phi_{\mathrm{C}}(\mathbf{k}_\alpha; \mathbf{r}) + \frac{1}{(2\pi)^{3/2}} \left[\bar{f}_\alpha^{(1)}(\mathbf{k}'_\alpha, \mathbf{k}_\alpha) \; \frac{e^{i\Theta_\alpha(r)}}{r} \right] \tag{11.14}$$

$$\chi^{(-)}(\alpha \mathbf{k}_\alpha; \mathbf{r}) \longrightarrow \phi_{\mathrm{C}}^*(-\mathbf{k}_\alpha; \mathbf{r}) + \frac{1}{(2\pi)^{3/2}} \left[\left(\bar{f}_\alpha^{(1)}(-\mathbf{k}'_\alpha, \mathbf{k}_\alpha) \right)^* \; \frac{e^{-i\Theta_\alpha(r)}}{r} \right]. \tag{11.15}$$

[1]Since we will be dealing with nuclear collisions, where the collision partners are charged, we use the general expressions that include Coulomb distortion. For pure short-range interaction, one replaces ϕ_{C} by the plane wave.

Table 11.1 Coulomb barriers for the systems discussed in the text.

Projectile:	^{1}H	^{4}He	^{7}Li	^{10}B	^{12}C	^{16}O
$V_{\rm B}$ (MeV):	8.50	17.26	23.10	38.50	49.48	66.72

The Gell-Mann Goldberger formula becomes particularly useful when the potential $\mathbb{V}^{(2)}$ is weak enough to be treated as a perturbation. One can then use the Distorted Waves Born Approximation (DWBA) for Eq. (11.12), replacing $|\,\psi_\alpha^{(+)}(0\mathbf{k}_0)\rangle \rightarrow |\chi^{(+)}(0\mathbf{k}_0)\rangle$. As an illustration, we discuss the application of the DWBA approximation to inelastic scattering in nucleus-nucleus collisions. In this case, the first term at the RHS of Eq. (11.12) vanishes and one gets

$$T^{\rm DWBA}_{\beta\mathbf{k}_\beta,0\mathbf{k}_0} = \langle\chi^{(-)}(\beta\mathbf{k}_\beta)|\,V^{(2)}_{\beta 0}\,|\chi^{(+)}(0\mathbf{k}_0)\rangle\,. \tag{11.16}$$

We consider the inelastic cross section of the first excited state of the ^{154}Sm target, in collisions with several light nuclear projectiles. Since ^{154}Sm is a highly deformed nucleus, its low lying spectrum consists of the rotational band of the ground state, with spins and parities $I^\pi = 0^+, 2^+, 4^+, 6^+, \cdots$. These states are modeled by the quantum symmetric top, with energies $E_I = \hbar^2\, I(I+1)/2\,\mathfrak{I}$. Owing to the large value of the momentum of inertia, $\mathfrak{I}$, the excitation energies are very low. The energy of the first excited state is only $E_{2^+} = 0.082$ MeV.

We have evaluated CC and DWBA cross sections for the first 2^+ state of ^{154}Sm, σ_{2^+}, in collisions with ^{1}H, ^{4}He, ^{7}Li, ^{10}B, ^{12}C and ^{16}O projectiles. To have meaningful comparisons of cross sections for different projectiles, in each case the calculations were performed at the energy of the Coulomb barrier. The barrier for each projectile-target combination was evaluated using the folding potentials of [Chamon *et al.* (2002)]. To simplify the discussion, we have neglected nuclear couplings and considered only the monopole and the quadrupole terms in the multipole expansion for the Coulomb interaction (see chapter 9). The CC calculations were performed within the space of the first channels with $I^\pi = 0^+, 2^+$ and 4^+. That is, the elastic channel and the two rotational channels with lowest excitation energies. We adopted the quadrupole deformation parameter $\beta_2 = 0.341$ of [Raman *et al.* (2001)], which was taken from experimental $B(E2)\uparrow$ values (see Sec. 9.3.3). The results are shown in figure 11.2. The CC and the DWBA cross sections are respectively represented by solid circles and open squares. Since the calculations were performed for the same target, the

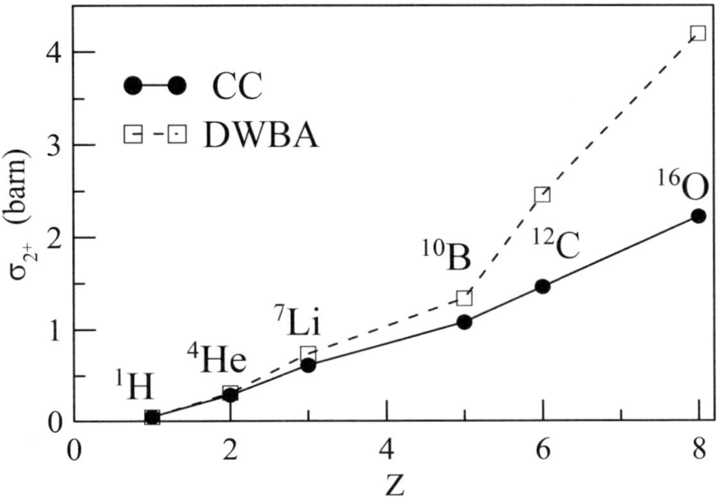

Fig. 11.2 Inelastic cross section for the lowest 2^+ state of ^{154}Sm, excited by several projectiles. In each case, the calculation is performed at the collision energy $E_{\rm c.m} = V_{\rm B}$. The solid circles represent the cross sections obtained from CC calculations whereas the open square are the corresponding cross sections given by the DWBA.

strength of the Coulomb quadrupole coupling depends only on the projectile's charge. For the lightest projectiles, ^{1}H, ^{4}He and ^{7}Li, the coupling is weak and the DWBA approximation is accurate. For heavier projectiles, the coupling is stronger and the DWBA becomes progressively worse.

11.1.3 *The Coupled Channel Born Approximation*

The Coupled Channel Born Approximation (CCBA) is an extension of the DWBA (see [Penny and Satchler (1964); Iano and Austern (1966)]), which may be very useful in some particular problems. Let us consider the situation where the channel space can be split into two partitions, A and B, such that the channels within the same partition are strongly coupled among themselves, whereas channels in different partitions are only weakly coupled. To develop the CCBA, one splits the interaction potential into two parts, as in Eq. (11.11), so that $\mathbb{V}^{(1)}$ is responsible for the strong couplings among channels of the same partition, but it does not couple channels in different partitions. This is done by the complementary term $\mathbb{V}^{(2)}$, which is assumed to be weak enough to be handled by perturbation theory. In this case, the channels in partition A, labeled α, and the channels in partition

B, labeled β, are approximately given by the separate sets of CC equations,

$$\left[E_{\alpha'} - K - V^{(1)}_{\alpha'\alpha'}\right] \left|\chi^{(\pm)}_{\alpha'}(0\mathbf{k}_0)\right\rangle = \sum_{\alpha} V^{(1)}_{\alpha'\alpha} \left|\chi^{(\pm)}_{\alpha}(0\mathbf{k}_0)\right\rangle \tag{11.17}$$

and

$$\left[E_{\beta'} - K - V^{(1)}_{\beta'\beta'}\right] \left|\chi^{(\pm)}_{\beta'}(\beta\mathbf{k}_\beta)\right\rangle = \sum_{\beta''} V^{(1)}_{\beta'\beta''} \left|\chi^{(\pm)}_{\beta''}(\beta\mathbf{k}_\beta)\right\rangle . \tag{11.18}$$

The T-matrix for the transition from the elastic channel, contained in the A partition, to a channel β in the B partition, is then given by

$$T^{\mathrm{CCBA}}_{\beta\mathbf{k}_\beta,0\mathbf{k}_0} = \sum_{\alpha',\beta'} \left\langle \chi^{(-)}_{\beta'}(\beta\mathbf{k}_\beta)\right| V^{(2)}_{\beta'\alpha'} \left|\chi^{(+)}_{\alpha'}(0\mathbf{k}_0)\right\rangle . \tag{11.19}$$

The CCBA approximation can be used in processes like transfer reactions involving a spherical heavy projectile and a highly deformed target. A concrete example is the transfer of an α-particle in the $^{16}O + {}^{154}Sm$ collision. Namely, the reaction: $^{16}O + {}^{154}Sm \rightarrow {}^{12}C + {}^{158}Gd$. As Fig. 11.2 indicates, the rotational channels of the $^{16}O + {}^{154}Sm$ system cannot be treated by the DWBA. Similarly, there are strong couplings among the rotational channels of the $^{12}C + {}^{158}Gd$ system, formed through the transfer of the α-particle. In this example, the channels of the $^{16}O + {}^{154}Sm$ system correspond to the A partition, whereas the B partition if the set of channels of the $^{12}C + {}^{158}Gd$ system. The distorted waves in each partition can be obtained by solving the corresponding set of CC equations (Eqs. (11.17) or Eqs. (11.18)) and the transfer T-matrix can be calculated by Eq. (11.19).

11.2 Semiclassical approximations

In many-body scattering several approximations use concepts of classical mechanics. Some of these approximations are discussed below.

11.2.1 *The classical trajectory approximation*

In the *classical trajectory approximation* (see e.g. [Alder *et al.* (1956); Alder and Winther (1975)]), also called *semiclassical approximation* and *impact parameter approximation*, the projectile-target relative motion is treated as a problem of classical mechanics whereas the excitation of the intrinsic degrees of freedom is described by full quantum mechanics. The semiclassical approximation can be used when two conditions are satisfied: (*a*) the wave length associated with the relative motion is small compared with the typical dimensions of the system; (*b*) the relative motion follows a trajectory determined by some average interaction, taken over the intrinsic states included in the channel expansion[2].

For simplicity, we consider here the scattering of spinless collision partners, interacting through a spherically symmetric potential. We assume that the projectile-target relative motion is not affected by the exchange of energy and angular momentum between the relative and the intrinsic degrees of freedom, so that the system follows a classical trajectory determined by the average interaction $\overline{V}(r)$. Thus, for a collision with impact parameter b, the system follows the trajectory schematically represented in figure 1.3 of chapter 1, and the time development of the radial coordinate is given by Eq. (1.37),

$$\frac{dr_b}{dt} = \mp v\sqrt{1 - \frac{b^2}{r_b^2} - \frac{\overline{V}(r)}{E}}.$$

Above, the minus and plus signs apply respectively for the incoming and outgoing branches of the trajectory. Solving this equation (see section 1.4), one obtains the function $r_b(t)$. The coupling potential acting on the intrinsic states then becomes a function of the time:

$$\mathbb{V}(r,\xi) \to \mathbb{V}(r_b(t),\xi) \to \mathbb{V}(\xi,t). \tag{11.20}$$

Within the semiclassical approximation, the intrinsic wave function, $\psi_b(\xi,t)$, is the solution of the time-dependent Schrödinger equation,

$$\Big[h + \mathbb{V}(\xi,t)\Big]\,\psi_b(\xi,t) = i\hbar\,\frac{\partial\psi_b(\xi,t)}{\partial t}. \tag{11.21}$$

To find $\psi_b(\xi,t)$, one first makes the channel expansion

$$\psi_b(\xi,t) = \sum_\alpha a_\alpha(b,t)\,\varphi_\alpha(\xi)\;e^{-i\varepsilon_\alpha t/\hbar}, \tag{11.22}$$

where

$$h\,\varphi_\alpha(\xi) = \varepsilon_\alpha\;\varphi_\alpha(\xi), \tag{11.23}$$

and truncate Eq. (11.22) after N terms ($\alpha = 0,\cdots,N-1$). The next step is to insert Eq. (11.22) into Eq. (11.21) and take scalar product of both sides of the resulting equation with each of the intrinsic states. In this way, one gets the semiclassical version of the coupled channel equations of chapter 9,

$$i\hbar\;\dot a_\alpha(b,t) = \sum_{\alpha'} e^{i(\varepsilon_\alpha - \varepsilon_{\alpha'})t/\hbar}\;V_{\alpha\alpha'}(t)\;a_{\alpha'}(b,t), \tag{11.24}$$

[2]A quantitative discussion of the validity of the semiclassical method can be found in section 8.6 of [Child (1994)].

where dots denote derivatives with respect to the time. Above, $\alpha, \alpha' = 0, ..., N-1$ and the matrix-elements of the potential are

$$V_{\alpha\alpha'}(t) = \int d\xi \; \varphi^*_\alpha(\xi) \; \mathbb{V}(\xi, t) \; \varphi_{\alpha'}(\xi). \tag{11.25}$$

Usually, the time scale is chosen so that the system is at closest approach at $t = 0$. Thus, the coupled differential equations should be solved integrating over time, starting from the initial conditions

$$a_\alpha(b, t \to -\infty) = \delta_{\alpha,0}. \tag{11.26}$$

For short-range interactions, $r_b(t)$ is outside the reach of the potential for large values of $|t|$. Thus, $\dot{a}_\alpha(b,t)$ vanish and the amplitudes remain constant. For long-range interactions, the diagonal matrix-elements, $V_{\alpha\alpha}(t)$, are not negligible, even at asymptotic distances. Thus, the amplitudes do not reach stationary values. However, the set of equations decouples, provided that the off-diagonal channel-coupling interaction vanishes asymptotically. In this way, the phase of each amplitude keeps changing but its modulus assumes a constant value. In both cases, the probability that the system emerges in an intrinsic state α is

$$P_\alpha(b) = \left| a_\alpha(b, t \to \infty) \right|^2. \tag{11.27}$$

The impact parameter is related to the scattering angle through the deflection function, $\Theta(b)$. On the other hand, the impact parameter corresponding to a given observation angle is its inverse function, $b(\Theta)$. When $b(\Theta)$ is single-valued, the cross section for the excitation of channel α can be approximated as

$$\frac{d\sigma_\alpha(\theta)}{d\Omega} \simeq \frac{d\sigma_{\rm el}(\theta)}{d\Omega} \times P_\alpha(b(\theta)). \tag{11.28}$$

The elastic cross section appearing above can be calculated by the classical expression,

$$\frac{d\sigma_{\rm el}(\Theta)}{d\Omega} = \frac{b(\Theta)}{\sin\Theta} \; \frac{db(\Theta)}{d\Theta}, \tag{11.29}$$

setting $\Theta = \theta$, or can be evaluated by quantum mechanics, neglecting channel coupling.

Frequently, the inverse function $b(\Theta)$ is not single-valued. This means that the observation angle can be reached by different classical trajectories, with different impact parameters. In such cases, the contributions from the trajectories interfere and the cross section has to be evaluated using the techniques discussed in chapter 5.

The total cross section for channel α can be obtained integrating over impact parameters. For the spherical potentials considered in the present section, it is given by

$$\sigma_\alpha = 2\pi \int b\, db\; P_\alpha(b) = 2\pi \int b\, db\; \Big| a_\alpha(b, t \to \infty) \Big|^2. \tag{11.30}$$

11.2.1.1 *Calculation of the classical trajectory*

Different approximations have been used to calculate the classical trajectory. In principle, the trajectory is given by the classical equations of motion for the Hamiltonian[3]

$$\mathcal{H} = \frac{\mathbf{P}^2}{2\mu} + \overline{V}(r), \tag{11.31}$$

with the average potential determined consistently with the solution of the quantum mechanical equation for the intrinsic wave function. That is, $\overline{V}(r) = \langle \psi | \mathbb{V} | \psi \rangle$. However, frequently one uses approximate trajectories.

The simplest approximation for the classical trajectory is to assume that it is not affected by the interaction, so that the projectile follows a straight line. This corresponds to scattering at forward angles, which dominates collisions at high energies. It can also be used at lower energies in collisions with large impact parameters. For an incident beam along the z-axis and setting $t = 0$ at closest approach, we have

$$\mathbf{r}_b(t) = \mathbf{b} + z\,\hat{\mathbf{z}} = \mathbf{b} + vt\,\hat{\mathbf{z}} \quad \to r_b(t) = \sqrt{b^2 + v^2\, t^2}. \tag{11.32}$$

This approximation has been used in several calculations in atomic and nuclear collisions (see [Grande *et al.* (1996)] and [Bertulani and Canto (1992)], for example).

The situation is quite different in large angle scattering, where the trajectory cannot be approximated by a straight line. One important example is Coulomb excitation in nucleus-nucleus collision at sub-barrier energies. In these collisions the Coulomb field may lead to inelastic scattering, exciting collective (vibrational or rotational) states in the projectile and/or the target. In this energy regime, the nuclear interaction can be ignored, since the classical turning point is located outside its range. In this way,

[3]Since we are assuming that the interaction is spherically symmetric, the average potential depends only on the projectile-target distance, r.

the projectile-target relative motion reduces to a Rutherford trajectory. In this case it is convenient to change from t to a new variable, w, for which the projectile-target distance is given analytically. This transformation is represented by

$$t = \frac{a}{v\,[\varepsilon \sinh w + w]} \qquad \text{and} \qquad r = a\,[\varepsilon \cosh w + 1]\,. \tag{11.33}$$

In the above equations (see chapter 3), a is half the distance of closest approach in a head-on collision, η is the Sommerfeld parameter and ε is the excentricity of the hyperbolic classical orbit. This change of variable leads to simple expressions for the relevant quantities in the scattering problem. Several applications of the semiclassical method to Coulomb excitation are discussed in detail in Alder and Winther's book [Alder and Winther (1975)].

At energies above the Coulomb barrier the nuclear interaction may be relevant. It is then necessary to take into account the effects of the nuclear forces both in the classical trajectory and in the excitation of collective states [Canto *et al.* (1990); Marta *et al.* (2008)]. This situation will be illustrated in an example presented later in this section.

To finish the discussion of the trajectory in the semiclassical method, we should mention a procedure used to improve the calculations when the excitation energies are not much smaller than the collision energies. In such cases, the trajectory for the incident energy, E, may be significantly different from the one corresponding to the relative energy in some excited channel α, E_α. The semiclassical calculation of the cross section for channel α can then be improved if one uses the symmetrized trajectory, corresponding to the average energy [Alder and Winther (1975)], $E_{\rm av} = \sqrt{E\,E_\alpha}$.

11.2.1.2 *Additional approximations in the semiclassical method*

The semiclassical cross sections may become much simpler in some limiting cases discussed in this sub-section.

The sudden limit

First we consider the sudden limit, in which the excitation energies in all channels included in the calculation are negligible. More precisely, this condition means

$$\tau\,\omega_{\alpha\alpha'} \approx 0, \quad \text{for any } \alpha. \tag{11.34}$$

Above, $\omega_{\alpha\alpha'} = (E_\alpha - E_{\alpha'})/\hbar$ and τ is the collision time, determined by the condition that $r_b(\tau)$ be larger than the range of the channel-coupling interaction.

To get to the sudden approximation, it is convenient to formulate the problem using the interaction picture [Merzbacher (1998); Messiah (1999)]. First, we write the Hamiltonian as the sum $H = H_0 + H'$, where H_0 is some model Hamiltonian and $H' \equiv H - H_0$ is the remaining part of H. This approach is useful when the time-evolution operator associated with H_0, $U_0(t, t_0)$, is well known. The states and operators in the Schrödinger and Interaction representations are related by linear transformations involving the propagator $U_0(t, t_0)$, which usually can be written as[4]

$$U_0(t, t_0) = e^{-iH_0(t-t_0)/\hbar}. \tag{11.35}$$

States ($|\psi\rangle$) and operators (A) in the Schrödinger picture become respectively (objects in the interaction picture are denoted by tildes)

$$\left|\tilde{\psi}\right\rangle = U_0^\dagger \left|\psi\right\rangle \quad \text{and} \quad \tilde{A} = U_0^\dagger \, A \, U_0. \tag{11.36}$$

It can be easily checked that the time evolution of the wave function in the interaction picture is given by the equation

$$i\hbar \frac{d}{dt} |\tilde{\psi}(t)\rangle = \tilde{H}' \, |\tilde{\psi}(t)\rangle, \tag{11.37}$$

where

$$\tilde{H}' = U_0^\dagger \, H' \, U_0. \tag{11.38}$$

Equivalently, the time evolution of $\tilde{\psi}$ can be given by the propagator $U'(t, t_0)$,

$$|\tilde{\psi}(t)\rangle = U'(t, t_0) \; |\tilde{\psi}(t_0)\rangle, \tag{11.39}$$

which will be discussed below.

Now we use this representation in the semiclassical approximation. Here, h plays the role of the Hamiltonian H_0 and V corresponds to the complementary Hamiltonian, H'. To convert the wave function of Eq. (11.22) to the interaction picture, we apply the operator $U_0^\dagger$ on it. We get

$$|\tilde{\psi}_b(t)\rangle = \sum_\alpha a_\alpha(b, t) \, |\varphi_\alpha\rangle. \tag{11.40}$$

[4]This expression is valid if the Hamiltonian H_0 does not depend explicitly on the time.

The final amplitudes $a_\alpha(t \to \infty)$, which are used to evaluate cross sections, are given by the scalar products

$$a_\alpha(b, t \to \infty) = \langle \varphi_\alpha | \tilde{\psi}_b(\infty) \rangle = \langle \varphi_\alpha | U'(-\infty, \infty) | \varphi_0 \rangle. \tag{11.41}$$

Thus, to evaluate the final amplitudes, we need the propagator $U'(t, t_0)$ in the $\{t_0 \to -\infty; t \to \infty\}$ limit. This propagator satisfies an integral equation that can be expressed as the infinite series [Alder and Winther (1975)]

$$U'(\infty, -\infty) = 1 + \frac{1}{i\hbar} \int_{-\infty}^{\infty} dt\, \tilde{V}(t) + \left(\frac{1}{i\hbar}\right)^2 \int_{-\infty}^{\infty} dt\, \tilde{V}(t) \int_{-\infty}^{t} dt'\, \tilde{V}(t') + \cdots. \tag{11.42}$$

This series cannot be summed because the operators $\tilde{V}(t')$ and $\tilde{V}(t)$ do not commute at different times t and t'. However, for commuting operators the upper limit of all integrals can be extended to infinity, with the contribution of the nth order divided by $n!$. In this way, the propagator reduces to the simple exponential,

$$U'(\infty, -\infty) = \exp\left[-\frac{i}{\hbar} \int_{-\infty}^{\infty} dt\, \tilde{V}(t)\right]. \tag{11.43}$$

It is straightforward to see that this result is valid in the sudden limit. In this case, the eingenvalues of H_0 are very small, in the sense that $\varepsilon_\alpha t \approx 0$ for any t contributing to the integral in the above equation. Thus, one can set $U_0 \simeq 1$, so that $\tilde{V}(t) \to V$ and

$$\left[\tilde{V}(t), \tilde{V}(t')\right]_- \to [V, V]_- = 0. \tag{11.44}$$

Eq. (11.43) can then be applied and the final amplitudes are given by

$$a_\alpha(b, t \to \infty) = \left\langle \varphi_\alpha \middle| \exp\left[-\frac{1}{\hbar} \int_{-\infty}^{\infty} dt\, V(t)\right] \middle| \varphi_0 \right\rangle, \tag{11.45}$$

or, in the representation of the intrinsic coordinates,

$$a_\alpha(b, t \to \infty) = \int d\xi\ \varphi_\alpha^*(\xi)\ \exp\left[-\frac{i}{\hbar} \int_{-\infty}^{\infty} dt\, V(\xi, t)\right] \varphi_0(\xi). \tag{11.46}$$

Using the straight line approximation for the trajectory, as in Eq. (11.32), and changing the time variable, $t \to z/v \Longrightarrow dt \to dz/v$, we get

$$a_\alpha(b, t \to \infty) = \int d\xi\ \varphi_\alpha^*(\xi)\ e^{i\, \hat{\chi}(\xi, b)}\, \varphi_0(\xi), \tag{11.47}$$

where the operator

$$\hat{\chi}(\xi, b) = \frac{1}{\hbar v} \int_{-\infty}^{\infty} V(\xi, |\mathbf{b} + z\hat{\mathbf{z}}|)\ dz, \tag{11.48}$$

acting on the intrinsic space, is a generalization of the eikonal phase of Eq. (5.118).

These results are very useful. They lead to very simple expressions when the sudden limit is appropriate. Furthermore, small deviations from the sudden limit can be taken into account in an approximate way, through the introduction of adiabaticity corrections [Alder and Winther (1975)].

The Born approximation in the semiclassical method

Now we consider the perturbative limit to lowest order, which corresponds to the first Born approximation in the semiclassical method. This approximation can be used when the amplitude in the elastic channel stays close to one and the excitation probabilities remain small along the collision. Since the Born approximation does not conserve unitarity, applying this approximation to problems with stronger interactions may lead to probabilities exceeding one.

If the above conditions are met, one can discard the first equation in Eq. (11.24) and set $a_0(t) = 1$ in the remaining ones (for $a_{\alpha\neq 0}(t)$). One can also neglect in the equation for $a_\alpha(t)$ the couplings with other inelastic channels. In this way, the coupled channel set reduces to the decoupled equations

$$i\hbar \frac{d}{dt} a_\alpha(b,t) = e^{i\omega_{\alpha 0}\, t}\, V_{\alpha 0}(t), \tag{11.49}$$

where we have introduced the relative frequencies

$$\omega_{\alpha 0} = \left(E_\alpha - E_0\right)/\hbar. \tag{11.50}$$

The above equation can be immediately integrated and the result is

$$a_\alpha(b,t) = -\frac{i}{\hbar} \int_{-\infty}^{t} dt'\; e^{i\omega_{\alpha 0}\, t'}\; V_{\alpha 0}(t'). \tag{11.51}$$

The excitation probabilities are then given by

$$P_\alpha(b) = \left|a_\alpha(b,\infty)\right|^2 = \frac{2\pi}{\hbar^2}\, \left|\mathcal{V}_{\alpha 0}\left(\omega_{\alpha 0}\right)\right|^2, \tag{11.52}$$

where $\mathcal{V}_{\alpha 0}\left(\omega_{\alpha 0}\right)$ is the Fourier transform of the time-dependent coupling interaction

$$\mathcal{V}_{\alpha 0}\left(\omega_{\alpha 0}\right) = \frac{1}{\sqrt{2\pi}} \int_{-\infty}^{\infty} dt\; e^{i\omega_{\alpha 0} t}\; V_{\alpha 0}(t). \tag{11.53}$$

In many cases, the Born approximation can be used to evaluate the excitation probability for collisions with large angular momenta (large impact parameters) and breaks down for small angular momenta. We illustrate this situation with an example in nuclear physics. We consider the breakup of ^{11}Li projectiles in collisions with a ^{208}Pb target. ^{11}Li is a very neutron rich nucleus, close to the neutron drip line[5]. The ^{11}Li nucleus is usually depicted as a ^{9}Li tightly bound core, surrounded by a halo of two weakly bound neutron. The binding energy of the halo to the core is only 0.3 MeV, which is much smaller than the separation energy of nucleons in typical nuclei (of the order of 10 MeV). In a collision with a heavy target, the strong dipole component of the Coulomb field tends to separate the charged core from the halo, leading to the Coulomb breakup process: $^{11}\text{Li} \rightarrow {}^{9}\text{Li} + 2n$.

Figure 11.3 shows the impact parameter dependence of the probability for the $^{11}\text{Li}+^{208}\text{Pb} \rightarrow {}^{9}\text{Li}+2n+^{208}\text{Pb}$ breakup reaction. The beam energy is $E_{\text{lab}} = 100$ MeV $\cdot A$. The solid line corresponds to the solution of the semiclassical coupled channel equations, whereas the dashed line is the Born approximation to the problem. The breakup probabilities are given only for $b > 10$ fm because in collisions with smaller impact parameters the incident flux is completely absorbed by the imaginary part of the optical potential. The results were obtained in [Bertulani and Canto (1992)], treating the continuum by the semiclassical version of the CDCC method (see section 9.4). Since ^{11}Li has no excited bound states, the elastic is the only bound channel in the calculation. Thus, the breakup probability is given by the relation, $P_{\text{bu}}(b) = 1 - |a_0(b,\infty)|^2$. It is clear that the Born approximation (dashed line) reproduces quite well the results of the semiclassical coupled channel calculation (solid line) for large impact parameters. However, it becomes very inaccurate at the smaller impact parameters, between 10 and 20 fm.

[5]For a given number of nucleons, A, there is a particular neutron-proton combination, $\{Z_s, N_s\}$, that gives maximal binding energy. The *stability line* corresponds to the line joining the cells of maximal stability over the periodic table. Frequently, nuclear reactions lead to isotopes that are not on the stability line. If they are close to it, they change their proton-to-neutron ratios through successive β-decays, until they reach the stability line. However, isotopes too far from the stability line can reduce the number of nucleons of the excessive kind through direct nucleon emission. The regions of the periodic table where these processes can take place are separated from the regions of beta-instability by the *drip lines*.

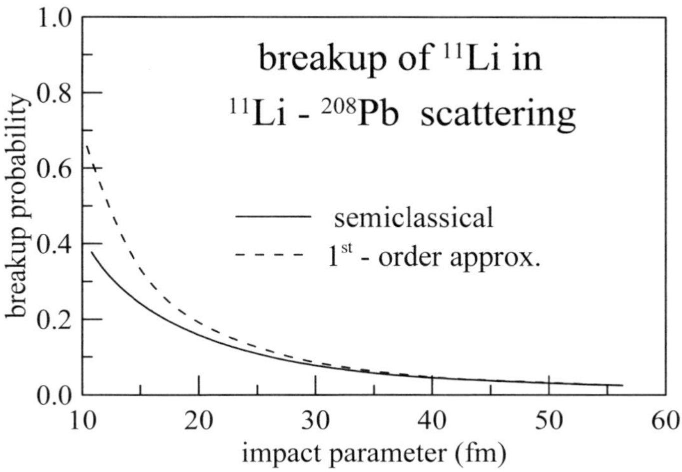

Fig. 11.3 The semiclassical probability of ^{11}Li breakup as a function of the impact parameter. The solid line was obtained solving the semiclassical coupled channel equations whereas the dashed line is the Born approximation to the problem. For details, see [Bertulani and Canto (1992)] and the text.

11.2.1.3 *Applications of the semiclassical method*

The semiclassical method has been applied in a wide variety of collisions in Molecular, Atomic and Nuclear Physics. In this section we give a few examples.

Our first example is a nuclear collision involving a small number of excited bound channels. We consider the scattering of ^{40}Ar projectiles from ^{160}Dy. In this case the projectile is spherical while the target is a highly deformed nucleus, which has several rotational states with low excitation energies. They are members of the rotational band of the ground state, having the spins and parities: $J^\pi = 0^+, 2^+, 4^+, 6^+, 8^+, 10^+$ and 12^+. These states are strongly coupled by the electric quadrupole interaction and should be included in the semiclassical coupled channel calculation. This has been done in [Canto *et al.* (1990)]. In figure 11.4 we show the inelastic probabilities for excitation of the four lowest rotational states of ^{160}Dy, bombarded by ^{40}Ar projectiles with energies in the $E_{\text{c.m.}} = 110 - 150$ MeV range. The probabilities were calculated at the scattering angle of 180 degrees. The figure exhibits results of a full quantum mechanical coupled channel calculation [Rhoades-Brown *et al.* (1981)] (filled circles), where the probabilities were obtained dividing the calculated cross section at this angle

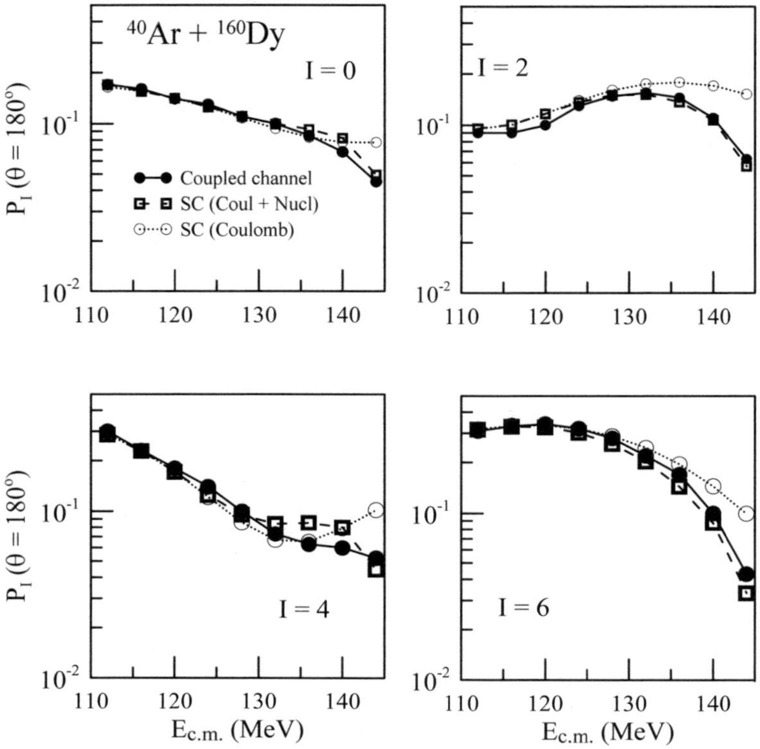

Fig. 11.4 Inelastic cross section for the excitation of the rotational band of ^{160}Dy in ^{40}Ar + ^{160}Dy scattering at $\theta = 180°$. For details, see the text.

by the corresponding Rutherford cross section. The figure also shows the results of two semiclassical coupled channel calculations. The first (open circles) neglects all nuclear effects. It is pure Coulomb excitation, as discussed in [Alder and Winther (1975)]. The second (filled squares) takes into account nuclear effects both in the trajectory and in the coupling matrix elements. Comparing the three calculations, we conclude that the semiclassical calculations for all the states are in good agreement with the quantum mechanical calculations at low energies. At the higher energies, however, the accuracy of the semiclassical calculation neglecting nuclear effects (open circles) is not very good. On the other hand, the semiclassical calculations with nuclear effects remains in good agreement with the ones based on full quantum mechanics.

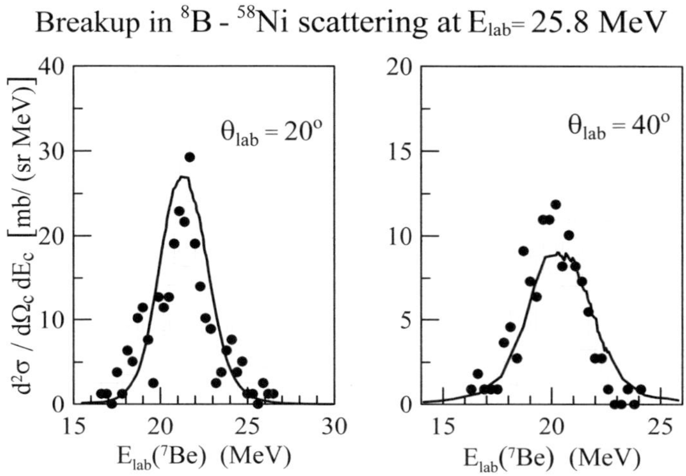

Fig. 11.5 Excitation function of the ^{7}Be fragment following ^{8}B breakup in the ^{8}B+^{58}Ni collision. The data are from [Kolata *et al.* (2001)] and the semiclassical coupled channel calculation if from [Marta *et al.* (2008)]. For details, see the text.

Our second example is also from nuclear physics. We consider the semiclassical description of the breakup reaction: $^8\mathrm{B} + ^{58}\mathrm{Ni} \rightarrow ^7\mathrm{Be} + p + ^{58}\mathrm{Ni}$, which has been measured by Kolata *et al.* [Kolata *et al.* (2001)]. In section 9.4 we showed a comparison of these data with the predictions of the CDCC calculation of Diaz-Torres and Thompson [Tostevin *et al.* (2001)]. The agreement between theory and experiment was shown to be very good (see figure 9.4). Now we compare the data with results of the semiclassical coupled channel calculation of [Marta *et al.* (2008)]. This comparison is made in figure 11.5, which shows excitation functions of the ^{7}Be fragment following the breakup of ^{8}B at the angles $\theta = 20^\circ$ and $\theta = 40^\circ$. In the calculations, the continuum spectrum of ^{8}B was discretized according to the procedure described in section 9.4, using the basis of [Tostevin *et al.* (2001)] (the same one used in figure 9.4). We see that predictions of the semiclassical model are in very good agreement with the data. The agreement is as good as that obtained with the full quantum mechanical calculations shown in figure 9.4.

We now discuss an application in atomic physics. We consider the electronic stopping of H^+ projectiles (protons) passing through a target of neutral hydrogen. The reduction of the projectile velocity results from three different processes: (*i*) inelastic scattering of the H^0 electron to excited orbits, (*ii*) ionization of the H^0 target and (*iii*) capture of the target's electron by the projectile. Although capture is the dominant process for incident protons with low energies, its contribution can be neglected at bombarding energies $E_{\rm lab} > 100$ keV. Thus, one can perform semiclassical calculations using a basis of electronic states in the H^0 target, including bound (inelastic scattering) and continuum (ionization) states. In this energy range, it is not necessary to include states of the electron bound to the projectile. Calculations along these lines were performed by Schiwietz and Grande [Schiwietz and Grande (1992)]. The discretization method and the basis of bound and continuum states used in the calculations were given in [Schiwietz (1990)]. Solving the semiclassical coupled channel equations they evaluated the excitation and ionization probabilities given by

$$P_{\alpha\neq 0}(b) = |a_{\alpha\neq 0}(b,\infty)|^2 \quad \text{and} \quad \frac{dP_\varepsilon(b)}{d\varepsilon} = \sum_n |a_\varepsilon(b,\infty)|^2 , \qquad (11.54)$$

where n stands for the set of quantum number which, together with ε, specifies the continuum states of the electron in the H^0 target. With these quantities, one can determine the average impact-parameter dependent Q-value, which is used in the calculations of the stoping power,

$$\overline{Q}(b) = \sum_\alpha P_\alpha(b)\ (E_\alpha - E_0) + \int_0^\infty d\varepsilon\ \frac{dP_\varepsilon(b)}{d\varepsilon}\ (\varepsilon + I_B) , \qquad (11.55)$$

where I_B is the binding energy of the electron in the initial state.

The inelastic (summed over intrinsic excitations) and the ionization cross sections are then given by

$$\sigma_{\rm inel} = 2\pi \sum_{\alpha\neq 0} \int b\ db\ P_\alpha(b), \qquad (11.56)$$

$$\sigma_{\rm ion} = 2\pi \int d\varepsilon \int b\ db\ \frac{dP_\varepsilon(b)}{d\varepsilon} . \qquad (11.57)$$

and the total reaction cross section contribution to stopping of the protons are

$$\sigma_{\rm tot} = \sigma_{\rm inel} + \sigma_{\rm ion} . \qquad (11.58)$$

In the calculations, the continuum states were normalized per square root of the energy, in a.u., as in the method discussed by Cowan [Cowan (1981)].

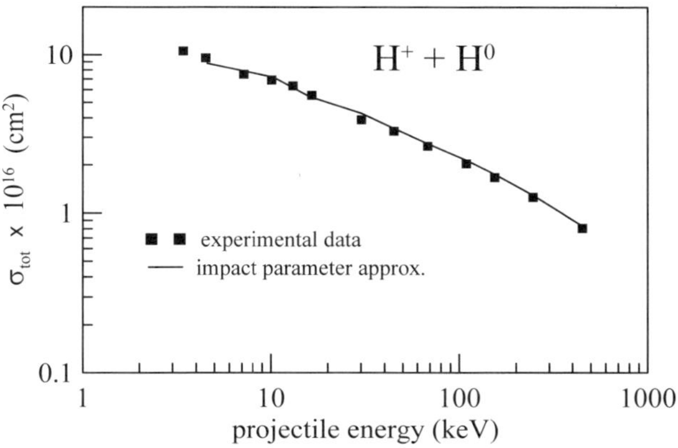

Fig. 11.6 Total reaction cross section in the collision of protons with atomic hydrogen, as a function of the bombarding energy. The details of the semiclassical calculation and of the data are given in [Schiwietz and Grande (1992)].

In figure 11.6 we show the total reaction cross section in collisions of protons with an atomic hydrogen target. The experimental data were obtained summing several cross sections, as described in [Schiwietz and Grande (1992)]. The theoretical curve was also taken from [Schiwietz and Grande (1992)]. The authors performed a semiclassical coupled channel calculation, as described above. The agreement between theory and experiment is very good, even at bombarding energies below 100 keV, where it was not supposed to work very well. In this energy range, the total reaction cross section is dominated by capture, which is not explicitly included in the model. Besides the capture process, one would expect also that the distortion of the single center electronic orbitals around the target would influence the ionization cross section. This process can be handled satisfactorily through the Continuum Distorted Wave Eikonal Initial States (CDW-EIS) [Fainstain *et al.* (1991)] (for an application see also [Montanari *et al.* (2010)]). It is likely that the success of the model of [Schiwietz and Grande (1992)] at low energies resides in the fact that the total reaction cross section is a very inclusive quantity. Using the unitarity of the semiclassical approximation, the total reaction probability for a given impact parameter can be evaluated through the relation $P_{\rm tot}(b) = 1 - |a_0(b,\infty)|^2$. This quantity does not depend strongly on how the flux removed from the elastic channel is distributed over the particular channels involved in the calculation.

11.2.2 *The classical-limit S-matrix*

The semiclassical approximation to the S-matrix in many-body scattering is valid when $\hbar$ is much smaller than the classical action involved in the scattering process. The techniques used in this case are similar to the ones employed in potential scattering, discussed in chapter 5. Usually, this approximation is called the *Classical-Limit S-Matrix* (CLSM). There are two ways to derive the CLSM. The first is based on the path integral approach and has been developed by Petchukas [Pechukas (1969a,b)] and by Miller [Miller (1970a,b, 1974)]. The second, which can be viewed as a multidimensional generalization of the WKB method, has been developed by Marcus [Marcus (1970, 1971, 1972)] and Connor [Connor and Marcus (1971)]. Here, we use Feynmann's Path Integral formulation of Quantum Mechanics [Feynman and Hibbs (1965)] and follow the work of [Miller (1974)].

11.2.2.1 *The semiclassical propagator*

In the Schrödinger picture of Quantum Mechanics, a state $|\psi(t_1)\rangle$ changes to a state $|\psi(t_2)\rangle$, as the time evolves from t_1 to t_2, according to the relation

$$|\psi(t_2)\rangle = e^{-iH(t_2-t_1)/\hbar}\,|\psi(t_1)\rangle\,, \tag{11.59}$$

where H is the system's Hamiltonian[6]. In this way, the probability amplitude of the position eigenstate $|q_1\rangle$ evolving to a new position eingenstate $|q_2\rangle$ as $t_1 \to t_2$ is given by the propagator

$$K(q_2t_2, q_1t_1) = \langle q_2|\; e^{-iH(t_2-t_1)/\hbar}\; |q_1\rangle\,. \tag{11.60}$$

In the path integral approach, the propagator of Eq. (11.60) is given by

$$K(q_2t_2, q_1t_1) = \int \mathcal{D}\left[q(t)\right]\; e^{i\Phi_{q[t]}(q_2,q_1)/\hbar}, \tag{11.61}$$

where $q[t]$ stands for the arbitrary trajectories connecting the point q_1 at $t = t_1$ with the point q_2 at $t = t_2$, and $\mathcal{D}\left[q(t)\right]$ is the volume element in the integration over paths[7]. In Eq. (11.61),

$$\Phi_{q[t]}(q_2, q_1) = \int \mathcal{L}_{q[t]}(q, \dot{q}, t)\, dt \tag{11.62}$$

---

[6]We are assuming that the Hamiltonian does not depend explicitly on time.

[7]The integration over paths may be performed dividing the time interval in a mesh $\{t_i\} \equiv \{t, t+\Delta t, t+2\Delta t, \cdots, t'\}$ and integrating over each $q(t_i)$. The desired result is then obtained taking the $\Delta t \to 0$. For a detailed discussion, see [Feynman and Hibbs (1965); Pechukas (1969a)].

is the action [Goldstein (1980)] evaluated along the particular trajectory $q[t]$.

In the classical limit, $\Phi_{q[t]}(q_2, q_1)/\hbar \gg 1$. Thus, the phase in the integrand of Eq. (11.61) oscillates wildly as the trajectory is varied around $q[t]$, producing strong cancellations of the contributions from neighboring trajectories. Therefore, the only important region in the integration over paths is the one where the action is stationary. That is, the neighborhood of the trajectories $\bar{q}[t]$ for which the variation of the action vanishes to first order. This is the condition leading to the Euler-Lagrance equations of motion [Goldstein (1980)]. Thus, the stationary phase occurs for the classical trajectory[8], $\bar{q}[t]$ (possibly more than one), connecting the space-time points $\{q_1, t_1\}$ and $\{q_2, t_2\}$. Using a delta-function like approximation for the path integral, the propagator reduces to the exponential

$$K(q_2, q_1) \to e^{i\bar{\Phi}(q_2,q_1)/\hbar}, \tag{11.63}$$

where

$$\bar{\Phi}(q_2, q_1) \equiv \Phi_{\bar{q}[t]}(q_2, q_1) = \int_{t_1}^{t_2} dt \; \mathcal{L}_{\bar{q}[t]}\,(q, \dot{q})\,. \tag{11.64}$$

However, the delta-function approximation is not appropriate. The vicinity of the classical trajectory gives a finite contribution to the propagator, leading to a slowly varying pre-exponential factor $A(q_2, q_1)$. Thus, the semiclassical propagator should be written as

$$K(q_2, q_1) = A(q_2, q_1)\; e^{i\bar{\Phi}(q_2,q_1)/\hbar}. \tag{11.65}$$

Owing to the unitarity of the propagator, $A(q_2, q_1)$ can be obtained from the phase factor through the relation [Miller (1970a)]

$$A(q_2, q_1) = \left| \left(\frac{1}{2\pi\hbar}\right) \frac{\partial^2 \bar{\Phi}(q_2, q_1)}{\partial q_2\, \partial q_1} \right|^{1/2}. \tag{11.66}$$

Although this result has been obtained for a single canonic pair $\{q, p\}$ it can easily be extended to any number of degrees of freedom.

It is convenient to write the classical action in terms of the momentum variable. This can be done through the Legendre transformation

$$\mathcal{L}(q, \dot{q}) = p\, \dot{q} - H(p, q), \tag{11.67}$$

where

$$p = \frac{\partial \mathcal{L}(q, \dot{q})}{\partial \dot{q}}. \tag{11.68}$$

[8]The classical trajectory and quantities associated with it will be denoted by an overline.

Using Eq. (11.67) and taking into account energy conservation, the classical action becomes

$$\bar{\Phi}_E(q_2, q_1) = -E\,(t_2 - t_1) + \int_{t_1}^{t_2} dt\; p\,\dot{q}, \tag{11.69}$$

where $\dot{q} \equiv \dot{q}(t)$ and $p \equiv p(t)$ are the values of the velocity and the momentum of the system at time t, along a trajectory that keeps the energy with the constant value E. They have the initial values $\{q_1, p_1\}$ and final values $\{q_2, p_2\}$.

Scattering problems involve the propagator in the momentum representation. In this case, the generator of the canonical transformation leading from the initial momentum (p_1) to the final momentum (p_2) keeping the energy constant is [Miller (1970a); Goldstein (1980)]

$$\begin{aligned}\bar{\Phi}_E(p_2, p_1) &= -\,(p_2 q_2 - p_1 q_1) + \bar{\Phi}_E(q_2, q_1)\\ &= -E\,(t_2 - t_1) - (p_2 q_2 - p_1 q_1) + \int_{t_1}^{t_2} dt\; p\,\dot{q},\end{aligned} \tag{11.70}$$

or, carrying out the integration by parts,

$$\bar{\Phi}_E(p_2, p_1) = -E\,(t_2 - t_1) - \int_{t_1}^{t_2} dt\; \dot{p}\,q. \tag{11.71}$$

The propagator in momentum space then becomes [Miller (1970a)],

$$\begin{aligned}K(p_2, p_1) = &\left|\left(\frac{1}{2\pi\hbar}\right) \frac{\partial^2 \bar{\Phi}(p_2, p_1)}{\partial p_2\, \partial p_1}\right|^{1/2}\\ &\times \exp\left[-i\left(E\,(t_2 - t_1) + \int_{t_1}^{t_2} dt\; \dot{p}\,q\right)\right].\end{aligned} \tag{11.72}$$

To derive the CLSM, we use the above propagator in the time dependent definition of the scattering operator (Eq. (4.106))

$$S_{p_2,p_1} = \lim_{\substack{t_1 \to -\infty \\ t_2 \to +\infty}} \left[\; \langle \phi(p_2)|\, e^{iH_0 t_2/\hbar}\, e^{-iH(t_2 - t_1)/\hbar}\, e^{-iH_0 t_1/\hbar}\, |\phi(p_1)\rangle \;\right],$$

or[9]

$$S_{p_1,p_1} = e^{iE(t_2 - t_1)/\hbar}\; \langle \phi(p_2)|\; e^{-iH(t_2 - t_1)/\hbar}\; |\phi(p_1)\rangle\,.$$

We get

$$S^{\rm CLSM}_{p_2,p_1} = \left|\left(\frac{1}{2\pi\hbar}\right) \frac{\partial^2 \bar{\Phi}(p_2, p_1)}{\partial p_2\, \partial p_1}\right|^{1/2} \times\; \exp\left[\frac{i}{\hbar}\, \Phi\,(p_2, p_1)\right], \tag{11.73}$$

[9] To simplify the notation, we have omitted the indication that we are taking the $t_1 \to -\infty, t_2 \to +\infty$ limits.

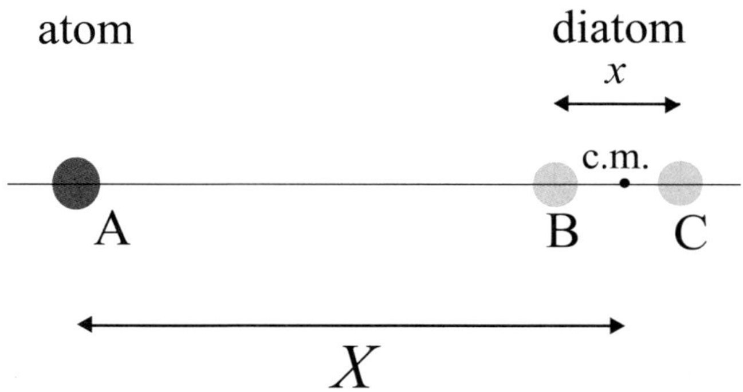

Fig. 11.7 Schematic representation of a collinear atom-diatom collision. The distance between the atom (A) and the center of mass of the diatom (B+C) is the collision coordinate, X. The distance between the two atoms of the diatom (B and C) is the intrinsic coordinate, x.

with

$$\bar{\Phi}\,(p_2, p_1) = -\int_{-\infty}^{\infty} dt\; \dot{p}\, q. \tag{11.74}$$

The relevant applications of the CLSM are for systems with intrinsic degrees of freedom. In such cases p represents the set of all generalized momenta $\{p^{(1)}, \cdots, p^{(n)}\}$. The pre-exponential factor of Eq. (11.73) then becomes a determinant [Miller (1970a)] and Eq. (11.74) changes as

$$\bar{\Phi}\,(p_2, p_1) = -\sum_{j=1}^{n} \int_{-\infty}^{\infty} dt\; \dot{p}^{(j)}\, q^{(j)}. \tag{11.75}$$

11.2.2.2 *Applications of the CLSM*

The CLSM has been applied to several problems in chemical dynamics (for a review, see [Sun (1998)]), atomic [Miller (1974)] and nuclear physics [Massmann and Rasmussen (1975); Donangelo *et al.* (1976); Guidry *et al.* (1976)]. To illustrate the use of this method, we describe its application to atom-diatom inelastic scattering, within the collinear approximation [Miller (1970b, 1974)]. The situation is illustrated in figure 11.7. Owing to the collinear approximation, X and x represent vectors in one dimension. The dynamics is governed by the Hamiltonian

$$H(X, P\,; x, p) = \frac{P^2}{2\mu} + h(x, p) + V(X, x), \tag{11.76}$$

where P and p are respectively the momenta conjugated to the coordinates X and x. In Eq. (11.76), $h(x,p)$ is the intrinsic Hamiltonian,

$$h(x,p) = \frac{p^2}{2m} + v(x), \tag{11.77}$$

where $v(x)$ is the interaction potential between the two atoms that constitute the diatom (B+C), m is the reduced mass associated with relative motion (B-C) inside the diatom and μ is the reduced mass of the atom-diatom system. The intrinsic states of the diatom satisfy the equation

$$h\,\varphi_n(x) = \varepsilon_n\,\varphi_n(x). \tag{11.78}$$

The collision is initiated with the diatom in its ground state ($n = 1$) and we want to evaluate the S-matrix for transitions to inelastic states $\varphi_n(x)$. In this case, the intrinsic coordinate is not appropriate to apply the CLSM. Since it is not cyclic, its conjugate momentum, p, is not a constant of motion of h. To deal with this situation, we should carry out a canonic transformation leading from the original intrinsic variables, x and p, to the action-angle variables of the oscillator, n and q. The action variable is given by the transformation [Miller (1970b, 1974)],

$$\left(n + \frac{1}{2}\right)\pi = \frac{\sqrt{2\,m}}{\hbar}\int_{x_<}^{x_>} dx'\sqrt{\varepsilon - v\left(x'\right)}, \tag{11.79}$$

where $x_<$ and $x_>$ are respectively the inner and the outer turning points in the B-C relative motion within the diatom, with the intrinsic energy

$$\varepsilon = \frac{p^2}{2m} + v(x), \tag{11.80}$$

and the variable q is

$$q = \frac{\sqrt{2\,m}}{\hbar\,n'(\varepsilon)}\int_{x_<}^{x_>} \frac{dx'}{\sqrt{\varepsilon - v\left(x'\right)}}. \tag{11.81}$$

The system Hamiltonian can be written in terms of the new coordinates as

$$H\left(X, P\,; q, n\right) = \frac{P^2}{2\mu} + \varepsilon(n) + V\left(X, x(q,n)\right). \tag{11.82}$$

Above, $\varepsilon(n)$ is the inverse function of $n(\varepsilon)$ and $x(q,n)$ is the original coordinate expressed as a function of the new coordinate and its conjugate momentum. For some particular vibrational potentials, like the harmonic oscillator and the Morse potential, the transformation can be obtained analytically [Miller (1970b, 1971)]. In other cases, it is necessary to resort to numerical procedures.

The S-matrix for the transition from the elastic channel, here labeled by n_1, to the inelastic channel, labelled by n_2, is given by Eq. (11.73), with the generalized phase of Eq. (11.75). In this case, the coordinates and momenta are: $q^{(1)} = X, q^{(2)} = q; p^{(1)} = P, p^{(2)} = n$, and Eq. (11.73) can be written

$$\bar{\Phi}\left(P_2\, n_2; P_1\, n_1\right) = -\int_{-\infty}^{\infty} \left[\dot{P}(t)\, X(t) + \dot{n}(t)\, q(t)\right]. \tag{11.83}$$

The above integral should be evaluated along the classical trajectory leading from the initial momenta $\{P_1, n_1\}$ to the final momenta $\{P_2, n_2\}$, where

$$P_1 = -\sqrt{2\mu\left[E - \varepsilon(n_1)\right]} \quad \text{and} \quad P_2 = \sqrt{2\mu\left[E - \varepsilon(n_1)\right]}. \tag{11.84}$$

The pre-exponential factor in Eq. (11.73) becomes

$$\left|\left(\frac{1}{2\pi\hbar}\right)\frac{\partial^2\bar{\Phi}(p_2, p_1)}{\partial p_2\,\partial p_1}\right|^{1/2} \longrightarrow \left|\left(\frac{1}{2\pi\hbar}\right)\frac{\partial^2\bar{\Phi}\left(P_2\, n_2; P_1\, n_1\right)}{\partial n_2\,\partial n_1}\right|^{1/2}. \tag{11.85}$$

The calculation of this factor can be simplified using the derivative relation of the phase [Miller (1974)],

$$\frac{\partial\bar{\Phi}\left(P_2\, n_2; P_1\, n_1\right)}{\partial n_1} = q_1 + X_1\frac{\partial P_1}{\partial n_1} \equiv \tilde{q}_1. \tag{11.86}$$

Using this result we can write

$$\frac{\partial^2\bar{\Phi}\left(P_2\, n_2; P_1\, n_1\right)}{\partial n_2\,\partial n_1} = \left[\left(\frac{\partial n_2}{\partial\tilde{q}_1}\right)_{n_1}\right]^{-1} \tag{11.87}$$

and the S-matrix becomes

$$S_{n_2,n_1}(E) = \left[\left(\frac{\partial n_2}{\partial\tilde{q}_1}\right)_{n_1}\right]^{-1}. \tag{11.88}$$

There is however a practical difficulty to calculate the phase and the pre-exponential factor of Eqs. (11.85) and (11.87). The integral of Eq. (11.83) should be evaluated along the classical trajectory starting with initial momenta P_1 and n_1 and and initial coordinates X_1 and q_1. The initial atom-diatom distance, X_1, should be large enough for the potential $V(X, x(q, n))$ to be negligible. The particular choice of X_1 does not affect the final value of n, provided that $V(X_1, x) \simeq 0$. On the other hand, the final momentum depends critically on q_1, which is not known. All we know is that it should

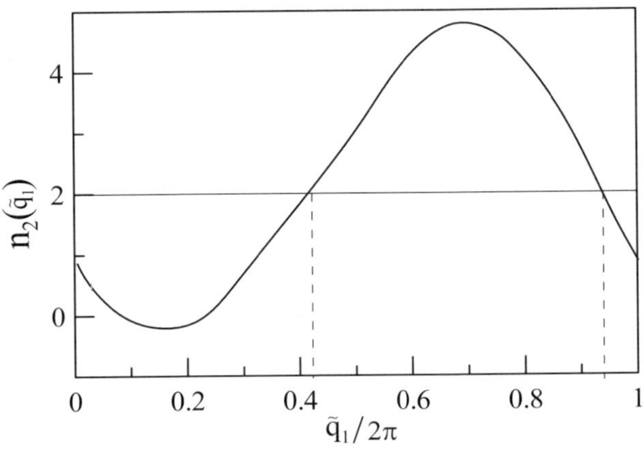

Fig. 11.8 The final value of the action variable as a function of the initial variable $\tilde{q}_1$. The figure is from [Miller (1970b)].

lead to the desired final value of the action variable (n_2). In order to determine q_1, or equivalently, $\tilde{q}_1$, it is necessary to solve the equations of motion for several initial conditions, constructing in this way the function $n_2(\tilde{q}_1)$, and then find the value (or values) of $\tilde{q}_1$ leading to the desired final momentum. This is illustrated in figure 11.8, from [Miller (1970b)], which shows the final value of the action variable as a function of $\tilde{q}_1$. The horizontal line indicates the excitation to the first excited state of the diatom, $n_2 = 2$. The vertical dashed lines indicate two values of $\tilde{q}_1$ that lead to this final momentm, n_2. In this particular case, the S-matrix is given by the sum of two independent contributions coming from the two trajectories. However, in other situations the two contributions cannot be evaluated separately and it is necessary to go beyond the stationary phase approximation, This is achieved through a generalization of the Uniform approximation to collisions with intrinsic degrees of freedom (see e.g. [Massmann and Rasmussen (1975)]).

Another extension of the method is necessary to evaluate the S-matrix for transitions to excited states that are predicted by quantum mechanics but vanish in the classical limit. This occurs when no trajectory leads to the required final value of the action variable. In the example of figure 11.8 it corresponds to any value of n above the maximum of the curve, like $n_2 = 6$. In this case, it is necessary to do an analytical continuation of the time to the complex plane [Miller (1974); Guidry *et al.* (1976)].

11.2.3 *Glauber approximation*

In section 5.2.1 we pointed out that the eikonal phase has the convenient property of depending linearly on the interaction potential. This makes the eikonal approximation a very powerful method to study collisions involving composite particles. A simple example is the scattering of protons from deuterons [Glauber (1955)]. This approach, which was presented in detail by Glauber is a series of lectures [Glauber (1959)], is known as Glauber approximation and it is discussed below.

Let us consider the collision of a structureless projectile with a target made of N particle, e.g. a nucleon-nucleus collision. In the center of mass reference frame, the system is described by the projectile-target separation vector, $\mathbf{r}$, and the position vector of the target particles, $\mathbf{x}_i$ $(i = 1, ..., N)$. The system's Hamiltonian can be written as

$$H(\mathbf{r}; \mathbf{x}_i) = H_0(\mathbf{r}; \mathbf{x}_i) + \sum_i V_i(\mathbf{r} - \mathbf{x}_i), \tag{11.89}$$

where

$$H_0(\mathbf{r}; \mathbf{x}_i) = -\hbar^2 \frac{\nabla_{\mathbf{r}}^2}{2\mu} + h(\mathbf{x}_1, ..., \mathbf{x}_N) \tag{11.90}$$

is the sum of the projectile-target kinetic energy with the intrinsic hamiltonian of the target, h. The many-body wave function of the system $\Psi(\mathbf{r}; \mathbf{x}_i)$ is given by the channel expansion

$$\Psi(\mathbf{r}; \mathbf{x}_i) = \sum_\alpha \psi_\alpha(\mathbf{r})\, \varphi_\alpha(\mathbf{x}_1, ..., \mathbf{x}_N), \tag{11.91}$$

with the intrinsic states satisfying the Schrödinger equation

$$h(\mathbf{x}_1, ..., \mathbf{x}_N)\, \varphi_\alpha(\mathbf{x}_1, ..., \mathbf{x}_N) = \varepsilon_\alpha\, \varphi_\alpha(\mathbf{x}_1, ..., \mathbf{x}_N). \tag{11.92}$$

Here, the projectile target interaction is given by the potential

$$V(\mathbf{r}) = \sum_{i=1}^{N} V_i(\mathbf{r} - \mathbf{x}_i). \tag{11.93}$$

First we assume that the target particles are frozen at the positions $\mathbf{r}_1, \mathbf{r}_2, ..., \mathbf{r}_N$, so that the total potential depends on these vectors parametrically. Then, since the eikonal phase depends linearly on the interaction, we can write[10] (see 5.2.1),

$$\chi_{\text{tot}}(\mathbf{b}, \{\mathbf{x}_i\}) = \sum_{i=1}^{N} \chi_i(\mathbf{r} - \boldsymbol{\rho}_i), \tag{11.94}$$

[10]Note that the interaction potential between the projectile and different target particles may be different. For example, in the case of nucleon-nucleus scattering, different interactions must be used for the protons and the neutrons in the target. For this reason, the phase associated with the i-th particle carries the index i.

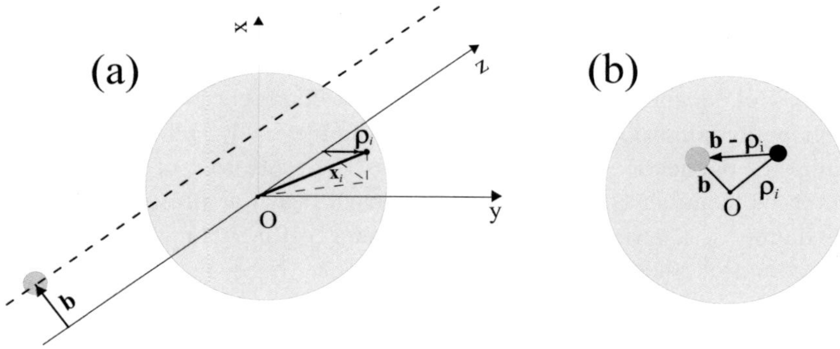

Fig. 11.9 Schematic representation of a collision of a structureless projectile with a many-body target. The figure shows the projectile approaching the target with impact parameter $\mathbf{b}$ and the i-th target particle. This particle's position vector is $\mathbf{x}_i$ and its component perpendicular to the z-axis is $\boldsymbol{\rho}_i$. Panels (a) and (b) are representations in three dimensions and on the x-y plane.

where the symbol $\{\mathbf{x}_i\}$ in the argument of the total phase means that it depends on the vector position of each target particle. Above, $\boldsymbol{\rho}_i$ stands for the component of $\mathbf{x}_i$ perpendicular to the incident beam. That is, if the incident beam is parallel to the z-axis and $\hat{\mathbf{z}}$ is the unit vector along this direction, $\boldsymbol{\rho}_i$ is the vector on the x-y plane

$$\boldsymbol{\rho}_i = \mathbf{x}_i - (\mathbf{x}_i \cdot \hat{\mathbf{z}})\, \hat{\mathbf{z}}. \tag{11.95}$$

This is illustrated in figure 11.9 in a three dimension (panel (a)) view and on a projection on the x-y plane (panel (b)). The contribution from particle i to the total phase is (see Eq. (5.118))

$$\chi_i\left(\mathbf{r} - \boldsymbol{\rho}_i\right) = -\frac{1}{\hbar v}\int_{-\infty}^{\infty} dz\; V\left(\mathbf{b} - \boldsymbol{\rho}_i, z\right). \tag{11.96}$$

The scattering amplitude is then given by the trivial generalization of Eq. (5.122):

$$f(\hat{\mathbf{k}}', \hat{\mathbf{k}}) = -\frac{ik}{2\pi}\int d^2\mathbf{b}\; e^{-i\mathbf{q}\cdot\mathbf{b}}\left[e^{i\,\chi_{\mathrm{tot}}(\mathbf{b},\{\mathbf{x}_i\})} - 1\right]. \tag{11.97}$$

Now we consider the realistic situation where the positions of target nucleons are dynamic variables. In this case the potentials appearing in Eqs. (11.93) and (11.96) become operators acting both on $\mathbf{r}$ and the on the intrinsic degrees of freedom. The collision then becomes a coupled channel problem, involving different intrinsic states. Labeling the channels by α, the scattering amplitude becomes

$$f_{\alpha,0}(\hat{\mathbf{k}}'_\alpha, \hat{\mathbf{k}}_\mathbf{0}) = -\frac{ik}{2\pi} \int d^2\mathbf{b}\; e^{-i\mathbf{q}\cdot\mathbf{b}} \int d^3\mathbf{x}_1\, d^3\mathbf{x}_2 ...\, d^3\mathbf{x}_N\; \varphi^*_\alpha\left(\mathbf{x}_1, ..., \mathbf{x}_N\right)$$
$$\times \left[e^{i\,\chi_{\rm tot}(\mathbf{b},\{\mathbf{x}_i\})} - 1\right] \varphi_0\left(\mathbf{x}_1, ..., \mathbf{x}_N\right). \qquad (11.98)$$

In particular, the elastic scattering amplitude is

$$f_{0,0}(\hat{\mathbf{k}}'_\alpha, \hat{\mathbf{k}}_\mathbf{0}) = -\frac{ik}{2\pi} \int d^2\mathbf{b}\; e^{-i\mathbf{q}\cdot\mathbf{b}} \int d^3\mathbf{x}_1\, d^3\mathbf{x}_2 ...\, d^3\mathbf{x}_N$$
$$\times \left|\varphi_0\left(\mathbf{x}_1, ..., \mathbf{x}_N\right)\right|^2 \left[e^{i\,\chi_{\rm tot}(\mathbf{b},\{\mathbf{x}_i\})} - 1\right]. \qquad (11.99)$$

Eq. (11.99) contains all the coupling effects on the elastic scattering.

We remark that Eq. (11.98) is a framework for the development of useful non-perturbative approximations to the coupled-channel equations, valid within the high energy limit. It can be easily checked that it contains channel coupling effects to all orders. This method, known as the *Dynamic Eikonal Approximation* (DEA), has been fully developed and applied in elastic scattering and breakup reactions by Baye *et al.* [Baye *et al.* (2005)] and by Goldstein *et al.* [Goldstein *et al.* (2006)]. An extension of a variant of this method to the relativistic domain was made by Bertulani [Bertulani (2005)]. Further development of the eikonal description of reactions were made by several authors. The divergent Coulomb effects in breakup reactions using the adiabatic or sudden approximation were remedied through the inclusion of an appropriate correction originally introduced by Margueron, Bonaccorso and Brink [Margueron *et al.* (2003)]. This correction consists in replacing the first-order term of the eikonal Coulomb phase by the first-order of the perturbation theory. The resulting eikonal theory, developed by Capel, Baye and Suzuki [Capel *et al.* (2008)], is not Coulomb-singular and compares quite well with the non-adiabatic Dynamic Eikonal Approximation. Another extention of the eikonal method to CC including the continuum, was made by K. Ogata *et al.* [Ogata *et al.* (2003)]. This method can be considered as a hybrid eikonal + CCDC version of the CCDC method.

Frequently, it is convenient to develop a one-channel description of elastic scattering containing the effects of couplings with other channels. This can be done through the introduction of optical and polarization potentials, as discussed in detail in chapter 10. In the present case, an approximation along these lines is to replace the expectation value of $\exp\left[i\,\chi_{\rm tot}(\mathbf{b}, \{\mathbf{x}_i\})\right]$, taken with respect to φ_0, by the expectation value of the total eikonal phase. That is

$$\left\langle e^{i\,\chi_{\rm tot}(\mathbf{b})}\right\rangle \to e^{i\,\langle\,\chi_{\rm tot}(\mathbf{b})\,\rangle}, \tag{11.100}$$

where we used the compact notation

$$\langle A(\mathbf{b})\rangle = \int d^3\mathbf{x}_1\, d^3\mathbf{x}_2 ...\, d^3\mathbf{x}_N\ \left|\varphi_\alpha\left(\mathbf{x}_1, ..., \mathbf{x}_N\right)\right|^2 A(\mathbf{b}, \{\mathbf{x}_i\}). \tag{11.101}$$

In order to keep the same elastic scattering amplitude of Glauber's theory, it is necessary to introduce a complementary polarization potential. Ignoring this latter potential, the amplitude is given by the so called *optical limit* of Glauber's theory, namely,

$$f_{0,0}^{\rm opt}(\hat{\mathbf{k}}_0', \hat{\mathbf{k}}_\mathbf{0}) = -\frac{ik}{2\pi}\int d^2\mathbf{b}\ e^{-i\mathbf{q}\cdot\mathbf{b}}\left[e^{i\,\langle\,\chi_{\rm tot}(\mathbf{b})\,\rangle} - 1\right]. \tag{11.102}$$

This prescription can be further simplified using directly the eikonal of the optical potential:

$$\langle\,\chi_{\rm tot}(\mathbf{b})\,\rangle = \frac{1}{\hbar v}\int_{-\infty}^{\infty} dz\ V_{\rm opt}\left(\mathbf{b}, z\right). \tag{11.103}$$

The optical limit of Glauber theory, expressed in Eq. (11.100), is an approximation. It becomes an inequality if $i\chi_{\rm tot}(\mathbf{b})$ is real, which would be the case if the potential is pure imaginary. This result comes under the name of Jensen's inequality, which rests on the convexity of the exponential function. In a nutshell, the Jensen inequality states [Johnson and Goebel (2000)]

$$\left\langle e^{i\,\chi_{\rm tot}(\mathbf{b})}\right\rangle \geq e^{i\,\langle\,\chi_{\rm tot}(\mathbf{b})\,\rangle}. \tag{11.104}$$

Accordingly,

$$\left\langle e^X\right\rangle \geq \langle 1 + X\rangle \tag{11.105}$$

Setting $X = i\chi_{\rm tot}(\mathbf{b}) - \langle i\chi_{\rm tot}(\mathbf{b})\rangle$, with $i\chi_{\rm tot}(\mathbf{b})$ real, we obtain the Jensen inequality, Eq. (11.104).

The inequality Eq. (11.104) has the consequence that the transmission coefficient that measures absorption, is smaller within the coupled channel theory than when calculated using the optical limit of Glauber theory. To be specific, we take for the optical potential the Lax form [Lax (1951)], namely, $V_{\rm opt} = t_{\rm NN}\,\rho$, and consider an energy high enough that the nucleon-nucleon T-matrix is purely imaginary. Then the interaction between the two nuclei will become imaginary (purely absorptive), and the corresponding elastic S-matrix element is real. The transmission coefficient defined as $\mathcal{T}(b) = 1 - |S(b)|^2$ determines the total reaction cross section, through

$$\sigma_{\mathrm{R}}^{\mathrm{opt}} = 2\pi \int_0^\infty b\, db\, \mathcal{T}(b). \tag{11.106}$$

If one calculates the same cross section within Glauber theory in the optical limit and uses the Jensen inequality, one necessarily arrives at,

$$\sigma_{\mathrm{R}}^{\mathrm{Glauber}} \leq \sigma_{\mathrm{R}}^{\mathrm{opt}}. \tag{11.107}$$

The equality usually approximately holds in cases where the matter inside the target is homogeneously distributed, as is the case in normal stable nuclei. In recent years nuclear research has produced and studied unstable nuclei with a large number of excess neutrons or protons. The studies revealed that some of these exotic nuclear species, such as ^{11}Li or ^{11}Be, have a cluster structure with a normal compact core plus a one- or two-neutron *halo* rotating around it. For such matter distribution the Glauber total reaction cross section may be 20% smaller than the optical one. If we parametrize these cross sections as $\pi R^2_{\mathrm{Glauber}}$ and πR^2_{opt}, respectively, then for a given set of data that is used to determine the radius of the exotic nucleus, the following inequality holds [Al-Khalili and Tostevin (1996); Al-Khalili *et al.* (1996)]

$$R_{\mathrm{Glauber}} \geq R_{\mathrm{opt}}. \tag{11.108}$$

The above relation was found to be very important to extract more reliable radii of these exotic nuclei. The values obtained for the halo nuclei were more than several percents larger than those extracted using the optical limit.

Exercises

(1) Write the semiclassical coupled channel equations (Eq. (11.24)) in terms of the variables w and ε, given by the transformation of Eq. (11.33).

(2) Consider the Coulomb exciation of multipolarity λ. The coupling interaction is of the form

$$\frac{V_0}{[r(t)]^{\lambda+1}}.$$

Evaluate approximately the excitation amplitude for channel-α using Eqs. (11.51) to (11.53), and show that it goes as

$$\exp\left[-2\ \omega_{\alpha,0}\ \frac{r(0)}{(\lambda+1)\ r''(0)}\right],$$

where $r(0)$ and $r''(0)$ are, respectively, the distance of closest approach and the radial acceleration at this distance[11].

(3) Use Eq. (11.28) to prove Eq. (11.30).

(4) The matrix-element of the dipole excitation of a state in channel α with wave vector $\mathbf{k}_\alpha$ is

$$V_{\alpha,0}\,\frac{1}{r^2}.$$

Use the first Born approximation to evaluate the transition amplitude for the corresponding excitation.

(5) Use the Jenson inequality to prove that the total reaction cross section in the Glauber limit satisfies the following

$$\langle\sigma_{\mathrm{R}}(\chi)\rangle \;\le\; \sigma(\langle\chi\rangle)$$

where $i\chi$ is the eikonal phase, taken to be real.

[11] *Hint*: expand $[r(t)]^{\lambda+1}$ around $t = 0$ (instant of closest approach) to second order in time, and evaluate the t-integral using contour integration

Chapter 12

Selected Topics in Many-body Scattering

The previous chapters dealt with the fundamentals of collision theory and several applications. There is a very large body of other topics which are not discussed and being of great relevance to much of the current research in the fields of atomic, molecular and nuclear physics, as well as other fields, these topics deserve full chapters if not entire books to cover. Since such detailed accounts are in fact abundantly available in the literature, we have decided to devote the last chapter of this book to give a very brief account of some of these topics. Our choice is entirely based on our own encounters with them on several occasions and for their own merit as being currently vigorously pursued and developed in several areas of research. The topics which are covered in this chapter are the concept of doorway states and resonances in nuclear reactions, the Feshbach resonances in atomic physics, the statistical theory of reactions, chaotic scattering, scattering in media and application to astrophysics. We also briefly discuss the multiple scattering formalism and the three-body problem. Finally, we briefly discuss the R-matrix Theory, which is frequently used to describe scattering at very low energy, as occurs in nucleosynthesis. We confess that the treatment is quite elementary and our aim is mostly to make the readers aware of the topics and direct them to the original literature and more specialized monographs and books dedicated to them.

12.1 Doorway resonances

An important modification of the reaction theory would be called for if a special closed channel dominates the coupling between the P and Q-subspaces. This situation could happen in nuclei, atomic clusters, and other many-body systems. Let us generalize the P-subspace to contain both the open channels, and this special closed channel, which we call the *Doorway* channel. To develop the theory in this situation[1], we introduce the doorway projection operator, D, and maintain the notation of P for the projection of the open channels. Thus, the generalized space, P', is the sum $P' = P + D$. The doorway projected wave function, could describe collective excitation in the system, which cannot be described by the compound nucleus ideas. The doorway, could be a nuclear quasi-molecule, which survives a short time before it collapses into the more complicated configurations of the compound nucleus. After performing the energy averaging to remove the statistical fluctuations due to the compound nucleus (CN), the *optical* S-matrix, would still exhibit a resonance-type modulation owing to the doorway. The doorway resonance width would necessarily contain two pieces: the width that describes the decay of the doorway to the open channels, the so-called *escape width*, $\Gamma^{\uparrow}_{\rm D}$, and the width that accounts for the decay of the doorway into the more complex CN configurations, the so-called *spreading* or *damping width*, $\Gamma^{\downarrow}_{\rm D}$.

For an analysis of the doorway dominated S-matrix, we introduce three projection operators: (*i*) the open channel projector, P, (*ii*) the doorway projector, D, and (*iii*) the compound nucleus projector, Q. These projectors have the usual properties of projection operators,

$$P + D + Q = 1, \quad P^2 = P, \quad Q^2 = Q, \quad D^2 = D. \tag{12.1}$$

and

$$PQ = QP = DP = PD = QD = DQ = 0. \tag{12.2}$$

Applying the projectors on the exact many-body Schrödinger equation, one gets the coupled equations[2],

$$[E - \mathbb{H}_{\rm PP}]\,|\Psi^{(+)}_{\rm P}\rangle = \mathbb{H}_{\rm PD}\;|\Psi_{\rm D}\rangle\; + \mathbb{H}_{\rm PQ}\;|\Psi_{\rm Q}\rangle\,, \tag{12.3}$$

$$[E - \mathbb{H}_{\rm DD}]\,|\Psi_{\rm D}\rangle = \mathbb{H}_{\rm DP}\;|\Psi^{(+)}_{\rm P}\rangle + \mathbb{H}_{\rm DQ}\;|\Psi_{\rm Q}\rangle\,, \tag{12.4}$$

$$[E - \mathbb{H}_{\rm QQ}]\,|\Psi_{\rm Q}\rangle = \mathbb{V}_{\rm QP}\;|\Psi^{(+)}_{\rm P}\rangle + \mathbb{H}_{\rm QD}\;|\Psi_{\rm D}\rangle\,. \tag{12.5}$$

Above and throughout this section we use the notation of section 10.1,

$$\mathbb{H}_{\rm AB} \equiv A\,\mathbb{H}\,B \quad \text{and} \quad |\Psi_{\rm A}\rangle \equiv A\,|\Psi\rangle\,, \tag{12.6}$$

with A and B standing for the any of the projectors P, D or Q.

[1]Our discussion of doorway states is based on [Feshbach (1992)].

[2]Since we are assuming that the subspaces D and Q contain only closed channels, their wave functions do not have the superscript $^{(+)}$.

The formal solution of the above three coupled equations, which is a bit cumbersome, will be postponed to a later stage of this section. First, we discuss the extreme doorway hypothesis, where there is no direct coupling between the P and the Q subspaces. In this case, the formation and decay of the compound nucleus take place through the doorway. This is achieved by setting,

$$\mathbb{H}_{\mathrm{PQ}} = \mathbb{H}_{\mathrm{QP}} = 0. \tag{12.7}$$

With the above, Eqs. (12.3) to (12.5) reduce to,

$$[E - \mathbb{H}_{\mathrm{PP}}]\,|\Psi_{\mathrm{P}}^{(+)}\rangle = \mathbb{H}_{\mathrm{PD}}\,|\Psi_{\mathrm{D}}\rangle\,, \tag{12.8}$$

$$[E - \mathbb{H}_{\mathrm{DD}}]\,|\Psi_{\mathrm{D}}\rangle = \mathbb{H}_{\mathrm{DP}}\,|\Psi_{\mathrm{P}}^{(+)}\rangle + \mathbb{H}_{\mathrm{DQ}}\,|\Psi_{\mathrm{Q}}\rangle\,, \tag{12.9}$$

$$[E - \mathbb{H}_{\mathrm{QQ}}]\,|\Psi_{\mathrm{Q}}\rangle = \mathbb{H}_{\mathrm{QD}}\,|\Psi_{\mathrm{D}}\rangle\,. \tag{12.10}$$

Since the Q-subspace couples exclusively to the D-subspace, Eq. (12.10) can be taken into account through the inclusion of an effective potential in Eq. (12.9), reducing the set to two coupled equations. Eq. (12.8) remains the same whereas Eq. (12.9) becomes

$$[E - \mathbb{H}_{\mathrm{DD}} - \mathcal{V}_{\mathrm{D}}^{(\mathrm{Q})}]\,|\Psi_{\mathrm{D}}\rangle = \mathbb{H}_{\mathrm{DP}}\,|\Psi_{\mathrm{P}}^{(+)}\rangle\,. \tag{12.11}$$

Above,

$$\mathcal{V}_{\mathrm{D}}^{(\mathrm{Q})} = \mathbb{H}_{\mathrm{DQ}}\,\mathbb{G}_{\mathrm{Q}}\,\mathbb{V}_{\mathrm{QP}} \tag{12.12}$$

is the contribution from the CN (Q-subspace) to the polarization potential acting on the D-subspace and $\mathbb{G}_{\mathrm{Q}}$ is the propagator in the Q-subspace[3],

$$\mathbb{G}_{\mathrm{Q}} = \frac{1}{E - \mathbb{H}_{\mathrm{QQ}}}. \tag{12.13}$$

For the effective doorway Hamiltonian $\mathbb{H}_{\mathrm{DD}} + \mathcal{V}_{\mathrm{D}}^{(\mathrm{Q})}$ to make any sense, we have to average out the poles in $\mathbb{G}_{\mathrm{Q}}$, as we did in section 10.3. This is accomplished by the energy averaging, which replaces this propagator by its average,

$$\mathbb{G}_{\mathrm{Q}} \rightarrow \overline{\mathbb{G}_{\mathrm{Q}}} = \frac{1}{E - \mathbb{H}_{\mathrm{QQ}} + i\,\frac{I}{2}}. \tag{12.14}$$

Using this propagator, one gets the energy averaged polarization potential,

$$\overline{\mathcal{V}_{\mathrm{D}}^{(\mathrm{Q})}} = \mathbb{H}_{\mathrm{DQ}}\,\overline{\mathbb{G}_{\mathrm{Q}}}\,\mathbb{H}_{\mathrm{QP}}. \tag{12.15}$$

[3]Since the partition Q contains only closed channels, it is not necessary to include the term $i\epsilon$ in the denominator of the corresponding Green's function.

In this way, Eqs. (12.9) and (12.11) become the set of coupled equations for the energy averaged wave functions (similarly to section 10.3),

$$\Big[E - \mathbb{H}_{\rm PP}\Big]\big|\overline{\Psi_{\rm P}^{(+)}}\big\rangle = \mathbb{H}_{\rm PD}\,\big|\overline{\Psi_{\rm D}}\big\rangle, \tag{12.16}$$

$$\Big[E - \mathbb{H}_{\rm DD} - \overline{\mathcal{V}_{\rm D}^{(\rm Q)}}\Big]\,\big|\overline{\Psi_{\rm D}}\big\rangle = \mathbb{H}_{\rm DP}\,\big|\overline{\Psi_{\rm P}^{(+)}}\big\rangle. \tag{12.17}$$

The formal solution of Eq. (12.16) is

$$\big|\,\overline{\Psi_{\rm P}^{(+)}}\,\big\rangle = \big|X_{\rm P}^{(+)}\big\rangle \;+\; \mathbb{G}_{\rm P}^{(+)}\,\mathbb{H}_{\rm PD}\,\big|\overline{\Psi_{\rm D}}\big\rangle. \tag{12.18}$$

Above,

$$\mathbb{G}_{\rm P}^{(+)} = \frac{1}{E - \mathbb{H}_{\rm PP} + i\epsilon} \tag{12.19}$$

is the propagator in the P-subspace and $X_{\rm P}^{(+)}$ is the scattering solution when the coupling with the D-subspace is switched off. It satisfies the equation

$$\Big[E - \mathbb{H}_{\rm PP}\Big]\big|X_{\rm P}^{(+)}\big\rangle = 0. \tag{12.20}$$

If the channels in the P-subspace are directly coupled among themselves, Eq. (12.20) corresponds to a set of coupled equations.

Inserting Eq. (12.18) into Eq. (12.17) and re-ordering the terms we get,

$$\Big[E - \mathbb{H}_{\rm DD} - \overline{\mathcal{V}_{\rm D}}\;\Big]\Big|\overline{\Psi_{\rm D}}\big\rangle = \mathbb{H}_{\rm DP}\,\big|X_{\rm P}^{(+)}\big\rangle, \tag{12.21}$$

where

$$\overline{\mathcal{V}_{\rm D}} = \overline{\mathcal{V}_{\rm D}^{(\rm Q)}} + \overline{\mathcal{V}_{\rm D}^{(\rm P)}} \tag{12.22}$$

is the total energy averaged polarization potential acting on the doorway space. Above, $\overline{\mathcal{V}_{\rm D}^{(\rm Q)}}$ is defined in Eq. (12.15) and $\overline{\mathcal{V}_{\rm D}^{(\rm P)}}$ is the polarization potential in the D-space arising from couplings with the open channels,

$$\overline{\mathcal{V}_{\rm D}^{(\rm P)}} = \mathbb{H}_{\rm DP}\,\mathbb{G}_{\rm P}^{(+)}\;\mathbb{H}_{\rm PD}. \tag{12.23}$$

The formal solution of Eq. (12.21) is

$$\big|\overline{\Psi_{\rm D}}\big\rangle = \overline{\mathcal{G}_{\rm D}}\;\mathbb{H}_{\rm DP}\;\big|X_{\rm P}^{(+)}\big\rangle\,, \tag{12.24}$$

where

$$\overline{\mathcal{G}_{\rm D}} = \frac{1}{E - \overline{\mathcal{H}_{\rm D}}} \tag{12.25}$$

is the Green's function associated with the effective energy averaged Hamiltonian in the D-subspace,

$$\overline{\mathcal{H}_{\rm D}} = \mathbb{H}_{\rm DD}\;+\overline{\mathcal{V}_{\rm D}^{(\rm Q)}} + \overline{\mathcal{V}_{\rm D}^{(\rm P)}}. \tag{12.26}$$

Now we evaluate the T-matrix between two states in the P-subspace. For this purpose, we use the two-potential formula for many-body scattering, discussed in section 9.2.1 (see Eq. (9.72)). We split the Hamiltonian in the P' space as

$$\mathbb{H}_{\mathrm{P'P'}} \equiv (P+D)\,\mathbb{H}\,(P+D) = \mathbb{H}_{\mathrm{PP}} + \mathbb{H}_{\mathrm{DD}} + \mathbb{H}_{\mathrm{PD}} + \mathbb{H}_{\mathrm{DP}} \tag{12.27}$$

and take $\mathbb{H}^{(1)} = \mathbb{H}_{\mathrm{PP}} + \mathbb{H}_{\mathrm{DD}}$ as the dominant Hamiltonian and $\mathbb{V}^{(2)} = \mathbb{H}_{\mathrm{PD}} + \mathbb{H}_{\mathrm{DP}}$ as the weaker part of the interaction. The T-matrix connecting the free initial state $\Phi\,(0\mathbf{k}_0)$ with the final state $\Phi\,(\alpha\mathbf{k}_\alpha)$ can be written

$$T_{\alpha\mathbf{k}_\alpha;\,0\mathbf{k}_0} = T^{\mathrm{P}}_{\alpha\mathbf{k}_\alpha;\,0\mathbf{k}_0} + T^{\mathrm{D}}_{\alpha\mathbf{k}_\alpha;\,0\mathbf{k}_0}. \tag{12.28}$$

The first term is the T-matrix when the coupling between the P- and the D-subspaces is switched off. It corresponds to direct transitions among the open channel in P and is given by[4]

$$T^{\mathrm{P}}_{\alpha\mathbf{k}_\alpha;\,0\mathbf{k}_0} = \big\langle \Phi(\alpha\mathbf{k}_\alpha) \big| \mathbb{V}_{\mathrm{PP}} \big| X^{(+)}_{\mathrm{P}}\,(0\mathbf{k}_0) \,\big\rangle. \tag{12.29}$$

The second term, which corresponds to transitions passing through intermediate doorway states, is[5]

$$\begin{aligned} T^{\mathrm{D}}_{\alpha\mathbf{k}_\alpha;\,0\mathbf{k}_0} &= \Big\langle X^{(-)}_{\mathrm{P},\alpha} \Big|\; \mathbb{H}_{\mathrm{PD}} + \mathbb{H}_{\mathrm{DP}} \;\Big|\; \overline{\Psi}^{(+)}\,(0\mathbf{k}_0) \,\Big\rangle, \\ &= \Big\langle X^{(-)}_{\mathrm{P},\alpha} \Big|\; \mathbb{H}_{\mathrm{PD}} \;\Big|\; \overline{\Psi_{\mathrm{D}}}\,(0\mathbf{k}_0) \,\Big\rangle, \end{aligned} \tag{12.30}$$

where $X^{(-)}_{\mathrm{P},\alpha}$ is the solution of Eq. (12.20) with incoming wave boundary condition, projected onto channel-α. Applying the operator $\mathbb{H}_{\mathrm{PD}}$ on both sides of Eq. (12.24) and using the result above, we get

$$T^{\mathrm{D}}_{\alpha\mathbf{k}_\alpha;\,0\mathbf{k}_0} = \Big\langle X^{(-)}_{\mathrm{P},\alpha} \Big|\; \mathbb{H}_{\mathrm{PD}}\, \overline{\mathcal{G}_{\mathrm{D}}}\; \mathbb{H}_{\mathrm{DP}} \Big| X^{(+)}_{\mathrm{P}} \Big\rangle. \tag{12.31}$$

Simple expressions can be derived in the case of a single doorway, when there is no direct couplings among the channels contained in the P-subspace. First, Eq. (12.31) can be simplified if one uses the spectral representation of $\overline{\mathcal{G}_{\mathrm{D}}}$. It can be written

$$\overline{\mathcal{G}_{\mathrm{D}}} = \big|\, D \,\big\rangle\; \frac{1}{(E - E_{\mathrm{D}}) + i\,\Gamma_{\mathrm{D}}/2}\; \big\langle\, \tilde{D} \,\big|, \tag{12.32}$$

[4]In Eq. (9.73) it was assumed that the potential $\mathbb{V}^{(1)}$ is diagonal in channel space. Thus, the first term in the two-potential formula would contribute exclusively to elastic scattering. Here we do not make this assumption.

[5]The second line of the equation below results from two properties of the coupling potentials. First, the projector D in $\mathbb{H}_{\mathrm{DP}} (\equiv D\mathbb{H}P)$ acts on the final state to the left, which has no component in the D-subspace. Thus, it does not contribute to the matrix element. Second, the projector D in $\mathbb{H}_{\mathrm{PD}} (\equiv P\mathbb{H}D)$ acts on the full scattering state at its right, eliminating its component in the P-subspace.

were $|D\rangle$ and $|\tilde{D}\rangle$ are respectively the eigenstate of $\overline{\mathcal{H}_{\rm D}}$ and its dual state[6], and $E_{\rm D}$ and $\Gamma_{\rm D}/2$ are the energy and total width of the doorway resonance, given by

$$E_{\rm D} = {\rm Re}\Big\{\langle\tilde{D}|\ \mathbb{H}_{\rm DD} + \overline{\mathcal{V}_{\rm D}^{\rm (Q)}} + \overline{\mathcal{V}_{\rm D}^{\rm (P)}}\ |D\rangle\Big\} \tag{12.33}$$

and

$$\Gamma_{\rm D} = -2\ {\rm Im}\big\{\langle\tilde{D}|\ \overline{\mathcal{V}_{\rm D}^{\rm (Q)}} + \overline{\mathcal{V}_{\rm D}^{\rm (P)}}\ \big|D\rangle\big\}. \tag{12.34}$$

The total width corresponds to a sum of contributions from couplings with the Q and the P-subspaces. That is

$$\Gamma_{\rm D} = \Gamma_{\rm D}^{\downarrow} + \Gamma_{\rm D}^{\uparrow}, \tag{12.35}$$

where

$$\Gamma_{\rm D}^{\downarrow} = -2\ {\rm Im}\big\{\langle\tilde{D}|\ \overline{\mathcal{V}_{\rm D}^{\rm (Q)}}\ \big|D\big\rangle\big\} \tag{12.36}$$

is the spreading width (decay to the CN states) and

$$\Gamma_{\rm D}^{\uparrow} = -2\ {\rm Im}\Big\{\langle\tilde{D}\,|\ \overline{\mathcal{V}_{\rm D}^{\rm (P)}}\ |\,D\rangle\Big\} \tag{12.37}$$

is the escape width (decay to open channels).

It is convenient to write the escape width as a sum of partial widths for the decay of the doorway state to each open channel. For this purpose, we use the explicit form of $\overline{\mathcal{V}_{\rm D}^{\rm (P)}}$ (Eq. (12.23)) and write separately the on-shell and off-shell parts of the Green's function,

$$\mathbb{G}_{\rm P}^{(+)} = \mathcal{P}\left\{\frac{1}{E - \mathbb{H}_{\rm PP}}\right\} - i\pi\ \delta(E - \mathbb{H}_{\rm PP}).$$

The imaginary part in Eq. (12.37) comes from the on-shell Green's function and we can write

$$\Gamma_{\rm D}^{\uparrow} = 2\pi\ \langle\tilde{D}\,|\ \mathbb{H}_{\rm DP}\ \delta(E - \mathbb{H}_{\rm PP})\ \mathbb{H}_{\rm PD}\ |\,D\rangle. \tag{12.38}$$

Inserting the identity

$$\mathbb{I} = \sum_{\alpha} \int d^3\mathbf{k}'\ \big|X_{{\rm P},\alpha}^{(+)}(\mathbf{k}')\big\rangle\ |\alpha)\ (\alpha|\ \big\langle X_{{\rm P},\alpha}^{(+)}(\mathbf{k}')\big|$$

between the operators above, we get

$$\Gamma_{\rm D}^{\uparrow} = \sum_{\alpha} \Gamma_{{\rm D},\alpha}^{\uparrow}, \tag{12.39}$$

[6]Since the effective Hamiltonian is complex, the spectral representation involves the eigestates and their dual counterparts (see chapter 6). In our notation, the latter is distinguished from the former by a tilde.

with[7]

$$\Gamma^{\uparrow}_{\rm D,\alpha} = 2\pi \int d^3\mathbf{k}' \left| \left\langle D \,\middle|\, \mathbb{H}_{\rm D\alpha} \,\middle|\, X^{(+)}_{\rm P,\alpha}(\mathbf{k}') \right\rangle \right|^2 \delta\left(E_k - E_{k'}\right). \tag{12.40}$$

The T-matrix of Eq. (12.31) can be written in terms of the doorway widths as

$$T^{\rm D}_{\alpha\mathbf{k}_\alpha;\,0\mathbf{k}_0} = \frac{\langle X^{(-)}_{\rm P,\alpha}\,(\alpha\mathbf{k}_\alpha) \,|\, \mathbb{H}_{\rm PD} |\, D\,\rangle \,\langle\, \tilde{D} \,|\, \mathbb{H}_{\rm DP}\, | X^{(+)}_{\rm P,0}\,(0\mathbf{k}_0)\,\rangle}{(E - E_{\rm D}) + i\,\Gamma_{\rm D}/2}. \tag{12.41}$$

The above equation is analogous to the one for long-lived resonances (Eq. (10.59)), derived in section 10.2.2. However, the decay widths involved are quite different. Here, they may be orders of magnitude larger.

Further simplifications can be made for time-reversal invariant spinless systems that conserves orbital angular momentum (see section 10.2.2). In this case, we can write

$$\langle X^{(-)}_{\rm P,\alpha}\,(\alpha\mathbf{k}_\alpha) \,|\, \mathbb{H}_{\rm PD} | D\,\rangle = \langle \tilde{D} \,|\, \mathbb{H}_{\rm DP}\, | X^{(+)}_{\rm P,\alpha}\,(0,-\mathbf{k}_0)\,\rangle^{\dagger}. \tag{12.42}$$

Performing partial-wave expansion as in section 4.6.1, one gets the partial wave-projected T-matrix[8],

$$T^{\rm D}_{\alpha l;0l} = \frac{1}{2\pi}\, e^{i\,(\delta_{\alpha l} + \delta_{0l})} \left[\frac{\sqrt{\Gamma^{\uparrow}_{\rm D,\alpha}(l)\;\Gamma^{\uparrow}_{\rm D,0}(l)}}{(E - E_{\rm D}) + i\,\Gamma_{\rm D}(l)/2}\right], \tag{12.43}$$

where δ_{0l} and $\delta_{\alpha l}$ are the phase shifts in the scattering within the P-subspace (couplings with D and Q switched off) at the $l^{\rm th}$ partial wave, for collisions initiated in the elastic channel and in channel α, respectively. This result is analogous to Eq. (10.68). The difference is that in the latter there is no doorway channel and the couplings are directly with the Q-subspace. For details, we refer to [Levin and Feshbach (1973)].

The doorway-dominated S-matrix element for elastic scattering, which is associated with $T_{0l;0l}$ $(= T^{\rm P}_{0l;0l} + T^{\rm D}_{0l;0l})$ is,

$$S_{00}(l) = S^{\rm P}_{00}(l) \left[1 - i\,\frac{\Gamma^{\uparrow}_{\rm D,0}(l)}{(E - E_{\rm D}) + i\,\Gamma_{\rm D}(l)/2}\right]. \tag{12.44}$$

[7]Below, we neglect the difference between the doorway state and its dual counterpart.

[8]When angular momentum projection is carried out, $\left\langle \overline{\Psi_{\rm D}} \right| \mathbb{H}_{\rm DP} \left| X^{(+)}_{\rm P,0}\,(0\mathbf{k}_0) \right\rangle$ gives rise to the radial integral

$$\int dr\, \varphi_{{\rm D}l}(r)\, V_{Dl;0l}\, w_{0l}(k_0, r),$$

where $\varphi_{{\rm D}l}(r)$ is the radial wave function of the doorway resonance, $V_{Dl;0l}$ is the angular momentum projected coupling potential and $w_{0l}(k_0, r)$ is the radial distorted wave associated with $X^{(+)}_{\rm P,0}\,(0\mathbf{k}_0)$. Since the doorway resonance is a closed channel and the Hamiltonian is hermitian, the first two factors in the integrand are real. The third factor can be written (see Eq. (2.42)) as $w_{0l}(k_0, r) = e^{i\,\delta_{0l}}\;\bar{w}_{0l}(k_0, r)$. For the real potential of the present case, $\bar{w}_{0l}(k_0, r)$ is a real function (see section 2.2). Thus, the above matrix element is given by a real integral multiplying the phase factor $e^{i\,\delta_{0l}}$.

Above, $S^{\text{P}}_{00}(l) = \exp[2i\,\delta_{0l}]$ is the partial-wave projected elastic S-matrix for the scattering within the P-subspace (couplings to D and Q switched off), which has modulus one.

Usually the doorway resonance affects one particular channel, say, the elastic channel. In such a case the escape width is just the decay width, $\Gamma^{\uparrow}_{\text{D}} = \Gamma^{\uparrow}_{\text{D},0}$. Further, the potential S-matrix is diagonal and unitary (considering one doorway or several isolated doorways), while the averaged, doorway-dominated S-matrix above is diagonal but not unitary, owing to the non-zero damping width. Therefore the transmission coefficient is,

$$\mathcal{T}_{00}(l) = 1 - |S_{00}(l)|^2\,. \tag{12.45}$$

Inserting Eq. (12.44) into Eq. (12.45) and splitting the total doorway width into its components, $\Gamma_{\text{D}} = \Gamma^{\downarrow}_{\text{D}} + \Gamma^{\uparrow}_{\text{D}}$, the above can be reduced to[9]

$$\mathcal{T}_{00}(l) = \frac{\Gamma^{\downarrow}_{\text{D}}\,\Gamma^{\uparrow}_{\text{D}}}{(E - E_{\text{D}})^2 + \Gamma^2_{\text{D}}/4}. \tag{12.46}$$

The transmission coefficient is symmetrical around $E = E_{\text{D}}$. This symmetry would be removed if the couplings $\mathbb{V}_{\text{PQ}}$ and $\mathbb{V}_{\text{QP}}$ were not zero. This happens in the case of an isobaric analog resonance [Toledo Piza and Kerman (1968)] excited in a (p, n) reaction. The S-matrix in this case can be written as,

$$S_{00}(l) = S^{\text{P}}_{00}(l)\,\left[1 - i\,\frac{e^{i\phi}\,\Gamma^{\uparrow}_{\text{D},0}(l)}{(E - E_{\text{D}}) + i\,\Gamma_{\text{D}}(l)/2}\right]. \tag{12.47}$$

Now, the situation is different: the modulus of $S^{\text{P}}_{00}(l)$ is smaller than one and there is an additional phase factor multiplying $\Gamma^{\uparrow}_{\text{D},0}(l)$. This factor comes from the fact that radial wave function $\bar{w}_{0l}(k_0, r)$ (see footnote on the previous page) is no longer real.

In more complicated situations, there are several doorway resonances and couplings of the P-subspace with both the D and the Q-subspaces. In such cases, it is convenient to eliminate the Q-subspace through the introduction of the energy averaged effective potential (see the development below),

$$\overline{\mathbb{V}^{(\text{Q})}} = \mathbb{V}\,Q\,\overline{G_{\text{Q}}}\,Q\,\mathbb{V}, \tag{12.48}$$

which carries the influence of the Q-subspace on the subspaces P and D. In Eq. (12.48), $\overline{G_{\text{Q}}}$ is the energy averaged propagator in the Q-space, defined in Eq. (12.13).

[9] Since $\mathbb{H}_{\text{PP}}$ is hermitian, $S^{\text{P}}_{00}(l)$ has modulus 1. Thus, it can be dropped in the calculation of the transmission factor.

To derive the effective equations in the subspaces P and Q, we use the energy averaged version of Eq. (12.5) to write,

$$\left|\overline{\Psi_{\rm Q}}\right\rangle = \overline{G_{\rm Q}}\; \mathbb{V}_{\rm QP} \left|\overline{\Psi_{\rm P}^{(+)}}\right\rangle + \overline{G_{\rm Q}}\; \mathbb{V}_{\rm QD} \left|\overline{\Psi_{\rm D}}\right\rangle, \tag{12.49}$$

and use this equation in Eqs. (12.3) and (12.4). Reordering the terms conveniently, the resulting coupled equations can be put in the form

$$\left[E - \mathcal{H}_{\rm P}^{\rm opt}\right] \left|\overline{\Psi_{\rm P}^{(+)}}\right\rangle = \mathcal{V}_{\rm PD} \left|\overline{\Psi_{\rm D}}\right\rangle, \tag{12.50}$$

$$\left[E - \mathcal{H}_{\rm D}^{\rm opt}\right] \left|\overline{\Psi_{\rm D}}\right\rangle = \mathcal{V}_{\rm DP} \left|\overline{\Psi_{\rm P}^{(+)}}\right\rangle. \tag{12.51}$$

Above, the effective Hamiltonians in subspaces P and D are,

$$\mathcal{H}_{\rm P}^{\rm opt} = K + h + \mathcal{V}_{\rm P}^{\rm opt} \quad \text{and} \quad \mathcal{H}_{\rm D}^{\rm opt} = K + h + \mathcal{V}_{\rm D}^{\rm opt},$$

with the optical potentials,

$$\mathcal{V}_{\rm P}^{\rm opt} \equiv P\,\left[\mathbb{V} + \overline{\mathbb{V}^{(\rm Q)}}\right]\, P = \mathbb{V}_{\rm PP} + \mathbb{V}_{\rm PQ}\, \overline{G_Q}\, \mathbb{V}_{\rm QP}, \tag{12.52}$$

$$\mathcal{V}_{\rm D}^{\rm opt} \equiv D\,\left[\mathbb{V} + \overline{\mathbb{V}^{(\rm Q)}}\right]\, D = \mathbb{V}_{\rm DD} + \mathbb{V}_{\rm DQ}\, \overline{G_Q}\, \mathbb{V}_{\rm QD}. \tag{12.53}$$

On the RHS of the above equations we identify two terms. The first corresponds to the channel potential plus couplings among the channels within the same subspace. The second accounts for the average influence of couplings with Q. In general, these potentials are complex (see chapter 10).

In Eqs. (12.50) and (12.51), $\mathcal{V}_{\rm PD}$ and $\mathcal{V}_{\rm DP}$ are the total couplings among channels in different subspaces. They are given by

$$\mathcal{V}_{\rm PD} = \mathbb{V}_{\rm PD} + P\, \overline{\mathbb{V}^{(\rm Q)}}\, D \tag{12.54}$$

$$\mathcal{V}_{\rm DP} = \mathbb{V}_{\rm DP} + D\, \overline{\mathbb{V}^{(\rm Q)}}\, P. \tag{12.55}$$

These potentials contain two terms. The first corresponds to all direct couplings of one channel in P with one channel in D. They represent transitions with a time-scale of the order of the doorway life time. On the other hand, the second term represents time delayed processes. In Eq. (12.54), it gives a transition from P to the Q, followed by propagation in Q, and the transition from Q to D. In Eq. (12.55), it gives a transition from D to the Q, followed by propagation in Q, and the transition from Q to P.

If the scattering process is not affected by doorways states, one drops the coupling potential $\mathcal{V}_{\rm PD}$ in Eq. (12.50). In this case, one is left with an equation involving channels in the P-subspace and the influence of the Q-subspace in taken into account through the inclusion of an energy averaged polarization potential (the optical potential), as discussed in section 10.3. The wave function, $X^{(+)}_{\text{opt}}$, then satisfies the decoupled equation,

$$\left[E - \mathcal{H}^{\text{opt}}_{\text{P}}\right] \left|X^{(+)}_{\text{opt}}(0, \mathbf{k}_0)\right\rangle = 0. \tag{12.56}$$

In general, the above is in fact a set of coupled equations within the P-subspace. Making the channel expansion,

$$\left|X^{(+)}_{\text{opt}}(0\mathbf{k}_0)\right\rangle = \sum_{\alpha'} \left|\chi^{(+)}_{\alpha'}(0\mathbf{k}_0)\right\rangle \left|\alpha'\right), \tag{12.57}$$

one gets the coupled equations,

$$\left[E_\alpha - K - \mathcal{V}^{\text{opt}}_{\alpha}\right] \left|\chi^{(+)}_{\alpha}(0\mathbf{k}_0)\right\rangle = \sum_{\alpha'} \mathcal{V}^{\text{opt}}_{\alpha,\alpha'} \left|\chi^{(+)}_{\alpha'}(0\mathbf{k}_0)\right\rangle, \tag{12.58}$$

for $\alpha, \alpha' = 0, 1, ..., N-1$, with the notations,

$$\mathcal{V}^{\text{opt}}_{\alpha,\alpha'} = \left(\alpha\right| \mathcal{V}^{\text{opt}}_{\text{P}} \left|\alpha'\right) \quad \text{and} \quad \mathcal{V}^{\text{opt}}_{\alpha} = \mathcal{V}^{\text{opt}}_{\alpha,\alpha}.$$

The corresponding T-matrix, which represents prompt transitions, are given by

$$T^{\text{P}}_{\alpha\mathbf{k}_\alpha,0\mathbf{k}_0} = \left\langle\Phi(\alpha\mathbf{k}_\alpha)\right| \mathcal{V}^{\text{opt}}_{\text{P}} \left|X^{(+)}_{\text{opt}}(0\mathbf{k}_0)\right\rangle \tag{12.59}$$

$$= \sum_{\alpha'} \left\langle\phi(\mathbf{k}_\alpha)\right| \mathcal{V}^{\text{opt}}_{\alpha,\alpha'} \left|\chi^{(+)}_{\alpha'}(0\mathbf{k}_0)\right\rangle. \tag{12.60}$$

Now we consider the more complicated situation where the collision dynamics is influenced by doorway states. Now we have to deal with the original equations (Eqs. (12.50) and (12.51)). For this purpose, we use the Gell-Mann & Goldberger relation. The potential is written as

$$\mathbb{V} = \mathbb{V}^{(1)} + \mathbb{V}^{(2)},$$

where

$$\mathbb{V}^{(1)} = \mathcal{V}^{\text{opt}}_{\text{P}} \tag{12.61}$$

is the dominant potential and

$$\mathbb{V}^{(2)} = \mathcal{V}_{\text{PD}} + \mathcal{V}_{\text{DP}} \tag{12.62}$$

is the complementary one, and use the Gell-Mann & Goldberger relation. The T-matrix is then given by

$$T_{\alpha\mathbf{k}_\alpha,0\mathbf{k}_0} = T^{\text{P}}_{\alpha\mathbf{k}_\alpha,0\mathbf{k}_0} + T^{\text{D}}_{\alpha\mathbf{k}_\alpha,0\mathbf{k}_0}, \tag{12.63}$$

where $T^{\text{P}}_{\alpha\mathbf{k}_\alpha,0\mathbf{k}_0}$ is the contribution from promp processes, given by Eq. (12.60), and $T^{\text{D}}_{\alpha\mathbf{k}_\alpha,0\mathbf{k}_0}$ represent transitions induced by the potential $\mathbb{V}^{(2)}$ in the space of waves distorted by $\mathbb{V}^{(1)}$. It is formally given by

$$T^{\text{D}}_{\alpha\mathbf{k}_\alpha,0\mathbf{k}_0} = \left\langle X^{(-)}_{\text{opt}}(\alpha\mathbf{k}_\alpha)\right| \mathcal{V}_{\text{PD}} + \mathcal{V}_{\text{DP}} \left|\overline{\Psi^{(+)}}(0, \mathbf{k}_0)\right\rangle.$$

First, we point out that the contribution from $\mathcal{V}_{\text{DP}}$ vanishes because it project on the left onto the subspace D and the final state is a channel in the P-subspace. On the other hand, $\mathcal{V}_{\text{DP}}$ eliminates the component of $\left|\overline{\Psi^{(+)}}(0\mathbf{k}_0)\right\rangle$ in the P-subspace. The above equation then becomes

$$T^{\text{D}}_{\alpha\mathbf{k}_\alpha,0\mathbf{k}_0} = \left\langle X^{(-)}_{\text{opt}}(\alpha\mathbf{k}_\alpha)\right| \mathcal{V}_{\text{PD}} \left|\overline{\Psi_{\text{D}}}(0\mathbf{k}_0)\right\rangle. \tag{12.64}$$

However, the above expression depends on the component of the exact wave function in the doorway subspace, $|\overline{\Psi_{\text{D}}}(0\mathbf{k}_0)\rangle$, which is not known. To cope with this situation, we resort to the LS equations corresponding to Eq. (12.50),

$$|\overline{\Psi^{(+)}}(0\mathbf{k}_0)\rangle = |\overline{X^{(+)}}(0\mathbf{k}_0)\rangle + G^{(+)}_{\text{opt}}\, \mathcal{V}_{\text{PD}}\, |\overline{\Psi_{\text{D}}}(0\mathbf{k}_0)\rangle\,, \tag{12.65}$$

where the optical Green's function in the P-subspace is given by

$$G^{(+)}_{\text{opt}} = \frac{1}{E - \mathcal{H}^{\text{opt}}_{\text{P}} + i\epsilon}. \tag{12.66}$$

Inserting Eq. (12.65) into Eq. (12.51) and reordering the terms, we get

$$\left[E - h - K - \overline{\mathcal{V}^{\text{tot}}_{\text{D}}}\right] |\overline{\Psi_{\text{D}}}(0,\mathbf{k}_0)\rangle = \mathcal{V}_{\text{DP}}\, |X^{(+)}_{\text{opt}}(0,\mathbf{k}_0)\rangle, \tag{12.67}$$

where $\overline{\mathcal{V}^{\text{tot}}_{\text{D}}}$ is the total potential in the Schrödinger equation for the wave function in the doorway space,

$$\begin{aligned}\overline{\mathcal{V}^{\text{tot}}_{\text{D}}} &= \mathcal{V}^{\text{opt}}_{\text{D}} + \mathcal{V}_{\text{DP}}\, G^{(+)}_{\text{opt}}\, \mathcal{V}_{\text{PD}} \\ &= \mathbb{V}_{\text{DD}} + \mathbb{V}_{\text{DQ}}\, \overline{G_Q}\, \mathbb{V}_{\text{QD}} + \mathcal{V}_{\text{DP}}\, G^{(+)}_{\text{opt}}\, \mathcal{V}_{\text{PD}}.\end{aligned} \tag{12.68}$$

Solving Eq. (12.67) for $|\overline{\Psi_{\text{D}}}(0,\mathbf{k}_0)\rangle$ we obtain,

$$|\overline{\Psi_{\text{D}}}(0,\mathbf{k}_0)\rangle = \overline{\mathcal{G}^{\text{full}}_{\text{D}}}\;\; \mathcal{V}_{\text{DP}}\, |X^{(+)}_{\text{opt}}(0,\mathbf{k}_0)\rangle, \tag{12.69}$$

with the full Green's function in the D-subspace given by

$$\overline{\mathcal{G}^{\text{full}}_{\text{D}}} = \frac{1}{E - h - K - \overline{\mathcal{V}^{\text{tot}}_{\text{D}}} + i\epsilon} \tag{12.70}$$

Now we evaluate $T^{\text{D}}_{\alpha\mathbf{k}_\alpha,0\mathbf{k}_0}$ using the above equation. We get

$$T^{\text{D}}_{\alpha\mathbf{k}_\alpha,0\mathbf{k}_0} = \left\langle X^{(-)}_{\text{opt}}(\alpha\mathbf{k}_\alpha)\right| \mathcal{V}_{\text{PD}}\;\, \overline{\mathcal{G}^{\text{full}}_{\text{D}}}\;\; \mathcal{V}_{\text{DP}} \left|X^{(+)}_{\text{opt}}(0,\mathbf{k}_0)\right\rangle. \tag{12.71}$$

Eq. (12.71) takes a simpler form in the case of M non-overlapping doorway resonances. In this case, we use the spectral representation of the full Green's function in the doorway space,

$$\overline{\mathcal{G}^{\text{full}}_{\text{D}}} = \sum_{m=1}^{M} |D_m\rangle\, \frac{1}{(E - E_{\text{D}_m}) + i\,\Gamma_{\text{D}_m}/2}\, \langle\tilde{D}_m| \tag{12.72}$$

and the T-matrix becomes

$$T^{\mathrm{D}}_{\alpha\mathbf{k}_\alpha,0\mathbf{k}_0} = \sum_{m=1}^{M} \langle X^{(-)}_{\mathrm{opt}}(\alpha\mathbf{k}_\alpha)\big|\, \mathcal{V}_{\mathrm{PD}} \,\big|\, D_m\rangle \frac{1}{(E - E_{\mathrm{D}_m}) + i\,\Gamma_{\mathrm{D}_m}/2}$$
$$\times \langle \tilde{D}_m \big|\, \mathcal{V}_{\mathrm{DP}} \,\big| X^{(+)}_{\mathrm{opt}}(0\mathbf{k}_0)\rangle. \tag{12.73}$$

Using the expansion of Eq. (12.57) for the distorted waves, the above equation takes the form

$$T^{\mathrm{D}}_{\alpha\mathbf{k}_\alpha,0\mathbf{k}_0} = \sum_{m,\beta,\beta'} \frac{g_{\beta,m}\;\tilde{g}_{m,\beta'}}{(E - E_{\mathrm{D}_m}) + i\,\Gamma_{\mathrm{D}_m}/2}, \tag{12.74}$$

where

$$g_{\beta,m} = \left\langle X^{(-)}_{\mathrm{opt},\beta}(\alpha\mathbf{k}_\alpha) \,\middle|\, \mathcal{V}_{\mathrm{PD}} \,\middle|\, D_m \right\rangle \tag{12.75}$$

and

$$\tilde{g}_{m,\beta'} = \left\langle \tilde{D}_m \,\middle|\, \mathcal{V}_{\mathrm{DP}} \,\middle| X^{(+)}_{\mathrm{opt},\beta'}(0\mathbf{k}_0) \right\rangle. \tag{12.76}$$

The complex form factors introduced above can be rewritten as a product of amplitude and a phase, just as in Eq. (10.85). However, here, the phase is complex and is not just the square root of the optical S-matrix.

The S-matrix associated to the T-matrix of Eq. (12.74) can be derived through the standard relation between the S and the T matrices. The elastic S-matrix is given by

$$S_{00} = S^{\mathrm{P}}_{00}\left[1 - i\sum_m e^{i\gamma_m}\, \frac{\Gamma^{\uparrow}_{\mathrm{D}_m,0}}{(E - E_{\mathrm{D}_m}) + i\,(\Gamma^{\uparrow}_{\mathrm{D}_m} + \Gamma^{\downarrow}_{\mathrm{D}_m})/2}\right] \tag{12.77}$$

where γ_m is the phase of the product $(g_{0,m}\,\tilde{g}_{m,0})$ and $\Gamma^{\uparrow}_{\mathrm{D}_m,0}$ is the partial width of the the doorway $|D_m\rangle$,

$$\Gamma^{\uparrow}_{\mathrm{D}_m,0} = 2\pi\,|g_{m,0}|^2, \tag{12.78}$$

with

$$\Gamma^{\uparrow}_{\mathrm{D}_m} = \sum_\alpha \Gamma^{\uparrow}_{\mathrm{D}_m,\alpha}. \tag{12.79}$$

Doorway resonances are important effects in nuclear reactions. One typical example is shown in figure 12.1. It is possible to excite a doorway resonance in the target or the projectile. In such a case, one speaks of an exit doorway. They are quite important both in nuclear physics and in atom clusters physics. In the following we discuss the reaction theory of the exit doorway.

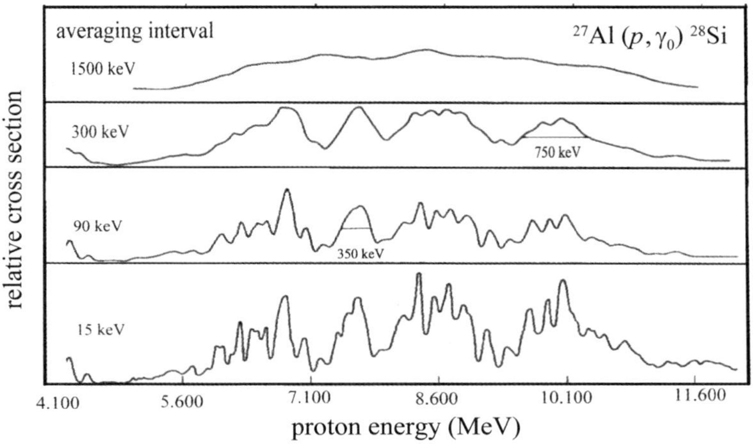

Fig. 12.1 An example of doorway resonances in the system $^{27}\mathrm{Al}(p,\gamma_0)^{28}\mathrm{Si}$. It corresponds to the results for some averaging intervals in figure 12 of [Singh *et al.* (1965)]. Notice that these resonances are directly related to the averaging interval employed. Wider doorways show up when use is made of a wider energy interval. One sees that at an energy interval of 300 keV, doorways with a total width of 750 keV are observed, while at an energy interval of 90 keV, doorways with half of the width are observed. The typical compound resonances in this system have a width of the order of 65 keV. For details see [Singh *et al.* (1965)].

12.2 Feshbach resonances in atom-atom collisions

In atomic gases at very low temperatures, where a great amount of experimental and theoretical work has been done in the last two decades in connection with Bose-Einstein condensation, the formation and manipulation of a compound atomic system is of a paramount importance [Inouye *et al.* (1998); Timmermans *et al.* (1999); Chin *et al.* (2010)]. The Feshbach, projection operator theory developed for the nuclear system is also applicable in this case. Here, the projection operator P projects onto two atoms in their ground states, while Q projects onto a quasi-molecular configuration. The energy of the system is very close to zero and an effective range expansion approach to the scattering amplitude is usually employed. As such, we have here a resonant scattering length and effective range. To see how this comes about, we have to indulge into the particularities of atom-atom interaction.

As we have seen in chapter 10, the long-range part of the atom-atom interaction is dominated by the dispersive Van der Waals force. This force supplies but a background potential contribution to the scattering. The interaction responsible for the coupling between the P- and Q-subspaces is the hyperfine electron-nuclear spin interaction. For the binary atom system, this interaction is,

$$V_{\mathrm{hf}} = \frac{\alpha_{\mathrm{hf}}}{\hbar^2}\left[\mathbf{i}_1 \cdot \mathbf{s}_1 + \mathbf{i}_2 \cdot \mathbf{s}_2\right], \tag{12.80}$$

where $\mathbf{i}$ and $\mathbf{s}$, are the nuclear and electronic spin vector operator, respectively, and α_{hf} is the hyperfine interaction strength. This interaction can be decomposed into a term that is diagonal in the P- and in the Q-subspaces, and a PQ coupling term. This is so, since we can write,

$$V_{\mathrm{hf}} = \frac{\alpha_{\mathrm{hf}}}{2\hbar^2}\left[\mathbf{I}_+ \cdot \mathbf{S}_+ + \mathbf{I}_- \cdot \mathbf{S}_-\right]. \tag{12.81}$$

Above, $\mathbf{I}_+ = \mathbf{i}_1 + \mathbf{i}_2$, $\mathbf{S}_+ = \mathbf{s}_1 + \mathbf{s}_2$ are the total nuclear and electronic spins, while $\mathbf{I}_- = \mathbf{i}_1 - \mathbf{i}_2$ and $\mathbf{S}_- = \mathbf{s}_1 - \mathbf{s}_2$ are the differences of the nuclear and electronic spins of the two atoms. The last term in Eq. (12.81) is responsible for the couplings between the continuum scattering states of the two atoms, which are contained in the P-subspace, and the closed quasi-molecular state contained in the Q-subspace [Timmermans *et al.* (1999)].

The solution of the scattering problem involving the effect of the closed channel quasi-molecular Q-subspace on the elastic channel contained in the P-subspace is already done in Chapter 10. We only give the final form of the S-matrix (Eq. (10.85)),

$$S = e^{2i\delta} = S^{(0)}\left[1 - \frac{\Gamma_r}{E - E_r - \Delta_r + i\Gamma_r/2}\right], \tag{12.82}$$

with, E_r, Δ_r and Γ_r representing respectively the energy, the shift, and width of the resonance, and $S^{(0)}$ being the background S-matrix arising from the scattering in the absence of the resonance.

At very low energies, only the s-wave contributes for the atom-atom scattering amplitude and we can write

$$f = \frac{1}{2ik}\left[S - 1\right] = \frac{1}{2ik}\left[e^{2i\delta} - 1\right]. \tag{12.83}$$

It is convenient to write the s-wave phase shift, δ, as (see Eqs. (10.93) - (10.95))

$$\delta = \delta^{(0)} + \delta^{(\mathrm{r})}, \tag{12.84}$$

where $\delta^{(0)}$ is a background phase shift and $\delta^{(\mathrm{r})}$ is the resonant term,

$$\delta^{(\mathrm{r})} = \tan^{-1}\left[\frac{\Gamma_r/2}{E_r + \Delta_r - E}\right]. \tag{12.85}$$

Inserting the S-matrix of Eq. (12.82) into Eq. (12.83), one gets the scattering amplitude,

$$f = \frac{1}{2ik}\left[e^{2i\,\delta^{(0)}}\,\frac{(E-\Delta_r-E_r)-i\Gamma_r/2}{(E-\Delta_r-E_r)+i\Gamma_r/2}\;-\;1\right]. \tag{12.86}$$

As we have shown in section 10.2.2, the resonance width, Γ_r, and the energy shift, Δ_r, arise respectively from the on-shell and from the off-shell parts of the Green's function in the P-subspace, and they are given by Eqs. (10.52) and (10.53). In the case of an isolated Feshbach Resonance, the width can be written[10],

$$\Gamma_r = 2\pi\int d^3\mathbf{k}'\;\left|\left\langle X_{\mathrm{Q}}\,|\,\mathbb{V}_{\mathrm{QP}}\,|\,X_{\mathrm{P}}^{(+)}(\mathbf{k}')\right\rangle\right|^2\;\delta(E_k-E_k'). \tag{12.87}$$

In Eq. (12.87), the wave function X_{Q} is the closed channel of a quasi-bound two atoms contained in the Q-space of the two atoms, and $X_{\mathrm{P}}(k')$ is the continuum two atoms state (both in their respective ground sate) contained in the P-subspace.

At very low energies, the matrix-element in the integrand does not depend on the orientation. Thus, performing the angular integration one gets[11]

$$\Gamma_r = \frac{8\pi^2\mu k}{\hbar^2}\;\left|\left\langle X_{\mathrm{Q}}\,|\,\mathbb{V}_{\mathrm{QP}}\,|\,X_{\mathrm{P}}(\mathbf{k})\right\rangle\right|^2,$$

or,

$$\Gamma_r = 2k\,\gamma. \tag{12.88}$$

Above, we have introduced the reduced width, γ, which is the slowly varying function of k,

$$\gamma = \frac{4\pi^2\mu}{\hbar^2}\;\left|\left\langle X_{\mathrm{Q}}\,|\,\mathbb{V}_{\mathrm{QP}}\,|\,X_{\mathrm{P}}(k)\right\rangle\right|^2. \tag{12.89}$$

The above behavior of the width of a resonance involving neutral particles, namely $\Gamma_r = 2k\gamma$, with γ being a slowly varying function of energy, is used extensively in neutron scattering from nuclei at low energies [Feshbach (1992)].

Frequently, low energy collisions are discussed in terms of the scattering length, defined as (see chapter 2),

$$a = -\lim_{k\to 0}\{f\}.$$

[10] As usual, we assume that the distorted waves are normalized according to the relation $\langle X^{(+)}(\mathbf{k})|X^{(+)}(\mathbf{k}')\rangle = \delta(\mathbf{k}-\mathbf{k}')$.

[11] The factor multiplying the modulus square of the matrix-element comes from the angular integration (4π) and from the conversion of the delta function in energy to a delta function in the wave number($\delta(E-E') = (k\mu/\hbar^2)\,\delta(k-k')$).

To evaluate the scattering length, we use Eq. (12.86). Since $\delta^{(0)}$ and Γ_r go to zero in the $k \to 0$ limit[12], we expand the function within square backers in Eq. (12.86) to first order in k. It is easy to show that the result can be put in the form

$$a = a^{(0)} - \frac{\gamma}{\varepsilon}, \tag{12.90}$$

where

$$a^{(0)} = - \lim_{k \to 0} \left\{ \frac{\delta^{(0)}}{k} \right\} \tag{12.91}$$

is the scattering length associated with the background scattering amplitude and

$$\varepsilon \equiv E_r + \Delta_r \tag{12.92}$$

is the so-called detuning parameter of the Feshbach resonance. This parameter corresponds to the energy of the quasi-bound two-particle system relative to the continuum level in the open elastic channel, whose threshold is $E = 0$. The modification of the effective scattering length which arises from the Feshbach resonance is contained in the reduced width, γ, and in the detuning parameter, ε. If the latter is made sufficiently small, the magnitude and sign of the effective scattering length suffer dramatic changes. Accordingly, the same happens with the atom-atom interaction in the dilute gas. The detuning parameter can be manipulated with the aid of the linear Zeeman effect through an applied magnetic field.

When a magnetic field, B, is applied to the system, the detuning parameter changes according to the equation,

$$\varepsilon = \varepsilon_0 + \Delta\mu\, B. \tag{12.93}$$

Above, ε_0 denotes the detuning parameter in the absence of the magnetic field and $\Delta\mu \equiv \mu^{(\mathrm{P})} - \mu^{(\mathrm{P})}$ is the difference between the values of the magnetic moments for the system in the configurations of two separated atoms and of the quasi-molecular state. Thus, one can measure the variation of the magnetic moment through the dependence of the detuning parameter on the applied magnetic field. Introducing the parameter,

$$B_0 = -\frac{\varepsilon_0}{\Delta\mu},$$

[12]The fact that Γ_r goes to zero in the $k \to 0$ limit is clear from Eq. 12.88), and the smooth behavior of the reduced width. Since we are assuming that the background potential does not support a bound state, the background phase-shift also goes to zero in this limit.

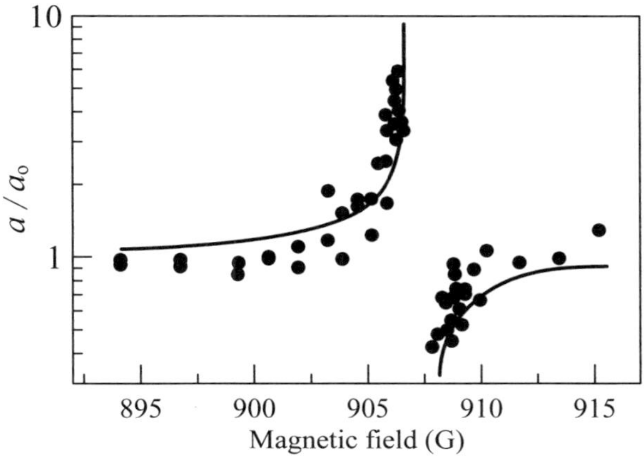

Fig. 12.2 Feshbach resonance in optically trapped Bose-Einstein condensates of Na atoms. The figure shows the scattering length normalized with respect to that of the background phase shift (Eq. (12.91)), as a function of the external magnetic field. The results are from the paper of Inouye *et al.* [Inouye *et al.* (1998)], where the details of the experiment can be found.

the B-dependent scattering length is given by the function,

$$a = a^{(0)} + \frac{\gamma/\Delta\mu}{B_0 - B} = a^{(0)}\left(1 + \frac{\Delta_{\rm B}}{B_0 - B}\right), \tag{12.94}$$

where we have introduced the width of the Feshbach resonance in terms of the magnetic field,

$$\Delta_{\rm B} = \frac{\gamma}{a^{(0)}\Delta\mu}. \tag{12.95}$$

In summary, in the context of cold atomic gases, the scattering length determines the strength of the effective, in-medium, atom-atom interaction. Therefore, by judiciously changing the value of ε, one can render the interaction amenable to manipulation; very large, very small, negative, or positive. This feature of the Feshbach resonance has proven to be of great importance in producing Bose-Einstein condensates of a variety of Bosonic atoms. In the case of Fermionic atoms, a condensate can be produced *only through the Feshbach resonance* which generates effective Bosonic entities.

An example of quasi-molecular Feshbach resonances is presented in figure 12.2. The strength of the magnetic field that leads to a vanishing detuning parameter is characterized by the discontinuity of the scattering length.

12.3 Statistical treatment of complex scattering: the compound nucleus

An important contribution to the nuclear reaction cross section at low energies comes from compound nucleus formation with subsequent decay. This cross section describes the process in which the projectile is absorbed by the target and the composite system lasts long enough time for the energy to be equilibrated. Eventually the compound nucleus decays, without any memory of how it was formed. In the following we give a detailed description of how such a process is described within the theoretical framework laid down by Feshbach. Before doing this we first give a heuristic derivation of the compound nucleus cross section, guided by the Bohr independence hypothesis which states that the formation and decay of the compound nucleus are two independent processes. This is in line with the statistical nature of the system. Using the Bohr hypothesis, we can write the partial wave cross section for the process[13] $\alpha \to \beta$ as [Tepel *et al.* (1973)],

$$\sigma_{\alpha\beta} = \sigma_0 \; \xi_\alpha \cdot \xi_\beta , \tag{12.96}$$

where σ_0 is the geometrical factor,

$$\sigma_0 = \frac{\pi}{k^2} \; (2l+1) \, . \tag{12.97}$$

Summing over β we get the transmission coefficient for the formation of the compound nucleus[14],

$$\mathcal{T}_\alpha = \frac{1}{\sigma_0} \sum_\beta \sigma_{\beta\alpha} . \tag{12.98}$$

Inserting the separable form of the cross section, we obtain

$$\mathcal{T}_\alpha = \xi_\alpha \sum_\beta \xi_\beta , \tag{12.99}$$

which can be solved for ξ in terms of $\mathcal{T}$,

$$\xi_\alpha = \mathcal{T}_\alpha \left(\sum_\gamma \mathcal{T}_\gamma \right)^{-1/2} . \tag{12.100}$$

[13]In this section it is convenient to consider collisions initiated in any channel ($\alpha = 0$ or not). Thus, the initial channel should be included in the label of the cross section, together with the final channel.

[14]In fact, quantities like $\sigma_{\beta\alpha}$, σ_0, ξ and $\mathcal{T}$ are functions of the orbital angular momentum, l. However, to simplify the notation, we do not indicate this dependence explicitly.

Using this result in Eq. (12.96), we get the well known Hauser-Feshbach formula for the partial cross section [Hauser and Feshbach (1952)],

$$\sigma_{\beta\alpha} = \sigma_0 \frac{\mathcal{T}_\alpha \, \mathcal{T}_\beta}{\sum_\gamma \mathcal{T}_\gamma}. \tag{12.101}$$

In deriving the above formula for the cross section, we allowed the elastic case to have the same structure as the non-elastic one. This is incorrect as the elastic process, described by $\sigma_{\alpha\alpha}$, involves correlations between the entrance and exit channels. The resulting cross section is in fact twice the one given by the Hauser-Feshbach one. We shall discuss this point in detail using Feshbach's theory, where the cross section are expressed in terms of the P and Q projectors.

An important concept in the theory and application of nuclear reactions is that of energy average. The wider the energy averaging interval is the less structure there is in the cross section. One example showing the effect of energy averaging on the experimental energy excitation function has been given in figure 12.1. We see clearly that more structure is removed as one widens the energy averaging interval. Models for the resulting substructure have been developed and are related to the doorway concept discussed in section 12.1. In the following we discuss the case where the energy averaging removes all the structure in the cross section and an optical model potential picture becomes necessary to account for part of the energy averaged cross section.

The treatment of Feshbach resonances presented in Chapter 10, hinges on the concept of *isolated resonances*, in the sense that the ratio of the average width of these resonances, $\overline{\Gamma_n}$, to that of the average spacing, $\overline{D_n} = \overline{E_n - E_{n-1}}$ is much smaller than unity,

$$\frac{\overline{\Gamma_n}}{\overline{D_n}} \ll 1. \tag{12.102}$$

The resulting cross section exhibits energy modulations with peaks corresponding to these resonances. One usually can read off the values of the widths directly from the data by performing an analysis using the unitary background $S^{(\mathrm{P})}$, plus sum-over poles expression of the S-matrix and the resulting background plus Breit-Wigner cross section (Eq. (12.77)).

The case involving the opposite limit,

$$\frac{\overline{\Gamma_n}}{\overline{D_n}} \gg 1, \tag{12.103}$$

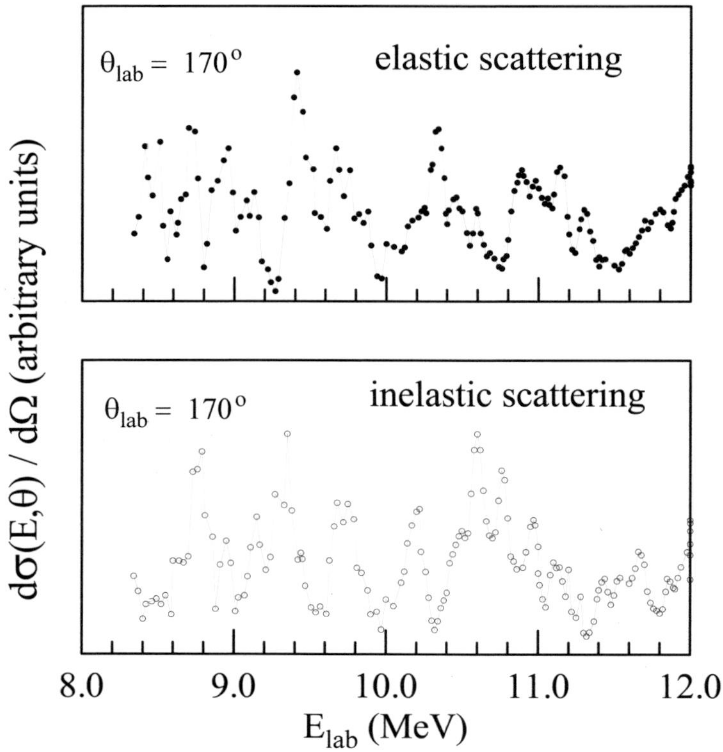

Fig. 12.3 Elastic and inelastic excitation functions in collisions of protons with ^{26}Mg [Häusser *et al.* (1964)]. For details, we refer to the original paper [Häusser *et al.* (1964)].

corresponds to what is known as *overlapping resonances*. When many overlapping Feshbach resonances are excited in a given scattering system, a statistical approach is more appropriate. One usually refers to such cases as chaotic scattering or, in nuclear physics, compound nuclear reactions. Typical cross sections exhibiting statistical fluctuations are shown in figure 12.3. It exhibits elastic (top panel) and inelastic (bottom panel) excitation functions for protons beams colliding with a ^{26}Mg target. The data are from [Häusser *et al.* (1964)]. Note that the energy oscillations of the two cross sections are out of phase. The maxima of the elastic excitation function do not occur at the same energies of the inelastic excitation functions. This rules out the possibility of doorway resonances. The oscillations are of statistical nature.

The treatment of overlapping Feshbach resonances in the compound nucleus requires the introduction of the concept of energy averaging, which leads naturally to the complex optical potential as the generating interaction of the now non-unitary background S-matrix (see section 10.3). This is necessary in order to find a representation of the S-matrix which can be used in the calculation of the energy averaged cross section composed of a sum of a background (the optical model contribution) plus a fluctuation part, with no interference term. This representation is called the *Optical Background Representation* (OBR) [Kawai *et al.* (1973)].

The first step needed to obtain the OBR of the S-matrix is the definition of the optical potential. It is the energy averaged effective potential arising from couplings with a subspace Q, which contains only closed channels. In this case, the polarization potential in the P-subspace is

$$\mathcal{V}_{\mathrm{P}}^{(\mathrm{Q})}(E) = \mathbb{V}_{\mathrm{PQ}}\, G_{\mathrm{Q}}\, \mathbb{V}_{\mathrm{QP}}, \tag{12.104}$$

with

$$G_{\mathrm{Q}} = \frac{1}{E - \mathbb{H}_{\mathrm{QQ}}}. \tag{12.105}$$

Note the important difference of the above from Eq. (10.14), where the Q channels may be opened. In that case, it was necessary to introduce $i\epsilon$ in the denominator, in order to guarantee outgoing boundary condition for the open channels. In the present case of closed compound nucleus channels, their wave functions are decaying and die out as $r \to \infty$.

The Feshbach potential above is ill-defined as it diverges when $E = \mathbb{H}_{\mathrm{QQ}}$. Accordingly, in order to find an entity which can be considered as a bona fid potential, one introduces an appropriate E-averaged version of $\mathcal{V}_{\mathrm{P}}^{(\mathrm{Q})}(E)$, denoted by $\overline{\mathcal{V}_{\mathrm{P}}^{(\mathrm{Q})}}(E)$. This is accomplished through an averaging function centered at the energy E with range I much larger than than the average spacing of the bound Q-states. Frequently, on uses a Lorentzian function, $\rho_I(E - E')$, whose integral over E' is unity. The E-averaged polarization potential is,

$$\overline{\mathcal{V}_{\mathrm{P}}^{(\mathrm{Q})}} = \int dE'\ \rho_I(E' - E)\ \mathcal{V}_{\mathrm{P}}^{(\mathrm{Q})}(E'), \tag{12.106}$$

with (calling $x = E' - E$)

$$\rho_I(x) = \frac{1}{2\pi}\left[\frac{I}{x^2 + I^2/4}\right]. \tag{12.107}$$

The integral over E' can be performed by contour integration and the result is,

$$\overline{\mathcal{V}_{\rm P}^{\rm (Q)}} = \mathbb{V}_{\rm PQ} \, \frac{1}{E - \mathbb{H}_{\rm QQ} + i\,I/2} \, \mathbb{V}_{\rm QP}. \tag{12.108}$$

The above potential operator is now a slowly varying function of energy owing to the presence of the factor I. The corresponding effective potential in the P-subspace,

$$\mathcal{V}^{\rm opt} = \mathbb{V}_{\rm PP} + \overline{\mathcal{V}_{\rm P}^{\rm (Q)}}, \tag{12.109}$$

is called the *optical potential* operator.

Note that whereas $\mathcal{V}_{\rm P}^{\rm (Q)}$ is Hermitian, the E-averaged polarization potential, $\overline{\mathcal{V}_{\rm P}^{\rm (Q)}}$, (and thus $\mathcal{V}^{\rm opt}$) is not. The imaginary part of the latter takes into account the loss of flux from the elastic channel for the formation of the compound nucleus. This type of absorption is different from that arising from couplings with open channels, since the flux lost to the CN will eventually reappear.

In order to derive the S-matrix, we go back to the effective equation for the P-projected wave function,

$$\left[E - \mathbb{H}_{\rm PP} - \mathcal{V}_{\rm P}^{\rm (Q)}\right] |\Psi_{\rm P}^{(+)}\rangle = 0. \tag{12.110}$$

We now add and subtract the E-averaged polarization potential operator, $\overline{\mathcal{V}_{\rm P}^{\rm (Q)}}$. The resulting equation can be cast in the following form,

$$\left[E - \mathcal{H}_{\rm P}^{\rm opt}\right] |\Psi_{\rm P}^{(+)}\rangle = \Delta\mathcal{V}_{\rm P} \, |\Psi_{\rm P}^{(+)}\rangle \, . \tag{12.111}$$

Above, we have introduced the optical Hamiltonian in the P-subspace,

$$\mathcal{H}^{\rm opt} = \mathbb{H}_{\rm PP} + \overline{\mathcal{V}_{\rm P}^{\rm (Q)}} = K + h + \mathcal{V}^{\rm opt}. \tag{12.112}$$

The remaining interaction in Eq. (12.111) is

$$\Delta\mathcal{V}_{\rm P} = \mathcal{V}_{\rm P}^{\rm (Q)} - \overline{\mathcal{V}_{\rm P}^{\rm (Q)}}, \tag{12.113}$$

or, using the explicit forms of $\mathcal{V}_{\rm P}^{\rm (Q)}$ (Eq. (12.104)) and $\overline{\mathcal{V}_{\rm P}^{\rm (Q)}}$ (Eq. (12.108)),

$$\begin{aligned} \Delta\mathcal{V}_{\rm P} &= \mathbb{V}_{\rm PQ} \left[\frac{1}{E - \mathbb{H}_{\rm QQ}} - \frac{1}{E - \mathbb{H}_{\rm QQ} + i\frac{I}{2}} \right] \mathbb{V}_{\rm QP} \\ &= \mathbb{V}_{\rm PQ} \left[\frac{i\,I/2}{(E - \mathbb{H}_{\rm QQ})\,(E - \mathbb{H}_{\rm QQ} + i\,I/2)} \right] \mathbb{V}_{\rm QP}. \end{aligned} \tag{12.114}$$

Now we introduce the complex coupling operators

$$\mathcal{U}_{\rm PQ} = \mathbb{V}_{\rm PQ} \left[\frac{i\,I/2}{E - \mathbb{H}_{\rm QQ} + i\,I/2} \right]^{1/2} \tag{12.115}$$

and

$$\mathcal{U}_{\rm QP} = \left[\frac{i\,I/2}{E - \mathbb{H}_{\rm QQ} + i\,I/2} \right]^{1/2} \mathbb{V}_{\rm QP}. \tag{12.116}$$

The potential $\Delta\mathcal{V}_{\rm P}$ can then be cast in the compact form

$$\Delta\mathcal{V}_{\rm P} = \mathcal{U}_{\rm PQ}\, G_{\rm Q}\, \mathcal{U}_{\rm QP}, \tag{12.117}$$

with $G_{\rm Q}$ given by Eq. (12.105), and Eq. (12.111) becomes,

$$[E - \mathcal{H}_{\rm P}^{\rm opt}]\, |\Psi_{\rm P}^{(+)}\rangle = \mathcal{U}_{\rm PQ}\, G_{\rm Q}\, \mathcal{U}_{\rm QP}\, |\Psi_{\rm P}^{(+)}\rangle\,. \tag{12.118}$$

The Lippmann-Schwinger equation corresponding to Eq. (12.118) is

$$|\Psi_{\rm P}^{(+)}\rangle = \left|X_{\rm opt}^{(+)}\right\rangle + G_{\rm opt}^{(+)}\, (\mathcal{U}_{\rm PQ}\, G_{\rm Q}\, \mathcal{U}_{\rm QP})\, |\Psi_{\rm P}^{(+)}\rangle\,, \tag{12.119}$$

where the optical wave function satisfies the equation

$$[E - \mathcal{H}_{\rm P}^{\rm opt}]\, \left|X_{\rm opt}^{(+)}\right\rangle = 0 \tag{12.120}$$

and the optical Green's function is

$$G_{\rm opt}^{(+)} = \frac{1}{E - \mathcal{H}_{\rm P}^{\rm opt} + i\epsilon}. \tag{12.121}$$

To evaluate the matrix-elements of the transition operator, $T_{\alpha\mathbf{k}_\alpha;0\mathbf{k}_0}$, we resort to the two-potential formula (see section 4.5), applied to many-body scattering (Eqs. (9.72) and (9.73)). Here, the dominant Hamiltonian and the complementary potential are respectively $\mathbb{H}^{(1)} = \mathcal{H}^{\rm opt}$ and $\mathbb{V}^{(2)} = \Delta\mathcal{V}_{\rm P}$. In this way, the optical wave functions play the role of distorted waves. The T-matrix is then given by the expression

$$T_{\alpha\mathbf{k}_\alpha;0\mathbf{k}_0} = T^{\rm opt}_{\alpha\mathbf{k}_\alpha;0\mathbf{k}_0} + T^{\rm fl}_{\alpha\mathbf{k}_\alpha;0\mathbf{k}_0}, \tag{12.122}$$

where

$$T^{\rm opt}_{\alpha\mathbf{k}_\alpha;0\mathbf{k}_0} = \left\langle \Phi(\alpha\mathbf{k}_\alpha) \right| \mathcal{V}^{\rm opt} \left| X_{\rm opt}^{(+)}(0\mathbf{k}_0) \right\rangle \tag{12.123}$$

is the *optical* T-matrix, associated with the optical Hamiltonian. For simplicity, we are assume that the optical potential is diagonal in channel space, so that the distorted waves reduces to a single channel. That is

$$\left|X_{\rm opt}^{(\pm)}(\alpha\mathbf{k}_\alpha)\right\rangle = \sum_{\alpha'} \left|\chi_{\alpha'}^{(\pm)}(\alpha\mathbf{k}_\alpha)\right\rangle\, |\alpha')\ \delta_{\alpha\alpha'} \quad\to\quad |\chi_\alpha^{(\pm)}(\alpha\mathbf{k}_\alpha)\rangle\, |\alpha)\,. \tag{12.124}$$

Therefore, with this assumption the optical component of the T-matrix gives only to elastic scattering.

The second term in Eq. (12.122) is the remaining part of the T-matrix, usually called the *fluctuation* T-matrix. It is given by

$$\begin{aligned} T^{\text{fl}}_{\alpha\mathbf{k}_\alpha;0\mathbf{k}_0} &= \left\langle X^{(-)}_{\text{opt}}(\alpha\mathbf{k}_\alpha)\middle| \Delta\mathcal{V}_{\text{P}} \middle|\Psi^{(+)}_{\text{P}}(0\mathbf{k}_0)\right\rangle \\ &= \left\langle X^{(-)}_{\text{opt}}(\alpha\mathbf{k}_\alpha)\middle| \mathcal{U}_{\text{PQ}}\, G_{\text{Q}}\, \mathcal{U}_{\text{QP}} \middle|\Psi^{(+)}_{\text{P}}(0\mathbf{k}_0)\right\rangle. \end{aligned} \tag{12.125}$$

In fact, Eq. (12.125) is not very useful because the wave function $\Psi^{(+)}_{\text{P}}(0\mathbf{k}_0)$ is not known. A better expression is obtained through the relation[15]

$$G_{\text{Q}}\, \mathcal{U}_{\text{QP}} \left|\Psi^{(+)}_{\text{P}}(0\mathbf{k}_0)\right\rangle = \mathcal{G}_{\text{Q}}\, \mathcal{U}_{\text{QP}} \left|X^{(+)}_{\text{opt}}(0\mathbf{k}_0)\right\rangle, \tag{12.126}$$

where $\mathcal{G}_{\text{Q}}$ is the full Green's function in the Q-subspace,

$$\mathcal{G}_{\text{Q}} = \frac{1}{E - \mathbb{H}_{\text{QQ}} - \mathcal{U}_{\text{QP}}\, G^{(+)}_{\text{opt}}\, \mathcal{U}_{\text{PQ}}}. \tag{12.127}$$

Using this result in Eq. (12.125), we get

$$T^{\text{fl}}_{\alpha\mathbf{k}_\alpha;0\mathbf{k}_0} = \left\langle X^{(-)}_{\text{opt}}(\alpha\mathbf{k}_\alpha)\middle| \mathcal{U}_{\text{PQ}}\, \mathcal{G}_{\text{Q}}\, \mathcal{U}_{\text{QP}} \middle|X^{(+)}_{\text{opt}}(0\mathbf{k}_0)\right\rangle. \tag{12.128}$$

The fluctuation T-matrix takes a more interesting form if we use the spectral representation of $\mathcal{G}_{\text{Q}}$,

$$\mathcal{G}_{\text{Q}} = \sum_n \frac{\left|\mathcal{X}_n\right\rangle \left\langle\tilde{\mathcal{X}}_n\right|}{E - E_n}, \tag{12.129}$$

where $\left|\mathcal{X}_n\right\rangle$ are the bound solutions of the equation,

$$\left[\mathbb{H}_{\text{QQ}} + \Delta\mathcal{V}_{\text{Q}}\right] \left|\mathcal{X}_n\right\rangle = E_n \left|\mathcal{X}_n\right\rangle. \tag{12.130}$$

Above, $\Delta\mathcal{V}_{\text{Q}}$ is the complex potential

$$\Delta\mathcal{V}_{\text{Q}} = \mathcal{U}_{\text{QP}}\, G^{(+)}_{\text{opt}}\, \mathcal{U}_{\text{PQ}} \tag{12.131}$$

and $\left|\tilde{\mathcal{X}}_n\right\rangle$ are states of the dual space. Owing to the complex nature of the potential, the orthogonality relations have the form

$$\left\langle\tilde{\mathcal{X}}_n\middle|\mathcal{X}_{n'}\right\rangle = \delta nn'. \tag{12.132}$$

Writing the complex energy eigenvalues as

$$E_n = \mathcal{E}_n - i\,\frac{\Gamma_n}{2}, \tag{12.133}$$

[15] This relation can be derived from the LS equation for $\left|\Psi^{(+)}_{\text{P}}(0\mathbf{k}_0)\right\rangle$ (see exercise 1, where hints for the derivation are given).

with

$$\mathcal{E}_n = \text{Re}\Big\{ \langle \tilde{\mathcal{X}}_\text{n} | \, \mathbb{H}_{\text{QQ}} + \Delta\mathcal{V}_\text{Q} | \mathcal{X}_\text{n} \rangle \Big\} \tag{12.134}$$

and

$$\Gamma_n = -2 \, \text{Im}\Big\{ \langle \tilde{\mathcal{X}}_n | \, \mathbb{H}_{\text{QQ}} + \Delta\mathcal{V}_\text{Q} | \mathcal{X}_n \rangle \Big\}, \tag{12.135}$$

the fluctuation T-matrix takes the form

$$T^{\text{fl}}_{\alpha\mathbf{k}_\alpha;0\mathbf{k}_0} = \sum_n \frac{\langle X^{(-)}_{\text{opt}}(\alpha\mathbf{k}_\alpha) | \, \mathcal{U}_{\text{PQ}} | \mathcal{X}_n \rangle \, \langle \tilde{\mathcal{X}}_n | \, \mathcal{U}_{\text{QP}} | X^{(+)}_{\text{opt}}(0\mathbf{k}_0) \rangle}{(E - \mathcal{E}_n) + i\,\Gamma_n/2}, \tag{12.136}$$

Owing to Eq. (12.122) and the usual relation between the S-matrix and the on-shell T-matrix: $S = 1 - 2\pi T$, we can write

$$S_{\alpha\mathbf{k}_\alpha;0\mathbf{k}_0} = S^{\text{opt}}_{\alpha\mathbf{k}_\alpha;0\mathbf{k}_0} + S^{\text{fl}}_{\alpha\mathbf{k}_\alpha;0\mathbf{k}_0}, \tag{12.137}$$

where S^{opt} is the optical S-matrix,

$$S^{\text{opt}}_{\alpha\mathbf{k}_\alpha;0\mathbf{k}_0} = \delta_{\alpha 0} - 2\pi i \, \langle \Phi(\alpha\mathbf{k}_\alpha) | \, \mathcal{V}^{\text{opt}} \, | X^{(+)}_{\text{opt}}(0\mathbf{k}_0) \rangle, \tag{12.138}$$

and S^{fl} is the fluctuation S-matrix. Using Eq. (12.136), it can be put in the form

$$S^{\text{fl}}_{\alpha\mathbf{k}_\alpha;0\mathbf{k}_0} = -\,i \sum_n \frac{g_{n0}\; g_{\alpha n}}{(E - \mathcal{E}_n) + i\,\Gamma_n/2}, \tag{12.139}$$

where

$$g_{n0} = \sqrt{2\pi} \, \langle X^{(-)}_{\text{opt}}(\alpha\mathbf{k}_\alpha) | \, \mathcal{U}_{\text{PQ}} | \mathcal{X}_n \rangle \equiv \sqrt{2\pi} \, \langle \chi^{(-)}_\alpha | \, \mathcal{U}_{\alpha\text{Q}} | \mathcal{X}_n \rangle \tag{12.140}$$

and

$$g_{\alpha n} = \sqrt{2\pi} \, \langle \tilde{\mathcal{X}}_n | \, \mathcal{U}_{\text{QP}} | X^{(+)}_{\text{opt}}(0\mathbf{k}_0) \rangle \equiv \sqrt{2\pi} \, \langle \tilde{\mathcal{X}}_n | \, \mathcal{U}_{\text{Q}0} | \chi^{(+)}_0 \rangle. \tag{12.141}$$

The matrix-elements g_{n0} and $g_{\alpha n}$ have the same structure and for elastic scattering they coincide. This can be proved using the time reversal symmetry, as done in section 10.2.2.

A convenient feature of the Optical Background Representation is that the energy-averages of the optical and the fluctuation terms in Eq. (12.137) have the following properties [Kawai *et al.* (1973)]:

$$\overline{S} \simeq S^{\text{opt}}, \tag{12.142}$$

$$\overline{S^{\text{opt}}} \simeq S^{\text{opt}}, \tag{12.143}$$

and

$$\overline{S^{\text{fl}}} \simeq 0. \tag{12.144}$$

These properties lead to an important conclusion about the modulus of the fluctuation S-matrix. Since the subspace Q contains only closed channels, the full S-matrix is unitary. Thus, we can write

$$SS^\dagger = (S^{\text{opt}} + S^{\text{fl}})\,(S^{\text{opt}} + S^{\text{fl}})^\dagger = 1. \tag{12.145}$$

Taking energy averages and using Eqs. (12.143) and (12.144), we get

$$S^{\text{opt}}S^{\text{opt}\dagger} + \overline{S^{\text{fl}}\,S^{\text{fl}\dagger}} = 1. \tag{12.146}$$

Reorganizing the terms and taking matrix-elements in the channel basis, the above equation becomes

$$\sum_\gamma \overline{S^{\text{fl}}_{\alpha,\gamma}\left(S^{\text{fl}}_{\beta,\gamma}\right)^*} = \delta_{\alpha\beta} - \sum_\gamma S^{\text{opt}}_{\alpha,\gamma}\left(S^{\text{opt}}_{\beta,\gamma}\right)^*. \tag{12.147}$$

In particular, for diagonal matrix-elements, we get

$$\sum_\gamma \overline{\left|S^{\text{fl}}_{\alpha,\gamma}\right|^2} = 1 - \sum_\gamma \left|S^{\text{opt}}_{\alpha,\gamma}\right|^2, \tag{12.148}$$

or

$$\mathcal{T}_\alpha(E_\alpha) = \sum_\gamma \overline{\left|S^{\text{fl}}_{\alpha,\gamma}\right|^2} = 1 - \sum_\gamma \left|S^{\text{opt}}_{\alpha,\gamma}\right|^2. \tag{12.149}$$

Above, $\mathcal{T}_\alpha(E_\alpha)$ is the optical transmission coefficient for a collision initiated in channel-α, with energy $E_\alpha = E - \varepsilon_\alpha$.

12.3.1 *Fluctuation cross section*

In this section we evaluate the energy-averaged cross section for channel β in a collision initiated in channel α. For simplicity, we consider the case of spinless collision partners and look at the contribution of a particular partial wave, say l, to the cross section.

First, we use Eqs. (12.137), (9.36), (9.29), (9.115) and the orthonormality relation of Legendre polynomials (Eq. (2.60)). The contribution from the lth partial-wave to the total (integrated over angle) cross section for this process then becomes

$$\sigma_{\beta\alpha}(l) = \frac{\pi}{k^2}\,(2l+1)\,\left|\left(S^{\text{opt}}_{\beta\alpha} - \delta_{\beta\alpha}\right) + S^{\text{fl}}_{\beta\alpha}\right|^2. \tag{12.150}$$

Now we take the energy average to get[16],

$$\overline{\sigma_{\beta\alpha}(l)} = \sigma^{\text{opt}}_{\beta\alpha}(l) + \overline{\sigma^{\text{fl}}_{\beta\alpha}(l)}, \tag{12.151}$$

[16]Owing to Eqs. (12.143) and (12.144), the interference between $(S^{\text{opt}}_{\beta\alpha} - 1)$ and $S^{\text{fl}}_{\beta\alpha}$ averages to zero.

where

$$\sigma^{\rm opt}_{\beta\alpha}(l) = \frac{\pi}{k^2}\,(2l+1)\,\left|S^{\rm opt}_{\beta\alpha}(l) - \delta_{\beta\alpha}\right|^2, \qquad (12.152)$$

is the optical cross section[17] and

$$\overline{\sigma^{\rm fl}_{\beta\alpha}(l)} = \frac{\pi}{k^2}\,(2l+1)\,\overline{S^{\rm fl}_{\beta\alpha}(l)\,S^{\rm fl\,*}_{\beta\alpha}(l)} \qquad (12.153)$$

is the so-called *fluctuation* cross section. The latter is to be related to the transmission coefficients of Eq. (12.149), for the theory to be consistent. This can be accomplished through the use of the explicit form of $S^{\rm fl}_{\beta\alpha}$, given in Eq. (12.139). We write

$$\overline{S^{\rm fl}_{\beta\alpha}(l)\,S^{\rm fl\,*}_{\beta\alpha}(l)} = \overline{\left[\sum_{n,m}\frac{g_{n\alpha}\,g_{\beta n}}{(E-\mathcal{E}_n)+i\Gamma_n/2}\,\frac{g^*_{m\alpha}\,g^*_{\beta m}}{(E-\mathcal{E}_m)-i\Gamma_m/2}\right]} \qquad (12.154)$$

Assuming that the matrix-elements $g_{n\alpha}$ have random phase, the contribution from $m \neq n$ resonances add to zero. Thus, only terms with $m = n$ survive and the above equation reduces to

$$\overline{S^{\rm fl}_{\beta\alpha}(l)\,S^{\rm fl\,*}_{\beta\alpha}(l)} = \overline{\left[\sum_{n}\frac{g_{n\alpha}\,g_{\beta n}\,g^*_{n\alpha}\,g^*_{\beta n}}{(E-\mathcal{E}_n)^2+\Gamma_n^2/4}\right]}. \qquad (12.155)$$

Now we perform the energy average explicitly. It is given by the integral

$$\overline{S^{\rm fl}_{\beta\alpha}(l)\,S^{\rm fl\,*}_{\beta\alpha}(l)} = \frac{1}{I}\int dE'\,\rho_I\,(E-E')\;S^{\rm fl}_{\beta\alpha}(E')\,S^{\rm fl\,*}_{\beta\alpha}(E'). \qquad (12.156)$$

For simplicity, we use a box average, which corresponds to the weight function

$$\begin{aligned}\rho_I\,(E-E') &= 1, \qquad \text{for } E+I/2 \geq E' \geq E-I/2,\\ &= 0, \qquad \text{otherwise.}\end{aligned} \qquad (12.157)$$

One should have in mind that although the averaging interval, I, contains a large number of resonances, it is much smaller than the resonance energies, $\mathcal{E}_n$, and small enough for the matrix-elements g_{ij} to be nearly constant within the $\{E-I/2; E+I/2\}$ interval. Thus, $\overline{S^{\rm fl}_{\beta\alpha}(l)\,S^{\rm fl\,*}_{\beta\alpha}(l)}$ can be written as

$$\overline{S^{\rm fl}_{\beta\alpha}(l)\,S^{\rm fl\,*}_{\beta\alpha}(l)} = \sum_n g_{n\alpha}(E)\,g_{\beta n}(E)\,g^*_{n\alpha}(E)\,g^*_{\beta n}(E)\;A_n, \qquad (12.158)$$

[17] If we assume that $\mathcal{H}^{\rm opt}$ is diagonal in channel space, the optical cross section gives only elastic scattering.

where A_n is the integral

$$A_n = \frac{1}{I}\int_{E-\mathcal{E}_n-I/2}^{E-\mathcal{E}_n+I/2}\frac{dx}{x^2+\Gamma_n^2/4} = \frac{2}{\Gamma_n I}\left[\tan^{-1}(2x/\Gamma_n)\right]_{E-\mathcal{E}_n-I/2}^{E-\mathcal{E}_n+I/2}. \qquad (12.159)$$

Typically, $E-\mathcal{E}_n \pm I/2$ is of the order of I, which is much larger than the widths Γ_n. Thus, the absolute values of the integration limits are much larger than one and we can approximate,

$$A_n \simeq \frac{2}{\Gamma_n I}\left[\tan^{-1}(\infty)-\tan^{-1}(-\infty)\right] = \frac{2\pi}{\Gamma_n I}. \qquad (12.160)$$

Now we approximate the resonance widths by their mean value, $\Gamma_n \to \Gamma$, and introduce the resonance average, denoted by $\langle\ \rangle_n$. This average is related with sums over n as

$$\langle\ \rangle_n = \frac{1}{N}\sum_n .$$

Above, $N = I/D$ is the number of resonances within the interval I and D is the average spacing between two consecutive resonances. Eq. (12.158) then becomes

$$\overline{S^{\rm fl}_{\beta\alpha}(l)\, S^{\rm fl\,*}_{\beta\alpha}(l)} \simeq \frac{2\pi}{D\Gamma}\left\langle g_{n\alpha}\, g_{\beta n}\, g^*_{n\alpha}\, g^*_{\beta n}\right\rangle_n. \qquad (12.161)$$

Assuming gaussian distributions for the couplings of the P and Q channels, we can use the relation (see [Agassi *et al.* (1975)], [Kawai *et al.* (1973)]),

$$\left\langle g_{n\alpha}\, g_{\beta n}\, g^*_{n\alpha}\, g^*_{\beta n}\right\rangle_n = \left\langle g_{n\alpha} g^*_{n\alpha}\right\rangle_n \left\langle g_{n\beta}\, g^*_{n\beta}\right\rangle_n + \left\langle g_{n\alpha}\, g^*_{n\beta}\right\rangle_n \left\langle g_{n\beta} g^*_{n\alpha}\right\rangle_n + \left\langle g_{n\alpha}\, g_{n\beta}\right\rangle_n \left\langle g^*_{n\beta} g^*_{n\alpha}\right\rangle_n. \qquad (12.162)$$

The last term in the above equation is usually set equal to zero in view of the fact that it contains averages of quantities containing large random phases. This situation is encountered in the compound nuclear reactions involving overlapping resonances. This simplifies Eq. (12.162). Introducing the notation of [Kawai *et al.* (1973)],

$$X_{\alpha\alpha'} = X^*_{\alpha'\alpha} = \langle g_{n\alpha}\, g^*_{n\alpha'}\rangle_n, \qquad (12.163)$$

it takes the form,

$$\left\langle g_{n\alpha}\, g_{\beta n}\, g^*_{n\alpha}\, g^*_{\beta n}\right\rangle_n = X_{\alpha\alpha}\, X_{\beta\beta} + X_{\alpha\beta}\, X_{\beta\alpha}. \qquad (12.164)$$

Thus, Eq. (12.161) can be written,

$$\overline{S^{\rm fl}_{\beta\alpha}(l)\, S^{\rm fl\,*}_{\beta\alpha}(l)} \simeq \frac{2\pi}{D\,\Gamma}\,\Big[X_{\alpha\alpha}\,X_{\beta\beta} + X_{\alpha\beta}\,X_{\beta\alpha}\Big]. \tag{12.165}$$

or,

$$\overline{S^{\rm fl}_{\beta\alpha}(l)\, S^{\rm fl\,*}_{\beta\alpha}(l)} \simeq Y_{\alpha\alpha}\,Y_{\beta\beta} + Y_{\alpha\beta}\,Y_{\beta\alpha}. \tag{12.166}$$

Above, we have introduced the new variables,

$$Y_{ij} = \sqrt{\frac{2\pi}{D\,\Gamma}}\;\; X_{ij} \qquad (i,j=\alpha,\beta). \tag{12.167}$$

Typical calculations of non-elastic processes involve many open channels. In such cases, the first term on the RSH of Eq. (12.166) is much larger the second [Kawai *et al.* (1973)]. Thus, to a first approximation, we can write

$$\overline{S^{\rm fl}_{\beta\neq\alpha}(l)\, S^{\rm fl\,*}_{\beta\neq\alpha}(l)} \simeq Y_{\alpha\alpha}\,Y_{\beta\beta}. \tag{12.168}$$

Inserting this result in Eq. (12.153), we get the contribution from the lth partial-wave to the cross section for non-elastic processes

$$\overline{\sigma^{\rm fl}_{\beta\neq\alpha}(l)} \simeq \frac{\pi}{k^2}\;(2l+1)\;\, Y_{\alpha\alpha}\,Y_{\beta\beta}. \tag{12.169}$$

Better approximations to include contributions from the term $Y_{\alpha\beta}Y_{\beta\alpha}$ can then be derived by iterative procedures.

On the other hand, for elastic scattering the two terms in Eq. (12.166) are identical. Thus, we can write

$$\overline{S^{\rm fl}_{\alpha\alpha}(l)\, S^{\rm fl\,*}_{\alpha\alpha}(l)} = 2\,Y_{\alpha\alpha}\,Y_{\alpha\alpha} \tag{12.170}$$

and the cross section becomes

$$\overline{\sigma^{\rm fl}_{\rm el}(l)} \equiv \overline{\sigma^{\rm fl}_{\alpha\alpha}(l)} = \frac{\pi}{k^2}\;(2l+1)\;\,[2\;Y_{\alpha\alpha}\,Y_{\beta\beta}]\,. \tag{12.171}$$

To derive the Hauser-Feshbach formula, we have to express $Y_{\alpha\alpha}$ and $Y_{\beta\beta}$ in terms of transmission factors[18] for the optical Hamiltonian. This can be

[18] In fact, for the general case of non-diagonal $S^{\rm opt}$, Y is non-diagonal, and the transmission factor becomes a transmission matrix. This matrix, which was first introduced by Satchler [Satchler (1963)], is defined by

$$\mathcal{T}_{\alpha,\beta} = \delta_{\alpha,\beta} - \sum_{\gamma} S^{\rm opt}_{\alpha,\gamma}\,S^{\rm opt\,*}_{\beta,\gamma}\,.$$

It can be derived from Eq. (12.147), in terms of the Y-matrices, giving the result, $\mathcal{T} = Y\,{\rm Tr}\{Y\} + Y^2$. This results in a modified Hauser-Feshbach formula extended to compound nucleus formation and decay in the presence of prompt, direct, reactions [Kawai *et al.* (1973)] (see also [Engelbrecht and Weidenmüller (1973)]).

achieved summing Eq. (12.166) over the final channel, β, and comparing the result with Eq. (12.149). We get[19]

$$\mathcal{T}_\alpha = \sum_\beta \overline{\left|S^{\text{fl}}_{\alpha,\beta}(l)\right|^2} \simeq Y_{\alpha\alpha} \sum_\beta Y_{\beta\beta} + \sum_\beta Y_{\alpha\beta}\, Y_{\beta\alpha}$$
$$= Y_{\alpha\alpha} \sum_\beta Y_{\beta\beta} + \left[Y^2\right]_{\alpha\alpha}. \tag{12.172}$$

Since the second term on the RHS of the above equation is much smaller than the first [Kawai *et al.* (1973)], we can approximate

$$\mathcal{T}_\alpha \simeq Y_{\alpha\alpha} \sum_\beta Y_{\beta\beta}. \tag{12.173}$$

Next, we sum over α to obtain

$$\sum_\alpha \mathcal{T}_\alpha = \Big(\sum_\gamma Y_{\gamma\gamma}\Big)^2 \longrightarrow \sum_\gamma Y_{\gamma\gamma} = \sqrt{\sum_\gamma \mathcal{T}_\gamma}, \tag{12.174}$$

and the above equation leads to

$$Y_{\alpha\alpha} = \frac{\mathcal{T}_\alpha}{\sqrt{\sum_\gamma \mathcal{T}_\gamma}}. \tag{12.175}$$

Using this result in the approximate equation for the non-elastic fluctuation cross section (Eq. (12.169)), we get the Hauser-Feshbach formula,

$$\overline{\sigma^{\text{fl}}_{\beta\neq\alpha}(l)} = \frac{\pi}{k^2}\,(2l+1)\,\frac{\mathcal{T}_\alpha\,\mathcal{T}_\beta}{\sum_\gamma \mathcal{T}_\gamma}. \tag{12.176}$$

Eq. (12.176), expresses in mathematical form the Bohr independence hypothesis, which states that the formation and decay of the compound nucleus are independent from each other. It reflects the fact that once the system is formed, it takes a long time to decay, and thus forgets how it was formed in the first place.

The fluctuation cross section for the compound elastic transition is derived in the same way. Inserting Eq. (12.175) in Eq. (12.171), we get

$$\overline{\sigma^{\text{fl}}_{\text{el}}(l)} = \frac{\pi}{k^2}\,(2l+1)\,\frac{2\,\mathcal{T}_\alpha^2}{\sum_\gamma \mathcal{T}_\gamma}. \tag{12.177}$$

This expression differs from the Hauser-Feshbach formula by a factor 2, which is called the *elastic enhancement factor*. In the next section we will show an evidence that this factor was verified experimentally.

[19]To keep the notation simple, we do not represent explicitly the l-dependence of the transmission coefficient.

12.3.2 Fluctuation part of angular distributions

The discussion of the fluctuation cross section and the resulting Hauser-Feshbach cross section have a direct bearing on the general characteristics of the compound nucleus cross section. In this section, we derive the contribution from compound nucleus formation (fluctuation term) to angular distributions.

For this purpose, we multiply Eq. (12.122) by the factor $\left(-4\pi^2\mu/\hbar^2\right)$, to obtain (see Eq. (4.64)),

$$f_{\beta,\alpha}\left(\mathbf{k}_\beta,\mathbf{k}_\alpha\right) = f^{\text{opt}}_{\alpha,\alpha}\left(\mathbf{k}'_\alpha,\mathbf{k}_\alpha\right)\,\delta_{\beta\alpha} + f^{\text{fl}}_{\beta,\alpha}\left(\mathbf{k}_\beta,\mathbf{k}_\alpha\right), \tag{12.178}$$

where $f^{\text{opt}}_{\alpha,\alpha}\left(\mathbf{k}'_\alpha,\mathbf{k}_\alpha\right)\delta_{\beta\alpha}$ and $f^{\text{fl}}_{\beta,\alpha}\left(\mathbf{k}_\beta,\mathbf{k}_\alpha\right)$ are the terms associated with the optical and the fluctuation parts of the T-matrix, respectively[20]. In the case of spinless collision partners, they are given by the partial-wave sums

$$f^{\text{opt}}_{\alpha,\alpha}\left(\mathbf{k}'_\alpha,\mathbf{k}_\alpha\right) = \frac{1}{2ik}\sum_l (2l+1)\;P_l(\cos\theta)\left[S^{\text{opt}}_{\alpha\alpha}(l)-1\right] \tag{12.179}$$

and

$$f^{\text{fl}}_{\beta,\alpha}\left(\mathbf{k}_\beta,\mathbf{k}_\alpha\right) = \frac{1}{2ik}\sum_l (2l+1)\;P_l(\cos\theta)\;S^{\text{fl}}_{\beta\alpha}(l). \tag{12.180}$$

The energy averaged cross section is given by

$$\overline{\left[\frac{d\sigma_{\beta\alpha}(\theta)}{d\Omega}\right]} = \overline{\left|f_{\beta,\alpha}\left(\mathbf{k}_\beta,\mathbf{k}_\alpha\right)\right|^2}, \tag{12.181}$$

where θ is the angle subtended by the vectors $\mathbf{k}_\beta$ and $\mathbf{k}_\alpha$.

Using Eqs. (12.178) to (12.180), and the properties of the OBR (Eqs. (12.142) to (12.144)), the interference of the two amplitudes vanish and one gets,

$$\overline{\left[\frac{d\sigma_{\beta\alpha}(\theta)}{d\Omega}\right]} = \frac{d\sigma^{\text{opt}}_{\beta\alpha}(\theta)}{d\Omega} + \overline{\left[\frac{d\sigma^{\text{fl}}_{\beta\alpha}(\theta)}{d\Omega}\right]}. \tag{12.182}$$

Above, $d\sigma^{\text{opt}}_{\beta\alpha}(\theta)/d\Omega$ is the cross section arising from the optical potential alone, which we assume that contributes only to elastic scattering, and the second term is the fluctuation cross section, given by

$$\overline{\left[\frac{d\sigma^{\text{fl}}_{\beta\alpha}(\theta)}{d\Omega}\right]} = \frac{1}{4k^2}\sum_{ll'}(2l+1)\;(2l'+1)\;\;P_l(\cos\theta)\,P_{l'}(\cos\theta)\;\overline{S^{\text{fl}}_{\beta\alpha}(l)\,S^{\text{fl}\,*}_{\beta\alpha}(l')}. \tag{12.183}$$

[20] As in the previous section, we are making the simplifying assumption that the optical Hamiltonian is diagonal in channel space.

The statistical property of $S^{\rm fl}_{\beta\alpha}(l)$ is now invoked and the following result is obtained,

$$\overline{S^{\rm fl}_{\beta\alpha}(l)\, S^{\rm fl\,*}_{\beta\alpha}(l')} = \overline{\left|S^{\rm fl}_{\beta\alpha}(l)\right|^2}\ \ \delta_{ll'}. \tag{12.184}$$

Eq. (12.183) then becomes

$$\overline{\left[\frac{d\sigma^{\rm fl}_{\beta\alpha}(\theta)}{d\Omega}\right]} = \frac{1}{4k^2}\sum_l (2l+1)^2\ P_l^2(\cos\theta)\ \overline{\left|S^{\rm fl}_{\beta\alpha}(l)\right|^2}. \tag{12.185}$$

Using the explicit form of $\overline{\left|S^{\rm fl}_{\beta\alpha}(l)\right|}$, one gets the fluctuation cross sections for the non-elastic and for the elatic processes. They are respectively given by (see Eqs. (12.176) and Eq. (12.177)),

$$\overline{\left[\frac{d\sigma_{\beta\neq\alpha}(\theta)}{d\Omega}\right]} = \frac{1}{4k^2}\sum_l\ (2l+1)^2\ P_l^2(\cos\theta)\ \ \left[\frac{\mathcal{T}_\alpha\,\mathcal{T}_\beta}{\sum_\gamma \mathcal{T}_\gamma}\right], \tag{12.186}$$

$$\overline{\left[\frac{d\sigma_{\rm el}(\theta)}{d\Omega}\right]} = \frac{1}{4k^2}\sum_l\ (2l+1)^2\ P_l^2(\cos\theta)\ \ \left[\frac{2\,\mathcal{T}_\alpha^2}{\sum_\gamma \mathcal{T}_\gamma}\right]. \tag{12.187}$$

An important property of the above cross sections is that they depend quadratically on Legendre polynomials[21]. Thus, angular distributions for processes resulting from compound nucleus formation are symmetric around $\theta = \pi/2$. In fact, this can be confirmed experimentally. An illustration is given in figure 12.4. The figure shows the angular dependence of the cross section for evaporation of an α-particle, following CN formation in ^{40}Ar + $^{\rm nat}$Se collisions. The data, from [Galin *et al.* (1974)], correspond to α-particles with energy in the range $20 \leq E_\alpha \leq 25$ MeV. Indeed, the data are clearly symmetric around $\theta = \pi/2$.

Figure 12.5 gives an experimental confirmation of the enhancement factor appearing in Eqs. (12.177) and (12.187). It shows angular distributions of compound elastic scattering for the $p+^{30}$Si system. The experimental cross section (solid circles) is compared with the Hauser-Feshbach formula (dashed line) and with the cross section of Eq. (12.187) (solid line), which includes the enhancement factor. The data and the theoretical curves are taken from [Kretschmer and Wangler (1978)]. Clearly, the data agrees with the theoretical curve that contains the enhancement factor.

[21]Since $P_l\,(\cos(\pi-\theta)) = (-)^l\, P_l(\cos\theta)$, the square of any Legendre polynomial (even or odd) is symmetrical around $\theta = \pi/2$. The angular distributions of Eqs. (12.186) and (12.187) must have the same property.

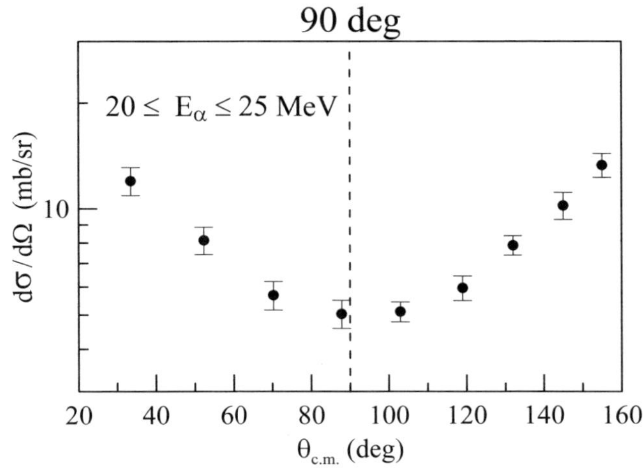

Fig. 12.4 Angular distribution of evaporated α-particles with energy in the range $20 \leq E_\alpha \leq 25$ MeV, following CN formation in the ^{40}Ar + ^{nat}Se collision. The data are from [Galin *et al.* (1974)], where the details of the experiment can be found.

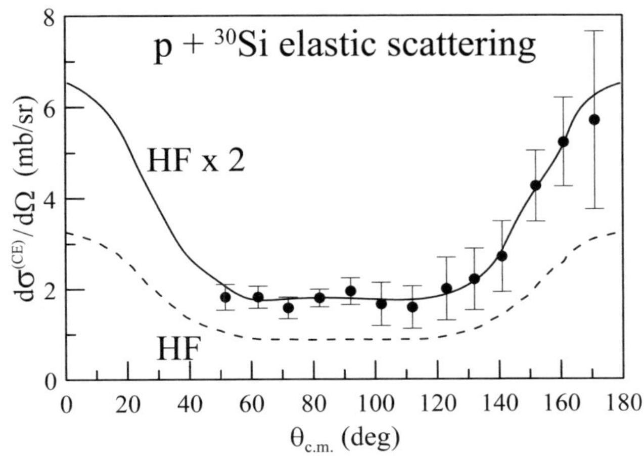

Fig. 12.5 Compound elastic angular distribution of protons in collisions with a ^{30}Si target [Kretschmer and Wangler (1978)]. For details of the experiment, including the technique to separate the CE cross section from direct elastic scattering are given in [Kretschmer and Wangler (1978)].

The statistical model has many applications in nuclear physics. One important example is in the study of heavy-ion fusion reactions. In these reactions, there is the formation of a highly excited compound nucleus, followed by the evaporation of light particles[22], like protons, neutrons and alpha-particles. A widely used experimental method to measure the fusion cross section consists in identifying the evaporation residues associated with each decay mode, through their de-excitation (like the emission of characteristic gamma-rays or alpha particles emitted by the residual nucleus). However, evaporation residues with very long half life cannot be identified in this way. Nevertheless, their contributions to the fusion cross section have to be taken into account. The statistical model provides the mean to achieve this goal. It gives the relative yield of each reaction channel and thus the contribution from undetected decay modes can be estimated through comparisons with those that have been measured. For this purpose, there are computer codes available like PACE [Gavron (1980)], CASCADE [Pühlhofer (1977)] and others.

To close this section, we briefly discuss the fluctuation cross section in the presence of a doorway. The results are similar to the ones derived in section 12.3.1, except for the important new feature of a doorway-modulated transmission coefficient (for detail we refer to [Feshbach (1992)]),

$$\mathcal{T}_\alpha = 1 - \left|S^{\rm opt}_{\beta\alpha}\right|^2 \times \left\{1 - \left[\frac{2\,\Gamma^{\uparrow}_{\rm D,\alpha}\,(E-E_{\rm D})\,\sin\gamma_{\rm D} - \Gamma_{\rm D}\,\cos\gamma_{\rm D} - \left(\Gamma^{\uparrow}_{\rm D,\alpha}\right)^2}{(E-E_{\rm D})^2 + \Gamma^2_{\rm D}/4}\right]\right\}, \tag{12.188}$$

where the total width of the doorway, $\Gamma_{\rm D}$, is,

$$\Gamma_{\rm D} = \Gamma^{\uparrow}_{\rm D} + \Gamma^{\downarrow}_{\rm D}. \tag{12.189}$$

At $E = E_{\rm D}$, the transmission coefficient becomes,

$$\mathcal{T}_\alpha = 1 - \left|S^{\rm opt}_{\beta\alpha}\right|^2 \times \left\{1 + \frac{4\,\cos\gamma_{\rm D}}{\Gamma_{\rm D}} + \left(\frac{2\,\Gamma^{\uparrow}_{\rm D,\alpha}}{\Gamma_{\rm D}}\right)^2\right\}, \tag{12.190}$$

while far away from the doorway resonance, we have the usual result,

$$\mathcal{T}_\alpha = 1 - \left|S^{\rm opt}_{\beta\alpha}\right|^2$$

Therefore, the transmission coefficient exhibits a dip at $E = E_{\rm D}$.

[22]This statement does not apply in the case of very heavy compound nuclei, which decay mainly through fission.

12.4 Chaotic scattering

Having dwelled on the statistical nature of reactions involving the population and decay of overlapping resonances, we are in a position to postulate that these reactions are examples of chaotic scattering in complex systems. As such, concepts borrowed from the theory of noise and from Random Matrix Theory become relevant. In this section we discuss a few applications of these theories in the context of compound nucleus reactions, though the ideas behind these theories are general enough to be be applicable to other systems such as very excited atoms in the presence of strong electric and/or magnetic fields, and to nano-systems such as quantum dots, the so-called artificial atoms. These latter devices are quantum wells containing several electrons. It has been shown recently that quantum dots behave exactly like the compound nucleus when the conductance of electrons is analyzed as a function of energy and/or magnetic fields [Alhassid (2000)]. In the following we briefly discuss these universal properties of chaotic many-body systems.

12.4.1 *Ericson fluctuations*

The regime of overlapping resonances requires for its description the introduction of statistical concepts such as energy averages and fluctuations. It is clear that this description leads to an average cross section which does not contain any direct reference to the underlying compound nuclear resonances. Thus, it leads to no information about resonance widths and time scales. This shortcoming can be remedied by the study of Ericson fluctuations [Ericson (1963)] or using the *number of maxima method* [Brink and Stephen (1963); Bizzeti and Maurenzig (1967)]. In this section we discuss the determination of average widths through Ericson's correlations.

We begin by introducing the S-matrix correlation function, given by[23]

$$C^{(\mathrm{S})}_{\beta\alpha}(E,\varepsilon) = \overline{\left[S^{\mathrm{fl}}_{\beta\alpha}(E+\varepsilon)\, S^{\mathrm{fl}\,*}_{\beta\alpha}(E)\right]}, \tag{12.191}$$

or, using the explicit form of $S^{\mathrm{fl}}_{\beta\alpha}$ (Eq. (12.139)),

$$C^{(\mathrm{S})}_{\beta\alpha}(E,\varepsilon) = \overline{\left[\sum_{n,m} \frac{g_{n\alpha}\, g_{\beta n}}{(E-\mathcal{E}_n)+i\Gamma_n/2}\; \frac{g^*_{m\alpha}\, g^*_{\beta m}}{(E+\varepsilon-\mathcal{E}_m)-i\Gamma_m/2}\right]}. \tag{12.192}$$

[23]Too keep the notation simple, in this section we omit the l-dependence of the S-matrix.

The above equation is analogous to Eq. (12.154), except for the energy shift ε in the denominator of the second fraction. Adopting the same approximations of section 12.3.1, namely that the g matrix-elements have random phases and are nearly energy independent, we get

$$C^{(S)}_{\beta\alpha}(E,\varepsilon) = \sum_n g_{n\alpha}\, g_{\beta n}\, g^*_{n\alpha}\, g^*_{\beta n}\, A_n. \tag{12.193}$$

This equation has the same form of Eq. (12.159) but here A_n is

$$A_n \simeq \frac{1}{I}\int_{-\infty}^{\infty} \frac{dx}{(x+\varepsilon+i\Gamma_n/2)\,(x-i\Gamma_n/2)}. \tag{12.194}$$

As in Eq. (12.159), the limits of integration have been extended to $\pm\infty$. This integral can be evaluated on the complex x-plane, closing the contour with the upper half-circle. It is then given by the residue of the pole at $x = i\Gamma_n/2$, and the result is

$$A_n \simeq \frac{2\pi}{I}\left[\frac{1}{\Gamma_n + i\,|\varepsilon|}\right] = \frac{2\pi}{\Gamma_n I}\left[\frac{1}{1+i\,|\varepsilon|\,/\Gamma_n}\right]. \tag{12.195}$$

Now we follow section 12.3.1. We express sums over n in terms of the n-averages and neglect variations of Γ_n ($\Gamma_n \simeq \Gamma$). Taking into account Eq. (12.161), we can write,

$$C^{(S)}_{\beta\alpha}(E,\varepsilon) = \overline{S^{\rm fl}_{\beta\alpha}(E)\; S^{\rm fl\,*}_{\beta\alpha}(E)} \times \left[\frac{1}{1+i\,|\varepsilon|\,/\Gamma_n}\right]. \tag{12.196}$$

Although this correlation function involves the resonance width, it is not a practical tool do to determine Γ because it is not an observable quantity. However, it helps us evaluating the Ericson correlation function, defined below.

Ericson's energy-correlation function is defined as

$$C^{(\sigma)}_{\beta\alpha}(E,\varepsilon) = \overline{\left[\sigma_{\beta\alpha}(E+\varepsilon)\,\sigma^*_{\beta\alpha}(E)\right]} - \left(\overline{\sigma_{\beta\alpha}(E)}\right)^2, \tag{12.197}$$

Looking at the contribution from the lth-partial wave to the cross section for spinless collision partners, and in the limit where the optical cross section does not contribute to the transition, we have

$$C^{(\sigma)}_{\beta\alpha}(E,\varepsilon) = \frac{\pi}{k^2}(2l+1)\,\overline{\left[S^{\rm fl}_{\beta\alpha}(E+\varepsilon)\; S^{\rm fl\,*}_{\beta\alpha}(E+\varepsilon)\;\; S^{\rm fl}_{\beta\alpha}(E)\; S^{\rm fl\,*}_{\beta\alpha}(E)\right]} - \left(\overline{\sigma^{\rm fl}_{\beta\alpha}(E)}\right)^2. \tag{12.198}$$

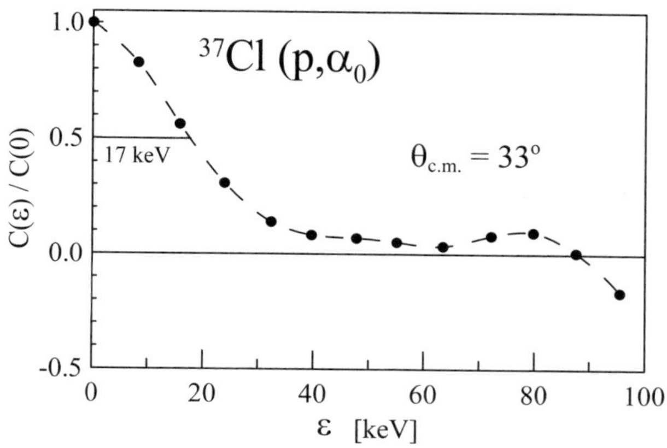

Fig. 12.6 Auto-correlation function for the ground-state level in the $^{37}\mathrm{Cl}(p,\alpha)$ reaction, at $\theta_{\mathrm{c.m.}} = 33^o$. The data is from [von Witsch *et al.* (1966)]. For detail, see the text.

Now we use properties of Gaussian distributions to reduce the above energy-average to energy-averages of products of the kind $S\,S^*$ (see [Kawai *et al.* (1973)] and appendix II of [Ericson (1963)]). We get

$$\begin{aligned} C^{(\sigma)}_{\beta\alpha}\,(E,\varepsilon) &= \frac{\pi}{k^2}\,(2l+1)\,\left|C^{(\mathrm{S})}_{\beta\alpha}\,(E,\varepsilon)\right|^2 \\ &= \left(\overline{\sigma^{\mathrm{fl}}_{\beta\alpha}(E)}\right)^2\,\frac{1}{1+\varepsilon^2/\Gamma^2}. \end{aligned} \tag{12.199}$$

The above equation has great practical importance. It allows the determination of the so-called *coherence energy*, Γ, from experimental values of the cross section over a fine mesh of collision energies. The coherence energy can be taken as an estimate of the average width of the resonances giving rise to the fluctuation cross section[24].

We finish this section with an illustration of the application of Ericson correlation functions to estimate the average width of the resonances giving rise to $\overline{\sigma^{\mathrm{fl}}_{\beta\alpha}}$. Figure 12.6 shows the auto-correlation function, $C(\varepsilon)$, for the $^{37}\mathrm{Cl}(p,\alpha)^{34}\mathrm{S}$ reaction, obtained from cross sections measured at $\theta_{\mathrm{c.m.}} = 33^o$. The data, taken from [von Witsch *et al.* (1966)], are normalized with respect to $C(\varepsilon = 0)$. This quantity corresponds to the correlation function of the

[24] One should keep in mind that the coherence energy gives a mean resonance width in a broad sense. Owing to the non-linear dependence of the correlation function on the width of each resonance, the sum over n cannot be expressed in terms of the mean width. Furthermore, the correlation function observed at a fixed angle involves many angular momenta, for which the resonance widths may be very different.

present section, normalized to its value at $\varepsilon = 0$. That is,

$$\frac{C(\varepsilon)}{C(0)} \simeq \frac{C^{(\sigma)}_{\beta 0)}(E,\varepsilon)}{C^{(\sigma)}_{\beta 0}(E,0)} = \frac{1}{1+\varepsilon^2/\Gamma^2}, \tag{12.200}$$

where β stands for the (p,α) channel, with ^{34}S in its ground state. In fact, Eq. (12.200) is only approximate because the expression for $C^{(\sigma)}_{\beta 0}(E,\varepsilon)$ was derived under the assumption that the contribution from direct processes to this channel is negligible, which is not quite true. The half width of the maximum at $\varepsilon = 0$ is the *coherence energy*, which gives an estimate of the resonance widths. From the figure, one gets $\Gamma \simeq 17$ keV, which corresponds to a half life of $\sim 4 \times 10^{-20}$ sec.

12.4.2 *Hanbury Brown-Twiss correlations*

Another feature of chaotic scattering can be found in the angular cross-correlation function[25] [Brink *et al.* (1964)],

$$C_{\beta\alpha}(\theta,\theta') = \frac{\overline{[\sigma_{\beta\alpha}(\theta)\,\sigma_{\beta\alpha}(\theta')]}}{\overline{[\sigma_{\beta\alpha}(\theta)]}\ \overline{[\sigma_{\beta\alpha}(\theta')]}} - 1, \tag{12.201}$$

where we used the short-hand notation

$$\sigma_{\beta\alpha}(\theta) \equiv \frac{d\sigma_{\beta\alpha}(\theta)}{d\Omega}.$$

As in the previous sections, overlines stand for energy-averages.

For simplicity, we restrict our discussion to non-elastic scattering and assume that the contribution of the optical amplitude to the cross section is neglible. In this way, the cross section is given exclusively by the fluctuation term.

The energy averaged cross sections appearing in the denominator of Eq. (12.201) have already been calculated and it is given by Eq. (12.185). Using the this result, the denominator of Eq. (12.201) becomes

$$D = \left[\frac{1}{4k^2}\sum_l (2l+1)^2\ P_l^2(\cos\theta)\ \overline{\left|S^{\rm fl}_{\beta\alpha}(l)\right|^2}\right] \times \left[\frac{1}{4k^2}\sum_l (2l+1)^2\ P_l^2(\cos\theta')\ \overline{\left|S^{\rm fl}_{\beta\alpha}(l)\right|^2}\right]. \tag{12.202}$$

[25] Since the cross-correlation function is defined only for the cross section, we will omit the upper script (σ) in $C^{(\sigma)}_{\beta\alpha}(\theta,\theta')$.

The numerator of Eq. (12.201) is more complicated. Combining Eq. (9.29) with the angular momentum expansion of the scattering amplitude (Eq. (9.114)), it can be cast in the form

$$N = \sum_{l_1 l_2 l_3 l_4} B_{l_1}(\theta)\, B^*_{l_2}(\theta)\, B_{l_3}(\theta')\, B^*_{l_4}(\theta)(\theta') \times \overline{S^{\text{fl}}_{\beta\alpha}(l_1)\, S^{\text{fl}\,*}_{\beta\alpha}(l_2)\, S^{\text{fl}}_{\beta\alpha}(l_3)\, S^{\text{fl}\,*}_{\beta\alpha}(l_4)}, \tag{12.203}$$

where

$$B_{l_i}(\theta) = \frac{1}{2ik}\,(2l_i + 1)\, P_{l_i}(\cos\theta), \tag{12.204}$$

with an analogous expression for $B_{l_i}(\theta')$.

Now we use the property of random matrixes [Rice (1954); Agassi *et al.* (1975)]

$$\overline{S(l_1)\, S^*(l_2)\, S(l_3)\, S^*(l_4)} = \overline{S(l_1)\, S^*(l_2)} \times \overline{S(l_3)\, S^*(l_4)}\ \delta_{l_1 l_2}\, \delta_{l_3 l_4} + \overline{S(l_1)\, S^*(l_4)} \times \overline{S(l_3)\, S^*(l_2)}\ \delta_{l_1 l_4}\, \delta_{l_3 l_2}.$$

This result greatly simplifies Eq. (12.203). Calling respectively N_1 and N_2 the contributions from the first and the second terms above, we can write

$$C_{\beta\alpha}(\theta, \theta') = \frac{N_1}{D} + \frac{N_2}{D} - 1. \tag{12.205}$$

It can be easily checked that N_1 is equal to the denominator, D (Eq. (12.202)), so that N_1/D cancels the last term in the above equation. Thus, the angular cross-correlation function reduces to

$$C_{\beta\alpha}(\theta, \theta') = \frac{N_2}{D}. \tag{12.206}$$

Now we calculate N_2. Using the explicit form of the coefficients $B_{l_i}(\theta)$ and $B_{l_j}(\theta')$ we get obtain,

$$N_2 = \left[\frac{1}{4k^2}\sum_l (2l+1)^2\ \ P_l(\cos\theta)\ P_l(\cos\theta')\ \overline{S^{\text{fl}}_{\beta\alpha}(l)\ S^{\text{fl}\,*}_{\beta\alpha}(l)}\right]^2. \tag{12.207}$$

To get a closed expression for $C^{(\sigma)}_{\beta\alpha}(\theta, \theta')$, we make a few approximations. First, we assume that there is a single non-elastic channel and that the absorption into this channel is given by a *sharp disc* of radius R ($\sim R_{\text{P}} + R_{\text{T}}$). That is

$$\begin{aligned} \overline{S^{\text{fl}}_{\beta\alpha}(l)\ S^{\text{fl}\,*}_{\beta\alpha}(l)} &= a, \qquad \text{for } l \le kR \\ &= 0, \qquad \text{for } l > kR, \end{aligned} \tag{12.208}$$

where a is a constant. Next, we assume that $kR \gg 1$, so that the angular momentum can be treated as a continuous variable: $l \to \lambda = l + 1/2$, and use the asymptotic form of the Legendre polynomials (Eq. (5.270))

$$P_l(\cos\theta) \quad \longrightarrow \quad P(\lambda, \cos\theta) \simeq \left(\frac{2}{\pi\lambda\sin\theta}\right)^{1/2} \cos(\lambda\theta - \pi/4). \qquad (12.209)$$

The above equation leads to two important results for our calculations. The first is that we can write

$$(2l+1)^2 \; P_l^2(\cos\theta) \simeq 4\lambda^2 \, \frac{2}{\pi\lambda\sin\theta} \; \cos^2(\lambda\theta - \pi/4)\,, \qquad (12.210)$$

or, using a standard trigonometrical relation,

$$(2l+1)^2 \; P_l^2(\cos\theta) \simeq \frac{4\lambda}{\pi\sin\theta} \; \left[1 + \cos(2\lambda\theta - \pi/2)\right]. \qquad (12.211)$$

For large angular momenta (and not too forward or backward angles) the second term above oscillates rapidly, so that it does not contribute to the partial-wave sum. Dropping this term and replacing the sum over l by the integration over λ, we get

$$\sum_l (2l+1)^2 \; P_l^2(\cos\theta) \simeq \frac{4}{\pi\sin\theta} \int_0^{kR} d\lambda \; \lambda = \frac{2}{\pi\sin\theta} \; (kR)^2\,. \qquad (12.212)$$

Using this result and Eq. (12.208) in Eq. (12.202), we get

$$D = \left(\frac{a^2}{4k^2}\right)^2 \; \frac{4}{\pi^2 \; \sin\theta \; \sin\theta'} \; (kR)^4\,. \qquad (12.213)$$

The second important consequence of Eq. (12.209) for our calculations is the approximate relation

$$(2l+1)^2 \; P_l(\cos\theta) \; P_l(\cos\theta') \simeq 4\lambda^2 \frac{1}{\pi\lambda\sqrt{\sin\theta\,\sin\theta'}} \left[2\,\cos(\lambda\theta - \pi/4)\,\cos(\lambda\theta' - \pi/4)\right]. \quad (12.214)$$

To evaluate N_2, we use the trigonometric relation,

$$2\,\cos(\lambda\theta - \pi/4)\,\cos(\lambda\theta' - \pi/4) = \cos\left[(\theta + \theta')\,\lambda - \pi/2\right] + \cos(\varphi\lambda)\,,$$

where we have introduced the notation

$$\varphi = \theta - \theta', \qquad (12.215)$$

and neglect the contribution from $\cos\left[(\theta+\theta')\,\lambda+\pi/2\right]$ to the partial-wave sum, because it oscillates rapidly. Replacing the sum over l by an integral over λ, we get

$$\sum_l (2l+1)^2\ P_l(\cos\theta)\ P_l(\cos\theta') \simeq \frac{4}{\pi\sqrt{\sin\theta\,\sin\theta'}}\int_0^{kR} d\lambda\ \lambda\ \cos(\lambda\varphi). \tag{12.216}$$

The above integral can be easily evaluated and one gets

$$\sum_l (2l+1)^2\ P_l^2(\cos\theta)\ P_l^2(\cos\theta') \simeq \frac{4}{\pi\sqrt{\sin\theta\,\sin\theta'}}\left[\frac{kR}{\varphi}\ \sin(\varphi kR) - \frac{2}{\varphi^2}\ \sin^2(\varphi kR/2)\right]. \tag{12.217}$$

Using this result and Eq. (12.208) in Eq. (12.207), we obtain

$$N_2 = \left(\frac{a^2}{4k^2}\right)^2 \frac{4}{\pi^2\ \sin\theta\ \sin\theta'}\left[\frac{2kR\ \sin(\varphi kR)}{\varphi} - \frac{4}{\varphi^2}\ \sin^2(\varphi kR/2)\right]^2. \tag{12.218}$$

Finally, using Eqs. (12.218) and (12.202) in Eq. (12.206), we get the approximate expression for the angular cross-correlation function

$$C(\varphi) \simeq \left\{2\ \frac{\sin(\varphi kR)}{\varphi kR} - \left[\frac{\sin(\varphi kR/2)}{\varphi kR/2}\right]^2\right\}^2. \tag{12.219}$$

Note that the cross-correlation function depends on the angles θ and θ' only through the angular difference φ. Furthermore, due to the simple approximation of Eq. (12.208), it does not depend on the channel indices. These facts lead us to adopt the simplified notation

$$C_{\alpha\beta}(\theta,\theta')\ \rightarrow\ C(\varphi).$$

The approximate angular cross-correlation function of Eq. (12.219), which attains the correct value of unity at $\varphi=0$, clearly exhibits a Fraunhofer type interference oscillations known in the literature as Hanbury-Brown and Twiss oscillations [Hanbury-Brown and Twiss (1954)]. This effect is very important in several branches of physics. It was originally used to determine angular sizes of astronomical radio sources [Hanbury-Brown and Twiss (1954)]. More recently, It has been used in high-energy physics, condensed matter and atomic physics applications[26]. Our result differ a bit from the one derived by Brink *et. al.* [Brink *et al.* (1964)],

$$C(\varphi) = \left[\frac{2\ J_1(\varphi kR)}{\varphi kR}\right]^2, \tag{12.220}$$

[26]For a review, we refer to [Baym (1998)].

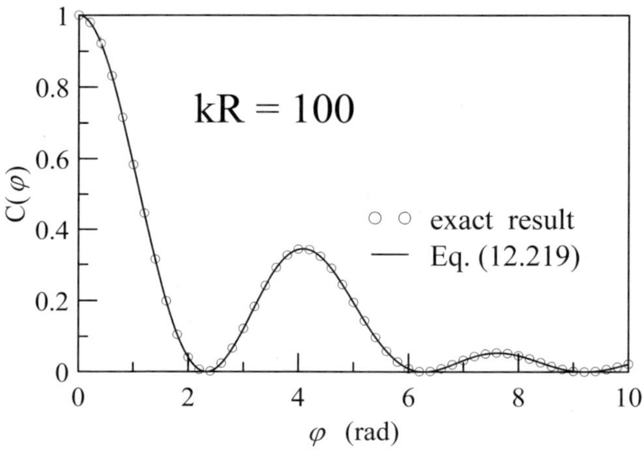

Fig. 12.7 Exact and approximate expressions for the angular cross correlation function. The solid line corresponds to the approximation of Eq. (12.219) whereas the open circles are exact results, obtained through the evaluation of the partial wave summation numerically, taking $\theta = \pi/4$ and $kR = 100$.

where $J_1(x)$ is the ordinary Bessel function of order 1. These authors also obtained a more approximate formula, given by,

$$C(\varphi) \simeq \left[\frac{\sin(\varphi k R)}{\varphi k R} \right]^2 . \tag{12.221}$$

In figure 12.7 we compare the approximate formula of Eq. (12.218) with the exact result. The latter was obtained by numerically summing over the angular momentum in Eqs. (12.202), (12.206) and (12.207). In the numerical calculation we took $\theta = \pi/4$ and $kR = 100$. The agreement is excellent. They all start at value of unity at $\varphi = 0$, and exhibit clear Fraunhofer diffraction oscillations. When used to extract the radius of the emitting source, the correlation function is constructed from the data using the defining equation. The value of the argument at which the correlation function attains the value 0.5, is, according to our result, $x = kR\ \Delta\varphi = 1.12$. This gives immediately the radius as

$$R = \frac{1.12}{k\ \Delta\varphi}.$$

The above treatment was made for the correlation among two neutral particle emitted from a random source. The case of charged particles, is more complicated and requires a more elaborate treatment. Further, in both neutral and charged particles angular correlations, final state interaction is important and must be taken into account. These extensions take us beyond our aim in this book, and we refer the reader to more dedicated publications [Baym (1998)].

12.5 Scattering in media and nuclear astrophysics

The discussion of scattering and reactions presented so far was restricted to particles in free space. In many applications, however, the collisions occur inside gases, or plasmas. In atomic physics, one deals with reaction rates in the gas of atoms or molecules. In astrophysics, the nuclear reactions which occur in stars, happen in the plasma of electron-stripped nuclear gases. The calculation of the reaction rates involve taking into account the statistical distribution of the energies of the reacting species[27].

In this section we present the essentials of the theory of reaction rates in gases and plasmas. The collision of particles is assumed to be mainly of two-body nature. The cross section of a given process is assumed to be not affected by the environment. The latter enters in determining the reaction rate for that process, R_{ij}, resulting from the collision of a particle of the species i with a particle of species j. It is given by

$$R_{ij} = \frac{N_i \, N_j}{1+\delta_{ij}} \, \big\langle v \; \sigma_{ij}(E) \big\rangle, \tag{12.222}$$

where N_i and N_j are the densities of the species i and j, with the Kronecker delta present to account for indistinguishability of identical species, and $\langle v \; \sigma_{ij}(E)\rangle$ is the thermal average

$$\big\langle v \; \sigma_{ij}(E(v))\big\rangle = \int d^3\mathbf{v} \;\, v \, P(v) \; \sigma_{ij}(E(v)). \tag{12.223}$$

Above, $E(v)$ is the kinetic energy of the relative motion associated with the velocity v and $P(v)$ is the thermal relative-velocity distribution. It is customary to employ the Maxwell-Boltzman form for $P(v)$, though other forms are sometimes used (see below). The normalized velocity distribution according to the Maxwell-Boltzman statistics is,

$$P(v) = \left(\frac{\mu}{2\pi k_{\rm B} T}\right)^{3/2} \; \exp\left[-\frac{\mu v^2/2}{k_{\rm B} T}\right] \tag{12.224}$$

where $k_{\rm B}$ is Boltzman's constant.

[27] A detailed description of the processes involved can be found, for example, in [Rolfs and Rodney (1988)].

With the above distribution, and knowing the dependence of the cross section on energy ($E = \mu v^2/2$), either from measurement or calculation, the integral appearing in the rate formula above can be evaluated either numerically or semi-analytically. Changing the variable of integration from v to E, we can rewrite the thermal average as,

$$\langle v\ \sigma_{ij}(E)\rangle = \left(\frac{8}{\pi\mu}\right)^{1/2} \left(\frac{1}{k_{\rm B}T}\right)^{3/2} \times\ I_{ij}(T), \tag{12.225}$$

where we have introduced the temperature-dependent energy integral,

$$I_{ij}(T) = \int_0^\infty dE\ E\ \sigma_{ij}(E)\ e^{-\,E/k_{\rm B}T}. \tag{12.226}$$

For charged particles, the cross section at low energies is dominated by barrier penetration. As an example, we consider the cross section for two protons in the interior of the Sun. The process $p+p \to d + e^+ + \nu_e$, is the principal reaction responsible for the burning of Hydrogen and the eventual formation of Helium. The temperature at the center of the Sun is about 1.5×10^7 Kelvin, which corresponds to an energy of 1.3 keV. At this energy the two protons, as they approach each other, encounter a Coulomb barrier. The potential is the sum of the Coulomb term,

$$V_{\rm C}(r) = \frac{1}{4\pi\epsilon_0}\ \frac{e^2}{r}, \tag{12.227}$$

where e is the absolute value of the electronic charge, with a short-range nuclear attraction. The latter can be approximated by the square well,

$$\begin{aligned} V_{\rm N}(r) &= -\,|V_0|\,, \qquad && \text{for } r \le R \\ &= \ 0, && \text{for } r > R. \end{aligned}$$

The nuclear potential is strong enough to overcome the Coulomb repulsion within its range, $R \sim R_{\rm P} + R_{\rm T}$. For $p+p$ collisions, $R_{\rm P} = R_{\rm T} = r_{\rm p} \sim 1$ fm, so that $R \simeq 2$ fm.

The potential barrier for the $p+p$ system is represented in figure 12.8. Its radius is $R_{\rm B} = R = 2$ fm and its height is $V_{\rm B} = V_{\rm C}\,(R_{\rm B}) = 720$ keV. For the fusion reaction to take place in a collision with energy E, the system must reach the internal region (shaded area). For this purpose, it has to tunnel through the barrier, as schematically represented in figure 12.8. Since the energy associated with the inner temperature of the sun, $E \sim 1.3$ keV, is much lower than the barrier height, the barrier is very thick. The classical turning point for this energy, given by the condition $V_{\rm C}\,(r_t) \sim 1.3$ keV, is $r_t\ \sim\ 1100$ fm $\gg R$. This leads to a vanishingly small tunneling

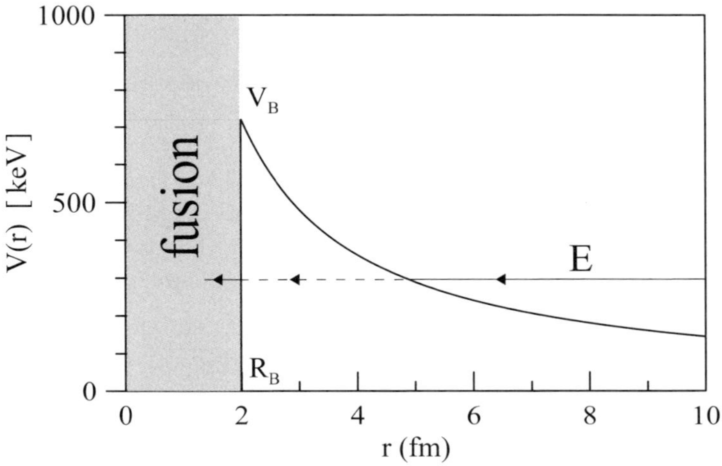

Fig. 12.8 The potential barrier for $p + p$ collisions.

probability. However, the reaction rate is appreciable because the densities of Eq. (12.222) are very high.

The tunneling probability can be evaluated by the WKB method (see section 5.2.2.5). It is given by

$$P = e^{-2\Phi}, \tag{12.228}$$

where

$$\Phi = \int_R^{r_t} dr\ \gamma(r). \tag{12.229}$$

Above

$$\gamma(r) = \frac{\sqrt{2\mu\ [V_C(r) - E]}}{\hbar} \tag{12.230}$$

is the absolute value of the local wave number, which is purely imaginary under the barrier.

For the Coulomb potential of Eq. (12.227), the integral can be carried out analytically and we get (see exercise 2),

$$\Phi = \eta\pi \times B(x), \tag{12.231}$$

where η is the Sommerfeld parameter defined in chapter 3, $x = R/r_t$ and $B(x)$ is the function

$$B(x) = \frac{2}{\pi}\left[\cos^{-1}\sqrt{x} - \sqrt{x(1-x)}\right]. \tag{12.232}$$

For all practical purposes x is very small. In the the case of $p+p$ collisions in the interior of the sun, for example, $x \sim 10^{-3}$. One immediately concludes that $B(x) \approx 1$ and the tunneling probability of Eq. (12.228) reduces to the so-called *Gamow factor*,

$$P \simeq e^{-2\eta\pi}. \tag{12.233}$$

The tunneling probability of the above equation carries an energy dependence through the Sommerfeld parameter. In the more general case of collisions of particles of charges $Z_1 e$ and $Z_2 e$, the Sommerfeld parameter is (see Eq. (3.3))

$$\eta = \frac{Z_1 Z_2 e^2}{4\pi\epsilon_0 \hbar v} = \frac{Z_1 Z_2 e^2}{4\pi\epsilon_0 \hbar c} \sqrt{\frac{\mu c^2}{2E}} \tag{12.234}$$

and the energy-dependent tunneling probability can be written as

$$P(E) = \exp\left[-b/\sqrt{E}\right], \tag{12.235}$$

with

$$b = 2\pi \left[\frac{Z_1 Z_2 e^2}{4\pi\epsilon_0 \hbar c}\right] \sqrt{\frac{\mu c^2}{2}}. \tag{12.236}$$

Accordingly, the cross section itself will contain an energy dependence of the above type. It is customary to define an astrophysical $\mathcal{S}$-factor where this barrier penetration factor is removed, defined as,

$$\mathcal{S}(E) = E\,\sigma(E)\,\exp[2\pi\eta] = E\,\sigma(E)\,\exp[b/\sqrt{E}]. \tag{12.237}$$

The factor E that multiplies the cross section is used to compensate the $1/k^2$ which appears in the latter. The resulting $\mathcal{S}$-factor is expected to show slow energy dependence even for reactions that occur at deeply sub-barrier energies, except in cases where a resonance dominates the cross section at the energies involved. With the introduction of $\mathcal{S}$, the integral of Eq. (12.226), which appears in the rate formula, becomes,

$$I(T) = \int_0^\infty dE\ \mathcal{S}(E)\ \exp\left[-\left(\frac{b}{\sqrt{E}} + \frac{E}{k_B T}\right)\right]. \tag{12.238}$$

For non-resonant reactions, the $\mathcal{S}$-factor is expected to be a slowly varying function of the energy. Further, the exponential factor in integrand of Eq. (12.238) exhibits a peak at the energy,

$$E_0 = \left[\frac{b\,k_B T}{2}\right]^{2/3}, \tag{12.239}$$

called the *Gamow energy*. A measure of the value of the integral can be obtained by extracting the $\mathcal{S}$-factor from the integral at $E = E_0$ and evaluating the resulting expression through an expansion to second order of the exponent. The resulting Gaussian integral is readily calculated to yield,

$$I(T) \simeq \mathcal{S}(E_0)\ \exp\left[-\left(\frac{b}{\sqrt{E_0}} + \frac{E_0}{k_{\rm B}T}\right)\right]\ \sqrt{\frac{8\pi}{3b}}\ E_0^{5/4}. \tag{12.240}$$

A simper expression can be obtained if we express b in terms of the Gamow energy. According to Eq. (12.239), we have

$$b = \left(\frac{2}{k_{\rm B}T}\right)\ E_0^{3/2} \tag{12.241}$$

and using this result, Eq. (12.240) takes the form

$$I(T) \simeq \mathcal{S}(E_0)\ E_0 \sqrt{\frac{4\pi}{\tau}}\ e^{-\tau}, \tag{12.242}$$

where

$$\tau = \frac{3\,E_0}{k_{\rm B}T} \tag{12.243}$$

is the argument of the exponential of Eq. (12.240).

A more precise treatment involves the introduction of an effective $\mathcal{S}$-factor which better represents the value of the integral. This factor is defined by the condition that it makes Eq. (12.242) exact. That is

$$I(T) = \mathcal{S}_{\rm eff}\ E_0 \sqrt{\frac{4\pi}{\tau}}\ e^{-\tau}. \tag{12.244}$$

Comparing this expression with Eq. (12.238) and changing the integration variable: $E \to x = E/k_{\rm B}T$, we get

$$\mathcal{S}_{\rm eff} = \frac{3}{2}\left[\frac{e^{\tau}}{\sqrt{\pi\tau}}\right]\int_0^\infty dx\ \mathcal{S}\left(x\,k_{\rm B}T\right)\ \exp\left[-\left(x + \frac{a}{\sqrt{x}}\right)\right], \tag{12.245}$$

with

$$a = \frac{b}{\sqrt{k_{\rm B}T}}.$$

The evaluation of the integral $I(T)$ or $\mathcal{S}_{\rm eff}$ relies on an expansion of $\mathcal{S}(E)$ around either the Gamow energy, E_0 or around $E = 0$. For the Gamow energy, we use the expansion,

$$\begin{aligned}\mathcal{S}(E) &= \sum_{n=0}^{\infty} S^{(n)}\left(E_0\right)\ \frac{1}{n!}\ \left(E - E_0\right)^n \\ &= \sum_{n=0}^{\infty} S^{(n)}\left(E_0\right)\ \frac{1}{n!}\ \sum_{m=0}^{\infty}(-)^n\binom{n}{m}\ E_0^m\ E^{n-m}.\end{aligned} \tag{12.246}$$

Above, we used the notation

$$\mathcal{S}^{(n)}(E_0) = \left[\frac{d^n \mathcal{S}(E)}{dE^n}\right]_{E=E_0}. \tag{12.247}$$

For the expansion around $E = 0$, we have just to replace $E_0 \to 0$ in the above equations. That is

$$\mathcal{S}(E) = \sum_{n=0}^{\infty} \mathcal{S}^{(n)}(0)\ \frac{1}{n!}\ E^n \tag{12.248}$$

and

$$\mathcal{S}^{(n)}(0) = \left[\frac{d^n \mathcal{S}(E)}{dE^n}\right]_{E=0}. \tag{12.249}$$

Using these expansions, the effective $\mathcal{S}$-factor becomes:

a) for the expansion around $E = E_0$ (with the notation $x_0 = E_0/k_B T$):

$$\mathcal{S}_{\text{eff}} = \frac{3}{2}\left[\frac{e^{\tau}}{\sqrt{\pi\tau}}\right] \sum_{n=0}^{\infty} \frac{(k_B T)^n}{n!}\ \mathcal{S}^{(n)}(E_0) \sum_{m=0}^{\infty} (-)^n \binom{n}{r}\ x_0^m\ f_{n-m}(a) \tag{12.250}$$

b) for the expansion around $E = 0$:

$$\mathcal{S}_{\text{eff}} = \frac{3}{2}\left[\frac{e^{\tau}}{\sqrt{\pi\tau}}\right] \sum_{n=0}^{\infty} \frac{(k_B T)^n}{n!}\ \mathcal{S}^{(n)}(0)\ f_n(a). \tag{12.251}$$

In the above equations, we have introduced the integrals

$$f_n(a) = \int_0^{\infty} dx\ x^n\ \exp\left[-\left(x + \frac{a}{\sqrt{x}}\right)\right]. \tag{12.252}$$

Using the asymptotic series expansion technique of section 3.3.1, it can be shown that,

$$f_0 = \frac{2}{3}\ (\pi\tau)^{1/2}\ e^{-\tau}\ \left[1 + \frac{5}{12\,\tau} - \frac{35}{2!\,(12)^2\,\tau^2} + \cdots\right]$$

$$f_1 = \frac{2}{3}\ (\pi\tau)^{1/2}\ e^{-\tau}\ \frac{\tau}{3}\ \left[1 + \frac{35}{12\,\tau} + \frac{385}{2!\,(12)^2\,\tau^2} + \cdots\right]$$

$$f_2 = \frac{2}{3}\ (\pi\tau)^{1/2} e^{-\tau}\ \left(\frac{\tau}{3}\right)^2\ \left[1 + \frac{89}{12\,\tau} + \frac{5005}{2!\,(12)^2\,\tau^2} + \cdots\right]$$

We thus find to lowest order in $1/\tau$,

$$\mathcal{S}_{\text{eff}} = \mathcal{S}(E_0)\left[1+\frac{1}{\tau}\left(\frac{5}{12}+\frac{5\,S^{(1)}(E_0)\,E_0}{2\,\mathcal{S}(E_0)}+\frac{S^{(2)}(E_0)\,E_0^2}{\mathcal{S}(E_0)}\right)\right], \quad (12.253)$$

for the expansion around $E = E_0$, and

$$\mathcal{S}_{\text{eff}} = \mathcal{S}(0)\left[1+\frac{5}{12\,\tau}+\frac{S^{(1)}(0)\,\left(E_0+\frac{35}{36}\,k_{\text{B}}T\right)}{\mathcal{S}(0)}\right.$$
$$\left.+\frac{\mathcal{S}^{(2)}(0)\,E_0}{\mathcal{S}(0)}\left(\frac{E_0}{2}+\frac{89}{72}\,k_{\text{B}}T\right)\right], \quad (12.254)$$

for the expansion around $E = 0$.

The above approximations of the effective $\mathcal{S}$-factor allows writing a closed expressions for the reaction rate formula, which is used in nucleosynthesis calculations. The rate formula in terms of the effective $\mathcal{S}$-factor for non-resonant reactions, is now the explicit function of temperature,

$$R_{i,j} = \left(\frac{N_i\,N_j}{1+\delta_{ij}}\right)\,\left(\frac{32\,E_0}{3\mu}\right)^{1/2}\,\frac{\exp\left(-\,3E_0/k_{\text{B}}T\right)}{k_{\text{B}}T}\,\,\mathcal{S}_{\text{eff}}^{(i,j)}(T). \quad (12.255)$$

Of course one can numerically evaluate the energy integral. However analytical expressions are always welcome as they show the explicit temperature dependence and may speed up the elemental production calculation in different stellar environments. The case of resonant reactions involves different type of approximations and depends to a large extent on whether the resonance is wider or narrower than the Gamow peak.

To get an idea about the parameters E_0 and τ in different stellar cases, and test the approximations developed above, we give the numerical values of these quantities,

$$E_0 = 5.0\left(\frac{2\mu}{N_0}\right)^{1/3}\,(Z_1Z_2)^{2/3}\,\,(k_{\text{B}}T)^{2/3}\,\,[\text{keV}] \quad (12.256)$$

and

$$\tau = 1.6\left(\frac{2\mu}{N_0}\right)^{1/3}\,\frac{(Z_1Z_2)^{2/3}}{(k_{\text{B}}T)^{1/3}}. \quad (12.257)$$

In the Sun's core the temperature is about 15 million degrees Kelvin. This gives $k_{\text{B}}T \simeq 1.3$ keV. Thus we get for the dominant process $p+p \to d+e^+ +\nu_e$, $E_0 = 5.9$ keV and $\tau = 5.9$. The value of τ is large enough to make the approximation for $\mathcal{S}_{\text{eff}}$ to first order in $1/\tau$ quite reasonable. For the purpose of comparison we give below a table for the three PP

Table 12.1 The Gamow energy for several processes that contribute to nucleosynthesis.

	T [10^7 Kelvin]	1.0	1.1	1.2	1.3	1.4	1.5
	$k_B T$ [keV]	0.86	0.95	1.03	1.12	1.21	1.29
PP-I							
	$p+p$	4.51	4.80	5.10	5.37	5.64	5.91
E_0 [keV]	$p+d$	4.96	5.29	5.60	5.91	6.21	6.50
	^{3}He $+^3$He	16.37	17.44	18.48	19.50	20.48	21.45
PP-II							
	^{3}He $+^4$He	17.10	18.22	19.31	20.37	21.40	22.41
E_0 [keV]	$p+^7$Li	11.29	12.04	12.75	13.45	14.14	14.80
PP-III							
E_0 [keV]	$p+^7$Be	13.68	14.58	15.45	16.30	17.12	17.93

reaction chains the lead to the formation of helium in the Sun; the PP-I, which involves the $p+p$, the $p+d$, and the ^{3}He + ^{3}He processes, the PP-II which involves $p+^7$ Li and ^{3}He $+^4$ He, and the PP-III, containing the reaction $p+^7$ Be. The Gamow energies for the above processes are given in table 12.1.

12.6 Multiple scattering

A microscopic framework to describe the scattering of composite systems is the multiple scattering theory. We discuss in this section the collision of a structureless projectile, with position vector $\mathbf{r}_1$, with a target composed of N particles with position vectors $\mathbf{r}_2, ..., \mathbf{r}_{N+1}$.

For simplicity, we consider the case of distinguishable spin zero particles. The projectile-target interaction can be written as

$$V(\mathbf{r}_1; \mathbf{r}_2, \cdots, \mathbf{r}_{N+1}) = \sum_{j=2}^{N+1} V^{(1j)}(\mathbf{r}_1 - \mathbf{r}_j) \, . \tag{12.258}$$

Calling h the intrinsic Hamiltonian of the target and K the kinetic energy operator of the projectile relative to the center of mass of the target, the full Hamiltonian of the projectile-target system can be written as,

$$H = K + h + V. \tag{12.259}$$

The scattering of the projectile from the target can best be formulated, as usual, using the LS equation for the T-matrix,

$$T = V + V\, G_0^{(+)}\, T. \tag{12.260}$$

Here, T is an operator in the $N+1$ particle Hilbert space and the non-interacting Green's function is

$$G_0^{(+)} = \frac{1}{E - K - h + i\varepsilon}. \tag{12.261}$$

We now introduce the two-body T-matrix of the pair ij in the space of $N+1$ particles, defined by the LS equation[28]

$$t^{(ij)} = V^{(ij)} + V^{(ij)}\, G_0^{(+)}\, t^{(ij)} = V^{(ij)} + t^{(ij)}\, G_0^{(+)}\, V^{(ij)}. \tag{12.262}$$

Note that $t^{(ij)}$ is different from the T-matrix for the collision of particle i with particle j in free space, which we will denote by $\bar{t}^{(ij)}$. This comes about from the fact that, contrary to the latter, $t^{(ij)}$ acts on the space of $N+1$ particles, due to the presence of the intrinsic target Hamiltonian in $G_0^{(+)}$.

The above two-body T-matrix is related to the T-matrix of our scattering problem by a LS-like equation. To derive this equation, we introduce an auxiliary T-matrix, defined by the equation

$$T^{(ij)} = V^{(ij)} + V^{(ij)}\, G_0^{(+)}\, T. \tag{12.263}$$

This matrix satisfies the relation,

$$T = \sum_{j=2}^{N+1} T^{(ij)} \qquad (\text{for } i = 1). \tag{12.264}$$

This can be trivially proved summing Eq. (12.263) over j and comparing the result with Eq. (12.260).

[28]Although the projectile is always labeled by $i = 1$, we leave it open because this general formulation is more convenient for the next section.

The next step is to express the potentials appearing in Eq. (12.258) in terms of the two-body T-matrix. From Eq. (12.262), we get,

$$V^{(ij)} = \left[1 + t^{(ij)}\, G_0^{(+)}\right]^{-1} t^{(ij)} \tag{12.265}$$

and, accordingly,

$$T^{(ij)} = \left[1 + t^{(ij)}\, G_0^{(+)}\right]^{-1} \left[t^{(ij)} + t^{(ij)}\, G_0^{(+)}\, T\right]. \tag{12.266}$$

Multiplying the above equation from the left by $\left[1 + t^{(ij)}\, G_0^{(+)}\right]$, re-ordering the terms and using Eq. (12.264), we obtain

$$T^{(ij)} = t^{(ij)} + t^{(ij)}\, G_0^{(+)} \sum_{k \neq j} T^{(ik)}. \tag{12.267}$$

Setting $i = 1$ and summing over j, one gets the equation for the exact T-matrix,

$$T = \sum_{j=2}^{N+1} \left[t^{(1j)} + t^{(1j)}\, G_0^{(+)} \sum_{k \neq j} T^{(1k)} \right]. \tag{12.268}$$

Iterating the above equation results in the multiple scattering series. The first term represents the single scattering approximation and can be used to define the corresponding optical potential for elastic scattering.

In practical calculations at high enough energies, one can approximate the $t^{(ij)}$ operator of Eq. (12.262) by its counterpart in free space, $\bar{t}^{(ij)}$ (*impulse approximation*). The latter is defined by a LS equation similar to Eq. (12.262) but replacing $G_0^{(+)}$ by the two-body Green's function in free space. That is, [29].

$$\bar{t}^{(ij)} = V^{(ij)} + V^{(ij)}\, g_{0\,(ij)}^{(+)}\, \bar{t}^{(ij)}, \tag{12.269}$$

with

$$g_{0\,(ij)}^{(+)} = \frac{1}{E - K_0 + i\varepsilon}, \tag{12.270}$$

where K_0 is the kinetic energy operator in the ij system. Note that $g_{0\,(ij)}^{(+)}$ is the inverse of $E - K_0$ whereas $G_0^{(+)}$ is the inverse of the $E - (h + K)$. The difference between the two kinetic energy operators is a term that goes as $1/N$. Thus, for large N and high enough energies ($E \gg h$) the difference

[29]For a more general presentation of this topic we refer to [Goldberger and Watson (2004)].

between the two Green's functions can be ignored. One can then use the impulse approximation, namely $t^{ij} \simeq \bar{t}^{ij}$.

The full multiple scattering series is generated by expanding Eq. (12.268) and taking the impulse approximation. That is,

$$T = \sum_{j=2}^{N+1} \bar{t}^{(1j)} + \sum_{j=2}^{N+1} \left(\bar{t}^{(1j)} G_0^{(+)} \sum_{k \neq j=2}^{N+1} \bar{t}^{(1k)} \right) + \cdots \tag{12.271}$$

The first order term of the above series corresponds to a single collision of the projectile with one particle in the target. The second order term represents processes where the impinging particle scatters off one of the particles in the target, propagates freely, and then suffers a second scattering from another particle in the target. Higher order terms are interpreted similarly, corresponding to successive collisions with a larger number of target particles. The ground state expectation value of the first order term of the multiple scattering series gives rise to the Lax or $t\rho$ optical potential, of wide use in nuclear and other many-body systems.

We should mention that a refined formulation of the multiple scattering theory involving fermions, such as the case in nuclear physics, must take into account the effect of the anti-symmetrization owing to to the identical nature of the projectile and the target constituent particles. This results in non-localities, short-range correlations and the so-called Pauli Blocking effects. These effects can be incorporated into the above formalism as was done in [Kerman *et al.* (1959)] (see also [Feshbach (1992)]). The extension of the multiple scattering theory to the collision between two nuclei can be found in, e.g., [Hussein *et al.* (1991)]. One way of investigating these modifications, in particular the short-range correlations, is through the study of the contribution of the second order term in the multiple scattering series [Kerman *et al.* (1959); Feshbach (1992)].

12.7 The three-body problem and Faddeev equations

In the previous section we laid the grounds for discussing few-particle scattering systems. This type of situation arises when calculating scattering observables for a particle colliding with a bound system of two particles or more. The three-body problem has received a great amount of attention in the literature, as they frequently appear in nuclear and atomic physics (e.g. the scattering of an atom from a diatomic molecule). For this reason and to give credit to the large body of literature on the subject, we develop in this section the fundamentals of the Faddeev formulation of the three-body scattering problem.

To derive the Faddeev equations we consider Eqs. (12.260) and (12.263), in the case of a three particle (labelled 1, 2, and 3) system, and write for the kinetic energy operator

$$K = \sum_{i=1}^{3} K_i.$$

Accordingly, the free Green's function is

$$G_0^{(+)} = \frac{1}{E - K_1 - K_2 - K_3 + i\varepsilon}.$$

To simplify the notation we write,

$$V^{(12)} = V_3, \quad V^{(13)} = V_2 \quad \text{and } V^{(23)} = V_1.$$

The corresponding auxiliary T-matrices of Eq. (12.263) are denoted by T_3, T_2, and T_1, respectively. Then the full T-matrix can be written as [Faddeev (1969)],

$$T = T_1 + T_2 + T_3. \tag{12.272}$$

The auxiliary T-matrices are related to the two-body T-matrices by Eq. (12.268). Using an analogous notation for the two-body T-matrices ($t^{(ij)} \equiv t_k$), Eq. (12.268) leads to the relations

$$T_1 = t_1 + t_1\ G_0\ [T_2 + T_3]\,, \tag{12.273}$$

$$T_2 = t_2 + t_2\ G_0\ [T_1 + T_3]\,, \tag{12.274}$$

$$T_3 = t_3 + T_3\ G_0\ [T_1 + T_2]\,. \tag{12.275}$$

The above set of three equations can be written in matrix form as follows,

$$\begin{pmatrix} T_1 \\ T_2 \\ T_3 \end{pmatrix} = \begin{pmatrix} t_1 \\ t_2 \\ t_3 \end{pmatrix} + \begin{pmatrix} 0 & t_1 & t_1 \\ t_2 & 0 & t_2 \\ 0 & t_3 & t_3 \end{pmatrix} G_0^{(+)} \begin{pmatrix} T_1 \\ T_2 \\ T_3 \end{pmatrix}. \tag{12.276}$$

Eq. (12.276) is a matrix form of the Faddeev equations. In several applications two of the particles are initially in a bound state. For such a case the above equations are modified as was originally done by Lovelace [Lovelace (1964)]. If, say, particles 2 and 3 are initially bound then the scattering with particle 1 will result in three processes. The elastic channel and two rearrangement channels, namely,

$$\begin{aligned} 1 + (2+3) &\to 1 + (2+3), \\ &\to 2 + (1+3), \\ &\to 3 + (1+2). \end{aligned}$$

Using the notation $\mathcal{T}_{i'i}$ for the T-matrix associated with the process

$$i + (j+k) \to i' + (j' + k'),$$

we write the LS equations,

$$\mathcal{T}_{11} = (V_2 + V_3) \ + \ (V_2 + V_3) \ G_1^{(+)} \ \mathcal{T}_{11} \tag{12.277}$$

$$\mathcal{T}_{21} = (V_1 + V_3) \ + \ (V_1 + V_3) \ G_1^{(+)} \ \mathcal{T}_{11} \tag{12.278}$$

$$\mathcal{T}_{31} = (V_1 + V_2) \ + \ (V_1 + V_2) \ G_1^{(+)} \ \mathcal{T}_{11}, \tag{12.279}$$

where

$$G_1^{(+)} = \frac{1}{E - K - V_1 + i\epsilon} \equiv \frac{1}{E - K - V^{(23)} + i\epsilon} \tag{12.280}$$

is the Green's function in which the incident particle is free (only the remaining two particle interact).

Eqs. (12.277) to (12.279) can be rewritten into a form similar to Faddeev's. To see this we define the auxiliary amplitudes,

$$\mathfrak{T}_1 = V_1 \ G_1^{(+)} \ \mathcal{T}_{11} \tag{12.281}$$

and, for $i = 2, 3$

$$\mathfrak{T}_i = V_i + V_i \ G_1^{(+)} \ \mathcal{T}_{11}. \tag{12.282}$$

Since $V_1 \, G_1^{(+)} = t_1 \, G_0^{(+)}$ and using the equation for $\mathcal{T}_{11}$ we can write,

$$\mathfrak{T}_1 = t_1 \ G_0^{(+)} \ (\mathfrak{T}_2 + \mathfrak{T}_3) \tag{12.283}$$

Similarly,

$$\mathfrak{T}_2 = t_2 + t_2 \ G_0^{(+)} \ (\mathfrak{T}_1 + \mathfrak{T}_3) \tag{12.284}$$

and

$$\mathfrak{T}_3 = t_3 + t_3 \ G_0^{(+)} \ (\mathfrak{T}_1 + \mathfrak{T}_2). \tag{12.285}$$

The above three equations are similar to the Faddeev equations except for absence of the inhomogeneous term in the equation for $\mathfrak{T}_1$.

In matrix form the Lovelace equations are,

$$\begin{pmatrix} \mathfrak{T}_1 \\ \mathfrak{T}_2 \\ \mathfrak{T}_3 \end{pmatrix} = \begin{pmatrix} 0 \\ t_2 \\ t_3 \end{pmatrix} + \begin{pmatrix} 0 & t_1 & t_2 \\ t_2 & 0 & t_2 \\ 0 & t_3 & t_3 \end{pmatrix} G_0^{(+)} \begin{pmatrix} \mathfrak{T}_1 \\ \mathfrak{T}_2 \\ \mathfrak{T}_3 \end{pmatrix} \tag{12.286}$$

The full T-matrices are given by the auxiliary T-matrix as

$$\mathcal{T}_{11} = \mathfrak{T}_1 + \mathfrak{T}_2,$$
$$\mathcal{T}_{21} = V_1 + \mathfrak{T}_1 + \mathfrak{T}_3,$$
$$\mathcal{T}_{31} = V_1 + \mathfrak{T}_1 + \mathfrak{T}_2.$$

It should be clear to the reader that the Faddeev equations or Lovelace equations have a compact kernel as defined in section 5.1.1. This makes the solution of the integral equations amenable to matrix inversion. The one issue that the Faddeev equations does not address fully is the issue of the long range Coulomb interaction. There are other formulations of the three particle scattering problem which rely on the wave functions. In particular we mention the contribution of Alt, Grassber and Sandhas [Alt *et al.* (1964)]. For more detailed discussion of the problem we refer the readers to more dedicated monographs (see, e.g. [Joachain (1983); Newton (2002)])

12.8 The R-matrix method

The calculation of resonance reactions relies on adopting the Schrodinger equation such as to contain both open and closed channels. In Feshbach theory, this is accomplished by using a decomposition of the Hilbert space through the introduction of projection operators.

An earlier approach to resonances was proposed by Kapur and Peierls [Kapur and Peierls (1938)] and an alternative method was introduced by Wigner [Wigner (1947a,b)] and developed further by Wigner and Eisenbud [Wigner and Eisenbud (1947)]. The latter is known as the R-matrix method. It uses the concept of internal and external regions in configuration space to solving the Schrödinger equation for resonance reactions. In this sense, the R-matrix is similar to the method of Kapur and Peierls. However, the boundary conditions of the former are easier to handle. The idea is to calculate a matrix, called the R-matrix, using the internal wave functions at his radius and its derivative. Then, matching the internal wave function with its asymptotic form, the S-matrix is expressed in terms of the R-matrix.

The R-matrix method can be used in two different ways. The first, usually called the *calculable R-matrix*, consists in using the R-matrix approach to solve the Schrödinger equation, either in single-channel or in coupled-channel problems. First, one solves the original equation for the projectile-target system using some bound state boundary condition. In this way, one finds a set of many-body states, which is used as a basis to expand the scattering wave function in the internal region. The R-matrix, which is the inverse of the logarithmic derivatives at the channel radii, is then calculated. Matching this wave functions with its asymptotic form at the channel radius, the scattering matrix is expressed in terms of the R-matrix. The second way to use the R-matrix method, usually called the *phenomenological R-matrix*, is to adopt the general form of the R-matrix simply as a parametrization, and determine the parameters fitting experimental data. These parameters have an important physical meaning. This approach is widely used to parametrize low energy cross section in nuclear astrophysics.

12.8.1 *The calculable R-matrix*

Now we introduce the concept of the R-matrix and show how it is related to the S-matrix in a scattering problem. For simplicity, we consider the simple case of a single channel, in a collision of spinless particles[30].

The decomposition of the physical space into internal and external regions, adopted in the R-matrix approach, can be put into mathematical form through a manipulation of the full Schrödinger equation. We use the formalism developed in Chapter 2. The full equation for a given angular momentum,

$$\Big[H - E\Big]\, \psi(Elm;\mathbf{r}) = 0, \tag{12.287}$$

has the solution

$$\psi(Elm;\mathbf{r}) = \frac{u_l(r)}{kr}\, Y_{lm}(\hat{r}). \tag{12.288}$$

The radial wave function satisfies the radial equation (Eq. (2.9))

$$-\frac{\hbar^2}{2\mu}\left[\frac{d^2}{dr^2} - \frac{l(l+1)}{r^2}\right] u_l(r) + V(r)\, u_l(r) = E\, u_l(r), \tag{12.289}$$

where, for simplicity, $V(r)$ is assumed to be a short-range potential. Thus, in the region external to the potential ($r > a$) the radial wave function has the asymptotic form

$$u_l^{\rm ext}(r) = \frac{i}{2}\left[\hat{h}_l^{(-)}(kr) - S_l\, \hat{h}_l^{(+)}(kr)\right], \tag{12.290}$$

[30] We follow section 4 of [Descouvemont (2003)].

where the incoming and outgoing Ricatti-Haenkel functions, $\hat{h}_l^{(-)}$ and $\hat{h}_l^{(+)}$, behave asymptotically as (see chapter 2),

$$\hat{h}_l^{(\pm)}(kr) \to e^{\pm i\,(kr - l\pi/2)}. \tag{12.291}$$

Now we introduce the set of radial wave functions, $\varphi_l^{(n)}(r)$, which are regular solutions of radial equation in the internal region,

$$-\frac{\hbar^2}{2\mu}\left[\frac{d^2}{dr^2} - \frac{l(l+1)}{r^2}\right] \varphi_l^{(n)}(r) + V(r)\,\varphi_l^{(n)}(r) = E_l^{(n)}\,\varphi_l^{(n)}(r) \quad (\text{for } r \le a), \tag{12.292}$$

with the boundary condition,

$$\left[\frac{d\varphi_l^{(n)}(r)}{dr}\right]_{r=a} = 0. \tag{12.293}$$

We remark that although the results presented in this section are all consistent, this boundary condition leads to mathematical difficulties (see sections 3.5 of [Descouvemont and Baye (2010)]). These difficulties can be eliminated by the R-matrix formulation of Bloch [Bloch (1957); Descouvemont and Baye (2010)].

The boundary condition for $\varphi_l^{(n)}(r)$ (Eq, (12.293)) together with the condition of being regular at the origin, establishes an eigenvalue equation, which has as solutions the set of energies $E_l^{(n)}$ and eigenfunctions $\varphi_l^{(n)}(r)$. These eigenfunctions are orthogonal and can be normalized as

$$\int_0^a dr\;\varphi_l^{(n)}(r)\;\varphi_l^{(n')}(r) = \delta_{nn'}. \tag{12.294}$$

In the R-matrix method, this set of eigenfunctions is used to expand the radial wave function in the internal region. That is, one writes

$$u_l(r) = \sum_n A_n\;\varphi_l^{(n)}(r) \tag{12.295}$$

and the coefficients of the expansion can be obtained taking the scalar product with each basis state and using the orthonormality relation. One gets

$$A_n = \int_0^a dr\;\varphi_l^{(n)}(r)\,u_l(r). \tag{12.296}$$

A more convenient expression can be obtained multiplying Eq. (12.289) by $\varphi_l^{(n)}(r)$, multiplying Eq. (12.292) by $u_l(r)$, and subtracting the results. One gets

$$\begin{aligned}\varphi_l^{(n)}(r)\;u_l''(r) - u_l(r)\;\varphi^{(n)\,\prime\prime}(r) &\equiv \frac{d}{dr}\left[\varphi_l^{(n)}(r)\,u_l'(r) - u_l(r)\,\varphi^{(n)\,\prime}(r)\right]\\ &= -\frac{2\mu}{\hbar^2}\left(E - E_l^{(n)}\right)\;\varphi_l^{(n)}(r)\;u_l(r),\end{aligned} \tag{12.297}$$

where primes and double primes denote respectively first and second derivatives with respect to r. Integrating both sides over r in the $\{0, a\}$ range and recalling that $u_l(0) = \varphi_l^{(n)}(0) = 0$ and $\varphi_l^{(n)\prime}(a) = 0$, we get

$$u_l'(a)\,\varphi_l^{(n)}(a) = \frac{2\mu}{\hbar^2}\left(E_l^{(n)} - E\right)\int_0^a dr\;\varphi_l^{(n)}(r)\,u_l(r). \tag{12.298}$$

Finally, comparing the above equation with Eq. (12.296), we get the expansion coefficient

$$A_n = \frac{\hbar^2}{2\mu}\left[\frac{\varphi_l^{(n)}(a)\;u_l'(a)}{E_l^{(n)} - E}\right]. \tag{12.299}$$

With the above form of the coefficients we can write the exact radial wave function as,

$$u_l(r) = \frac{\hbar^2}{2\mu a}\;(a\,u_l'(a))\sum_n \frac{\varphi_l^{(n)}(a)}{E_l^{(n)} - E}\;\varphi_l^{(n)}(r). \tag{12.300}$$

We define the R-matrix for the one-channel problem as the inverse of the logarithmic derivative of $u_l(r)$ evaluated at $r = a$,

$$R_l(E) = \frac{1}{\mathcal{L}_l} = \frac{1}{a}\left[\frac{u_l(a)}{u_l'(a)}\right]. \tag{12.301}$$

Using Eq. (12.300), one immediately gets

$$R_l(E) = \sum_n \frac{\gamma^{(n)}\,\gamma^{(n)}}{E_l^{(n)} - E}. \tag{12.302}$$

Above, we have introduced the reduced widths,

$$\gamma^{(n)} = \left(\frac{\hbar^2}{2\mu a}\right)^{1/2}\varphi_l^{(n)}(a). \tag{12.303}$$

Thus, the reduced widths of the resonances are determined by values of internal basis functions at the surface. Clearly they are dependent on the radius that defines the surface which is usually taken as a free parameter. Note that in the single-channel case Eq. (12.302) does not correspond to a true matrix. It reduces reduces to a function.

To use the R-matrix method to calculate cross sections, we must express the S-matrix elements in terms of the R-matrix. The required relation is readily obtained from the external wave function at the surface, Eq. (12.290), namely,

$$S_l = \frac{h_l^{(-)}(ka)}{h_l^{(+)}(ka)}\left[\frac{1 - \left(L_l^{(+)}\right)^*\,R_l(E)}{1 - L_l^{(+)}\,R_l(E)}\right]. \tag{12.304}$$

Above, we have introduced the logarithmic derivative of the outgoing Ricatti-Haenkel function at the surface radius,

$$L_l^{(+)} = \frac{ka}{h_l^{(+)}(ka)} \left[\frac{dh_l^{(+)}(\rho)}{d\rho}\right]_{\rho=ka}. \tag{12.305}$$

Although the R-matrix depends on radius parameter, a, the dependence of the S-matrix is partially canceled by the radial dependence of the Ricatti-Haenkel functions. Thus, in practical applications the cross sections are nearly independent of a. This completes the derivation of the R-matrix and its relation to the S-matrix in the case of a single channel.

In the general case of several open channels and charged collision partners with spin, calculating the R-matrix and and deriving its relation with the S-matrix are much harder tasks. Owing to the charge of the particles, the asymptotic wave function of Eq. (12.290) should be expressed in terms of Coulomb wave functions and the S-matrix associated with the short-range part of the potential. That is, the asymptotic radial wave function emerging in channel β in a collision initiated in channel α is

$$u^{\text{ext}\,(J)}_{\beta l_\beta,\alpha l_\alpha}(r) = \frac{i}{2}\left[H_l^{(-)}(k_\alpha r)\,\delta_{\beta\alpha} - \bar{S}^J_{\beta l_\beta,\alpha l_\alpha}\, H_l^{(+)}(k_\beta r)\right]. \tag{12.306}$$

We merely give the form of the R-matrix in terms of the channel reduced widths, $\gamma_\alpha^{(n)}$,

$$R_{\alpha,\beta}(E) = \sum_n \frac{\gamma_\alpha^{(n)}\,\gamma_\beta^{(n)}}{E^{(n)} - E}, \tag{12.307}$$

which is a natural extension of Eq. (12.302).

The relation between the S-matrix and the R-matrix is a bit more involved. We refer the reader to the review articles [Lane and Thomas (1958); Descouvemont and Baye (2010); Descouvemont (2003)] and to Burke's book [Burke (2011)] for a detailed derivation of these expressions.

Implementations of the R-matrix method as a tool to solve the couple channel problem can be found in [Haglund and Robson (1965); Buttle (1967)], for CC problems restricted to bound channels, and in [Druel *et al.* (2010)], for couplings with continuum states. The latter is a study the $d + {}^{58}$Ni collision, involving the breakup of the deuteron. In this study, the authors solve the CDCC equations by the R-matrix method, employing the Lagrange-mesh technique.

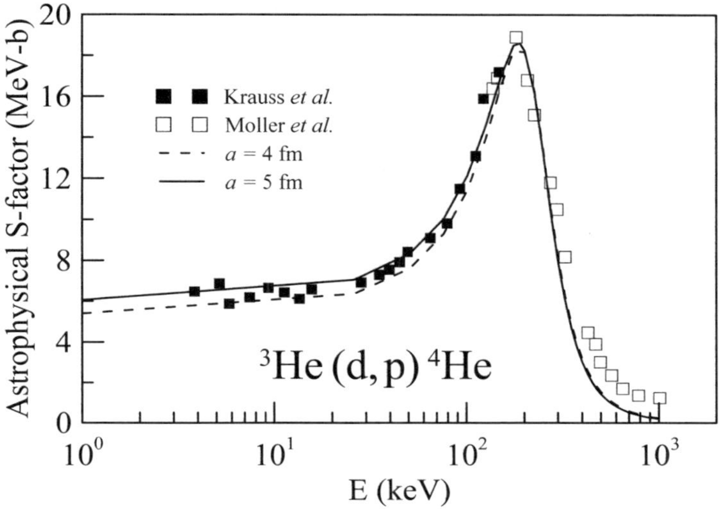

Fig. 12.9 R-matrix fits for the ^{3}He (p, d) ^{4}He, for two values of the surface radius, a. The data are from [Krauss *et al.* (1987)] (solid squares) and from [Möller and Besenbacher (1980)] (open squares). The calculations of the astrophysical S-fator, defined in Eq. (12.237), are taken from [Descouvemont (2003)].

12.8.2 *The phenomenological R-matrix*

The R-matrix is also used as a convenient parametrization to describe the low-energy scattering data. This approach is widely used to analyze low-energy scattering experiments and reactions of astrophysical interest (see, e.g., [Pellegriti *et al.* (2008); Mountford *et al.* (2012)]). In this variant, the R-matrix parameters are not calculated from any model, but considered as free parameters, and are fitted to experimental data. The one drawback of the method is the introduction of the surface parameter. However, the dependence of the cross sections on this parameter is not very strong, provided that a reasonable values is used.

An illustration of the phenomenological R-matrix analysis is presented in figure 12.9. It shows the astrophysical S-fator (see section 12.5) for the transfer reaction $^3\text{He} + \text{p} \to \text{d} +^4 \text{He}$. Experimental data in the energy range 5 − 1000 keV are compared with R-matrix fits obtained with two different values of the parameter a. The details of this study are given in [Descouvemont (2003)]. The fits are very good and they are very important to give estimates of $\mathcal{S}$ at very low collision energies, which are encountered in astrophysical processes and are very hard to measure in laboratories. The results for the surface parameters $a = 4$ fm and $a = 4$ are quite close, which indicates that the dependence of the astrophysical S-factor on this parameter is weak.

Exercises

(1) Prove Eq. (12.126),

$$G_{\mathrm{Q}}\, \mathcal{U}_{\mathrm{QP}} \left|\Psi_{\mathrm{P}}^{(+)}\right\rangle = \mathcal{G}_{\mathrm{Q}}\, \mathcal{U}_{\mathrm{QP}} \left|X_{\mathrm{opt}}^{(+)}\right\rangle.$$

For this purpose, follow the steps listed below:

a) use the LS for $\left|\Psi_{\mathrm{P}}^{(+)}(0\mathbf{k}_0)\right\rangle$ to show that the state defined as

$$\left|\, Y \,\right\rangle = (G_{\mathrm{Q}})^{1/2}\, \mathcal{U}_{\mathrm{QP}} \left|\Psi_{\mathrm{P}}^{(+)}\right\rangle$$

is related to the optical wave function through the equation

$$\left|\, Y \,\right\rangle = \left[\frac{1}{1 - G_{\mathrm{Q}}^{1/2}\, \mathcal{U}_{\mathrm{QP}}\, G_{\mathrm{opt}}^{(+)}\, \mathcal{U}_{\mathrm{PQ}}\, G_{\mathrm{Q}}^{1/2}}\right] G_{\mathrm{Q}}^{1/2}\, \mathcal{U}_{\mathrm{QP}} \left|X_{\mathrm{opt}}^{(+)}\right\rangle.$$

b) Derive Eq. (12.126) using the above result.

(2) Prove that the probability of tunneling through the potential barrier in proton-proton collision can be approximated by the expression

$$P(E) = \exp\left[-2\eta\pi \times B(x)\right],$$

with $x = R/r_t$, where R is the range of the nuclear attractive potential (approximated by a square well) and r_t is the turning point in a collision with energy E. Above, η is the Sommerfeld parameter and $B(x)$ is the function,

$$B(x) = \frac{2}{\pi}\left[\cos^{-1}\left(\sqrt{x}\right) - \sqrt{x(1-x)}\right].$$

(3) Show that when the optical potential cross section contributes to the transition, the energy correlation function becomes

$$C_{\beta\alpha}^{(\sigma)}(E,\varepsilon) = \left[2\, \sigma_{\beta\alpha}^{\mathrm{opt}}(E)\;\; \sigma_{\beta\alpha}^{\mathrm{fl}}(E) + \left(\sigma_{\beta\alpha}^{\mathrm{fl}}(E)\right)^2\right] \frac{1}{1 + \varepsilon^2/\Gamma^2}.$$

(4) Prove that the enhancement factor in the compound elastic cross section in the limit where the amplitudes $g_{n\alpha}$ are real (situation encountered at very low energies, where the resonances are isolated) is equal to 3.

(5) Derive Eq. (12.304) and find the connection between the width of the single resonance and the R-matrix amplitudes, $\gamma^{(n)}$.

Bibliography

Abramowitz, M. and Stegun, I. (1972). *Handbood of Mathematical Functions* (Dover, NY).

Abriola, D., Di Gregorio, D., Testoni, J. E., Etchegoyen, A., Etchegoyen, M. C., Fernández Niello, J. O., Ferrero, A. M. J., Gil, S., Macchiavelli, A. O., Pacheco, A. J. and Kittl, J. (1989). *Phys. Rev. C* **39**, p. 546.

Adhikari, S. K. (1986). *Am. J. Phys.* **54**, p. 362.

Adhikari, S. K. and Hussein, M. S. (2008). *Am. J. Phys.* **76**, p. 1108.

Agassi, D., Weidenmüller, H. A. and Mantzouranis, G. (1975). *Phys. Rep.* **22**, p. 145.

Aguiar, C. E., Zardi, F. and Vitturi, A. (1997). *Phys. Rev. C* **56**, p. 1511.

Airy, G. B. (1839). *Proc. Cambridge Phil. Soc.* **6**, p. 379.

Akyüz, O. and Winther, A. (1981). in R. A. Broglia, C. H. Dasso and R. A. Ricci (eds.), *Nuclear Structure of Heavy Ion Reaction* (North Holland), proc. E. Fermi Summer School of Physics.

Al-Khalili, J. S. and Tostevin, J. A. (1996). *Phys. Rev. Lett.* **76**, p. 3903.

Al-Khalili, J. S., Tostevin, J. A. and Thompson, I. J. (1996). *Phys. Rev. C* **54**, p. 1843.

Alder, K., Bohr, A., Huus, T., Mottelson, B. and Winther, A. (1956). *Rev. Mod. Phys.* **28**, p. 432.

Alder, K. and Winther, A. (1975). *Electromagnetic Excitations* (North-Holland, Amsterdam).

Alhassid, Y. (2000). *Rev. Mod. Phys.* **72**, p. 895.

Almeida, F. I. A. and Hussein, M. S. (1985). *Phys. Rev C* **31**, p. 2120.

Alt, E. and Sandhas, W. (1966). *Coulomb Interactions in Nuclear and Atomic Few-Body Problems* (Plenum Press).

Alt, E. O., Grassber, P. and Sandhas, W. (1964). *Nucl. Phys. B* **2**, p. 1967.

Amos, K., Dortmans, P. J., von Geramb, H. V., Karataglidis, S. and Raynal, J. (2000). *Adv. Nucl. Phys.* **25**, p. 275.

Amos, K., Karataglidis, S., van der Knijff, D., Canton, L., Pisent, G. and Svenne, J. P. (2005). *Phys. Rev. C* **72**, p. 064604.

Anantaraman, N., Toki, H. and Bertsch, G. F. (1983). *Nucl. Phys. A* **398**, p. 269.

Andrés, M. V., Gómez-Camacho, J. and Nagarajan, M. A. (1994). *Nucl. Phys. A* **579**, p. 273, ibid. A583 (1994) 817.

Austern, N. (1970). *Direct Nuclear Reaction Theory* (John Wiley).

Austern, N., Iseri, Y., Kamimura, M., Kawai, M., Rawitscher, G. and Yashiro, M. (1987). *Phys. Rep.* **154**, p. 125.

Balakrishnan, N., Kharchenko, V., Forrey, R. C. and Dalgarno, A. (1997). *Chem. Phys. Lett.* **280**, p. 5.

Baltz, A. J. and Bayman, B. F. (1982). *Phys. Rev. C* **26**, p. 1969.

Baltz, A. J., Glendenning, N. K., Kauffmann, S. K. and Pruess, K. (1979). *Nucl. Phys. A* **327**, p. 221.

Baltz, A. J., Hussein, M. S. and Carlson, B. V. (1984). *Phys. Rep.* **113**, p. 133.

Baltz, A. J., Kauffmann, S. K., Glendenning, N. K. and Pruess, K. (1978). *Phys. Rev. Lett.* **40**, p. 20.

Barata, J. C. A., Canto, L. F. and Hussein, M. S. (2011). *Braz. J. Phys.* **41**, p. 50.

Barlette, V. E., Leite, M. M. and Adhikari, S. K. (2000). *Eur. J. Phys.* **21**, p. 435.

Barlette, V. E., Leite, M. M. and Adhikari, S. K. (2001). *Am. J. Phys.* **69**, p. 1010.

Bartnitzky, G., Blazevic, A., Bohlen, H. G., Casandjian, J. M., Chartier, M., Clement, H., Gebauer, B., Gilibert, A., Kirchner, T., Khoa, D. T., Lepine-Szily, A., Mittig, W., von Ortzen, W., Ostrowski, A. N., Roussel-Chomaz, P., Siegler, J., Wilpert, M. and Wilpert, T. (1996). *Phys. Lett. B* **365**, p. 23.

Bass, R. (1977). *Phys. Rev. Lett.* **39**, p. 265.

Baye, D. (2004). *Phys. Rev. C* **70**, p. 015801.

Baye, D., Capel, P. and Goldstein, G. (2005). *Phys. Rev. Lett.* **95**, p. 082502.

Baym, G. (1998). *Acta. Physica Polonica B* **29**, p. 1839.

Becchetti, F. D. and Greenless, G. W. (1969). *Phys. Rev.* **182**, p. 745.

Berry, M. V. (1966). *Proc. Phys. Soc.* **98**, p. 479.

Berry, M. V. (1969). *J. Phys. B* **2**, p. 381.

Berry, M. V. and Mount, E. V. (1972). *Rep. Prog. Phys.* **35**, p. 315.

Bertsch, G., Borysowicz, J., McManus, H. and Love, W. G. (1977). *Nucl. Phys. A* **284**, p. 399.

Bertulani, C. A. (2005). *Phys. Rev. Lett.* **94**, p. 072701.

Bertulani, C. A. and Canto, L. F. (1992). *Nucl. Phys. A* **539**, p. 163.

Bizzeti, P. G. and Maurenzig, P. R. (1967). *Il Nuovo Cimento* **XLVII B**, p. 29.

Blatt, J. M. and Weisskopf, V. F. (1991). *Theoretical Nuclear Physics* (Dover Publications, Inc.).

Bloch, C. (1957). *Nucl. Phys.* **4**, p. 503.

Blocki, J., Randrup, J., Swiatecki, W. J. and Tsang, C. F. (1977). *Ann. Phys. (NY)* **105**, p. 427.

Bohlen, H. G., Stiliaris, E., Gebauer, B., von Ortzen, W., Wilpert, M., Wilpert, T., Ostrowski, A., Khoa, D. T., Demyanova, A. S. and A., O. A. (1993). *Z. Phys. A* **346**, p. 189.

Bohr, A. and Mottelson, B. R. (1998). *Nuclear Structure*, Vol. II (World Scientific).

Bohr, N. (1948). *K. Dan. Vidensk. Selllsk. Mat. Fys. Medd.* **18**, p. No. 8.

Brandan, M. E. and Satchler, G. R. (1997). *Phys. Rep.* **285**, p. 143.

Bransden, B. H. and Joachain, C. J. (1983). *Physics of Atoms and Molecules* (Longman).

Bray, I. (2002). *Phys. Rev. Lett.* **89**, p. 273201.

Bray, I. and Stelbovics, A. T. (1992). *Phys. Rev. C* **46**, p. 6995.

Breitschaft, A. M. S., Canto, L. F., Schechter, H., Hussein, M. S. and Moniz, E. J. (1995). *Ann. Phys. (NY)* **243**, p. 420.

Bremmer, H. (1951). *Commun. Pure Appl. Math.* **4**, p. 105.

Brieva, F. A. and Rook, J. R. (1977a). *Nucl. Phys. A* **291**, p. 299.

Brieva, F. A. and Rook, J. R. (1977b). *Nucl. Phys. A* **291**, p. 317.

Brieva, F. A. and Rook, J. R. (1978). *Nucl. Phys. A* **297**, p. 206.

Brink, D. M. (1965). *Nuclear Forces* (Pergamon Press).

Brink, D. M. (1966). in *Proc. Intl. School of Physics 'Enrico Fermi', course 36*, p. 247.

Brink, D. M. (1985). *Semi-classical Methods for Nucleus-Nucleus Scattering* (Cambridge University Press, Cambridge, England).

Brink, D. M. and Canto, L. F. (1977). *Nucl. Phys. A* **279**, p. 85.

Brink, D. M. and Satchler, G. R. (1971). *Angular Momentum* (Clarendon Press - Oxford).

Brink, D. M. and Stancu, F. (1978). *Nucl. Phys. A* **299**, p. 321.

Brink, D. M. and Stephen, R. O. (1963). *Phys. Lett.* **5**, p. 77.

Brink, D. M., Stephen, R. O. and Tanner, N. W. (1964). *Nucl. Phys.* **54**, p. 577.

Brink, D. M. and Takigawa, N. (1977). *Phys. Rev. C* **47**, p. R2470.

Broglia, R. A. and Winther, A. (1991). *Heavy Ion Reactions*, Vol. I (Addison-Wesley).

Buck, B., Stamp, A. and Hodgson, P. (1963). *Phil. Mag.* **8**, p. 1805.

Buck, U. (1974). *Rev. Mod. Phys.* **46**, p. 369.

Buck, U., Hoppe, H. O. and Huisken, H., F. Pauly (1973). *Discuss. Faraday Soc.* **55**, p. 185.

Burke, P. G. (2011). *R-Matrix Theory of Atomic Collisions*, Vol. 61 (Springer Series on Atomic Optical and Plasma Physics).

Buttle, P. G. A. (1967). *Phys. Rev.* **160**, p. 719.

Calogero, F. (1967). *The Variable Phase Approach to Potential Scattering* (Academic Press).

Cândido-Ribeiro, M. A., Chamon, L. C., Pereira, D., Hussein, M. S. and Galetti, D. (1997). *Phys. Rev. Lett.* **78**, p. 3270.

Canto, L. F., Donangelo, R. and Hussein, M. S. (1991). *Nucl. Phys. A* **529**, p. 243.

Canto, L. F., Donangelo, R., Hussein, M. S. and Lotti, P. (1995). *Nucl. Phys. A* **589**, p. 117.

Canto, L. F., Donangelo, R., Hussein, M. S. and Pato, M. P. (1992). *Nucl. Phys. A* **542**, p. 131.

Canto, L. F., Donangelo, R., Rasmussen, J. O., Ring, P. and Stoyer, M. A. (1990). *Phys. Lett. B* **248**, p. 10.

Canto, L. F., Gomes, P. R. S., Donangelo, R. and Hussein, M. S. (2006). *Phys. Rep.* **424**, p. 1.

Canton, L., Pisent, G., Svenne, J. P., van der Knijff, D., Amos, K. and Karataglidis, S. (2005). *Phys. Rev. Lett.* **94**, p. 122503.

Capel, P., Baye, D. and Suzuki, Y. (2008). *Phys. Rev. C* **78**, p. 054602.

Casimir, H. B. G. and Polder, D. (1948). *Phys. Rev.* **73**, p. 360.

Chamon, L. C. (2010). *private communication* .

Chamon, L. C., Carlson, B. V., Gasques, L. R., Pereira, D., De Conti, C., Alvarez, M. A. G., Hussein, M. S., Cândido Ribeiro, M. A., Rossi Jr., E. S. and Silva, C. P. (2002). *Phys. Rev. C* **66**, p. 014610.

Chamon, L. C., Pereira, D., Hussein, M. S., Cândido-Ribeiro, M. A. and Galetti, D. (1997). *Phys. Rev. Lett.* **79**, p. 5218.

Chester, C., Friedman, B. and Ursell, F. (1957). *Proc. Camb. Phil. Soc.* **53**, p. 599.

Child, M. S. (1994). *Molecular collision theory* (Dover Publications, Inc.).

Chin, C., Grimm, R., Julienne, P. and Tiesinga, E. (2010). *Rev. Mod. Phys.* **82**, p. 1225.

Christensen, P. R. and Winther, A. (1976). *Phys. Lett. B* **65**, p. 19.

Churchill, R. (1974). *Complex Variables and Applications* (Princeton University Press, Princeton, New Jersey).

Coester, F. (1964). *Phys. Rev.* **133**, p. B1516.

Cole, R. K., Waddell, C. N., Dittman, R. R. and Sandhu, H. S. (1966). *Nucl. Phys.* **75**, p. 241.

Colton, D. and Kress, R. (1998). *Inverse Acoustic and Electromagnetic Scattering Theory* (Springer).

Connor, J. N. L. and Marcus, R. A. (1971). *J. Chem. Phys.* **55**, p. 5636.

Cowan, R. D. (1981). *The Theory of Atomic Structure and Spectra* (Londonc s).

Cramer, J. G., DeVries, R. M., Goldberg, D. A., Zisman, M. S. and Maguire, C. F. (1976). *Phys. Rev C* **14**, p. 2158.

Cuyt, A., Petersen, V., Verdonk, B., Waadeland, H. and Jones, W. B. (2008). *Handbook of continued fractions for special functions* (Springer).

Dalgarno, A. (1967). *Adv. Chem. Phys.* **12**, p. 143.

Dalgarno, A. and Davidson, W. D. (1966). *Adv. At. Mol. Opt. Phys.* **2**, p. 1.

Dasgupta, M., Hinde, D. J., Diaz-Torres, A., Bouriquet, B., Low, C. I., Milburn, C. J. and Newton, J. O. (2007). *Phys. Rev. Lett.* **99**, p. 192701.

Dasso, C. H. and Pollarolo, G. (2003). *Phys. Rev C* **68**, p. 054604.

Descouvemont, P. (2003). *Theoretical Models for Nuclear Astrophysics* (Nova Science Pub. Inc.).

Descouvemont, P. and Baye, D. (2010). *Rep. Prog. Phys.* **73**, p. 1.

Diaz-Torres, A. and Thompson, I. J. (2002). *Phys. Rev. C* **65**, p. 024606.

Dollard, J. (1964). *J. Math. Phys.* **5**, p. 729.

Donangelo, R., Canto, L. F. and Hussein, M. S. (1979). *Nucl. Phys. A* **320**, p. 422.

Donangelo, R., Guidry, M. W., Boisson, J. P. and Rasmussen, J. O. (1976). *Phys. Lett. B* **64**, p. 377.

Druel, T., Baye, D., Descouvemont, P. and Sparenberg, J. M. (2010). *Nucl. Phys. A* **845**, p. 88.

Eberly, J. H. (1965). *Am. J. Phys.* **33**, p. 771.

Edmonds, A. (1974). *Angular Momentum in Quantum Mechanics* (Princeton University Press, Princeton).

Engelbrecht, C. A. and Weidenmüller, H. A. (1973). *Phys. Rev. C* **8**, p. 859.

Erb, K. and Bromley, D. A. (1984). *Treatise on Heavy Ion Science* (Plenum Press, N.Y.).
Ericson, T. (1963). *Ann. Phys. (NY)* **23**, p. 390.
Esbensen, H. and Mişicu, S. (2007). *Phys. Rev. C* **76**, p. 054609.
Faddeev, L. D. (1969). *Trudy MIAN SSSR* **61**, p. 1.
Fainstain, P. D., Ponce, V. H. and Rivarola, R. D. (1991). *J. Phys. B: At. Mol. Opt. Phys.* **24**, p. 3091.
Feshbach, H. (1962). *Ann. Phys. (NY)* **19**, p. 287, ibid. 5 (1958) 357.
Feshbach, H. (1985). *Ann. Phys. (NY)* **165**, p. 398.
Feshbach, H. (1992). *Theoretical Nuclear Physics - Nuclear Reactions.*
Feynman, R. P. and Hibbs, A. R. (1965). *Quantum Mechanics and Path Integrals* (McGraw-Hill, NY).
Firk, F. W. K., Lynn, J. E. and Moxton, M. C. (1960) (Kingston Conference).
Flugge, S. (1999). *Practical Quantum Mechanics* (Springer-Verlag, Berlin Heildelberg).
Ford, K. W. and Wheeler, J. A. (1959). *Ann. Phys. (NY)* **7**, p. 259.
Formanek, J. (1976). *Am. J. Phys.* **44**, p. 778.
Frahn, W. (1966). *Nucl. Phys. A* **75**, p. 577.
Frahn, W. E. (1971). *Phys. Rev. Lett.* **26**, p. 568.
Frahn, W. E. (1985). *Diffractive processes in Nuclear Physics* (Oxford University Press, Oxford).
Frahn, W. E. and Venter, R. H. (1963). *Ann. Phys. (NY)* **24**, p. 243.
Fraser, P., Amos, K., Karataglidis, S., Canton, L., Pisent, G. and Svenne, J. P. (2008). *Eur. Phys. J. A* **35**, p. 69.
Friedman, W. A. and Goebel, C. J. (1977). *Ann. Phys. (NY)* **104**, p. 145.
Friedrich, H. (1981). *Phys. Rep.* **74**, p. 210.
Friedrich, H. (1998). *Theoretical Atomic Physics*, 2nd edn. (Springer-Verlag).
Froelich, P., Jonsell, S., Saenz, A., Zygelman, B. and Dalgarno, A. (2000). *Phys. Rev. Lett.* **84**, p. 4577.
Fröman, N. and Fröman, P. (1965). *JWKB APPROXIMATION: Contributions to the theory* (North-Holland, Amsterdam).
Fuller, R. C. (1975a). *Phys. Rev. C* **12**, p. 1561.
Fuller, R. C. (1975b). *Phys. Lett. B* **57**, p. 217.
Galin, J., Gatty, B., Guerrau, D., Rousset, C., Schlotthauer-Voos, U. C. and Tarrago, X. (1974). *Phys. Rev. C* **9**, p. 1113.
Gavron, A. (1980). *Phys. Rev. C* **21**, p. 230.
Glauber, R. J. (1955). *Phys. Rev.* **100**, p. 242.
Glauber, R. J. (1959). *Lectures in Theoretical Physics, Vol. 1* (University of Colorado Press, Boulder).
Glover, C. W. (1985). *Phys. Rev. C* **31**, p. 1.
Goldberger, M. L. and Watson, K. M. (2004). *Collision Theory* (Dover Books on Physics).
Goldstein, G., Baye, D. and Capel, P. (2006). *Phys. Rev. C* **73**, p. 024602.
Goldstein, H. (1980). *Classical Mechanics*, 2nd edn. (Addison-Wesley).
Gómez-Camacho, J., Andrés, M. V. and Nagarajan, M. A. (1994). *Nucl. Phys. A* **580**, p. 156.

Gontchar, I., Hinde, D., Dasgupta, M. and Newton, J. (2004). *Phys. Rev. C* **69**, p. 024610.

Gradshteyn, I. and Ryzhik, I. (1965). *Table of Integrals, Series, and Products*, 4th edn. (Academic Press).

Grande, P. L., Schiwietz, G., Sigaud, G. M. and Montenegro, E. C. (1996). *Phys. Rev. A* **54**, p. 2983.

Griffin, J. J. and Wheeler, J. A. (1957). *Phys. Rev.* **108**, p. 311.

Guidry, M. W., Massmann, H., Donangelo, R. and Rasmussen, J. O. (1976). *Nucl. Phys. A* **274**, p. 183.

Haglund, M. E. and Robson, D. (1965). *Phys. Lett.* **14**, p. 225.

Hanbury-Brown, R. and Twiss, R. G. (1954). *Phil. Mag.* **45**, p. 663.

Hauser, W. and Feshbach, H. (1952). *Phys. Rev.* **87**, p. 336.

Häusser, O., von Brentano, P. and Mayer-Kuckuk, T. (1964). *Phys. Lett.* **12**, p. 226.

Hill, D. L. and Wheeler, J. A. (1953). *Phys. Rev.* **89**, p. 1102.

Hodgson, P. E. (1967). *Ann. Rev. Nucl. Sci.* **17**, p. 1.

Holdeman, J. T. and Thaler, R. (1965). *Phys. Rev.* **139**, p. B1186.

Huby, R. (1954). *Proc. Phys. Soc. (London) A* **67**, p. 1113.

Hussein, M. S., Chen, Y. T. and Almeida, F. I. A. (1984). *Am. J. Phys.* **52**, p. 650.

Hussein, M. S., Li, W. and Wüster, S. (2008). *J. Phys. A: Math. and Theor.* **41**, p. 475207.

Hussein, M. S. and McVoy, K. W. (1984). *Rep. Prog. Part. Nucl. Phys.* **12**, p. 103.

Hussein, M. S. and Moniz, E. J. (1984). *Phys. Rev. C* **29**, p. 2054.

Hussein, M. S., Pato, M. P., Canto, L. F. and Donangelo, R. (1993). *Phys. Rev. C* **47**, p. 2398.

Hussein, M. S., Rego, R. A. and Bertulani, C. A. (1991). *Phys. Rep.* **201**, p. 279.

Iano, P. J. and Austern, N. (1966). *Phys. Rev.* **151**, p. 853.

Ibraheem, A. A. and Bonaccorso, A. (2005). *Nucl. Phys. A* **748**, p. 414.

Ichikawa, T., Hagino, K. and Iwamoto, A. (2007a). *Phys. Rev. C* **75**, p. 057603.

Ichikawa, T., Hagino, K. and Iwamoto, A. (2007b). *Phys. Rev. C* **75**, p. 064612.

Ichikawa, T., Hagino, K. and Iwamoto, A. (2009). *Phys. Rev. Lett.* **103**, p. 202701.

Inouye, S., Andrews, M. R., Stenger, J., Miesner, H.-J., Stamper-Kurn, D. M. and Ketterle, W. (1998). *Nature* **392**, p. 151.

Jackson, D. F. and Johnson, R. C. (1974). *Phys. Lett. B* **49**, p. 249.

Jackson, J. D. (1975). *Classical Electrodynamics* (John Wiley & Sons, Inc.).

Jiang, C. L., Esbensen, H., Rehm, K. E., Back, B. B., Janssens, R. V. F., Caggiano, J. A., Collon, P., Greene, J., Heinz, A. M., Henderson, D. J., Nishinaka, I., Pennington, T. O. and Seweryniak, D. (2002). *Phys. Rev. Lett.* **89**, p. 052701.

Jiang, C. L., Rehm, K. E., Janssens, R. V. F., Esbensen, H., Ahmad, I., Back, B. B., Collon, P., Davids, C. N., Greene, J. P., Henderson, D. J., Mukherjee, G., Pardo, R. C., Paul, M., Pennington, T. O., Seweryniak, D., Sinha, S. and Zhou, Z. (2004). *Phys. Rev. Lett.* **93**, p. 012701.

Joachain, C. J. (1983). *Quantum Collision Theory* (North Holland).

Johnson, R. C. and Goebel, C. J. (2000). *Phys. Rev. C* **62**, p. 027603.

Kamal, A. N. (1984). *Am. J. Phys.* **52**, p. 46.

Kapur, P. L. and Peierls, R. E. (1938). *Proc. Roy. Soc. (London) A* **166**, p. 277.

Kawai, M., Kerman, A. K. and McVoy, K. W. (1973). *Ann. Phys. (N.Y.)* **75**, p. 156.

Kemble, E. C. (1935). *Phys. Rev.* **48**, p. 549.

Kerman, A. K., McManus, H. and Thaler, R. M. (1959). *Ann. Phys. (NY)* **8**, p. 551.

Khoa, D. T., Khan, E., Còlo, G. and van Giai, N. (2002). *Nucl. Phys. A* **706**, p. 61.

Khoa, D. T., von Ortzen, W., Bohlen, H. G. and Ohkubo, S. (2007). *J. Phys. G: Nucl. Part. Phys.* **34**, p. R111.

Kim, Y. J., Amos, K., Karataglidis, S. and Richter, W. A. (2008). *Nucl. Phys. A* **807**, p. 119.

Kolata, J. J., Guimarães, V., Peterson, D., Santi, P., White-Stevens, R. H., Vincent, S. M., Becchetti, F. D., Lee, M. Y., O'Donnell, T. W., Roberts, D. A. and Zimmerman, J. A. (2001). *Phys. Rev. C* **63**, p. 024616.

Komineas, S. and Papanicolau, N. (2008). *New J. Phys.* **10**, p. 043021.

Kondo, Y., Sugiyama, Y., Tomita, Y., Yamanouchi, Y., Ikezoe, H., Ideno, K., Hamada, S., Sugimitsu, T., Hijiya, M. and Fujita, H. (1996). *Phys. Lett. B* **365**, p. 17.

Krappe, H. J., Nix, J. R. and Sierk, A. J. (1979a). *Phys. Rev. Lett.* **42**, p. 215.

Krappe, H. J., Nix, J. R. and Sierk, A. J. (1979b). *Phys. Rev. C* **20**, p. 992.

Krauss, A., Becker, H. W., Trautvetter, H.-P., Rolfs, C. and Brand, K. (1987). *Nucl. Phys. A* **465**, p. 150.

Kretschmer, W. and Wangler, M. (1978). *Phys. Rev. Lett.* **41**, p. 1244.

Kuo, T. T. S. and Brown, G. E. (1966). *Nucl. Phys.* **85**, p. 40.

Lane, A. (1962). *Nucl. Phys.* **35**, p. 676.

Lane, A. M. and Thomas, R. G. (1958). *Rev. Mod. Phys.* **30**, p. 257.

Langer, R. E. (1934). *Bull. Amer. Math. Soc.* **42**, p. 545.

Langer, R. E. (1937). *Phys. Rev.* **51**, p. 669.

Langsdorj Jr., A., Lane, R. O. and Monahan, J. E. (1957). *Phys. Rev.* **107**, p. 1077.

Lapidus, I. R. (1992). *Am. J. Phys.* **50**, p. 45.

Lax, M. (1951). *Rev. Mod. Phys.* **23**, p. 287.

Levin, F. S. and Feshbach, H. (1973). *Reaction Dynamics* (Gordon and Breach, Science Publishers, Inc., New York & London & Paris).

Levinson, N. (1949). *Kgl. Danske Videnskab. Selskab, Mat.-fys. Medd.* **25**, p. 9.

Love, W. G., Terasawa, T. and Satchler, G. R. (1977). *Phys. Rev. Lett.* **39**, p. 6.

Lovelace, C. (1964). *Phys. Rev.* **135**, p. B1225.

Lubian, J. (2010). Private communication, .

Lubian, J., Correa, T., Aguilera, E. F., Canto, L. F., Gomez-Camacho, A., Quiroz, E. M. and Gomes, P. R. S. (2009). *Phys. Rev C* **79**, p. 064605.

Lubian, J. and Nunes, F. M. (2007). *J. Phys. G: Nucl. Part. Phys.* **34**, p. 513.

Macias, A. and Riera, A. (1982). *Phys. Rep.* **90**, p. 299.

Maitra, N. T. and Heller, E. J. (1996). *Phys. Rev. A* **54**, p. 4763.

Marcus, R. A. (1970). *Chem. Phys. Lett.* **7**, p. 525.

Marcus, R. A. (1971). *J. Chem. Phys.* **54**, p. 3965.
Marcus, R. A. (1972). *J. Chem. Phys.* **56**, p. 311.
Margueron, J., Bonaccorso, A. and Brink, D. M. (2003). *Nucl. Phys. A* **720**, p. 337.
Marta, H. D., Canto, L. F. and Donangelo, R. (2008). *Phys. Rev C* **78**, p. 034612.
Marta, H. D., Canto, L. F., Donangelo, R. and Lotti, P. (2002). *Phys. Rev. C* **66**, p. 024605.
Massmann, H. and Rasmussen, J. O. (1975). *Nucl. Phys. A* **243**, p. 155.
Matsumoto, T., Hiyama, E., Ogata, K., Iseri, Y., Kamimura, M., Chiba, S. and Yahiro, M. (2004). *Phys. Rev. C* **70**, p. 061601(R).
Mayer, H., Hall, J. R., Jacobs, W. W., Schwandt, P. and Singh, P. P. (1981a). *Phys. Rev. C* **24**, p. 1782.
Mayer, H., Schwandt, P., Jacobs, W. W. and Hall, J. R. (1983). *Phys. Rev. C* **27**, p. 459.
Mayer, H., Schwandt, P., Moake, G. L. and Singh, P. P. (1981b). *Phys. Rev. C* **23**, p. 616.
McDowell, M. and Coleman, C. (1970). *Introduction to the Theory of Ion-Atom collisions* (North-Holland).
McVoy, K. W. and Brandan, M. E. (1992). *Nucl. Phys. A* , 542, p. 295.
Melkanoff, M., Sawada, T. and Raynal, J. (1966). *Methods in Computational Physics - vol. 6* (Academic Press (NY)).
Merzbacher, E. (1970). *Quantum Mechanics* (John Wiley & sons, Inc.).
Merzbacher, E. (1998). *Quantum Mechanics*, 3rd edn. (John Wiley & sons, Inc.).
Messiah, A. (1999). *Quantum Mechanics* (Dover).
Mişicu, S. and Esbensen, H. (2006). *Phys. Rev. Lett.* **96**, p. 112701.
Mişicu, S. and Esbensen, H. (2007). *Phys. Rev. C* **75**, p. 034606.
Miller, W. H. (1970a). *J. Chem. Phys.* **53**, p. 1949.
Miller, W. H. (1970b). *J. Chem. Phys.* **53**, p. 3578.
Miller, W. H. (1971). *J. Chem. Phys.* **55**, p. 3150.
Miller, W. H. (1974). *Adv. Chem. Phys.* **25**, p. 69.
Möller, W. and Besenbacher, F. (1980). *Nucl. Instrum. Methods* **168**, p. 111.
Montanari, C. C., Montenegro, E. C. and Miraglia, J. E. (2010). *J. Phys. B: At. Mol. Opt. Phys.* **43**, p. 165201.
Morse, P. and Feshbach, H. (1953). *Methods of Theoretical Physics* (McGraw-Hill, New York).
Mountford, D. J., Murphy, A. S., Achouri, N. L., Angulo, C., Brown, J. R., Davinson, T., Oliveira Santos, F. d., Séréville, N. d., Descouvemont, P., Kamalou, O., Laird, A. M., Pittman, S. T., Ujic, P. and Woods, P. J. (2012). *Phys. Rev. C* **85**, p. 022801 (R).
Nagarajan, M. A., Mahaux, C. C. and Satchler, G. R. (1985). *Phys. Rev. Lett.* **54**, p. 1136.
Newton, R. (2002). *Scattering Theory of Waves and Particles*, 2nd edn. (Courier Dover Publications).
Nunes, F. M. and Thompson, I. J. (1998). *Phys. Rev.* **C57**, p. R2818.
Nunes, F. M. and Thompson, I. J. (1999). *Phys. Rev. C* **59**, p. 2652.
Nussenzveig, H. (1977). *Scientific American* **236**, p. 116.

Nussenzveig, H. (1992). *Diffraction Effects in Semiclassical Scattering* (Cambridge University Press, Cambridge).

Nutt, G. (1968). *J. Math. Phys* **9**, p. 796.

Ogata, K., Yahiro, M., Iseri, Y., Matsumoto, T. and Kamimura, M. (2003). *Phys. Rev. C* **68**, p. 06409.

Park, J. T., Aldag, J. E. and George, J. M. (1975). *Phys. Rev. Lett.* **34**, p. 1253.

Park, J. T., Aldag, J. E., George, J. M. and Peacher, J. L. (1976). *Phys. Rev. A* **14**, p. 608.

Pechukas, P. (1969a). *Phys. Rev.* **181**, p. 166.

Pechukas, P. (1969b). *Phys. Rev.* **181**, p. 174.

Pellegriti, M. G., Achouri, N. L., Angulo, C., Angélique, J. C., Berthoumieux, E., Casarejos, E., Couder, M., Davinson, T., Ghag, C., Murphy, A. S., Orr, N. A., Ray, I., Stefan, I. G. and Descouvemont, P. (2008). *Phys. Lett. B* **659**, p. 864.

Penny, S. K. and Satchler, G. R. (1964). *Nucl. Phys.* **53**, p. 145.

Perey, F. and Buck, B. (1962). *Nucl. Phys.* **32**, p. 353.

Pires, K. C. C., Lichtenthäler, R., Lépine-Szily, A., Guimarães, V., Faria, P. N., Barioni, A., Mendes Junior, D. R., Morcelle, V., Pampa Condori, R., Morais, M. C., Zamora, J. C., Crema, E., Moro, A. M., Rodrígues-Gallardo, M., Assunção, M., Shorto, J. M. B. and Mukherjee, S. (2011). *Phys. Rev. C* **83**, p. 064603.

Press, W., Teukolsky, S., Vettering, W. and Flannery, B. (1992). *Numerical Recepes in Fortran*, 2nd edn. (Cambridge University Press).

Pühlhofer, F. (1977). *Nucl. Phys. A* **280**, p. 267.

Put, L. W. and Paans, A. M. J. (1974). *Phys. Lett. B* **49**, p. 266.

Put, L. W. and Paans, A. M. J. (1977). *Nucl. Phys. A* **291**, p. 93.

Raman, S., Nestor Jr., C. W. and Tikkanen, P. (2001). *Atomic and Nuclear Data Tables* **78**, p. 1.

Randrup, J. (1975). Report LBL-4302, Lawrence Berkeley Laboratory.

Randrup, J. (1976). *Nucl. Phys. A* **259**, p. 253.

Randrup, J. and Swiatecki, W. J. (1984). *Nucl. Phys. A* **429**, p. 105.

Raynal, J. (1996). in *Proceedings of the Specialist's Meeting on the Nucleon Nucleus Optical Model up to 200 MeV* (Bruyères-Le-Chatel, France).

Rhoades-Brown, M., Macfarlane, M. and Pieper, S. (1980). *Phys. Rev. C* **21**, p. 2417.

Rhoades-Brown, M. J., Donangelo, R., Guidry, M. W. and Neese, R. E. (1981). *Phys. Rev. C* **24**, p. 2747.

Rice, S. O. (1954). *Selected Papers on Noise and Stochastic Processes* (Dover, New York).

Rich, A. (1981). *Rev. Mod. Phys.* **53**, p. 127.

Ring, P. and Schuck, P. (1980). *The Nuclear Many-body Problem* (Springer-Verlag).

Rodberg, L. and Thaler, R. (1967). *Introduction to the Quantum Theory of Scattering* (Academic Press, N.Y.).

Rodrígues-Gallardo, M., Arias, J. M., Gómez-Camacho, J., Johnson, R. C., Moro, A. M., Thompson, I. J. and Tostevin, J. A. (2008). *Phys. Rev. C* **77**, p. 064609.

Rodrígues-Gallardo, M., Arias, J. M., Gómez-Camacho, J., Moro, A. M., Thompson, I. J. and Tostevin, J. A. (2009). *Phys. Rev. C* **80**, p. 051601(R).

Rolfs, C. and Rodney, W. S. (1988). *Cauldrons in the Cosmos* (University of Chicago press, Chicago).

Roman, P. (1965). *Advanced Quantum Mechanics* (Addison-Wesley).

Rutherford, E. (1911). *Phil. Mag.* **XXI**, p. 669.

Sakuragi, Y., Yahiro, M. and Kamimura, M. (1983). *Prog. Theor. Phys.* **70**, p. 1047.

Sakuragi, Y., Yahiro, M. and Kamimura, M. (1986). *Prog. Theor. Phys. Suppl.* **89**, p. 136.

Satchler, G. R. (1963). *Phys. Lett.* **7**, p. 55.

Satchler, G. R. (1983). *Direct Nuclear Reactions* (Oxford University Press).

Satchler, G. R. and Love, W. G. (1979). *Phys. Rep.* **55**, p. 183.

Satchler, G. R., Nagarajan, M. A., Liley, J. S. and Thompson, I. J. (1987). *Ann. Phys. (NY)* **178**, p. 110.

Scadron, M., Weinberg, S. and Wright, J. (1964). *Phys. Rev.* **135**, p. B202.

Schiwietz, G. (1990). *Phys. Rev. A* **42**, p. 296.

Schiwietz, G. and Grande, P. L. (1992). *Nucl. Instrum. Methods Phys. Res. B* **69**, p. 10.

Schultz, H. and Wiebicke, H. (1966). *Phys. Lett.* **21**, p. 190.

Schwinger, J. (1961). *Proc. Natl. Acad. Sci. U.S.* **47**, p. 122.

Segrè, E. (1977). *Nuclei & Particles* (Addison-Wesley).

Seyler, R. G. and Blanchard, C. H. (1963). *Phys. Rev.* **131**, p. 355.

Shulgina, N. B., Jonson, B. and Zhukov, M. V. (2009). *Nucl. Phys. A* **825**, p. 175.

Singh, P. P., Segel, R. E., L., M.-S., Hanna, S. S. and Allas, R. G. (1965). *Nucl. Phys.* **65**, p. 577.

Sinha, B. (1975). *Phys. Rep.* **20**, p. 1.

Stefanini, A., Montagnoli, G., Silvestri, S., Corradi, L., Courtin, S., Fioretto, E., Guiot, B., Haas, F., Lebhertz, D., Mason, P., Scarlassara, F. and Szilner, S. (2009). *Phys. Lett. B* **679**, p. 95.

Stiliaris, E., Bohlen, H. G., Fröbrich, P., Gebauer, B., Kolbert, D., von Ortzen, W., Wilpert, M. and Wilpert, T. (1989). *Phys. Lett. B* **223**, p. 291.

Strayer, M. R., Cusson, R. Y., Stoecker, H., Maruhn, J. A. and Greiner, W. (1983). *Phys. Rev. C* **28**, p. 228.

Sun, X. (1998). *Semiclassical Approximations in Chemical Dynamics*, Ph.D. thesis, Department of Chemistry, University of California, Berkeley, Berkeley, USA.

Swiatecki, W. J. (1980). *Prog. Part. Nucl. Phys.* **4**, p. 383.

Swiatecki, W. J. (1981). *Phys. Scripta* **42**, p. 113.

Takigawa, N., Kuratani, M. and Sagawa, H. (1993). *Phys. Rev. C* **47**, p. R2470.

Tamura, T. (1965). *Rev. Mod. Phys.* **37**, p. 679.

Taylor, J. (2006). *Scattering Theory* (Dover Publications, Inc.).

Teichmann, T. (1951). *Phys. Rev.* **83**, p. 141.

Tepel, J. W., Hoffmann, H. M. and Weidenmüller, H. A. (1973). *Phys. Lett. B* **49**, p. 1.

Thaler, R. M. (1959). *Phys. Rev.* **114**, p. 827.

Thomas, R. (1958). *Rev. Mod. Phys.* **30**, p. 257.

Thompson, I. J. (1988). *Comput. Phys. Rep.* **7**, p. 167.

Thompson, I. J., Nagarajan, M. A., Lilley, J. S. and Smithson, M. J. (1989). *Nucl. Phys. A* **505**, p. 84.

Thorn, C. E., LeVine, M. J., Kolata, J. J., Flaum, C., Bond, P. D. and Sens, J. C. (1977). *Phys. Rev. Lett.* **38**, p. 384.

Timmermans, E., Tommasini, P., Hussein, M. S. and Kerman, A. K. (1999). *Phys. Rep.* **315**, p. 199.

Toledo Piza, A. F. R. and Kerman, A. K. (1968). *Ann. Phys. (NY)* **48**, p. 173.

Tostevin, J. A., Nunes, F. M. and Thompson, I. J. (2001). *Phys. Rev. C* **63**, p. 024617.

Ueda, M., Pato, M. P., Hussein, M. S. and Takigawa, N. (1999). *Nucl. Phys. A* **648**, p. 229.

van Haeringen, H. (1976). *J. Math. Phys* **17**, p. 995.

Volkov, A. (1965). *Nucl. Phys.* **74**, p. 33.

von Witsch, W., von Brentano, P., Mayer-Kuckuk, T. and Richter, A. (1966). *Nucl. Phys.* **80**, p. 394.

Wallace, R. J. (1973). *Ann. Phys. (N.Y.)* **78**, p. 190.

Watson, G. N. (1966). *A treatise on the theory of Bessel Functions*, 2nd edn. (Cambridge University Press, Cambridge).

Weise, H. P., Mittmann, H. U., Ding, A. and Henglein, H. (1971). *Z. Naturforsch. A* **26**, p. 1122.

Wheeler, J. (1937). *Phys. Rev.* **52**, p. 1083.

Wigner, E. P. (1947a). *Phys. Rev.* **70**, p. 15.

Wigner, E. P. (1947b). *Phys. Rev.* **70**, p. 606.

Wigner, E. P. and Eisenbud, L. (1947). *Phys. Rev.* **72**, p. 29.

Wilmore, D. and Hodgson, P. E. (1964). *Nucl. Phys.* **55**, p. 673.

Zagrevaev, V. and Greiner, W. (2005). *J. Phys. G: Nucl. Part. Phys.* **31**, p. 825.

Index